Annual Review of Ecology, Evolution, and Systematics

Editorial Committee (2006)

Michael J. Foote, University of Chicago
Douglas J. Futuyma, State University of New York, Stony Brook
C. Drew Harvell, Cornell University
Lars O. Hedin, Princeton University
Mark Kirkpatrick, University of Texas, Austin
Douglas W. Schemske, Michigan State University
H. Bradley Shaffer, University of California, Davis
Daniel Simberloff, University of Tennessee

Responsible for the Organization of Volume 37 (Editorial Committee, 2004)

Michael J. Foote
Douglas J. Futuyma
C. Drew Harvell
Lars Hedin
Michael J. Ryan
Douglas W. Schemske
H. Bradley Shaffer
Daniel Simberloff
Stephen Palumbi (Guest)
Barry Sinervo (Guest)

Production Editor: Roselyn Lowe-Webb
Managing Editor: Veronica Dakota Padilla
Bibliographic Quality Control: Mary A. Glass
Electronic Content Coordinator: Suzanne K. Moses
Illustration Editors: Eliza K. Jewett/Glenda Lee

Annual Review of Ecology, Evolution, and Systematics

Volume 37, 2006

Douglas J. Futuyma, *Editor*
State University of New York, Stony Brook

H. Bradley Shaffer, *Associate Editor*
University of California, Davis

Daniel Simberloff, *Associate Editor*
University of Tennessee

www.annualreviews.org • science@annualreviews.org • 650-493-4400

Annual Reviews
4139 El Camino Way • P.O. Box 10139 • Palo Alto, California 94303-0139

Annual Reviews
Palo Alto, California, USA

COPYRIGHT © 2006 BY ANNUAL REVIEWS, PALO ALTO, CALIFORNIA, USA. ALL RIGHTS RESERVED. The appearance of the code at the bottom of the first page of an article in this serial indicates the copyright owner's consent that copies of the article may be made for personal or internal use, or for the personal or internal use of specific clients. This consent is given on the condition that the copier pay the stated per-copy fee of $20.00 per article through the Copyright Clearance Center, Inc. (222 Rosewood Drive, Danvers, MA 01923) for copying beyond that permitted by Section 107 or 108 of the U.S. Copyright Law. The per-copy fee of $20.00 per article also applies to the copying, under the stated conditions, of articles published in any *Annual Review* serial before January 1, 1978. Individual readers, and nonprofit libraries acting for them, are permitted to make a single copy of an article without charge for use in research or teaching. This consent does not extend to other kinds of copying, such as copying for general distribution, for advertising or promotional purposes, for creating new collective works, or for resale. For such uses, written permission is required. Write to Permissions Dept., Annual Reviews, 4139 El Camino Way, P.O. Box 10139, Palo Alto, CA 94303-0139 USA.

International Standard Serial Number: 1543-592X
International Standard Book Number: 0-8243-1437-9
Library of Congress Catalog Card Number: 71-135616

All Annual Reviews and publication titles are registered trademarks of Annual Reviews.

⊗ The paper used in this publication meets the minimum requirements of American National Standards for Information Sciences—Permanence of Paper for Printed Library Materials, ANSI Z39.48-1992.

Annual Reviews and the Editors of its publications assume no responsibility for the statements expressed by the contributors to this *Annual Review*.

TYPESET BY TECHBOOKS, FALLS CHURCH, VIRGINIA
PRINTED AND BOUND BY FRIESENS CORPORATION, ALTONA, MANITOBA, CANADA

Contents

Annual Review of
Ecology, Evolution,
and Systematics

Volume 37, 2006

Birth-Death Models in Macroevolution
  Sean Nee 1

The Posterior and the Prior in Bayesian Phylogenetics
  Michael E. Alfaro and Mark T. Holder 19

Unifying and Testing Models of Sexual Selection
  Hanna Kokko, Michael D. Jennions, and Robert Brooks 43

Genetic Polymorphism in Heterogeneous Environments: The Age
of Genomics
  Philip W. Hedrick 67

Ecological Effects of Invasive Arthropod Generalist Predators
  William E. Snyder and Edward W. Evans 95

The Evolution of Genetic Architecture
  Thomas F. Hansen 123

The Major Histocompatibility Complex, Sexual Selection,
and Mate Choice
  Manfred Milinski 159

Some Evolutionary Consequences of Being a Tree
  Rémy J. Petit and Arndt Hampe 187

Late Quaternary Extinctions: State of the Debate
  Paul L. Koch and Anthony D. Barnosky 215

Innate Immunity, Environmental Drivers, and Disease Ecology of
Marine and Freshwater Invertebrates
  Laura D. Mydlarz, Laura E. Jones, and C. Drew Harvell 251

Experimental Methods for Measuring Gene Interactions
  Jeffery P. Demuth and Michael J. Wade 289

Corridors for Conservation: Integrating Pattern and Process
  Cheryl-Lesley B. Chetkiewicz, Colleen Cassady St. Clair, and Mark S. Boyce 317

The Population Biology of Large Brown Seaweeds: Ecological Consequences of Multiphase Life Histories in Dynamic Coastal Environments
David R. Schiel and Michael S. Foster 343

Living on the Edge of Two Changing Worlds: Forecasting the Responses of Rocky Intertidal Ecosystems to Climate Change
Brian Helmuth, Nova Mieszkowska, Pippa Moore, and Stephen J. Hawkins 373

Has Vicariance or Dispersal Been the Predominant Biogeographic Force in Madagascar? Only Time Will Tell
Anne D. Yoder and Michael D. Nowak 405

Limits to the Adaptive Potential of Small Populations
Yvonne Willi, Josh Van Buskirk, and Ary A. Hoffmann 433

Resource Exchange in the Rhizosphere: Molecular Tools and the Microbial Perspective
Zoe G. Cardon and Daniel J. Gage 459

The Role of Hybridization in the Evolution of Reef Corals
Bette L. Willis, Madeleine J.H. van Oppen, David J. Miller, Steve V. Vollmer, and David J. Ayre 489

The New Bioinformatics: Integrating Ecological Data from the Gene to the Biosphere
Matthew B. Jones, Mark P. Schildhauer, O.J. Reichman, and Shawn Bowers 519

Incorporating Molecular Evolution into Phylogenetic Analysis, and a New Compilation of Conserved Polymerase Chain Reaction Primers for Animal Mitochondrial DNA
Chris Simon, Thomas R. Buckley, Francesco Frati, James B. Stewart, and Andrew T. Beckenbach 545

The Developmental, Physiological, Neural, and Genetical Causes and Consequences of Frequency-Dependent Selection in the Wild
Barry Sinervo and Ryan Calsbeek 581

Carbon-Nitrogen Interactions in Terrestrial Ecosystems in Response to Rising Atmospheric Carbon Dioxide
Peter B. Reich, Bruce A. Hungate, and Yiqi Luo 611

Ecological and Evolutionary Responses to Recent Climate Change
Camille Parmesan 637

Indexes

Cumulative Index of Contributing Authors, Volumes 33–37 671

Cumulative Index of Chapter Titles, Volumes 33–37 674

Errata

An online log of corrections to *Annual Review of Ecology, Evolution, and Systematics* chapters (if any, 1997 to the present) may be found at
http://ecolsys.annualreviews.org/errata.shtml

Related Articles

From the *Annual Review of Earth and Planetary Sciences*, Volume 34 (2006)

Phanerozoic Biodiversity Mass Extinctions
Richard K. Bambach

Explaining the Cambrian "Explosion" of Animals
Charles R. Marshall

Dates and Rates: Temporal Resolution in the Deep Time Stratigraphic Record
Douglas H. Erwin

The General Circulation of the Atmosphere
Tapio Schneider

From the *Annual Review of Entomology*, Volume 51 (2006)

Botanical Insecticides, Deterrents, and Repellents in Modern Agriculture and an Increasingly Regulated World
Murray B. Isman

Plant Chemistry and Natural Enemy Fitness: Effects on Herbivore and Natural Enemy Interactions
Paul J. Ode

Apparent Competition, Quantitative Food Webs, and the Structure of Phytophagous Insect Communities
F.J. Frank van Veen, Rebecca J. Morris, and H. Charles J. Godfray

Host Plant Selection by Aphids: Behavioral, Evolutionary, and Applied Perspectives
Glen Powell, Colin R. Tosh, and Jim Hardie

The Ecological Significance of Tallgrass Prairie Arthropods
Matt R. Whiles and Ralph E. Charlton

Biogeographic Areas and Transition Zones of Latin America and the Caribbean Islands Based on Panbiogeographic and Cladistic Analyses of the Entomofauna
Juan J. Morrone

Conflict Resolution in Insect Societies
Francis L.W. Ratnieks, Kevin R. Foster, and Tom Wenseleers

Assessing Risks of Releasing Exotic Biological Control Agents of Arthropod Pests
J. C. van Lenteren, J. Bale, F. Bigler, H.M.T. Hokkanen, and A.J.M. Loomans

Plant-Mediated Interactions Between Pathogenic Microorganisms and Herbivorous Arthropods
Michael J. Stout, Jennifer S. Thaler, and Bart P.H.J. Thomma

From the *Annual Review of Environment and Resources*, Volume 31 (2006)

Abrupt Change in Earth's Climate System
Jonathan T. Overpeck and Julia E. Cole

Integrated Regional Changes in Arctic Climate Feedbacks: Implications for the Global Climate System
A. David McGuire, F.S. Chapin III, John E. Walsh, and Christian Wirth

Global Marine Biodiversity Trends
Enric Sala and Nancy Knowlton

Biodiversity Conservation Planning Tools: Present Status and Challenges for the Future
Sahotra Sarkar, Robert L. Pressey, Daniel P. Faith, Christopher R. Margules, Trevon Fuller, David M. Stoms, Alexander Moffett, Kerrie A. Wilson, Kristen J. Williams, Paul H. Williams, and Sandy Andelman

From the *Annual Review of Microbiology*, Volume 60 (2006)

Streamlining and Simplification of Microbial Genome Architecture
Michael Lynch

Subterfuge and Manipulation: Type III Effector Proteins of Phytopathogenic Bacteria
Sarah R. Grant, Emily J. Fisher, Jeff H. Chang, Beth M. Mole, and Jeffery L. Dangl

Origins of Mutations Under Selection: The Adaptive Mutation Controversy
John R. Roth, Elisabeth Kugelberg, Andrew B. Reams, Eric Kofoid, and Dan I. Andersson

Understanding Microbial Metabolism
Diana M. Downs

From the *Annual Review of Phytopathology*, Volume 44 (2006)

The Current and Future Dynamics of Disease in Plant Communities
Jeremy J. Burdon, Peter H. Thrall, and Lars Ericson

Significance of Inducible Defense-related Proteins in Infected Plants
L.C. van Loon, M. Rep, and C.M.J. Pieterse

Coexistence of Related Pathogen Species on Arable Crops in Space and Time
Bruce D.L. Fitt, Yong-Hu Huang, Frank van den Bosch, and Jonathan S. West

Biology of Flower-Infecting Fungi
Henry K. Ngugi and Harald Scherm

The Dawn of Fungal Pathogen Genomics
Jin-Rong Xu, You-Liang Peng, Martin B. Dickman, and Amir Sharon

Climate Change Effects on Plant Disease: Genomes to Ecosystems
K.A. Garrett, S.P. Dendy, E.E. Frank, M N. Rouse, and S.E. Travers

Annual Reviews is a nonprofit scientific publisher established to promote the advancement of the sciences. Beginning in 1932 with the *Annual Review of Biochemistry*, the Company has pursued as its principal function the publication of high-quality, reasonably priced *Annual Review* volumes. The volumes are organized by Editors and Editorial Committees who invite qualified authors to contribute critical articles reviewing significant developments within each major discipline. The Editor-in-Chief invites those interested in serving as future Editorial Committee members to communicate directly with him. Annual Reviews is administered by a Board of Directors, whose members serve without compensation.

2006 Board of Directors, Annual Reviews

Richard N. Zare, *Chairman of Annual Reviews, Marguerite Blake Wilbur Professor of Chemistry, Stanford University*
John I. Brauman, *J.G. Jackson–C.J. Wood Professor of Chemistry, Stanford University*
Peter F. Carpenter, *Founder, Mission and Values Institute, Atherton, California*
Sandra M. Faber, *Professor of Astronomy and Astronomer at Lick Observatory, University of California at Santa Cruz*
Susan T. Fiske, *Professor of Psychology, Princeton University*
Eugene Garfield, *Publisher, The Scientist*
Samuel Gubins, *President and Editor-in-Chief, Annual Reviews*
Steven E. Hyman, *Provost, Harvard University*
Daniel E. Koshland Jr., *Professor of Biochemistry, University of California at Berkeley*
Joshua Lederberg, *University Professor, The Rockefeller University*
Sharon R. Long, *Professor of Biological Sciences, Stanford University*
J. Boyce Nute, *Palo Alto, California*
Michael E. Peskin, *Professor of Theoretical Physics, Stanford Linear Accelerator Center*
Harriet A. Zuckerman, *Vice President, The Andrew W. Mellon Foundation*

Management of Annual Reviews

Samuel Gubins, President and Editor-in-Chief
Richard L. Burke, Director for Production
Paul J. Calvi Jr., Director of Information Technology
Steven J. Castro, Chief Financial Officer and Director of Marketing & Sales
Jeanne M. Kunz, Human Resources Manager and Secretary to the Board

Annual Reviews of

Anthropology
Astronomy and Astrophysics
Biochemistry
Biomedical Engineering
Biophysics and Biomolecular Structure
Cell and Developmental Biology
Clinical Psychology
Earth and Planetary Sciences
Ecology, Evolution, and Systematics
Entomology
Environment and Resources

Fluid Mechanics
Genetics
Genomics and Human Genetics
Immunology
Law and Social Science
Materials Research
Medicine
Microbiology
Neuroscience
Nuclear and Particle Science
Nutrition
Pathology: Mechanisms of Disease

Pharmacology and Toxicology
Physical Chemistry
Physiology
Phytopathology
Plant Biology
Political Science
Psychology
Public Health
Sociology

SPECIAL PUBLICATIONS
Excitement and Fascination of Science, Vols. 1, 2, 3, and 4

Birth-Death Models in Macroevolution

Sean Nee

Institute of Evolutionary Biology, School of Biological Sciences, University of Edinburgh, West Mains Road, Edinburgh, EH9 3JT, United Kingdom; email: sean.nee@ed.ac.uk

Key Words

fossil record, paleobiology, phylogenetics

Abstract

Birth-death models, and their subsets—the pure birth and pure death models—have a long history of use for informing thinking about macroevolutionary patterns. Here we illustrate with examples the wide range of questions they have been used to address, including estimating and comparing rates of diversification of clades, investigating the "shapes" of clades, and some rather surprising uses such as estimating speciation rates from data that are not resolved below the level of the genus. The raw data for inference can be the fossil record or the molecular phylogeny of a clade, and we explore the similarites and differences in the behavior of the birth-death models when applied to these different forms of data.

INTRODUCTION

Mathematical modeling of the dynamical process of speciation and extinction—more generally "birth-death" models—can be used to address many questions in macroevolution. For example, a simple mathematical model of the growth of a clade through speciation can be used to ask the straightforward question, What is the rate of speciation in the clade? This may be asked, for example, to discover if there was anything unusual about the rate of speciation during the Cambrian explosion (Lieberman 2001), the explosion referring, of course, to body plans. Birth-death models can also be used to address less obvious questions in macroevolution. How bad would the fossil record have to be to explain the discrepancy between the fossil ages of major radiations and their molecular ages (Foote et al. 1999)? For another example, species numbers were very low at both the start of the Cambrian and the period after the Late Permian mass extinction. In the subsequent radiations the number of higher taxa appearing was much higher in the former than the latter—why? (Nee & May 1997) Is it "strange" to have so many animal phyla with a very small number of species (Strathman & Slatkin 1983)?

The modern period of the use of birth-death processes to inform thinking about macroevolution began with the work of Raup and colleagues in 1973 (Raup et al. 1973) when computers were just starting to become a readily used tool. They modelled a scenario in which the species in a clade had equal probabilities of either speciating to create a second species or going extinct, and the overall clade size was kept roughly constant. They discovered that this purely random process could produce apparent trends and patterns resembling those in the fossil record, such as "adaptive radiations." As Maynard Smith put it (Maynard Smith 1989), they showed that "...it is fatally easy to read a pattern into stochastically generated data." This is obviously an important realization and familiarity with the "patterns" that random processes create is an essential piece of a scientist's mental furniture.

We can extend the use of such models back further to 1924 to the work of the statistician Yule (1924) who modelled a clade growing according to the pure birth process in which extinction does not occur (see below). The Willis in his paper's title ("A mathematical theory of evolution based on the conclusions of Dr. J.C. Willis, FRS") was an expert on angiosperms who had formed what were perceived to be anti-Darwinian views on the distributions of numbers of subtaxa per taxon, e.g., the numbers of species in genera. The paper illustrates the sorts of taxon sizes that are generated by the random birth process—and, so, shows us the kind of patterns that random processes can generate. However, Yule went further and used his model for the second important use to which such models are put—parameter estimation. He used the birth process to estimate the rate of cladogenesis of the angiosperms. We will see more contemporary examples below.

Branching processes (Gilinsky & Good 1991) and purely statistical models of random walks (e.g., Cornette & Lieberman 2004)—as opposed to models with explicit births and deaths—are also models that have informed macroevolutionary thinking, although their mathematical structure differs from that of the models to be discussed here. But they are little used compared to birth-death models in the technical sense and will not be discussed here. Before leaving them entirely, we will note one interesting result. Apart from the past 75 Myr, the diversity fluctuations seen in Sepkoski's celebrated 540 Myr fossil compendium (Sepkoski 1982) cannot be distinguished from a random walk (Cornette & Lieberman 2004). The past 75 Myr have seen diversity climb higher than random expectations and this is probably not an artifact of a better fossil record, the so-called pull of the recent (Jablonski et al. 2003).

Phylogenetic information is central to inferences about macroevolution as we are interested in the fates of *clades*, i.e., groups of related organisms. After the work of Raup and his colleagues, the second wave of interest in birth-death models and macroevolutionary inference was stimulated by the increasing availability of molecular phylogenies (Hey 1992, Nee et al. 1992). Fossil family trees describe the times of appearance of taxa as well as the genealogical relationships among them. So, too, molecular phylogenies also come equipped with a temporal dimension provided by the molecular "clock." Increasingly sophisticated statistical advances allow the dating (possibly only relative) of the nodes in a molecular phylogeny even in the absence of a universal metronomic molecular time piece (e.g., Aris-Brosou & Yang 2003, Sanderson 1997). Of course, dated molecular phylogenies can be the subject of controversy (e.g., Aris-Brosou & Yang 2003, Benton & Ayala 2003), but such controversy is premised on their existence. In this second phase of interest in birth-death models the emphasis has been primarily on using these models for statistical inference rather than familiarizing us with the sort of patterns that can be generated by random processes.

We will now methodically go though the various kinds of birth-death models that have been used and see the sorts of macroevolutionary inferences that have been drawn from them as well as discuss their general properties. The mathematics of these models will not be presented—this has been done many times before. An important source of original results for the general birth-death process with varying birth and death rates is Kendall (1948). Useful summaries of various results, as well as a few new results, can be found in Raup (1985), Foote (2001), and Nee et al. (1994). Statistical analysis of macroevolutionary data is not addressed here because such analyses are constantly evolving and frequently need to be tailored for the idiosyncracies of particular data sets. The studies discussed in this review are the best guides to how to perform such analyses.

THE PURE BIRTH AND PURE DEATH MODELS

The Pure Birth Model

In this model each species has a constant probability, b, of producing a new species at each point in time and extinction never occurs. In this case, starting with $N(0)$ species at the start of the process, the number of species in the clade, $N(t)$, is expected to grow exponentially over time t:

$$N(t) = N(0)e^{bt}. \qquad 1.$$

Of course, this model may be applied to other entities in addition to species—higher taxa for example. It immediately follows from the exponential growth that a

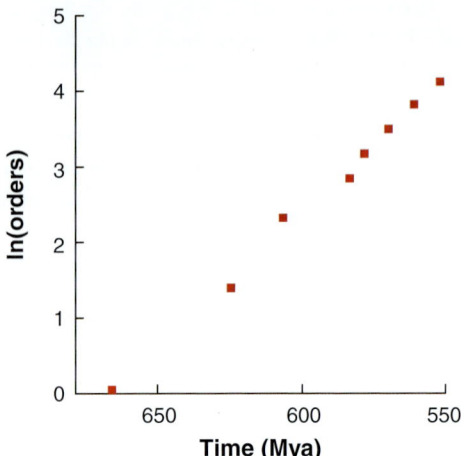

Figure 1
Cumulative increase in the natural logarithm of the number of metazoan orders over the Vendian and Lower Cambrian. The use of the natural logarithm allows one to assess the fit of the birth model from the linearity of the plot and estimate the rate of diversification from the slope of the points. Data from Sepkoski (1978).

plot of the logarithm of the number of species against time should be linear and the slope of the plot provides an estimate of b, the per-capita birth rate, as long as it is understood that this particular calculation assumes natural logarithms. Equation 1 can be interpreted as a purely deterministic model of exponential growth, as it was in Stanley (1975) and Sepkoski (1978). It is also the expected number of species of the stochastic model (e.g., Nee 2001).

Figure 1 shows the increase in the number of metazoan orders as we cross into the Cambrian (Sepkoski 1978) and shows that, for this data set at least, ordinal diversification can be well-described by a model of exponential growth. The rate of growth, b, can be roughly computed from the Figure as 0.033/Myr, although Sepkoski was more interested in the constancy of the diversification rate rather than its actual value.

Figure 2 illustrates the same sort of analysis carried out on a molecular phylogeny (Baldwin & Sanderson 1998). The phylogeny is of the plants in the Hawaiian silversword alliance, a group with remarkable morphological diversity (trees, shrubs, vines, etc.) and habitat heterogeneity. Although there is an apparent slight curvature suggesting a slight increase in diversification rate over time, this is not so dramatic as to obviate the use of the pure birth model to estimate speciation rates as 0.56 ± 0.17/Myr. This is a remarkably high rate, exceeding that of continental radiations. For example, the Old World monkeys had the highest speciation rate of the four main groups of primate—0.342/Myr—the other groups being New World monkeys, Madagascar primates, and apes (Purvis et al. 1995). The latter three groups had similar rates of diversification that were statistically significantly smaller than that of the Old World monkeys. Based on fossil data, Alroy reports that origination rates of mammals during the Paleocene averaged 0.43/Myr, but this was accompanied by high extinction rates resulting in a rather slower net accumulation rate (Alroy 1999). To make these numbers meaningful consider that a clade diversifying at a rate of 0.5/Myr would grow to $\approx$150 after 10 Myr and $\approx$270,000 after 25 Myr.

Paleontologists will be aware that Sepkoski and others did not imagine they were simply estimating origination rates. In fact, they were interested in net rates, r, of

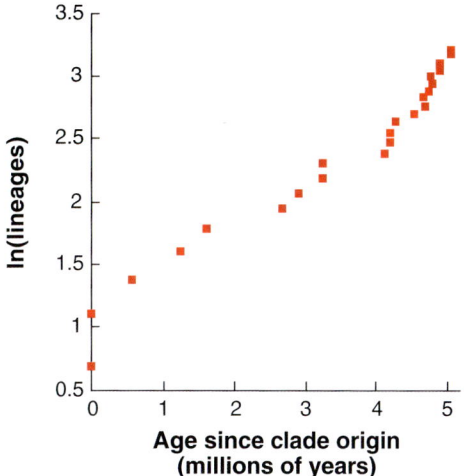

Figure 2

Cumulative increase in the logarithm of the number of lineages in a dated molecular phylogeny of the Hawaiian silversword alliance. Data from Baldwin & Sanderson (1998).

diversification, origination minus extinction, $b–d$, so the b in Equation 1 should be interpreted as this rate r for paleontological studies. We have included macroevolutionary paleontological inference in this section simply because of the interest in a single parameter.

However, it is not possible to have this more general interpretation of the parameter when dealing with molecular phylogenies. As discussed in the next section, introducing the death rate d qualitatively changes the behavior of the model and the apparent growth of the clade through time is no longer log-linear. However, although it is unrealistic to assume that no extinction occurs, this is not actually relevant. If, as in the case of the silversword alliance diversification discussed above, the model is appropriate for the data, and the rough log-linearity of **Figure 2** says that it is, then it is appropriate for estimation of a diversification rate even if, technically, this is purely a speciation rate.

So far, we have seen the pure birth model used to make inferences about the rate of diversification and its constancy, as evidenced by log-linearity. Other questions can be addressed. For example, Ribera and colleagues used molecular phylogenies to investigate whether there exist differences in diversification rates of aquatic beetle clades that inhabit running versus stagnant water bodies (Ribera et al. 2001). As can be seen from the roughly parallel slopes of the two clades in **Figure 3** there is no substantial difference. A comparison based on fossil data suggests balanoid barnacles diversified more rapidly than molluscs (Stanley & Newman 1980). (Of course, the next step is to investigate possible causes of such differences—or lack thereof—but that is beyond the scope of this review.)

Many authors have suggested that speciation rates may have increased during the Pleistocene (roughly two million to 10,000 years ago) because of recurring ice

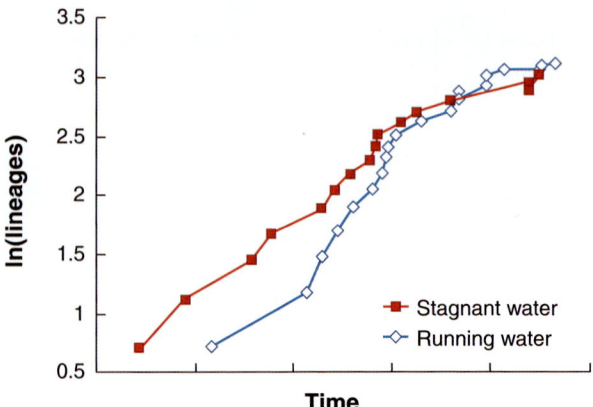

Figure 3
Cumulative increase in the logarithm of the number of lineages in a molecular phylogeny of two species of aquatic beetle, one living in stagnant water and the other in running water. The time axis is in units of genetic distance and is not labeled because it is not calibrated. Data from Ribera et al. (2001).

ages (e.g., Hewitt 1999). Though a molecular phylogenetic study of birds (Zink & Slowinski 1995) actually suggested that rates declined, another such study of North American tiger beetles (Barraclough & Vogler 2002) found some, albeit weak, evidence for a Pleistocene increase. Much stronger molecular evidence for an increase is found for Holarctic damselflies, which also display a geographical signal: a more northerly clade experienced a 24-fold increase in the rate of cladogenesis compared to a sixfold increase in a more southerly clade (Turgeon et al. 2005).

In fact, the primate analyses of Purvis et al. (1995) and the damselfly analysis of Turgeon et al. (2005) used a birth-death model (see below) but the best estimate for the death rate was zero, so they are discussed in this section. A best estimate of $d = 0$ is frequently the case (e.g., Losos & Schluter 2000). It is unclear why this is the case.

The Pure Death Model

The pure death model (Raup 1975, van Valen 1973) is only useful for paleontological data. Suppose we have data on the longevities of a number of species. On the assumption that at any point in time each species has the same probability of going extinct, d, then we expect an exponential decay in the number of species that survive to age t, $N(t)$:

$$N(t) = N(0)e^{-dt}.$$

A semilog plot of the number of species surviving to age t is the analogue of our previous semilog plots and, once again, the linearity of the plot informs us of the adequacy of the model and the slope of the plot informs us of the extinction rate d.

On the basis of such analyses, van Valen proposed an evolutionary law (Van Valen 1973). In the words of Raup (Raup 1975, p. 82), "... within an ecologically homogenous taxonomic group, extinction occurs at a stochastically constant rate." To explain this law, van Valen proposed a Red Queen view of evolution in which evolutionary advances in one species are seen as environmental deterioration by others, leading to perpetual coevolution and, somehow, constant rates of extinction. However dubious the merits of the model (Maynard Smith 1989), there remains the empirical question of the constancy of extinction rates. To not observe such constancy would be surprising—and interesting—as it would imply either that species senesce, or the converse, that the longer a species survives, the more likely it is to survive even longer. I think we would expect the future lifetime of a species to be independent of how long it has been around already, and that is what the pure death process postulates.

THE BIRTH-DEATH MODEL

We now introduce extinction by allowing species to not only speciate, with constant rate b, but to also die, with constant rate d. This introduces no radically new features for the interpretation of paleontological data: Clades are expected to grow exponentially at a rate $b-d$, the net rate of diversification, and this is the interpretation that paleontologists have always put on the single parameter of the model. But adding extinction has a dramatic impact on the interpretation of molecular phylogenies.

We saw in the previous section that the semilog representation of the growth of a molecular phylogeny is expected to be linear under a pure birth process. **Figure 4** illustrates that this is no longer the case when the extinction rate is nonzero. In fact, over much of the history of the clade the plot is expected to be linear with a slope of $b-d$, just as it would be with paleontological data, but as we approach the present the slope is expected to increase, asymptotically approaching b, thus creating the illusion of an accelerating rate of cladogenesis. This happens because the molecular phylogeny is based entirely on data from extant species, and species that have originated more recently in the past have had less time to go extinct. One consequence of this behavior is that a clade that we might naively think to be in rude health, with an accelerating rate of diversification, is, in fact, subject to high rates of extinction. This is discussed in the context of the phylogeny of plethodontid salamanders in Nee et al. (1995).

Intuition, based on this behavior of the lineage-through-time plot, suggests that we have two items of information about the growth of the clade (b, $b-d$) and this

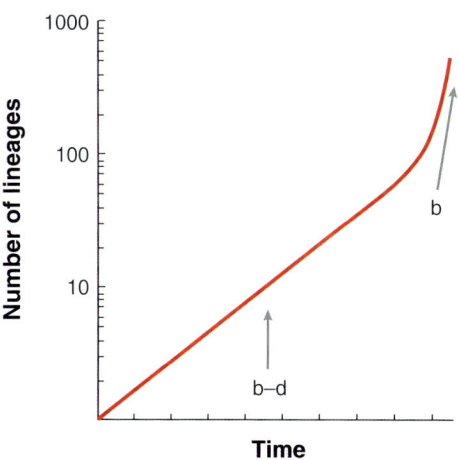

Figure 4

Theoretical plot showing the expected cumulative increase in the logarithm of the number of lineages in a molecular phylogeny growing according to a birth-death process. Theory from Nee et al. (1994).

suggests the possibility that we may be able to estimate speciation and extinction rates separately. This is, in fact, the case (Nee et al. 1994).

The stochastic theory is such that it is more natural to consider the composite parameters b–d and d/b. The first is, of course, the net rate of diversification and the second controls the degree to which the growth of the clade departs in its behavior from that of a pure birth process—visually, the larger d/b the greater degree of curvature in **Figure 4**. A crucial probability relevant for both paleontological and molecular phylogenetic data is the probability that a lineage that arose at some time in the past still exists at time t later, i.e., it has at least one progeny. This is given by:

$$\Pr(t) = \frac{1 - \frac{d}{b}}{1 - \frac{d}{b} e^{-(b-d)t}}. \qquad 2.$$

Probability Equation 2 is central because a lineage will not appear in a molecular phylogeny unless it has at least one extant descendant. The importance of the probability in paleontology is that it forms the basis of cohort analysis of, say, the survival of genera through time, which can be used to estimate origination and extinction rates of the species (Raup 1975, 1978; Foote 1988, 2001). Foote's analysis returned a high speciation rate for Cambrian trilobites of 0.4/Myr.

So, we have two surprises from the birth-death model. First, we can estimate speciation and extinction rates from molecular phylogenies even though they do not contain information from extinct species. Second, we can estimate per-species speciation and extinction rates even from fossil data that are not resolved to a level below that of, say, the genus.

Van Valen's empirical data in support of his law included data on the survivorship of taxa higher than species. Raup (1975) made the observation that semilog survivorship plots are not expected to be linear under a birth-death model if the taxa are higher than species: $\log N(t)$, $N(t)$ from Equation 2 as $N(0)$ $\Pr t$ is no longer a linear function of time. This underlies the nonlinearity of **Figure 4** as well.

Equation 2 forms the basis of an interesting argument that life arose repeatedly on Earth, even though all current life is descended from a single ancestor (Raup & Valentine 1983). I invite the reader to consult this admirably speculative paper for details.

Although it is possible in principle to estimate both composite parameters (and so, if you wish, recover b and d), in fact our estimates of b–d are generally much more precise than that of d/b. This is not obvious from Equation 2. This lack of obviousness is good, because it is not necessarily true! If, for example, we had as our data the fact that after a time t one genus was extinct and one other extant, then a maximum likelihood analysis will return an estimate of $d/b = 0.5$ and any number b–d from very small to infinity so the probability in Equation 2 becomes 1/2. As soon as the exponential term in Equation 2 gets small then any value of b–d will generate an estimated probability of 0.5, which is what it should be on the basis of these data. It is only as we accumulate more data stretched over time that the parameter b–d becomes of crucial importance in fitting our estimated probability to the data. The imprecision of our estimates of *d/b* compared to *b–d* was noted in paleontology by Foote (1988) in a study of trilobite genera survivorship and by Nee et al. in a study of *Drosophila* (Nee et al. 1995). A related fact is that although it is possible in principle to detect mass extinctions in molecular phylogenies, which is remarkable, the signal is expected to be weak in the kinds of data sets we can realistically expect to observe (Harvey et al. 1994).

THE MORAN PROCESS

The previous models are suitable for radiations that have not hit any limits to diversity. At the other extreme, we require a model for clades that have reached a plateau, as, for example, metazoan orders appear to have done in the Ordovician and Silurian and remained at the plateau for about 200 million years (Sepkoski 1978). A useful model for this was first introduced in population genetics by Moran (1958): at each point in time, each lineage (or higher taxon) has a probability of going extinct and when a lineage does go extinct it is replaced by the progeny of another lineage chosen at random. So in this model the clade is kept at a constant size deterministically. There are many analytical results available for this process in the population genetics literature, particularly that branch known as coalescence theory (e.g., Hudson 1990), and these tell us what a molecular phylogeny of the extant members of a clade that has arisen from this process should look like. As might be, to some extent, anticipated from the behavior of the birth-death process, the Moran process generates a clade with an apparently accelerating rate of cladogenesis over its entire history.

There is no substantial difference between the Moran process and one of the simulation algorithms studied by, among others, Raup et al. (1973) and Gould et al. (1977), nor between this and the algorithm employed by Sepkoski (1978) and Sepkoski & Kendrick (1993), once the plateau has been reached for some time. (In Sepkoski 1978 the clade grows logistically to the plateau.) In the simulations of Raup et al. (1973) and Gould et al. (1977), they decided on a ceiling diversity and set $b = d$ at this ceiling. When the diversity dropped below the ceiling, they set $b > d$ to get it back there: similarly, they set $b < d$ when diversity rises above.

The Moran process was introduced into the study of molecular trees by Hey (1992), although not identified as such. In an analysis of eight phylogenetic trees (of salamanders, for example), Hey found that a pure birth process model ($d = 0$) provided a statistically superior description of the data than the Moran model ($d = b$) and argued on this basis for the merits of the pure birth process model as a model of macroevolution. However, Nee (2001) argues on methodological grounds that the Moran process should remain an integral component of our arsenal.

Random Walk

There is a very big difference between the Moran process and a birth-death process in which $b = d$ and these rates are kept constant. In this model, the total size of the clade follows a random walk. This model has been studied extensively in the context of fossil data with particular reference to the shape of clades (Gould et al. 1977, Kitchell

& MacLeod 1988, Uhen 1996), and not at all, to my knowledge, with reference to molecular data. This is understandable when we realize that extinction is inevitable for clades following this random walk: In the fossil record, clades come and go but in molecular trees of extant species, clades only come.

THE SHAPES OF CLADES

Paleontologists have long been interested in the shapes of clades, i.e., their diversity paths as they first appear in the fossil record, reach their peak diversity level, and ultimately dwindle to extinction (see **Figure 5**). So, for example, are clades symmetrical, with their rise to peak diversity being the mirror image of their decline to extinction? This is what would be expected from a birth-death process with $b = d$: A random walk is symmetrical with respect to time. As mentioned in the Introduction, the ability of birth-death simulation processes to produce clades that look like actual ones as they appear in the fossil record was one of the first surprising results in the modern period of the use of these models for macroevolutionary inference (Gould et al. 1977, Raup et al. 1973, Stanley et al. 1981).

A remarkable result was obtained by Gould et al. (1987). In a study of eight major groups of marine invertebrates they found that clades arising early in time (Cambrian and Ordovician) tended to be bottom-heavy (as in **Figure 5**), whereas those arising in later periods tended, on average, to be symmetrical. This means that the temporal direction of diversification has a signal and that diversification is asymmetrical with respect to time, unlike the simple random walk model. They developed macroevolutionary hypotheses for this phenomenon in terms of adaptive radiations into empty niche space.

A simulation study of birth-death processes casted doubt on this result by noting that the average bottom-heaviness Gould et al. observed in the early period—as measured by a center of gravity, CG, a statistic they had devised—was quite likely

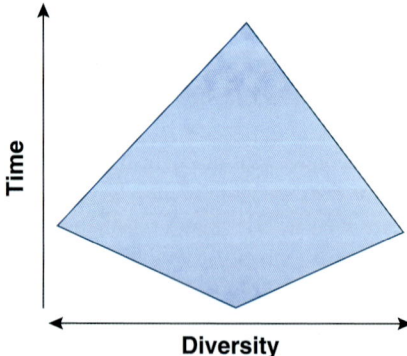

Figure 5

A spindle diagram showing the increase and decline of a clade over time. The width of the diagram at any point in time indicates its diversity at that time. The particular clade shown here is bottom heavy.

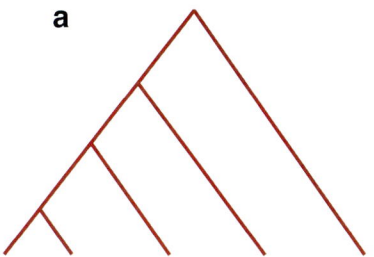 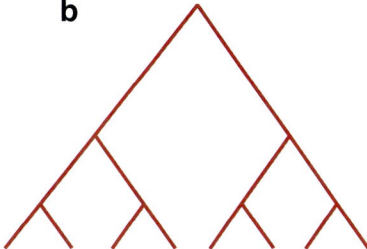

Figure 6

Two phylogenetic topologies. Typology *a* is completely unbalanced and typology *b* is completely balanced.

to arise by chance alone (Kitchell & MacLeod 1988). This critique is not entirely definitive, however. It is true that the original paper did not do a statistical analysis of the CG difference between the early and later periods—an omission remedied by this paper. But the paper also included a regression analysis of the CG of mammalian genera as a function of time since the base of the Tertiary ($\approx$65 Mya) and observed a highly significant tendency for clades to become more top-heavy over time.

Alas, this result seems extremely sensitive to the data used. In a reanalysis using an updated data set, exactly the opposite pattern is found (Uhen 1996). Uhen concluded that it is pointless to generate macroevolutionary speculations for observations that are clearly highly volatile as new data are accumulated. It is a shame that Gould's arrow of time joins the 26-million-year extinction periodicity and its associated Death Star (Muller 1988) in the bin of exciting paleontological might-have-beens.

The analogue of the analysis of clade shape in the study of trees of extant species is the investigation of tree balance (see **Figure 6**). A variety of statistics and null models for studying tree balance have been suggested (Mooers & Heard 1997). There is really only one null model that is meaningful in a macroevolutionary context, however. This is commonly referred to as the "equal rates Markov" (ERM) model. Thus far we have not discussed tree topology. Under the pure birth process, each lineage is as likely to give rise to a new lineage as any other, so if we are tracking topology as well as internode intervals in the growing tree, whenever a new lineage appears it arises from an existing lineage that is chosen at random. This is the usual description of the ERM model. Analysis of tree shape consists of (*a*) defining a tree balance statistic—several are defined by Kirkpatrick & Slatkin (1993); (*b*) calculating the distribution of the statistic under the ERM model, and (*c*) comparing the statistic's value in a real tree with this distribution to see if it is unusual.

The general result that has emerged is that real trees are more unbalanced than expected under the ERM model (Mooers & Heard 1997, Pinelis 2003) and the conclusion is that there is heterogeneity among lineages in their propensity to diversify.

Though the birth process requires the ERM model of topology, the model itself is not particularly wedded to the birth process: It is simply an algorithm for generating tree topologies under the hypothesis of lineage equivalence. If the tree has a temporal dimension (so far in this topology discussion we have not assumed that it does), we can do a different sort of analysis that is clearly tied to the birth process and, perhaps, reveals more than the summary-statistic approach to tree balance studies.

Figure 7

We ask the question: For each lineage that crosses line *a* in the phylogeny, how many daughter lineages does it have at the time of line *b*?

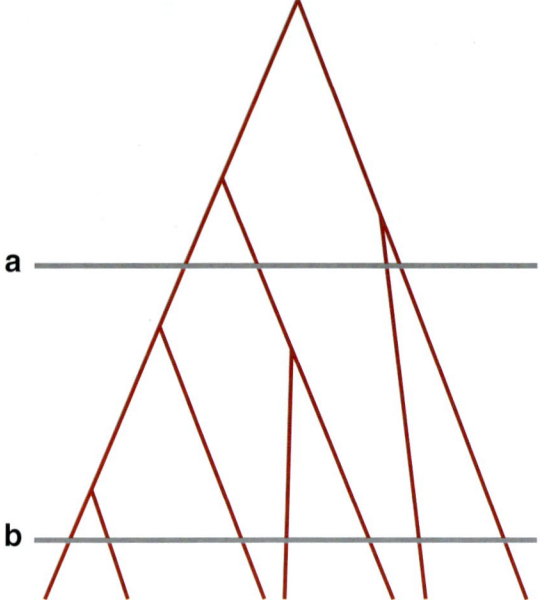

Consider **Figure 7**. If the tree has grown according to the pure birth process, then the number of daughter lineages for each parental lineage is expected to have a one parameter geometric distribution. This result is very general and does not require, for example, that the birth rate be constant—it can change arbitrarily (Nee et al. 1994). All that matters is that it be the same for each lineage at each point in time.

Figure 8 illustrates the results of such an analysis carried out on Sibley and Ahlquist's molecular phylogeny of birds. As is readily apparent the Passeri (song birds) and Ciconiiformes (shore birds, waders, flamingos, gulls, and such like) were identified as statistical outliers from the fitted geometric distribution. One can speculate as to why these groups radiated so exceptionally: It is surely no coincidence that the Ciconiiform radiation occurred at the time of the breakup of Gondwanaland (Cotgreave & Harvey 1994). Analyzing these two radiations on their own displays an excellent fit to a pure birth process (Nee et al. 1992). This kind of analysis also identified Old World monkeys as a significant radiation, in the sense of being anomalous under the null model, as did our previous analyses of rates of cladogenesis.

What we have just done, of course, is look at the numbers of subtaxa per taxon, and this kind of analysis is meaningful for both molecular and fossil data. It has been known since the work of Willis in the 1920s that the frequency distributions answering such questions—how many species per genus or how many families per order, etc., are of the "hollow curve" variety, with a mode of monotypic taxa and a long tail, such as seen in **Figure 8**. Sepkoski (Sepkoski 1978) was interested in such distributions in the context of the question of whether or not studies of diversity that are only resolved to, say, ordinal level are informative about diversification at the species level. He concluded that they are.

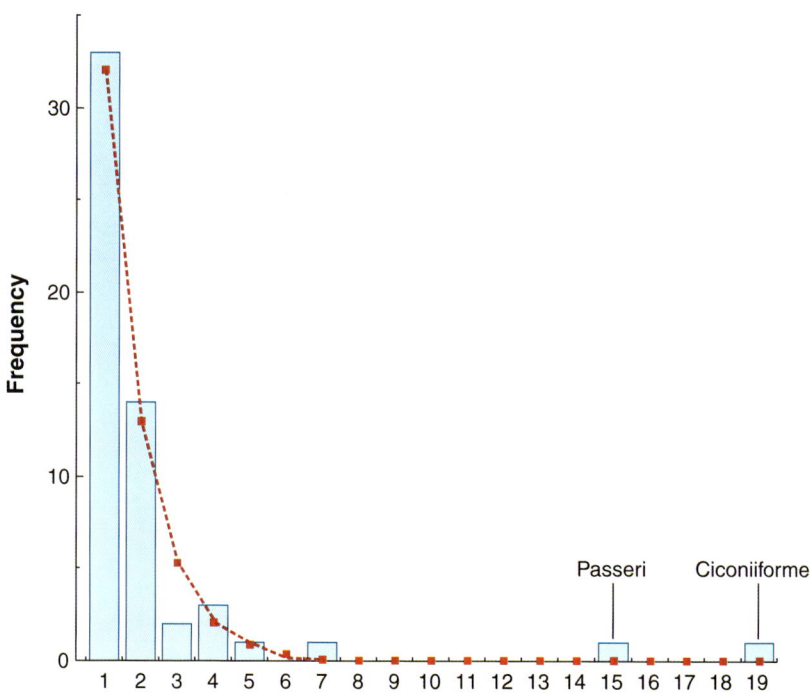

Figure 8
Frequency distribution of the number of daughter lineages per parental lineage for the sort of analysis described in **Figure 7** applied to actual data: in this case the Sibley and Ahlquist phylogeny of the birds. For details see Nee et al. (1992).

SAMPLING AND PARAPHYLY

So far, we have been implicitly assuming that we have all the members of a clade in our analysis. Though there has been great progress in producing more-or-less complete phylogenies of large groups with a temproal dimension (Bininda-Emonds et al. 1999, Purvis 1995), it is often going to be the case that molecular phylogenies will only be a sample of a clade. Paleontological data are often incomplete as well and the exclusion of a lineage and all its descendants from a clade on the basis of the possession of some characteristic is known as paraphyly.

Paraphyly stalks the compendia of fossil data and much thought has been given to how it may bias results. Stung by criticism that the large numbers of paraphyletic taxa in his compendium (Sepkoski 1982) constituted "taxonomic noise" (Smith & Patterson 1988), Sepkoski & Kendrick carried out simulations of evolutionary history using every sort of model discussed so far and threw in the odd mass extinction as well (Sepkoski & Kendrick 1993). They then sampled taxa from the known evolutionary history and asked how readily they allowed that history to be inferred. They concluded that although monophyletic taxa, as expected, perform best, nonetheless paraphyletic taxa do indeed contain a good signal.

On the basis of intuition, it has been thought that paraphyly may make clades appear more bottom-heavy than they actually are (Gould et al. 1987, Smith 1994). In fact, simulation shows that paraphyly is more likely to make clades appear top-heavy

(Uhen 1996). Although Uhen does discuss why this is the case, I am unable to distill his explanation into a succinct, intuitive morsel and encourage any interested reader to consult his paper.

Molecular phylogenies that are based on simply a random sample of a clade that has grown according to the birth or birth-death process will give the misleading impression that the rate of cladogenesis has been slowing down, so the upward curvature in **Figure 4** becomes downward instead (Nee et al. 1994). This effect arises because lineages that have arisen in the recent past are likely to have fewer progeny than older lineages and are, as a result, less likely to have any progeny lineages in the sample. This effect is particularly striking in the phylogenetic trees of viruses that have been spreading exponentially, such as HIV, which have effectively a star-burst phylogeny (e.g., Nee et al. 1995, Slatkin & Hudson 1991) (remember that, for viruses, decades are a macroevolutionary timescale). Sampling has no qualitative effect on trees that have grown according to the Moran process.

CONCLUSION

We will finish off where we began, with a visit to some less obvious uses of birth-death processes for macroevolutionary inference. The two papers to be mentioned here both exploit the concept of evolutionary history, which, in the present context, simply means the sum total of the branch lengths in the phylogeny of a clade (Nee & May 1997). Foote et al. (1999) coupled a birth model—to generate an evolutionary history potentially being laid down in rock—with a stochastic model of actual preservation. The purpose was to ask about the probability that eutherian mammals arose much earlier than the fossil record suggests. They concluded that it is highly unlikely that they arose very long before they actually left a trace of their existence, contrary to molecular evidence.

Nee & May (1997) asked how much evolutionary history would remain in a clade after a mass extinction using birth-death models for analytical results and simulations. They concluded that a surprising amount could be preserved—in a sound-bite, 80% of the evolutionary history of a clade could remain after the extinction of 95% of the species. Essentially this is because the deep branches of the clade are likely to survive even a substantial "pruning of the twigs." Erwin suggested that this could account for why the biodiversity recovery after the Late Permian mass extinction produced so few higher taxa such as phyla: They were mainly likely to have survived and maintained their position in niche space.

Birth-death models have a distinguished and continuing history in macroevolutionary inference. Occasionally, improved models are recommended. These are usually premised on the idea that more realistic models are to be preferred or that a model that better fits observed patterns is a step forward. In my opinion, there is no substitute for models that have a simple interpretation and lead to easy inference, and whose failure is often of more interest than the ability to fit an observed pattern. I expect the sort of models discussed here to maintain their pre-eminence for a long time to come.

LITERATURE CITED

Alroy J. 1999. The fossil record of North American mammals: evidence for a Paleocene evolutionary radiation. *Syst. Biol.* 48:107–18

Aris-Brosou S, Yang Z. 2003. Bayesian models of episodic evolution support a Late Precambrian explosive diversification of the Metazoa. *Mol. Biol. Evol.* 20:1947–54

Baldwin BG, Sanderson MJ. 1998. Age and rate of diversification of the Hawaiian silversword alliance (Compositae). *Proc. Natl. Acad. Sci. USA* 95:9402–6

Barraclough T, Vogler AP. 2002. Recent diversification rates in North American tiger beetles estimated from a dated mtDNA phylogenetic tree. *Mol. Biol. Evol.* 19:1706–16

Benton MJ, Ayala FJ. 2003. Dating the tree of life. *Science* 300:1698–700

Bininda-Emonds ORP, Gittleman JL, Purvis A. 1999. Building large trees by combining phylogenetic information: a complete phylogeny of the extant Carnivora (Mammalia). *Biol. Rev. Camb. Philos. Soc.* 74:143–75

Cornette JL, Lieberman BS. 2004. Random walks in the history of life. *Proc. Natl. Acad. Sci. USA* 101:187–91

Cotgreave P, Harvey PH. 1994. Associations among biogeography, phylogeny and bird species diversity. *Biodivers. Lett.* 2:46–55

Foote M. 1988. Survivorship analysis of Cambrian and Ordovician trilobites. *Paleobiology* 14:258–71

Foote M. 2001. Evolutionary rates and the age distributions of living and extinct taxa. In *Evolutionary Patterns—Growth, Form and Tempo in the Fossil Record*, ed. JBC Jackson, S Lidgard, FK McKinney, pp. 245–95. Chicago: Univ. Chicago Press

Foote M, Hunter JP, Janis CM, Sepkoski JJ Jr. 1999. Evolutionary and preservational constraints on origins of biological groups: divergence times of eutherian mammals. *Science* 283:1310–14

Gilinsky NL, Good IJ. 1991. Probabilities of origination, persistence and extinction of families of marine invertebrate life. *Paleobiology* 17:145–66

Gould SJ, Gilinsky NL, German RZ. 1987. Asymmetry of lineages and the direction of evolutionary time. *Science* 236:1437–41

Gould SJ, Raup DM, Sepkoski JJ Jr, Schopf TJM, Simberloff DS. 1977. The shape of evolution: a comparison of real and random clades. *Paleobiology* 3:23–40

Harvey PH, May RM, Nee S. 1994. Phylogenies without fossils. *Evolution* 48:523–29

Hewitt GM. 1999. Post-glacial recolonization of European biota. *Biol. J. Linn. Soc.* 68:87–112

Hey J. 1992. Using phylogenetic trees to study speciation and extinction. *Evolution* 46:627–40

Hudson RR. 1990. Gene genealogies and the coalescent process. In *Oxford Surveys in Evolutionary Biology*, ed. D Futuyma, J Antonovics, 7:1–49. Oxford, UK: Oxford Univ. Press

Jablonski D, Roy K, Valentine JW, Price RM, Anderson PS. 2003. The impact of the pull of the recent on the history of marine diversity. *Science* 300:1133–35

Kendall DG. 1948. On the generalized birth-and-death process. *Ann. Math. Stat.* 19:1–15

Kirkpatrick M, Slatkin M. 1993. Searching for evolutionary pattern in the shape of a phylogenetic tree. *Evolution* 47:1171–81

Kitchell JK, MacLeod N. 1988. Macroevolutionary interpretations of symmetry and synchroneity in the fossil record. *Science* 240:1190–93

Lieberman BS. 2001. A test of whether rates of speciation were unusually high during the Cambrian radiation. *Proc. R. Soc. London Ser. B* 268:1707–14

Losos JB, Schluter D. 2000. Analysis of an evolutionary species-area relationship. *Nature* 408:847–50

Maynard Smith J. 1989. The causes of extinction. *Philos. Trans. R. Soc. London Ser. B* 325:241–52

Mooers AO, Heard SB. 1997. Inferring evolutionary process from phylogenetic tree shape. *Q. Rev. Biol.* 72:31–55

Moran PAP. 1958. Random processes in genetics. *Proc. Camb. Philos. Soc.* 54:60–71

Muller R. 1988. *Nemesis: The Death Star*. New York: Weidenfeld & Nicolson

Nee S. 2001. Inferring speciation rates from phylogenies. *Evolution* 55:661–68

Nee S, Holmes EC, May RM, Harvey PH. 1995. Estimating extinction from molecular phylogenies. In *Extinction Rates*, ed. JL Lawton, RM May, pp. 164–82. Oxford, UK: Oxford Univ. Press

Nee S, May RM. 1997. Extinction and the loss of evolutionary history. *Science* 278:692–94

Nee S, May RM, Harvey PH. 1994. The reconstructed evolutionary process. *Philos. Trans. R. Soc. London Ser. B* 344:305–11

Nee S, Mooers AO, Harvey PH. 1992. Tempo and mode of evolution revealed from molecular phylogenies. *Proc. Natl. Acad. Sci. USA* 89:8322–26

Pinelis I. 2003. Evolutionary models of phylogenetic trees. *Proc. R. Soc. London Ser. B* 270:1425–31

Purvis A. 1995. A composite estimate of primate phylogeny. *Philos. Trans. R. Soc. London Ser. B* 348:405–21

Purvis A, Nee S, Harvey PH. 1995. Macroevolutionary inferences from primate phylogeny. *Proc. R. Soc. London Ser. B* 260:329–33

Raup DM. 1975. Taxonomic survivorship curves and Van Valen's law. *Paleobiology* 1:82–96

Raup DM. 1978. Cohort analysis of generic survivorship. *Paleobiology* 4:1–15

Raup DM. 1985. Mathematical models of cladogenesis. *Paleobiology* 11:42–52

Raup DM, Gould SJ, Schopf TJM, Simberloff DS. 1973. Stochastic models of phylogeny and the evolution of diversity. *J. Geol.* 81:525–42

Raup DM, Valentine JW. 1983. Multiple origins of life. *Proc. Natl. Acad. Sci. USA* 80:2981–84

Ribera I, Barraclough T, Vogler AP. 2001. The effect of habitat type on speciation rates and range movements in aquatic beetles: inferences from species-level phylogenies. *Mol. Ecol.* 10:721–35

Sanderson MJ. 1997. A nonparametric approach to estimating divergence times in the absence of rate constancy. *Mol. Biol. Evol.* 14:1218–31

Sepkoski JJ Jr. 1978. A kinetic model of Phanerozoic taxonomic diversity I. Analysis of marine orders. *Paleobiology* 4:223–51

Sepkoski JJ Jr. 1982. A compendium of fossil marine animal families. *Milwaukee Public Mus. Contrib. Biol. Geol.* 51:1–125

Sepkoski JJ Jr, Kendrick DC. 1993. Numerical experiments with model monophyletic and paraphyletic taxa. *Paleobiology* 19:168–84

Slatkin M, Hudson RR. 1991. Pairwise comparison of mitochondrial DNA sequences in stable and exponentially growing populations. *Genetics* 129:555–62

Smith AB. 1994. *Systematics and the Fossil Record*. Oxford, UK: Blackwell Sci.

Smith AB, Patterson C. 1988. The influence of taxonomic method on the perception of patterns of evolution. *Evol. Biol.* 23:127–216

Stanley SM. 1975. A theory of evolution above the species level. *Proc. Natl. Acad. Sci. USA* 72:646–50

Stanley SM, Newman WA. 1980. Competitive exclusion in evolutionary time: the case of the acorn barnacles. *Paleobiology* 6:173–83

Stanley SM, Signor PW III, Lidgard S, Karr AF. 1981. Natural clades differ from "random" clades: simulations and analyses. *Paleobiology* 7:115–27

Strathman RR, Slatkin M. 1983. The improbability of animal phyla with few species. *Paleobiology* 9:97–106

Turgeon J, Stoks R, Thum RA, Brown JM, McPeek MA. 2005. Simultaneous Quaternary radiations of three damselfly clades across the Holarctic. *Am. Nat.* 165:E78–107

Uhen MD. 1996. An evaluation of clade-shape statistics using simulations and extinct families of mammals. *Paleobiology* 22:8–22

Van Valen L. 1973. A new evolutionary law. *Evol. Theory* 1:1–30

Yule GU. 1924. A mathematical theory of evolution based on the conclusions of Dr. J.C. Willis, FRS. *Philos. Trans. R. Soc. London Ser. B* 213:21–87

Zink RM, Slowinski JB. 1995. Evidence from molecular systematics for decreased avian diversification in the Pleistocene epoch. *Proc. Natl. Acad. Sci. USA* 92:5832–35

The Posterior and the Prior in Bayesian Phylogenetics

Michael E. Alfaro[1] and Mark T. Holder[2]

[1]School of Biological Sciences, Washington State University, Pullman, Washington 99164-4236; email: alfaro@wsu.edu

[2]School of Computational Science, Florida State University, Tallahassee, Florida 32306-4120; email: mholder@scs.fsu.edu

Key Words

bootstrap, Markov chain Monte Carlo, PP, prior probability, phylogenetic confidence

Abstract

Bayesian analysis has enjoyed explosive growth in phylogenetics over the past five years. Accompanying this popularity has been increased focus on the meaning of the posterior probability (PP) and the role of the prior in phylogenetic inference. Here we discuss the behavior of the PP in Bayesian and frequentist terms and its relationship to parametric and nonparametric bootstrapping. We also review the use of priors in phylogenetics and the issues surrounding the specification of informative and minimally informative prior distributions.

INTRODUCTION

Over the past five years, Bayesian analysis has established itself as a major methodological innovation in the field of phylogenetics (Huelsenbeck et al. 2001, Lewis 2001). However, serious criticisms have been leveled against the method since near its inception (e.g., Suzuki et al. 2002). In this review we focus on two controversial aspects of Bayesian inference: (*a*) the use of posterior probabilities (PPs) as measures of clade support and (*b*) the specification of priors distributions. In particular we discuss the PP in relation to the bootstrap. Although the two support measures are superficially similar, empirical observations, simulation studies, and theoretical work clearly show they are not (and should not be) identical.

Bayesian phylogenetic analyses are conducted using a simulation technique known as Markov chain Monte Carlo (MCMC). Verifying that the MCMC analysis has been sufficiently thorough is clearly a major concern for practitioners, but coverage of the MCMC convergence issues is beyond the scope of this review.

Applying Bayes' Theorem to Phylogenetics

Bayesian inference focuses on a quantity known as the posterior probability (PP, also denoted as Pr [*Tree*|*Data*]), which is the probability that a tree is the true tree based on the prior beliefs and the likelihood. If we use θ to represent the parameters of a model, then

$$\Pr(Tree|Data) = \frac{\int \Pr[Data|Tree, \theta] \times \Pr[Tree, \theta] d\theta}{\Pr[Data]}. \qquad 1.$$

The PP is calculated by integrating the PP density over all possible parameters of a model. This density is the product of the likelihood of the tree (Pr [*Data*|*Tree*, θ], the probability of the data given the tree and model) and the prior probability density of the tree and model (Pr[*Tree*, θ], simply referred to as the prior) divided by the probability of the data (Pr[*Data*]).

The prior encapsulates knowledge from sources other than the data at hand, such as information from earlier studies. The likelihood used in Bayesian calculations is a measure of how well a hypothesis fits the observed data. In maximum likelihood inference, trees (or other hypotheses) are chosen solely on the basis of the likelihood function. In contrast, Bayesians choose the most probable tree, which is a function of both the likelihood and the prior. The same models of character evolution used in maximum likelihood can be used in the Bayesian framework simply by specifying prior distributions for the model parameters.

The PP cannot be calculated analytically because it requires a complex integration over the model parameters (branch lengths, rates, base frequencies) and a summation over all possible trees to calculate Pr[*Data*] (the denominator of Equation 1). Thus phylogeneticists must use MCMC to approximate the PP (for a review of MCMC in phylogenetics, see Yang 2005). In MCMC, a so-called walk through tree/model space is simulated, and trees are periodically sampled. These samples approximate draws from the PP distribution. We can manipulate the MCMC samples to infer trees (and support for the trees).

Rise of Bayesian Phylogenetic Methods

The development of modern Bayesian applications in phylogenetics had multiple origins in the mid-1990s, with work by Rannala and Yang (Rannala & Yang 1996, Yang & Rannala 1997), Li (1996; Li et al. 2000), and Mau, Newton, Larget, and Simon (Larget & Simon 1999, Mau 1996, Mau & Newton 1997, Mau et al. 1999, Newton et al. 1999, Simon & Larget 2001). Yang (2005) compares these approaches in his recent review. The rapid acceptance of Bayesian phylogenetics can be explained by several factors: the availability of user-friendly software implementations, MRBAYES (Huelsenbeck & Ronquist 2001, Ronquist & Huelsenbeck 2003) and BAMBE (Simon & Larget 2001); the incorporation of complex models of character evolution; the perceived computational advantages of Bayesian MCMC over maximum likelihood; the capacity to accommodate uncertainty in model parameter values, topologies, and ancestral state reconstructions (e.g., Lutzoni et al. 2001, Pagel & Lutzoni 2002); and the ability to simultaneously infer trees and support for clades.

BAYESIAN POSTERIOR PROBABILITIES AS MEASURES OF CLADE SUPPORT

Although PP statements can be made about any parameter of the model, systematists are usually interested in support for trees or clades. In current practice, workers decide on models of evolution (see review in Sullivan & Joyce 2005), specify prior distributions, perform an MCMC run, and produce a majority-rule consensus tree of all the sampled topologies. The numbers on the branches of the consensus tree (the proportion of the MCMC samples that contain a particular split) approximate the PP of the split. Because all the MCMC samples are used in calculating the PP, support for some clades may come from trees that are poorly supported. The MCMC samples can also be summarized using tree-based approaches, which find the 95% credible set of trees and summarize the topological similarities of the trees within this set (Holmes 2003, 2005).

Early developers of Bayesian phylogenetic methods (Larget & Simon 1999, Rannala & Yang 1996, Yang & Rannala 1997) noted that the PP of a clade was often higher than the associated bootstrap proportion (BP). Subsequent empirical studies confirmed this as a general phenomenon, and the first major controversy of Bayesian phylogenetics emerged (Buckley et al. 2002, Karol et al. 2001, Leache & Reeder 2002, Miller & Andrew 2005, Miller et al. 2002, Murphy et al. 2001, Streelman et al. 2002, Whittingham et al. 2002, Wilcox et al. 2002). On the heels of these studies came a small flurry of simulations and theoretical papers that sought to explain the statistical behavior of the PP, especially in relation to the BP (Alfaro et al. 2003, Cummings et al. 2003, Douady et al. 2003, Erixon et al. 2003, Miller & Andrew 2005, Simmons et al. 2004, Suzuki et al. 2002). Some authors suggested the BP was too conservative and argued for the PP as the preferred support measure (Murphy et al. 2001, Wilcox et al. 2002). They noted that PPs of 95% were generally associated with bootstrap values of 70% or greater and suggested this result was in agreement with earlier work arguing that BPs of 70% actually indicated confidence levels of 95%

(Hillis & Bull 1993; but see Efron et al. 1996, Felsenstein & Kishino 1993, Newton 1996). Others (e.g., Buckley et al. 2002) suggested that a less-conservative PP might be more susceptible to model misspecification than the bootstrap. Consensus has yet to emerge on the merit of using the PP as a support measure in phylogenetics. In part this is because the results of these studies appear to be contradictory. Some of the controversy arises from the unfortunate notion that Bayesian methods should be judged by their frequentist behavior (Huelsenbeck & Rannala 2004).

Frequentist and Bayesian Statistics

Most biologists have been trained in frequentist statistical approaches and are relatively unfamiliar with Bayesian methods. Fundamentally, the approaches differ in their treatment of parameters: The frequentist approach has fixed but unknown values, whereas the Bayesian approach has random variables. Hypothesis testing and model-selection strategies also differ between these frameworks (e.g., Gelman et al. 2000; for phylogenetic applications, see Bollback & Huelsenbeck 2001, Huelsenbeck et al. 2001). In frequentist statistics, the decision to accept or reject a null hypothesis involves ascertaining whether an observed summary of the data (a test statistic) falls within the tails of the null distribution. In contrast, in Bayesian statistics, probabilities are assigned directly to hypotheses themselves, and no hypothesis is designated the null. Both Bayesian and frequentist approaches are defensible, but because their underlying approaches differ, tests in different frameworks may sometimes produce different results (Holmes 2005).

Studied Properties of the Posterior Probability of a Clade

Simulation studies and real-data analyses have been used to study several properties of the PP of a clade. Two possible interpretations of a support probability for an observed clade are apparent:

1. The Bayesian perspective is the probability that the clade is present on the true tree.
2. The frequentist perspective is 1.0 minus the p-value. If we take the null hypothesis to be that any particular clade is not in the tree, then the p-value is the probability a clade would be supported at least as strongly by the data even though the clade is not present in the true tree.

Despite the relevance of the former (Bayesian) interpretation to PPs, studies in theoretical systematics have evaluated how well the posterior from a Bayesian MCMC serves to estimate both of these properties (see **Figure 1**). We discuss several properties of the PP that have been studied in theoretical papers below.

Probability that the clade is correct. In simulation studies many data sets can be produced, and the true tree is known. Thus one can test how well the PP from a Bayesian analysis estimates the actual probability that a clade is correct (Hillis & Bull 1993). Ninety-five percent of all clades assigned a PP of 0.95 should be found on

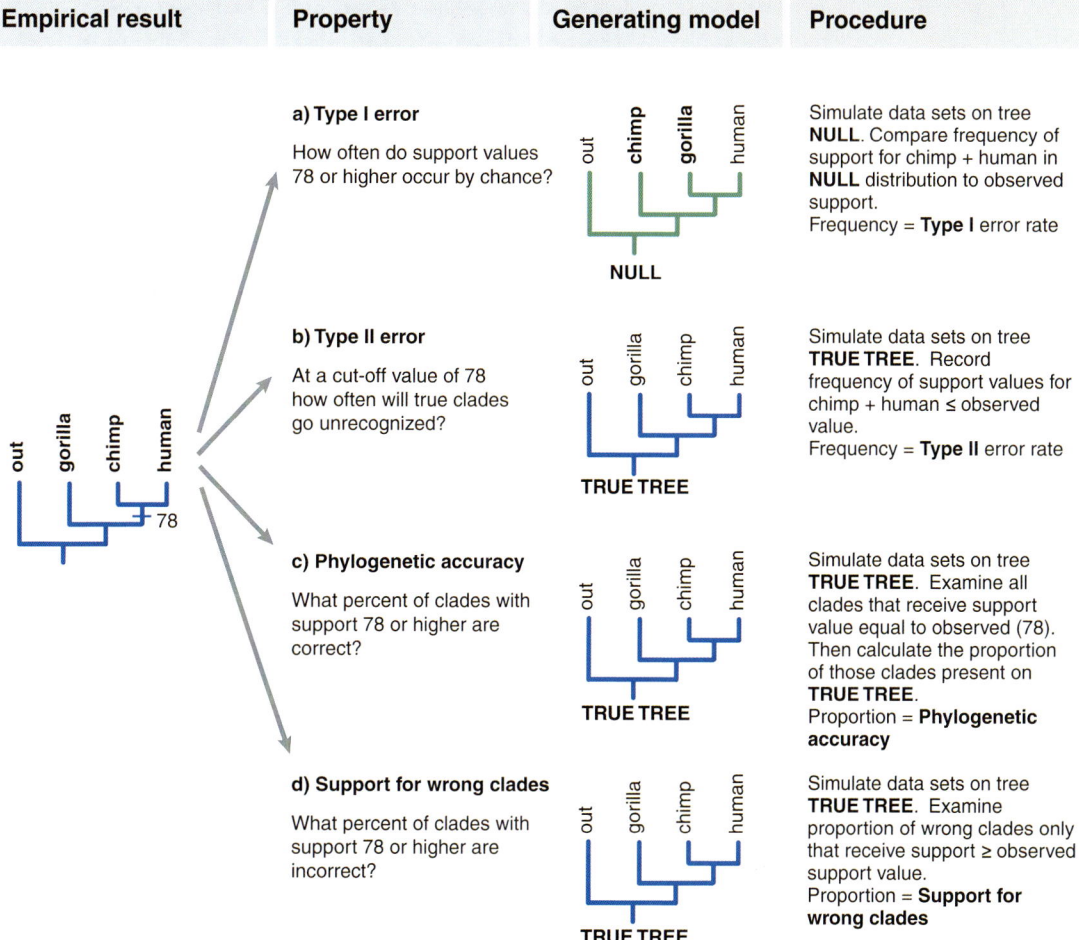

Figure 1

Four properties of phylogenetic support methods. The meaning of a 78% bootstrap proportion (BP) or posterior probability (PP) attached to the clade human + chimp depends on statistical context. (*a*) Type I error rate determination requires a null model in which the clade does not appear. (*b*) Type II error measures the rate of falsely rejecting human + chimp when it is true. (*c*) Phylogenetic accuracy (Hillis & Bull 1993) measures the probability that clades with a given support value are correct under generating conditions when the clade appears. (*d*) Support for wrong clades measures the strength of support assigned to clades that are not present under the generating conditions. Panels *a* and *b* are cornerstones of frequentist statistical testing. Panel *c* has become a common benchmark for phylogenetic confidence methods, although statements about accuracy are dependent on the generating conditions. Panel *d* is sometimes measured in studies comparing the BP with the PP.

the true tree. Theoretically, the PP is an ideal estimator, provided that the model is correct and the parameters for the simulation (including the tree shape) are drawn from the same probability distribution that is used as a prior. However, because PPs in phylogenetics are estimated with MCMC, inaccurate PPs may be obtained if the chain fails to reach the stationary distribution. This has not been shown to be a problem with real phylogenetic data. Simulation studies under relatively simple conditions confirm that MCMC performs well (Alfaro & Huelsenbeck 2006, Huelsenbeck & Rannala 2004). When the model is known and the prior is taken seriously, the PP perfectly matches phylogenetic accuracy (Huelsenbeck & Rannala 2004). Several simulation studies have compared phylogenetic accuracy to the PP using a fixed tree. Given the true model and a fairly easy tree to estimate, the PP is a conservative measure of phylogenetic accuracy, although it is less conservative than the nonparametric BP (Alfaro et al. 2003, Wilcox et al. 2002).

A number of studies have demonstrated that PP is sensitive to violations to the model (Buckley 2002, Douady et al. 2003, Erixon et al. 2003, Huelsenbeck & Rannala 2004, Waddell et al. 2002). Interestingly there seems to be little downside to slight overparameterization of the model, although the expectation is that all the parameters should have increased variance (Cunningham et al. 1998). In contrast, underparameterization can lead to greatly inflated estimates of the PP (Erixon et al. 2003, Huelsenbeck & Rannala 2004, Lemmon & Moriarty 2004, Suzuki et al. 2002). In light of these results, it appears prudent to use sufficiently complex models in performing Bayesian MCMC and err on the side of complexity rather than simplicity. The results about the relatively minor ill effects of overparameterization come from analyses in which the analysis model is slightly too parameter-rich; these conclusions do not imply that one should avoid model-selection techniques and always prefer the most-complex model available.

Posterior probability interpreted as a function of the p-value. Assessing the probability of a Type I error (i.e., rejecting the null hypothesis when it is true) forms the cornerstone of most frequentist statistical tests. Phylogenetic analyses fit naturally into the estimation or model-selection frameworks, whereas p-values are integral to hypothesis testing. To convert clade support statements into a hypothesis test, one must choose a threshold (critical value); support higher than the threshold is taken to mean the rejection of the null hypothesis that the tree lacks the clade. In phylogenetics, assigning a p-value of 0.05 to a branch should mean that the corresponding clade is only seen with greater support from the data in 5% of data sets drawn from trees that do not contain the clade (the null model).

It may seem intuitive to treat the one minus the PP of a clade as a p-value. However, practitioners working with real data generally do not interpret PPs (or bootstrap values for that matter) attached to nodes on a phylogenetic tree as indications of Type I error. Nevertheless, theoretical studies that generate data from star trees and focus only on the incidence of strongly supported clades are implicitly examining how well Bayesian procedures produce p-values. Frequentist p-values are usually calculated under the least-favorable conditions—i.e., under a null model that is most likely to lead to a false rejection. In testing the significance of a clade, for example, the null hypothesis is the maximum likelihood tree constrained to not have that clade. Thus frequentist p-values are defined with respect to Type I error and lead to conservative inference because they are based on a worst-case scenario. In contrast, Bayesian methodology does not place a premium on controlling Type I error. Thus, unsurprisingly, converting Bayesian PPs into a frequentist test can result in inflated Type I error rates (Suzuki et al. 2002). When the model of evolution is severely misspecified, the PP leads to dramatically high rates of Type I error (Buckley 2002, Suzuki et al. 2002). Type II error (failing to reject an incorrect hypothesis) for a given cut-off value appears to be lower for the PP than for the BP under simple conditions (Alfaro et al. 2003).

Predictor of the best empirical estimate. If we examine the behavior of methods on real data, we do not have the luxury of knowing the true tree. One can use the best tree from an empirical study as a proxy for the true tree. Tree inference procedures can be studied by subsampling taxa and/or characters from the original matrix, analyzing these subsets, and comparing the results with the analysis of the full data set (Simmons et al. 2004, Taylor & Piel 2004). Because the true generating model is unknown, one interpretation of these studies is that they show the effects of biologically relevant model misspecification.

Testing clade support based on a data subset approximates the probability that the clade will be found in the full data analysis and gives systematists a sense of whether they can interpret a support measure as "the probability of the clade being found if we collected a much larger data set." Bootstrapping or jackknifing significantly underestimates this probability, whereas the Bayesian PP significantly overestimates it (Simmons et al. 2004, Taylor & Piel 2004). If the conditions in these studies are phylogenetically representative, these results suggest a clade that receives an 85% bootstrap or jackknife value is more likely to appear after more-extensive data collection than one that receives an 85% PP. Strongly supported nodes (PP > 95%), however, appear to be only slightly inflated (Taylor & Piel 2004)—they do not appear to be as extremely overestimated as some simulation studies (Suzuki et al. 2002) indicate.

Correlation with the bootstrap proportion. Some studies have simply quantified the correlation coefficient of the PP and the BP, typically calculated across all clades in an analysis. The PP shows moderate correlation with the nonparametric BP, although the strength of the relationship can vary considerably depending on the simulation conditions (Alfaro et al. 2003, Cummings et al. 2003, Douady et al. 2003). Importantly, even in cases where the two measures differ substantially, a staggeringly large number of splits will have received support equal to zero. Thus the overall correlation between the PP and BP is generally high.

The Posterior Probability and the Bootstrap Proportion

The nonparametric bootstrap has a long and controversial history in phylogenetics (Berry & Gascuel 1996; Efron et al. 1996; Felsenstein 1985; Felsenstein & Kishino

1993; Hillis & Bull 1993; Li & Zharkikh 1994; Newton 1996; Sanderson 1995; Sanderson & Wojciechowski 2000; Zharkikh & Li 1992, 1995). Despite the controversy surrounding the meaning and interpretation of the BP, the procedure is easy to describe: Pseudoreplicate data matrices are created from the original data by sampling the columns of site patterns, and the optimal tree for the pseudoreplicate is found. The process is repeated hundreds of times, and support for a clade is expressed as the proportion of times the clade appeared in the pseudoreplicates (**Figure 2***a*).

Several papers have suggested there are theoretical reasons to expect the PP and the BP to be the same (Cummings et al. 2003, Erixon et al. 2003, Randle et al. 2005). A key difference between the methods, however, is that bootstrapping generates new data sets (by subsampling the original matrix), whereas Bayesian MCMC treats the data matrix and relies on a model to account for sampling variability. Given the similarities between the methods, we predict maximum likelihood and Bayesian analysis of any given bootstrap pseudoreplicate to be strongly congruent. In fact, when PPs are calculated for the bootstrap pseudoreplicates rather than the original matrix, the BP and the averaged PP are similar (Erixon et al. 2003, Waddell et al. 2002).

In a highly cited paper, Efron et al. (1996) state that the BP can be interpreted as an approximate PP if a noninformative prior is used on site patterns. Although we do not dispute this interpretation, this unconstrained model of site patterns differs substantially from the typical Bayesian phylogenetic models. Therefore, Efron et al.'s (1996) statement does not imply that the BP and the PP obtained from a MRBAYES run should be the same (see also Efron 2003). In the following section we elaborate on the differences between the Bayesian perspective on the BP and a typical Bayesian phylogenetic analysis.

Figure 2

Three perspectives of the bootstrap. (*a*) Typically, bootstrapping is performed by sampling columns of the data matrix (site patterns), finding the optimal tree for the pseudoreplicates, and recording the proportion of times a clade of interest appears. (*b*) The multinomial bootstrap mimics the resampling procedure in panel *a* by parameterizing the frequency of occurrence of the site patterns. Data sets can be simulated by drawing site patterns from this model, producing results that are identical to sampling with replacement from the original data matrix. (*c*) A nonhypothesis testing form of the parametric bootstrap would involve fitting the best tree, instead of a null or constraint tree as in the SOWH test (Goldman et al. 2000), and model of sequence evolution to the original data. New data sets can then be simulated under this model and analyzed, and the presence or absence of clades can be assessed. Note that procedures in panels *b* and *c* may yield different bootstrap proportions (BPs). Although they are both parametric procedures, the underlying models are quite different. Bayesian Markov chain Monte Carlo (MCMC) makes use of models that are similar to those used in parametric bootstrapping with the addition of prior probabilities on model parameters. Thus the Bayesian bootstrap posterior probability (PP) and the Bayesian MCMC PP may also differ owing to the difference in the underlying models (Svennblad et al. 2006).

a Bootstrap

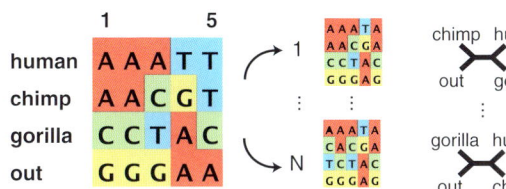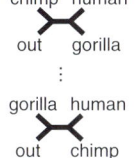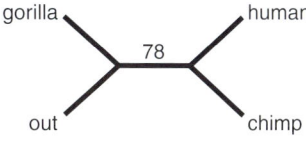

1 Resample from original matrix   **2** Optimize   **3** Calculate support

b Multinomial bootstrap

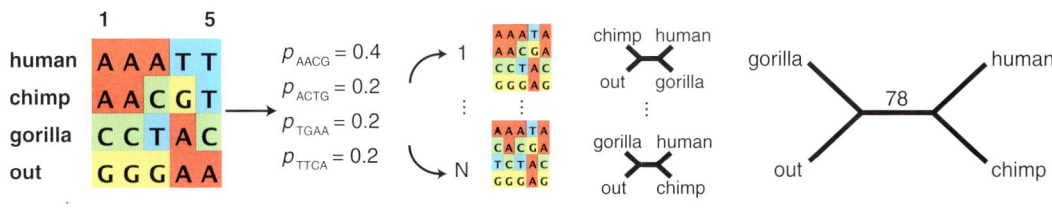

1 Estimate frequency of data site patterns   **2** Simulate   **3** Optimize   **4** Calculate support

c Phylogenetic parametric bootstrap

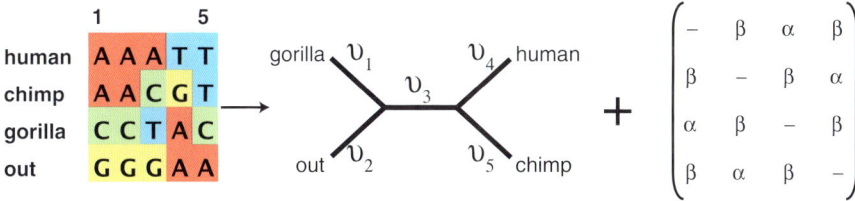

1 Fit probabilistic model to original data

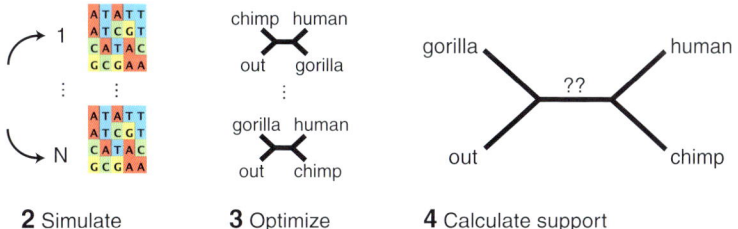

2 Simulate   **3** Optimize   **4** Calculate support

The Parameters of the Nonparametric Bootstrap

Parametric bootstrapping involves estimating parameters of a model from the data, then generating data sets according to this fitted model. The typical description of nonparameteric bootstrapping (i.e., resampling the original data with replacement data) sounds like a different procedure. It is possible to describe the nonparametric bootstrap as a procedure of estimating parameters from the original data under a peculiar model, then generating new data from this model (and parameters)—this is exactly the same strategy taken in parametric bootstrapping. The peculiar model is the unconstrained model of Goldman (1993). No tree is assumed in this model; instead one simply estimates the frequencies of every site pattern from the original matrix.

The unconstrained model gets its name from the fact that it enforces no relationships between the frequencies of different character types. In essence, the only assumptions made are that the sites are independent draws from the same underlying distribution—the multinomial distribution with expected character frequencies exactly matching the proportions seen in the observed data (**Figure 2b**) (Holmes 2003, 2005). Simulating from this multinomial distribution is identical to resampling the original data with replacement data because the probability of a site pattern generated at any position in the matrix is simply the proportion of sites in the original matrix that displayed that pattern. Describing the nonparametric bootstrapping as fitting the unconstrained model and then sampling from it may seem awkward and unnecessarily formal, but this description is the key to understanding Efron et al.'s (1996) Bayesian view of the BP.

A Bayesian View of the Nonparametric Bootstrap

Efron et al.'s (1996) Bayesian interpretation of the nonparametric bootstrap is framed in the context of estimating the means of normal distributions. However, one can apply the argument to phylogenetics using the unconstrained model with uninformative prior probability on the parameters. (Efron et al., however, do not explicitly describe this prior; see Svennblad et al. 2006 for possible implementations).

Phylogenies cannot be inferred using the unconstrained model because trees are not parameters of the model—the only parameters are the expected frequencies of each site pattern. The link between phylogenetic inference and this unconstrained model is made by associating every point in the space of pattern frequencies with a tree. Every possible data matrix can be represented as a vector of site pattern frequencies (where elements indicate the proportion of columns from the original matrix that shows a particular site pattern). If we view the elements of the vector as coordinates in multidimensional space, then each data set maps to a point in the space of possible pattern frequencies (Kim 2000). The choice of an optimality criterion (e.g., specification of a maximum likelihood model) induces a partition of this space. Each tree corresponds to a region of the parameter space. Any data set that displays pattern frequencies within this region will favor the tree. Thus, the data matrix that a researcher collects can be summarized as a location in the space of possible pattern frequencies, and tree searching is an attempt to determine which tree's region contains the observed data location.

Of course, the presence of sampling error implies that the observed pattern frequencies will deviate from the so-called true pattern frequencies (i.e., those one would obtained from infinitely long sequences). Assuming just the unconstrained model and taking the partitioning of pattern frequency space as a given, a Bayesian assessment of confidence attempts to estimate the probability that the true pattern frequencies are within a particular tree's region of parameter space. Efron et al.'s (1996) Bayesian interpretation of the nonparametric BP is then the probability that the true pattern frequency falls within the region of parameter space that would favor a given tree. Thus, the nonparametric BP can be considered an approximation to a PP, although only when using a multinomial distribution over site pattern frequencies.

The Parametric Bootstrap

In the phylogenetic parametric bootstrap, the parameters are not site pattern frequencies but instead are components of a probabilistic model of sequence evolution—substitution rate parameters, branch lengths, and base frequencies. In the nonparametric bootstrap, new data sets are generated using the multinomial distribution centered at the observed data. In the parametric bootstrap, new data sets are generated from a phylogenetic model that has been displaced from the observed pattern frequencies to the closest point on a "model manifold," to use Kim's (2000) terminology. Note that closeness here is measured by the Kullback-Leibler distance, which is a form of likelihood ratio between a phylogenetic model and the unconstrained model (**Figure 2c**). In the hypothesis testing form of parametric bootstrapping [e.g., the SOWH test (Goldman et al. 2000)], the pattern frequencies are moved to the closest model manifold for a tree that does not have the clade of interest. Data are simulated using a multinomial distribution (as in the nonparametric bootstrap), but the parameters of the multinomial (the probabilities assigned to each site pattern) are not free parameters—they are determined by the phylogenetic model. Given these differences, we do not expect parametric bootstrap and multinomial bootstrap support calculated from the same data to always correspond. Although they are both bootstrap procedures, the underlying sampling distributions are not the same. A similar case can be made for an expectation of differences between the nonparametric bootstrap and Bayesian MCMC, even though both may be viewed as PPs (see below).

Contrasting the Bayesian Interpretation of the Bootstrap with Markov Chain Monte Carlo

Although one could interpret the nonparametric BP as a PP derived from a unconstrained model (as discussed above), this PP should not always be equal to the PP from Bayesian MCMC. This is because Bayesian MCMC more closely resembles the parametric bootstrap in that both procedures fully exploit a model of character evolution as the estimate support. Thus, even with a Bayesian interpretation of the nonparametric BP, we should not expect the BP and the Bayesian MCMC PP to always be the same because the underlying models are quite different. Svennblad et al. (2006) provide a more in-depth, analytical discussion.

Summary of the Discussion of Posterior Probabilities

The proliferation of Bayesian studies demands that phylogeneticists carefully consider the statistical meaning of the posterior in Bayesian and frequentist contexts. A Bayesian perspective provides the natural framework for interpreting the PP as the probability that the clade is correct given the prior, the data, and the model of evolution. We believe this view has utility in phylogenetic systematics, and PPs are a natural and useful way for comparing alternative hypotheses. However, even within the Bayesian framework, phylogeneticists must be aware of several factors that can influence the PP. Most important among these is the sensitivity of the PP to model misspecification. Assessments of model adequacy (Bollback 2002) and model averaging procedures (Huelsenbeck et al. 2004, Lewis et al. 2005) should accompany Bayesian analyses so that model-sensitive inferences can be identified. Fortunately, the flexibility of the Bayesian framework and MCMC computations allows complex models to be implemented in a fairly straightforward fashion (Castoe et al. 2004, Nylander et al. 2004).

It is unwise to automatically use the PP as a guide to rates of Type I error because the Bayesian procedures have not been constructed with frequentist performance in mind (though it is theoretically possible to define Bayesian models and priors that give good frequentist performance). Similarly, one should not expect PPs to correspond well with nonparametric BPs any more than one would expect the p-values from a parametric and a nonparametric test to be the same (Svennblad et al. 2006).

ISSUES SURROUNDING THE PRIOR IN BAYESIAN PHYLOGENETICS

Prior probability distributions (referred to as priors hereafter) are often a main focus of debates between Bayesians and advocates of other forms of statistical inference. The use of explicit prior distributions is unique to the Bayesian perspective, and the elicitation of priors is clearly a difficult and subjective step in an analysis. Even within the Bayesian school, practitioners disagree about the specification of priors when an obvious, uncontroversial prior is not available.

The Subjective Bayesian Approach

In subjective Bayesian analysis, prior probabilities are assigned according to the researcher's degree of belief in different hypotheses (or parameter values). Advocates of subjective Bayesianism (de Finetti 1974, 1975; Goldstein 2006; Lindley 2000) usually cite the coherence of the methodology as one of its strengths. If the priors accurately reflect one's initial beliefs, and the model is good, then the PP distribution is the ideal guide to the researcher: Conclusions based on the posterior will be neither too conservative nor too liberal. The primary objection to the subjective Bayesian approach is that, although analyses to update personal beliefs may be useful in many situations, the scientific process demands objective statements about the data and hypotheses. Explicitly incorporating one's personal beliefs makes the analysis unnecessarily limited to the scientist publishing the data and makes the results possibly unconvincing to other researchers.

Objective Bayesian Analyses

Objective Bayesian analysis has been a major focus of statistics in the past few decades. Simply put, the goals of an objective Bayesian analysis are to allow the data (via the likelihood) to dominate the conclusions that are drawn (for a readable review, see Berger 2004). Objective Bayesian priors are often called noninformative priors; however, minimally informative may be a more-accurate description (i.e., all probability distributions are informative about some hypotheses). Priors are chosen to have a minimal effect on the parameter of interest, and interpreting a prior as a statement of one's beliefs is shunned (Bernardo 1997).

Objective Bayesian priors (e.g., Bernardo 1979) are not always uniform (flat) priors; instead they take the form of the model's likelihood function into account to allow the data to dominate the analysis. Objective Bayesian priors have a number of desirable properties: Many avoid problems that arise from priors that are sensitive to arbitrary choices (such as how to parameterize a model), and they often result in analyses that behave well from a frequentist viewpoint (Bayarri & Berger 2004). Unfortunately, they are not used in phylogenetics because the likelihood function for character data on trees cannot be expressed in a tractable form. Instead, priors in Bayesian phylogenetics are usually chosen to be vague over a wide range of parameter values, in the hope that such priors will have minimal effect on the results of the analysis. Some refer to this kind of approach as pseudo-Bayes because the priors are neither truly subjective nor objective (Berger 2004). Despite the lack of a formal justification, many Bayesians regard it as an effective way of analyzing real data (Berger 2004).

Overview of the Use of Priors in Bayesian Phylogenetics

All calculations in a Bayesian analysis are performed using the joint probability density of all parameters. Systematists are usually interested in tree topologies, but prior probabilities must be specified for every aspect of the model. In practice, systematists specify the joint prior probability density by multiplying the prior densities for several groups of parameters. This is equivalent to treating these parameter groups as if they were independent of each other. When parameters are tightly correlated (such as the proportion of invariants sites and the parameter governing the amount of rate heterogeneity among varying sites), the assumption of independence is probably unwarranted, but is still generally made.

Priors on the Model of Sequence Evolution

In many cases, priors on parameters of the sequence evolution model may be of less concern than tree and edge-length priors. Sequence evolution parameters are estimated from all sites in the data matrix and thus tend to have sharp likelihood profiles and are relatively insensitive to the choice of prior. Yang et al. (1995) noted that, in the context of maximum likelihood, the parameters of the sequence evolution model were often similar from tree to tree. If this general pattern holds, then even poorly chosen priors on the parameters of the sequence model may have little effect on tree inference because all trees may be affected in similar ways.

Conversely, increases in model complexity should lead to a flattening of likelihood profiles because the data are used to estimate more parameters. If this happens, prior influence may become a concern, especially in regions of parameter space where models are relatively insensitive to parameter change (Felsenstein 2004, Rannala 2002, Zwickl & Holder 2004).

Priors on the branching process. Trees represent the results of real biological processes (e.g., replication, anagenesis, cladogenesis); thus it seems natural to model the process of forming trees itself—to place a prior on different tree shapes according to models of how trees are generated. As Yang et al. (1995) note, probability distributions over tree shapes were employed in some of the earliest attempts to use likelihood techniques to reconstruct genealogies (Edwards & Cavalli-Sforza 1963, Thompson 1975). The Yule process models tree formation by repeated division (speciation) without lineage extinction, whereas the birth-death process includes lineage extinction. Priors on both processes do not place uniform probability on all trees, but instead place them on all labeled histories. Historically, birth-death priors have been used in Bayesian phylogenetics (Rannala & Yang 1996), although nearly all recent analyses use a uniform prior on topologies. The effect of choosing different tree priors on inference remains uninvestigated.

Informative topological priors. At the outset of any new investigation, most systematists have some prior expectation for tree shapes because few groups remain completely unstudied in terms of their evolutionary history. Therefore, a default uniform prior on topologies is usually not appropriate. Effectively, the failure to accommodate relevant prior information may produce a conservative analysis, especially if the tree agrees with the previous estimates (the reported clade support will probably be lower than an estimate using an informative prior).

Instead of ignoring prior information, one could attempt to include it in the analysis. One approach is to use a super matrix, wherein all previous data are included and reanalyzed. Alternatively, one could use measures of clade support from previous studies as the topological prior in the analysis of new data. Ideally, one would use the joint PP distribution of trees and all parameters from the first analysis directly as the prior for the subsequent analysis. Unfortunately, this is not practical for two reasons: First, the computer memory required to represent this distribution is generally prohibitive, and second, sufficiently accurate estimates of the probability density are not available. MCMC sampling may provide reasonable estimates of quantities such as clade PPs, but a detailed view of the entire parameter space of trees and models of character evolution is beyond our grasp. Ronquist et al. (2004) have pioneered an approach to accommodate measures of clade support from one analysis into a second analysis. From a tree with estimates of branch support (ideally PPs of the clades), they propose two sets of factors that can be used to approximate the entire joint distribution on trees. These approximations appear to capture the relative probabilities of trees that are good (close to optimal). This implies that if the final analysis recovers a tree similar to previous results, then the approximate topology prior has probably mimicked the ideal prior—the posterior distribution of trees from the previous analysis. One outstanding problem is how to deal with partial overlap of characters among the input trees (if some of them are based, in part, on the same sequence data). Downweighting the trees to account for redundant use of data is in order, but it is difficult to give precise guidelines on how such weighting should be done (Ronquist et al. 2004)

Clade Priors and Tree Priors

The prior probabilities of clades are usually not specified as independent inputs to the Bayesian inference. In practice, a prior on different topologies is specified, and this prior induces a prior distribution on splits. The most common prior in real analysis is a uniform prior over all topologies or over all labeled histories. A broader class of priors on tree topology includes label-invariant distributions (Steel & Penny 1993, Steel & Pickett 2006). In this class of distributions, the tree shape alone determines the prior probability on a topology. Any prior that seeks to be uninformative about the relationships among a group of taxa belongs to the class of label-invariant priors. Steel & Pickett (2006) prove it is impossible for any leaf-invariant prior to imply a uniform prior on all possible clades (assuming that analysis includes over four taxa for rooted inference or over five taxa for unrooted trees). This property arises from the compatibility of clades of differing sizes with differing numbers of topologies.

Pickett and Randle clearly view this property as a flaw in Bayesian phylogenetic analysis and as a reason to distrust clade PP: "[W]hen all trees are treated as equally probable, a priori, Bayesian clade support distributions are affected by priors that fail to model ignorance accurately" (Picket & Randle 2005, Randle & Pickett 2006). However, unequal priors on clades of different sizes do not imply a problem with clade PPs—if the model and prior are correct, the PP will accurately reflect the probability that both exist in the true tree. The unequal prior arises from an intrinsic relationship between clades and trees: Smaller clades are compatible with more trees than larger clades. Thus the phenomenon reflects a truth about probability statements on bifurcating trees (and not about Bayesian analyses, per se). As an analogy, let us consider the problem of estimating the number of beans in a jar. We could assign a prior probability to every possible value, but it would not be reasonable to expect this prior to assign equal probabilities to hypotheses such as "the number of beans is divisible by two" and "the number is divisible by three." Although it may seem desirable to express a priori ignorance about every conceivable hypothesis, it is not possible to do this and make coherent probability statements. For large trees, all clades have a low prior, and empirically, only modest amounts of data appear necessary to result in high PPs, even for clades that are assigned relatively low prior probabilities (Brandley et al. 2006).

Priors on Branches

Most Bayesian phylogenetic analyses follow the tradition of maximum likelihood methods and treat each branch length as an independent parameter. In practice, a single prior distribution is applied to each branch in the tree (implicitly assuming independence of branch lengths). A disadvantage of treating each branch length as an independent parameter is that it loses some information relevant to the phylogenetic inference. For example, assuming that the sampled tips are contemporaneous and the rate of evolution does not vary wildly over short timescales, branch lengths should be close to ultrametric. Models of tree generation can be used to specify that prior distributions are branch lengths that conform to the molecular clock hypothesis.

Suchard et al. (2001) employed a hierarchical approach to the assignment of priors on branch lengths that partially accommodates the information implied by contemporaneous taxonomic sampling. In their implementation, the prior probability for a particular edge had an exponential distribution. However, instead of specifying a fixed value for the mean of this exponential distribution, they treat the mean of the branch length prior as a type of parameter (a hyperparameter). A hyperprior distribution is assigned to the mean of the branch length prior, producing a model that is hierarchical—some parameters only affect predictions about the data that will be seen (and hence affect the likelihood), whereas hyperparameters only affect the probability density of other parameters. In the context of a joint prior over branch lengths, this hierarchical approach effectively expresses the intuition that the branch lengths will be roughly the same scale as each other.

Most practitioners are interested in the topology of the tree returned by a phylogenetic analysis, not in the other values inferred for other aspects of the models of evolution used during the inference. Despite the fact that the topology is the focal parameter, the prior distributions placed on other nuisance parameters can affect the topology inferred or the strength of support for aspects of the tree (Felsenstein 2004, Yang 2005, Zwickl & Holder 2004). The prior distribution placed on branch lengths warrants particular scrutiny for several reasons: There is no unambiguous upper limit on branch lengths; the length of internal branches is correlated with the difference in likelihoods between different topologies; and the prior distribution applied to branches influences 2N-3 factors (the prior densities of each branch) that are multiplied to contribute to the prior probability density.

Branch lengths are usually expressed in terms of expected number of changes per site. In this parameterization, branch lengths are unbounded parameters. Clearly enormous branch lengths are not realistic, but it is difficult to justify any particular upper limit on the branch length. As Felsenstein (2004) points out, the likelihood does not go to zero as branch lengths become infinite. As all the branches in the tree become large, the likelihood approaches a number determined by the equilibrium state frequencies of the model. This likelihood is usually extremely small, but not zero. Thus, the posterior distribution may be particularly sensitive to the upper bound placed on prior used for the branch length. Of course, the prior on branch lengths does not need to be a distribution that requires the specification of an upper limit (e.g., the default branch length in version 3.1 of MRBAYES is an exponential distribution). The use of a prior that puts decreasing density on higher branch length is justifiable because the likelihood is more informative about short branch lengths—it requires less data to discriminate between branch lengths of 0.01 and 0.02 than to discriminate between lengths of 1.01 and 1.02.

The potential for the branch length prior to affect clade probabilities is apparent when one considers extreme priors. If a strong prior were used to force all internal branch lengths to be infinitesimal, then all trees would have similar likelihoods (similar to a star tree's likelihood). A prior emphasizing longer internal branch lengths may make Bayesian phylogenetic analyses too liberal (Yang & Rannala 2005). This causes concern in light of the tendency to use priors with unrealistically high means for branch lengths. Yang and Rannala give several suggestions for making Bayesian analyses more conservative, in particular using priors with small internal means.

Interestingly, Yang & Rannala (2005) also note that if one uses informative priors on topologies, one can reasonably make the branch length prior specifically tailored to the branch in question. If one wishes to express the prior belief that a clade is particularly unlikely, then one can also reasonably assume that if this is wrong (and the clade does exist), it is not separated by the other taxa by a long branch. (If it were, then the clade probably would have been recovered in previous analyses.)

Inferring Trees from Star Trees

Interestingly, data sets simulated from an unresolved model tree occasionally show strong support of one resolution of the tree (Cummings et al. 2003, Lewis et al. 2005, Suzuki et al. 2002). The effect can be particularly pronounced if the model of character evolution used to generate the data is more complex than the model used for inference (Suzuki et al. 2002). Simulations from star trees can be used to detect a bias in a method, such as long-branch attraction (Bruno & Halpern 1999). However, in some cases (Cummings et al. 2003, Lewis et al. 2005, Suzuki et al. 2002), the effect occurs even if external branches in the simulations are equal length and there is no bias for one tree over another. Given that Bayesian methods are consistent (if the assumed model is accurate and the prior does not rule out the true tree), one might expect the effect to decrease as more data are collected. Interestingly, it does not (Lewis et al. 2005). At first this result seems paradoxical. However, Bayesian theory guarantees the posterior will accurately reflect the uncertainty about a hypothesis that is true or false (eventually leading to complete confidence in or rejection of the hypothesis as more data is collected) (Cummings et al. 2003, Lewis et al. 2005). Theory does not guarantee that conditions on the border of multiple hypotheses (multiple trees in this case) will return equal support for all bordering hypotheses, even as the amount of data increases.

On one level, these results are not too troubling. First, we do not expect true hard polytomies to occur frequently. Even small internal branches can be recovered if there is sufficient data, so the spurious support given to essentially infinite data is confined to hard polytomies. Second, the percentage of data sets showing strong support for a resolution is not high (unless the model of character evolution is also misspecified). Third, most researchers are accustomed to using moderate α-levels in frequentist statistics. If one views a polytomy as a null hypothesis, then it is not surprising that one occasionally rejects a null inappropriately. We do not advocate interpreting Bayesian PPs in these frequentist terms. Instead, we merely want to point out that most practitioners accept some false positives from a statistical method.

Conversely, returning strong support for nonexistent clades is clearly an undesirable property. Even if false resolutions only occasionally receive strong support, a large tree could contain multiple spurious groupings.

Ideally, if data came from a polytomy, an inference method would return equal support for all possible resolutions. Because of sampling error, this ideal cannot be achieved in all cases. The frequencies of data patterns generated by a star tree are similar to those generated from a tree with a short internal branch. The frequencies are so similar that it is not feasible to discriminate between them, given a small sample of data. This fundamental reality means that if one were to engineer an analysis that avoids incorrectly resolved polytomies, there would have to be some loss of power (i.e., an inability to detect short branches when they exist).

Priors for More Conservative Bayesian Markov Chain Monte Carlo

Given that Bayesian methods (and other highly parametric approaches) tend to have high power, it seems appropriate to consider modifications of standard Bayesian approaches to make the results more conservative. Lewis et al. (2005) propose considering unresolved trees during the MCMC analysis and even using priors that prefer unresolved trees over more-resolved trees. Favoring polytomies may be done for the sake of conservatism, not realism. Thus, we may favor polytomies because errors that return unresolved trees are less troublesome than incorrectly supporting a false group. The consideration of polytomies during the MCMC requires a new set of proposals utilizing reversible-jump MCMC (Green 1995) to switch between models (trees in this case) that differ in the number of parameters they contain (for details, see Lewis et al. 2005). Simulations and real-data analyses indicate that considering polytomies during the MCMC does not dramatically affect the power of the analysis to detect short branches. Yang & Rannala (2005) discuss several approaches to using priors that favor short internal branches as a means of making Bayesian analysis. An ad hoc approach could also be used to mimic the behavior of a conservative Bayesian analysis (e.g., one could collapse branches that are shorter than some cut-off value before summarizing the trees sampled during the MCMC).

Summary of the Discussion of Priors

The requirement that practitioners of Bayesian methods must specify prior probability statements for every aspect of the model may seem daunting. Prior specification merits careful consideration, but for many aspects of the model, the data speak strongly. In these cases, a broad range of priors will lead to similar conclusions. Recent work suggests that the use of priors favoring short branch lengths, or polytomies, may prove to be an effective technique for allowing users to make Bayesian support statements conservative (without sacrificing too much power). Improving the feasibility of inference under complex models of character evolution will undoubtedly be one of the most important contributions of Bayesian phylogenetics to the field of systematics. Using sound priors with these parameter-rich models is particularly important because many possible parameter values will be compatible with the data.

Specifying priors that have no effect on the analysis is not an attainable goal. Thus if one wishes to utilize minimally informative priors, it is prudent to examine analyses using multiple combinations of priors to demonstrate that the priors are not having an unexpectedly strong effect on the conclusions.

CONCLUSIONS

Perhaps the greatest justification for the continued use and development of Bayesian methodology is the potential it offers as a model-based framework for analyzing multigene or genomic data sets. Bayesian methods are not, however, a panacea, and workers must familiarize themselves with the strengths and limitations of the approach when interpreting their phylogenetic analysis (reviewed in Huelsenbeck et al. 2002). Bayesian phylogenetics does not always produce statistics that have a good frequentist interpretation. Research into the development of probability distributions on tree space (Holmes 2005) has the potential to bridge this difference to some degree by improving the basis of all phylogenetic statistical tests. Beyond the issues we discuss here, additional work is also needed to understand the adequacy of current Bayesian MCMC samplers for yielding reliable PP under difficult tree conditions (Mossel & Vigoda 2005). Continued efforts in the areas of phylogenetic model development, model selection, and model adequacy are also necessary to ensure the adequate performance of Bayesian methods on the increasingly complex data sets of the future.

ACKNOWLEDGMENTS

Paul Lewis provided helpful discussions of several issues, in particular the use of hierarchical priors on edge lengths. Susan Holmes clarified several points regarding the Bayesian interpretation of the nonparametric bootstrap and statistical incoherency in phylogenetics. Devin Drown and Jessica Lynch Alfaro provided helpful comments on an earlier version of the manuscript. NSF DEB-0445453 (M.E.A.) provided funding for this project. M.T.H. was funded by a CIPRES grant (NSF EF 0331495 to David Swofford).

LITERATURE CITED

Alfaro ME, Huelsenbeck JP. 2006. Comparative performance of Bayesian and AIC-based measures of phylogenetic model uncertainty. *Syst. Biol.* 55:89–96

Alfaro ME, Zoller S, Lutzoni F. 2003. Bayes or bootstrap? A simulation study comparing the performance of Bayesian Markov chain Monte Carlo sampling and bootstrapping in assessing phylogenetic confidence. *Mol. Biol. Evol.* 20:255–66

Bayarri MJ, Berger JO. 2004. The interplay of Bayesian and frequentist analysis. *Stat. Sci.* 19:58–80

Berger JO. 2004. The case for objective Bayesian analysis. *Bayesian Anal.* 1:1–17

Bernardo JM. 1979. Reference posterior distributions for Bayesian inference. *J. R. Stat. Soc. Ser. B* 41:113–47

Bernardo JM. 1997. Non-informative priors do not exist: a dialogue with José M. Bernardo. *J. Stat. Plan. Inference* 65:159–89

Berry V, Gascuel O. 1996. On the interpretation of bootstrap trees: appropriate threshold of clade selection and induced gain. *Mol. Biol. Evol.* 13:999–1011

Bollback JP. 2002. Bayesian model adequacy and choice in phylogenetics. *Mol. Biol. Evol.* 19:1171–80

Bollback JP, Huelsenbeck JP. 2001. Phylogeny, genome evolution, and host specificity of single-stranded RNA bacteriophage (family Leviviridae). *J. Mol. Evol.* 52:117–28

Brandley MC, Leache AD, Warren DL, McGuire JA. 2006. Are unequal clade priors problematic for Bayesian phylogenetics? *Syst. Biol.* 55:138–46

Bruno WJ, Halpern AL. 1999. Topological bias and inconsistency of maximum likelihood using wrong models. *Mol. Biol. Evol.* 16:564–66

Buckley TR. 2002. Model misspecification and probabilistic tests of topology: evidence from empirical data sets. *Syst. Biol.* 51:509–23

Buckley TR, Arensburger P, Simon C, Chambers GK. 2002. Combined data, Bayesian phylogenetics, and the origin of the New Zealand cicada genera. *Syst. Biol.* 51:4–18

Castoe T, Doan T, Parkinson C. 2004. Data partitions and complex models in Bayesian analysis: the phylogeny of gymnophthalmid lizards. *Syst. Biol.* 53:448–69

Cummings MP, Handley SA, Myers DS, Reed DL, Rokas A, Winka K. 2003. Comparing bootstrap and posterior probability values in the four-taxon case. *Syst. Biol.* 52:477–87

Cunningham CW, Zhu H, Hillis DM. 1998. Best-fit maximum-likelihood models for phylogenetic inference: empirical tests with known phylogenies. *Evolution* 52:978–87

de Finetti B. 1974. *Theory of Probability*, Vol. 1. Chichester: Wiley

de Finetti B. 1975. *Theory of Probability*, Vol. 2. Chichester: Wiley

Douady CJ, Delsuc F, Boucher Y, Doolittle WF, Douzery EJP. 2003. Comparison of Bayesian and maximum likelihood bootstrap measures of phylogenetic reliability. *Mol. Biol. Evol.* 20:248–54

Edwards AWF, Cavalli-Sforza LL. 1963. The reconstruction of evolution. *Heredity* 18:553

Efron B. 2003. Second thoughts on the bootstrap. *Stat. Sci.* 18:135–40

Efron B, Halloran E, Holmes S. 1996. Bootstrap confidence levels for phylogenetic trees. *Proc. Natl. Acad. Sci. USA* 93:13429–34

Erixon P, Svennblad B, Britton T. 2003. Reliability of Bayesian posterior probabilities and bootstrap frequencies in phylogenetics. *Syst. Biol.* 52:665–73

Felsenstein J. 1985. Confidence limits on phylogenies: an approach using the bootstrap. *Evolution* 39:783–91

Felsenstein J. 2004. *Inferring Phylogenies*. Sunderland, MA: Sinauer

Felsenstein J, Kishino H. 1993. Is there something wrong with the bootstrap on phylogenies? A reply to Hillis and Bull. *Syst. Biol.* 42:193–200

Gascuel O, ed. 2005. *Mathematics of Evolution and Phylogeny*. Oxford, UK: Oxford Univ. Press

Gelman A, Carline JB, Stern HS, Rubin DB. 2000. *Bayesian Data Analysis*. Boca Raton, FL: Chapman & Hall. 526 pp.

Goldman N. 1993. Statistical tests of models of DNA substitution. *J. Mol. Evol.* 36:182–98

Goldman N, Anderson JP, Rodrigo AG. 2000. Likelihood-based tests of topologies in phylogenetics. *Syst. Biol.* 49:652–70

Goldstein M. 2006. Subjective Bayesian analysis: principles and practice. *Bayesian Anal.* 1(3):403–20

Green PJ. 1995. Reversible jump Markov chain Monte Carlo computation and Bayesian model determination. *Biometrika* 82:711–32

Hillis DM, Bull JJ. 1993. An empirical test of bootstrapping as a method for assessing confidence in phylogenetic analysis. *Syst. Biol.* 42:182–92

Holmes S. 2003. Bootstrapping phylogenetic trees: theory and methods. *Stat. Sci.* 18:241–55

Holmes S. 2005. Statistical approach to tests involving phylogenies. See Gascuel 2005, pp. 91–120

Huelsenbeck JP, Larget B, Alfaro ME. 2004. Bayesian phylogenetic model selection using reversible jump Markov chain Monte Carlo. *Mol. Biol. Evol.* 2004:1123–33

Huelsenbeck JP, Larget B, Miller RE, Ronquist F. 2002. Potantial applications and pitfalls of Bayesian inference of phylogeny. *Syst. Biol.* 51:673–88

Huelsenbeck JP, Rannala B. 2004. Frequentist properties of Bayesian posterior probabilities of phylogenetic trees under simple and complex substitution models. *Syst. Biol.* 53:904–13

Huelsenbeck JP, Ronquist F. 2001. MRBAYES: Bayesian inference of phylogenetic trees. *Bioinformatics* 17:754–55

Huelsenbeck JP, Ronquist F, Nielsen R, Bollback JP. 2001. Bayesian inference of phylogeny and its impact on evolutionary biology. *Science* 294:2310–14

Karol KG, McCourt RM, Cimino MT, Delwiche CF. 2001. The closest living relatives of land plants. *Science* 294:2351–53

Kim J. 2000. Slicing hyperdimensional oranges: the geometry of phylogenetic estimation. *Mol. Phylogenet. Evol.* 17:58–75

Larget B, Simon DL. 1999. Markov chain Monte Carlo algorithms for the Bayesian analysis of phylogenetic trees. *Mol. Biol. Evol.* 16:750–59

Leache AD, Reeder TW. 2002. Molecular systematics of the Eastern Fence Lizard (*Sceloporus undulatus*): a comparison of parsimony, likelihood, and Bayesian approaches. *Syst. Biol.* 51:44–68

Lemmon AR, Moriarty EC. 2004. The importance of proper model assumption in Bayesian phylogenetics. *Syst. Biol.* 53:265–77

Lewis PO. 2001. Phylogenetic systematics turns over a new leaf. *Trends Ecol. Evol.* 16:30–37

Lewis PO, Holder MT, Holsinger KE. 2005. Polytomies and Bayesian phylogenetic inference. *Syst. Biol.* 54:241–53

Li S, Pearl D, Doss H. 2000. Phylogenetic tree construction using Markov chain Monte Carlo. *J. Am. Stat. Assoc.* 95:493–508

Li W. 1996. *Phylogenetic tree construction using Markov chain Monte Carlo*. PhD thesis. Ohio State Univ., Columbus

Li W, Zharkikh A. 1994. What is the bootstrap technique? *Syst. Biol.* 43:424–30

Lindley DV. 2000. The philosophy of statistics. *Statistician* 49:293–337

Lutzoni F, Pagel M, Reeb V. 2001. Major fungal lineages are derived from lichen symbiotic ancestors. *Nature* 411:937–40

Mau B. 1996. *Bayesian phylogenetic inference via Markov chain Monte Carlo methods*. PhD thesis. Univ. Wis., Madison. 106 pp.

Mau B, Newton MA. 1997. Phylogenetic inference for binary data on dendrograms using Markov chain Monte Carlo. *J. Comput. Graph. Stat.* 6:122–31

Mau B, Newton MA, Larget B. 1999. Bayesian phylogenetic inference via Markov chain Monte Carlo methods. *Biometrika* 55:1–12

Miller AN, Andrew NMA. 2005. Multi-gene phylogenies indicate ascomal wall morphology is a better predictor of phylogenetic relationships than ascospore morphology in the sordariales (ascomycota, fungi). *Mol. Phylogenet. Evol.* 35:60–75

Miller RE, Buckley TR, Manos P. 2002. An examination of the monophyly of morning glory taxa using Bayesian phylogenetic inference. *Syst. Biol.* 51:740–53

Mossel E, Vigoda E. 2005. Phylogenetic MCMC algorithms are misleading on mixtures of trees. *Science* 309:2207–9

Murphy WJ, Eizirik E, O'Brien SJ, Madsen O, Scally M, et al. 2001. Resolution of the early placental mammal radiation using Bayesian phylogenetics. *Science* 294:2348–51

Newton MA. 1996. Bootstrapping phylogenies: large deviations and dispersion effects. *Biometrika* 83:315–28

Newton MA, Mau B, Larget B. 1999. Markov chain Monte Carlo for the Bayesian analysis of evolutionary trees from aligned molecular sequences. In *Statistics in Molecular Biology and Genetics*, ed. F Seiller-Moseiwitch, TP Speed, M Waterman, 33:143–62. Hayward, CA: Monogr. Ser. Inst. Math. Stat.

Nylander J, Ronquist F, Huelsenbeck JP. 2004. Bayesian phylogenetic analysis of combined data. *Syst. Biol.* 53:47–67

Pagel M, Lutzoni F. 2002. Accounting for phylogenetic uncertainty in comparative studies of evolution and adaptation. In *Biological Evolution and Statisitcal Physics*, ed. M Laessig, A Valleriani, pp. 151–64. Berlin: Springer-Verlag

Pickett KM, Randle CP. 2005. Strange Bayes indeed: uniform topological priors. *Mol. Phylogenet. Evol.* 34:203–11

Randle CP, Mort ME, Crawford DJ. 2005. Bayesian inference of phylogenetics revisited: developments and concerns. *Taxon* 54:9–15

Randle CP, Pickett KM. 2006. Are nonuniform clade priors important in Bayesian phylogenetic analysis? A response to Brandley et al. *Syst. Biol.* 55:147–50

Rannala B. 2002. Identifiability of parameters in MCMC Bayesian inference of phylogeny. *Syst. Biol.* 51:754–60

Rannala B, Yang Z. 1996. Probability distribution of molecular evolutionary trees: a new method of phylogenetic inference. *J. Mol. Evol.* 43:304–11

Ronquist F, Huelsenbeck JP. 2003. MRBAYES 3: Bayesian phylogenetic inference under mixed models. *Bioinformatics* 19:1572–74

Ronquist F, Huelsenbeck JP, Britton T. 2004. Bayesian supertrees. In *Phylogenetic Supertrees: Combining Information to Reveal the Tree of Life*, ed. ORP Bininda-Emonds, pp. 193–224. Amsterdam: Kluwer

Sanderson MJ. 1995. Objections to bootstrapping phylogenies: a critique. *Syst. Biol.* 44:299–320

Sanderson MJ, Wojciechowski MF. 2000. Improved bootstrap confidence limits in large-scale phylogenies, with an example from the Neo-Astragalus (Leguminosae). *Syst. Biol.* 49:671–85

Simmons M, Pickett K, Miya M. 2004. How meaningful are Bayesian support values? *Mol. Biol. Evol.* 21:188–99

Simon D, Larget B. 2000. Bayesian analysis in molecular biology and evolution (BAMBE), version 2.03 beta. Dept. Math. Comput. Sci., Duquesne Univ., Pittsburgh

Steel M, Pickett KM. 2006. On the impossibility of uniform priors on clades. *Mol. Phylogenet. Evol.* 39:585–86

Steel MA, Penny D. 1993. Distributions of tree comparison metrics: some new results. *Syst. Biol.* 42:126–41

Streelman JT, Alfaro ME, Westneat MW, Bellwood DR, Karl SA. 2002. Evolutionary history of the parrotfishes: biogeography, ecomorphology, and comparative diversity. *Evolution* 56:961–71

Suchard MA, Weiss RE, Sinsheimer JS. 2001. Bayesian selection of continuous-time Markov chain evolutionary models. *Mol. Biol. Evol.* 18:1001–13

Sullivan J, Joyce P. 2005. Model selection in phylogenetics. *Annu. Rev. Ecol. Evol. Syst.* 36:445–66

Suzuki Y, Glazko GV, Nei M. 2002. Overcredibility of molecular phylogenies obtained by Bayesian phylogenetics. *Proc. Natl. Acad. Sci. USA* 99:16138–43

Svennblad B, Erixon P, Oxelman B, Britton T. 2006. Fundamental differences between the methods of maximum likelihood and maximum posterior probability in phylogenetics. *Syst. Biol.* 55:116–21

Taylor D, Piel W. 2004. An assessment of accuracy, error, and conflict with support values from genome-scale phylogenetic data. *Mol. Biol. Evol.* 21:1534–37

Thompson EA. 1975. *Human Evolutionary Trees*. Cambridge, UK: Cambridge Univ. Press

Waddell PJ, Kishino H, Ota R. 2002. Very fast algorithms for evaluating the stability of ML and Bayesian phylogenetic trees from sequence data. *Genome Inform.* 13:82–92

Whittingham LA, Slikas B, Winkler DW, Sheldon FH. 2002. Phylogeny of the tree swallow genus, *Tachycineta* (Aves: Hirundinidae), by Bayesian analysis of mitochondrial DNA sequences. *Mol. Phylogenet. Evol.* 22:430–41

Wilcox T, Zwick D, Heath T, Hillis D. 2002. Phylogenetic relationships of the dwarf boas and a comparison of Bayesian and bootstrap measures of phylogenetic support. *Mol. Phylogenet. Evol.* 25:361–71

Yang Z. 2005. Bayesian inference in molecular phylogenetics. See Gascuel 2005, pp. 63–90

Yang Z, Goldman N, Friday A. 1995. Maximum likelihood tree from DNA sequences: a peculiar statistical estimation problem. *Syst. Biol.* 44: 384–99

Yang Z, Rannala B. 1997. Bayesian phylogenetic inference using DNA sequences: a Markov chain Monte Carlo method. *Mol. Biol. Evol.* 14:717–24

Yang Z, Rannala B. 2005. Branch-length prior influences Bayesian posterior probability of phylogeny. *Syst. Biol.* 54:455–70

Zharkikh A, Li W. 1992. Statistical properties of bootstrap estimation of phylogenetic variability from nucleotide sequences. I. Four taxa with a molecular clock. *Mol. Biol. Evol.* 9:1119–47

Zharkikh A, Li W. 1995. Estimation of confidence in phylogeny: the complete-and-partial bootstrap technique. *Mol. Phylogenet. Evol.* 4:44–63

Zwickl D, Holder MT. 2004. Model parameterization, prior distributions, and the general time-reversible model in Bayesian phylogenetics. *Syst. Biol.* 53:877–88

Unifying and Testing Models of Sexual Selection

Hanna Kokko,[1] Michael D. Jennions,[2] and Robert Brooks[3]

[1]Department of Biological and Environmental Science, University of Helsinki, FIN–00014 Helsinki, Finland; email: hanna.kokko@helsinki.fi

[2]School of Botany and Zoology, Australian National University, Canberra ACT 0200, Australia; email: michael.jennions@anu.edu.au

[3]School of Biological, Earth, and Environmental Sciences, University of New South Wales, Sydney NSW 2052, Australia; email: rob.brooks@unsw.edu.au

Key Words

anisogamy, female choice, mate choice, sex roles, sexual conflict

Abstract

Sexual reproduction is associated with the evolution of anisogamy and sperm-producing males and egg-laying females. The ensuing competition for mates has led to sexual selection and coevolution of the sexes. Mathematical models are extensively used to test the plausibility of different complicated scenarios for the evolution of sexual traits. Unfortunately, the diversity of models is now itself equally bewildering. Here we clarify some of the current debate by reviewing evolutionary explanations for the relationship between anisogamy, potential reproductive rates, parental care, sex roles, and mate choice. We review the benefits females might gain by mating with certain males rather than others. We also consider other forms of selection that can make females mate nonrandomly. One way empiricists can contribute to resolving theoretical disputes is to quantify the cost of expressing mating biases in the appropriate life-history currency.

INTRODUCTION

Sexual selection occurs when individuals compete for access to gametes of the opposite sex (Andersson 1994). The kinds of traits it generates are so flamboyant and even devious that sexual selection, despite a component of natural selection, is often elevated to the status of the so-called other force shaping the evolution of organisms.

Sexual selection is the result of interactions between individuals of the same sex, opposite sexes, or both. Males usually produce enough sperm over their lifetime to inseminate many females. Their reproductive success depends primarily on how many females they mate with, the fecundity of these females, and the proportion of each female's eggs they fertilize. Males are therefore under strong sexual selection to improve all three aspects of reproductive success. In addition to the number of available gametes, the payoff of mating can also vary depending on qualitative differences among gametes. Mate choice is additionally under natural selection: It can increase access to material resources required for reproduction (e.g., nesting sites) or other aspects of fitness (although this too can be considered sexual selection, e.g., if the mate phenotype determines access to these resources). Given this complexity, verbal arguments about how sexual selection influences both sexes can be convoluted and are certainly insufficient.

Mathematical modeling has played a major role in the development of the field. It clarifies verbal arguments, exposes hidden assumptions, and tests whether the stated outcomes actually emerge from the starting premises. Unfortunately, this has led to a proliferation of superficially identical models, making it difficult to see the forest for the trees. Claims and counterclaims about whether specific processes can occur (e.g., compare Kirkpatrick 1986 and Grafen 1990a) have caused some biologists to argue that theoretical models make inconsistent and often transitory claims about what is possible in the real world (Eberhard 2005). In reality, if models reach different conclusions, they must be making different assumptions. Differences should therefore be considered illuminating. It is the empiricist's task to test assumptions directly or to argue whether they are biologically plausible. The view that modelers are divorced from reality is strengthened when theoretical models are used to make blanket statements. For example, Maynard Smith (1977) claimed that paternity cannot have an effect on paternal care because similar reductions of paternity would occur in future broods. Subsequent empirical studies tend to find, however, that care increases with a greater share of paternity (Sheldon 2002), and more recent models have shown this to be the theoretical expectation when paternity varies between broods (Houston & McNamara 2002).

The sexual-selection literature is enormous, covering a gamut of specialized topics (e.g., fighting assessment, signaling, sperm competition, intragenomic conflict, sex ratio skew, and differential allocation). Instead of attempting a comprehensive review of every available model, we begin with the foundations of sexual selection (anisogamy and parental roles) and then focus on the single most-debated effect of sexual selection: the evolution of mate choice for elaborate traits. We hope to show that the field could benefit from conceptual unification of several key areas that have developed in partial isolation.

Sexual selection: selection generated by differential access to opposite-sex gametes (or mates)

Anisogamy: difference in the size of the gametes that forms the basis of the definition of males and females

SEXUAL SELECTION AND SEX ROLES: IS YOUR TEXTBOOK RIGHT?

Sex and the fusion of gametes are prerequisites for sexual selection. The evolution of sexual reproduction per se is beyond the scope of our review, but it is interesting to note that sexual selection might help maintain sexual reproduction (Agrawal 2001, Siller 2001). Once there is sexual reproduction with two different mating types, distinct reproductive strategies can evolve. Typically one type provides offspring with fewer material benefits. The simplest case of unequal investment is a difference in gamete size (anisogamy). The sex producing smaller gametes is, by definition, the male. This is why seahorses that become pregnant after insemination are still referred to as males: They produced the smaller gametes (Berglund & Rosenqvist 2003).

Although anisogamy is sometimes argued to have evolved to ensure uniparental inheritance of cytoplasmic elements (Randerson & Hurst 2001), its evolution is more-widely attributed to disruptive selection. Two main assumptions are required: (*a*) Smaller gametes can be produced in larger numbers, and (*b*) larger zygotes with more resources survive better (Parker et al. 1972). Intermediate-sized gametes fall in between these two advantages, and, assuming suitable survival functions, disruptive selection arises when it is best to produce either small searcher or large resource-provider gametes (Bulmer & Parker 2002). In many species, this leads to individuals that are exclusively male or female and specialize in producing a single gamete type, although limited mate availability can also select for simultaneous hermaphrodites (Puurtinen & Kaitala 2002).

Even if the total biomass of gametes produced by each sex is equal, anisogamy implies that abundant sperm compete for far less numerous eggs. This competition creates a race among males to ensure their sperm locate eggs and fertilize them before those of rivals: Sexual selection is born. In external fertilizers, it selects for mechanisms to optimize the timing and quantity of sperm released (Levitan 2004) and for improved sperm efficiency in locating and entering eggs. If males are mobile, it selects for improvements in their ability to track down females; if females are mobile, it selects for males that increase the ease with which females locate them and the likelihood that they mate with them. When mating encounters allow sperm from several males to have the potential to fertilize eggs—multiple mating in internal fertilizers (Jennions & Petrie 2000) or simultaneous sperm release by adjacent males in external fertilizers (Byrne et al. 2003)—selection favors traits of males or their sperm that increase the prospects of fertilizing more eggs than their rivals (Simmons 2001). All these traits evolve through sexual selection initiated by anisogamy.

The relative size difference between sperm and eggs is large, but, in absolute terms, the difference in gamete-production costs can be trivial compared with those of other reproductive activities, such as lactation by female mammals or intense calling by male frogs. Still, there are good reasons to believe that anisogamy fundamentally influences sexual dimorphism in these traits. The standard textbook explanation for why females care for offspring and males seek out additional mates runs as follows: Anisogamy allows males to invest initially less per reproductive event. They therefore have a higher potential reproductive rate (PRR); hence they gain more than

> **Potential reproductive rate (PRR):** the highest possible reproduction rate for an individual given the constraints of speed of gamete production and parental care and assuming unlimited access to opposite-sex mates

Operational sex ratio (OSR): the ratio of male versus female individuals that are available for mating at any given time

females from deserting their current offspring if there is a trade-off between caring for current young and finding new mates. This amplifies the initial asymmetry: Males that provide less parental investment are selected to provide even less, and females compensate by increasing their investment (Clutton-Brock & Parker 1992, Emlen & Oring 1977, Maynard Smith 1977, Trivers 1972, Williams 1966). The result is a male-biased operational sex ratio (OSR), which forces males to compete even more intensely for mates and allows females to divert investment from sexual competition into providing more care for their offspring. It also permits females to be choosy, as they constantly encounter eager males. Subsequent models have refined this general theme, for example, by pointing out that choosiness is not a perfect mirror image of competition for mates, as choosiness depends on variation in the quality of potential mates (Johnstone et al. 1996, Owens & Thompson 1994).

Queller (1997), in a paper largely ignored at the time, questioned the logic of the standard explanation for asymmetries in parental care. His insight was that it is inconsistent to claim that the higher PRR of males allows them to gain more than females from searching for additional mates. The PRR is only a potential rate. In a population with a male-biased OSR, few males ever achieve anything near their potential. More generally, each offspring in diploid species has one father and one mother (Fisher 1930). This so-called Fisher condition (sensu Houston & McNamara 2005) guarantees that the number of offspring produced by males cannot exceed that produced by females. Anisogamy may lead to an asymmetry in PRR, but this cannot explain why females provide more care than males (see also Kokko & Jennions 2003).

Queller (1997) pointed out two ways to escape this dilemma. First, multiple mating and sperm competition create uncertainty about paternity for males, diminishing the expected fitness gain of caring for young. In females, uncertainty about maternity only occurs in rare circumstances such as egg dumping (Tallamy 2005). Second, a subset of males may be consistently more likely to mate than others if their traits are favored by sexual selection. Under a male-biased OSR with some males consistently more successful than others, the future reproductive rate of a breeding male is higher than that of the average male. Because average male reproductive rates equal average female reproductive rates, a successful male reproduces faster than the average breeding female. If success is nonrandom, the costs of caring owing to lost mating opportunities are therefore higher for males than for females. Interestingly, all these points have been made by those using PRR to predict the direction of sexual selection (e.g., Parker & Simmons 1996). However, before Queller (1997), they were never fully integrated with the evolution of parental care. (A final possibility is that the adult sex ratio is female-biased so that males must have a higher mating rate. To be a distinct explanation, however, this bias cannot be a result of sexual selection.)

In summary, anisogamy does not lead to different male-female parental roles because differential investment per gamete creates a sexual asymmetry in potential mating rates. The causal route is instead that anisogamy generates reproductive competition among males. The resultant sexual selection, including that owing to sperm competition, leads to uncertainty of paternity as well as predictable differences in reproductive rates between females and a subset of reproductively successful males.

These two factors make it more likely that males will pursue additional mating opportunities rather than invest in their current offspring.

Interestingly, Queller's insights coincided with growing recognition that models should be self-consistent. It is trivially true that models must be internally consistent (the books must balance; for example, the Fisher condition must hold so that every offspring has the correct number of parents), but there is also a growing trend to ensure that models of specific phenomena are also consistent with the wider system in which they are embedded (Houston & McNamara 2005). The way to ensure this is to avoid fixing parameters and instead let them emerge from the model. For example, recent models of caring do not fix the value of carer-offspring relatedness or caring payoffs. Instead these values change as a result of the population-level pattern of desertion, mortality, and migration (e.g., Härdling et al. 2003, Houston & McNamara 2002, Houston et al. 2005, Webb et al. 1999). Problems with consistency are nevertheless often overlooked, as illustrated by some recent models purporting to fix the problem (Wade & Shuster 2002) that in fact do not (Houston & McNamara 2005).

WHO CHOOSES AND WHO COMPETES?

Isogamous species tend to be unicellular; only multicellular species are anisogamous (Randerson & Hurst 2001). This makes testing whether the evolution of anisogamy promotes divergent sex roles empirically intractable: It is difficult to envisage how sex roles could develop or be measured in unicellular species independently of gamete size. Fortunately, anisogamy does not always produce classical sex roles. Competing for mates and being choosy are not mutually exclusive alternatives, and several species of birds, fish, frogs, and insects show either permanent or context-dependent sex-role reversal. Studies of these species provide an empirical testing ground to clarify what factors other than anisogamy drive sexual differences in choosiness, competitiveness, and mate-searching behavior (e.g., Barlow 2005, Simmons & Kvarnemo 2006).

Prior to Queller (1997), models simply assumed that the effect of relative parental investment on the PRR and OSR would predict which sex competed for mates (Clutton-Brock & Parker 1992, Parker & Simmons 1996) and, by extension, that the other sex was choosy. The insight that factors such as variance in mate quality also influence these predictions (Johnstone et al. 1996, Owens & Thompson 1994), as well as the requirement of self-consistency, has prompted new research into sex-role reversal and choosiness by both sexes (Kokko & Johnstone 2002, Kokko & Monaghan 2001). These models refrain from assuming that a higher male PRR and male-biased OSR equate to greater competition among males. Instead, these models employ a life-history approach to calculate the sex-specific fitness consequences of varying the investment into competition and/or choosiness.

The predictions are similar to those of earlier models, but there are subtle differences. The most important is that the cost of breeding can be seen as a more-fundamental factor than either the OSR or PRR in determining which sex competes more intensely for mates (Kokko & Monaghan 2001). This is best illustrated by the finding that the OSR at which the intensity of competition is greater in males than females need not be unity. Despite a female-biased OSR, males may still compete more intensely for mates if breeding is costlier for females (Kokko & Monaghan 2001, equation 4). Even so, the familiar relationships between the OSR, PRR, and competition for mates are often still present. The cost of breeding increases if there is a greater rate of adult mortality while breeding and/or if it takes longer to breed, and the OSR is influenced by the effect of breeding on adult mortality and the PRR by the time taken to re-enter the pool of potential mates. Therefore, a relatively higher breeding cost for females, unless balanced by greater nonbreeding male mortality, is associated with a male-biased OSR and higher male PRR. The statement that the costs of breeding are pivotal in determining which sex competes does not preclude other factors as better predictors in specific cases. If, for example, the adult sex ratio is female-biased for reasons unrelated to breeding costs, then females might still compete for males even if breeding was costlier for them.

The value of these more-recent models depends on perspective. On one hand, they make fewer implicit assumptions, thus providing a more-integrated picture of how sex roles, choosiness, and parental investment coevolve. Investigating additional components also increases predictive power, such as predicting the switch point in species that show conditional sex-role reversal (e.g., Forsgren et al. 2004, Simmons & Kvarnemo 2006). On the the other hand, if everything interacts with everything, it is difficult to make general predictions that are amenable to empirical tests because so many parameters have to be measured. But that is life.

There is always a tension between general heuristic models and specific models of particular systems. A clear trend exists in sexual-selection modeling toward more-inclusive models that consider future mating prospects in a self-consistent way and include feedback between choice (by both sexes), parental care, and mating effort (Houston et al. 2005, McNamara et al. 1999). Instead of fixing parameters, these models allow interactions between individuals to generate parameters (sex ratio, availability of mates) that can change over time (e.g., Alonzo & Warner 2000). Often these models are more complex than their predecessors, but there is a good reason. Houston et al. (2005) have argued that earlier models sometimes oversimplified to the point of being misleading. Empiricists need to recognize that modeling, similar to empirical data collection, progresses. Conclusions are made on the basis of the information and techniques available at the time. The influential models of Maynard Smith (1977) alerted us all to conflicts inherent in parental care, even if they do not satisfy current standards of consistency (Houston & McNamara 2005).

UNIFICATION OF MODELS OF THE EVOLUTION OF MATE CHOICE

Lande (1981) offered a breakthrough genetic model of the potential for runaway selection of male ornaments. In so doing he fleshed out an idea briefly sketched by Fisher (1930) a half-century earlier. Another quarter-century of subsequent modeling has led to a bewildering variety of evolutionary scenarios of how and when mate choice will evolve. There are still unresolved disputes as to the number of distinct processes involved in mate-choice evolution and their relative importance.

Mate choice is the outcome of the inherent propensity of an individual to mate more readily with certain phenotypes of the opposite sex (i.e., mating preference or bias) and the extent to which an individual engages in mate sampling before deciding to mate (i.e., choosiness). Here we pool these two aspects of choice, simply noting that preferences might be cost-free, whereas choosiness is less likely to be so.

The evolutionary trajectory of mate choice, similar to that of any other trait, depends on genetic variation in and the strength of direct selection on choice, and indirect selection that relies on genetically correlated traits. The strength of selection is proportional to the covariance between fitness and a trait. Fuller et al. (2005) recently provided a simplified multivariate, quantitative genetic framework to discuss the various evolutionary forces acting on mate choice, and we follow their approach here. In a few cases we have made relationships more explicit (e.g., in the absence of indirect selection and mutation, two forces of direct selection on a trait must be opposite and equal at equilibrium). Pomiankowski & Iwasa (2001) provide a more-rigorous mathematical review of these models.

The change in the mean values of a set of traits after a generation of selection ($\Delta \bar{z}$) can be expressed as

$$\Delta \bar{z} = \mathbf{G}\boldsymbol{\beta} + \mathbf{u}. \qquad 1.$$

Here $\mathbf{G}$ is the matrix of genetic variances in and covariances among a set of traits (reviewed in Steppan et al. 2002). The vector of selection gradients, $\boldsymbol{\beta}$, describes the intensity of direct selection on each trait, whereas $\mathbf{u}$ describes the extent to which each trait's mean values change owing to mutation (Lande 1979, Lande & Arnold 1983). Theory suggests, and empirical evidence shows, that mutations contribute more to the genetic variance of some traits than others. This is partly a result of the number of genes influencing a trait and the timing of their expression (Houle et al. 1996, Rifkin et al. 2005). This should alert us to the fact that $\mathbf{u}$ and $\mathbf{G}$ are not independent. Phenotypic change in traits (e.g., age when expressed) and their underlying genetic basis (e.g., the number of genes that influence their expression) also mean that neither $\mathbf{u}$ nor $\mathbf{G}$ is necessarily constant (Steppan et al. 2002).

Equation 1 can be used to review the evolution of three traits: a male sexual display (d), a female mating preference (p), and residual fitness or viability (v). The term v is difficult to quantify empirically: It is the naturally selected fitness left over after variation in fecundity solely attributable to mate quality or number is excluded. It cannot be measured, unless assumptions are made about the function relating investment in d to v: By how much is v reduced as d increases? The implicit assumption that the value of v is independent of the sex in which the relevant genes are expressed is also likely to be invalid in many cases (Fedorka & Mousseau 2004).

Fuller et al. (2005) decompose $\boldsymbol{\beta}$ into sexual selection (β_S) and natural selection (β_N). By Fuller et al.'s definition (which we follow here), sexual selection covers any selection that arises as a result of variation in the number and/or quality of mates.

Direct benefits of mate choice: an increase in a fitness component (e.g., survival or fecundity) of a choosy individual that is not based on the genetic quality of offspring

The initial equation when expanded is

$$\begin{pmatrix} \Delta \bar{d} \\ \Delta \bar{p} \\ \Delta \bar{v} \end{pmatrix} = \begin{pmatrix} V_d & C_{dp} & C_{dv} \\ C_{dp} & V_p & C_{pv} \\ C_{dv} & C_{pv} & V_v \end{pmatrix} \times \left(\begin{bmatrix} \beta_{Nd} \\ \beta_{Np} \\ \beta_{Nv} \end{bmatrix} + \begin{bmatrix} \beta_{Sd} \\ \beta_{Sp} \\ \beta_{Sv} \end{bmatrix} \right) + \begin{pmatrix} u_d \\ u_p \\ u_v \end{pmatrix}. \quad 2.$$

We can now discuss the various scenarios for the evolution of a mating preference, p, in terms of different selective forces and how these are influenced by additive genetic variance (V) in each trait and genetic covariances (C) between traits.

Direct Selection: The Believable Trinity of Sensory Bias, Direct Benefits, and Sexual Conflict

The least-controversial forms of selection on mate choice occur when there is direct selection (where β_{Np} or $\beta_{Sp} \neq 0$) and additive genetic variation in choice ($V_p > 0$). When choice is favored by direct natural selection ($\beta_{Np} > 0$), the resultant preference is not an adaptation to ensure chosen males confer benefits; rather it is an incidental effect of selection in another context. Ignoring mutation, at equilibrium the preference is under natural selection ($\beta_{Np} = 0$), and there is no sexual selection on it. In contrast, the display trait evolves (if $V_d > 0$) until there is an equilibrium between sexual and natural selection on it ($\beta_{Sd} = -\beta_{Nd}$). Sexual-selection theory refers to this process as sensory bias or sensory exploitation (Endler & Basolo 1998). Many empirical examples have been proposed to illustrate sensory bias (e.g., Garcia & Ramirez 2005), but, as Fuller et al. (2005) point out, critical tests are usually lacking. One key aim should be to demonstrate $\beta_{Sp} \equiv 0$ by showing that the value of d does not indicate whether a male confers greater benefits or reduced mating costs to females (remember that $\equiv$ means no values of the trait experience selection, as opposed to a local selective optimum having been reached). Fuller et al. (2005) suggest three empirical tests:

1. Breeding experiments to see if male phenotype predicts female reproductive performance. Many such studies show that $|\beta_{Sp}| > 0$ (e.g., Friberg & Arnqvist 2003, Head et al. 2005).
2. Selection experiments in which females are either randomly assigned mates or allowed to choose (given $\beta_{Np} = 0$ and $\beta_{Sp} \equiv 0$, there should be no decline in the preference as there are no sexually selected costs to random mating). Again, available studies generally suggest $|\beta_{Sp}| > 0$ (see Hosken & Tregenza 2005).
3. Artificial selection for a female trait that might influence mate choice (e.g., Endler et al. 2001). Because strong selection on almost any trait is likely to influence mating patterns, and hence mate choice (Wiley & Poston 1996), the value of this approach depends critically on whether experiments mimic a plausible force of natural selection and help to understand the proximate basis of mating preferences.

Another uncontroversial way in which a mating preference can evolve is through direct sexual selection on choice ($\beta_{Sp} > 0$, $V_p > 0$). In standard terminology, this is defined as natural selection for direct benefits (e.g., Hoelzer 1989, Price et al.

1993). It assumes an initially positive phenotypic correlation between a male display, d, and the benefit of mating with him, owing to a direct increase in female fitness or offspring survival. For example, males in better condition might provide better parental care or be less likely to transfer venereal diseases. Assuming $V_d > 0$ and again ignoring **u**, d evolves until natural selection opposes sexual selection, $\beta_{Sd} = -\beta_{Nd}$. The mating preference similarly evolves under sexual selection, until the benefit of stronger preferences is countered by their costs owing to, for example, the energetics of mate searching ($\beta_{Sp} = -\beta_{Np}$). In some cases, however, one can also imagine a preference that has no naturally selected cost ($\beta_{Np} = 0$) initially evolving because $\beta_{Sp} > 0$, until $\beta_{Sp} = 0$ because the benefit originally signaled by d is eventually exactly countered by a fitness cost that d itself imposes on females (e.g., long-tailed males provide less parental care) (Price et al. 1993; see also Kokko 1998b).

Sexual-conflict theory is all about the diverse range of direct costs that a male can impose on females while attempting to increase his own fitness (Arnqvist & Rowe 2005). Let us consider a situation where mate choice has already evolved: Females prefer males with larger d. How do direct costs influence the subsequent evolution of the mating preference? There could be sexual selection to diminish preferences if attractive males impose greater direct costs on females than less-attractive males ($\beta_{Sp} < 0$). There has long been circumstantial evidence for such costs in birds where females contribute relatively more to parental care if their mate is attractive (Møller & Jennions 2001). Recent experimental studies report that females assigned attractive males (Head et al. 2005) or males with phenotypes correlated with attractiveness (Friberg & Arnqvist 2003, Pitnick & García-González 2002) can have decreased lifetime offspring production. This begs a question: If some males impose direct costs, why prefer them? Logically, one seeks out other forces counteracting the preference reduction. One possibility is that females gain genetic benefits for their offspring, which offset the direct costs these harmful males impose (Hosken & Tregenza 2005, Kokko 2005). But what about compensatory direct benefits?

One major area of sexual conflict regards mating rates (Arnqvist & Rowe 2005). The optimal number of matings is almost certainly lower for a female than for a male (Arnqvist & Nilsson 2000). This can create sexual selection to elevate mating preferences ($\beta_{Sp} > 0$) to reduce the number of times females mate (Gavrilets et al. 2001; for an updated review, see Rowe et al. 2005). Such cost minimization results in some males mating more often than others, owing to female traits. This is, by definition, a form of mate choice (Wiley & Poston 1996) involving a mating preference/resistance that selects for attractive, seductive, or persistent males (Holland & Rice 1998). If stronger mating preferences help reduce the direct cost of excessive mating, they can in principle counter the selection on females to avoid mating with males with larger values of d that impose greater direct costs per mating. Although not explicitly modeled, the difference between the cost of rejecting and accepting a male could increase with d. This would make it cheaper to mate with preferred males (larger d) than to reject them. (Obviously some males must be cheaper to reject than accept, otherwise females would accept all males.)

There could also be selection to mate less readily under certain conditions, e.g., when predators are common so that the benefit of choosiness is a reduced mating

G-matrix: a table of genetic variances (along the diagonal) and covariances (elsewhere) between traits

rate ($\beta_p > 0$). This creates sexual selection on males if some traits increase the rate of encounter with females in their receptive state. Putting aside sexual conflict over mating rates and differential costs of rejection, however, current mathematical models suggest females should avoid mating with males that impose greater costs per mating unless there are genetic benefits. [We have so far ignored mutations **u**, but they play a large role in this argument (Kokko 2005).]

Under sexual conflict, the equilibrium situation is still one where sexual selection to increase p is countered by natural selection ($\beta_{Sp} = -\beta_{Np}$). Natural selection to reduce p arises, for example, if overly resistant females take longer to acquire a mate and therefore breed at suboptimal times (e.g., Backwell & Passmore 1996). In sum, in the above three scenarios, there is no net-directional selection on p at equilibrium. This reflects opposing sexual and natural selection in sexual conflict and direct benefit models. Although sexual conflict has recently been elevated in status as an explanation for mate choice, the only real distinction between it and direct benefit models is that the main reason not to reduce p is the cost of an increased mating rate. Sexual-conflict models may be seen as distinctive because they alone involve male-imposed costs. However, given direct benefits to matings with certain males, one could, without torturing the language, equally well state that p is maintained through the cost of mating with males with smaller displays that provide fewer benefits. The sensory bias scenario differs from the other two, however, because there is stabilizing natural selection on p at equilibrium.

Indirect Selection: The Devil Is in the Details

In the section above, aside from noting that traits only evolve if $\mathbf{V} > 0$, we studiously ignored the G-matrix and mutations. In effect, we prevented a genetic covariance between traits developing. Is this realistic? Can the effect of mutations be ignored? Most importantly, if genetic correlations between traits evolve, do equilibrium states stay the same? The answers are no, no, and no.

Given additive genetic variation ($V_p > 0$, $V_d > 0$), genes for a display and those for a preference associate more often than expected because choosy females mate more often with males with better displays. This gametic phase disequilibrium results in a positive genetic covariance ($C_{dp} > 0$) between d and p (Lande 1981). Whenever $p > 0$, as a result of drift or sensory bias (Payne & Pagel 2000), mate choice results in direct sexual selection on display d ($\beta_{Sd} > 0$) and indirect selection on preferences p because $C_{dp} > 0$. As p increases, so does β_{Sd}; a positive feedback loop is established, driving an increase in p. Lande (1981) formally showed that rapidly increasing values of p and d evolve if $C_{dp} > V_d$, but if $C_{dp} < V_d$, d only evolves to match p.

Indirect selection on p is often referred to as the Fisher process and the rapid coevolution of exaggerated display and choice as the Fisherian runaway (Andersson 1994). This is in recognition of Fisher (1930) who first outlined this process in a verbal model to account for Darwin's claim that female aesthetic sensibilities drive the evolution of male courtship traits. In the absence of any naturally selected cost of p ($\beta_{Np} \equiv 0$), a line of equilibrium emerges: For any given level of p, there is a matching expression of d; $\beta_{Sp} \equiv 0$ because all males are of equal fitness. The sexually selected benefit of a larger display is perfectly balanced by natural selection owing to the mortality cost it imposes ($\beta_{Sd} = -\beta_{Nd}$). Females gain nothing by producing more-attractive sons with higher mating success, as these sons have lower survivorship. We discuss this further below to consider what happens when preferences are costly ($\beta_{Np} \neq 0$).

Mate-choice models are most controversial when we consider the third trait of residual fitness, or viability (v), that covaries positively with male display ($C_{dv} > 0$). Phenotypic viability (measured by proxies such as lifespan, immunocompetence, or physiological condition) may still not be positively correlated with sexual display (Grafen 1990a, Hunt et al. 2004b, Kokko 2001, Tomkins et al. 2004). Depending on the life-history schedule and effect of display on mating success (Kokko 1998a), males with the highest residual viability may invest disproportionately in display and end up in the worst condition with the lowest survivorship (e.g., Hunt et al. 2004a). Life-history schedules also determine whether display increases with age (Kokko 1998a, Proulx et al. 2002).

The most-likely cause of a positive relationship, $C_{dv} > 0$, is condition-dependent expression of sexual displays (Tomkins et al. 2004, Zahavi 1977). Numerous studies have shown that sexual display is sensitive to male condition (Bondurianksy & Rowe 2005, but see also Cotton et al. 2004). Condition refers to a male's state prior to development of the display and is therefore equivalent to residual fitness. Fuller et al. (2005) suggested that V_d can be zero. One can indeed imagine a scenario where the gene for d simply codes for condition-determined expression. Quantitative genetics has, however, its basis in partitioning phenotypic variation among genetic and environmental sources. Genes for conditions that covary with the expression of d are thus synonymous with additive genetic variation in d. If condition is heritable, a father-son regression will show that their display sizes covary, which, excluding effects of a shared environment, is generally interpreted as $V_d > 0$. Although not widely discussed, if condition-dependent display and viability have a near-identical genetic basis (pleiotropy), they could be considered as one trait [although whether genotype by environment interactions strongly affect allocation decisions about investment in d is still undetermined (Greenfield & Rodriguez 2004)]. How d can honestly signal v to females is, of course, a separate question (Getty 2006). So too is the question of what maintains additive genetic variation in v (V_v), i.e., lek paradox (Kirkpatrick & Ryan 1991, Rowe & Houle 1996). We return to the role of mutations **u** below.

Given positive V_d, V_p, C_{dp}, and C_{dv}, there is also a genetic correlation between p and v ($C_{pv} > 0$). Natural selection on viability ($\beta_{Nv} > 0$) therefore generates additional indirect selection on mate choice: Choosy females produce fitter offspring and not just more-attractive sons. This is commonly referred to as the good-gene viability indicator, or handicap model of sexual selection. It was first verbally modeled by Zahavi (1975, 1977). After some initial failures, this process has been successfully modeled many times (e.g., Grafen 1990a, Iwasa et al. 1991, Iwasa & Pomiankowski 1999, Pomiankowski 1987b).

> **Lek paradox:** the problem of maintaining sufficient genetic variation in a trait subject to directional natural and sexual selection to sustain costly nonrandom mating

Indirect Selection: The Difficult Second Album

The role of indirect selection in the evolution of mate choice has sparked considerable debate. First, many questioned whether certain processes could work in principle [e.g., compare Kirkpatrick (1986) with Grafen (1990b) debating Zahavi's (1975) handicap ideas]. Usually differences arose because models made different structural assumptions, for example, about the relationship between the expression of genes for v and d (see Eshel et al. 2000, 2002), or violated assumptions implicit in other models, for example, the assumption that V_v remains positive. In other cases, minor modifications to the original models were required. For example, early models of the Fisher process concluded that costly preferences cannot be maintained (e.g., Kirkpatrick 1985), but subsequent work showed this no longer applies if there is biased mutation on a complex male display (Pomiankowski et al. 1991). Mutations **u** can therefore make a qualitative difference, even in (by far the most-common type of) models that ignore their effect on the G-matrix and therefore do not directly help answer the question of how $\mathbf{V} > 0$ can be maintained.

Second, many question the importance of indirect selection as a major force. Indirect benefits of mate choice are probably small. A chain of events are required for choosy females to obtain genetic benefits (Kirkpatrick & Barton 1997), so the rewards are minimal if even one process does not work as required. For example, erosion of additive genetic variation in fitness (lower V_v), inaccurate choice (lower C_{dp}), unreliable signals of viability (lower C_{dv}), and weaker sexual selection on male display (lower β_{Sd}) all reduce the benefits of choosiness. The key empirical question, therefore, becomes, how strong is direct selection on choosiness? If it is weak, then small, indirect genetic benefits may still suffice to maintain preferences that, by virtue of being cheap to express, have significant consequences on trait evolution in the opposite sex. Unfortunately, the costs of choosiness are largely unknown (Houle & Kondrashov 2002), although some researchers have recently attempted to compare the relative strength of direct costs and indirect benefits (Arnqvist & Kirkpatrick 2005), as well as to quantify precisely mate-sampling costs (Byers et al. 2005).

Third, confusion exists about whether different processes should or can be distinguished. Specifically, there is a long history of trying to dichotomize traits into those that evolve or are maintained solely through Fisherian or good-genes processes. This has led to a quixotic attempt to empirically test these as alternatives, generating controversy where none exists. In the section below, we outline why we believe there is a unitary process that can be accommodated in a single set of models. This view is neither new nor ours alone. To quote Pomiankowski & Iwasa (2001), "[t]he two hypotheses are really extremes on a continuum... In the real world, male traits involved in mate choice simply do not fall into two types, Fisher and handicap traits" (p. 216).

Indirect benefits of mate choice: an increase in the fitness of a choosy individual based on the genetic quality of offspring

A UNITARY MODEL OF INDIRECT BENEFITS

We have recently been involved in debating whether processes for the evolution of mate choice that rely on benefits from producing attractive sons are fundamentally different from those from producing more-viable offspring (Cameron et al. 2003, Eshel et al. 2000, Fuller et al. 2005, Kokko et al. 2002). We first outline the arguments others offer for such a division before presenting the case for a unified view. As a modeling exercise, one can certainly eliminate certain indirect benefits to clarify an argument, in the same way as a researcher controls for confounding factors known to influence the outcome of an experiment. On the basis of such work, we know, for example, that the handicap process works even if choosy females do not produce sons with higher mating success (Grafen 1990a,b).

It is surprisingly difficult to find an unambiguous definition of the Fisher process in the literature. One criterion is that choosy females produce more-attractive sons owing to linkage disequilibrium (C_{dp}) (e.g., Fuller et al. 2005, Mead & Arnold 2004). Some authors focus on self-reinforcing selection on p and d (e.g., Heisler et al. 1987, Takahashi 1997) or, more narrowly, on the possibility of open-ended runaway (Fisher's runaway process sensu Iwasa & Pomiankowski 1995 and Day 2000). Others provide a definition where the only forces on male display are natural selection for reduction and female choice for exaggeration (e.g., van Doorn et al. 2004; implicit in Shuster & Wade 2003, p. 185). All these features arise in a classical Fisherian model such as Lande (1981), but which are essential?

Linkage disequilibrium, self-reinforcement, and runaway cannot all define the Fisher process as they do not always co-occur. Linkage disequilibrium, per se, does not always predict self-reinforcing selection as the male trait does not evolve if p is below a critical threshold value (Pomiankowski 1988) and runaway only occurs if $C_{dp} > V_d$ (Lande 1981). What of the requirement that there is only natural selection against and sexual selection for the male display? Reading Fisher's (1930) original contribution provides some insight. He allowed a positive correlation between a male trait and viability to initiate the spread of the female preference, although the relationship eventually turned negative during the coevolutionary process.

Following Fisher's lead (see also Kirkpatrick & Ryan 1991), here we define the Fisher process as simply referring to cases where female preferences experience indirect selection caused by offspring inheriting their father's attractiveness (the so-called sexy sons). There is no reference to any correlation between male display and other fitness traits, although models that address the Fisher process on its own usually analyze cases where male display reduces survival (Lande 1981). There is also no statement about whether choice is costly. Weatherhead & Robertson (1979) originally used the term sexy son in a context of costly choice, but the term is now more widely applied.

Cameron et al. (2003) argue that we must formally distinguish Fisherian and good-genes processes because a costly female preference can never be maintained solely by the indirect benefit of producing sexy sons (see also Kirkpatrick & Ryan 1991, Pomiankowski 1987a). Their model shows that equilibrium is only reached when female choice and male display are at their naturally selected optima. Thus the Fisher process is unstable and will not be seen in nature. As Pomiankowski et al. (1991) first noted, however, p and d lie above their naturally selected optimum at equilibrium if there is a negative mutational bias on the male trait. Cameron et al.'s (2003) insight is that in the *absence* of female choice, deleterious mutations push d below its naturally selected optimum. This implies that natural selection favors males with larger ornaments. As attractive males are more viable at this stage, their evolution cannot be considered a Fisher process, but it is rather a version of good genes.

Our interpretation differs for the following reason: One must distinguish between a process and a model that exemplifies its evolution. Of necessity, every model makes various assumptions that typically make it one of many possible examples of how a process might run. Relaxing model assumptions one by one is why we usually find many published models addressing the same topic. Cameron et al. (2003) make two important assumptions: (*a*) The male display d can take on any value, positive or negative, and (*b*) mutation pressure is constant and negative for all values of d. What happens if we change these assumptions and consider a male trait that can only have non-negative values (e.g., crest height) with a naturally selected optimum at $d = 0$ (absence of crest)? This simple assumption yields a model with an equilibrium where female choice is costly and the male trait is subject to a destructive mutational bias (i.e., mutations tend to interfere negatively with the development of a complex ornament). However, in the absence of female choice, the net mutation bias is necessarily positive: Traces of an ornament that is not selected for can be created by mutation, thus $d > 0$ (see **Supplemental Appendix**; follow the Supplemental Material link from the Annual Reviews home page at **http://www.annualreviews.org**).

> **Fisher-Zahavi process:** coevolution of female preferences and male displays influenced by the fact that females mating with preferred fathers produce sexy sons

This example is neither more nor less general than that of Cameron et al. (2003). Our model differs because it considers a case where natural selection only ever favors smaller traits, and consequently natural and sexual selection on d are never in the same direction. If the Fisher process cannot lead to trait exaggeration unless elements of good genes are added, this example of a stable costly preference at equilibrium should be impossible.

The issue reduces to one of definitions. Many models investigate the Fisher process when there is mutational input or display is positively correlated with viability-related fitness (e.g., Eshel et al. 2000, Kokko et al. 2002, Pomiankowski et al. 1991). Cameron et al. (2003) appear to limit the definition of a Fisher process to a purer context where these factors are excluded. It is illustrative to compare the implicit definition of Cameron et al. (2003) with that of Shuster & Wade (2003, pp. 182–85). Both envisage a female preference for a trait that is originally related to viability but no longer is after preference evolution. The interpretations are exact opposites of each other: Shuster & Wade use the same example to reject the idea of good genes in favor of Fisher. Definitions cannot be imposed, but we recommend the general usage and consider a model to incorporate the Fisher process if it helps to delimit the conditions under which the production of sexy sons influences female-choice evolution (Parker 2006). The term Fisher-Zahavi process appears appropriate when additionally emphasizing the possibility that male displays covary with viability-related traits (see Eshel et al. 2000).

At equilibrium, there is always a choice among males who offer different indirect and/or direct benefits. The net effect of mating with a male can be quantified as the sum of its offsprings' reproductive values, including attractiveness of male offspring (Kokko et al. 2002, McNamara et al. 2003). This argument has been incorrectly criticized for ignoring C_{dp} (Fuller et al. 2005); if this critique were true, the method could not conceivably produce cases where females prefer less-viable males yet preferences still spread (Kokko et al. 2002). Reproductive-value calculations include C_{dp} implicitly, as choosy females are more likely to mate with a certain subset of males in each generation. Reproductive value also depends on the environment. Small changes in environmental conditions can lead to different phenotypic relationships between major fitness traits, such as display and longevity, without any change in the underlying genetic architecture (Kokko et al. 2002). This is because male attractiveness is often subject to life-history trade-offs (Grafen 1990b, Kokko 2001), and males can adaptively adjust their current investment in attractiveness in response to environmental challenges (Badyaev & Qvarnström 2002, Candolin 2000, Hunt et al. 2004a).

It might be impossible to distinguish empirically between a pure Fisher process and cases where display and residual fitness v are genetically correlated. A partial solution is to look only at the viability of daughters, but this assumes that the fitness effects of genes are not sex-dependent, although recent work suggests otherwise (Chippindale et al. 2001, Fedorka & Mousseau 2004). This is one purely pragmatic reason why the Fisher process (a benefit of producing sexy sons, usually because $C_{dp} > 0$) and the effects of a correlation between male display and other fitness-related traits ($C_{dv} > 0$) are best seen as part of a continuum of outcomes of a unitary evolutionary process (Kokko et al. 2002, Pomiankowski & Iwasa 2001; compare Brooks 2000 and Kokko 2001).

ON PHENOTYPIC AND QUANTITATIVE GENETIC MODELS

Sometimes the conclusions of different sexual-selection models seem related to the modeling technique used. Fuller et al. (2005), Shuster & Wade (2003) and others have claimed that phenotypic models are inappropriate to model aspects of sexual selection that depend on genetic covariances. This is an unfortunate criticism: Methods that appear to have ignored a term may simply have let it emerge from the model, rather than giving it a fixed value. Phenotypic models preferentially emphasize that mate choice changes the types of offspring produced (Kokko et al. 2002, McNamara et al. 2003), instead of assuming a fixed $C_{dp} > 0$.

Models can differ in the simplifying assumptions made, but they are, ultimately, special cases of the same mathematical framework (Hammerstein 1996, Page & Nowak 2002, Waxman & Gavrilets 2005). The same assumptions (e.g., biased mutation) are expressed in different ways depending on the modeling technique. The apparent effect of modeling techniques on conclusions only arises because researchers using different approaches tend to make different assumptions. All models are approximations. We know, for example, that the G-matrix must ultimately evolve in response to mutations **u**, but this is typically ignored in quantitative genetic models. Empiricists prefer to ignore **u**'s immediate effect as well, as the G-matrix is easier to estimate assuming **u** = 0. This is convenient, but because **u** is important for long-term evolution (Iwasa et al. 1991, Pomiankowski et al. 1991), the outcome may feature a mismatch between short- and long-term evolutionary arguments (Pigliucci & Schlichting 1997). We prefer the current plurality in modeling, as each modeling tool has some disadvantages, making it more difficult to capture some features of reality than others. Using one approach can lead only to the neglect of some real-world issues. To quote the psychologist Abraham Maslow, "[i]f the only tool you have is a hammer, you tend to see every problem as a nail."

IDEAS ON TESTING THESE IDEAS

The prediction that choosy females produce sexy sons is critical to all models of indirect selection on choice. If inheritance is genetic, $C_{dp} > 0$, although genetic preferences for displays transmitted culturally from father to son can evolve in principle (Ihara et al. 2003, McNamara et al. 2003). Several empirical studies have used artificial selection on d to show that p evolves as a correlated response (e.g., Brooks & Couldridge 1999, Wilkinson & Reillo 1994), but results are difficult to interpret. C_{dp} is maintained by nonrandom mating, so it should break down rapidly during artificial selection using forced random matings, yet correlated responses to selection may occur owing to sexual selection within selected lines in which mate choice can occur (Gray & Cade 1999).

Within-generation estimates of C_{dp} are valuable (e.g., Ritchie et al. 2005). We suspect future estimates will show that at equilibrium C_{dp} is small both in species where mate choice is strong and unconstrained and in species where females can rarely exercise their preferences freely. For example, the unanimity of choice in lekking birds means there is little phenotypic variation in choice, and thus little scope for current additive genetic variation. Similarly, the requirement of social monogamy in most passerines means that many females mate with an available male rather than the one they would prefer, so d and p (measured as observed choice) barely covary (Qvarnström et al. 2006). Blows' (1999) study of the evolution of C_{dp} after experimental hybridization of *Drosophila serrata* and *D. birchii* illustrates our point. Hybridization disrupted the linkage disequilibrium underlying C_{dp}, resulting in a relatively small correlation of 0.39 at generation 5. Choice built the correlation up to 1.02 at generation 37, but C_{dp} then plunged to –0.04 by generation 56, possibly because much of the variation in choice and display present at and before generation 37 had been fixed by generation 56.

A conceptual issue emerges: Must we only consider the extant set of females when calculating C_{dp}? Let us assume all peahens prefer the peacock with the largest train; hence $C_{dp} = 0$. There is no Fisher process evident, but $C_{dp} > 0$ is reintroduced if a mutation causes a peahen to mate randomly. As her sons are unsuccessful, this mutant soon is eliminated. Should we not distinguish between cases where $C_{dp} = 0$ because females show a near-unanimous preference for the sexiest males and a case where $C_{dp} = 0$ because females mate randomly with respect to the value of d? This highlights the difficulty of arguing about long-term evolution using short-term measures of selection (Hammerstein 1996, Pigliucci & Schlichting 1997). Empirical progress clearly requires a combination of single-generation, multigeneration, and clever experimental approaches.

The most-fundamental tests of mate-choice models are whether (*a*) female choice maximizes the sum of direct and indirect benefits; (*b*) direct selection results in females choosing males that do not maximize their fitness when compared with other mates, but this behavior is still adaptive because the mechanism driving choice maximizes net-female fitness (e.g., by reducing the cost of excess mating); or (*c*) female choice is maladaptive (Gustafsson & Qvarnström 2006). The extent to which (*b*) occurs may depend on how easily females can modify the effect of traits valuable in a nonmating context that might adversely influence mate choice (Houle & Kondrashov 2002) or on whether females can make themselves unavailable to males by becoming sexually unreceptive (Rowe et al. 2005). In turn, (*c*) is more likely if male-female coevolution is at a stage where the necessary mutations have not arisen for females to adjust their behavior to maximize their fitness, given the male types to which they are currently exposed. Empirical tests of these scenarios are daunting (Hunt et al. 2004b). It is difficult to produce an all-inclusive measure of offspring fitness and merge it with parental-lifetime reproductive success, and even this is only a proxy for fitness (Brommer et al. 2004).

What We Tend to Forget to Measure

To summarize, all models agree that mate choice can evolve by a combination of direct and/or indirect benefits. In the latter case, the selection on female preferences is weak and easily overridden by the costs of choice. It is important to remember, however, that even small changes in female behavior (which cost little) can generate strong selection when a male's fitness depends primarily on his mating success (Alatalo et al. 1998, Reynolds & Gross 1990).

There is an important way to phrase this prediction that is rarely used in the literature: Preferences can exist despite substantial costs of expression if they yield large direct benefits, whereas preferences that evolve in the absence of direct benefits should be cheap to express (unless there is substantial genetic variation in overall fitness). Although we do not discuss the lek paradox extensively, new multivariate methods indicate it is far from resolved (see Blows & Hoffmann 2005). In the case of a practically cost-free behavior that creates mating biases, the consequences for male-trait evolution can still be dramatic. Unfortunately, showing there are no measurable costs in the appropriate life-history currency is probably even more demanding than measuring indirect benefits. It may, however, still be possible to rank preferences in terms of their estimated costs and to test the prediction that costs are smaller where we presume only indirect benefits than in cases where direct benefits of choice have been firmly established. The challenge is there.

SUMMARY POINTS

1. The multitude of mathematical models in the field of sexual selection can appear confusing, but the diversity of models, similar to the diversity of experimental approaches, is necessary for healthy progress toward understanding which assumptions are crucial for an argument. There is a growing trend toward more-inclusive models that make fewer implicit assumptions and avoid inconsistencies generated by too many fixed parameters.

2. The importance of self-consistent modeling is highlighted by recent reassessment of the way parental sex roles follow from anisogamy. Earlier explanations based on PRRs are at odds with the fact that total reproduction by the male population cannot exceed that by the female population. Instead, parental roles can follow from anisogamy because it causes sperm competition and sexual selection, which feed back to the evolution of parental roles.

3. Mate-choice evolution can be influenced by direct and indirect benefits. These can all be merged into a simple, quantitative genetic framework, including cases in which superfluous mating is costly to females. Care should be taken, however, when arguing about long-term evolution using short-term measures of selection.

4. The recent debate regarding the distinctness of the Fisher process from the good-genes hypothesis boils down to definitions. We suggest using the term Fisher-Zahavi process to help remember that indirect benefits can include both improved viability and greater attractiveness of offspring.

5. Detecting indirect benefits is often challenging as the lek paradox (in its univariate or multivariate form) predicts these to be small; however, costs of expressing preferences can be minute too. A central prediction emerges that requires future tests: Costs of preference expression should be small in cases of indirect benefits, whereas they can be larger if direct benefits are involved.

ACKNOWLEDGMENTS

Many people have helped us to formulate our ideas more clearly, and we are particularly indebted to Göran Arnqvist, Pat Backwell, Troy Day, Matt Hall, David Houle, John McNamara, Locke Rowe, and Geoff Parker; we would like to add many others. The frank comment by a *Larus* sp. individual in Toronto must also be mentioned here. Funding was provided by the Academy of Finland and by the Australian Research Council.

LITERATURE CITED

Agrawal AF. 2001. Sexual selection and the maintenance of sexual reproduction. *Nature* 411:692–95

Alatalo RV, Kotiaho J, Mappes J, Parri S. 1998. Mate choice for offspring performance: major benefits or minor costs? *Proc. R. Soc. London Ser. B* 265:2297–301

Alonzo SH, Warner RR. 2000. Female choice, conflict between the sexes and the evolution of male alternative reproductive behaviours. *Evol. Ecol. Res.* 2:149–70

Andersson M. 1994. *Sexual Selection*. Princeton, NJ: Princeton Univ. Press

Arnqvist G, Kirkpatrick M. 2005. The evolution of infidelity in socially monogamous passerines: the strength of direct and indirect selection on extrapair copulation behaviour in females. *Am. Nat.* 165:S26–37

Arnqvist G, Nilsson T. 2000. The evolution of polyandry: multiple mating and female fitness in insects. *Anim. Behav.* 60:145–64

Arnqvist G, Rowe L. 2005. *Sexual Conflict*. Princeton, NJ: Princeton Univ. Press

Backwell PRY, Passmore NI. 1996. Time constraints and multiple choice criteria in the sampling behaviour and mate choice of the fiddler crab, *Uca annulipes*. *Behav. Ecol. Sociobiol.* 38:407–16

Badyaev AV, Qvarnström A. 2002. Putting sexual traits into the context of an organism: a life-history perspective in studies of sexual selection. *Auk* 119:301–10

Barlow GW. 2005. How do we decide that a species is sex-role reversed? *Q. Rev. Biol.* 80:28–35

Berglund A, Rosenqvist G. 2003. Sex role reversal in pipefish. *Adv. Study Behav.* 32:131–67

Blows MW. 1999. Evolution of the genetic covariance between male and female components of mate recognition: an experimental test. *Proc. R. Soc. London Ser. B* 266:2169–74

Blows MW, Hoffmann AA. 2005. A reassessment of genetic limits to evolutionary change. *Ecology* 86:1371–84

Bonduriansky R, Rowe L. 2005. Sexual selection, genetic architecture, and the condition dependence of body shape in the sexually dimorphic fly *Prochyliza xanthostoma* (Piophilidae). *Evolution* 59:138–51

Brommer JE, Gustafsson L, Pietiäinen H, Merilä J. 2004. Single-generation estimates of individual fitness as proxies for long-term genetic contribution. *Am. Nat.* 163:505–17

Brooks R. 2000. Negative genetic correlation between male sexual attractiveness and survival. *Nature* 406:67–70

Brooks R, Couldridge V. 1999. Multiple sexual ornaments coevolve with multiple mating preferences. *Am. Nat.* 154:37–45

Bulmer MG, Parker GA. 2002. The evolution of anisogamy: a game-theoretic approach. *Proc. R. Soc. London Ser. B* 269:2381–88

Byers JA, Wiseman PA, Jones L, Roffe TJ. 2005. A large cost of female mate sampling in pronghorn. *Am. Nat.* 166:661–68

Byrne PG, Simmons LW, Roberts JD. 2003. Sperm competition and the evolution of gamete morphology in frogs. *Proc. R. Soc. London Ser. B* 270:2079–86

Cameron E, Day T, Rowe R. 2003. Sexual conflict and indirect benefits. *J. Evol. Biol.* 16:1055–60

Candolin U. 2000. Changes in expression and honesty of sexual signalling over the reproductive lifetime of sticklebacks. *Proc. R. Soc. London Ser. B* 267:2425–30

Chippindale AK, Gibson JR, Rice WR. 2001. Negative genetic correlation for adult fitness between sexes reveals ontogenetic conflict in *Drosophila*. *Proc. Natl. Acad. Sci. USA* 98:1671–75

Clutton-Brock TH, Parker GA. 1992. Potential reproductive rates and the operation of sexual selection. *Q. Rev. Biol.* 67:437–56

Cotton S, Fowler K, Pomiankowski A. 2004. Do sexual ornaments demonstrate heightened condition-dependent expression as predicted by the handicap hypothesis? *Proc. R. Soc. London Ser. B* 271:771–83

Day T. 2000. Sexual selection and the evolution of costly female preferences: spatial effects. *Evolution* 54:715–30

Eberhard WG. 2005. Evolutionary conflicts of interest: Are female sexual decisions different? *Am. Nat.* 165:S19–25

Emlen ST, Oring LW. 1977. Ecology, sexual selection, and the evolution of mating systems. *Science* 197:215–23

Endler JA, Basolo A. 1998. Sensory ecology, receiver biases and sexual selection. *Trends Ecol. Evol.* 13:415–20

Endler JA, Basolo A, Glowacki S, Zerr J. 2001. Variation in response to artifical selection for light sensitivity in guppies, *Poecilia reticulata*. *Am. Nat.* 158:36–48

Eshel I, Sansone E, Jacobs F. 2002. A long-term genetic model for the evolution of sexual preference: the theories of Fisher and Zahavi re-examined. *J. Math. Biol.* 45:1–21

Eshel I, Volovik I, Sansone E. 2000. On Fisher-Zahavi's handicapped sexy son. *Evol. Ecol. Res.* 2:509–23

Fedorka KM, Mousseau TA. 2004. Female mating bias results in conflicting sex-specific offspring fitness. *Nature* 429:65–67

Fisher RA. 1930. *The Genetical Theory of Natural Selection*. Oxford, UK: Oxford Univ. Press

Forsgren E, Amundsen T, Borg AA, Bjelvenmark J. 2004. Unusually dynamic sex roles in a fish. *Nature* 429:551–54

Friberg U, Arnqvist G. 2003. Fitness effects of female mate choice: Preferred males are detrimental for *Drosophila melanogaster* females. *J. Evol. Biol.* 16:797–811

Fuller RC, Houle D, Travis J. 2005. Sensory bias as an explanation for the evolution of mate preferences. *Am. Nat.* 166:437–46

Garcia CM, Ramirez E. 2005. Evidence that sensory traps can evolve into honest signals. *Nature* 434:501–5

Gavrilets S, Arnqvist G, Friberg U. 2001. The evolution of female mate choice by sexual conflict. *Proc. R. Soc. London Ser. B* 268:531–39

Getty T. 2006. Sexually selected signals are not similar to sports handicaps. *Trends Ecol. Evol.* 21:83–88

Grafen A. 1990a. Biological signals as handicaps. *J. Theor. Biol.* 144:517–46

Grafen A. 1990b. Sexual selection unhandicapped by the Fisher process. *J. Theor. Biol.* 144:473–516

Gray DA, Cade WH. 1999. Correlated-response-to-selection experiments designed to test for a genetic correlation between female preferences and male traits yield biased results. *Anim. Behav.* 58:1325–27

Greenfield MD, Rodriguez RL. 2004. Genotype-environment interactions and the reliability of mating signals. *Anim. Behav.* 68:1461–68

Gustafsson L, Qvarnström A. 2006. A test of the "sexy son" hypothesis: sons of polygynous collared flycatchers do not inherit their father's mating status. *Am. Nat.* 167:297–302

Hammerstein P. 1996. Darwinian adaptation, population genetics and the streetcar theory of evolution. *J. Math. Biol.* 34:511–32

Härdling R, Kokko H, Arnold KE. 2003. Dynamics of the caring family. *Am. Nat.* 161:395–412

Head ML, Hunt J, Jennions MD, Brooks RC. 2005. The indirect benefits of mating with attractive males outweigh the direct costs. *PLoS Biol.* 3:289–94

Heisler L, Andersson M, Arnold SJ, Boake CR, Borgia G, et al. 1987. Evolution of mating preferences and sexually selected traits. In *Sexual Selection: Testing the Alternatives*, ed. JW Bradbury, M Andersson, pp. 97–118. Chichester, UK: Wiley

Hoelzer GA. 1989. The good parent process of sexual selection. *Anim. Behav.* 38:1067–78

Holland B, Rice WR. 1998. Perspective: chase-away sexual selection: antagonistic seduction versus resistance. *Evolution* 52:1–7

Hosken DJ, Tregenza T. 2005. Evolution: Do bad husbands make good fathers? *Curr. Biol.* 15:836–38

Houle D, Kondrashov AS. 2002. Coevolution of costly mate choice and condition-dependent display of good genes. *Proc. R. Soc. London Ser. B* 269:97–104

Houle D, Morikawa B, Lynch M. 1996. Comparing mutational variabilities. *Genetics* 143:1467–83

Houston AI, McNamara JM. 2002. A self-consistent approach to paternity and parental effort. *Philos. Trans. R. Soc. London Ser. B* 357:351–62

Houston AI, McNamara JM. 2005. John Maynard Smith and the importance of consistency in evolutionary game theory. *Biol. Philos.* 20:933–50

Houston AI, Szekely T, McNamara JM. 2005. Conflict between parents over care. *Trends Ecol. Evol.* 20:33–38

Hunt J, Brooks R, Jennions MD, Smith MJ, Bentsen CL, Bussière LF. 2004a. High-quality male field crickets invest heavily in sexual display but die young. *Nature* 432:1024–27

Hunt J, Bussière LF, Jennions MD, Brooks R. 2004b. What is genetic quality? *Trends Ecol. Evol.* 19:329–33

Ihara Y, Aoki K, Feldman MW. 2003. Runaway sexual selection with paternal transmission of the male trait and gene-culture determination of the female preference. *Theor. Popul. Biol.* 63:53–62

Iwasa Y, Pomiankowski A. 1995. Continual change in mate preferences. *Nature* 377:420–22

Iwasa Y, Pomiankowski A. 1999. Good parent and good genes models of handicap evolution. *J. Theor. Biol.* 200:97–109

Iwasa Y, Pomiankowski A, Nee S. 1991. The evolution of costly mate preferences. II. The "handicap" principle. *Evolution* 45:1431–42

Jennions MD, Petrie M. 2000. Why do females mate multiply? A review of the genetic benefits. *Biol. Rev.* 75:21–64

Johnstone RA, Reynolds JD, Deutsch JC. 1996. Mutual mate choice and sex differences in choosiness. *Evolution* 50:1382–91

Kirkpatrick M. 1985. Evolution of female choice and male parental investment in polygynous species: the demise of the sexy son. *Am. Nat.* 125:788–810

Kirkpatrick M. 1986. The handicap mechanism of sexual selection does not work. *Am. Nat.* 127:222–40

Kirkpatrick M, Barton NH. 1997. The strength of indirect selection on female mating preferences. *Proc. Natl. Acad. Sci. USA* 94:1282–86

Kirkpatrick M, Ryan MJ. 1991. The evolution of mating preferences and the paradox of the lek. *Nature* 350:33–38

Kokko H. 1998a. Good genes, old age and life history trade-offs. *Evol. Ecol.* 12:739–50

Kokko H. 1998b. Should advertising parental care be honest? *Proc. R. Soc. London Ser. B* 265:1871–78

Kokko H. 2001. Fisherian and 'good genes' benefits of mate choice: how (not) to distinguish between them. *Ecol. Lett.* 4:322–26

Kokko H. 2005. Treat 'em mean, keep 'em (sometimes) keen: evolution of female preferences for dominant and coercive males. *Evol. Ecol.* 19:123–35

Kokko H, Brooks R, McNamara JM, Houston AI. 2002. The sexual selection continuum. *Proc. R. Soc. London Ser. B* 269:1331–40

Kokko H, Jennions M. 2003. It takes two to tango. *Trends Ecol. Evol.* 18:103–4

Kokko H, Johnstone RA. 2002. Why is mutual mate choice not the norm? Operational sex ratios, sex roles, and the evolution of sexually dimorphic and monomorphic signalling. *Philos. Trans. R. Soc. London Ser. B* 357:319–30

Kokko H, Monaghan P. 2001. Predicting the direction of sexual selection. *Ecol. Lett.* 4:159–65

Lande R. 1979. Quantitative genetical analysis of multivariate evolution, applied to brain:body size allometry. *Evolution* 33:402–16

Lande R. 1981. Models of speciation by sexual selection on polygenic traits. *Proc. Natl. Acad. Sci. USA* 78:3721–25

Lande R, Arnold SJ. 1983. The measurement of selection on correlated characters. *Evolution* 37:1210–26

Levitan DR. 2004. Density-dependent sexual selection in external fertilizers: variances in male and female fertilization success along the continuum from sperm limitation to sexual conflict in the sea urchin *Strongylocentrotus franciscanus*. *Am. Nat.* 164:298–309

Maynard Smith J. 1977. Parental investment: a prospective analysis. *Anim. Behav.* 25:1–9

McNamara JM, Gasson CE, Houston AI. 1999. Incorporating rules for responding into evolutionary games. *Nature* 401:368–71

McNamara JM, Houston AI, Marques dos Santos M, Kokko H, Brooks R. 2003. Quantifying male attractiveness. *Proc. R. Soc. London Ser. B* 270:1925–32

Mead LS, Arnold SJ. 2004. Quantitative genetic models of sexual selection. *Trends Ecol. Evol.* 19:264–71

Møller AP, Jennions MD. 2001. How important are direct fitness benefits of sexual selection? *Naturwissenschaften* 88:401–15

Owens IPF, Thompson DBA. 1994. Sex-differences, sex-ratios and sex-roles. *Proc. R. Soc. London Ser. B* 258:93–99

Page KM, Nowak MA. 2002. Unifying evolutionary dynamics. *J. Theor. Biol.* 219:93–98

Parker GA. 2006. Sexual conflict over mating and fertilization: an overview. *Philos. Trans. R. Soc. London Ser. B* 361:235–59

Parker GA, Simmons LW. 1996. Parental investment and the control of sexual selection: predicting the direction of sexual competition. *Proc. R. Soc. London Ser. B* 263:315–21

Parker GA, Smith VGF, Baker RR. 1972. The origin and evolution of gamete dimorphism and the male-female phenomenon. *J. Theor. Biol.* 36:181–98

Payne RJH, Pagel M. 2000. Inferring the origins of state-dependent courtship traits. *Am. Nat.* 157:42–50

Pigliucci M, Schlichting CD. 1997. On the limits of quantitative genetics for the study of phenotypic evolution. *Acta Biotheor.* 45:143–60

Pitnick S, García-González F. 2002. Harm to females increases with male body size in *Drosophila melanogaster*. *Proc. R. Soc. London Ser. B* 269:1821–28

Pomiankowski A. 1987a. The costs of choice in sexual selection. *J. Theor. Biol.* 128:195–218

Pomiankowski A. 1987b. The 'handicap principle' does work—sometimes. *Proc. R. Soc. London Ser. B* 127:123–45

Pomiankowski A. 1988. The evolution of female mate preferences for mate genetic quality. *Oxford Surv. Evol. Biol.* 5:136–84

Pomiankowski A, Iwasa Y. 2001. How does mate choice contribute to exaggeration and diversity in sexual characters? In *Economics in Nature: Social Dilemmas, Mate Choice and Biological Markets*, ed. R Noë, J van Hooff, P Hammerstein, pp. 203–20. Cambridge, UK: Cambridge Univ. Press

Pomiankowski A, Iwasa Y, Nee S. 1991. The evolution of costly mate preferences. 1. Fisher and biased mutation. *Evolution* 45:1422–30

Price T, Schluter D, Heckman NE. 1993. Sexual selection when the female directly benefits. *Biol. J. Linn. Soc. London* 48:187–211

Proulx SR, Day T, Rowe L. 2002. Older males signal more reliably. *Proc. R. Soc. London Ser. B* 269:2291–99

Puurtinen M, Kaitala V. 2002. Mate-search efficiency can determine the evolution of separate sexes and the stability of hermaphroditism in animals. *Am. Nat.* 160:645–60

Queller DC. 1997. Why do females care more than males? *Proc. R. Soc. London Ser. B* 264:1555–57

Qvarnström A, Brommer JE, Gustafsson L. 2006. Testing the genetics underlying the co-evolution of mate choice and ornament in the wild. *Nature* 441:84–86

Randerson JP, Hurst LD. 2001. The uncertain evolution of the sexes. *Trends Ecol. Evol.* 16:571–79

Reynolds JD, Gross MR. 1990. Costs and benefits of female mate choice: Is there a lek paradox? *Am. Nat.* 136:230–43

Rifkin SA, Houle D, Kim J, White KP. 2005. A mutation accumulation assay reveals a broad capacity for rapid evolution of gene expression. *Nature* 438:220–23

Ritchie MG, Saarikettu M, Hoikkala A. 2005. Variation, but no covariance, in female preference functions and male song in a natural population of *Drosophila montana*. *Anim. Behav.* 70:849–54

Rowe L, Cameron E, Day T. 2005. Escalation, retreat, and female indifference as alternative outcomes of sexually antagonistic coevolution. *Am. Nat.* 165:S5–18

Rowe L, Houle D. 1996. The lek paradox and the capture of genetic variance by condition dependent traits. *Proc. R. Soc. London Ser. B* 263:1415–21

Sheldon BC. 2002. Relating paternity to parental care. *Philos. Trans. R. Soc. London Ser. B* 357:341–50

Shuster SM, Wade MJ. 2003. *Mating Systems and Strategies*. Princeton, NJ: Princeton Univ. Press

Siller S. 2001. Sexual selection and the maintenance of sex. *Nature* 411:689–92

Simmons LW. 2001. *Sperm Competition and Its Evolutionary Consequences in the Insects*. Princeton, NJ: Princeton Univ. Press

Simmons LW, Kvarnemo C. 2006. Costs of breeding and their effects on the direction of sexual selection. *Proc. R. Soc. London Ser. B*. 273:465–70

Steppan SJ, Phillips PC, Houle D. 2002. Comparative quantitative genetics: evolution of the G matrix. *Trends Ecol. Evol.* 17:320–27

Takahasi K. 1997. Models of selective mating and the initiation of the Fisherian process. *Proc. R. Soc. London Ser. B* 264:839–44

Tallamy DW. 2005. Egg dumping in insects. *Annu. Rev. Entomol.* 50:347–70

Tomkins J, Radwan J, Kotiaho JS, Tregenza T. 2004. Genic capture and resolving the lek paradox. *Trends Ecol. Evol.* 19:323–28

Trivers R. 1972. Parental investment and sexual selection. In *Sexual Selection and the Descent of Man*, ed. B Campbell, pp. 136–79. Chicago: Aldine

van Doorn GS, Dieckmann U, Weissing FJ. 2004. Sympatric speciation by sexual selection: a critical reevaluation. *Am. Nat.* 163:709–25

Wade MJ, Shuster SM. 2002. The evolution of parental care in the context of sexual selection: a critical reassessment of parental investment theory. *Am. Nat.* 160:285–92

Waxman D, Gavrilets S. 2005. 20 questions on adaptive dynamics. *J. Evol. Biol.* 18:1139–54

Weatherhead PJ, Robertson RJ. 1979. Offspring quality and the polygyny threshold: "the sexy son hypothesis." *Am. Nat.* 113:201–8

Webb JN, Houston AI, McNamara JM, Székely T. 1999. Multiple patterns of parental care. *Anim. Behav.* 58:983–93

Wiley RH, Poston J. 1996. Indirect mate choice, competition for mates, and coevolution of the sexes. *Evolution* 50:1371–81

Wilkinson GS, Reillo PR. 1994. Female choice response to artificial selection on an exaggerated male trait in a stalk-eyed fly. *Proc. R. Soc. London Ser. B* 255:1–6

Williams GC. 1966. *Adaptation and Natural Selection: A Critique of Some Current Thought*. Princeton, NJ: Princeton Univ. Press

Zahavi A. 1975. Mate selection—a selection for the handicap. *J. Theor. Biol.* 53:205–14

Zahavi A. 1977. The cost of honesty (further remarks on the handicap principle). *J. Theor. Biol.* 67:603–5

Genetic Polymorphism in Heterogeneous Environments: The Age of Genomics

Philip W. Hedrick

School of Life Sciences, Arizona State University, Tempe, Arizona 85287-4501;
email: philip.hedrick@asu.edu

Key Words

balancing selection, melanism, MHC, pesticide resistance, tests for selection

Abstract

The selective mechanisms for maintaining polymorphism in natural populations has been the subject of theory, experiments, and review over the past half century. Advances in molecular genetic techniques have provided new insight into many examples of balancing selection. In addition, new theoretical developments demonstrate how diversifying selection over environments may maintain polymorphism. Tests for balancing selection in the current generation, the recent past, and the distant past provide a comprehensive approach for evaluating selective impacts. In particular, sequenced-based tests provide new ways to evaluate the long-term impact of selection on particular genes and the overall genome in natural populations. Overall, there appear to be many loci exhibiting the signal of adaptive directional selection from genomic scans, but the present evidence suggests that the proportion of loci where polymorphism is maintained by environmental heterogeneity is low. However, as more molecular genetic details become available, more examples of polymorphism maintained by selection in heterogeneous environments may be found.

INTRODUCTION

In 1976 and 1986, I wrote reviews (Hedrick 1986, Hedrick et al. 1976; see also Felsenstein 1976) discussing the maintenance of genetic polymorphism in heterogeneous (variable) environments. At that time, most of the well-known examples of genetic polymorphism that appeared to be maintained by diversifying selection were either visible or allozyme polymorphisms. Here, I revisit some of these examples, particularly ones for which there have been recent developments, such as relevant DNA sequence information, and I discuss some new examples. In my previous reviews, theory suggested that maintenance of polymorphism was possible in variable environments but that it was most likely when selection varied in space and/or there was limited gene flow or habitat selection. Here, I discuss some new theoretical developments related to these ideas and some other theory related to selection in variable environments.

Overall, the topics in this review are somewhat changed from my previous reviews. In my earlier reviews, an important issue was the evaluation of experimental studies, mainly in *Drosophila*, examining genetic perturbations, environmental perturbations, and habitat selection, but in recent years, the emphasis on these studies has declined. A new focus has been the emphasis on using DNA sequence data to identify and understand the impact of selection, particularly favorable directional selection, on specific genes. If there is favorable selection for different alleles in different environments, then diversifying selection may occur. I discuss some of this research and how it can be used to identify genes under diversifying selection.

GENERAL EXAMPLES

How important is variable selection in space and/or time in maintaining genetic polymorphism at specific genes? Below, I discuss some of best examples, but the overall number for which there is detailed information is limited. There are a substantial number of examples for which directional selection has been or is important (note that several other examples are discussed as illustrations in the section on theory). Within a given species, present evidence suggests that polymorphism is maintained by diversifying selection at only a few genes. And, except for some genes related to pathogen resistance, these appear to be maintained only in the given species. If polymorphisms are in related species, researchers have often found a separate origin for the polymorphism. In other words, many of these polymorphisms are relatively recent and not maintained over the formation of new species.

I begin with comments about recent investigations in some well-known, or particularly relevant, examples of genetic polymorphism in heterogeneous environments. Because of space limitations, I do not discuss shell polymorphism in the land snail *Cepaea nemoralis* (Cook 1998, Jones et al. 1977) or the extensive experimental work in microorganisms (except for one example in *Pseudomonas fluorescens* below), but as entry into this literature, I recommend the following: Suiter et al. (2003), Elena & Lenski (2003), Zhong et al. (2004), Dykhuizen & Dean (2004), Brisson & Dykhuizen (2004), MacLean (2005), Barrett et al. (2005).

Color Polymorphism

Melanism in *Biston betularia*. A classic example of a genetic polymorphism influenced by diversifying selection is industrial melanism in the peppered moth, *Biston betularia* (Cook 2003, Grant 2005, Kettlewell 1973, Majerus 1998). Industrial melanics were first noticed in England in the mid-nineteenth century and appear to have rapidly spread from a single source. At a particular location, the increase from a low frequency to a frequency greater than 90% often took only several decades. The main selective agent appears to have been differential predation by birds on polluted and unpolluted resting backgrounds, and extensive gene flow significantly influenced the spread and distribution of melanics (Cook 2003). The overall case for natural selection acting on the rise and fall of melanics in peppered moths is indisputable.

Since the introduction of clean air legislation in the 1960s in Great Britain, the frequency of dominant melanic moths has declined. For example, at Caldy Common, the frequency of melanics declined from greater than 90% in 1960 to below 5% by 2002 (**Figure 1**) (Grant 2005). Also given in **Figure 1** is the very similar expected decrease of the melanic phenotype, assuming a 15.3% selective disadvantage for melanics [that is, the fitnesses for the dominant melanics and recessive typicals are 0.847 and 1.0, respectively (Grant et al. 1996)].

B. betularia is also found in North America, and in some areas the frequency of melanics once was quite high (Grant et al. 1996). In the 1960s, clean air legislation also began to result in better air quality in North America, and the frequency of melanics dropped (Grant & Wiseman 2002). Presumably, similar reversals in selection pressures that occurred on both continents were responsible for this remarkable parallel decline. On the basis of crosses between American and British moths, researchers concluded that the melanic variants from the two continents are alleles at the same locus (Grant 2004). When the gene that determines melanism in *B. betularia* is discovered, then DNA data should be able to determine the cause and relationship

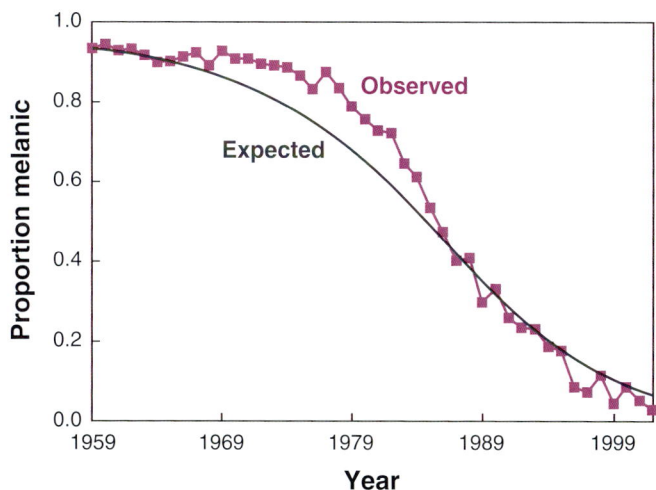

Figure 1

The observed decline in the frequency of melanics from 1959 to 2002 at Caldy Common in England (*magenta squares*) (Grant 2005) and the expected decline when there is 15.3% selection against the melanics (*dark gray line*).

of melanism in different populations, the number of different melanic mutants, their age, what amino acid changes cause melanism, and related factors.

Melanism in mice. From studies of coat color mutants in house mice, researchers identified a number of genes that can cause melanism. However, several molecularly documented cases of melanism in wild vertebrate populations appear to be the result of variation in the melancortin-1 receptor gene (*Mc1r*), including in birds (Mundy 2005), in mammals (Eizirik et al. 2003, Nachman et al. 2003), and in reptiles (Rosenblum et al. 2004).

Nachman et al. (2003) presented a detailed examination of melanism that is due to *Mc1r* variation in the rock pocket mouse (*Chaetodipus intermedius*), a light-colored mouse that generally lives on light-colored granite rocks but has melanic forms that live on black lava in several restricted sites in the southwestern United States. **Figure 2** shows the frequencies of the normal recessive and dominant melanic forms from a 35-km transect in southwestern Arizona (Hoekstra et al. 2004). Here, the frequency of melanics is highly concordant with substrate color: High frequencies of melanics occurred in the center of this transect, which has approximately 10 km of black

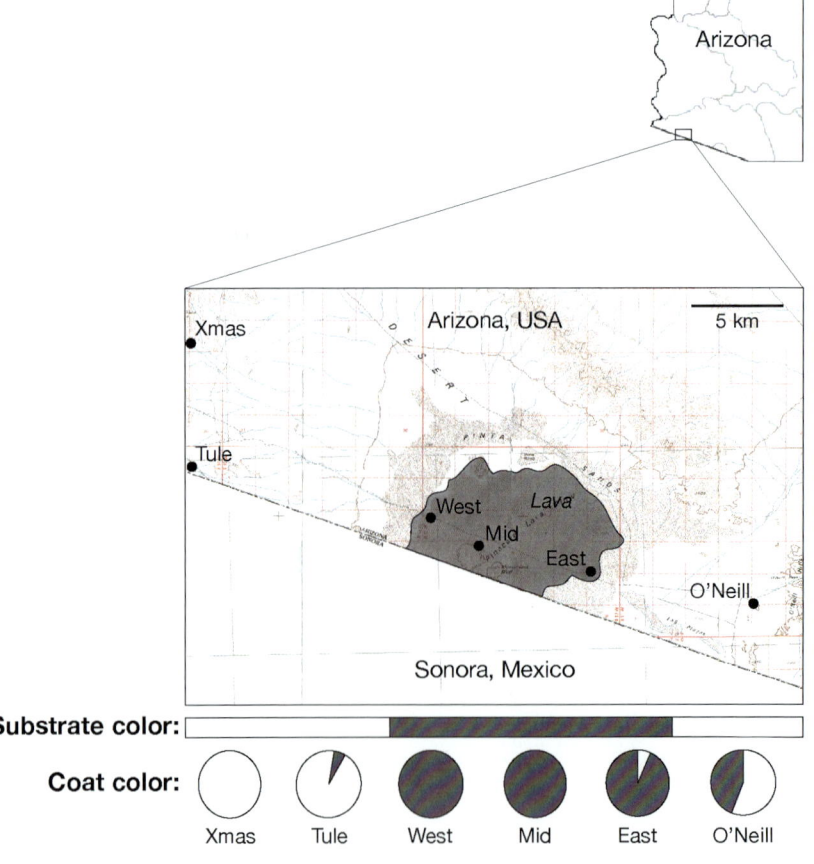

Figure 2

Six sampling sites (three on dark volcanic rock and three on light-colored substrate) and coat color frequencies (in pie diagrams) in rock pocket mice across a transect in the Sonoran desert (Hoekstra et al. 2004).

lava, and lower frequencies of melanics occurred on the light-colored substrate sites at either end of the transect. Overall, 95% of the mice were melanic on the three lava sites, and 23% of the mice were melanic in the three sites with light-colored substrate (most of these melanics were from the O'Neill sample, which, based on mtDNA sequence data, had the most gene exchange with the lava sites). There was no association of mtDNA variation and background color in these samples, and estimates of gene flow for neutral mtDNA markers were substantial.

Investigation of molecular variation in the *Mc1r* gene, which is known to have variants that produce dark-colored house mice, was found to correlate nearly completely with the light and melanic phenotypes. The melanic and normal alleles were found to differ by four amino acid differences (Nachman et al. 2003), and the nucleotide diversity for the melanic alleles was 1/20th that for light alleles. The lower variation among the melanic alleles is consistent with the expected pattern if selection has recently increased its frequency. Although variants at *Mc1r* appear to determine melanism patterns in this instance, in another population with melanics in New Mexico, *Mc1r* was not the gene responsible.

Recently, Hoekstra et al. (2006) identified a single nucleotide mutation in *Mc1r* that is a major determinant of the derived light color in a recent (<6,000 years old) "beach mouse" subspecies of *Peromyscus polionotus*. Matching of background color with pelage coloration for concealment appears to be the selective mechanism. However, in other beach mice, the light color does not appear to be determined by *Mc1r*. A single amino acid change in another gene (*SLC24A5*) has been shown to have a major influence on the difference in skin color between African and European human populations (Lamason et al. 2005). In this case, there is strong evidence for the recent evolution of the light color variant, such as very low heterozygosity for linked genes on haplotypes of the light color variant. These studies of the genetics of color differences within species are providing some of the most convincing examples of strong adaptive selection in different environments.

Pesticide Resistance

Some of the best-documented examples of selection in heterogeneous environments are those resulting from recent human changes in the environment, such as the use of chemicals to control pests (McKenzie 1996). The genetic basis of insecticide resistance may be the result of many genes (particularly in laboratory-selected strains), of mutants at a single or a few genes (ffrench-Constant et al. 2004, Oakeshott et al. 2003), or of expansion of gene families (Labbe et al. 2005, Ranson et al. 2002). Because the molecular basis of many of these genes is known, detailed genetic and evolutionary understanding is possible.

Insects. Resistance to some insecticides among mosquitoes that are vectors for diseases such as malaria (*Anopheles gambiae*) and West Nile virus (*Culex pipiens*) are the result of single amino acid substitutions. For example, a single nucleotide change, GGC (glycine) to AGC (serine) at codon position 119 in the gene for the enzyme acetylcholin-esterase (*ace*-1), in *C. pipiens* results in insensitivity to organophosphates

(Weill et al. 2003). A complete lack of variation within samples among resistant haplotypes suggests that they have originated and spread quite recently. In addition, the mutated amino acid is in the active "gorge" of the enzyme, and the same catalytic properties and insecticidal sensitivity were present in transfected mosquitoes (Weill et al. 2003).

Instead of undergoing a qualitative change, some insects become resistant to insecticides by producing large amounts of detoxifying enzymes. For example, the overtranscription of a single allele at a single cytochrome P450 gene in *Drosophila melanogaster* has been shown to confer worldwide resistance to DDT (Daborn et al. 2002). In this case, microarray analysis of all the P450 genes identified the specific overtranscribed gene [although further analysis (Pedra et al. 2004) suggested that the resistance is more complicated], and subsequent genetic and sequence investigation demonstrated that the single resistance allele resulted from the insertion of an *Accord* transposable element in the 5′ end of the gene. A survey of 34 populations worldwide found a complete correspondence of *Accord* insertion and insecticide resistance (Catania et al. 2004). Another example of an insertion of a transposable element adaptive to organophosphate pesticides by truncation of a gene has recently been documented (Aminetzach et al. 2005).

Warfarin resistance in rats. Warfarin and related anticoagulants were introduced in the early 1950s to control rodent pests, such as the Norway or brown rat, *Rattus norvegicus*, a widespread agricultural and domestic pest. However, resistance to warfarin evolved quickly, and resistant rats (both brown and black) and house mice are now found throughout the United States and Europe (Pelz et al. 2005). Resistance in brown rats is the result of a dominant allele R at an autosomal locus Rw, and the resistant animals are generally heterozygotes for R and for the wild-type, susceptible (S) allele. Warfarin inhibits blood coagulation by repressing the vitamin K reductase reaction (VKOR). Rw has recently been identified as the gene *VKORC1* encoding for a protein in the VKOR pathway (Rost et al. 2004). Missense mutations in this protein have been found in warfarin-resistant humans and rats and have reduced enzyme activity.

In an area along the England-Wales border, RR homozygotes had low viability because of a 20-fold increase in vitamin K requirement (Greaves et al. 1977). When warfarin was applied, a net heterozygote advantage resulted, and the relative survivals of genotypes RR, RS, and SS were estimated as 0.37, 1.0, and 0.68, respectively. In one population, the frequency of R was stable and polymorphic over nearly a decade. When warfarin was no longer applied in another population, the frequency of R declined very quickly, a result that illustrated the cost of maintaining the resistance allele.

Kohn et al. (2000, 2003) attempted to locate the warfarin resistance gene (Rw) and characterize variation at closely linked markers in German rat populations, which have varying levels of warfarin resistance. They determined that there has been a single and recent origin of warfarin resistance in these populations. On the basis of indirect genetic evidence, researchers determined that heterozygote advantage is not present in the German populations perhaps because these resistant rats do not suffer vitamin K deficiency, as documented in other populations (Markussen et al. 2003).

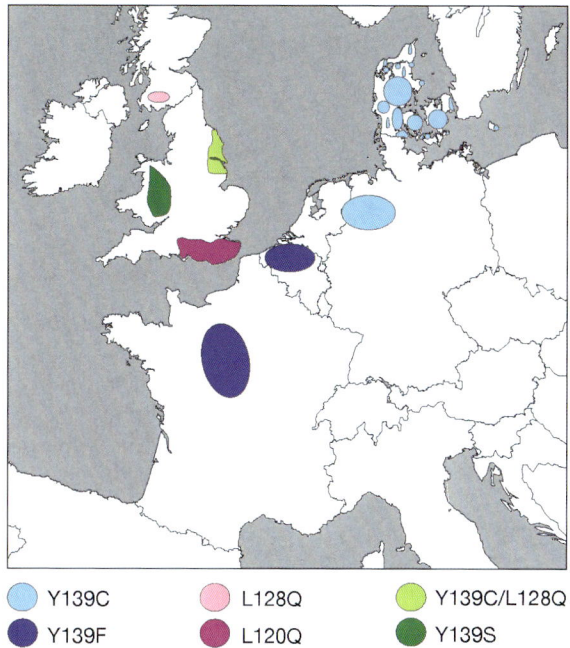

Figure 3
Geographic distribution of warfarin-resistant brown rats in Europe (Pelz et al. 2005). The areas where the six different resistant mutants have been found are indicated by different colors (indicated by ancestral amino acid, amino acid site, and derived amino acid).

In other words, the net heterozygote advantage for warfarin resistance appears to be present for some resistance alleles (populations) and not for others.

Recently, Pelz et al. (2005) have characterized *VKORC1* resistant variants in wild brown rat populations over much of Europe (**Figure 3**). They found eight different mutations, five of which affected only the two amino acids at positions 128 and 139. The mutant Y139F was found in both France and Belgium populations, and another mutant Y139C was found in both Germany and Denmark populations. The variant previously found to exhibit heterozygote advantage along the England-Wales border was a third variant, Y139S, at the same amino acid position. Because new amino acids are found at position 139 for so many resistance alleles, Pelz et al. speculated that Y139 may be part of the warfarin binding site of *VKORC1*. Because these variants have reached substantial frequency in only a few decades, they likely are not the result of new mutants, but rather were already present in these populations at low frequency.

Pathogen Resistance

Major histocompatibility complex (MHC). The MHC genes are part of the immune system in vertebrates, and differential selection through resistance to pathogens is widely thought to be the basis of their high genetic variation (Edwards & Hedrick 1998, Hedrick & Kim 2000, Hughes 1999, Meyer & Thomson 2001, Piertney & Oliver 2006). I provide a brief introduction here because I use MHC genes below as an example of balancing selection. Variation in the genes of the human MHC, known as *HLA* genes, have been the subject of intensive study for many years because of their role in acceptance or rejection of transplanted organs, in many autoimmune diseases, and in recognition of pathogens. With sequence-determined alleles in worldwide human surveys, genes *HLA-A*, *HLA-B*, and *HLA-DR1* have 243, 499, and 321 alleles, respectively (Garrigan & Hedrick 2003), with *HLA-B* being the most variable gene in the human genome (Mungall et al. 2003).

In addition, most of the nucleotide variation is in functionally important parts of the MHC genes and results in amino acid variation. For example, Hedrick et al. (1991) determined the amino acid heterozygosity for the 366 (*HLA-A*) or 363 (*HLA-B*) amino acid sites in humans. There were a number of individual amino acid sites that have heterozygosities greater than 50% for both genes, and most of these amino acid sites appear, based on the three-dimensional structure of the HLA molecule, to have important functions involved with initiating the immune response. Similarly, high heterozygosity for functional amino acid sites has been found for other *HLA* genes (Salamon et al. 1999) and for MHC genes in the Arabian oryx (Hedrick et al. 2000) and wolves (Hedrick et al. 2002).

CCR5 and disease resistance. An apparent adaptive selection example is the 32-base pair coding sequence deletion of the human chemokine receptor 5, *CCR5-$\Delta 32$*. From population surveys, homozygotes for the deletion are resistant to HIV infection, and heterozygotes have a delayed onset of AIDS (Samson et al. 1996). This variant has a quite nonuniform geographical distribution, with a relative high frequency, particularly for a null allele, in northern Europe (16% in some areas) and a lower frequency in southern Europe (4% in Greece), and it has not been found in African and Asian samples (Novembre et al. 2005).

Using theoretical approaches and the observed amount of linkage disequilibrium at two nearby microsatellite loci, the age of the deletion was estimated at only about 700 years old (Stephens et al. 1998) and was conjectured to result in resistance to the plague (Stephens et al. 1998) or smallpox (Galvani & Slatkin 2003). However, Hummel et al. (2005) examined a sample of 2900-year-old Bronze Age skeletons and found that the frequency of the deletion (11.8%) was at similar levels as that found in a modern German sample (9.2%). If the *CCR5-$\Delta 32$* mutant provided protection from the plague, then one would predict that the frequency of this allele would be reduced in plague victims compared with a control group. Hummel et al. (2005) also found no evidence of a reduction in deletion frequency in a sample of fourteenth-century plague victims (14.3%) when compared with a contemporaneous sample of famine victims (12.5%).

Sabeti et al. (2005) used human genomic diversity data to suggest that aspects of the *CCR5-$\Delta 32$* data, including linkage disequilibrium, are not significantly different from other variants at the locus or throughout the human genome. They also used an improved genetic map of closely linked microsatellite loci and 32 SNP (single nucleotide polymorphism) markers and estimated deletion age to be 5075 (3150–7800, 95% confidence interval), consistent with a date in the Bronze Age.

These studies use two approaches that hopefully will become more widespread in other situations, ancient DNA and detailed genomic data. First, dating the deletion to at least 3000 years ago with the Bronze Age samples showed that the deletion was present long before plague was an important human pathogen and likely before smallpox was widespread. Nonetheless, this variant was probably adaptive in conferring resistance to general infectious diseases that activate the immune system response provided by CCR5.

CCR5-Δ32 may be unusual for a selected variant because it results in a nonfunctional allele with a selective advantage. Usually loss of gene function would be under purifying selection, but because HIV and other pathogens use CCR5 as a gateway to initiate infection, loss of function appears advantageous. In red-capped mangabeys, there is a 24-bp deletion at CCR5 that has reached a frequency of 87% (Chen et al. 1998), and CCR5 is silenced altogether in sooty mangabeys, the natural hosts of SIV (simian immunodeficiency virus) (Veazey et al. 2003). In other words, without any major cost to resistance (de Silva & Stumpf 2004) that would result in a balanced selection in the presence of such a pathogen, adaptive selection could eventually result in fixation of *CCR5-Δ32*.

Malaria-resistant variants in humans. Malaria kills more than one million children each year in Africa alone and is the strongest selective pressure in recent human history (Kwiatkowski 2005). As a result, selective protection from malaria by sickle cell, thalassemia, G6PD, Duffy, and many other host variants provide some of the clearest examples of adaptive variation and diversifying selection for pathogen resistance. Genomic studies have demonstrated that selection for malarial resistance is strong, up to 10%, and that variants conferring resistance to malaria are recent (Hamblin et al. 2002, Ohashi et al. 2004, Saunders et al. 2005, Wood et al. 2005), generally less than 5000 years old, consistent with the proposed timing of malaria as an important human disease. Often the resistant variants are in different populations, probably owing in part to their recent independent mutation origin. However, Williams et al. (2005) provided evidence of negative epistasis between thalassemia and sickle cell and suggested that because there is no known cost to α^+-thalassemia, this is a potential explanation for the failure of the thalassemia variant to increase or become fixed in African populations.

Sickle cell hemoglobin provides the classic example of balanced polymorphism, with heterozygotes having a greater resistance to malaria than homozygotes, but sickle cell homozygotes suffer a cost from sickle cell anemia. A third allele *C* at this locus, which has a different single amino acid substitution at the same amino acid position, is in substantial frequency in several west African populations and has been shown to confer higher protection to malaria than sickle cell (Modiano et al. 2001). With estimated relative fitness from this data set, Hedrick (2004) showed that if *C* is introduced by mutation or gene flow, it will always increase and eventually become fixed, although the rate of increase of *C* is a function of the initial frequency of *S*. This outcome occurs primarily because genotype *CC* has the highest estimated relative fitness of any genotype. Overall, it appears that *C* has been slowly increasing in this west African population because of the presence of *S* and that now it is poised to rapidly increase to a high frequency within the next 50 generations, eliminating allele *S* and sickle cell anemia and going to fixation.

***R* (disease-resistant) genes in plants.** In recent years, there has been extensive research examining *R* (disease-resistant) genes in plants (Bergelson et al. 2001, Mauricio et al. 2003, Meyers et al. 2005, Rose et al. 2004, Xiao et al. 2004), which has provided evidence for balancing selection. For example, Stahl et al. (1999) showed that resistance and susceptibility in the highly selfing plant *Arabidopsis thaliana* to the bacterial pathogen *Pseudomonas* is the result of molecular variation at the gene *Rpm1*. In a worldwide survey of *A. thaliana* accessions, they found that all susceptible plants were genotype *rr* and that all resistant plants were *RR*, except for one resistant accession from Kazakhstan that was heterozygous *Rr*. The overall frequency of the resistant *R* allele was estimated to be 0.52 throughout the range of the species. A fitness cost for the presence of this resistance allele in the absence of pathogens has been shown (Tian et al. 2003), providing a counter to the advantage of this allele for disease resistance and an explanation for the polymorphism of this allele. In general, researchers believe that to maintain polymorphism in a gene-for-gene system there needs to be a cost to the resistant host allele and a cost for the virulent pathogen allele, or they will go to fixation.

Allozyme Variation

Diversifying selection was often given as an important basis for the maintenance of genetic variation at allozymes, and discussion of these data was included in my previous reviews. Eanes (1999) provided an extensive review and evaluation of most of the subsequent data and the well-known case studies (see also Schmidt & Rand 2001, Storz & Dubach 2004, Véliz et al. 2004, Watt et al. 2003). Studies have continued to examine functional differences in allozyme molecules and, more recently, the amino acid basis for allozyme alleles (for a summary of these data for *Drosophila melanogaster* allozymes, see Eanes 1999). Recently, analysis of North American latitudinal clines for a number of allozymes in *D. melanogaster* showed that the derived allele is the one increasing with latitude, consistent with the model of an ancestral African species adapting to temperate climates (Sezgin et al. 2004).

Perhaps the most studied allozyme example is variation in the *Adh* locus in *D. melanogaster*, which provided the first detailed examination of DNA variation for an allozyme polymorphism (Kreitman 1983). But even for this extensively studied and cited example, there are still questions about the agent of selection, the selective importance of differences between the *F* (Fast) and *S* (Slow) variants other than the amino acid substitution, and the selective interaction of different loci or sites (Eanes 1999). Further, examination of DNA sequence data illustrated that the *Adh F* haplotypes had much less variation than the *S* haplotypes and showed little divergence from them (Kreitman 1983). As a result, it appears that the *F* haplotypes have increased in frequency only recently (the *Adh* polymorphism is also absent in closely related species).

Examination of *S* haplotypes in samples from Zimbabwe *D. melanogaster* (Begun et al. 1999) demonstrated that there was as much variation within this group as in the worldwide sample (Kreitman 1983) that included *F* haplotypes (see also Veuille et al. 1998). In other words, the excess of variation seen around the amino acid difference between the *F* and *S* haplotypes (Kreitman & Hudson 1991) may have predated the generation and the increase of the *F* haplotypes. Surprisingly, this instance of high, localized variation correlated with a known functional polymorphism remains the only such example in *D. melanogaster*. This suggests either that there are few such balanced polymorphisms in this species or that recombination generally has obscured this signal of balancing selection.

THEORETICAL MODELS

The first theoretical paper that showed that diversifying selection over different habitats could maintain genetic polymorphism was that of Levene (1953). The important and continuing impact of Levene's letter to the editors in *American Naturalist*, which was just over two pages long, is illustrated by its citation nearly 600 times (**Figure 4**), starting with a number of citations in the influential 1955 volume of the Cold Springs Harbor Symposium on Quantitative Biology on "Population Genetics: The Nature and Causes of Genetic Variability in Populations." The number of citations continued to increase over time, but in the late 1980s and early 1990s, there was a dip in citation number. In contemporary literature, Levene's paper continues to be widely cited, with an average of 20 citations per year over the past decade. However, the type of citations has changed, with many early manuscripts extending the theory developed by Levene and later manuscripts stating that this type of selection is a potentially important explanation for the observation of genetic variation in a given species.

As an introduction to this section, I discuss the intuition of the evolutionary biologists that first considered diversifying selection and maintenance of polymorphism. In 1990, I wrote to Howard Levene regarding his short paper on polymorphism when there is variable selection in space (Levene 1953). In his response, Levene wrote that

> Dobzhansky asked me during a conversation whether polymorphism would be easier to achieve in a divided population. His intuitive feeling was that it would. When I told him the results in the letter he was very pleased, and said, "Fine, go ahead and send it

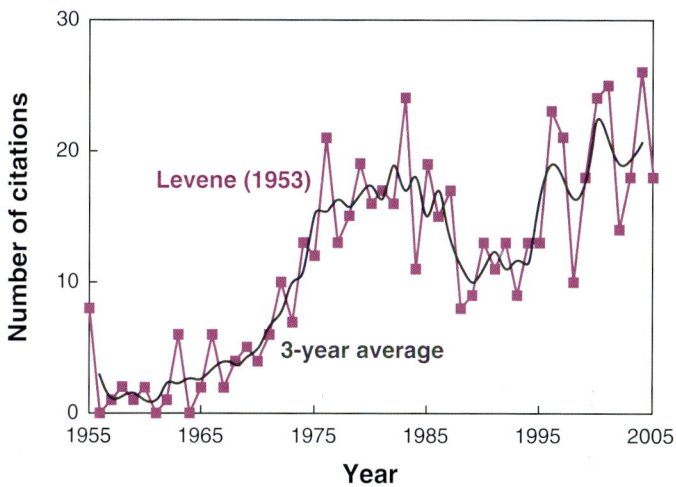

Figure 4

The number of annual citations for the classic article by Levene (1953) (*magenta squares*) over the past 50 years and the three-year running average of citation number (*dark gray line*).

as a letter. But of course, you will now go into it in detail." I doubted I would and said the only really interesting question was how about selection differing by time instead of location. But then that was done by Everett Dempster. I don't remember what the other model with variable population size was and obviously I never did anything with it.

In addition, in the summary of Levene (1953), he stated that

> The model proposed is obviously not realistic: however, if it is modified by supposing that individuals move preferentially to niches they are better fitted for, or that there is a tendency for mating to occur within a niche rather than at random over the whole population, conditions will be more favorable for equilibrium, so that in a sense we are considering the worst possible case.

From the theoretical models developed since Levene's, we know that diversifying selection over either time or space generally does result in broader conditions for maintenance of polymorphism than when fitnesses are constant over environments. However, the conditions are generally more robust when there is spatial variation in selection rather than temporal variation in selection (Frank & Slatkin 1990, Hedrick 1986, Hedrick et al. 1976). Intuitively, if the environment varies over time—for example, one year is wet and the next year is dry—then every individual must endure every different environment whether or not they are genetically adapted to it. However, if the environment varies over space—for example, one area is wet and another dry—then only part of the population encounters a particular environment at a given time. In other words, when there is spatial variation in selection, some genotypes and a proportion of their descendants that are well adapted in a particular environment may not even encounter environments to which they are not adapted.

It is again important to emphasize that variable selection over space or time is not sufficient for the maintenance of polymorphism (Hedrick 1986, Levene 1953) because of the often-stated assumption that genotype-environment interaction will result in maintenance of genetic variation. Further, when there are more than two alleles and/or more than one locus and multiple environments, the conditions are not necessarily more robust.

For example, Prout & Savolainen (1996) examined the potential for stable polymorphism in the leafhopper, *Mocydia crocea*, using the fitness estimates of six leafhopper morphs on 20 host plants (Müller 1987). Exploring different starting frequencies of the gametes involved in determining the phenotypes, they found that variation at one gene was always fixed, while a two-allele polymorphism was maintained at the other locus. In other words, the conclusion that variation in fitness values of different leafhopper morphs on different host plants could explain the genetic variation (Müller 1987) does not appear to be true. However, they did find that each allele was favored on at least one host plant, and if gene flow was small between host plant species, then there would be a global polymorphism for all alleles.

For quantitative characters, the conditions for maintenance of genetic variation vary with genotype-environment interaction and sometimes do not result in the maintenance of variation (Via & Lande 1987), whereas other models do (Gillespie & Turelli

1989). Because of space constraints, I do not examine the maintenance of quantitative genetic variation by diversifying selection (see Byers 2005, Turelli & Barton 2004).

Spatial Models

Levene (1953) introduced the first theoretical model for examining the impact of diversifying selection in space. With random mating (and soft selection), the conditions that Levene derived depend on the harmonic mean of the heterozygotes being larger than the harmonic mean of the homozygotes. When there is limited gene flow and/or habitat selection, this further increases the proportion of the population that escapes the environment to which it is not adapted.

One of the best examples of spatial structure in the environment resulting in maintenance of genetic polymorphism is in the common bacterium *Pseudomonas fluorescens*, which evolves rapidly under novel environmental conditions and generates a variety of mutants (Rainey & Travisano 1998). In a homogeneous environment (constantly shaken microcosm), the original morph, smooth, is the only detectable one. But in a spatially heterogeneous environment (no shaking of microcosm) at least three morphs, smooth, wrinkly-spreader, and fuzzy-spreader, were found in substantial frequency in all replicate microcosms. Further, transfer of cultures from the heterogeneous environment to the homogeneous environment resulted in loss of about three quarters of the variation in just two weeks, and subsequent transfer back to the heterogeneous environment quickly restored the diversity level. Additional study of this system has examined the evolution of fitness trade-offs for these mutants, the genetic basis of the mutants, and other aspects of the system (Brockhurst et al. 2004, MacLean et al. 2004, Spiers et al. 2002).

When there is limited gene flow and differential selection between populations, the situation may become somewhat complicated (Hedrick 1995a, Kawecki & Ebert 2004, Lenormand 2002). As an example, on several islands in Lake Erie, the water snake, *Nerodia sipedon*, has both banded forms, as found on the mainland, and gray, unbanded forms thought to be a balance between selection against the banded forms on flat limestone rocks of the island because of bird predation and gene flow to the island from the banded mainland populations. The inheritance of banding appears to be determined primarily by variation at a single locus (King 1993), with the banded form dominant over the unbanded. King & Lawson (1995) estimated that the selective advantage (s) of unbanded individuals on the islands was between 0.11 and 0.28, and from variation in frequencies of allozyme alleles among the islands, they obtained an estimate of gene flow of $m = 0.01$.

Using these values of s and assuming that $m = 0.01$, the expected allele frequency of the unbanded allele on the islands can be calculated (Hedrick 2005). If $s = 0.11$, then there is an unstable equilibrium at 0.101 and a stable one at 0.890, and if $s = 0.28$, then there is an unstable equilibrium at 0.037 and a stable one at 0.953. For both levels of selection, the stable equilibrium frequencies are somewhat higher than the frequencies observed [0.53 to 0.86 on different islands, with a mean of 0.73 (King & Lawson 1995)]. This difference may be because s is lower or m higher than estimated, or perhaps the populations have not yet come to an equilibrium state. Overall, this case study illustrates how the various evolutionary factors can be estimated and then put in a theoretical context to explain the observed patterns of genetic variation. It also illustrates that even for this relatively simple situation, the findings may be complicated.

Genotype-specific habitat selection such that individuals prefer niches in which they have higher fitness results in broader theoretical conditions for a polymorphism (Hedrick 1990). Further, estimates of average habitat selection (Hedrick 1990) based on *Drosophila* habitat choice data also provided much broader conditions for polymorphism. A single gene appears responsible for the polymorphism in bill size in the African finch *Pyrenestes* (Smith 1993), and, in this case, bill size and feeding preference are correlated, suggesting that the maintenance of genetic variation in this random mating population is related to habitat selection. In addition, sex-dependent habitat selection by itself can sometimes enhance the maintenance of polymorphism (Hedrick 1993). Recently, Ravigné et al. (2004) evaluated protected polymorphism for habitat selection under different models of population regulation.

Temporal Models

The first considerations of the maintenance of polymorphism from temporally variable selection were by Dempster (1955) and Haldane & Jayakar (1963). The conditions for an equilibrium for two alleles in diploids were that the geometric mean fitness of the heterozygote needs to be larger than the geometric mean fitness of the homozygotes (Haldane & Jayakar 1963), and the same conditions were shown to apply to multiple alleles (Turelli 1981). Even though the conditions for a stable polymorphism in an infinite population are unrelated to environmental pattern (Gillespie 1973), the distribution of allele frequencies is strongly affected by the autocorrelation between subsequent environments (Hedrick 1976). Recently, a temporal selection approach was used to model maintenance of multiple alleles at a MHC locus in which multiple pathogens were present or absent in different generations (Hedrick 2002).

Although individuals generally cannot avoid unfavorable environments when selection varies in time, some organisms may be able to avoid an environment for which they are not adapted because they can exist in a life stage that does not encounter the effect of the environment. For example, some plants have extensive seed pools, and the seeds that do not germinate do not experience many of the environmental effects encountered by the seeds that do germinate. Likewise, insects or other animals that undergo diapause can avoid environments encountered by individuals that develop without diapause. Such models have been investigated in an ecological context to determine their influence on the coexistence of competing species. This storage effect allows each of two competing species to survive environments for which they are not adapted by having a life-history stage that avoids the unfavorable environment.

These ecological models have been extended to show that the conditions for maintenance of genetic variation can be greater in fluctuating environments in which there is the opportunity for genotypes to escape unfavorable environments via the storage mechanism (Ellner & Hairston 1994, Hedrick 1995b). An example of temporal selection variation and seed dormancy is provided by flower color variation in a small annual plant of the Mohave Desert, *Linanthus parryae*. This variation was the focus of debate among early evolutionary geneticists about what factors were important in determining the pattern of genetic variation and whether it was consistent with genetic drift (Epling & Dobzhansky 1942, Wright 1943) or explained by natural selection (Epling et al. 1960). Schemske & Bierzychudek (2001) reexamined this classic example in a polymorphic population where the frequency of plants with blue flowers ranged from 9% to 15.9%. For the four years with lower seed numbers (1988, 1989, 1992, and 1993), blue-flowered plants had many more seeds than did white-flowered plants, whereas for the three years with higher seed numbers, the opposite was true.

Turelli et al. (2001) developed theory to explain this situation. They determined the conditions for a stable polymorphism with variable selection and variable contributions to the seed bank and showed that a genotype whose arithmetic and geometric means are both less than unity can persist if its relative fitness is greater than unity in years with high reproduction (high contribution to the seed bank). In fact, they found that the higher fitness of the blue-flowered plants is balanced by a lower contribution of blue-flowered plants to the seed bank.

TESTS FOR BALANCING SELECTION

In the past few years, with the availability of multiple genome sequences for a number of organisms (particularly in humans), there has been an intensive search for genomic regions exhibiting a signal of adaptive (positive Darwinian) selection (Akey et al. 2004, Bustamante et al. 2005, Nielsen et al. 2005, Wang et al. 2006). Many of the regions identified have undergone a selective sweep because of favorable directional selection, as indicated by the signal of a high nonsynonymous to synonymous substitution ratio, or by low genetic variation and/or high linkage disequilibrium in closely linked regions. There appear to be many genomic regions with a history of adaptive selection, but only a small proportion of these indicate a genetic signal consistent with balancing selection. However, the smaller region thought to exhibit a balancing selection signal, compared with a selective sweep, may make detection beyond the resolution of most genomic scans. Further, alleles identified under directional selection may be part of balancing selection at a gene, and further examination of specific alleles may be necessary to clarify the nature of selection (Wang et al. 2006).

Recent studies in plants indicate that many loci reflect balancing or positive selection, although this may be an overestimate because demographic effects may not be accounted for and the examined loci are not a random sample (Wright & Gaut 2004). Genomic regions of *Arabidopsis thaliana* with high levels of intraspecific variability were specifically examined by Cork & Purugganan (2005), who identified several loci with other signals of balancing selection. Of course, identifying the basis of the fitness and phenotypic differences between the genotypes for these genes is necessary to understand their adaptive effects.

The MHC genes have become a paradigm for balancing selection, so much so that neutralists have often used them as an example. Garrigan & Hedrick (2003) examined the evidence for balancing selection at the MHC and for heuristic reasons divided the evidence as that in the current generation, that over the history of populations (recent past), and that over the history of species (distant past). Evidence in the current generation is produced in that generation and can be lost in a generation, given that selection is not present. The generation of a balancing selection signal in the recent past is primarily determined by selection in combination with factors such as genetic drift, gene flow, and recombination and is generated or lost over tens to thousands of generations, depending on the influence of these factors. The signal of selection in the distant past is primarily determined by mutation and takes many thousands or millions of generations to generate or to lose (Garrigan & Hedrick 2003). Although the extent of evidence is less for other loci than it is for MHC genes, this division based on timescale and evolutionary factors is useful in evaluating evidence for balancing selection.

Current Generation

Identifying balancing selection in the current generation has been the classical approach, and it uses deviations from Hardy-Weinberg proportions, Mendelian proportions, or random mating proportions, as well as associations of specific genotypes and fitness when exposed to a given environment. As an illustration of the last phenomenon, Arkush et al. (2002) examined survival of endangered winter-run chinook salmon from known MHC genetic matings (one parent was heterozygous and the other homozygous) when exposed to three different important salmonid pathogens. For example, for the 10 families, each with 60 progeny, exposed to the infectious hematopoietic necrosis virus, 245 of 299 (81.9%) heterozygotes survived, and 228 out of 305 (74.8%) homozygotes survived. The estimated selection against homozygotes was 8.8%.

Recent Past

Since the development of the neutral theory nearly 40 years ago, a number of tests for evidence of balancing selection in the recent past have been developed. These include the Ewens-Watterson test (Ewens 1972, Watterson 1978) of allele frequency distribution, more linkage disequilibrium and less genetic variation in the particular region than for neutral markers, and geographic differentiation different from that found for neutral markers. For example, **Table 1** gives the estimate of F, the Hardy-Weinberg level of homozygosity for *HLA-A* for 12 different human populations (Garrigan & Hedrick 2003). Using the Ewens-Watterson test, three of these populations have less Hardy-Weinberg homozygosity than expected under neutrality, given the sample size and number of alleles.

Lewontin & Krakauer (1973) developed a test to examine the variation over populations for given loci. In theory, all loci undergo similar effects of genetic drift and gene flow, producing an expected variance over populations, whereas differential selection over populations increases this expected variance, and the same balancing selection over populations decreases this variance. In recent years, versions of this test have been used to identify outlier loci, that is, loci that have a significantly higher (or lower) F_{ST} than expected and therefore may be selectively important (for reviews, see Beaumont 2005, Luikart et al. 2003, Storz 2005).

Table 1 Ewens-Watterson's homozygosity *F* and Tajima's *D* statistics, and their probabilities (significant values are given in boldface), calculated for a global sample of *HLA-A* sequences, given the observed sample size (2N) and observed number of alleles (*k*) (Garrigan & Hedrick 2003)

Population	2N	k	F	Prob (F)	D	Prob (D)
Ainu	100	9	0.182	0.06	3.143	<**0.01**
Australian	367	7	0.327	0.24	3.736	<**0.01**
Chinese	298	22	0.140	0.48	2.356	**0.01**
French	248	20	0.162	0.60	2.440	<**0.01**
Havasupai	244	3	0.401	**0.03**	4.049	<**0.01**
Kagui	100	4	0.356	0.09	3.254	<**0.01**
Molucca	52	7	0.214	0.09	2.514	<**0.01**
Omani	236	27	0.100	0.42	2.498	<**0.01**
PNG-Lowland	94	6	0.336	0.32	2.447	<**0.01**
PNG-Highland	188	5	0.634	0.75	1.013	0.13
Zapotec	137	6	0.235	**0.02**	1.976	**0.01**
Zulu	200	23	0.069	<**0.01**	2.441	<**0.01**
Global	2164	45	0.126	0.73	3.691	<**0.01**

Distant Past

Finally, for evidence of balancing selection that may be from the distant past, tests used to detect these signals (Bamshad & Wooding 2003) are more recent and are based on the distribution of sequenced alleles and/or the level of sequence variability, such as the tests of Tajima (1989), Fu & Li (1993), and Fay & Wu (2000), or the comparative sequence divergence and/or variability between different classes of mutation, such as the ratios of nonsynonymous to synonymous substitutions and the tests of Hudson et al. (1987) and McDonald & Kreitman (1991). For example, Tajima's *D* was calculated in the same populations for which we examined *F* (**Table 1**), and 11 of 12 populations examined are significant using Tajima's test (Garrigan & Hedrick 2003). This significance occurs because Tajima's test specifically examines the extent of sequence divergence between the alleles, and the *HLA-A* alleles are very divergent from each other, whereas the Ewens-Watterson test does not incorporate sequence data. As a result, the sequence-based test of Tajima has much greater statistical power than does the allele frequency–based Ewens-Watterson test.

One of the most widely used sequence-based tests is that of McDonald & Kreitman (1991), a simple statistical test to examine synonymous and nonsynonymous sites in the coding region of a gene within and between species. If the observed variation is neutral, then the rate of substitution between species and the amount of variation within species are both a function of the mutation rate, and the ratio of nonsynonymous (replacement) to synonymous (silent) fixed differences between species should be the same as the ratio of nonsynonymous to synonymous polymorphisms within species.

Table 2 The number of nonsynonymous (N) and synonymous (S) substitutions for fixed (F) differences between species and polymorphism (P) within species, and the ratio N/S, for *Adh* (McDonald & Kreitman 1991) and *G6pd* (Eanes et al. 1993) in *Drosophila* and for *G6pd* (Verrelli et al. 2002) and *HLA-A* (Garrigan & Hedrick 2003) in humans and chimpanzees

	Drosophila				Human			
	Adh		*G6pd*		*G6pd*		*HLA-A*	
	F	P	F	P	F	P	F	P
N	7	2	21	2	0	5	0	76
S	17	42	26	36	44	23	0	49
Ratio	0.41	0.05	0.81	0.06	0.00	0.28	—	1.61

McDonald & Kreitman (1991) applied the test to data from the coding region of *Adh* from *Drosophila* species (**Table 2**), and the ratio of nonsynonymous to synonymous fixed differences was 7/17 = 0.41, whereas the ratio of nonsynonymous to synonymous polymorphisms was only 2/42 = 0.05. They suggested that the excess of nonsynonymous substitutions results from fixation of selectively advantageous mutations [see a more extreme situation in **Table 2** for the *G6pd* gene in *Drosophila* (Eanes et al. 1993)]. **Table 2** also gives data for two human genes, *G6pd* and *HLA-B*, known to be under positive selection. Although the pattern for the human *G6pd* gene suggested strong, historical, purifying selection, polymorphic nonsynonymous variants are in high frequency, contrary to the prediction of low frequency under purifying selection (Verrelli et al. 2002). For *HLA-B* (and three other *HLA* genes), an extreme pattern was found, reflective of very strong balancing selection (Garrigan & Hedrick 2003). That is, there were more nonsynonymous polymorphic sites than polymorphic synonymous sites, the polymorphic nonsynonymous variants are often in intermediate frequency, and there were no fixed differences between humans and chimpanzees, either for nonsynonymous or synonymous variants.

If balancing selection maintains variation for a long period, then the same polymorphism may be present in different related species. This phenomenon, known as trans-species polymorphism (Klein 1987), is a common feature for MHC genes (Garrigan & Hedrick 2003, Klein et al. 1998), and, as a result, often the most similar MHC sequence is not in the same species but in a related species. A dramatic illustration of the long-term impact of balancing selection is in artiodactyls (sheep, goats, cows, etc.), where balancing selection appears to have maintained allelic lineages for over 20 million years (Gutiérrez-Espeleta et al. 2001). Takahata (1990) showed that balancing selection results in a phylogenetic pattern similar to that of neutrality but on a much longer timescale [see discussion of self-incompatibility alleles in plants that have similar patterns in Hughes (1999) and Castric & Vekemans (2004)].

How common is trans-species polymorphism for other loci? An initial screen of sequences from humans and chimps found very little trans-species polymorphism (Asthana et al. 2004), and there are very few other documented examples (Hughes

1999, Wuif et al. 2004). In other words, balancing selection generally is not strong enough, nor does it act long enough to maintain polymorphism over species.

CONCLUSIONS

Recent genetic information has provided details on the basis of some of the examples of selection in heterogeneous environments. The next few years should provide many more molecular details of known examples and provide potentially new candidate loci to be examined further. There appear to be many loci exhibiting the signal of adaptive directional selection from genomic scans, but overall the proportion of loci where polymorphism is maintained by environmental heterogeneity is low. Future investigations should provide a general picture of how selection has shaped the history of different parts of the genome and indicate genes for further detailed investigation of selection in heterogeneous environments.

LITERATURE CITED

Akey JM, Eberle MA, Rieder MJ, Carlson CS, Shriver MD, et al. 2004. Population history and natural selection shape patterns of genetic variation in 132 genes. *PLoS Biol.* 2:1591–99

Aminetzach YT, Macpherson JM, Petrov DA. 2005. Pesticide resistance via transposition-mediated adaptive gene truncation in *Drosophila*. *Science* 309:764–67

Arkush KD, Giese AR, Mendonca HL, McBride AM, Marty GD, Hedrick PW. 2002. Resistance to three pathogens in the endangered winter-run chinook salmon (*Oncorhynchus tshawytscha*): effects of inbreeding and major histocompatibility complex genotypes. *Can. J. Fish. Aquat. Sci.* 59:966–75

> This large-scale experiment examined MHC resistance in endangered salmon to three different pathogens.

Asthana S, Schmidt S, Sunyaev S. 2004. A limited role for balancing selection. *Trends Genet.* 21:30–32

Bamshad M, Wooding SP. 2003. Signatures of natural selection in the human genome. *Nat. Rev. Genet.* 4:99–111

Barrett RDH, MacLean RC, Bell G. 2005. Experimental evolution of *Pseudomonas fluorescens* in simple and complex environments. *Am. Nat.* 166:470–80

Beaumont MA. 2005. Adaptation and speciation: what can F_{st} tell us? *Trends Ecol. Evol.* 20:435–40

Begun DJ, Betancourt AJ, Langley CH, Stephan W. 1999. Is the fast/slow allozyme variation at the *Adh* locus of *Drosophila melanogaster* an ancient balanced polymorphism? *Mol. Biol. Evol.* 16:1816–19

Bergelson J, Fretman M, Stahl EA, Tian D. 2001. Evolutionary dynamics of plant R-genes. *Science* 292:2281–85

Brisson D, Dykhuizen DE. 2004. *ospC* diversity in *Borrelia burgdorferi*: different hosts are different niches. *Genetics* 168:713–22

Brockhurst MA, Rainey PB, Buckling A. 2004. The effect of spatial heterogeneity and parasites on the evolution of host diversity. *Proc. R. Soc. London Ser. B* 271:107–11

Bustamante CD, Fledel-Alon A, Williamson S, Nielsen R, Hubisz MT, et al. 2005. Natural selection on protein-coding genes in the human genome. *Nature* 437:1153–57

Byers DL. 2005. Evolution in heterogeneous environments and the potential of maintenance of genetic variation in traits of adaptive significance. *Genetica* 123:107–24

Castric V, Vekemans X. 2004. Plant self-incompatibility in natural populations: a critical assessment of recent theoretical and empirical advances. *Mol. Ecol.* 13:2873–89

Catania F, Kauer MO, Daborn PJ, Yen JL, Ffrench-Constant RH, Schlötterer C. 2004. World-wide survey of an *Accord* insertion and its association with DDT resistance in *Drosophila melanogaster*. *Mol. Ecol.* 13:2491–504

Chen Z, Kwon D, Jin Z, Monard S, Telfer P, et al. 1998. Natural infection of a homozygous Δ24 CCR5 red-capped mangabey with an R2b-tropic simian immunodeficiency virus. *J. Exp. Med.* 188:2057–65

Cook LM. 1998. A two-stage model for *Cepaea* polymorphism. *Philos. Trans. R. Soc. London Ser. B* 353:1577–93

Cook LM. 2003. The rise and fall of the *Carbonaria* form of the peppered moth. *Q. Rev. Biol.* 78:399–417

Cork JM, Purugganan MD. 2005. High-diversity genes in the *Arabidopsis* genome. *Genetics* 170:1897–911

Daborn PJ, Yen JL, Bogwitz MR, Le Goff G, Feil E, et al. 2002. A single P450 allele associated with insecticide resistance in *Drosophila*. *Science* 297:2253–56

Dempster E. 1955. Maintenance of genetic heterogeneity. *Cold Spring Harbor Symp. Quant. Biol.* 20:25–32

de Silva E, Stumpf MPH. 2004. HIV and the CCR5-Δ32 resistance allele. *FEMS Microbiol. Lett.* 241:1–12

Dykhuizen DE, Dean AM. 2004. Evolution of specialists in an experimental microcosm. *Genetics* 167:2015–26

Eanes WF. 1999. Analysis of selection on enzyme polymorphisms. *Annu. Rev. Ecol. Syst.* 30:301–26

Eanes WF, Kirchner M, Yoon Y. 1993. Evidence for adaptive evolution of the G6PD gene in the *Drosophila melanogaster* and *D. simulans* lineages. *Proc. Natl. Acad. Sci. USA* 90:7475–79

Edwards S, Hedrick PW. 1998. Evolution and ecology of MHC molecules: from genomics to sexual selection. *Trends Ecol. Evol.* 13:305–11

Eizirik E, Yukhi N, Johnson WE, Menotti-Raymond M, Hannah SS, O'Brien SJ. 2003. Molecular genetics and evolution of melanism in the cat family. *Curr. Biol.* 13:448–53

Elena SF, Lenski RE. 2003. Evolution experiments with microorganisms: the dynamics and genetic bases of adaptation. *Nat. Rev. Genet.* 4:457–69

Ellner S, Hairston NG. 1994. Role of overlapping generations in maintaining genetic variation in a fluctuation environment. *Am. Nat.* 14:403–17

Epling C, Dobzhansky T. 1942. Genetics of natural populations. IV. Microgeographic races in *Linanthus parryae*. *Genetics* 27:317–32

Epling C, Lewis H, Ball FM. 1960. The breeding group and seed storage: a study in population dynamics. *Evolution* 14:238–55

Ewens WJ. 1972. The sampling theory of selectively neutral alleles. *Theor. Popul. Biol.* 3:87–112

Fay JC, Wu CI. 2000. Hitchhiking under positive Darwinian selection. *Genetics* 155:1405–13

Felsenstein J. 1976. The theoretical population genetics of variable selection and migration. *Annu. Rev. Genet.* 10:253–80

ffrench-Constant RH, Daborn PJ, Le Goff G. 2004. The genetics and genomics of insecticide resistance. *Trends Genet.* 20:167–70

Frank SA, Slatkin M. 1990. Evolution in a variable environment. *Am. Nat.* 136:244–60

Fu YX, Li WH. 1993. Maximum likelihood estimation of population parameters. *Genetics* 134:1261–70

Galvani AP, Slatkin M. 2003. Evaluating plague and smallpox as historical selective pressures for the *CCR5-Δ32* HIV-resistance allele. *Proc. Natl. Acad. Sci. USA* 100:15276–79

Garrigan D, Hedrick PW. 2003. Perspective: detecting adaptive molecular evolution, lessons from the MHC. *Evolution* 57:1707–22

Gillespie JH. 1973. Natural selection with varying selection coefficients—a haploid model. *Genet. Res.* 21:115–20

Gillespie JH, Turelli M. 1989. Genotype-environment interactions and the maintenance of polygenic variation. *Genetics* 121:129–38

Grant BS. 2004. Allelic melanism in American and British peppered moths. *J. Hered.* 95:97–102

Grant BS. 2005. Industrial melanism. In *Encyclopedia of Life Sciences*, pp. 1–9. Chichester, UK: Wiley. http://www.els.net/. doi:10.1038/npg.els.0004.150

Grant BS, Owen DF, Clarke CA. 1996. Parallel rise and fall of melanic peppered moths in America and Britain. *J. Hered.* 87:351–57

Grant BS, Wiseman LL. 2002. Recent history of melanism in American peppered moths. *J. Hered.* 93:86–90

Greaves JH, Redfern R, Ayres PB, Gill JE. 1977. Warfarin resistance: a balanced polymorphism in the Norway rat. *Genet. Res.* 30:257–63

Gutiérrez-Espeleta GA, Hedrick PW, Kalinowski ST, Garrigan D, Boyce WM. 2001. Is the decline of desert bighorn sheep from infectious disease the result of low MHC variation? *Heredity* 86:1–13

Haldane JBS, Jayakar SD. 1963. Polymorphism due to selection of varying direction. *J. Genet.* 58:237–42

Hamblin MT, Thompson EE, Di Rienzo A. 2002. Complex signatures of natural selection at the Duffy blood group locus. *Am. J. Hum. Genet.* 70:169–83

Hedrick PW. 1976. Genetic variation in a heterogeneous environment. II. Temporal heterogeneity and directional selection. *Genetics* 84:145–50

Hedrick PW. 1986. Genetic polymorphism in heterogeneous environments: a decade later. *Annu. Rev. Ecol. Syst.* 17:535–66

Hedrick PW. 1990. Genotypic-specific habitat selection: a new model and its application. *Heredity* 65:145–49

> Perspective on the strongest example of balancing selection, MHC, and the gain and loss of balancing selection signals.

Hedrick PW. 1993. Sex-dependent habitat selection and genetic polymorphism. *Am. Nat.* 141:491–500

Hedrick PW. 1995a. Gene flow and genetic restoration: the Florida panther as a case study. *Conserv. Biol.* 9:996–1007

Hedrick PW. 1995b. Genetic polymorphism in a temporally varying environment: effects of delayed germination or diapause. *Heredity* 75:164–70

Hedrick PW. 1998. Maintenance of genetic polymorphism: spatial selection and self-fertilization. *Am. Nat.* 152:145–50

Hedrick PW. 2002. Pathogen resistance and genetic variation at MHC loci. *Evolution* 56:1902–8

Hedrick PW. 2004. Estimation of relative fitnesses from relative risk data and the predicted future of hemoglobin alleles S and C. *J. Evol. Biol.* 17:221–24

Hedrick PW. 2005. *Genetics of Populations*. Boston: Jones & Bartlett. 3rd ed.

Hedrick PW, Ginevan ME, Ewing EP. 1976. Genetic polymorphism in heterogeneous environments. *Annu. Rev. Ecol. Syst.* 7:1–32

Hedrick PW, Kim T. 2000. Genetics of complex polymorphisms: parasites and maintenance of MHC variation. In *Evolutionary Genetics: From Molecules to Morphology*, ed. R Singh, C Krimbas, pp. 204–34. New York: Cambridge Univ. Press

Hedrick PW, Lee RN, Garrigan D. 2002. MHC variation in red wolves: evidence for ancestry from coyotes and balancing selection. *Mol. Ecol.* 11:1905–13

Hedrick PW, Parker KM, Gutierrez-Espeleta G, Rattink A, Lievers K. 2000. Major histocompatibility complex variation in the Arabian oryx. *Evolution* 54:2145–51

Hedrick PW, Whittam TS, Parham P. 1991. Heterozygosity at individual amino acid sites: extremely high levels for *HLA-A* and *-B* genes. *Proc. Natl. Acad. Sci. USA* 88:5897–901

Hoekstra HE, Drumm KE, Nachman MW. 2004. Ecological genetics of adaptive color polymorphism in pocket mice: geographic variation in selected and neutral genes. *Evolution* 58:1329–41

Hoekstra HE, Hirschmann RJ, Bundey RA, Insel PA, Crossland JP. 2006. A single amino acid mutation contributes to adaptive beach mouse color pattern. *Science* 313:101–4

Hudson RR, Kreitman M, Aguade M. 1987. A test of neutral molecular evolution based on nucleotide data. *Genetics* 116:153–59

Hughes AL. 1999. *Adaptive Evolution of Genes and Genomes*. New York: Oxford Univ. Press

Hummel S, Schmidt D, Kremeyer B, Herrmann B, Oppermann M. 2005. Detection of the *CCR5-Δ32* HIV resistance gene in Bronze Age skeletons. *Genes Immun.* 6:371–74

Jones JS, Leith BH, Rawlings P. 1977. Polymorphism in *Cepaea*: a problem with too many solutions. *Annu. Rev. Ecol. Syst.* 8:109–43

Kawecki TJ, Ebert D. 2004. Conceptual issues in local adaptation. *Ecol. Lett.* 7:1225–41

Kettlewell B. 1973. *The Evolution of Melanism*. London: Oxford Univ. Press

Report on the detection of the HIV-resistant null deletion *CCR5-Δ32* in 2900 Bronze Age skeletons.

King RB. 1993. Color pattern variation in Lake Erie water snakes: inheritance. *Can. J. Zool.* 71:1985–90

King RB, Lawson R. 1995. Color-pattern variation in Lake Erie water snakes: the role of gene flow. *Evolution* 49:885–96

Klein J. 1987. Origin of major histocompatibility complex polymorphism: the trans-species hypothesis. *Hum. Immunol.* 19:155–62

Klein J, Sato A, Nagl S, O'hUigin C. 1998. Molecular trans-species polymorphism. *Annu. Rev. Ecol. Syst.* 29:1–21

Kohn MH, Pelz HJ, Wayne RK. 2000. Natural selection mapping of the warfarin-resistance gene. *Proc. Natl. Acad. Sci. USA* 97:7911–15

Kohn MH, Pelz HJ, Wayne RK. 2003. Locus-specific genetic differentiation at *Rw* among warfarin-resistant rat (*Rattus norvegicus*) populations. *Genetics* 164:1055–70

Kreitman M. 1983. Nucleotide polymorphism at the *alcohol dehydrogenase* locus of *Drosophila melanogaster*. *Nature* 304:411–17

Kreitman M, Hudson RR. 1991. Inferring the evolutionary histories of the *Adh* and *Adh-dup* loci in *Drosophila melanogaster* from patterns of polymorphism and divergence. *Genetics* 127:565–82

Kwiatkowski DP. 2005. How Malaria has affected the human genome and what human genetics can teach us about malaria. *Am. J. Hum. Genet.* 77:171–92

Labbe P, LeNormand T, Raymond M. 2005. On the worldwide spread of an insecticide resistance gene: a role for local selection. *J. Evol. Biol.* 18:1471–84

Lamason RL, Mohideen MAPK, Mest JR, Wong AC, Norton HL, et al. 2005. SLC24A5, a putative cation exchanger, affects pigmentation in zebrafish and humans. *Science* 210:1782–86

Lenormand T. 2002. Gene flow and the limits to natural selection. *Trends Ecol. Evol.* 17:183–89

Levene H. 1953. Genetic equilibrium when more than one ecological niche is available. *Am. Nat.* 87:331–33

Lewontin RC, Krakauer J. 1973. Distribution of gene frequency as a test of the theory of the selective neutrality of polymorphism. *Genetics* 74:175–95

Luikart G, England PR, Tallmon D, Jordan S, Taberlet P. 2003. The power and promise of population genomics: from genotyping to genome typing. *Nat. Rev. Genet.* 4:981–94

MacLean RC. 2005. Adaptive radiation in microbial microcosms. *J. Evol. Biol.* 18:1376–86

MacLean RC, Bell G, Rainey PB. 2004. The evolution of a pleiotropic fitness tradeoff in *Pseudomonas fluorescens*. *Proc. Natl. Acad. Sci. USA* 101:8072–77

Majerus MEN. 1998. *Melanism: Evolution in Action*. Oxford: Oxford Univ. Press

Markussen MDK, Heiberg AC, Nielsen R, Leirs H. 2003. Vitamin K requirement in Danish anticoagulant-resistant Norway rats (*Rattus norvegicus*). *Pest Manag. Sci.* 59:913–20

Mauricio R, Stahl EA, Korves T, Tian D, Kreitman M, Bergelson J. 2003. Natural selection for polymorphism in the disease resistance gene *Rps2* of *Arabidopsis thaliana*. *Genetics* 163:735–46

Review of the many human genes that confer resistance to malaria, the strongest selective factor in recent human history.

Genetic variants at the SLC24A5 gene result in lighter pigmentation in both zebrafish and humans.

McDonald J, Kreitman M. 1991. Adaptive protein evolution at the *Adh* locus in *Drosophila*. *Nature* 351:652–54

McKenzie JA. 1996. *Ecological and Evolutionary Aspects of Insecticide Resistance*. New York: Academic

Meyer D, Thomson G. 2001. How selection shapes variation of the human major histocompatibility complex: a review. *Ann. Hum. Genet.* 68:1–26

Meyers BC, Kaushik S, Nandety RS. 2005. Evolving disease resistance genes. *Curr. Opin. Plant Biol.* 8:128–34

Modiano D, Luoni G, Sirima BS, Simporé J, Verra F, et al. 2001. Haemoglobin C protects against clinical *Plasmodium falciparum* malaria. *Nature* 414:305–8

Müller HJ. 1987. Uber die Vitalität der Larvenformen der Jasside *Mocydia crocea* (H.-S.) (Homoptera Auchenorrhycha) und ihre ökologische Bedeutung. *Zool. Jahr. Syst.* 114:105–29

Mundy NI. 2005. A window on the genetics of evolution: MC1R and plumage colouration in birds. *Proc. R. Soc. London Ser. B* 272:1633–40

Mungall AJ, Palmer SA, Sims SK, Edwards CA, Ashurst JL, et al. 2003. The DNA sequence and analysis of human chromosome 6. *Nature* 425:805–11

Nachman MW, Hoekstra HE, D'Agostino SL. 2003. The genetic basis of adaptive melanism in pocket mice. *Proc. Natl. Acad. Sci. USA* 100:5268–73

> Adaptive melanism in pocket mice living on dark lava results from amino acid changes in the *Mc1r* gene.

Nielsen R, Bustamante C, Clark AG, Glanowski S, Sackton TB, et al. 2005. A scan for positively selected genes in the genomes of humans and chimpanzees. *PLoS Biol.* 3:976–85

Novembre J, Galvani AP, Slatkin M. 2005. The geographic spread of the CCR5 Δ32 HIV-resistance allele. *PLoS Biol.* 3:1954–62

Oakeshott JG, Home I, Sutherland TD, Russett RJ. 2003. The genomics of insecticide resistance. *Genome Biol.* 4(1):202

Ohashi J, Naka I, Patarapotikul J, Hananantachai H, Brittenham G, et al. 2004. Extended linkage disequilibrium surrounding the hemoglobin E variant due to malarial selection. *Am. J. Hum. Genet.* 74:1198–208

Pedra JHF, McIntyre LM, Scharf ME, Pittendrigh BR. 2004. Genome-wide transcription profile of field- and laboratory-selected dichlorodiphenytrichloroethane (DDT)-resistant *Drosophila*. *Proc. Natl. Acad. Sci. USA* 101:7034–39

Pelz HJ, Rost S, Hünerberg M, Fregin C, Heiberg AC, et al. 2005. The genetic basis of resistance to anticoagulants in rodents. *Genetics* 170:1839–47

> Resistance to pesticides in rats is the result of a number of independent mutations at the *VKORC1* gene.

Piertney SB, Oliver MK. 2006. The evolutionary ecology of the major histocompatibility complex. *Heredity* 96:7–21

Prout T, Savolainen O. 1996. Genotype-by-environment interaction is not sufficient to maintain variation: Levene and leafhopper. *Am. Nat.* 148:930–36

Rainey PB, Travisano M. 1998. Adaptive radiation in a heterogeneous environment. *Nature* 394:69–72

Ranson H, Claudianos C, Ortelli F, Abgrall C, Hemingway J, et al. 2002. Evolution of supergene families associated with insecticide resistance. *Science* 298:179–81

Ravigné V, Olivieri I, Dieckman U. 2004. Implications of habitat choice for protected polymorphisms. *Evol. Ecol. Res.* 6:125–45

Rose LE, Bittner-Eddy PD, Langley CH, Holub EB, Michelmore RW, Beynon JL. 2004. The maintenance of extreme amino acid diversity at the disease resistance gene, RPP13, in *Arabidopsis thaliana*. *Genetics* 166:1517–27

Rosenblum EB, Hoekstra HE, Nachman MW. 2004. Adaptive reptile color variation and the evolution of the MC1R gene. *Evolution* 58:1794–808

Rost S, Fregin A, Ivaskevicius V, Conzelmann E, Hörtnagel K, et al. 2004. Mutations in VKORC1 cause warfarin resistance and multiple coagulation factor deficiency type 2. *Nature* 427:537–41

Sabeti PC, Walsh E, Schaffner SF, Varilly P, Fry B, et al. 2005. The case for selection at *CCR5-Δ32*. *PLoS Biol.* 3:1963–69

Salamon H, Klitz W, Easteal S, Gao X, Erlich HA, et al. 1999. Evolution of HLA class II molecules: allelic and amino acid site variability across populations. *Genetics* 152:393–400

Samson M, Libert F, Doranz BJ, Rucher J, Liesnard C, et al. 1996. Resistance to HIV-1 infection in caucasian individuals bearing mutant alleles of the CCR-5 chemokine receptor gene. *Nature* 382:722–25

Saunders MA, Slatkin M, Garner C, Hammer MF, Nachman MW. 2005. The extent of linkage disequilibrium caused by selection on G6PD in humans. *Genetics* 171:1219–29

Schemske DW, Bierzychudek P. 2001. Perspective: evolution of flower color in the desert annual *Linanthus parryae*: Wright revisited. *Evolution* 55:1269–82

Schmidt PS, Rand DM. 2001. Adaptive maintenance of genetic polymorphism in an intertidal barnacle: habitat-and-life-stage specific survivorship of MPI genotypes. *Evolution* 55:1336–44

Sezgin E, Duvernell DD, Matzkin LM, Duan Y, Zhu CT, et al. 2004. Single-locus latitudinal clines in metabolic genes and derived alleles in *Drosophila melanogaster*. *Genetics* 168:923–31

For a number of allozyme variants in North America, the derived allele increases with increasing latitude.

Smith TB. 1993. Disruptive selection and the genetic basis of bill size polymorphism in the African finch *Pyrenestes*. *Nature* 363:619–20

Spiers AJ, Kahn SG, Bohannon J, Tavisano M, Rainey PB. 2002. Adaptive divergence in experimental populations of *Pseudomonas fluorescens*. I. Genetic and phenotypic bases of wrinkly spreader fitness. *Genetics* 161:33–46

Stahl EA, Dwyer G, Mauricio R, Kreitman M, Bergelson J. 1999. Dynamics of disease resistance polymorphism at the *Rpm1* locus of *Arabidopsis*. *Nature* 400:667–71

Stephens JC, Reich DE, Goldstein DB, Shin HD, Smith MW, et al. 1998. Dating the origin of the *CCR5-Δ32* AIDS-resistance allele by the coalescence of haplotypes. *Am. J. Hum. Genet.* 62:1507–15

Storz JF. 2005. Using genome scans of DNA polymorphism to infer adaptive population divergence. *Mol. Ecol.* 14:671–88

Storz JF, Dubach JM. 2004. Natural selection drives altitudinal divergence at the albumin locus in deer mice, *Peromyscus maniculatus*. *Evolution* 58:1342–52

Suiter AM, Bänziger O, Dean AM. 2003. Fitness consequences of a regulatory polymorphism in a seasonal environment. *Proc. Natl. Acad. Sci. USA* 100:12782–86

Tajima F. 1989. Statistical method for testing the neutral mutation hypothesis by DNA polymorphism. *Genetics* 123:585–95

Takahata N. 1990. A simple genealogical structure of strongly balanced allelic lines and trans-species evolution of polymorphism. *Proc. Natl. Acad. Sci. USA* 87:2419–23

Tian D, Traw MB, Chen JQ, Kreitman M, Bergelson J. 2003. Fitness costs of R-gene-mediated resistance in *Arabidopsis thaliana*. *Nature* 423:74–77

Turelli M. 1981. Temporally varying selection on multiple alleles. A diffusion analysis. *J. Math. Biol.* 13:115–29

Turelli M, Barton NH. 2004. Polygenic variation maintained by balancing selection: pleiotropy, sex-dependent allelic effects and G × E interactions. *Genetics* 166:1053–79

Turelli M, Schemske DW, Bierzychudek P. 2001. Stable two-allele polymorphisms maintained by fluctuation fitnesses and seed banks: protecting the blues in *Linanthus parryae*. *Evolution* 55:1283–98

Veazey R, Ling BH, Pandrea I, McClure H, Lackner A, Marz P. 2003. Decrease CCR5 expression on CD4$^+$ T cells of SIV-infected sooty mangabeys. *AIDS Res. Hum. Retroviruses* 19:227–33

Véliz D, Bourget E, Bernatchez L. 2004. Regional variation in the spatial scale of selection at MPI* and GPI* in the acorn barnacle *Semibalanus balanoides* (Crustacea). *J. Evol. Biol.* 17:953–66

Verrelli BC, McDonald JH, Argyropoulos G, Destro-Bisol G, Froment A, et al. 2002. Evidence for balancing selection from nucleotide sequence analyses of human G6PD. *Am. J. Hum. Genet.* 71:1112–18

Veuille M, Bénassi V, Aulard S, Depaulis F. 1998. Allele-specific population structure of *Drosophila melanogaster* alcohol dehydrogenase at the molecular level. *Genetics* 149:971–81

Via S, Lande R. 1987. Evolution of genetic variability in a spatially heterogeneous environment: effects of genotype-environment interaction. *Genet. Res.* 49:147–56

Wang ET, Kodama G, Baldi P, Moyzis RK. 2006. Global landscape of recent inferred Darwinian selection for *Homo sapiens*. *Proc. Natl. Acad. Sci. USA* 103:135–40

Watt WB, Wheat CW, Meyer EH, Martin JF. 2003. Adaptation at specific loci. VII. Natural selection, dispersal and the diversity of molecular-functional variation patterns among butterfly species complexes (*Colias*: Lepidoptera: Pieridae). *Mol. Ecol.* 12:1265–75

Watterson GA. 1978. The homozygosity test of neutrality. *Genetics* 88:405–17

Weill R, Lutfalla G, Mogensen K, Chandre F, Berthomieu A, et al. 2003. Insecticide resistance in mosquito vectors. *Nature* 423:136–37

Williams TN, Mwang TW, Wambua S, Peto TEA, Weatherall DJ, et al. 2005. Negative epistasis between the malaria-protective effects of α^+-thalassemia and the sickle cell trait. *Nat. Genet.* 37:1253–57

Wood ET, Stover DA, Slatkin M, Nachman MW, Hammer MF. 2005. The β-globin recombinational hotspot reduces the effects of strong selection around HbC, a recently arisen mutation providing resistance to malaria. *Am. J. Hum. Genet.* 77:637–42

Wright S. 1943. An analysis of local variability of flower color in *Linanthus parryae*. *Genetics* 28:139–56

Wright SI, Gaut BS. 2004. Molecular population genetics and the search for adaptive evolution in plants. *Mol. Biol. Evol.* 22:506–19

Wuif C, Zhao K, Innan H, Nordborg M. 2004. The probability and chromosomal extent of trans-specific polymorphism. *Genetics* 168:2362–72

Xiao S, Emerson B, Ratanasut K, Patrick E, O'Neill C, et al. 2004. Origin and maintenance of a broad-spectrum disease resistance locus in *Arabidopsis*. *Mol. Biol. Evol.* 21:1661–72

Zhong S, Khodursky A, Dykhuizen DE, Dean AM. 2004. Evolutionary genomics of ecological specialization. *Proc. Natl. Acad. Sci. USA* 101:11719–24

Ecological Effects of Invasive Arthropod Generalist Predators

William E. Snyder[1] and Edward W. Evans[2]

[1]Department of Entomology, Washington State University, Pullman, Washington 99164; email: wesnyder@wsu.edu

[2]Department of Biology, Utah State University, Logan, Utah 84322; email: ewevans@biology.usu.edu

Key Words

ant, beetle, crab, crayfish, praying mantis, wasp

Abstract

Arthropod generalist predators (AGP) are widespread and abundant in both aquatic and terrestrial ecosystems. They feed upon herbivores, detritivores, and predators, and also on plant material and detritus. In turn, AGP serve as prey for larger predators. Several prominent AGP have become invasive when moved by humans beyond their native range. With complex trophic roles, AGP have diverse effects on other species in their introduced ranges. The invaders displace similar native species, primarily through competition, intraguild predation, transmission of disease, and escape from predation and/or parasites. Invasive AGP often reach higher densities and/or biomass than the native predators they replace, sometimes strengthening herbivore regulation when invasive AGP feed on key herbivores, but sometimes weakening herbivore suppression when they eat key predators. The complexity and unpredictability of ecological effects of invasive AGP underscores the high risk of adverse consequences of intentional introductions of these species (e.g., for biological control or aquaculture).

INTRODUCTION

Many arthropods are predatory, and of these predators many species are quite generalized in their feeding. Predatory arthropods have received much attention from ecologists recently, because of their complex roles in community dynamics. While some theories of trophic control envision distinct trophic levels, arthropod generalist predators (AGP) can blur these distinctions. Many AGP feed not only upon herbivores, but also upon other predators and detritivores; many AGP feed also on plants and/or detritus, and in turn are fed upon by larger predators. Thus, AGP link multiple trophic levels. Although these links can be strong, with AGP exerting "keystone" effects that dictate community structure (Paine 1966), in many other cases AGP have numerous, relatively weak but pervasive interconnections with other community members (Polis 1991). Thus, AGP have sometimes been found to disrupt the top-down regulation of herbivores, generally through strong intraguild predation (Polis et al. 1989), but in other cases these predators provide some of our clearest evidence for strong top-down regulation of herbivores (Halaj & Wise 2001, Schmitz et al. 2000). Intensification of human transport and commerce around the world has led to widespread movement of species outside of their native range (Mack et al. 2000), including many AGP species. Because of the complex trophic role of AGP, these invaders can have particularly widespread impacts on the communities they invade (**Figure 1**).

Here, we review the growing literature documenting the ecological impacts of invasive AGP. We follow the convention of considering any taxon as invasive that not only becomes established, but also spreads readily in its new range (Elton 1958). We exclude commensal species (e.g., the common house spider *Achaearanea tepidariorum*) that fit our ecological criteria but are unable to survive without direct human influence. We also exclude specialist predators that attack one or just a few key prey species. The inclusion of nonanimal foods (e.g., honeydew, plant material, detritus) as a major portion of the diet did not exclude a species in this review, if that species also attacks a broad array of animal prey species. For many invasive taxa of AGP, excellent reviews exist that recount the course of invasion and/or list species of AGP that appear to be invasive (**Table 1**). Some provide particularly complete compilations of observed feeding relationships between invasive AGP and native species (**Table 1**). We focus instead on experimental results that demonstrate the ecological impact of invasive AGP. Necessarily, we concentrate on a few well-studied species (**Table 1**). While experimental results for any single species are incomplete, we review and discuss the clear patterns that emerge from the experimental studies across these taxa.

ARRIVAL AND SPREAD

Currently invasive AGP have taken a variety of routes to their introduced ranges, but in all cases initial transport is known, or thought, to be by humans. In several cases, humans have intentionally introduced exotic arthropod generalists in search of economic gain. For example, ladybird beetles have been widely introduced worldwide for biological control (e.g., Obrycki et al. 2000). Gordon (1985) documented

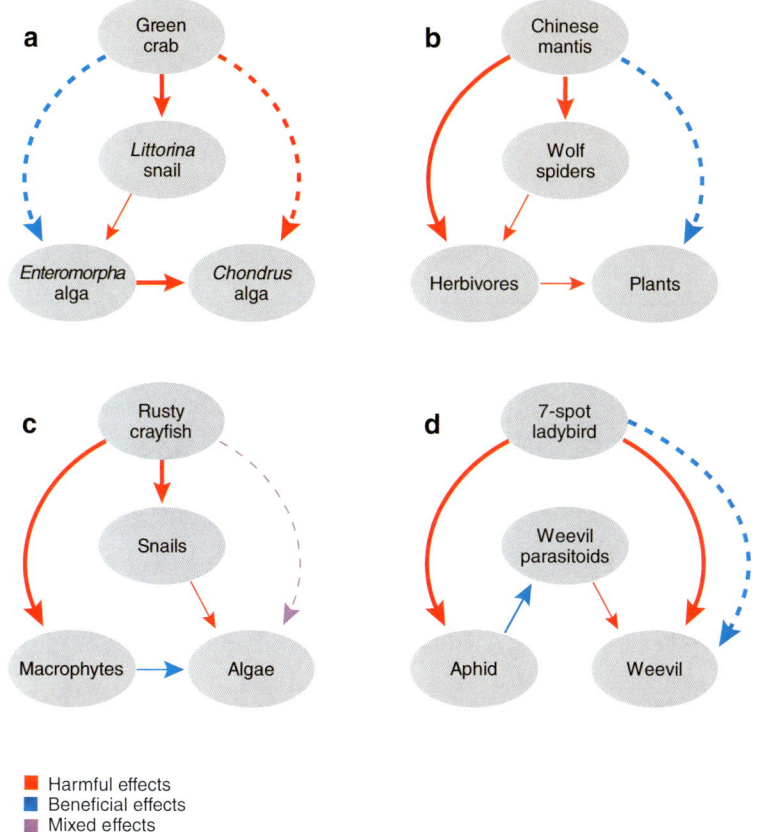

Figure 1

Impact of invasive arthropod generalist predators (AGP) on species interactions within four communities. Solid lines indicate direct interactions, whereas dashed lines indicate indirect effects. Colors indicate whether the interaction is to the benefit or harm of the species at the arrow's point; thick lines show interactions that have strengthened following AGP invasion, thin lines show interactions that have weakened. (*a*) On the east coast of North America the invasive green crab feeds heavily upon and decreases densities of *Littorina* snails, decreasing snail herbivory on the competitively dominant alga *Enteromophora* and allowing this species to outcompete the alga *Chondrus* (Lubchenco 1978). Thus, green crab invasion indirectly benefits *Enteromophora*, and indirectly harms *Chondrus*. (*b*) In Delaware (United States) old fields, wolf spiders leave areas inhabited by the invasive Chinese mantis, with lower spider densities presumably benefiting herbivores and thus indirectly harming plants. However, mantids feed directly on herbivores as well, so that their net impact is herbivore reduction leading to enhanced plant growth (Moran et al. 1996). (*c*) In Trout Lake, Wisconsin (United States), invasive rusty crayfish prey upon and reduce densities of snails, reducing snail herbivory of algae (Lodge et al. 1994). However, crayfish also damage macrophytes and remove this important substrate for algal growth, counteracting any indirect benefit of crayfish for algae. (*d*) In alfalfa fields in Utah (United States), invasive seven-spot ladybird beetles both suppress alfalfa weevil through direct predation and weaken weevil suppression by eating the aphids that provide food (honeydew) for weevil parasitoids, simultaneously enhancing and weakening weevil control (Evans & England 1996).

Table 1 Distributions of well-studied invasive arthropod generalist predators including their habitat, native and invasive ranges, and representative taxon-specific reviews

Key species	Common name	Habitat	Native to	Invasive in	Recent, taxon-specific review
Crustacea: Decapoda					
Portunidae					
Carcinus maenas	Green crab	Marine	Europe	Asia, Australia, North America, South America, South Africa	Carlton & Cohen (2003); Thresher et al. (2003)
Grapsidae					
Eriocheir sinensis	Mitten crab	Marine/freshwater	Asia	Europe, North America	
Cambaridae					
Orconectes rusticus	Rusty crayfish	Freshwater	North America	North America	Lodge et al. (2000)
Pacifastacus leniusculus	Signal crayfish	Freshwater	North America	Asia, Europe, North America	"
Procambarus clarkii	Red swamp crayfish	Freshwater	North America	Africa, Asia, Europe, North America	"
Insecta: Coleoptera					
Carabidae					
Pterostichus melanarius	None	Terrestrial	Europe	North America	
Coccinellidae					
Coccinella septempunctata	Seven-spot ladybird beetle	Terrestrial	Europe, Asia	North America	Obrycki et al. (2000)
Harmonia axyridis	Multicolored Asian ladybird beetle	Terrestrial	Asia	Europe, North America	Koch (2003); Pervez & Omkar (2006)
Insecta: Hymenoptera					
Formicidae					
Linepithema humile	Argentine ant	Terrestrial	South America	Africa, Asia, Australia, North America	Holway et al. (2002); Ness & Bronstein (2004)
Solenopsis invicta	Red imported fire ant	Terrestrial	South America	North America, Australia, New Zealand	"
Vespidae					
Polistes dominulus	European paper wasp	Terrestrial	Asia, Europe	North & South America	
Vespula germanica	German wasp	Terrestrial	Europe	North America, Australia, New Zealand	Beggs (2001); Spradberry & Maywald (1992)
Vespula vulgaris	European common wasp	Terrestrial	Europe	North America, Australia, New Zealand	Beggs (2001)
Insecta: Mantodea					
Mantidae					
Tenodera sinensis	Chinese mantis	Terrestrial	Asia	North America	Fagan et al. (2002)

introductions of 179 ladybird species into North America alone. Of these at least 18 have become established (Obrycki et al. 2000), and two in particular, *Coccinella septempunctata* and *Harmonia axyridis*, are generalists that have become highly invasive. Species in one other major taxon of invasive arthropod predator, crayfish, have also been widely, intentionally introduced for economic reasons (primarily for aquaculture, but also as bait or pets) (Lodge et al. 2000). Once moved, crayfish are then either released intentionally into natural systems, escape from culture, or both (Lodge et al. 2000).

Shipping traffic is responsible for the majority of accidental AGP introductions. For example, the ground beetle *Pterostichus melanarius*, a European native that has invaded a wide swath of North America, is believed to have arrived in soil ballast dumped from ships (Niemelä et al. 1997). Similarly, the European green crab, *Carcinus maenas* (and its cryptic congener *C. aestuarii*), native to Europe, and the Chinese mitten crab, *Eriocheir sinensis*, native to Asia, have been transported in ballast water (Carlton & Cohen 2003, Herborg et al. 2003). The wasp *Polistes dominulus* is thought to have established nests in shipping crates in its native Europe and then moved with the freight to North America (Hathaway 1986). Interestingly, despite repeated, extensive deliberate releases of the ladybird beetles *Coccinella septempunctata* and *Harmonia axyridis* into North America for aphid biological control, both species became established near shipping ports, where these successful colonists may have arrived as stowaways on cargo ships (Day et al. 1994). Movement of ornamental plants constitutes the second main avenue for accidental introduction of AGP among continents. Three praying mantid species native to Europe and/or Asia, *Tenodera sinensis*, *T. angustipennis*, and *Mantis religiosa*, are all believed to have reached eastern North America on nursery stock (Hurd 1999). Movement of intact soil is particularly crucial for the successful initial establishment of soil-nesting social Hymenoptera, because it allows the transport of functional social groups (Hee et al. 2000). Such transport is believed to be responsible for the movement of two particularly damaging invasive ant species, the red imported fire ant *Solenopsis invicta* and the Argentine Ant *Linepithema humile* (Hee et al. 2000, Holway et al. 2002), and also the movement of the German wasp *Vespula germanica* from Europe to elsewhere in the world (Spradberry & Maywald 1992).

Upon reaching the introduced range, some AGP readily disperse on their own. Although ladybird beetles introduced for biological control are often collected at establishment sites and distributed intentionally, adults of these beetles are winged and have impressive overland dispersal abilities (Hodek et al. 1993). *Coccinella septempunctata*, for example, arrived in Utah (United States) on its own in the early 1990s, and quickly became widespread and abundant (Evans 2004). Invasive crabs have mobile larval stages that can allow rapid, widespread distribution in the exotic range (Herborg et al. 2005, Yamada et al. 2005). However, for other predator taxa, human-mediated dispersal appears crucial. The mantis *Tenodera sinensis*, native to Asia but widely distributed in eastern North America, appears to be limited by barriers as seemingly insignificant as paved roads and thus depends on humans for its movement (Bartley 1982, Hurd 1999). Invasive crayfish can disperse within single river systems but rely on humans to release them into disjunct waterways (Lodge et al.

2000). For example, the rusty crayfish (*Orconectes rusticus*) can slowly disperse on its own through streams, but new populations are strongly associated with the intensity of human use (Puth & Allen 2004), suggesting that human transport remains the most efficient means of spread. Observed spread of the red imported and Argentine ants in their invasive ranges is too rapid to be due to natural spread and instead appears to be human-mediated (Holway et al. 2002, Suarez et al. 2001, Ward et al. 2005).

FACTORS AFFECTING INVASION

Many factors may affect the success of AGP invasions. We consider here three major potential influences: genetic bottlenecks, propagule pressure, and habitat disturbance.

Genetic Bottlenecks

Owing to small founder populations, genetic diversity for invasive species may often be low, yet low diversity clearly does not prevent success in invasion (Mack et al. 2000). Indeed, for the red imported fire ant and Argentine ant, low initial genetic diversity fosters invasion success (Holway et al. 2002). Both ant species possess a chemical kin recognition system that in the native range mediates conflicts between colonies (Holway et al. 1998, Ross et al. 1996). However, with the low initial genetic diversity typical of invasion by these species, ants are too similar genetically for soldiers to differentiate between members of their own versus a different colony (Ross et al. 1996, Tsutsui et al. 2000). With intraspecific aggression thereby eliminated, ants achieve far higher densities than are seen in the native range. One notes, however, that other invasive ant species, and indeed some populations of these same species, remain intraspecifically aggressive yet nonetheless are successful invaders (Holway et al. 2002). For other invasive AGP, founder populations appear to be relatively diverse. Indeed, promotion of high genetic diversity has often been an explicit goal in intentional introductions for biological control, as it is generally considered important for widespread establishment (Roush 1990). Hence it is interesting that no relationship is apparent between genetic diversity and successful invasion among ladybirds (Krafsur et al. 2005). German wasps invading Australia appear to be the result of more than two invasion events (Goodisman et al. 2001), suggesting some initial genetic diversity, whereas invasive populations of the wasp *P. dominulus* show genetic diversity similar to that in the native range (Johnson & Starks 2004). Thus it appears that, unlike in multiqueen forms of the invasive red imported fire and Argentine ants, low initial genetic diversity is not a prerequisite for successful invasion by these other AGP, even for the social wasps.

Propagule Pressure

For AGP, as for other taxa (Allendorf & Lundquist 2003), invasion success reflects not just performance once in the introduced range, but also the number of opportunities to become established in the first place (i.e., higher "propagule pressure" makes successful invasion more likely). A recent study suggests that propagule number is the most predictive factor in determining the likelihood of particular ant species becoming established in the United States, with those species discovered and intercepted in quarantined shipments also the most likely to become established eventually (Suarez et al. 2005). As another example, the invasive ladybird beetles *H. axyridis* and *C. septempunctata* both were released intentionally many times in eastern North America for classical biological control, but apparently each species established successfully only once (Day et al. 1994). Unfortunately, for most accidentally introduced AGP, almost nothing is known about the number, if any, of unsuccessful propagules that may have preceded successful establishment. Circumstantial evidence suggests that repeated reintroduction may be crucial for other taxa in addition to ants and ladybirds; for example, establishment of invasive green crabs throughout much of the world corresponds with greater shipping traffic, which presumably would have increased the frequency of propagule delivery (Carlton & Cohen 2003). Indeed, examination of ballast water reveals diverse populations of would-be AGP invaders, suggesting that large numbers of propagules are released every time a ship dumps its ballast (Carlton & Geller 1993).

Disturbance

For several invasive AGP, habitat disturbance facilitates invasion. For example, river plains that are naturally disturbed by periodic flooding are the native habitats of both red imported fire and Argentine ants, apparently predisposing these ants to take advantage of human-caused disturbance in their invasive ranges (Holway et al. 2002, Zettler et al. 2004). Argentine ants can invade even undisturbed habitats once densities build along disturbed edges (Suarez et al. 1998). Similarly, invasive praying mantids in the northeastern United States occur only in old fields and cannot survive succession to the native forest (W.E. Snyder, personal observation). It is unclear whether this reliance on old fields derives from structural characteristics or net productivity; marshy areas dominated by grasses and low-growing forbs appear to be the only less-disturbed habitats the mantids use. Invasive ladybirds rapidly come to dominate heavily managed agricultural habitats (e.g., Colunga-Garcia & Gage 1998, Elliott et al. 1996), but these are the only habitats where the ecology of invasive ladybirds has been investigated in detail. Though invasive crayfish readily invade relatively undisturbed, naturally diverse communities (Lodge et al. 2000), even for these taxa disturbance can play a facilitative role. For example, the invasive red swamp crayfish (*Procambarus clarkii*) is more tolerant of pollution than are native European crayfish species; this trait apparently contributes to the dominance of the invader in some areas (Gil-Sánchez & Alba-Tercedor 2002). Similarly, the crayfish *Cherax destructor*, which is currently invading southwestern Australia following translocation from elsewhere on the continent, is taking advantage of high salinity following human water management schemes to speed its invasion—the locally native crayfish are unable to withstand high salinity conditions (Beatty et al. 2005).

ECOLOGICAL IMPACTS

Here, we track the complex ecological effects of invasive AGP. We follow their impacts from the top of the food web to the bottom, moving from predators of invasive AGP, to intraguild interactions among invasive AGP and other species, and finally to the cascading impacts of invasive AGP on herbivores and plants. Often, these effects are magnified relative to the native AGP that the invasives replace, because the abundance and/or biomass of invasive predators so often is greater than was achieved by the replaced native AGP (Alyokhin & Sewell 2004, Armstrong & Stamp 2003, Beggs 2001, Grosholz et al. 2000, Holway et al. 2002, Wilson et al. 2004).

Additional Prey for Native Predators

Typically, native AGP are fed upon by larger, often vertebrate predators (e.g., Polis 1991). However, we found no clear cases in which invasions of AGP have increased food resources for higher predators. In Spain, it is uncertain whether addition of the exotic red swamp crayfish to local streams that lack native crayfish has bolstered densities of otters (*Lutra lutra*) and other vertebrate predators (Beja 1996). The red swamp crayfish now contributes roughly 20% of otter energy intake, but this occurs primarily in summer when other prey are also abundant; the crayfish likely depress densities of these other prey for otter through competition and predation (Beja 1996, Correia 2001). Similarly, on the west coast of North America predatory seabirds feed heavily upon invasive European green crabs, but nonetheless seabird densities did not increase as green crab densities rose dramatically (Grosholz et al. 2000). Indeed, abundant novel AGP can be detrimental to native predators higher in the food web. In California (United States), for example, coastal horned lizards (*Phrynosoma coronatum*), which are specialist predators of ants, decline following invasion by Argentine ants (Suarez & Case 2002). More ant biomass is present following invasion, but the lizards apparently do not recognize Argentine ants as prey. Because native ants are displaced by Argentine ants, the lizards are left with few acceptable prey. Similarly, the Texas horned lizard (*Phrynosoma cornutum*) in Texas (United States) has declined following invasion by the red imported fire ant (Donaldson et al. 1994), perhaps because red imported fire ants are also avoided as prey.

Displacement of Native Predators

Although some invaders may join communities of native AGP with little upset (Burger et al. 2001, Niemelä et al. 1997), it is striking how often invasion by AGP results in precipitous decline, and often complete exclusion, of related native AGP species. For example, invasion of North America by the ladybird beetle *Coccinella septempunctata* was followed by rapid, marked declines in several native species, perhaps most dramatically for *C. novemnotata* and *C. transversoguttata* (Alyokhin & Sewell 2004, Elliott et al. 1996, Wheeler & Hoebeke 1995). Unfortunately, nearly all such information on abundance of native and exotic ladybirds is from agricultural systems, and so it is unclear whether equally dramatic declines of native species have occurred in less-disturbed habitats. Intriguingly, native ladybirds reappeared in agricultural fields with artificially induced aphid outbreaks; thus, the native species may persist in sizable numbers in other habitats (Evans 2004). Similarly, the invasive red swamp crayfish in Spain extirpates entirely the native white-clawed crayfish (*Austropotamobius pallipes*) from heavily disturbed and polluted, lower watercourses (Gil-Sánchez & Alba-Tercedor 2002). However, fast moving higher-altitude streams are unsuitable for the invader and serve as a refuge for the native (Gil-Sánchez & Alba-Tercedor 2002). In yet another similar case, in California (United States), while the Argentine ant extirpates most native ant species from riparian zones, dry upland habitats (where Argentine ants are not successful) serve as refuges for native ants (Holway 2002).

The replacement of one invasive AGP by another (arriving later) is a recurrent theme. In Trout Lake in Wisconsin (United States), the native crayfish *Orconectes virilis* was displaced initially following invasion by the nonnative *Orconectes propinquus* (Lodge et al. 1986). The two species persisted together until the late 1970s when a second nonnative, the rusty crayfish, was released into the lake. This newcomer rapidly drove both *O. virilis* and *O. propinquus* to near extinction (Lodge et al. 1986). Similarly, guilds of North American ladybird beetles were invaded first by the European species *C. septempunctata*. A second exotic species, *Harmonia axyridis*, arrived several years later and now is replacing *C. septempunctata* (Alyokhin & Sewell 2004, Brown & Miller 1998). A similar progression has been documented for the invasive green crab, which in some habitats is being replaced by the more recently arriving shore crab *Hemigrapsus sanguineus* (Lohrer & Whitlatch 2002). The German wasp first invaded the honeydew-rich beech forests of New Zealand but later was displaced by invading common wasps (*Vespula vulgaris*) (Beggs 2001). The mechanism driving the wasp-species replacement was unusual—German wasps are strongly attracted to fermenting honeydew and become intoxicated and lethargic, reducing their foraging intensity relative to that of the teetotaling common wasps (Beggs 2001). Similar successive waves of replacement have been reported for predaceous marine amphipods in Europe (Dick & Platvoet 2000). For mantids in eastern North America the order of invasion is not known, but the invasive praying mantis *Tenodera sinensis* is a key predator maintaining low densities of another exotic mantis, *T. angustipennis* (Snyder & Hurd 1995).

Perhaps because the displacement of native AGP is such a dramatic and pervasive result of AGP invasion, this is the impact that has been most frequently studied experimentally. We next review the diverse mechanisms thought to underlie these species replacements.

Exploitative competition. For several invasive AGP, greater efficiency of resource use may be a key to species replacement. This is the "R*" principle": The species that drives the key limiting resource(s) to the lowest level will dominate competitively in that environment, as the low levels of that resource are insufficient to support other species (Tilman 1982). This may occur, for example, in the displacement of native ladybirds by the invasive *Coccinella septempunctata* from alfalfa fields in western North America (Evans 2004). Dispersing adults of native ladybirds appear particularly sensitive in their habitat selection to local aphid density and hence have largely abandoned alfalfa fields in which predation by *C. septempunctata* now prevents aphids from reaching high numbers. Alyokin & Sewell (2004) report a similar scenario following invasion by the ladybird *H. axyridis* in Maine potato crops. Reduced prey abundance following invasion has been reported also for ants (Holway et al. 2002), wasps (Beggs 2001) and crayfish (Lodge et al. 2000).

Other cases exist in which invasive AGP use shared resources more efficiently than do native AGP. An example concerns the wasp *Polistes dominulus* invading North America from Europe. By controlling experimentally the energy intake of foragers of *P. dominulus* and its native congener *P. fuscatus*, Armstrong & Stamp (2003) demonstrated that the invasive wasp produced 2.5 times more offspring from the same available food, than the native wasp. At least three characteristics of the invader led to this result. First, *P. dominulus* is smaller that *P. fuscatus*, so production of each new wasp requires less energy from foragers in the preceding generation (Armstrong & Stamp 2003). Second, *P. dominulus* nests within human structures (e.g., under the eaves of houses) and thus is able to put less protein and energy into nest construction and correspondingly more into producing offspring (Curtis et al. 2005). Finally, *P. dominulus* refrains from intra- or interspecific interference competition with other wasps, saving additional energy for reproduction (Gamboa et al. 2002).

Introduced AGP also may prevail over native competitors because of unusually broad diets. For example, the ladybird *C. septempunctata* may readily exploit alternative prey such as alfalfa weevil larvae, thereby enabling it to persist in alfalfa fields even when aphid density is low and native ladybirds disperse (Evans 2004). *Harmonia axyridis*, another spectacularly invasive predator of aphids in North America, has the unusual ability to prey upon other ladybirds (and other aphid predators), and this may enhance its success as an invader (Cottrell & Yeargan 1998, Michaud 2002). Similarly, the broad diet of the German wasp promotes this species' invasion of Australia and replacement of the native wasp *Polistes humilis*. Kasper et al. (2004) used molecular methods to characterize the gut contents of both species (the wasps masticate their prey, rendering many soft-bodied prey unrecognizable). In contrast to *P. humilis*, German wasps attacked many other prey in addition to Lepidoptera. Kasper et al. (2004) hypothesize that such diet supplementation is enabling the German wasp to exclude *P. humilis* by driving Lepidoptera densities to levels too low to sustain the native species. In several cases, carbohydrates from sources other than prey can support high densities of invasive AGP. In beech forests in New Zealand, huge numbers of a native scale insect (*Ultracoelostoma assimile* Maskell) produce copious honeydew. The common wasp feeds heavily on the honeydew. Thus maintained at abnormally high densities, the wasps take over 99% of Lepidoptera and spiders (Beggs 2001). The red-imported fire ant will "tend" aphids for honeydew. In southeastern North America, this ant tends the abundant cotton aphids, *Aphis gossypii*, on cultivated cotton, *Gossypium hirsutum* (Kaplan & Eubanks 2005). The aphids support high ant densities, thereby exacerbating the negative effect of the ants on other predators (Kaplan & Eubanks 2005). Consumption of honeydew also contributes to high densities for many other invasive ant species [reviewed by Holway et al. (2002)].

Predator-predator aggression. In nearly all cases of invasion by AGP, predator-predator aggression has been suggested to play some role. Among AGP, aggressive encounters are strongly influenced by relative body size, with larger predators defeating (and often consuming) smaller ones (Polis et al. 1989). In several cases, larger exotic AGP competitively dominate over smaller native species [examples include exotic versus native crayfish (Gherardi & Cioni 2004), crabs (Grosholz et al. 2000), and ground beetles (Prasad & Snyder 2006)]. Perhaps more surprising are cases where smaller (or similarly sized) invasive AGP nonetheless aggressively dominate native species. For example, despite being larger than the invasive crayfish *O. propinquus* and *O. rusticus*, the displaced native *O. virilis* is submissive to these species in competition over both food (Hill & Lodge 1999) and shelter (Garvey et al. 1994, Hill & Lodge 1994). Another example comes from the invasive ladybird beetle *Harmonia axyridis*. The eggs of ladybirds are preyed upon by all other stages, but eggs of *H. axyridis* are distinctive in their chemical protection against predation by other ladybirds (Cottrell 2004, Sato & Dixon 2004). Ladybird larvae aggressively prey on each other as well as on eggs. Larvae of *H. axyridis* possess both a relatively strongly developed chemical defense system, and strongly adherent tarsi, both of which apparently contribute to this species' ability to win contests with other larvae even when the *H. axyridis* larva is smaller than its adversary (Snyder et al. 2004; Yasuda et al. 2001, 2004). For invasive ants, greater colony size, rather than the size of individuals, can yield victory in contests with natives. Human & Gordon (1996) and Holway (1999) examined interactions between naturally occurring native ants and Argentine ants from colonies translocated experimentally to areas ahead of the invasion front. Individually, Argentine ant foragers often lost battles with native foragers. However, with colonies larger than those of the natives because of reduced intraspecific aggression, Argentine ants inevitably dominated the resource through sheer force of numbers. The red imported fire ant also appears to use superior numbers, rather than individual superiority in aggressive encounters, to overwhelm and displace native ants from food (Morrison 2000). Competitive superiority of Argentine ants over native ants disappears when intraspecific aggressiveness among Argentine ants is experimentally restored (Holway & Suarez 2004).

Relative immunity from shared natural enemies. In a number of cases, invasive AGP leave specialist natural enemies behind when they invade a new region. In their native range red imported fire ants dramatically curtail foraging in the presence of phorid flies that parasitize the ants, but these flies were not introduced along with the ants (Feener 1981, Orr et al. 1995). Native fire ants suffer competitively against invasive fire ants because the natives are attacked by phorid species that do not readily switch to attacking red imported fire ants (Morrison 1999). In a study of parasitism by Strepsiptera, the native North American wasp *P. fuscatus* experienced 16–18% parasitism, whereas the invasive species *P. dominulus* was never found to be parasitized (Pickett & Wenzel 2000). In an interesting departure from this general theme, however, Hoogendoorn & Heimpel (2002) suggest that because the relatively unsuitable host *H. axyridis* serves as a sink for wasted eggs of a shared braconid parasitoid, the presence of this invasive ladybird may in fact promote higher rather than lower numbers of the native North American ladybird *Coleomegilla maculata*.

Often, exotic AGP aggressively drive native species from shelter, heightening exposure of the natives to shared predators. For example, the rusty crayfish facilitates its own invasion by dominating rock-strewn areas of lakebeds (Hill & Lodge 1994). These habitats offer greatest protection from vertebrate predators (fish, herons), and once driven from these refugia native crayfish are subject to increased predation by visually searching predators during the day (Hill & Lodge 1994). Another invasive crayfish, the signal crayfish (*Pacifastacus leniusculus*) similarly outcompetes Atlantic salmon (*Salmo salar*), bullhead (*Cottus gobio* L.) and stone loach (*Noemacheilus barbatulus* L.) in Britain (Griffiths et al. 2004, Guan & Wiles 1997), and sculpin (*Cottus beldingi*) in western North America (Light 2005), for shelter from predators. The European green crab drives Dungeness crab (*Cancer magister*) from shelter, enhancing the risk of predation by vertebrates for this "delightful with butter" native (McDonald et al. 2001). A similar example is the displacement of an endemic New Zealand spider (*Latrodectus katipo*) by the exotic spider *Steatoda capensis* (Hann 1990). The invader may be superior in rapidly colonizing vacant coastal areas following storm damage. Once established as web-occupying adults, individuals of the invasive spider prevail in antagonistic encounters with the later arriving immatures of *L. katipo* seeking web sites.

The importance of predator-predator interactions in AGP invasion is apparent when indigenous predators retard invasion by AGP [a form of "biotic resistance" (Elton 1958)]. Accidentally introduced to the port city of Osaka, the two-spotted ladybird *Adalia bipunctata* has been slow to expand its distribution in Japan (Sakuratani et al. 2000). As elsewhere, *H. axyridis* in its native Japan is an aggressive intraguild predator of eggs and larvae of other ladybirds (Yasuda & Shinya 1997). *Adalia* is particularly vulnerable to intraguild predation by *H. axyridis*, at least in simplified laboratory settings, and this has been linked to *Adalia*'s failure to establish where *H. axyridis* is native (Kajita et al. 2006a). Similarly, DeRivera et al. (2005) link the failure of the green crab to expand its range down the southeastern coast of North America to this species' susceptibility to intraguild predation by the native blue crab, *Callinectes sapidus*. Blue crabs grow increasingly common further south along the coast, and DeRivera et al. (2005) found that predation rates of tethered green crabs were dramatically higher at sites closer to the southern edge of the crab's invasion front than at other sites throughout the invasive range. In contrast, where the green crab has successfully invaded along the west coast of North America, the green crab is an intraguild predator of the native shore crab, *Hemigrapsus oregonensis* (Grosholz et al. 2000). Similarly, Gruner (2005) reported increased density of up to 80-fold of a theridiid spider, *Achaearanea* cf. *riparia*, in Hawaii following experimental exclusion of predatory birds. Apparently, the guild of generalist birds prevents this established but uncommon nonnative spider from becoming invasive.

True parasites also may be lost when AGP invade. In its native range, the green crab is attacked by a taxonomically diverse community of parasites. However, these crabs generally enter their invasive range as larvae free of parasites. Torchin et al. (2001) compared parasite loads and size of green crabs across their native range in

Europe to those in their invasive range in North America, South Africa, and Australia. While more than 90% of crabs in the native range were parasitized by at least one parasite species, less than 10% of invasive crabs were parasitized. Consequently, green crabs in their invasive range were on average 1.3 times larger than those in Europe (Torchin et al. 2001). Similarly, both Argentine and imported fire ants have also lost parasitic *Wolbachia* bacteria during introduction outside the native range (Shoemaker et al. 2000, Reuter et al. 2005). Recent studies of ladybirds similarly indicate that *H. axyridis* is less susceptible to the endemic entomopathogen *Beauveria bassiana* than is the native North American ladybird *Olla v.-nigrum* (Cottrell & Shapiro-Ilan 2003).

Introduction of new pathogens. We know of only one case in the literature where this mechanism appears important. Exotic crayfish brought to Europe for aquaculture ventures came with a genetically distinct strain of the crayfish plague (the fungal pathogen *Aphanomyces astaci*). While the exotic signal crayfish is relatively immune to the ill effects of the fungal strains brought with them from the native range, native European crayfish are quite susceptible to the novel (from their perspective) strain (Lilley et al. 1997). Collapse of native crayfish stocks owing to plague often leads humans to introduce exotic crayfish species to restore these fisheries, reinforcing the spread of both exotic crayfish and exotic plague strains (Lodge et al. 2000).

Disrupted mating systems. Although exotic AGP and natives sometimes interbreed, the role of interbreeding in species replacement appears minor. Invading rusty crayfish interbreed with an earlier invader, *Orconectes propinquus* (Perry et al. 2001). Because hybrid crayfish suffer no obvious fitness costs, interbreeding alone would not lead to the disapearance of *O. propinquus* genes. Similarly, the red imported fire ant in southeastern North America interbreeds with a fire ant that had invaded earlier, *Solenopsis richteri*. However, because red imported fire ants are competitively superior to both pure and hybrid *S. richteri*, this interbreeding likely has little effect on the eventual displacement of *S. richteri* (Shoemaker et al. 1996). We found no other clear examples of intermating and genetic dilution as a factor speeding invasion by AGP. Among ladybirds, for example, the invader *C. septempunctata* readily copulates with the native North American *C. transversoguttata*, but females of neither species produce fertile eggs from such couplings (E.W. Evans, unpublished data). Nonetheless, such interspecific matings allow for transmission of sexually transmitted diseases, and invasive species such as *H. axyridis* (Japanese populations of which harbor male-killing bacteria at high incidence) could negatively affect native ladybirds by such means (Majerus 1997).

Behavioral and Plastic Responses by Natives

Prey species often have developed complex behavioral and developmentally plastic responses that reduce their risk of being preyed upon (Benard 2004, Schmitz et al. 2004, Werner & Peacor 2003). Invasion by exotic AGP species, with which the natives have not coevolved, may circumvent these antipredator responses, and in several cases improper antipredator behavior by natives appears to exacerbate the negative effects of invasive AGP. Ineffective antipredator behavior of the green crab facilitates its replacement by another invading crab, *Hemigrapsus sanguineus*. While juvenile *H. sanguineus* actively move away from larger crabs, which are aggressive cannibals/intraguild predators, juvenile green crabs instead remain inactive and rely upon protective coloration. This passive defense is ineffective against other crabs, which are tactile hunters (Lohrer & Whitlatch 2002). In contrast, *H. sanguineus* juveniles escape attack from larger conspecifics and also avoid intraguild predation by green crabs. In an unusual twist, the exotic rusty crayfish has more effective antipredator behavior against native predatory fish than does the native crayfish species, *O. virilis*. *O. virilis* swims when encountering fish predators and is easily captured by the fish, whereas the rusty crayfish approaches fish aggressively with claws raised and generally escapes fish predation (Garvey et al. 1994).

What is perhaps more surprising, however, is how quickly native species may adopt effective antipredator behavior, or initiate phenotypic changes, in response to invasive AGP. Shell thickness of the snail *Littorina obtusata* increased from 50% to 80% across different populations within several decades following invasion of green crabs, perhaps as a rapid evolutionary response to these voracious snail predators (Seeley 1986). However, chemical cues from green crabs alone induce snail shell thickening (Trussell & Smith 2000), and the snails also reduce activity (and thereby risk of predation) in the presence of green crabs (Trussell et al. 2003). Similar induced-defense responses may occur in other green crab prey (Appleton & Palmer 1988, Leonard et al. 1999). Tadpoles of the red-legged frog, *Rana aurora*, recognize the invasive red swamp crayfish as a threat and spend more time safely hidden within a refuge (Pearl et al. 2003), as does the rare fish species the Little Colorado spinedace (*Lepidomeda vittata*) in the presence of the introduced crayfish *Orconectes virilis* (Bryan et al. 2002). Ladybirds refrain from ovipositing when encountering chemical cues associated with the tracks and frass of conspecifics or other ladybird species that might act as intraguild predators (Agarwala et al. 2003, Hemptinne et al. 2001, Růžička 2001). Such responses may contribute to negative effects in interactions of invasive and native ladybirds (Kajita et al. 2006b). For all of these species, however, reduced foraging activity and/or phenotypic responses come at a tradeoff with other physiological needs, and thus AGP invasion likely remains costly to the prey (Bryan et al. 2002, Palmer 1992, Pearl et al. 2003, Trussell & Nicklin 2002).

Alteration of Top-Down Control

In several cases, exotic AGP have weakened control by herbivores. Several clear examples come from biological control systems. The herbivorous mite *Tetranychus lintearius*, introduced into western North America from Europe to attack the invasive weed gorse, *Ulex europaeus*, instead is itself controlled by the invasive predatory mite *Phytoseiulus persimilis* (Pratt et al. 2003). Similarly, the ground beetle *Pterostichus melanarius*, native to Europe and invasive in North America, disrupts suppression of the pea aphid, *Acyrthosiphon pisum*, in cultivated alfalfa (*Medicago sativum*) fields in Wisconsin (United States) (Snyder & Ives 2001). The beetles feed heavily on aphids when alfalfa plants are short following cutting, but as plants regrow the aphids are increasingly able to escape predation by poor-climbing *P. melanarius*. However, the immobile pupae of a key parasitoid of the aphids, *Aphidius ervi*, remain susceptible to ground beetle predation. By culling parasitoids from the aphid population, *P. melanarius* can exacerbate pea aphid outbreaks later in the cutting cycle. Interestingly, another invasive predator, the ladybird beetle *H. axyridis*, appears able to ameliorate any negative effects of *P. melanarius* on aphid control when both predators are present, because larvae of the ladybird selectively prey upon aphids rather than parasitoid pupae (Snyder & Ives 2003).

However, invasive AGP usually have complex impacts on other species, to the benefit of some species while harming others. In a now-classic paper, Lubchenco (1978) described how the snail *Littorina littorea* influences plant community structure in sheltered, shoreline tidal pools along the northeastern coast of the United States (**Figure 1a**). When abundant the snails allow the competitively inferior alga *Chondrus* to dominate, because snails avoid eating this alga species and prefer to consume instead the competitively dominant alga *Enteromorpha*. However, when green crabs invade pools they suppress snail densities and allow *Enteromorpha* to escape competition with *Chondrus*. Clearly, invasion by green crabs has changed the trophic structure of these tidal pool communities, but the absence of data predating crab invasion prevents more detailed assessment. It is clear, however, that predators other than green crabs exert strong top-down control in adjacent intertidal zones where green crabs are rare (Menge 1976).

A similar but terrestrial example for altered top-down regulation following invasion comes from old fields in Delaware (United States). Moran et al. (1996) manipulated densities of the invasive mantid *Tenodera sinensis* in relatively large, open plots surrounded only by a sticky-trap barrier, and measured mantid impacts on other arthropods and, indirectly, on plants (the two-month study encompassed the main period of growth for mantids, and also for most old-field plants) (**Figure 1b**). Overall, predator biomass did not differ between plots with and without mantids, because wolf spiders, the dominant large native predators prior to mantid invasion, emigrated at relatively high rates from plots with mantids. The spiders apparently recognize the intraguild predation risk that *T. sinensis* poses and leave areas inhabited by mantids to negate this risk (Moran et al. 1996, Wilder & Rypstra 2004). Despite reduced spider densities, the greater voracity and larger adult size of *T. sinensis* led to reduced densities of many taxa of herbivores, indirectly enhancing plant growth. Fagan & Hurd (1994) found that an invasive mantid species in Delaware (United States), *Mantis religiosa*, reduced total arthropod biomass by 45%, but densities of some herbivores (mirid bugs) nonetheless increased even as densities of others (crickets and Homoptera other than aphids) declined dramatically. Other experiments with invasive mantids have reported a similar mix of positive and negative, direct and indirect effects on other species (Fagan et al. 2002).

Another example was recorded following a crayfish invasion. Lodge et al. (1994) conducted an exclusion/inclosure experiment within Trout Lake in Wisconsin (United States), in which the invasive rusty crayfish was either caged at natural postinvasion densities or excluded from large field pens (crayfish exclusion most closely matched the much lower crayfish densities typical of the preinvasion community) (**Figure 1c**). Over the next 3 months, rusty crayfish reduced submerged macrophyte standing biomass ca. 90%, partly through direct feeding but more typically by laceration of plant tissue during hunting. However, rusty crayfish also reduced snail densities by 99%, leading to a per-surface-area increase in algae. Despite the initiation of this trophic cascade, crayfish did not affect total algal densities, because macrophytes removed by crayfish no longer provided a growth surface for algae (Lodge et al. 1994). The impact of the rusty crayfish on Trout Lake was followed over the initial 20 years of crayfish invasion (Wilson et al. 2004), revealing that results from the field experiment quite accurately predict longer term impacts of rusty crayfish. As the invasion progressed, total snail densities in the lake decreased 99.9%, macrophyte biomass decreased up to 75%, and macrophyte species richness declined up to 90%. Impacts of invasion not recorded over the short-scale field experiment, but quite evident in the long-term, were a dramatic increase in total crayfish density (up to 18-fold), and less dramatic decreases in the densities of bluegill (*Lepomis macrochirus*) and pumpkinseed (*L. gibbosus*) sunfish (presumably through competition with rusty crayfish). Nyström et al. (2001) report nearly identical community effects of the invasive signal crayfish on European pond communities, and Smart et al. (2002) report a dramatic reduction in macrophyte biomass and cover in Lake Naivasha in Kenya following the intentional release for aquaculture of the red swamp crayfish.

Several other examples of complex impacts of invasive AGP on other community members can be found in the literature. The invasive ladybird *C. septempunctata* both directly reduces survivorship of alfalfa weevils through predation, and indirectly increases survivorship through removal of aphids (and honeydew) that enhance weevil parasitism (**Figure 1d**) (Evans & England 1996). Grosholz et al. (2000) reported that over a 9-year period spanning green crab invasion, several species of small clams (preferred prey of green crabs but apparently less so of native crabs) declined fivefold (see also Walton et al. 2002). However, substrate-dwelling polychaete worms and tube-building crustaceans increased dramatically, to densities up to 100 times preinvasion levels. The effect on substrate-dwellers appeared to be indirect; apparently, the burrowing activity of clams, combined perhaps with competition for shared prey, otherwise limits polychaete and tube-crustacean densities (Grosholz et al. 2000).

Invasional Meltdown or Reconstruction?

Recently, ecologists have grown concerned that invasion by one exotic species will hasten invasion by subsequent species, a process called "invasional meltdown" (Simberloff & Von Holle 1999). Invasive AGP provide good examples. Ants are often mutualists with other species, for example tending Homoptera for honeydew or transporting plant seeds. In a few cases, invasive ants have established mutualistic relationships with other invasive species, presumably hastening invasion by both (nicely reviewed by Holway et al. 2002 and Ness & Bronstein 2004). For example, the red imported fire ant tends and protects colonies of the invasive scale insect *Antonina graminis* (Helms & Vinson 2003). Honeydew from the scales provides up to 48% of colony energy needs, while scale survivorship is poor when ants are absent, such that ants and scales are likely facilitating one another's invasion. Similarly, because invasive ants are generally smaller than the native ants they replace, and thus are relatively poor seed dispersers, ant invasions can speed replacement of native by invasive plant species, when native plants depend upon ant-mediated seed dispersal (Ness et al. 2004). In fragmented native habitats nested within a largely urban landscape, abundances of the invasive Argentine ant and of nonnative Isopoda, Dermaptera, and Blattaria all are correlated positively with one another and inversely with patch size (Bolger et al. 2000), although it is unclear whether exotics are facilitating one another's invasion or simply responding in kind to the same environmental factors. In another example, exotic northern pike, *Esox lucieus*, and largemouth bass, *Micropterus salmoides*, two large predatory fish species, co-occur with the also exotic red swamp crayfish in some Spanish lakes. The fish and crayfish are sympatric in the native range (the southeastern United States), and there is some evidence that without the red swamp crayfish as prey, the invasive fish would not persist in sufficient numbers to have widespread negative effects on native species (Geiger et al. 2005). The European green crab preferentially preys upon native clams in the genus *Nutricola* along coastal California (Grosholz 2005). With the native clams in decline following green crab invasion, the exotic clam *Gemma gemma*, long present but at low densities, is released from competition and becomes invasive; green crabs selectively prey upon the native clams rather than *G. gemma*.

However, there are also several cases in which an invasive AGP is reducing rather than promoting the harm caused by another invasive species, as top-down control pathways important in the native range are "reconstructed" (**Figure 2**). For example, the soybean aphid (*Aphis glycine*), recently introduced accidentally from its native Asia into the midwestern United States, has become a key pest of cultivated soybean, *Glycine max* (Rutledge et al. 2004). The aphid invasion has some attributes of an invasional meltdown—the obligate winter plant host of the aphid is the invasive plant buckthorn (common buckthorn, *Rhamnus cathartica*, and glossy buckthorn, *Frangula alnus*), already well-established in the region. However, a key predator of the soybean aphid in Asia, the invasive ladybird beetle *H. axyridis*, had already invaded the region, and *H. axryridis* contributes substantial biological control of the soybean aphid (**Figure 2a**; Rutledge et al. 2004). In a similar example from an aquatic system, in Japan both largemouth bass and the red swamp crayfish, which are sympatric in their native range (southeastern United States), are widely invasive. Red swamp crayfish strongly reduce densities of native macrophytes of several species, but harm to native macrophytes is greatly reduced where bass are present and can control crayfish densities (**Figure 2b**; Maezono et al. 2005). Thus, any attempt to extirpate bass in Japan would carry the risk of endangering native aquatic macrophytes as crayfish densities inevitably rise. These studies highlight the risk of removing from communities exotic species with multichannel trophic connections to both native and other exotic species (Zavaleta et al. 2001).

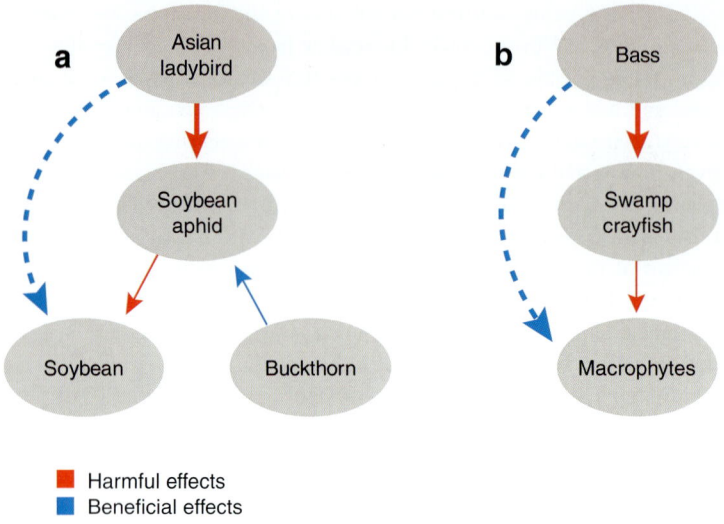

Figure 2

Restoration of top-down control in translocated communities including invasive arthropod generalist predators. (*a*) When the soybean aphid was accidentally introduced into the midwestern United States, both its summer host, the nonnative crop plant soybean, and its overwintering host, the invasive plant buckthorn, were already present to facilitate aphid invasion. However, the multicolored Asian ladybird beetle, a key predator of soybean aphid in their shared native range, was also already established, and the ladybird contributes significantly to soybean aphid suppression (Rutledge et al. 2004). (*b*) In Japan, bass, an invasive fish, are key predators of the invasive red swamp crayfish. Where both invasive largemouth bass and crayfish are present, native macrophytes are spared the otherwise devastating effects of red swamp crayfish (Maezono et al. 2005). Once integrated into communities, removal of exotic AGP often would have both beneficial and harmful impacts on other species (Zavaleta et al. 2001). Lines as in **Figure 1**.

CONCLUSIONS

Human activity is central in promoting invasion of AGP. Invasive AGP have widespread and varied ecological impacts, many of which stem from displacement of similar native predators. Such displacement arises by various mechanisms; these include both exploitative and interference competition, predation, escape from natural enemies, disruption of mating systems, and introduction of pathogens. Invasive AGP may weaken or strengthen top-down control, and they may contribute either to invasional meltdown or to prevention of further damage from other invaders. Predicting impacts is challenging because invasive AGP join an often diverse array of natural enemies that are already interacting among themselves in many complex ways (Sih et al. 1998). Given the large potential for adverse effects of invasive AGP, however, intentional introductions of such species (e.g., for biological control or aquaculture) should be discontinued. Unfortunately, further accidental introductions are all but inevitable, and they alone are likely to pose many management challenges for applied ecologists in the years to come.

FUTURE ISSUES

1. **Using molecular diet analysis to determine diet breadth and predator impact.** For many invasive arthropod predators, prey identity is difficult to establish because prey fragments are almost nonexistent in species that are fluid feeders (Holway et al. 2002), or because predation events are rare or difficult to observe (e.g., Johnson et al. 2005). However recent advances in the molecular identification of prey remains using PCR now allows a full assessment of predator diet breadth (Symondson 2002). Molecular identification of prey is extremely useful in making direct comparisons between the diets of native versus introduced AGP (e.g., Kasper et al. 2004) and also in identifying the impact of invasive predators upon key endangered prey species (Sheppard et al. 2004).

2. **Experimental studies that consider multiple mechanisms of invasion success.** Studies of invasive species often consider mechanisms of invasive success in isolation from one another. However, as demonstrated in this review, many attributes of a given invasive species likely contribute to its invasion success. Often missing is a better understanding of the interaction among, and relative importance of, multiple mechanisms of species replacement. There is circumstantial evidence that multiple mechanisms synergistically feed one another, through complex webs of interacting effects. For example, the green crab escapes its parasites during invasion (Torchin et al. 2001), leading to greater body size and thus competitive advantage in aggressive contests with natives (Grosholz et al. 2000). In turn, when natives lose aggressive contests and are driven from shelter, the natives then have a higher risk of being preyed upon by shared predators (McDonald et al. 2001).

3. **Invasive predators and altered predator biodiversity.** Recently, there has been growing interest in the role of predator species richness in determining the strength of herbivore suppression. For predators, herbivore suppression sometimes strengthens, but sometimes weakens, with the inclusion of more predator species (Ives et al. 2005). Invasive AGP alter predator diversity either by adding new species to communities or by changing species composition when they entirely displace native predators. Because invasive AGP tend to be particularly active in predator-predator interference (this review), replacement of native predators by invasive AGP might be expected to generally weaken herbivore suppression (e.g., Finke & Denno 2004). However, it is interesting that invasive AGP are members of several communities where the experimental manipulation of predator species diversity has revealed the strongest herbivore suppression when predator diversity is greatest (Cardinale et al. 2003, Snyder et al. 2006).

ACKNOWLEDGMENTS

We thank N. Davidson, D. Finke, Y. Kajita, R. Ramirez, G. Snyder, S. Steffan, and C. Straub for comments that improved this manuscript. W. Snyder was supported by grant #2004–01215 from the National Research Initiative of the USDA Cooperative Research, Education and Extension Service (CSREES).

LITERATURE CITED

Agarwala BK, Yasuda H, Kajita Y. 2003. Effect of conspecific and heterospecific feces on foraging and oviposition of two predatory ladybirds: role of fecal cues in predator avoidance. *J. Chem. Ecol.* 29:357–76

Allendorf FW, Lundquist LL. 2003. Population biology and invasive species. *Conserv. Biol.* 17:24–29

Alyokhin A, Sewell G. 2004. Changes in a lady beetle community following the establishment of three alien species. *Biol. Invasions* 6:463–71

Appleton RD, Palmer AR. 1988. Water-borne stimuli released by crabs and damaged prey induce more predator-resistant shells in a marine gastropod. *Proc. Natl. Acad. Sci. USA* 85:4387–91

Armstrong TR, Stamp NE. 2003. Colony productivity and foundress behavior of a native versus an invasive social wasp. *Ecol. Entomol.* 28:635–44

Bartley JA. 1982. Movement patterns of adult male and female mantids (*Tenodera sinensis* Saussure). *Environ. Entomol.* 11:1108–11

Beatty S, Morgan D, Gill H. 2005. Role of life history in the colonization of Western Autralian aquatic systems by the introduced crayfish *Cherax destructor* Clark, 1936. *Hydrobiologia* 549:219–37

Beggs J. 2001. The ecological consequences of social wasps (*Vespula* spp.) invading an ecosystem that has an abundant carbohydrate source. *Biol. Conserv.* 99:17–28

Beja PR. 1996. An analysis of otter *Lutra lutra* predation on introduced American crayfish *Procambarus clarkii* in Iberian streams. *J. Appl. Ecol.* 33:1156–70

Benard MF. 2004. Predator-induced phenotypic plasticity in organisms with complex life histories. *Annu. Rev. Ecol. Evol. Syst.* 35:651–73

Bolger DT, Suarez AV, Crooks KR, Morrison SA, Case TJ. 2000. Arthropods in urban habitat fragments in southern California: area, age and edge effects. *Ecol. Appl.* 10:1230–48

Brown MW, Miller SS. 1998. Coccinellidae in apple orchards of eastern West Virginia and the impact of invasion by *Harmonia axyridis* (Coleoptera: Coccinellidae). *Entomol. News* 109:136–42

Bryan SD, Robinson AT, Sweetser MG. 2002. Behavioral responses of a small native fish to multiple introduced predators. *Environ. Biol. Fishes* 63:49–56

Burger JC, Patten MA, Prentice TR, Redak RA. 2001. Evidence for spider community resilience to invasion by non-native spiders. *Biol. Conserv.* 98:241–49

Cardinale BJ, Harvey CT, Gross K, Ives AR. 2003. Biodiversity and biocontrol: emergent impacts of a multi-enemy assemblage on pest suppression and crop yield in an agroecosystem. *Ecol. Lett.* 6:857–65

Carlton JT, Cohen AN. 2003. Episodic global dispersal in shallow water marine organisms: the case history of the European shore crabs *Carcinus maenus* and *C. aesturarii*. *J. Biogeogr.* 30:1809–20

Carlton JT, Geller JB. 1993. Ecological roulette: the global transport of nonindigenous marine organisms. *Science* 261:78–82

Colunga-Garcia M, Gage SH. 1998. Arrival, establishment, and habitat use of the multicolored Asian lady beetle (Coleoptera: Coccinellidae) in a Michigan landscape. *Environ. Entomol.* 27:1574–80

Correia AM. 2001. Seasonal and interspecific evaluation of predation by mammals and birds on the introduced red swamp crayfish *Procambarus clarkii* (Crustacea, Cambaridae) in a freshwater marsh (Portugal). *J. Zool.* 255:533–41

Cottrell TE. 2004. Suitability of exotic and native lady beetle eggs (Coleoptera: Coccinellidae) for development of lady beetle larvae. *Biol. Control* 31:362–71

Cottrell TE, Shapiro-Ilan DI. 2003. Susceptibility of a native and an exotic lady beetle (Coleoptera: Coccinellidae) to *Beauveria bassiana*. *J. Invertebr. Pathol.* 84:137–44

Cottrell TE, Yeargan KV. 1998. Intraguild predation between an introduced lady beetle, *Harmonia axyridis* (Coleoptera: Coccinellidae), and a native lady beetle, *Coleomegilla maculata* (Coleoptera: Coccinellidae). *J. Kans. Entomol. Soc.* 71:159–63

Curtis TR, Aponte Y, Stamp NE. 2005. Nest paper absorbency, toughness, and protein concentration of a native vs an invasive social wasp. *J. Chem. Ecol.* 31:1089–1100

Day WH, Prokrym DR, Ellis DR, Chianese RJ. 1994. The known distribution of the predator *Propylea quattuordecimpunctata* (Coleoptera: Coccinellidae) in the United States, and thoughts on the origin of this species and 5 other exotic lady beetles in eastern North America. *Entomol. News* 105:244–56

DeRivera CE, Ruiz GM, Hines AH, Jivoff P. 2005. Biotic resistance to invasion: native predator limits abundance and distribution of an introduced crab. *Ecology* 86:3364–76

Dick JTA, Platvoet D. 2000. Invading predatory crustacean *Dikerogammarus villosus* eliminates both native and exotic species. *Proc. R. Soc. London Ser. B* 267:977–83

Donaldson W, Price AH, Morse J. 1994. The current status and future prospects of the Texas horned lizard (*Phrynosoma cornutum*) in Texas. *Tex. J. Sci.* 46:97–113

Elliott NC, Kieckhefer R, Kauffman W. 1996. Effects of an invading coccinellid on native coccinellids in an agricultural landscape. *Oecologia* 105:537–44

Elton C. 1958. *The Ecology of Invasions by Plants and Animals*. London, UK: Methuen

Evans EW. 2004. Habitat displacement of North American ladybirds by an introduced species. *Ecology* 85:637–47

Evans EW, England S. 1996. Indirect interactions in biological control of insects: pests and natural enemies in alfalfa. *Ecol. Appl.* 6:920–30

Fagan WF, Hurd LE. 1994. Hatch density variation of a generalist arthropod predator: population consequences and community impact. *Ecology* 75:2022–32

Fagan WF, Moran MD, Rango JJ, Hurd LE. 2002. Community effects of praying mantids: a meta-analysis of the influences of species identity and experimental design. *Ecol. Entomol.* 27:385–95

Feener DH. 1981. Competition between ant species: outcome controlled by parasitic flies. *Science* 214:815–17

Finke DL, Denno RF. 2004. Predator diversity dampens trophic cascades. *Nature* 429:407–10

Gamboa GJ, Greig EI, Thom MC. 2002. The comparative biology of two sympatric paper wasps, the native *Polistes fuscatus* and the invasive *Polistes dominulus*. *Insect. Soc.* 49:45–49

Garvey JE, Stein RA, Thomas HM. 1994. Assessing how fish predation and interspecific prey competition influence a crayfish assemblage. *Ecology* 75:532–47

Geiger W, Paloma P, Baltanas A, Montes C. 2005. Impact of an introduced Crustacean on the trophic webs of Mediterranean wetlands. *Biol. Invasions* 7:49–73

Gherardi F, Cioni A. 2004. Agonism and interference competition in freshwater decapods. *Behaviour* 141:1297–324

Gil-Sánchez JM, Alba-Tercedor J. 2002. Ecology of native and introduced crayfishes *Austropotamobius pallipes* and *Procambarus clarkii* in southern Spain and implications for conservation of the native species. *Biol. Conserv.* 105:75–80

Goodisman MAD, Matthews RW, Crozier RH. 2001. Hierarchical genetic structure of the introduced *Vespula germanica* in Australia. *Mol. Ecol.* 10:1423–32

Gordon RD. 1985. The Coccinellidae (Coleoptera) of America north of Mexico. *J. NY Entomol. Soc.* 93:1–912

Griffiths SW, Collen P, Armstrong JD. 2004. Competition for shelter among overwintering signal crayfish and juvenile Atlantic salmon. *J. Fish Biol.* 65:436–47

Grosholz ED. 2005. Recent biological invasion may hasten invasional meltdown by accelerating historical introductions. *Proc. Natl. Acad. Sci. USA* 102:1088–91

Grosholz ED, Ruiz GM, Dean CA, Shirley KA, Maron JL, Connors PG. 2000. The impacts of a nonindigenous marine predator in a California bay. *Ecology* 81:1206–24

Gruner DS. 2005. Biotic resistance to an invasive spider conferred by generalist insectivorous birds on Hawai'i Island. *Biol. Invasions* 7:541–46

Guan RZ, Wiles PR. 1997. Ecological impact of introduced crayfish on benthic fishes in a British lowland river. *Conserv. Biol.* 11:641–47

Halaj J, Wise DH. 2001. Terrestrial trophic cascades: how much do they trickle? *Am. Nat.* 157:262–81

Hann SW. 1990. Evidence for the displacement of an endemic New Zealand spider, *Latrodectus kapito* Powell by the South African species, *Steatoda capensis* Hann (Araneae: Theridiidae). *NZ J. Zool.* 17:295–308

Hathaway MA. 1986. A new paper wasp in North America. *Pest Manag.* 5:32–34

Hee JJ, Holway DA, Suarez AV, Case TJ. 2000. Role of propagule size in the success of incipient colonies of the invasive Argentine ant. *Conserv. Biol.* 14:559–63

Helms KR, Vinson SB. 2003. Apparent facilitation of an invasive mealybug by an invasive ant. *Insect. Soc.* 50:403–4

Hemptinne JL, Lognay G, Doumbia M, Dixon AFG. 2001. Chemical nature and persistence of the oviposition deterring pheromone in the tracks of the larvae of the two spot ladybird, *Adalia bipunctata* (Coleoptera: Coccinellidae). *Chemoecology* 11:43–47

Herborg M, Rushton SP, Clare AS, Bentley MG. 2003. Spread of the Chinese mitten crab (*Eriocheir sinensis* H. Milne Edwards) in Continental Europe: analysis of a historical data set. *Hydrobiologia* 503:21–28

Herborg M, Rushton SP, Clare AS, Bentley MG. 2005. The invasion of the Chinese mitten crab (*Eriocheir sinensis*) in the United Kingdom and its comparison to continental Europe. *Biol. Invasions* 7:959–68

Hill AM, Lodge DM. 1994. Diel changes in resource demand: competition and species replacement among crayfishes. *Ecology* 75:2118–26

Hill AM, Lodge DM. 1999. Replacement of resident crayfishes by an exotic crayfish: the roles of competition and predation. *Ecol. Appl.* 9:678–90

Hodek I, Iperti G, Hodková M. 1993. Long-distance flights in Coccinellidae (Coleoptera). *Eur. J. Entomol.* 90:403–14

Holway DA. 1999. Competitive mechanisms underlying the displacement of native ants by the invasive Argentine ant. *Ecology* 80:238–51

Holway DA. 2002. Role of abiotic factors in governing susceptibility to invasion: a test with Argentine ants. *Ecology* 83:1610–19

Holway DA, Lach L, Suarez AV, Tsutsui ND, Case TJ. 2002. The causes and consequences of ant invasions. *Annu. Rev. Ecol. Evol. Syst.* 33:181–233

Holway DA, Suarez AV. 2004. Colony-structure variation and interspecific competitive ability in the invasive Argentine ant. *Oecologia* 138:216–22

Holway DA, Suarez AV, Case TJ. 1998. Loss of intraspecific aggression in the success of a widespread invasive social insect. *Science* 282:949–52

Hoogendoorn M, Heimpel GE. 2002. Indirect interactions between an introduced and a native ladybird beetle species mediated by a shared parasitoid. *Biol. Control* 25:224–30

Human KG, Gordon DM. 1996. Exploitation and interference competition between the invasive Argentine ant, *Linepithema humile*, and native ant species. *Oecologia* 105:405–12

Hurd LE. 1999. Ecology of praying mantids. In *The Praying Mantids*, ed. FR Prete, H Wells, PH Wells, LE Hurd, pp. 43–60. Baltimore: Johns Hopkins Univ. Press

Ives AR, Cardinale BJ, Snyder WE. 2005. A synthesis of subdisciplines: predator-prey interactions, and biodiversity and ecosystem functioning. *Ecol. Lett.* 8:102–16

Johnson MT, Follett PA, Taylor AD, Jones VP. 2005. Impacts of biological control and invasive species on a nontarget native Hawaiian insect. *Oecologia* 142:529–40

Johnson RN, Starks PT. 2004. A surprising level of genetic diversity in an invasive wasp: *Polistes dominulus* in the Northeastern United States. *Environ. Entomol.* 97:732–37

Kajita Y, Takano F, Yasuda H, Evans EW. 2006a. Interactions between introduced and native predatory ladybirds (Coleoptera, Coccinellidae): factors influencing the success of species introductions. *Ecol. Entomol.* 31:58–67

Kajita Y, Yasuda H, Evans EW. 2006b. Effects of native ladybirds on oviposition of the exotic species, *Adalia bipunctata* (Coleoptera: Coccinellidae) in Japan. *Appl. Entomol. Zool.* 41:57–61

Kaplan I, Eubanks MD. 2005. Aphids alter the community-wide impact of fire ants. *Ecology* 86:1640–49

Kasper ML, Reeson AF, Cooper SJB, Perry KD, Austin AD. 2004. Assessment of prey overlap between a native (*Polistes humilis*) and an introduced (*Vespula germanica*) social wasp using morphology and phylogenetic analyses of 16s rDNA. *Mol. Ecol.* 13:2037–48

Koch RL. 2003. The multicolored Asian lady beetle, *Harmonia axyridis*: a review of its biology, uses in biological control, and nontarget impacts. *J. Insect Sci.* 3:1–16

Krafsur ES, Obrycki JJ, Harwood JD. 2005. Comparative genetic studies of native and introduced Coccinellidae in North America. *Eur. J. Entomol.* 102:469–74

Leonard GH, Bertness MD, Yund PO. 1999. Crab predation, waterborne cues, and inducible defenses in the blue mussel, *Mytilus edulis*. *Ecology* 80:1–14

Light T. 2005. Behavioral effects of invaders: alien crayfish and native sculpin in a California stream. *Biol. Invasions* 7:353–67

Lilley JH, Cerenius L, Soderhall K. 1997. RAPD evidence for the origin of crayfish plague outbreaks in Britain. *Aquaculture* 157:181–85

Lodge DM, Kershner MW, Aloi JE. 1994. Effects of an omnivorous crayfish (*Orconectes rusticus*) on a freshwater littoral food web. *Ecology* 75:1265–81

Lodge DM, Kratz TM, Capelli GM. 1986. Long-term dynamics of three crayfish species in Trout Lake, Wisconsin. *Can. J. Fish. Aquat. Sci.* 43:993–98

Lodge DM, Taylor CA, Holdich DM, Skurdal J. 2000. Nonindigenous crayfishes threaten North American freshwater biodiversity: lessons from Europe. *Fisheries* 25:7–148

Lohrer AM, Whitlatch RB. 2002. Interactions among aliens: apparent replacement of one exotic species by another. *Ecology* 83:719–32

Lubchenco J. 1978. Plant species diversity in a marine intertidal community: importance of herbivore food preference and algal competitive abilities. *Am. Nat.* 112:23–39

Mack RN, Simberloff D, Lonsdale WM, Evans H, Clout M, Bazzaz FA. 2000. Biotic invasions: causes, epidemiology, global consequences, and control. *Ecol. Appl.* 10:689–710

Maezono M, Koboyashi R, Kusahara M, Miyashita T. 2005. Direct and indirect effects of exotic bass and bluegill on exotic and native organisms in farm ponds. *Ecol. Appl.* 15:638–50

Majerus MEN. 1997. Interspecific hybridization in ladybirds (Col.: Coccinellidae). *Entomol. Rec.* 109:11–23

McDonald PS, Jensen GC, Armstrong DA. 2001. The competitive and predatory impacts of the nonindigenous crab *Carcinus maenas* (L.) on early benthic phase Dungeness crab *Cancer magister* Dana. *J. Exp. Mar. Biol. Ecol.* 258:39–54

Menge BA. 1976. Organization of the New England rocky intertidal community: role of predation, competition and environmental heterogeneity. *Ecol. Monogr.* 46:355–93

Michaud JP. 2002. Invasion of the Florida citrus ecosystem by *Harmonia axyridis* (Coleoptera: Coccinellidae) and asymmetric competition with a native species, *Cycloneda sanguinea*. *Environ. Entomol.* 31:827–35

Moran MD, Rooney TP, Hurd LE. 1996. Top-down cascade from a bitrophic predator in an old-field community. *Ecology* 77:2219–27

Morrison LW. 1999. Indirect effects of phorid fly parasitoids on the mechanisms of interspecific competition among ants. *Oecologia* 121:113–22

Morrison LW. 2000. Mechanisms of interspecific competition among an invasive and two native fire ants. *Oikos* 90:238–52

Ness JH, Bronstein JL. 2004. The effects of invasive ants on prospective ant mutualists. *Biol. Invasions* 6:445–61

Ness JH, Bronstein JL, Andersen AN, Holland JN. 2004. Ant body size predicts dispersal distance of ant-adapted seeds: implications of small-ant invasions. *Ecology* 85:1244–50

Niemelä J, Spence JR, Carcamo H. 1997. Establishment and interactions of carabid populations: an experiment with native and introduced species. *Ecography* 20:643–52

Nyström P, Svensson O, Lardner B, Brönmark C, Granéli W. 2001. The influence of multiple introduced predators on a littoral pond community. *Ecology* 82:1023–39

Obrycki JJ, Elliott NC, Giles KL. 2000. Coccinellid introductions: potential for and evaluation of nontarget effects. In *Nontarget Effects of Biological Control*, ed. PA Follett, JJ Duan, pp. 127–45. Boston, MA: Kluwer Acad.

Orr MR, Seike SH, Benson WW, Gilbert LE. 1995. Flies suppress fire ants. *Nature* 373:292–93

Paine RT. 1966. Food-web complexity and species diversity. *Am. Nat.* 100:65–75

Palmer AR. 1992. Calcification in marine mollusks: how costly is it? *Proc. Natl. Acad. Sci. USA* 89:1379–82

Pearl CA, Adams MJ, Schuytema GS, Nebeker AV. 2003. Behavioral responses of anuran larvae to chemical cues of native and introduced predators in the Pacific Northwestern United States. *J. Herpetol.* 37:572–76

Perry WL, Feder JL, Dwyer G, Lodge DM. 2001. Hybrid zone dynamics and species replacement between *Orconectes* crayfishes in a northern Wisconsin lake. *Evolution* 55:1153–66

Pervez A, Omkar. 2006. Ecology and biological control application of multicoloured Asian ladybird, *Harmonia axyridis*: a review. *Biocontrol Sci. Technol.* 16:111–28

Pickett KM, Wenzel JW. 2000. High productivity in haplometric colonies of the introduced paper wasp *Polistes dominulus* (Hymenoptera: Vespidae; Polistinae). *J. NY Entomol. Soc.* 108:314–25

Polis GA. 1991. Complex trophic interactions in deserts: an empirical critique of food-web theory. *Am. Nat.* 138:123–55

Polis GA, Myers CA, Holt RD. 1989. The ecology and evolution of intraguild predation: potential competitors that eat each other. *Annu. Rev. Ecol. Syst.* 20:297–330

Prasad RP, Snyder WE. 2006. Diverse trait-mediated interactions in a multi-predator, multi-prey community. *Ecology* 87:1131–37

Pratt PD, Coombs EM, Croft BA. 2003. Predation by phytoseiid mites on *Tetranychus lintearius* (Acari: Tetranychidae), an established weed biological control agent of gorse (*Ulex europaeus*). *Biol. Control* 26:40–47

Puth LM, Allen TFH. 2004. Potential corridors for the rusty crayfish, *Orconectes rusticus*, in northern Wisconsin (USA) lakes: lessons for exotic invasions. *Landscape Ecol.* 20:567–77

Reuter M, Pederson JS, Keller L. 2005. Loss of Wolbachia infection during colonization in the invasive Argentine ant *Linepithema humile*. *Heredity* 94:364–69

Ross KG, Vargo EL, Keller L. 1996. Social evolution in a new environment: the case of introduced fire ants. *Proc. Natl. Acad. Sci. USA* 93:3021–25

Roush RT. 1990. Genetic variation in natural enemies: critical issues for colonization in biological control. In *Critical Issues in Biological Control*, ed. M Mackauer, LE Ehler, J Roland, pp. 263–88. Andover, Hants, UK: Intercept

Rutledge CE, O'Neil RJ, Fox TB, Landis DA. 2004. Soybean aphid predators and their use in integrated pest management. *Ann. Entomol. Soc. Am.* 97:240–48

Růžička Z. 2001. Oviposition responses of aphidophagous coccinellids to tracks of ladybird (Coleopterea: Coccinellidae) and lacewing (Neuroptera: Chrysopidae) larvae. *Eur. J. Entomol.* 98:183–88

Sakuratani Y, Matumoto Y, Oka M, Kubo T, Fujii A, et al. 2000. Life history of *Adalia bipunctata* (Coleoptera, Coccinellidae) in Japan. *Eur. J. Entomol.* 97:555–58

Sato S, Dixon AFG. 2004. Effect of intraguild predation on the survival and development of three species of aphidophagous ladybirds: consequences for invasive species. *Agric. For. Entomol.* 6:21–24

Schmitz OJ, Hamback PA, Beckerman AP. 2000. Trophic cascades in terrestrial systems: a review of the effect of predator removals on plants. *Am. Nat.* 155:141–53

Schmitz OJ, Krivan V, Ovadia O. 2004. Trophic cascades: the primacy of trait-mediated indirect interactions. *Ecol. Lett.* 7:153–63

Seeley RH. 1986. Intense natural selection caused a rapid morphological transition in a living marine snail. *Proc. Natl. Acad. Sci. USA* 83:6897–901

Sheppard SK, Henneman ML, Memmott J, Symondson WOC. 2004. Infiltration by alien predators into invertebrate food webs in Hawaii: a molecular approach. *Mol. Ecol.* 13:2077–88

Shoemaker DD, Ross KG, Arnold ML. 1996. Genetic structure and evolution of a fire ant hybrid zone. *Evolution* 50:1958–76

Shoemaker DD, Ross KG, Keller L, Vargo EL, Werren JH. 2000. Wolbachia infections in native and introduced populations of fire ants (*Solenopsis* spp.). *Insect Mol. Biol.* 9:661–73

Sih A, Englund G, Wooster D. 1998. Emergent impacts of multiple predators on prey. *Trends Ecol. Evol.* 13:350–55

Simberloff D, Von Holle B. 1999. Positive interactions of nonindigenous species: invasional meltdown? *Biol. Invasions* 1:21–32

Smart AC, Harper DM, Malaisse F, Schmitz S, Coley S, de Beauregard AG. 2002. Feeding of the exotic Louisiana red swamp crayfish, *Procambarus clarkii* (Crustacea, Decapoda), in an African tropical lake: Lake Naivasha, Kenya. *Hydrobiologia* 488:129–42

Snyder WE, Clevenger GM, Eigenbrode SD. 2004. Intraguild predation and successful invasion by introduced ladybird beetles. *Oecologia* 140:559–65

Snyder WE, Hurd LE. 1995. Egg hatch phenology and intraguild predation between two mantid species. *Oecologia* 104:496–500

Snyder WE, Ives AR. 2001. Generalist predators disrupt biological control by a specialist parasitoid. *Ecology* 82:705–16

Snyder WE, Ives AR. 2003. Interactions between specialist and generalist natural enemies: parasitoids, predators, and pea aphid biocontrol. *Ecology* 84:91–107

Snyder WE, Snyder GB, Finke DL, Straub CS. 2006. Predator biodiversity strengthens herbivore suppression. *Ecol. Lett.* In press

Spradbery JP, Maywald GF. 1992. The distribution of the European or German wasp, *Vespula germanica* (F.) (Hymenoptera: Vespidae) in Australia: Past, present and future. *Aust. J. Zool.* 40:495–510

Suarez AV, Bolger DT, Case TJ. 1998. Effects of fragmentation and invasion on native ant communities in coastal Southern California. *Ecology* 79:2041–56

Suarez AV, Case TJ. 2002. Bottom-up effects on persistence of a specialist predator: ant invasions and horned lizards. *Ecol. Appl.* 12:291–98

Suarez AV, Holway DA, Case TJ. 2001. Patterns of spread in biological invasions dominated by long-distance jump dispersal: insights from Argentine ants. *Proc. Natl. Acad. Sci. USA* 98:1095–1100

Suarez AV, Holway DA, Ward PS. 2005. The role of opportunity in the unintentional introduction of nonnative ants. *Proc. Natl. Acad. Sci. USA* 102:17032–35

Symondson WOC. 2002. Molecular identification of prey in predator diets. *Mol. Ecol.* 11:627–41

Thresher R, Proctor C, Ruiz G, Gurney R, MacKinnon C, et al. 2003. Invasion dynamics of the European shore crab, *Carcinus maenas*, in Australia. *Mar. Biol.* 142:867–76

Tilman D. 1982. *Resource Competition and Community Structure*. Princeton, NJ: Princeton Univ. Press

Torchin ME, Lafferty KD, Kuris AM. 2001. Release from parasites as natural enemies: increased performance of a globally introduced marine crab. *Biol. Invasions* 3:333–45

Trussell GC, Ewanchuk PJ, Bertness MD. 2003. Trait-mediated effects in rocky intertidal food chains: predator risk cues alter prey feeding rates. *Ecology* 84:629–40

Trussell GC, Nicklin MO. 2002. Cue sensitivity, inducible defense and trade-offs in a marine snail. *Ecology* 83:1635–47

Trussell GC, Smith LD. 2000. Induced defenses in response to an invading crab predator: An explanation of historical and geographic phenotypic change. *Proc. Natl. Acad. Sci. USA* 97:2123–27

Tsutsui ND, Suarez AV, Holway DA, Case TJ. 2000. Reduced genetic variation and the success of an invasive species. *Proc. Natl. Acad. Sci. USA* 97:5948–53

Walton WC, MacKinnon C, Rodriguez LF, Proctor C, Ruiz GM. 2002. Effect of an invasive crab upon a marine fishery: green crab, *Carcinus maenas*, predation upon a venerid clam, *Katelysia scalarina*, in Tasmania (Australia). *J. Exp. Mar. Biol. Ecol.* 272:171–89

Ward DF, Harris RJ, Stanley MC. 2005. Human-mediated range expansion of Argentine ants *Linepithema humile* (Hymenoptera: Formicidae) in New Zealand. *Sociobiology* 45:401–7

Werner EE, Peacor SD. 2003. A review of trait-mediated indirect interactions in ecological communities. *Ecology* 84:1083–1100

Wheeler AG, Hoebeke ER. 1995. *Coccinella novemnotata* in northeastern North America: historical occurrence and current status (Coleoptera: Coccinellidae). *Proc. Entomol. Soc. Wash.* 97:701–16

Wilder SM, Rypstra AL. 2004. Chemical cues from an introduced predator (Mantodea, Mantidae) reduce the movement and foraging of a native wolf spider (Araneae, Lycosidae) in the laboratory. *Environ. Entomol.* 33:1032–36

Wilson KA, Magnuson JJ, Lodge DM, Hill AM, Kratz TK, et al. 2004. A long-term rusty crayfish (*Orconectes rusticus*) invasion: dispersal patterns and community change in a north temperate lake. *Can. J. Fish. Aquat. Sci.* 61:2255–66

Yamada SB, Dumbauld BR, Kalin A, Hunt CE, Figlar-Barnes R, Randall A. 2005. Growth and persistence of a recent invader *Carcinus maenas* in estuaries of the northeastern Pacific. *Biol. Invasions* 7:309–21

Yasuda H, Evans EW, Kajita Y, Urakawa K, Takizawa T. 2004. Asymmetric larval interactions between introduced and indigenous ladybirds in North America. *Oecologia* 141:722–31

Yasuda H, Kikuchi T, Kindlmann P, Sato S. 2001. Relationships between attack and escape rates, cannibalism, and intraguild predation in larvae of two predatory ladybirds. *J. Insect Behav.* 14:373–83

Yasuda H, Shinya K. 1997. Cannibalism and interspecific predation in two predatory ladybirds in relation to prey abundance in the field. *Entomophaga* 42:153–63

Zavaleta ES, Hobbs RJ, Mooney HA. 2001. Viewing invasive species removal in a whole-ecosystem context. *Trends Ecol. Evol.* 16:454–59

Zettler JA, Taylor MD, Allen CR, Spira TP. 2004. Consequences of forest clear-cuts for native and indigenous ants (Hymenoptera: Formicidae). *Ann. Entomol. Soc. Am.* 97:513–18

The Evolution of Genetic Architecture

Thomas F. Hansen[1,2]

[1]Department of Biology, Center for Ecological and Evolutionary Synthesis, University of Oslo, 0316 Oslo, Norway; email: thomas.hansen@bio.uio.no

[2]Department of Biological Sciences, Florida State University, Tallahassee, Florida 32306

Key Words

canalization, epistasis, evolvability, genotype-phenotype map, pleiotropy

Abstract

Genetic architecture, the structure of the mapping from genotype to phenotype, determines the variational properties of the phenotype and is instrumental in understanding its evolutionary potential. Throughout most of the history of evolutionary biology, genetic architecture has been treated as a given set of parameters and not as a set of dynamic variables. The past decade has seen renewed interest in incorporating the genotype-phenotype map as a dynamical part of population genetics. This has been aided by several conceptual advances. I review these developments with emphasis on recent theoretical work on the evolution of genetic architecture and evolvability.

INTRODUCTION

Genetic architecture refers to the pattern of genetic effects that build and control a given phenotypic character and its variational properties. A description of genetic architecture may include statements about gene and allele number, the distribution of allelic and mutational effects, and patterns of pleiotropy, dominance, and epistasis. Despite the obvious complexity of the developmental processes that underlie the genetic architecture, the vast majority of evolutionary theory is built around a few very simple models of the genotype-phenotype relationship. These are models of additive unconstrained gene action, including the polygenic models of quantitative genetics, and the simple one-allele, one-trait models used in the study of adaptation. At least for R.A. Fisher, these models were not merely a matter of simplifying convenience. Fisher presented some ingenious arguments to justify his focus on genes in isolation (Fisher 1930). One of these was that in a large, panmictic, recombining population, all combinations of alleles would occur, and the evolutionary effect of an allele could be found by averaging over all the genotype combinations in which it participates. Fisher then defined additive effects in terms of these averages. The underlying assumption is not that the genetic architecture is simple, but rather that complex gene interactions can be averaged and treated like statistical noise. Similarly, he argued that the high dimensionality of biological organisms meant that there would always be a way to achieve a goal and improve the organism. Thus, there is always an advantageous mutation, and if the population is large enough, the advantageous mutation will arise and become fixed. In effect, Fisher used the very complexity of organismal architecture to argue that we could treat genes and gene effects in a manner resembling mass action in statistical mechanics where particulars average out and general laws emerge.

Here I will argue that many of the complexities of genetic architecture are evolutionarily relevant and should not be treated as noise. I will not do this by challenging Fisher's statistical-mechanics philosophy and replacing it with some unsatisfactory listing of particulars, but rather by arguing that many complexities can have systematic influences on evolutionary dynamics, which we should aim to model and measure.

Classical population genetics tends to treat genetic architecture as a set of invariant parameters and not as evolutionary variables. The Neodarwinian paradigm more or less defined evolution as change in allele frequencies and left little room for the evolution of allelic effects. Concepts such as genetic canalization, the evolution of reduced genetic variability (Waddington 1942), and genetic assimilation—evolutionary responses based on environmentally induced variation (Waddington 1953)—were difficult to accommodate into this framework. Work on the subject was pursued by a few people (e.g., Rendel 1967, Schmalhausen 1949, Waddington 1957), and was largely empirical (for review see Moreno 1994; Scharloo 1991). Canalization has, however, finally been given a solid population genetic interpretation in terms of evolution of reduced gene effects through epistatic interactions with an evolving genetic background (Wagner et al. 1997), and is currently the focus of considerable theoretical and empirical interest (de Visser et al. 2003, Dworkin 2005a, Flatt 2005, Gibson & Dworkin 2004, Gibson & Wagner 2000, Rutherford 2000, Wagner 2005).

Indeed, the past one or two decades have seen a renewed interest in incorporating development and more realistic representations of the genotype-phenotype map into population genetics. This interest is evident in a number of recent edited volumes (Hallgrímsson & Hall 2005, Pigliucci & Preston 2004, Schlosser & Wagner 2004, Wagner 2001, Wolf et al. 2000), and may be motivated by advances in evolutionary developmental biology and interest in the evolution of evolvability (Gerhart & Kirschner 1997, Kauffman 1993, Maynard Smith & Szathmáry 1995, Raff 1996, Wagner & Altenberg 1996, Weiss & Buchanan 2004).

Understanding genetic architecture is important for many biological questions, including: speciation, the evolution of sex and recombination, the survival of small populations, inbreeding, understanding diseases, animal and plant breeding, and understanding the processes and genetics of adaptation and population differences. There is, however, one overarching reason why genetic architecture is essential in understanding evolutionary theory, and that is because it describes or determines the variational properties of characters, and thus their evolutionary potential. Understanding the evolution of evolvability requires understanding the evolution of genetic architecture.

In this review I concentrate on recent theoretical and conceptual developments. There is an enormous empirical literature relevant to genetic architecture, but mostly without good connection to theory; consequently I discuss empirical work only in so far as it connects with or exemplifies conceptual issues. Traditionally, most research on genetic architecture has been concerned with population concepts, such as genetic variances and covariances, but I focus my attention to the biological underpinnings of genetic architecture as embodied in the genotype-phenotype map. This is the most fundamental level, and the "architectures" of population variance or population differences are less clear-cut because they are influenced both by genotype effects and genotype frequencies. The goal is disentangling these and understanding how underlying genetic architecture influences evolution.

GENETIC ARCHITECTURE: DEFINITIONS AND DISTINCTIONS

Genetic architecture can be studied on many different levels. We can ask about the genetic basis for the differences between species or populations, how many genes are involved, whether dominance and epistasis are important, etc. These differences can be studied by line-cross analysis or by quantitative trait loci (QTL) analysis. We can also ask about the architecture of segregating variation in a population. The classical quantitative genetics model, developed by Fisher and others, was based on a decomposition of the phenotype into contributions from different genes and their interactions (Lynch & Walsh 1998). These are defined as regression parameters on the presence of the particular allele combinations, so that the dominance and epistatic effects are interaction terms in the regression. The total phenotypic variance can then be partitioned into contributions from the various terms, including the additive genetic variance, which is the sum of the variances in additive effects, and a series of dominance and epistatic variance components stemming from the interaction terms

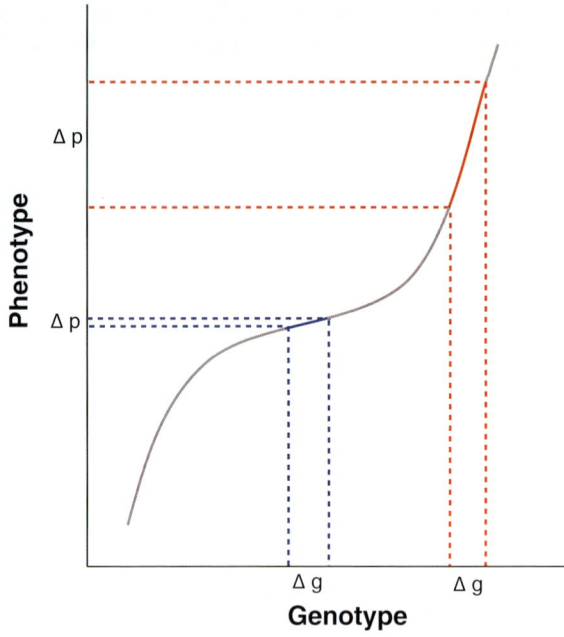

Figure 1
The genotype-phenotype map and canalization: The mapping from genotype to phenotype is a mathematical function that assigns a phenotypic change to each change in genotype. In the flat (canalized) region a given genetic change (Δg) has a small effect (Δp), whereas in the steep (decanalized) region the same change has a large effect. Canalization is evolution toward flat regions, whereas decanalization is evolution toward steep regions.

in the model. This decomposition also generalizes to multiple traits. In this case, the additive genetic variance is replaced with a variance matrix, the G-matrix, which contains the additive genetic variances and covariances between the traits.

We can also study genetic architecture on the level of the genotype-phenotype map, the relationship between individual genotypes and their phenotype (**Figure 1**). It is useful here to recall G.P. Wagner's (1996, Wagner & Altenberg 1996) distinction between variance and variability. Variability has a technical meaning as propensity or disposition to vary. This propensity is conceptually independent of allele frequencies, variances, and other population parameters; it depends on the genotype-phenotype map and mutation rates. The distinction between variation and variability is related to the difference between the G-matrix describing segregating variation, and the mutational M-matrix, describing new (additive) variance and covariance that arise by mutation each generation (e.g., Pigliucci 2004).

Until recently, relatively little attention was paid to the distinction between population parameters and parameters describing the genotype-phenotype map. We may suspect that these were often confused or thought to measure similar things. The development of an explicit conceptual distinction between population or statistical notions of genetic architecture on one hand, and functional/physiological/biological notions on the other was a great step forward. An important contribution was made by Cheverud & Routman (1995), who developed an explicit model of "physiological" epistasis defined without regard to allele frequencies and showed how this physiological epistasis differed from the Fisherian notion of statistical epistasis and even contributed to the additive genetic variance. The Fisherian regression model minimizes the statistical influence of gene interactions (Moreno 1994, Templeton 2000, Whitlock et al. 1995) and, more seriously, the interaction effects in the model miss important distinctions between different types of dominance and epistasis, such as between overdominance and partial dominance, or between positive and negative epistasis (Hansen & Wagner 2001a).

Cheverud & Routman's (1995) model was, however, based on an arbitrary, and not very operational, parameterization of genetic effects. Hansen & Wagner (2001a) generalized this approach by giving an operational definition of a genetic effect as the phenotypic effect of a substitution of an allele (or set of alleles) into a given reference genotype. They then showed how any particular or average genotype could be used as a reference point and how one could translate from one reference genotype to another (see also Barton & Turelli 2004). Genetic effects are not only relative to a reference genotype, but also to a scale of measurement, which will depend on the trait in question. For fitness, which is inherently on a ratio scale, multiplicative interactions are fundamental, and dominance and epistasis are most properly seen as deviances from multiplicative effects (G.P. Wagner, personal communication).

Epistasis occurs when the effect of a gene, or genotype, is different in two different genetic backgrounds (Wagner et al. 1998). The term epistasis hides a vast diversity of complex gene interactions (Rice 2000), and to make progress it is necessary to identify patterns and types of gene interactions that are evolutionarily relevant. One important distinction in this respect is between directional and nondirectional gene interactions (Hansen & Wagner 2001a). Epistasis is said to be directional if genes systematically modify each other in particular patterns or directions in morphospace. Nondirectional epistasis occurs when genes interact, but there are no systematic patterns to the interactions. We can also think of directional epistasis as curvature in the genotype-phenotype map (**Figure 1**; Rice 1998, 2000, 2002). Directional dominance can also occur when allelic effects are systematically altered in relation to the effect of the allele they are combined with. I later show that positive and negative gene interactions can have very different evolutionary consequences. Positive directional gene interactions occur when genes systematically reinforce each other's effects in a particular direction. Negative directional gene interactions occur when genes systematically diminish each other's effect. Directional epistasis for fitness is well recognized in classical population genetics as important in determining the mutation load (e.g., Charlesworth 1990a). Somewhat counterintuitively, negative epistasis for fitness on a multiplicative scale is often known as synergistic epistasis, because it means that deleterious mutations are worse when occurring together. Another important distinction is between epistatic interactions that change the order of effects of a set of alleles (or genotypes), and epistatic interactions that only change the magnitude of effects. Weinreich et al. (2005) referred to the case where the order of fitness effects of a set of alleles is different in different genetic backgrounds as sign epistasis. Obviously, sign epistasis

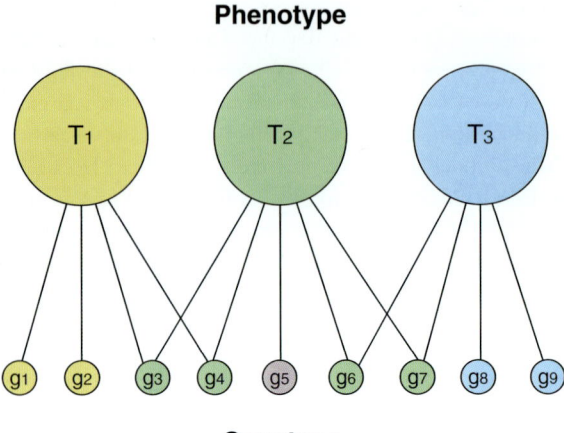

Figure 2

Pleiotropy in the genotype-phenotype map: Nine genes (g_1 to g_9) affect three traits (T_1 to T_3). The genes g_3, g_4, g_6, and g_7 affect more than one character and are said to be pleiotropic. Traits T_1 and T_3 are the most autonomous, as half the genes affecting them have no pleiotropic effects. Trait T_2 is less autonomous with only one fifth of its genes being without pleiotropic effects, but this trait has a larger mutational target size, as it is affected by more genes (five) than the others (four). Traits T_1 and T_3 are independent of each other in that they share no genes. Note, however, that a change in T_1 could affect T_2 through g_3 and g_4, and this could lead to compensatory changes in g_6 and g_7, which would affect T_3.

can lead to dramatic qualitative changes of evolutionary dynamics. Phillips et al. (2000) provide further discussion of epistatic nomenclature.

Pleiotropy is the other main factor in the genotype-phenotype relationship (**Figure 2**). A gene is said to be pleiotropic if it affects more than one character. Together with linkage disequilibrium—the statistical associations between alleles at different loci–pleiotropy is the underlying cause of genetic covariation between characters on the population level. There is, however, no simple relationship between genetic covariance and pleiotropy. Typically there will be a variety of different pleiotropic effects that may cancel or overwhelm each other, as when variation in acquisition of resources hides variation in the allocation of resources (Charlesworth 1990b; Houle 1991). Thus, even mutational covariances cannot fully describe pleiotropy. It is also important to remember that the pattern of pleiotropy depends entirely upon the delineation of characters, and from the theoretical point of view it may be more precise to investigate the dimensionality of mutations and alleles (Wagner & Mezey 2000).

GENETIC ARCHITECTURE AND EVOLVABILITY

What is Evolvability?

Literally, evolvability means ability to evolve. As reviewed by Schlichting & Murren (2004), numerous more formal definitions of evolvability have been put forward. In my opinion, none of these captures the, admittedly varied, usage of the term very well. Hence, I venture a new formulation:

Evolvability is the ability of the genetic system to produce and maintain potentially adaptive genetic variants.

There are three things I emphasize with this definition. First, I follow Wagner & Altenberg (1996) in defining evolvability as a property of the genotype-phenotype map (the genetic system) and not as a population property. Evolvability is a disposition and thus related to variability more than to variation. Often we measure evolvability on the population level, for example, by using mean-scaled additive genetic variances (Hansen et al. 2003b, Houle 1992), but this should be seen as a measure or instantiation of evolvability and not as a definition. For clarity we may use the term population evolvability in these cases. Second, not only the production of variation is important, but also the ability to maintain variation. We want to include the capability for storing and transmitting genetic variation as a part of what makes a system evolvable. Finally, and contrary to Schlichting & Murren (2004), I view the literature on evolvability as almost exclusively concerned with the potential for adaptive evolution, and I suggest that we do not want to include the capacity for producing unconditionally deleterious mutations as a part of a system's evolvability.

With this definition in mind, I review the main aspects of genetic architecture that are important in determining the ability to produce and maintain potentially adaptive variants.

Autonomy

The existence of quasi-independent characters with the potential for evolutionary autonomy is one of the core assumptions of evolutionary biology (Lewontin 1978, Wagner 2001). In their influential review of evolvability, Wagner & Altenberg (1996) drew attention to structural features of the genotype-phenotype map that facilitate autonomy. They argued that the evolvability of a complex system depends on solving the problem of unbounded pleiotropy. If every part depends on every other part, it becomes more and more difficult to find potential advantageous changes when the number of parts is increasing. This is the essence of Fisher's (1930) geometric model of evolution, where he showed that the probability that a new mutation would bring the genotype closer to a given optimum was a decreasing function of the number of traits affected by the mutation. Wagner & Altenberg suggested that this problem could be solved through modularity of the genotype-phenotype map. If the map was divided into modules with relatively few genes having pleiotropic effects across modules, then each module would have evolutionary autonomy (**Figure 2**).

The evolutionary autonomy of a character can be quantified as a conditional evolvability, its evolvability (however defined) when other aspects of the organism are kept fixed (Hansen 2003; Hansen et al. 2003a). The conditional evolvability of a character may be only a fraction of its unconditional evolvability, as a lot of the variation and variability may stem from alleles and mutations with severe pleiotropic constraints, raising the possibility that the apparent evolvability of most quantitative characters may be illusory (Hansen & Houle 2004). In support of this many studies have found that most genetic variation is concentrated along a few axes in morphospace (Björklund 1996; Blows & Hoffmann 2005). On the other hand, although genetic correlations are generally prevalent, they rarely equal one (Roff 1996). There are examples of large responses to artificial selection with relatively small fitness costs (Weber 1996), and one very careful study showed that there exist a very large number of underlying genetic dimensions for *Drosophila* wing morphology (Mezey & Houle 2005).

Variational autonomy is not just a question of presence or absence of pleiotropy. It depends on patterns of variability of the pleiotropic effects (**Figure 2**). If two characters are affected by the same genes in the same way, they have no evolutionary autonomy, but if the genes vary in their pleiotropic effects, for example, if one set of genes have positive effects on both characters, while another set of genes have positive effects on one character and negative effects on the other, then the two characters are largely autonomous despite all genes being pleiotropic. In fact, the two characters may have higher conditional evolvabilities in this situation because they each have a larger mutational target size. Hansen (2003) found that conditional evolvability was typically maximized at intermediate degrees of pleiotropy in several different classes of models and hypothesized that evolvability was generally maximized by genetic architectures with maximal variation in pleiotropic effects. Welch & Waxman (2003) also found that increased modularity does not necessarily increase the rate of adaptation. Baatz & Wagner (1997), however, found that hidden pleiotropic effects (i.e., pleiotropic effects that cancel to generate zero genetic correlation) could reduce the long-term conditional evolvability of a character, presumably through the evolution of covariance or higher cross moments between the traits owing to gene-frequency changes.

Pleiotropy also affects the maintenance of genetic variation in mutation-selection balance (e.g., Wagner 1989). Although less genetic variance is maintained when the genes underlying a character have pleiotropic effects, the pleiotropy also increases the mutational target size per character so as to maintain more variation per character (Hansen 2003). Still, individual genes with many pleiotropic effects will show less molecular variation (Waxman & Peck 1998).

Mutability

Mutation is the ultimate source of variation and thus sets a fundamental upper limit to evolvability. Many large-scale mutation-accumulation experiments have revealed, however, that surprisingly large amounts of phenotypic variation are generated each generation for most quantitative characters (Houle et al. 1996, Lynch et al. 1999). Combined with the general observation of abundant segregating genetic variation (e.g., Houle 1992), this observation has led to the notion that high evolvability is the rule, and that populations are rarely mutation limited. As discussed above, however, the focus on variation in individual characters ignores pleiotropic constraints, leaving levels of conditional evolvability poorly understood (e.g., Blows & Hoffmann 2005, Hansen & Houle 2004, Galis 1999). The degree of pleiotropic constraints on mutational variation is little explored and also likely to be more severe than pleiotropic constraints on standing variation because mutational variation is not yet filtered by selection.

In interpreting mutational variability we must distinguish between rates and effects. Traits may differ in their mutational variability owing to differences in the effects of the mutations that occur (i.e., reflecting different degrees of canalization), or because of the rate at which the mutations appear. Although the molecular mutation rate differs between different organisms and different regions of the genome (Drake et al. 1998), the most important determinant of the rate of mutation affecting a phenotypic trait is the mutational target size, the number, size, and mutability of the genes and regulatory elements that affect a trait (Houle 1998). Houle (1992, 1998) has argued that mutational target size can explain the generally higher population evolvability of life-history traits and fitness components as compared to morphological traits.

Experimental evidence for the importance of mutation rates for evolvability comes from de Visser et al. (1999), who compared rates of adaptation in different mutator strains of *Escherichia coli* with orders-of-magnitude different mutation rates, and found evidence of mutation limitation on evolvability in small, poorly adapted populations, but not in large or well-adapted populations.

Coordination

The evolvability of a complex system depends crucially on its potential for coordinated variability (Riedl 1977). Such coordination arises through development, or through the functional coordination of the organism. The general principle is that simple, nonobstructive changes in developmental processes can lead to large changes in phenotype, which may have a higher likelihood of being advantageous because they are structured by the developmental process. Examples include changes in time or rate of development as in classical cases of heterochronic and allometric change (e.g., Gould 2002).

Developmental modularity is one source of coordinated change through co-option. Co-option consists of employing or expressing one part of the organism—a gene, a gene network, a developmental program, an organ, a physiological response, or a behavior—in a new context. Evidence of evolution by co-option is well documented (e.g., Gerhart & Kirschner 1997, Raff 1996). For example, the widespread observation of genetic programs or modules expressed in different tissues and at different stages of development is evidence of co-option (von Dassow & Munro 1999).

Gene Interaction

Although the response to selection is generally thought to depend only on the additive genetic variance, epistasis may have dramatic effects on the response to selection even over a few generations (Carter et al. 2005, Hansen & Wagner 2001a, Hansen et al. 2006; **Figure 3**), because epistasis makes gene effects become evolvable, and directional epistasis enables rapid changes in additive effects, additive variances, and evolvability. In contrast, Fisherian statistical definitions of gene effects implicitly assume nondirectional epistasis, which has hardly any effect on the response from standing variation (**Figure 3**). Estimates of epistatic variance components are mixtures of positive and negative epistasis, as well as different types of sign epistasis, etc. They are therefore not suitable in predicting evolvability and evolutionary dynamics.

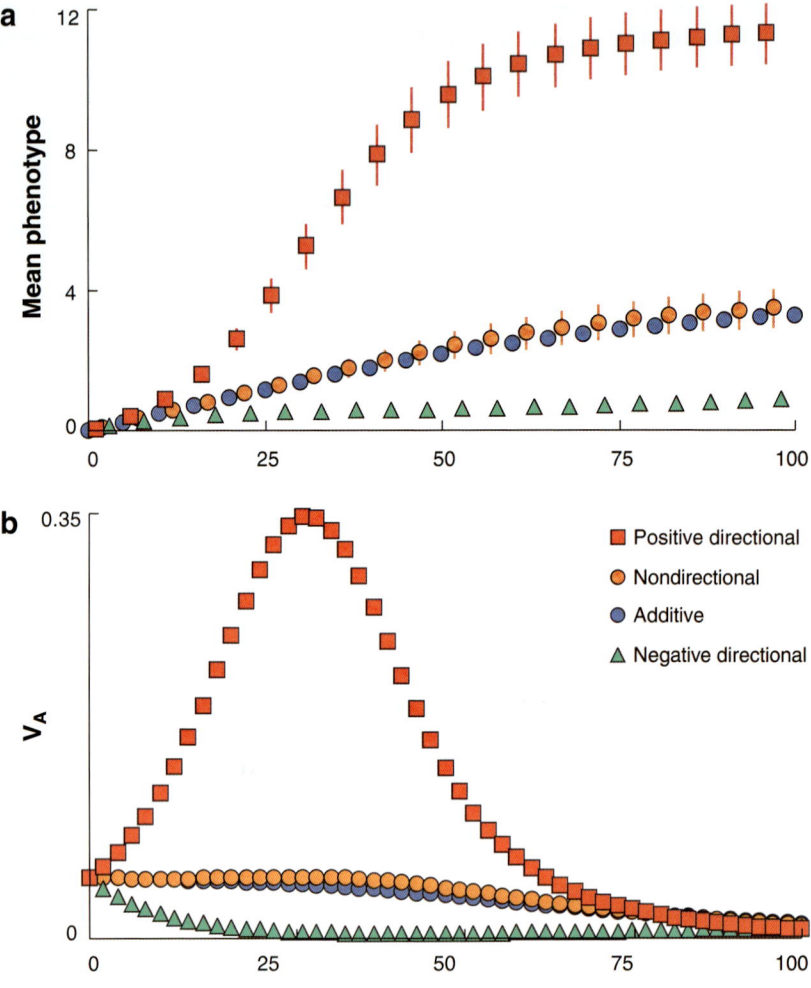

Figure 3
Individual-based simulations illustrating the effects of epistasis on the response to selection from standing genetic variation. Panel (*a*) shows the mean phenotype and (*b*) shows the additive genetic variance as averages over 100 populations subject to linear directional selection. Bars in (*a*) indicate one standard deviation over the ensemble. Four different genetic architectures are shown: pure additive, positive directional epistasis, negative directional epistasis, and nondirectional epistasis. Each simulated population consists of 1000 individuals, and there are 20 segregating loci connected with normally distributed epistasis with mean epistatic effects being 1, 0, or -1 depending on treatment. Modified from Carter et al. (2005).

An influence of patterns of gene interactions on evolvability also appears in other models. For example, analyses of Boolean gene networks indicate that evolvability is maximized at intermediate connectivity between genes (Frank 1999, Kauffman 1993).

A substantial theoretical and empirical literature has developed on how evolvability can increase during a population bottleneck by epistatic variance being converted into additive variance by genetic drift (see Barton & Turelli 2004 for an overview and general analysis). The changes in additive genetic variance due to drift are mainly the result of random evolution of additive effects of alleles caused by their epistatic interaction with a randomly changing genetic background. There is thus no direct relationship between changes in epistatic and additive variance components, and the term conversion is a misnomer. Barton & Turelli (2004) argued that these changes are unlikely to be important. Furthermore, not just drift, but any means of change in the genetic background may cause the evolution of additive effects when epistasis is present. If the epistasis is directional, systematic changes owing to selection will have more powerful effects on the additive genetic variance. For example, Bradshaw & Holzapfel (2000) and Bradshaw et al. (2005) evoked drift to explain increased levels of additive genetic variance for photoperiodism and developmental time in recently colonized northern populations of pitcher-plant mosquitoes (*Wyeomia smithii*). Selection on a directional epistatic architecture is a more plausible alternative.

Epistasis also affects the maintenance of variation. Hermisson et al. (2003) showed that a multilinear epistatic architecture, i.e., an architecture where gene effects are linear functions of the effects of other loci, always maintains less genetic variation in mutation-selection balance than what is maintained by an additive architecture, and sometimes much less. The mutation load, however, may increase or decrease depending on the directionality of epistasis (Hansen & Wagner 2001b, Kondrashov 1988). Sign epistasis has the potential for generating protected polymorphisms and can thus sometimes maintain large amounts of genetic variation in the absence of mutation (e.g., Gavrilets 1993, Gimelfarb 1989). Overdominance for fitness (heterozygote superiority) is a similarly powerful mechanism for maintenance of variation (e.g., Turelli & Barton 2004), but appears to be rare (Charlesworth & Charlesworth 1987).

Hidden and Cryptic Variation

One reason why epistatic genetic architectures tend to maintain less genetic variation under stabilizing selection is because epistatic interactions allow the evolution of reduced gene effects (canalization). Although this leads to less variation expressed at the phenotypic level it also allows for the accumulation of cryptic molecular variation hidden from selection. If cryptic variation can be revealed, it may fuel rapid adaptation. A genetic system can hide genetic variation through negative linkage disequilibrium, where alleles with opposite effects on a trait occur together to produce a gamete with a small total effect, or through canalization, where the effects of alleles are reduced by the genetic background through epistasis.

In a sexually recombining population it is difficult to build up much variation in linkage disequilibrium unless the loci are very strongly linked (Bürger 2000, Hastings 1989). Linkage disequilibrium may, however, be very important if recombination is infrequent, such as in cyclical parthenogens where the life history alternates between parthenogenetic and sexual phases (Lynch & Gabriel 1983). Lynch and coworkers (Deng & Lynch 1996, Pfrender & Lynch 2000) studied the accumulation and release of genetic variance in a cyclically parthenogenic population of *Daphnia pulex*. During the parthenogenetic phase hidden genetic variance accumulates as clones with successful genotypes increase in frequency. During the sexual phase this hidden variation is released leading to increased population evolvability. They also observed genetic slippage, as sexual reproduction reduced fitness components by breaking up coadapted gene complexes. Although the genetic slippage is a fitness drawback to sexual reproduction, one can easily imagine that the increased evolvability could be beneficial if the resting eggs produced from sex are dispersed to hatch in a new environment, or if parasites or predators are adapting to the more successful clones. The maintenance of sexual reproduction in such life histories is a candidate example for adaptive evolvability.

The idea that genetic variation may get canalized under stabilizing selection and released under directional selection or under stress originated with Waddington (1953, 1957). I discuss canalization in a later section and here focus on the release of variation. Hermisson & Wagner (2004) provided a general theoretical analysis of how cryptic variation can be released during a genetic or environmental shift. They showed that a release of genetic variation is expected to happen even without canalization of the wild type. This counterintuitive result can be understood through their metaphor of the moving rug. Cryptic variation can be likened to potentially deleterious mutations accumulating like dirt under a rug. During a genetic or environmental perturbation, the rug is changed to expose the mutations. This can happen either by shrinking the rug (classical decanalization, directional epistasis) or simply by moving the rug at random (nondirectional epistasis), which also exposes cryptic variation. Note that this result undermines some of the classical evidence for adaptive canalization of the wild type, which is based on observing more genetic variation after a perturbation.

Evolutionary capacitors are mechanisms that reveal cryptic variation in a reversible fashion. The discovery that they may be common is one of the most exciting recent developments in evolutionary genetics. It started with Rutherford & Lindquist's (1998) study of the heat-shock protein Hsp90. If *Drosophila* Hsp90 protein is rendered ineffective by mutation or chemical inhibition, large amounts of cryptic genetic variation are revealed in a variety of traits. Based on the idea that Hsp90 acts as a specialized chaperone for signal-transducing proteins, Rutherford & Lindquist proposed that Hsp90 normally acts to buffer variation in signal-transduction pathways, but during times of stress Hsp90, as a heat-shock protein, may be titrated from its normal functions by binding to denatured protein, so that cryptic variation in the signal-transduction pathways is revealed. They proposed that this is an adaptation to specifically increase evolvability during periods of stress. It was later shown that Hsp90 acts the same way in *Arabidopsis* (Queitsch et al. 2002), and similar mechanisms were discovered involving heat-shock proteins in bacteria (Fares et al. 2002) and prions in yeast (Masel & Bergman 2003, True & Lindquist 2000). It is also plausible that general stress can act as a capacitor to reveal variation (Badyaev 2005, Rutherford 2000). Theoretical work has revealed many possible mechanisms for capacitance (Bergman & Siegal 2003, Eshel & Matessi 1998, Hansen et al. 2000, Masel 2005, Masel & Bergman 2003). In particular, Bergman & Siegal (2003) found that knocking out any gene in a gene network would tend to reveal variation [as expected from Hermisson & Wagner's (2004) general analysis]. The crucial and still open question is whether there exist systems that are specifically adapted for revealing variation in times of need.

Robustness may be a general feature of genetic systems that generates cryptic variation (Wagner 2005). It has been argued that the genotype-phenotype map is usually overdetermined in the sense that there are more genes than traits, and that this implies the existence of large neutral networks or neutral spaces of equivalent genotypes (Gavrilets 2004, Kauffman 1993, Schuster et al. 1994, Wagner 2005, Weiss & Fullerton 2000). Cryptic variation may accumulate in neutral spaces, and neutral drift leads to possibilities for population divergence and the evolution of novelty.

Genome Dynamics

The dynamic genome of eukaryotes is a great source of evolvability extending far beyond classical recombination. The fact that genes generally function independently of their position in the genome allows a number of genomic changes such as gene or genome duplication, gene transposition, gene conversion, and inversion, which may all generate novel possibilities. Gene duplication in particular is an important mechanism for generating new genes that may be free to evolve novel functions (Ohno 1970, Raff 1996) or to increase specialization of gene function (Force et al. 1999).

Of course, recombination of genetic material both within and between genomes is an important mechanism of evolvability allowing the decoupling of good from bad genetic material, as well as the rapid generation of good gene combination. On the other hand, recombination also acts against evolvability by precluding the maintenance of coadapted gene complexes.

The Major Transitions

Changes in genetic architecture and evolvability come from changes in the development or structure of the organism. The most fundamental changes involve major reorganizations of the organism. Maynard Smith & Szathmáry (1995) discussed the major transitions in evolution, which they described as major changes in how the organism transmits genetic information. Examples include the evolution of multicellularity, which vastly increased the potential for specialization and reduced the need for genetic specification by building the organism of repeated units, and the evolution of sex and recombination, which allows genes to adapt more easily without inferring with each other, and allowed large increases in genome complexity. Dennett (1995) makes much the same point when he eloquently describes the emergence of evolutionary cranes, which are biological devices that ease the building of adaptations in "design space." Increases in evolvability happen when new cranes, such as sex, multicellularity, a nervous system, a new sensory modality, or communication system, emerge to generate qualitatively new variational possibilities.

Major transitions or novelties may originate as historical contingencies, and are thus not easily accessible to population-genetic analysis. Their maintenance, however, is a question of population genetics, as evidenced by the huge literature on the maintenance of sexual reproduction, but I will not develop this important topic here.

EVOLUTION OF GENETIC ARCHITECTURE: GENERAL PRINCIPLES

Modes of Change

Beyond changes in genotype frequencies, we can identify three genetic mechanisms for the evolution of genetic architecture. The first is through epistatic interactions with an evolving genetic background. Epistasis means that gene or genotype effects depend on the genetic background, and when the background changes because of selection, drift, mutation, or any other mechanism, gene effects will change. The character of these changes depends on the type of epistasis involved. For there to be a systematic relationship between changes in the phenotype and changes in gene effects and evolvability, the epistasis must be directional. Directional epistasis is the basis of canalization and the evolution of evolvability. Directional dominance may have similar effects.

Genetic architecture can also evolve through heritable allelic effects. If there is a correlation between the effects of alleles and the mutations they can generate, then the fixation of an allele will alter the mutational-effects distribution, and thus the variability of the trait. This can include changes in pleiotropy and epistasis. If, for example, an allele that has gained or lost a regulatory element is fixed, then most subsequent mutations to this locus will inherit this particular gain or loss. There is thus a correlation between the effects of the allele and its mutations. The generality of this scenario is supported by a high degree of modularity in the regulatory region of most genes (Stern 2000). This modularity means that many mutations will affect only a part of the spatiotemporal expression patterns of the gene, so as to set up a strong correlation between the effects of alleles and their mutations. For example, if the allele gains a regulatory element that expresses the gene in a novel context, the gene may gain novel pleiotropic effects, and these pleiotropic effects will be inherited by all further mutations of this allele that do not interfere with the new regulatory element. A possible example is discussed below.

Just like genotype effects can evolve through epistatic interactions with a changing genetic background, they can evolve through genotype-by-environment interactions with a changing environment. This may be particularly important when there are systematic biotic changes in the environment or when there are feedback mechanisms between genetic and environmental changes, as when the environment consists of other individuals (Lynch 1987, Wolf 2003).

Modes of Selection

Because genetic architecture is a function of general organismal development and structure, it can be affected by basically any evolutionary change in the organism. It is thus naive to expect any single force of selection to dominate or to think that genetic architecture evolves as a unit. Instead we must look for the effects of selection on the individual elements of architecture and be open to different answers in different cases.

We may recognize three main classes of hypotheses for the evolution of variational properties (de Visser et al. 2003). The first class is the adaptation hypotheses, which posits that variational properties are directly influenced by selection on variation or variability. The two main possibilities are that the genotype-phenotype map could be adapted to increase evolvability, or that it could be adapted to increase robustness toward genetic disturbances (mutation, recombination, hybridization, etc.). These could well be group-level adaptations.

The second class is the intrinsic hypotheses, which posit that variational properties are intrinsic features of the organism not under direct selection. As the genotype-phenotype map is fundamentally linked to organismal design, it is likely to be strongly affected by indirect selection stemming from a variety of sources. Thus, we face the likely prospect that many or most features of genetic architecture are not primary adaptations related to variability, but instead indirect effects of selection on characters themselves, as opposed to selection on the variational properties of characters.

The third class is the congruence hypotheses, which are intermediate between the adaptation and the intrinsic hypotheses. Here we start with the premise that genetic and environmental sources of variability are often affected by the same functional pathways in the organism, such that selection to structure (usually to reduce) environmental variation will lead to correlated changes in the structure of the genetic architecture. Thus organismal architecture is not adapted to structure genetic variation, but it is adapted to structure environmental variation. The generic example of a congruence hypothesis is Haldane's view that dominance evolved owing to selection on the wild-type allele to produce excess enzyme to cover for unusual environmental circumstances. The consequence is that the wild-type allele also evolved the capacity to cover for alleles with reduced activity; dominance evolves, dominance is selected, but dominance is not an adaptation for genetic robustness.

The shape of the fitness function, the mapping from phenotype to fitness, is instrumental in understanding the evolution of variational properties. A fundamental result to bear in mind is that variation is favored when the fitness function is convex (positive second derivative), and that variation is selected against when the fitness function is concave (negative second derivative) (Layzer 1980). Hence, stabilizing selection reduces variation and disruptive selection increases variation, whereas directional selection may favor or disfavor variation depending on the exact curvature of the fitness function. The direct selection pressure on genetic correlations is determined by the second cross derivatives of the fitness function so that if selection on trait 1 leads to an increased sensitivity of fitness to trait 2, then there is selection for positive correlation between the two traits (Layzer 1980).

Note that the evolutionary quantitative genetics concept of stabilizing selection as a concave peak in the adaptive landscape (e.g., Lande & Arnold 1983) that I use here is different from the stabilizing-selection concept of Waddington (1957) and Schmalhausen (1949), who used the term narrowly to refer to intrinsic selection for developmental stability. Similarly, the term canalizing selection is sometimes used explicitly to refer to selection to reduce allelic effects, as opposed to selection on gene frequencies (Wagner et al. 1997). A more formal way of understanding the effects of selection on the genotype-phenotype map, and making notions such as canalizing selection precise, is to decompose selection into different gradients with interpretable effects. Rice (1998, 2002) has developed a very general system of this sort. He used Taylor expansions to write the mapping from genotype (or any set of underlying variables) to phenotype to fitness as a set of terms, each consisting of factors being fitness gradients and derivatives of the genotype-phenotype map. One of these terms can be taken to represent canalizing selection in the sense that, if the fitness function is quadratic, it describes a force that moves the population toward a point of minimum curvature of the genotype-phenotype map, and thus toward maximum robustness. Hermisson et al. (2003) obtained a similar result from a different decomposition.

The classical approach to studying selection on genetic architecture is through modifier models. A modifier is a hypothetical gene that does not have a direct effect on the trait in question, but can alter the effect of other genes on the trait, or affect other aspects of architecture such as mutation or recombination rates. A general result from this type of model is that selection on the modifier is on the second order of, and thus much weaker than, selection on the direct effects (e.g., Proulx & Phillips 2005). The use of very few loci may, however, give a misleading picture of the opportunity for selection on variability. The more loci that are affected by the modifier, the larger the opportunity is for selection (Proulx & Phillips 2005), and the more modifiers there are, the larger the evolvability. The concept of a modifier of gene effects is also problematic in that the symmetry of epistasis (Hansen & Wagner 2001a) implies that a modifier must have direct effects on the phenotype for some state of the modified locus. The evolution of modifiers will usually be determined by their direct effects. For these reasons, it may be more instructive to focus on sets of mutually interacting loci.

EVOLUTION OF THE GENOTYPE-PHENOTYPE MAP

Canalization and Genetic Assimilation

The general hypothesis that emerged from the classical work on canalization (Rendel 1967, Scharloo 1991, Waddington 1957) was that wild types, i.e., the common genotypes found in natural populations, were typically canalized in the sense that they had evolved robustness toward genetic and environmental perturbances. The evidence is based on the common observation that perturbed individuals, such as hybrids, carriers of large-effect mutations, or stressed individuals, are often more variable than the wild type (Flatt 2005, Scharloo 1991; but see Hermisson & Wagner 2004). As illustrated in **Figure 1**, this can be explained with a genotype-phenotype map that is flat in the region of the wild type and steeper away from the wild type (Moreno 1994, Rendel 1967). Perturbances could then shift the genotype away from the flat, canalized region into the steeper region where variation can be expressed. If we think of the genotype in **Figure 1** as an underlying physiological variable (e.g., Rice 1998), this model can also explain genetic assimilation, because environmental perturbations may push the physiological variable into the steeper regions where phenotypic differences between genotypes will be amplified and can thus be selected upon.

Does Stabilizing Selection Favor Canalization?

The question remains, why should we expect to see the wild type in the robust regions of the genotype-phenotype map? Waddington (1957) proposed that this was caused by selection against variation. Because stabilizing selection is generally expected to be the most common mode of selection, it has the potential to explain the general phenomenon. There is a body of recent theory addressing this question (Azevedo et al. 2006; Eshel & Matessi 1998; Gavrilets & Hastings 1994; Hermisson & Wagner 2004; Hermisson et al. 2003; Kawecki 2000; Proulx & Phillips 2005; Rice 1998, 2002; Siegal & Bergman 2002; van Nimwegen et al. 1999; A. Wagner 1996; Wagner et al.

1997). Although all studies find that genetic canalization is possible under stabilizing selection, they differ on its universality, power, and exact mechanisms.

In general, stabilizing selection for an optimum phenotype controlled by a polygenic system allows evolution to choose among many equivalent genotypes. A number of researchers have found that selection in such a situation tends to favor the more robust genotypes in the flatter regions of the genotype-phenotype map or into the interior of the neutral space (e.g., Rice 1998, 2002; van Nimwegen et al. 1999; A. Wagner 1996). Hermisson et al. (2003), however, found that complex epistatic interactions often introduce constraints that limit optimal robustness. In an analysis of the multilinear epistatic model under stabilizing selection, they found a general tendency to evolve to reduce additive genetic variation, but mutational variability is not generally minimized. The problem is that, in a complex epistatic system, the canalization of different loci may conflict with each other and with selection to keep the mean at optimum. Some loci, typically those with high mutation rates, may become canalized, but at the price of decanalizing other loci. This predicts a negative correlation between mutation rates and mutation effects if adaptive canalization has taken place.

Wagner et al. (1997) found that, near mutation-selection equilibrium, genetic canalization was relatively independent of the strengths of stabilizing selection, as stronger selection removed variation thereby reducing the opportunity for selection on genetic canalization. In general, the opportunity for selection on genetic canalization is strongly dependent on the mutational target size. Studies of two-locus models have typically found that genetic canalization is implausible, as selection becomes too weak (e.g., Nowak et al. 1997, Proulx & Phillips 2005, Wagner 1999). It has been argued that there is much less opportunity for direct selection for genetic canalization than there is for environmental canalization because there is much less mutational than environmental variation (e.g., Wagner 2005). If true, this would make the congruence scenario much more likely than the adaptive scenario. This neglects, however, to take into account that selection acts on segregating alleles and not only on new mutations, and the variation generated by segregating alleles is usually comparable in magnitude to the environmental components of variation (e.g., Houle 1992). Thus, the opportunities for genetic and environmental canalization may be comparable. The likelihood of the adaptive scenario may depend more on the commonality of mechanisms that can canalize many loci at once. Selection for canalization of individual loci is weak, but selection for the canalization of large systems of loci is stronger. Similarly, the opportunity for environmental canalization may depend on the commonality of mechanisms that simultaneously provide robustness toward many sources of environmental disturbance.

These results leave us with a complex picture of the effects of stabilizing selection on genetic canalization. A direct effect of selection is theoretically possible, but depends on the details of genetic and organismal architecture. One general insight is that canalization may be less dependent on strength of stabilizing selection than it is on mutation rates and target sizes (Hermisson et al. 2003, Houle 1998, Proulx & Phillips 2005, Wagner et al. 1997). The models also suggest some plausible alternatives involving indirect selection based on environmental robustness or developmental stability. Genetic canalization may also be effective on deleterious variation generated by segregation or migration load (Proulx & Phillips 2005).

Environmental Canalization and the Congruence Hypothesis

Stabilizing selection favors increased robustness toward environmental noise, and genetic mechanisms that reduce such noise can be favored and lead to the evolution of environmental canalization. If there is a genetic link between the mechanisms leading to environmental canalization and genetic canalization, the latter can evolve due to indirect selection. This is known as the congruence hypothesis for the evolution of genetic canalization, which has been found plausible by a number of theoreticians (Ancel & Fontana 2000, de Visser et al. 2003, Gavrilets & Hastings 1994, Nowak et al. 1997, Wagner et al. 1997).

A link between genetic and environmental robustness is plausible, because genetic and environmental disturbances may often affect the same functional pathways in the organism, and any increase in the robustness of a pathway leads to both genetic and environmental canalization. Evidence for a correlation between environmental and genetic robustness has been found in studies of RNA folding (Ancel & Fontana 2000), in the effects of heat-shock proteins (Rutherford 2000), and in effects of P-element insertions in *Drosophila* (Stearns et al. 1995). Particularly striking is the recent finding by Rifkin et al. (2005) of a strong correlation between mutational and environmental variance in gene expression in a panel of mutation accumulation lines of *Drosophila*. Dworkin (2005b), however, did not find a link between environmental and genetic canalization in *Drosophila* bristles.

A variant of the congruence hypothesis is that genetic canalization evolves as a side effect of selection for developmental stability (e.g., Gavrilets & Hastings 1994; Siegal & Bergman 2002). While theoretically plausible, a number of recent studies failed to find a link between canalization and measures of developmental stability such as fluctuating asymmetry (Debat et al. 2000, Dworkin 2005a, Milton et al. 2003, Pélabon et al. 2004, Rutherford 2000).

Does Directional Selection Favor Decanalization?

The other part of the classical theory of canalization is that directional selection away from the canalized wild type should lead to decanalization and increased evolvability. A decanalizing effect of directional selection has been suggested by a number of theoreticians (e.g., Layzer 1980, Rice 1998, Wagner 1996, Wagner et al. 1997). The results of Carter et al. (2005) and Hansen et al. (2006), however, show that this depends on the directionality of epistasis (**Figure 3**). Positive directional epistasis indeed leads to decanalization under directional selection, but negative directional epistasis causes canalization. The exact curvature of the fitness landscape is also likely to be important.

Prolonged response to directional selection may not be frequent in nature, and an important, but neglected, area of theoretical research is investigating the effects of fluctuating directional selection, or moving optima, on the evolution of genetic architecture. To my knowledge, Kawecki's (2000) simulation study is the only contribution. Kawecki considered the evolution of a modifier of the effects of a set of other loci that additively determines a trait under different forms of fluctuating selection. He found that a canalizing allele was favored when the period of fluctuation was short, but that polymorphisms or advantages to decanalization may result when the period becomes longer. Given the results on directional selection, we expect that the details of genetic architecture, such as the directionality of epistasis would be important, and further investigations of different genetic architectures under fluctuating selection are called for. I note that fluctuating selection can favor increased mutation rates through mutation-rate modifiers hitchhiking with the beneficial alleles they create (e.g., Ishii et al. 1989). See also Bürger & Gimelfarb (2002) and Jones et al. (2004) for analysis of the effects of fluctuating optima on the maintenance of genetic variation in additive architectures.

A recent theoretical study addresses the effects of disruptive selection on the evolution of genetic architecture. Kopp & Hermisson (2006) showed with a modifier approach that disruptive selection generated by intraspecific competition favors the evolution of an asymmetric genetic architecture where most of the effects are concentrated on a few major loci.

Evolution of Pleiotropy

Pleiotropy can evolve through differential epistatic modification of different traits (Cheverud 2001; Cheverud et al. 1997, 2004; Wolf et al. 2005). Although this process still lacks dynamical analysis, its operation can be extrapolated from the results on the evolution of single trait architecture. If the epistatic modifications of the effects of a gene on different traits are the same (i.e., not differential), then pleiotropy will not change. If epistasis is differential, then changes in pleiotropy will depend on the details of how epistasis acts on the different trait effects, and in particular the directionalities of this epistasis. Thus, we need to look for differential directional epistasis. For a precise understanding of the dynamics of pleiotropy we must distinguish between trait-specific epistatic effects and cross epistasis, where changes in gene effects on one trait modify effects of other genes on another trait (Hansen & Wagner 2001a, Wagner & Mezey 2000).

Pleiotropy can also evolve through heritable allelic changes in modular regulatory sites. There are no established models or examples of this, but the possibility may be illustrated with the gene *Bab2*, which affects abdominal pigmentation and trichome patterning in *Drosophila* (Gompel & Carroll 2003). In *D. melanogaster Bab2* represses melanic pigmentation and trichomes, and explains much of the segregating variation in abdominal pigmentation, and presumably also in trichome patterns, making it likely that the two are pleiotropically correlated. In some other species, however, *Bab2* has by various mechanisms lost its effects on either one or the other of the two traits. Presumably pleiotropy is also lost and further changes in the expression pattern of *Bab2* will affect only one of the traits.

The classic question on the evolution of pleiotropy is whether selection for a particular functional relationship between traits can lead to a corresponding functional organization of the pleiotropic effects (e.g., Cheverud 1996, Riedl 1977, G.P. Wagner 1996)? For example, can selection strengthen pleiotropy between two traits that are functionally interdependent, and can it decouple two traits that are functionally unrelated? Cheverud (1982) argued that the G-matrix would evolve to match the shape of the fitness landscape. G.P. Wagner (1996), however, argued that stabilizing selection is not a very powerful mechanism for the evolution of pleiotropic integration and/or decoupling, as it would favor canalization in all directions. This is underscored by the observation that canalization is relatively unaffected by the strength of stabilizing selection (Wagner et al. 1997). Instead, G.P. Wagner (1996) suggested that fluctuating directional selection on one character combined with stabilizing selection on another character could be a powerful mechanism for the evolution of pleiotropic decoupling. These suggestions need be investigated with formal models. The sensitivity of the evolution of gene effects to directional epistasis suggests that these hypotheses may depend on genetic detail. It seems plausible, however, that a combination of directional and stabilizing selection may favor alleles with heritable loss of pleiotropic effects, and thus support Wagner's hypothesis. At the population level, joint directional selection on two characters can increase their genetic correlation when there are alleles with different pleiotropic effects in the population (Slatkin & Frank 1990).

In general, pleiotropy is determined by organismal architecture and may evolve by intrinsic mechanisms (Houle 2001, Klingenberg 2005). An understanding of the evolution of pleiotropy by intrinsic mechanisms requires an understanding of the development of the specific phenotypes in question. The continued progress of research in developmental biology will prove illuminating.

Evolution of Gene Interactions

Epistatic gene interactions are not just a cause of the evolution of individual gene effects, they are also themselves evolvable according to similar principles. A particular epistatic interaction can be modified by higher-order epistasis (Hansen & Wagner 2001a, Hansen et al. 2006, Wagner et al. 1998), and epistatic interactions can evolve through inherited allelic interaction effects, as when an allele loses or gains the ability to physiologically modify another locus.

Hansen et al. (2006) studied the evolution of epistatic interactions under long-term directional selection. The evolution of pairwise interactions is influenced by the directionality of third-order interactions and, in general, the evolution of m-order interactions is influenced by the directionality of m+1-order interactions. In addition to this there was a strong inherent tendency for positive epistatic interactions to weaken and for negative epistatic interactions to strengthen when epistasis is measured with reference to the mean genotype of the evolving population. A tendency toward the evolution of negative epistasis has also been found in simulations based on several specific models (Azevedo et al. 2006; Wilke & Adami 2001). Hansen et al. (2006) identified three types of nonadditive quasi-equilibrium architectures that, although not strictly stable, could be maintained for an extended time: (*a*) nondirectional epistatic architectures, (*b*) canalized architectures with strong erratic epistasis, and (*c*) near-additive architectures where epistasis gets weaker.

Hermisson et al. (2003) studied the evolution of pairwise epistatic interactions under a balance between mutation and stabilizing selection and found that directional epistasis measured with reference to the mean genotype tends to disappear. If some directional epistasis remains at equilibrium, it will act as a constraint and keep the population mean away from the optimum. Liberman & Feldman (2005) found that stronger epistasis could evolve in a two-locus system with a modifier locus under conditions of polymorphic equilibrium for the two loci. The key to this result appears to be that mean fitness was an increasing function of the epistasis parameter. In general, however, we have shown that epistasis does not necessarily evolve to maximize fitness or evolvability (Carter et al. 2005, Hansen et al. 2006, Hermisson et al. 2003).

Dating back to Wright's shifting-balance theory there has been a lot of interest in the evolution of coadapted gene complexes, combinations of alleles from different loci that work well together. A supergene can only be effectively maintained segregating in a population under strong linkage, and few examples are known. There is, however, good evidence for coadapted genomes from crosses between isolated populations (e.g., Fenster & Galloway 2000). In fact, postzygotic isolation is generally thought to result from genetic incompatibilities between isolated genomes. Such coadaptation may result from the fixation of new mutations that work well with, or at least are compatible with, the genetic background. Randomness in the order of appearance and fixation of new mutations will typically make the genetic architectures of isolated populations diverge (Mani & Clarke 1990). In this sense, we expect coadapted, integrated genomes to evolve.

Evolution of Dominance

Mutations with deleterious effects on fitness generally have a degree of recessivity, and various hypotheses to explain this phenomenon have been with us almost since the beginning of population genetics. Fisher, Haldane, and Wright took different positions that may be described as adaptive, congruent, and intrinsic (de Visser et al. 2003). Fisher's view that dominance is an adaptation to minimize the deleterious effects of mutations is now viewed as problematic on both theoretical and empirical grounds. Theoretically, it has long been argued that the strength of selection on dominance modifiers is very weak (e.g., Proulx & Phillips 2005), and Fisher's model has not been able to explain why mutations with small effects should be close to additive, whereas those with large effects should be recessive (e.g., Phadnis & Fry 2005). Orr (1991) made the telling observation that dominance also occurs in species of *Chlamydomonas* that spend almost all of their life cycle in a haploid state.

Haldane, and to some extent Wright, held a congruence view where dominance was seen as a safety factor against disturbances. However, a model of dominance as an intrinsic consequence of enzyme biochemistry became the favored explanation when Kacser & Burns (1981) showed that dominance is inherent in models of metabolic flux, and that these models can also explain why dominance is correlated with the homozygous effect of the mutation (for review see

Keightley 1996). The Kacser-Burns theory is, however, limited in that it applies only to flux in linear pathways under Michelis-Menten kinetics, and cannot explain the general phenomenon (e.g., Bagheri-Chaichian & Wagner 2004, Omholt et al. 2000, Phadnis & Fry 2005). Furthermore, Bagheri-Chaichian et al. (2003) point out that epistasis makes dominance evolvable in the same sense as it makes gene effects evolvable, and thus they challenge the notion that any particular pattern of dominance is inevitable.

Genetic Architecture: Intrinsic or Adapted?

Most theory on the evolution of genetic architecture focuses on adaptive or congruent scenarios, as the selection pressures are well defined. The intrinsic scenario is harder to represent in a population-genetics framework, as almost any complex relationship between genotype and phenotype would lead to changes in genetic architecture under selection on the phenotype and the character of the changes would depend on the details of the system. We may recognize two classes of models of intrinsic phenomena. One type are general, but very abstract, models that look for intrinsic properties such as order or robustness (e.g., Kauffman 1993). These may provide some general insight, but do not yield much in terms of testable hypotheses or measurable parameters in empirical systems. The other class of models are highly specific representations of particular physiological or developmental genotype-phenotype maps, as for instance in models of metabolic or gene-regulatory networks. The problem with these models is that their results are highly system specific.

Models of metabolic networks have been used to predict patterns of dominance (see above) and epistasis (Bagheri-Chaichian & Wagner 2004, Bagheri-Chaichian et al. 2003, Keightley 1989, Szathmáry 1993, Wagner et al. 1998). One general insight from these models is that both dominance and epistasis are likely to be ubiquitous. Similar results emerge from models of gene-regulatory networks (Omholt et al. 2000). It is, however, doubtful how far any particular patterns should be extrapolated. For example, Szathmáry (1993) showed that mutations acting on different enzymes in a linear unsaturated pathway would act antagonistically on flux, but were likely to act synergistically on the pool size of metabolites, as long as these were not downstream from both mutated enzymes. He then argued that we should expect antagonistic epistasis in prokaryotes, because their fitness is determined by flux, but synergistic epistasis in eukaryotes, because fitness here would be determined by pool sizes. This argument ignores the complexity of physiology where metabolic pathways normally involve feedback loops and forks and the fact that the mapping from genotype to phenotype would usually consist of many physiological or developmental steps. A metabolic reaction is but one step in the process. It therefore appears difficult to predict specific epistatic properties of most genotype-phenotype maps from metabolic first principles.

Certain generic properties, such as robustness, may, however, exist. Wagner (2005) argued that there are many equivalent solutions to any given biological problem and that evolution tends to find solutions supported by many possible genotypes just by chance. These solutions also tend to be robust toward mutations, because many mutations are other genotypes coding for the same solution, particularly because equivalent solutions tend to form connected networks in genotype space (e.g., Schuster et al. 1994). Thus, robustness may be intrinsic to many biological systems. Many researchers have also found robustness in specific systems (e.g., Alon et al. 1999; Ancel & Fontana 2000; Barkai & Leibler 1997; A. Wagner 1996, 2000, 2005; Wagner & Stadler 1999; von Dassow et al. 2000).

Evolution of Genome Architecture

The number and type of genes affecting a character is the basis of its genetic architecture, as it will determine its mutational target size, the number of possible gene interactions, and the potential for specialization and refinement of the character's variational properties. New genes affecting a character appear by recruitment (co-option) or by gene duplication. These two events have different consequences for genetic architecture; recruitment tends to increase the integration of the genotype-phenotype map, while duplication tends to favor parcellation.

A recruited gene may normally come with heritable pleiotropic effects on other characters (its original function) and may act to increase the complexity and pleiotropy of the genotype-phenotype map. In contrast, a gene duplication may produce a new gene that is similar to an old gene, and have the same pleiotropic and epistatic effects. Importantly, gene duplication can also generate genes that are more specialized through subfunctionalization, where each duplicate loses different parts of the original functions (e.g., by losing different regulatory regions). Force et al. (1999) showed that subfunctionalization is a likely outcome of a gene duplication event. A subfunctionalized gene may conduct only a subset of the functions of the old gene and may thus be less burdened by pleiotropic and epistatic constraints. Subfuctionalization may lead to the evolution of modularity (Force et al. 2005) and may indeed be a way to evolve new autonomous evolutionary characters through parcellation in the sense of Wagner & Altenberg (1996).

Wagner (1994) showed that gene duplication is most likely to be nondeleterious if either a single gene or the entire genetic system is duplicated. Whole-genome duplications may qualify as major transitions in evolution producing material for increase in genome complexity through specialization and novelty.

Lynch & Conery (2003) argued that the retention of gene duplications, as well as mobile genetic elements and introns, were more likely in small- or moderately sized populations, because selection against weak deleterious effects, which may be the most common immediate effect of a duplication, is less efficient in a small population. Based on this they suggested that that the increase in genome complexity among eukaryotes could be explained as a result of lower effective population sizes. As larger organisms generally have smaller population sizes, large size may be the origin of complexity, rather than the other way around.

There are many other evolvable aspects of genome architecture, including mutation rates (Drake et al. 1998) and recombination rates (Barton 1995), that I will not address owing to space constraints. I note that directional epistasis is also instrumental in understanding the evolution of recombination rates (Barton 1995; Otto & Feldman 1997).

EMPIRICAL CHALLENGES

Theory and Experiment

Although there is a huge empirical literature describing genetic architectures in many systems on many different levels, most of this is poorly connected to theory. This situation has a number of causes. The first is that the statistical models of classical quantitative genetics were set up in a way that did not facilitate either the empirical or the theoretical study of gene interactions. A second problem is that most of the dynamic theory is based on models that lack parameters that are experimentally measurable even in principle. The situation resembles that described by Gavrilets (2003) for models of speciation, where there are a number of highly specific, arbitrary, and often confusing, simulation studies, but a lack of simple general analytical results that can guide empirical work. Thus, a major challenge for the field is achieving a better connection between theory and experiment.

One fundamental problem with many models and representations of gene interaction is that they lack an adequate representation of the phenotype. Genetic architecture is about the mapping from genotype to phenotype, and some arbitrary assignment of fitness to genotypes does not help us understand how this mapping evolves. A haphazard treatment of the phenotype combined with detailed modeling of the genetic system may also introduce conceptual biases. For example, could it be that the notions of overdetermined phenotypes and neutral spaces of equivalent genotypes are artifacts of models with simple one-dimensional phenotypes and complex genotypes? One avenue to connect genotype and phenotype goes through a better understanding of gene-regulatory networks underlying specific phenotypes. The rapid accumulation of data on gene networks will make it possible to assemble realistic dynamic models of the variational properties of phenotypes. These can then be studied from the vantage point of interesting theoretical entities such as the directionality of epistasis or the extent of variation in pleiotropy. Another interesting possibility might be to use artificial selection on phenotypes with well-understood underlying gene networks to understand what parts are changing and whether those parts would be the same in replicate experiments. Beldade et al. (2002) provide a nice example.

Epistasis

Both the theoretical and the empirical study of epistasis have suffered badly from the statistical definitions of epistasis as variance components. Empirically, epistatic variance components are difficult to estimate, and when they are estimated they tend to be small (e.g., Whitlock et al. 1995). Small epistatic variance components are, however, compatible with strong epistasis in the underlying genotype-phenotype map (Cheverud & Routman 1995, Keightley 1989, Moreno 1994, Templeton 2000). A more serious problem is that epistatic variance components cannot be used to distinguish between different types of epistasis (Hansen & Wagner 2001a). Different types of epistasis can have profoundly different evolutionary effects, and we need methods that can identify and measure the relevant aspects of epistasis, directional epistasis and sign epistasis in particular.

On the surface, QTL analysis is a promising avenue to estimate directional epistasis, sign epistasis, and pleiotropic effects (Carlborg & Haley 2004, Mackay 2001, Zeng 2005). Epistasis is often found in QTL studies. A telling example is Dilda & Mackay's (2002) finding of many epistatic QTLs for *Drosophila* bristle number, the classic textbook example of an additive character. There are, however, serious shortcomings with current methodology for estimating QTL effects. The use of significance thresholds for QTL detection biases the results toward the detection of too few loci with too large effects (Beavis 1998). When it comes to epistasis this bias is extreme, as there is less power for the detection of interaction effects. In other words, only the extreme tail of the epistasis distribution is studied, and we have little guarantee that this is representative of the general pattern of epistasis, which may be dominated by numerous weak effects. Developing methods for accurately estimating the number, mean, and variance of epistatic (and additive) effects from QTL data is a major statistical challenge.

With these caveats in mind there is at present little evidence pointing to strong directional epistasis. Most QTL data display nondirectional epistasis (e.g., Weber et al. 2001). Due to the importance of directional epistasis for the mutation load and the maintenance of sexual reproduction (Hansen & Wagner 2001b, Kondrashov 1988), many studies have looked for evidence of synergistic epistasis among deleterious fitness mutations (de Visser et al. 1997a, b; Elena & Lenski 1997, 2001; Lynch et al. 1999; Mukai 1969; Remold & Lenski 2004; Whitlock & Bourguet 2000), but the results are equivocal. In any case, this literature illustrates how good experimental work can be motivated by theoretical questions, but we need to study directional epistasis in more traits than fitness.

The study of canalization also has methodological problems (Dworkin 2005a). It is now clear that a simple increase in genetic variation after a perturbation is not necessarily evidence for canalization of the wild type (Hermisson & Wagner 2004). Better evidence, looking directly at changes in genotype effects, is necessary. Stearns & Kawecki (1994; Stearns et al. 1995), studied canalization by looking at the effects of inserted P-elements on the variation of different traits in *Drosophila*. Across life-history traits they found a negative correlation between the genetic variation induced by the insertions and the fitness sensitivity of the traits, and interpreted this as evidence of increased canalization under stronger selection. Apart from the difficulty that theory predicts canalization to be relatively insensitive to the strength of selection, this is also problematic in that the traits differ in the mutational target size. Houle (1998) argued that differences in mutational target size are more likely to explain the pattern. Future studies must distinguish differences in gene effects from differences in mutational target sizes.

Pleiotropy

Evolutionary developmental biology has provided an enormous amount of data on character variability. The theoretical framework used to interpret this data has, however, been largely verbal and focused on the issue of modularity. The effects of pleiotropy are, however, more refined than its mere absence or presence. It is necessary to describe the pleiotropic effects in more detail. One important issue is studying the signs and degree of variation in pleiotropic effects. Tentative evidence for variation in pleiotropic effects come from Phillips et al. (2001), who found differences in estimated G-matrices for *Drosophila* wing characters across lines inbred from the same base population. It is, however, unclear how much of the differences were due to sampling error. Several researchers have studied patterns of pleiotropy and the degree of modularity between characters with QTL data (Cheverud 1996; Cheverud et al. 1997, 2004; Hall et al. 2006; Juenger et al. 2005 Mezey et al. 2000; Weber et al. 2001; Wolf et al. 2005).

An important challenge is estimating the underlying dimensionality of characters, so as to better understand what absolute pleiotropic constraints exist. This may in principle be approached by estimating the dimensionality of G- or M-matrices, but there are formidable statistical difficulties that must be overcome (Houle et al. 2002, Mezey & Houle 2003).

A final question of theoretical interest concerns the degree of pleiotropic constraint on new mutations. There is a need for focus on the quality rather than just the quantity of new mutations (Hansen 2003, Hansen & Houle 2004).

CONCLUSIONS

Fisher's focus on additive effects was a very successful research strategy that clarified the first principles of character evolution. His treatment of gene interactions, however, was designed to minimize and hide their effects. The result was that the modern synthesis was left without tools to study the evolution of genetic architecture and consequently without tools to study the evolution of evolvability. This situation is now improving owing to several conceptual advances. These include the notion that gene effects, and not just gene frequencies, are evolutionary variables, as well as the distinctions between variation and variability and between statistical and biological effects. Particularly important is the emerging understanding that the statistical notion of gene interaction is not equivalent to the way genes interact in the genotype-phenotype map and that different types of interaction have very different evolutionary consequences. To a first approximation, the evolution of evolvability depends on the directionality of epistasis.

We still need much more theoretical work on the dynamic evolution of genetic architecture under different selection regimes. Another theoretical challenge is providing a better connection to experiment through the identification of measurable parameters and general analytical results. On the empirical side, the general challenge is connecting the masses of genetic data to interesting phenotypes. I thus support Houle's (2001) call for a science of phenomics.

ACKNOWLEDGMENTS

Thanks to J. Alvarez-Castro, J. Fierst, D. Futuyma, J. Hermisson, D. Houle, A. Labra, C. Pélabon, J. Pienaar, S. Proulx, M. Tomaiuolo, and G.P. Wagner for discussions and helpful comments on the draft, and to A.J.R. Carter and A. Labra for help with figures. This work was supported by NSF grants #0344417 and #0444157.

LITERATURE CITED

Alon U, Surette MG, Barkai N, Leibler S. 1999. Robustness in bacterial chemotaxis. *Nature* 397:168–71

Ancel LW, Fontana W. 2000. Plasticity, evolvability, and modularity in RNA. *J. Exp. Zool.* 288:242–83

Azevedo RBR, Lohaus R, Srinivasan S, Dang KK, Burch CL. 2006. Sexual reproduction selects for robustness and negative epistasis in artifical gene networks. *Nature* 440:87–90

Baatz M, Wagner GP. 1997. Adaptive inertia caused by hidden pleiotropic effects. *Theor. Popul. Biol.* 51:49–66

Badyaev AV. 2005. Stress-induced variation in evolution: from behavioral plasticity to genetic assimilation. *Proc. R. Soc. London Ser. B* 272:877–86

Bagheri-Chaichian H, Hermisson J, Vaisnys JR, Wagner GP. 2003. Effects of epistasis on phenotypic robustness in metabolic pathways. *Math. Biosci.* 184:27–51

Bagheri-Chaichian H, Wagner GP. 2004. The evolution of dominance in metabolic pathways. *Genetics* 168:1713–35

Barkai N, Leibler S. 1997. Robustness in simple biochemical networks. *Nature* 387:913–17

Barton NH. 1995. A general model for the evolution of recombination. *Genet. Res.* 65:123–44

Barton NH, Turelli M. 2004. Effects of genetic drift on variance components under a general model of epistasis. *Evolution* 58:2111–32

Beavis WD. 1998. QTL analyses: power, precision, and accuracy. In *Molecular Dissection of Complex Traits*, ed. AH Patterson, pp. 145–62. Boca Raton, FL: CRC

Beldade P, Brakefield PM, Long AD. 2002. Contribution of *Distal-less* to quantitative variation in butterfly eyespots. *Nature* 415:315–18

Bergman A, Siegal ML. 2003. Evolutionary capacitance as a general feature of complex gene networks. *Nature* 424:549–52

Björklund M. 1996. The importance of evolutionary constraints in ecological time scales. *Evol. Ecol.* 10:423–31

Blows MW, Hoffmann AA. 2005. A reassessment of genetic limits to evolutionary change. *Ecology* 86:1371–84

Bradshaw WE, Haggerty BP, Holzapfel CM. 2005. Epistasis underlying a fitness trait within a natural population of the pitcher-plant mosquito, *Wyeomyia smithii*. *Genetics* 169:485–88

Bradshaw WE, Holzapfel CM. 2000. The evolution of genetic architectures and the divergence of natural populations. See Wolf et al. 2000, pp. 245–63

Bürger R. 2000. *The Mathematical Theory of Selection, Recombination, and Mutation*. Chichester, UK: Wiley

Bürger R, Gimelfarb A. 2002. Fluctuating environments and the role of mutation in maintaining quantitative genetic variation. *Genet. Res.* 80:31–46

Carlborg Ö, Haley C. 2004. Epistasis: too often neglected in complex trait studies. *Nat. Rev. Genet.* 5:618–25

Carter AJR, Hermisson J, Hansen TF. 2005. The role of epistatic gene interactions in the response to selection and the evolution of evolvability. *Theor. Popul. Biol.* 68:179–96

Charlesworth B. 1990a. Optimization models, quantitative genetics, and mutation. *Evolution* 44:520–38

Charlesworth B. 1990b. Mutation-selection balance and the evolutionary advantage of sex and recombination. *Genet. Res.* 55:199–221

Charlesworth D, Charlesworth B. 1987. Inbreeding depression and its evolutionary consequences. *Annu. Rev. Ecol. Syst.* 18:237–68

Cheverud JM. 1982. Phenotypic, genetic, and environmental morphological integration in the cranium. *Evolution* 36:499–516

Cheverud JM. 1996. Developmental integration and the evolution of pleiotropy. *Am. Zool.* 36:44–50

Cheverud JM. 2001. The genetic architecture of pleiotropic relations and differential epistasis. See Wagner 2001, pp. 411–33

Cheverud JM, Ehrich TH, Vaughn TT, Koreishi SF, Linsey RB, Pletscher LS. 2004. Pleiotropic effects on mandibular morphology II: Differential epistasis and genetic variation in morphologial integration. *J. Exp. Zool. B* 302:424–35

Cheverud JM, Routman EJ. 1995. Epistasis and its contribution to genetic variance components. *Genetics* 139:1455–61

Cheverud JM, Routman EJ, Irschick DJ. 1997. Pleiotropic effects of individual gene loci on mandibular morphology. *Evolution* 51:2006–16

Debat V, Alibert P, David P, Paradis E, Auffray JC. 2000. Independence between developmental stability and canalization in the skull of the house mouse. *Proc. R. Soc. London Ser. B* 267:423–30

Deng HW, Lynch M. 1996. Change of genetic architecture in response to sex. *Genetics* 143:203–12

Dennett DC. 1995. *Darwin's Dangerous Idea: Evolution and the Meanings of Life*. New York: Simon & Schuster

de Visser JAGM, Hermisson J, Wagner GP, Ancel Meyers LA, Bagheri-Chaichian H, et al. 2003. Perspective: evolution and detection of genetic robustness. *Evolution* 57:1959–72

de Visser JAGM, Hoekstra RF, van den Ende H. 1997a. An experimental test for synergistic epistasis in *Chlamydomonas*. *Genetics* 145:815–19

de Visser JAGM, Hoekstra RF, van den Ende H. 1997b. Test of interaction between genetic markers that affect fitness in *Aspergillus niger*. *Evolution* 51:1499–505

de Visser JAGM, Zeyl CW, Gerrish PJ, Blanchard JL, Lenski RE. 1999. Diminishing returns from mutation supply rate in asexual populations. *Science* 283:404–6

Dilda CL, Mackay TFC. 2002. The genetic architecture of *Drosophila* sensory bristle number. *Genetics* 162:1655–74

Drake J, Charlesworth B, Charlesworth D, Crow JF. 1998. Rates of spontaneous mutation. *Genetics* 148:1667–86

Dworkin I. 2005a. Canalization, cryptic variation, and developmental buffering: A critical examination and analytical perspective. See Hallgrímsson & Hall 2005, pp. 131–58

Dworkin I. 2005b. A study of canalization and developmental stability in the sternopleural bristle system of *Drosophila melanogaster*. *Evolution* 59:1500–9

Elena SF, Lenski RE. 1997. Test of synergistic interactions among deleterious mutations in bacteria. *Nature* 390:395–98

Elena SF, Lenski RE. 2001. Epistasis between new mutations and genetic background and a test of genetic canalization. *Evolution* 55:1746–52

Eshel I, Matessi C. 1998. Canalization, genetic assimilation, and preadaptation: A quantitative genetic model. *Genetics* 149:2119–33

Fares MA, Ruiz-Gonzalez MX, Moya A, Elena SF, Barrio E. 2002. Endosymbiotic bacteria - GroEL buffers against deleterious mutations. *Nature* 417:398

Fenster CB, Galloway LF. 2000. Population differentiation in an annual legume: genetic architecture. *Evolution* 54:1157–72

Fisher RA. 1930. *The Genetical Theory of Natural Selection*. Oxford, UK: Oxford Univ. Press

Flatt T. 2005. The evolutionary genetics of canalization. *Q. Rev. Biol.* 80:287–316

Force A, Cresko WA, Pickett FB, Proulx SR, Amemiya C, Lynch M. 2005. The origin of subfunctions and modular gene regulation. *Genetics* 170:433–46

Force A, Lynch M, Pickett FB, Amores A, Yan YL, Postlethwait J. 1999. Preservation of duplicate genes by complementary degenerative mutations. *Genetics* 151:1531–45

Frank SA. 1999. Population and quantitative genetics for regulatory networks. *J. Theor. Biol.* 197:281–94

Galis F. 1999. Why do almost all mammals have seven cervical vertebrae? Developmental constraints, Hox genes, and cancer. *J. Exp. Zool.* 285:19–26

Gavrilets S. 1993. Equilibria in an epistatic viability model under arbitrary strength of selection. *J. Math. Biol.* 31:397–410

Gavrilets S. 2003. Models of speciation: What have we learned in 40 years? *Evolution* 57:2197–215

Gavrilets S. 2004. *Fitness Landscapes and the Origin of Species*. (*Monogr. Popul. Biol. Ser.*), ed. SA Levin, H Horn. Princeton, NJ: Princeton Univ. Press. 476 pp.

Gavrilets S, Hastings A. 1994. A quantitative-genetic model for selection on developmental noise. *Evolution* 48:1478–86

Gerhart J, Kirschner M. 1997. *Cells, Embryos and Evolution: Towards a Cellular and Developmental Understanding of Phenotypic Variation and Evolutionary Adaptability*. Malden, MA: Blackwell

Gibson G, Dworkin I. 2004. Uncovering cryptic genetic variation. *Nat. Rev. Genet.* 5:681–90

Gibson G, Wagner GP. 2000. Canalization in evolutionary genetics: a stabilizing theory? *Bioessays* 22:372–80

Gimelfarb A. 1989. Genotypic variation for a quantitative character maintained under stabilizing selection without mutations: epistasis. *Genetics* 123:217–27

Gompel N, Carroll SB. 2003. Genetic mechanisms and constraints governing the evolution of correlated traits in drosophilid flies. *Nature* 424:931–35

Gould SJ. 2002. *The Structure of Evolutionary Theory*. Cambridge, MA: Belknap

Hall MC, Basten CJ, Willis JH. 2006. Pleiotropic quantitative trait loci contribute to population divergence in traits associated with life-history variation in *Mimulus guttatus*. *Genetics* 172:1829–44

Hallgrímsson B, Hall BK, eds. 2005. *Variation: a Central Concept in Biology*. Oxford, UK: Academic

Hansen TF. 2003. Is modularity necessary for evolvability? Remarks on the relationship between pleiotropy and evolvability. *Biosystems* 69:83–94

Hansen TF, Álvarez-Castro JM, Carter AJR, Hermisson J, Wagner GP. 2006. Evolution of genetic architecture under directional selection. *Evolution*. 60:1523–36

Hansen TF, Armbruster WS, Carlson ML, Pélabon C. 2003a. Evolvability and genetic constraint in *Dalechampia* blossoms: Genetic correlations and conditional evolvability. *J. Exp. Zool. B* 296:23–39

Hansen TF, Chiu CH, Carter AJR. 2000. Gene conversion may aid adaptive peak shifts. *J. Theor. Biol.* 207:495–511

Hansen TF, Houle D. 2004. Evolvability, stabilizing selection, and the problem of stasis. See Pigliucci & Preston 2004, pp. 130–50

Hansen TF, Pelabon C, Armbruster WS, Carlson ML. 2003b. Evolvability and genetic constraint in *Dalechampia* blossoms: Components of variance and measures of evolvability. *J. Evol. Biol.* 16:754–65

Hansen TF, Wagner GP. 2001a. Modeling genetic architecture: a multilinear model of gene interaction. *Theor. Popul. Biol.* 59:61–86

Hansen TF, Wagner GP. 2001b. Epistasis and the mutation load: a measurement-theoretical approach. *Genetics* 158:477–85

Hastings A. 1989. Linkage disequilibrium and genetic variances under mutation—selection balance. *Genetics* 121:857–60

Hermisson J, Hansen TF, Wagner GP. 2003. Epistasis in polygenic traits and the evolution of genetic architecture under stabilizing selection. *Am. Nat.* 161:708–34

Hermisson J, Wagner GP. 2004. The population genetic theory of hidden variation and genetic robustness. *Genetics* 168:2271–84

Houle D. 1991. Genetic covariance of fitness correlates: What genetic correlations are made of and why it matters. *Evolution* 45:630–48

Houle D. 1992. Comparing evolvability and variability of quantitative traits. *Genetics* 130:195–204

Houle D. 1998. How should we explain variation in the genetic variance of traits? *Genetica* 102/103:241–53

Houle D. 2001. Characters as the units of evolutionary change. See Wagner 2001, pp. 109–40

Houle D, Mezey J, Galpern P. 2002. Interpretation of the results of common principal component analyses. *Evolution* 56:433–40

Houle D, Morikawa B, Lynch M. 1996. Comparing mutational variabilities. *Genetics* 143:1467–83

Ishii K, Matsuda H, Iwasa Y, Sasaki A. 1989. Evolutionary stable mutation rate in a periodically changing environment. *Genetics* 121:163–74

Jones AG, Arnold SJ, Burger R. 2004. Evolution and stability of the G-matrix on a landscape with a moving optimum. *Evolution* 58:1639–54

Juenger T, Perez-Perez JM, Bernal S, Micol JL. 2005. Quantitative trait loci mapping of floral and leaf morphology traits in *Arabidopsis thaliana*: evidence for modular genetic architecture. *Evol. Devel.* 7:259–71

Kacser H, Burns JA. 1981. The molecular basis of dominance. *Genetics* 97:639–66

Kauffman SA. 1993. *The Origins of Order. Self-Organization and Selection in Evolution*. New York: Oxford Univ. Press

Kawecki TJ. 2000. The evolution of genetic canalization under fluctuating selection. *Evolution* 54:1–12

Keightley PD. 1989. Models of quantitative variation of flux in metabolic pathways. *Genetics* 121:869–76

Keightley PD. 1996. A metabolic basis for dominance and recessivity. *Genetics* 143:621–25

Klingenberg CP. 2005. Developmental constraints, modules, and evolvability. See Hallgrímsson & Hall 2005, pp. 219–47

Kondrashov AS. 1988. Deleterious mutations and the evolution of sexual reproduction. *Nature* 336:435–40

Kopp M, Hermisson J. 2006. Evolution of genetic architecture under frequency-dependent disruptive selection. *Evolution*. 60:1537–50

Lande R, Arnold SJ. 1983. The measurement of selection on correlated characters. *Evolution* 37:1210–26

Layzer D. 1980. Genetic variation and progressive evolution. *Am. Nat.* 115:809–26

Lewontin RC. 1978. Adaptation. *Sci. Am.* 239:212–31

Liberman U, Feldman MW. 2005. On the evolution of epistasis I: Diploids under selection. *Theor. Popul. Biol.* 67:141–60

Lynch M. 1987. Evolution of intrafamilial interactions. *Proc. Natl. Acad. Sci. USA* 84:8507–11

Lynch M, Blanchard J, Houle D, Kibota T, Schultz S, et al. 1999. Perspective: Spontaneous deleterious mutation. *Evolution* 53:645–63

Lynch M, Conery JS. 2003. The origins of genome complexity. *Science* 302:1401–4

Lynch M, Gabriel W. 1983. Phenotypic evolution and parthenogenesis. *Am. Nat.* 122:745–64

Lynch M, Walsh B. 1998. *Genetics and Analysis of Quantitative Characters*. Sunderland, MA: Sinauer

Mackay TFC. 2001. The genetic architecture of quantitative traits. *Annu. Rev. Genet.* 35:303–39

Mani GS, Clarke FRS. 1990. Mutational order: a major stochastic process in evolution. *Proc. R. Soc. London Ser. B* 240:29–37

Masel J. 2005. Evolutionary capacitance may be favored by natural selection. *Genetics* 170:1359–71

Masel J, Bergman A. 2003. The evolution of the evolvability properties of the yeast prion [PSI$^+$]. *Evolution* 57:1498–1512

Maynard Smith J, Szathmáry E. 1995. *The Major Transitions in Evolution*. Oxford, UK: Freeman

Mezey JG, Cheverud JM, Wagner GP. 2000. Is the genotype-phenotype map modular?: a statistical approach using mouse quantitative trait loci. *Genetics* 156:305–11

Mezey JG, Houle D. 2003. Comparing G matrices: Are common principal components informative? *Genetics* 165:411–25

Mezey JG, Houle D. 2005. The dimensionality of genetic variation for wing shape in *Drosophila melanogaster*. *Evolution* 59:1027–38

Milton CC, Huynh B, Batterham P, Rutherford SL, Hoffmann AA. 2003. Quantitative trait symmetry independent of Hsp90 buffering: Distinct modes of genetic canalization and developmental stability. *Proc. Natl. Acad. Sci. USA* 100:13396–401

Moreno G. 1994. Genetic architecture, genetic behavior, and character evolution. *Annu. Rev. Ecol. Syst.* 25:31–45

Mukai T. 1969. The genetic structure of natural populations of *Drosophila melanogaster*. VII. Synergistic interaction of spontaneous mutant polygenes controlling viability. *Genetics* 61:749–61

Nowak MA, Boerlijst MC, Cooke J, Maynard Smith J. 1997. Evolution of genetic redundancy. *Nature* 388:167–71

Ohno S. 1970. *Evolution by Gene Duplication*. Heidelberg, Germany: Springer-Verlag

Omholt SW, Plahte E, Øyehaug L, Xiang KF. 2000. Gene regulatory networks generating the phenomena of additivity, dominance and epistasis. *Genetics* 155:969–80

Orr HA. 1991. A test of Fisher's theory of dominance. *Proc. Natl. Acad. Sci. USA* 88:11413–15

Otto SP, Feldman MW. 1997. Deleterious mutations, variable epistatic interactions, and the evolution of recombination. *Theor. Popul. Biol.* 51:134–47

Pélabon C, Carlson ML, Hansen TF, Yoccoz NG, Armbruster WS. 2004. Consequences of interpopulation crosses on developmental stability and canalization of floral traits in *Dalechampia scandens* (Euphorbiaceae). *J. Evol. Biol.* 17:19–32

Pfrender ME, Lynch M. 2000. Quantitative genetic variation in *Daphnia*: Temporal changes in genetic architecture. *Evolution* 54:1502–9

Phadnis N, Fry JD. 2005. Widespread correlations between dominance and homozygous effects of mutations: Implications for the theories of dominance. *Genetics* 171:385–92

Phillips PC, Otto SP, Whitlock MC. 2000. Beyond the average: the evolutionary importance of gene interactions and variability of epistatic effects. See Wolf et al. 2000, pp. 20–38

Phillips PC, Whitlock MC, Fowler K. 2001. Inbreeding changes the shape of the genetic covariance matrix in *Drosophila melanogaster*. *Genetics* 158:1137–45

Pigliucci M. 2004. Studying mutational effects on G-matrices. See Pigliucci & Preston 2004, pp. 231–48

Pigliucci M, Preston K, eds. 2004. *Phenotypic Integration: Studying the Ecology and Evolution of Complex Phenotypes*. New York: Oxford Univ. Press

Proulx SR, Phillips PC. 2005. The opportunity for canalization and the evolution of genetic networks. *Am. Nat.* 165:147–62

Queitsch C, Sangster TA, Lindquist S. 2002. Hsp90 as a capacitor of phenotypic variation. *Nature* 417:618–24

Raff RA. 1996. *The Shape of Life: Genes, Development, and the Evolution of Animal Form*. Chicago: Univ. Chicago Press

Remold SK, Lenski RE. 2004. Pervasive joint influence of epistasis and plasticity on mutational effectes in *Escherichia coli*. *Nat. Genet.* 36:423–26

Rendel R. 1967. *Canalization and Gene Control*. London: Logos

Rice SH. 1998. The evolution of canalization and the breaking of von Baer's laws: Modeling the evolution of development with epistasis. *Evolution* 52:647–56

Rice SH. 2000. The evolution of developmental interactions: Epistasis, canalization, and integration. See Wolf et al. 2000, pp. 82–98

Rice SH. 2002. A general population genetic theory for the evolution of developmental interactions. *Proc. Natl. Acad. Sci. USA* 99:15518–23

Riedl RJ. 1977. A systems-analytical approach to macroevolutionary phenomena. *Q. Rev. Biol.* 52:351–70

Rifkin SA, Houle D, Kim J, White KP. 2005. A mutation accumulation assay reveals a broad capacity for rapid evolution of gene expression. *Nature* 438:220–23

Roff DA. 1996. The evolution of genetic correlations: an analysis of patterns. *Evolution* 50:1392–1403

Rutherford SL. 2000. From genotype to phenotype: buffering mechanisms and the storage of genetic information. *BioEssays* 22:1095–105

Rutherford SL, Lindquist S. 1998. Hsp90 as a capacitor for morphological evolution. *Nature* 396:336–42

Scharloo W. 1991. Canalization: Genetic and developmental aspects. *Annu. Rev. Ecol. Syst.* 22:65–93

Schlichting CD, Murren CJ. 2004. Evolvability and the raw materials for adaptation. In *Plant Adaptation: Molecular Genetics and Ecology*, ed. QCB Cronk, J Whitton, RH Ree, IEP Taylor, pp. 18–29. Ottawa, Ont.: NRC Res. Press

Schlosser G, Wagner GP. 2004. *Modularity in Development and Evolution*. Chicago: Chicago Univ. Press

Schmalhausen II. 1949. *Factors of Evolution: The Theory of Stabilizing Selection*. Philadelphia: Blakiston

Schuster P, Fontana W, Stadler PF, Hofacker IL. 1994. From sequences to shapes and back - A case-study in RNA secondary structures. *Proc. R. Soc. London Ser. B* 255:279–84

Siegal ML, Bergman A. 2002. Waddington's canalization revisited: Developmental stability and evolution. *Proc. Natl. Acad. Sci. USA* 99:10528–32

Slatkin M, Frank SA. 1990. The quantitative genetic consequences of pleiotropy under stabilizing and directional selection. *Genetics* 125:207–13

Stearns SC, Kaiser M, Kawecki TJ. 1995. The differential genetic and environmental canalization of fitness components in *Drosophila* melanogaster. *J. Evol. Biol.* 8:539–57

Stearns SC, Kawecki TJ. 1994. Fitness sensitivity and the canalization of life-history traits. *Evolution* 48:1438–50

Stern D. 2000. Perspective: Evolutionary developmental biology and the problem of variation. *Evolution* 54:1079–91

Szathmáry E. 1993. Do deleterious mutations act synergistically? Metabolic control theory provides a partial answer. *Genetics* 133:127–32

Templeton AR. 2000. Epistasis and complex traits. See Wolf et al. 2000, pp. 41–57

True HL, Lindquist SL. 2000. A yeast prion provides a mechanism for genetic variation and phenotypic diversity. *Nature* 407: 477–83

Turelli M, Barton N. 2004. Polygenic variation maintained by balancing selection: pleiotropy, sex-dependent allelic effects and G x E interactions. *Genetics* 166:1053–79 van Nimwegen E, Crutchfield WY, Huynen M. 1999. Neutral evolution of mutational robustness. *Proc. Natl. Acad. Sci. USA* 96:9716–20

Von Dassow G, Meir E, Munro EM, Odell GM. 2000. The segment polarity network is a robust developmental module. *Nature* 406:188–92

Von Dassow G, Munro E. 1999. Modularity in animal development and evolution: elements of a conceptual framework for EvoDevo. *J. Exp. Zool.* 285:307–25

Waddington CH. 1942. Canalization of development and the inheritance of acquired characters. *Nature* 150:563–65

Waddington CH. 1953. Genetic assimilation of an acquired character. *Evolution* 7:118–26

Waddington CH. 1957. *The Strategy of the Genes: A Discussion of Some Aspects of Theoretical Biology*. London: Allen & Unwin

Wagner A. 1994. Evolution of gene networks by gene duplications: a mathematical model and its implications on genome organization. *Proc. Natl. Acad. Sci. USA* 91:4387–91

Wagner A. 1996. Does evolutionary plasticity evolve? *Evolution* 50:1008–23

Wagner A. 1999. Redundant gene functions and natural selection. *J. Evol. Biol.* 12:1–16

Wagner A. 2000. Robustness against mutations in genetic networks of yeast. *Nat. Genet.* 24:355–61

Wagner A. 2005. *Robustness and Evolvability in Living Systems*. Princeton, NJ: Princeton Univ. Press

Wagner A, Stadler PF. 1999. Viral RNA and evolved mutational robustness. *J. Exp. Zool.* 285:119–27

Wagner GP. 1989. Multivariate mutation-selection balance with constrained pleiotropic effects. *Genetics* 122:223–34

Wagner GP. 1996. Homologues, natural kinds and the evolution of modularity. *Am. Zool.* 36:36–43

Wagner GP, ed. 2001. *The Character Concept in Evolutionary Biology*. San Diego: Academic

Wagner GP, Altenberg L. 1996. Complex adaptations and evolution of evolvability. *Evolution* 50:967–76

Wagner GP, Booth G, Bagheri-Chaichian H. 1997. A population genetic theory of canalization. *Evolution* 51:329–47

Wagner GP, Laubichler MD, Bagheri-Chaichian H. 1998. Genetic measurement theory of epistatic effects. *Genetica* 102/103:569–80

Wagner GP, Mezey J. 2000. Modeling the evolution of genetic architecture: a continuum of alleles model with pairwise AxA epistasis. *J. Theor. Biol.* 203:163–75

Waxman D, Peck JR. 1998. Pleiotropy and the preservation of perfection. *Science* 279:1210–13

Weber KE. 1996. Large genetic change at small fitness cost in large populations of *Drosophila melanogaster* selected for wind tunnel flight: rethinking fitness surfaces. *Genetics* 144:205–13

Weber KE, Eisman R, Higgins S, Morey L, Patty A, et al. 2001. An analysis of polygenes affecting wing shape on chromosome 2 in *Drosophila melanogaster*. *Genetics* 159:1045–57

Weinreich DM, Watson RA, Chao L. 2005. Sign epistasis and genetic constraint on evolutionary trajectories. *Evolution* 59:1165–74

Weiss KM, Buchanan AV. 2004. *Genetics and the Logic of Evolution*. New York: Wiley

Weiss KM, Fullerton SM. 2000. Phenogenetic drift and the evolution of genotype-phenotype relationships. *Theor. Popul. Biol.* 57:187–95

Welch JJ, Waxman D. 2003. Modularity and the cost of complexity. *Evolution* 57:1723–34

Whitlock MC, Bourguet DB. 2000. Factors affecting the genetic load in *Drosophila*: synergistic epistasis and correlations among fitness components. *Evolution* 54:1654–60

Whitlock MC, Phillips PC, Moore FBG, Tonsor SJ. 1995. Multiple fitness peaks and epistasis. *Annu. Rev. Ecol. Syst.* 26:601–29

Wilke CO, Adami C. 2001. Interaction between directional epistasis and average mutational effects. *Proc. R. Soc. London Ser. B* 268:1469–74

Wolf JB. 2003. Genetic architecture and evolutionary constraint when the environment contains genes. *Proc. Natl. Acad. Sci. USA* 100:4655–60

Wolf JB, Broodie ED III, Wade MJ, eds. 2000. *Epistasis and the Evolutionary Process*. New York: Oxford Univ. Press

Wolf JB, Leamy LJ, Routman E, Cheverud JM. 2005. Epistatic pleiotropy and the genetic architecture of covariation within early- and late-developing skull trait complexes in mice. *Genetics* 171:683–94

Zeng ZB. 2005. QTL mapping and the genetic basis of adaptation: recent developments. *Genetica* 123:25–37

The Major Histocompatibility Complex, Sexual Selection, and Mate Choice

Manfred Milinski

Department of Evolutionary Ecology, Max Planck Institute of Limnology, D-24306 Plön, Germany; email: milinski@mpil-ploen.mpg.de

Key Words

evolution of sex, immune system, immunogenetic optimality, parasites, olfaction

Abstract

To maintain sexual reproduction, recombination of good genes through selective mate choice must achieve a twofold genetic benefit in each generation. "Fragrant" immune genes of the major histocompatibility complex (MHC) allow the choosy sex to complement her own set of alleles with a more or less diverse set of male alleles to reach an optimal number of different MHC alleles for the offspring. The optimal complement from the partner should include those MHC alleles that provide resistance against the current parasites, which could be revealed by the expression of costly secondary sexual characters. This maximizes resistance to ever-changing infectious diseases. Because the advantage of sex must be produced through recombination, assortative mating should combine currently advantageous MHC alleles. Preferring just MHC dissimilar mates is only a best-of-bad-job rule. MHC ligand peptides may be the natural "perfume" that reveals a potential partner's MHC genetics probably in all vertebrates. Perfumes may mimick MHC related signals.

THE EVOLUTIONARY PUZZLE OF SEXUAL REPRODUCTION

The existence of sex is an outstanding, still unsolved problem in evolutionary biology (Ridley 2004) or "the deepest mystery in all of biology" (Trivers 1985): We cannot explain why the great majority of animals and plants reproduce sexually. More precisely, we do not understand why males exist (Maynard Smith 1976). The conventional textbook explanation (Crow & Kimura 1965) states that sexual populations can react more quickly to environmental changes than do asexual populations. If two new mutations A and B would be advantageous to cope with recent environmental change and these occur as usual in different individuals, sexuals can combine them within one generation, whereas A and B have to occur sequentially in the same line in an asexual population, which takes time (Fisher 1930, Muller 1932). The problem is that mixed populations containing both reproductive modes will turn very quickly into asexual ones (**Figure 1**).

Imagine two females, a sexual and an asexual one with two offspring each. The sexual one has a son and a daughter, each with only half of the mother's genome, whereas the asexual one has two daughters, identical to their mother. Only daughters again produce two offspring each. Sons neither lay eggs nor give birth although they cost as much as daughters to be produced. Sons seem to be a waste. Because the asexual female has twice as many grandchildren as the sexual one, the proportion of asexual females doubles in each generation (**Figure 1**). Therefore, sexuals will be outcompeted within several generations (Maynard Smith 1976). Sex needs a strong short-term advantage to survive this competition. To favor sexual reproduction, "the environment has to change in such a way that the correlations between selectively relevant features of the environment change sign between generations" (Maynard Smith 1976), e.g., the environment would have to change from rainforest to desert between

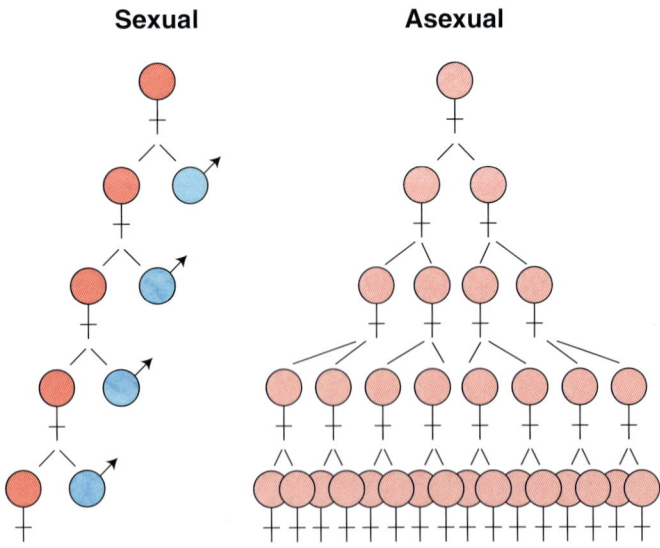

Figure 1

The disadvantage of sexual versus asexual reproduction. See text for details.

generations. "But it is hard to believe that the world is like that," wrote Maynard Smith (1976) and, after he had evaluated the existing hypotheses for the evolution of sex, he concluded: "I fear that the reader may find these models insubstantial and unsatisfactory. But they are the best we have." What may be the strong short-term advantage of sexual reproduction?

Muller's Ratchet and Kondrashov's Hypothesis

When a population consists entirely of asexually reproducing females, a fatal evolutionary mechanism drives the population slowly, but definitely, to extinction by a process called "Muller's ratchet" (Muller 1964). The asexual lines will assemble deleterious germline mutations, about one mutation per generation, until the genome is so degenerated after many generations that the population goes extinct. Mutations can be purged only by sexual reproduction: If two diploid organisms mate that carry one deleterious mutation each, one quarter of their offspring will be mutation free. However, the additive effect of several deleterious mutations seems insufficient to explain that producing some mutation-free offspring would compensate for the twofold efficiency advantage of asexual reproduction. Muller's ratchet is too weak and too slow: Sexuals would not survive the competition with asexuals.

Kondrashov (1988) rescued Muller's ratchet as a mechanism supporting sex by additional assumptions. (*a*) The deleterious mutation rate must be greater than one per generation. (*b*) Multiple mutations must have an increasingly damaging effect ("synergistic epistasis") giving rise to truncation selection under which individuals carrying more than some specific number of mutations will not reproduce at all. However, it is not obvious that a certain proportion of each sexually produced clutch remains without some offspring. Furthermore, most experiments have so far failed to demonstrate a predominance of synergistic epistasis among deleterious mutations (Bonhoeffer et al. 2004, de la Pena et al. 2000, de Visser et al. 1997, Elena 1999, Elena & Lenski 1997, Keightley & Eyre-Walker 2000, Sanjuan et al. 2004, West et al. 1998, Wloch et al. 2001). This suggests that a potential mutation clearance function alone is not likely to render sexual reproduction superior to asexual reproduction.

A Parasite "Red Queen" May Favor Sex if Selective Mate Choice Produces a Twofold Genetic Benefit in Each Generation

Sex is obviously an inefficient way to reproduce: A female throws away half of her genes (during meiosis), tries to find a male who has done similarly, and fills up what she has dropped with what she gets from him. If sex is superior to asexual reproduction, the offspring with the combination of the new and the retained half of genes should be about twice as fit as if she had either made just a copy of herself or chosen a mate randomly. Selective mate choice has to achieve a twofold genetic benefit in each generation compared to random mating. Is this feasible?

Infectious diseases appear in enormous numbers of species and individuals. They may be the selectively relevant features of the environment that change sign between generations in order to favor sexual reproduction. They may change so rapidly that for hosts "it takes all the running you can do, to keep in the same place," as the Red Queen said to Alice in Wonderland (Carroll 1872). The parasite Red Queen hypothesis (e.g., Bell 1982, Ebert & Hamilton 1996, Hamilton 1980, Hamilton et al. 1990, Jaenike 1978, Ladle 1992, Van Valen 1973) assumes that new combinations of genes for resistance are required in every generation to cope with the currently dominating parasites. A female is predicted to select the male possessing resistance genes that would, in combination with her own genes, provide the offspring with the best immune response against the current infectious diseases. This requires that (*a*) there is an enormous variation in resistance genes in the population (otherwise the female would not be able to find better genes than she has dropped), (*b*) she "knows" her immune genes, and (*c*) she can recognize the immune genes of potential mates and choose accordingly. These assumptions may appear too demanding to be fulfilled.

However, they seem to be fulfilled in vertebrates including humans. It is extremely difficult to find a genetically suitable donor for someone who needs transplantation. This is mainly a consequence of polymorphisms of genes within the major histocompatibility complex (MHC) (Horton et al. 2004). The MHC contains the most polymorphic gene loci known in vertebrates (e.g., Janeway et al. 2001). Although this is bad news for potential organ recipients and depicts another unsolved puzzle of evolutionary biology—we cannot explain how this unusual polymorphism is maintained—it is good news for the Red Queen hypothesis: There are hundreds of different immune alleles on the population market in various combinations of a few per individual that a female can potentially choose from. A number of studies have shown that both female and male mice prefer mates that have MHC alleles that differ from their own alleles and that they can choose by smell only (e.g., Egid & Brown 1989, Potts et al. 1991, Yamazaki et al. 1976, 1978; reviews in Penn & Potts 1999 and Penn 2002). This preference would result in offspring that are MHC heterozygous. MHC genes "smell" (actually their gene products or specific ligands smell) and can thus be literally sniffed out by a mouse. The fact that (inbred) mice prefer MHC-dissimilar mates proves that a female mouse "knows" her MHC alleles (and males do the same). All conditions required for the Red Queen hypothesis to work seem to be fulfilled, at least for mice. But what about other vertebrates? All vertebrates studied so far have a very similar MHC system (Cooper & Alder 2006, Kelley et al. 2005), and results similar to those from mice have been found in other vertebrates, e.g., in rats, fish, lizards, birds, and humans, as we will see.

A quick test of the Red Queen hypothesis, although correlative in nature, can be done in populations with both sexual and asexual lines. If sex provides offspring with higher resistance, we would expect that in mixed natural populations (*a*) asexually produced individuals are more often infected than sexually produced ones, and (*b*) the proportion of sexually reproducing individuals increases with the parasite pressure. (*a*) Moritz et al. (1991) measured the prevalence of the mite (*Geckobia* spp.) on both parthenogenetically and sexually produced geckos (*Heteronotia binoei*). A higher proportion of parthenogenetic than sexual geckos was infected in each of six mixed populations. (*b*) Lively (1987) found more often sexual reproduction in the parthenogenetic snail *Potamopyrgus antipodarum* with an increasing degree of trematode infection. Sex was rare in populations where parasites were absent but widespread where parasites were common. This is suggestive evidence. The latter example suggests that invertebrates, which do not possess an MHC system, might nevertheless have polymorphic immune genes and engage in mate choice that takes these genes into account, e.g., a highly polymorphic locus involved in histocompatibility has recently been characterized in the tunicate *Botryllus schlosseri* (DeTomaso et al. 2005). Furthermore, a mutation purging function, though not strong enough on its own (see above), might help the "parasite Red Queen" to favor sex (Howard & Lively 1994).

MAJOR HISTOCOMPATIBILITY COMPLEX GENETICS AND SEXUAL SELECTION

I will not try to add another comprehensive review to the list of existing excellent reviews (e.g., Apanius et al. 1997; Bernatchez & Landry 2003; Brown & Eklund 1994; Edwards & Hedrick 1998; Jordan & Bruford 1998; Mays & Hill 2004; Penn 2002; Penn & Potts 1998a,b, 1999; Piertney & Oliver 2006; Tregenza & Wedell 2000; Zelano & Edwards 2002; Ziegler et al. 2005). Instead I will try to answer a key question for understanding the evolution of sexual reproduction from an evolutionary ecologist's point of view after reviewing the most relevant features of the immune system, particularly the MHC: Does MHC-dependent mate choice lead to offspring that are about twice as resistant to the natural spectrum of infectious diseases as compared to random choice?

How Does the Major Histocompatibility Complex Help to Defeat Infectious Diseases?

When a cell of the body is infected by, e.g., a virus, it contains virus-derived proteins, which are degraded into small peptides of a length of about nine amino acids. Outside the cell cytotoxic T lymphocytes (T cells) are waiting, which are able to tell whether such a peptide is derived from foreign or self-proteins. Because mature T cells are usually tolerant to uninfected cells, they kill infected cells thereby eliminating the virus. However, virus-derived peptides have to be picked up within the cell, taken to the cell surface and displayed to T cells. This is done by MHC molecules, i.e., glycoproteins encoded in the large cluster of genes known as the MHC. Their most striking structural feature is a groove running across their outermost surface, in which a variety of short peptides can be bound.

There are two different classes of MHC molecules, known as class I and II, which deliver peptides from different cellular compartments to the surfaces of infected cells. Peptides from viruses are bound to MHC class I molecules and are recognized by cytotoxic T cells (which kill infected cells). The function of MHC class II molecules is to bind peptides from extracellular bacteria and larger parasites. These peptides are generated in the intracellular vesicles of macrophages, immature dendritic cells, B cells, and other antigen-presenting cells. MHC class II molecules present these peptides to other specific T cells, which activate B cells and thus stimulate the production of specific antibodies, i.e., induce a vaccination effect.

MHC molecules show great genetic variation in the population, and each (human) individual carries on her nine loci (Horton et al. 2004) 18 of the possible variants, i.e., alleles, if she is maximally heterozygous. Each of the resulting 18 different MHC molecules may bind different sets of peptides. The peptides that can bind to a given MHC molecule have the same (or very similar) amino acid residues at two or three particular positions along their sequence but can vary otherwise. The amino acids at these positions, the anchor residues, insert into pockets in the binding groove of the appropriate MHC molecule. Peptides binding to other MHC molecules may have other anchors. Different MHC molecules bind different sets of peptides (with different anchors). Having different MHC alleles and thus different MHC molecules increases the range of pathogen-derived peptides that can be bound and displayed to T cells at the cell surface (Janeway et al. 2001, Suri et al. 2003).

Is it Good to Be Major Histocompatibility Complex Heterozygous?

An individual can be infected by a wide variety of pathogens. Their proteins will not generally have many peptide sequences in common. If T cells are to be alerted to all possible infections, then an individual's MHC molecules must be able to bind to many different peptides. Although each parasite contains different proteins, which are degraded into different peptides, studies in various vertebrate taxa have shown that only one or a few different MHC alleles provide resistance to a specific parasite (e.g., Bonneaud et al. 2005, Briles et al. 1983, Grimholt et al. 2003, Harf & Sommer 2005, Hill 1999, Hill et al. 1991, Jeffery et al. 1999, Langefors et al. 2001, Lohm et al. 2002, Olsson et al. 2005, Paterson et al. 1998, Slade & McCallum 1992, reviews in Hedrick & Kim 1999, Jeffery & Bangham 2000, Tiwari & Terasaki 1985). This implies that each parasite is detected through only one or very few of its specific peptides. Thus, specific MHC alleles provide resistance to specific infectious diseases, suggesting that it seems to be a good strategy to produce offspring with as many different MHC alleles as possible, resulting in maximal heterozygosity.

Many of the studies that demonstrated or implied MHC-dependent mate choice found that the choosy sex prefers partners with somewhat dissimilar MHC alleles (e.g., Bonneaud et al. 2006; Egid & Brown 1989; Eklund et al. 1991; Freeman-Gallant et al. 2003; Landry et al. 2001; Ober et al. 1997; Olsson et al. 2003; Potts et al. 1991, 1994; Richardson et al. 2005; Wedekind & Füri 1997; Wedekind et al. 1995; Yamazaki et al. 1976, 1978). MHC disassortative mating may function to increase the resistance of offspring to infectious diseases by increasing their MHC heterozygosity (heterozygote advantage hypothesis; e.g., Apanius et al. 1997, Potts & Wakeland 1990) and/or it may operate to prevent kin-matings (inbreeding avoidance hypothesis; Brown & Eklund 1994, Potts et al. 1994), because this can have fitness benefits (Arkush et al. 2002, Meagher et al. 2000). Some correlative field studies supported the heterozygote advantage hypothesis (e.g., Carrington et al. 1999, Thursz et al. 1997) while others did not (e.g., Hill et al. 1991, Paterson et al. 1998). Also experimental studies beginning with Doherty & Zinkernagel (1975) provided ambiguous results as a recent meta-analysis showed (Penn 2002).

Often MHC homo- and heterozygous individuals are exposed in experiments to only one or very few infectious agents. Although increasing an individual's MHC heterozygosity by one or a few MHC alleles would increase the probability that a specific peptide of this agent is presented to T cells, there is no guarantee that this will happen and, given the great number of alternative alleles, it is not likely either. Just by chance the added MHC variant might be one (or one of the few) that codes for an appropriate MHC molecule. If, e.g., 50 alleles for one locus exist in the population, a randomly chosen homozygote would be resistant by chance in one of 50 cases, and a heterozygote with one randomly chosen additional allele would be resistant in two of 50 cases—not much to gain just from heterozygosity. This might be different if individuals are exposed simultaneously to many different parasites (e.g., Apanius et al. 1997, Hughes & Nei 1992). I have reviewed above that it needs the quality of an MHC molecule not just the quantity of different MHC molecules for specific peptide presentation. Or as De Boer et al. (2004) put it, "reflecting the mechanisms underlying immune protection, we assume that the fitness of an individual is directly related to the properties of the MHC alleles harbored by this individual. This is in contrast to earlier models, where the fitness of heterozygotes is supposed to be unrelated to the specificity of the MHC alleles."

Knowing which MHC alleles offer resistance against which infectious diseases should allow a prediction of which heterozygotes or homozygotes have higher resistance. Grimholt et al. (2003) evaluated the association between disease resistance and MHC class I and class II polymorphism in Atlantic salmon (*Salmo salar*), which possesses only one locus each of MHC class I and class II. As seems typical for fishes, class I and II loci are found on different chromosomes (Sato et al. 2000, Stet et al. 2003). Standardized disease challenge trials were performed either with salmon anaemia virus (ISAV), causing infectious salmon anaemia to which each fish from 25 families was singly exposed, or with the *Aeromonas salmonicida* bacteria, causing furunculosis to which each fish from 33 families was singly exposed. Of the many alleles in the population only very few class I and class II alleles provided resistance to one of the diseases. Interestingly, the best performing class II alleles were advantageous in heterozygotes but allowed for highest resistance as homozygotes. The best of these had a frequency of as high as 20% in the population. The latter finding had been shown already by Langefors et al. (2001) for *S. salar* experimentally exposed to *A. salmonicida*. Lohm et al. (2002) confirmed that resistance to furunculosis infection in Atlantic salmon is associated with specific MHC class IIB alleles rather than MHC heterozygosity.

In conclusion, a heterozygote advantage per se has not been found by a number of studies and it is not necessarily expected either. Furthermore, as recent models have shown, a heterozygote advantage on its own fails to explain the high degree of polymorphism of the MHC (De Boer et al. 2004) in contrast to predictions of earlier models. Mate choice just for dissimilar MHC alleles would not necessarily improve the resistance of offspring nor would it help to maintain MHC polymorphism in the population. Obviously, the highest resistance against all kinds of parasites would be achieved if each individual would possess several hundred loci and carry all MHC alleles that exist in the population. Before we explore this option we should discuss what could be gained by MHC-related mate choice if only two to nine MHC loci existed per individual.

The Red Queen Hypothesis Predicts Interaction Between Specific Parasites and Frequency of Specific Major Histocompatibility Complex Alleles

When a parasite P1 is common, those host individuals that happen to possess the MHC allele, say A1, which codes for an MHC molecule that can present a specific peptide of this parasite to their T cells, will be most resistant and thus produce more offspring than their less resistant competitors. Hence the next host generation preferentially consists of individuals with allele A1 and is therefore largely resistant against parasite P1. Now another parasite, P2, against which allele A2 would provide resistance, can invade, because A2 has a low frequency. Now those individuals, which happen to carry A2, are most resistant and leave more offspring than their competitors. Thus A2 becomes a high-frequency allele and P2 can no longer persist, etc. This kind of negative frequency-dependent selection can maintain a high degree of polymorphism of genes providing resistance, e.g., MHC genes (Bodmer 1972, Clarke & Kirby 1966, Ebert & Hamilton 1996, Hamilton 1980, Hamilton & Zuk 1982, Hamilton et al. 1990, Hedrick 2002, Howard & Lively 1994, Slade & McCallum 1992, Takahata & Nei 1990). This effect can be amplified by mate choice if individuals carrying the MHC allele that currently provides highest resistance are selected as mates. The above scenario describes the kind of battlefield dynamics between hosts and parasites, which allows the choosy sex to potentially achieve a double fitness gain of its offspring every new generation (e.g., Ebert & Hamilton 1996, Hamilton 1980, Hamilton et al. 1990, Jaenike 1978). However, not every rare MHC allele increases in frequency. Only those rare alleles that happen to provide resistance against an invading disease have this potential.

If frequency-dependent selection of certain rare MHC alleles by ever-changing parasites is the universal evolutionary mechanism that maintains both the MHC polymorphism and sexual reproduction in vertebrates, it should happen in every population in every generation. Conclusive evidence of this kind is still missing. There must be a two-step procedure. (*a*) One has to monitor both parasite and MHC allele frequencies in a natural population for several years. If one parasite has high impact in one year, a specific MHC allele that is found more often in individuals that are not infected by this parasite should increase in frequency until the next generation. Then another parasite may be found to invade and another MHC allele carried by noninfected individuals should increase until the third generation, whereas the previous fashionable allele should become rare again, etc. Although such evidence would be promising it would be only correlative. (*b*) In a second step one would have to test experimentally whether the MHC allele that increased in frequency upon invasion of a previously rare parasite provides resistance against this parasite. Individuals with and without the MHC allele in question from lab-bred sib-groups (to control for background genes at least on other chromosomes) should be exposed each to a natural dose of the parasite in question (twice to allow for a vaccination effect). If several such

MHC allele parasite causal relationships can be proven experimentally over several generations, the maintenance of the MHC polymorphism of that population may begin to become unraveled.

The first step recently has been done. So far, only two studies have attempted to look for frequency changes of MHC alleles in natural populations. Westerdahl et al. (2004) found significant frequency variation between years for one MHC class I allele in great reed warblers (*Acrocephalus arundinaceus*) (see also Westerdahl 2003). The authors argue that only those individuals possessing the allele were able to survive infection with avian malaria. To test whether changes in MHC allele frequency correlate with parasite prevalence in three-spined sticklebacks (*Gasterosteus aculeatus*) Wegner (2004) analyzed a time series of three consecutive generations of fish from three different populations. He found significant frequency changes in 7 out of 41 analyzed MHC class IIB alleles. Some 7 macroparasite species showed significant variation in prevalence within the three generations; 4 of the 7 MHC class II alleles were significantly associated as predicted to prevalence of at least one of the fluctuating parasite species against which they appeared to confer protection. Rare MHC alleles rose with simultaneous decreases of parasite prevalence and vice versa. Now evidence from testing the assumptions for the second step is needed.

Mate Choice for Major Histocompatibility Complex Alleles that Currently Provide Resistance?

When individuals with specific rare MHC alleles are resistant against the currently predominating parasite, a female's best choice would be to prefer a male with that allele. But how does she know who has this allele? Even if she would be able to detect each individual MHC allele by smell, she would have to know both the species of the current parasite and the identity of the MHC allele that provides resistance against it. The evolution of such abilities is hard to imagine. There are several approximations to this "gold standard rule." (*a*) Without any information, preferring MHC dissimilar mates might be a best-of-bad-job rule. More heterozygous offspring may not reliably be more resistant but may profit from a higher probability to carry the needed allele. Many studies found either that MHC-dissimilar mates were preferred, that dissimilar individuals were rated more positively (not necessarily as mates) or that offspring in a population were more heterozygous than would be expected from no mate choice, e.g., in mice (Egid & Brown 1989; Eklund et al. 1991; Potts et al. 1991, 1994; Yamazaki et al. 1976, 1978), rats (Brown et al. 1987), fish (Landry et al. 2001), lizards (Olsson et al. 2003), birds (Bonneaud et al. 2006, Freeman-Gallant et al. 2003, Richardson et al. 2005), and humans (Ober et al. 1997, Wedekind & Füri 1997, Wedekind et al. 1995). However, several other studies found no evidence for increased heterozygosis through potential mate choice.

(*b*) A rule that is closer to the gold standard would be to test whether a potential mate is infected (or just in poor condition). If a male is not infected it might carry the MHC alleles providing resistance against the currently dominating parasites. To detect a resistant mate "the methods used should have much in common with those of a physician checking eligibility for life insurance, i.e., unclothe the subject, weigh, listen, observe vital capacity, and take blood, urine, and fecal samples" (Hamilton & Zuk 1982). Indeed, mice recognize and avoid odors of parasitized conspecifics (Kavaliers & Colwell 1995, Kavaliers et al. 2005). Penn & Potts (1998b) review many examples of scent being used to detect illness including a list of body odors that physicians had used as diagnostic indicators for human diseases.

(c) Animals could choose mates by scrutiny of characters whose full expression is dependent on health and vigor (Hamilton & Zuk 1982). If a male bird can grow a peacock's tail with long feathers and bright colors, and display vigorously while producing an energetically costly song, it must be healthy and possess the alleles that provide resistance against the current parasites. This idea offered a solution to the long-standing problem of why females evolved a preference for males as mates that are handicapped by showy and costly epigamic characters (Darwin 1871, Fisher 1930). As Hamilton & Zuk (1982) suggest, such handicaps may reveal health and vigor and thus resistance against parasites to which the offspring of a choosy female will be exposed soon. Many experimental studies, not yet including the MHC, support the Hamilton-Zuk hypothesis showing that such male handicaps are indeed costly to produce, reveal health and vigor, and are important for female choice (e.g., Hill 1991, Houde & Torio 1992, Milinski & Bakker 1990, Møller 1990, Zuk et al. 1990). Furthermore, in ring-necked pheasants (*Phasianus colchicus*) an association was demonstrated between the MHC genotype of males and both variability in survival and expression of tarsal spur length (Von Schantz et al. 1996), a condition-dependent secondary sexual trait predicting fitness and attractiveness to females (Von Schantz et al. 1989). Ekblom et al. (2004) found in a lekking bird, the great snipe (*Gallinago media*), that females prefer males with certain MHC class IIB allele types. They speculate that these alleles may be linked to resistance to common parasites, as such males were larger.

The advantage that favors sexual reproduction and thus mate choice must be produced through recombination. Although being a dogma of evolutionary biology, this condition appears to be demanding when applied to MHC-related mate choice. The choosy sex should not only shop for good genes but also combine them with its own good genes. Assortative mating with respect to the possession of MHC alleles that provide resistance to several different harmful parasites currently present would be expected. To allow for a short-term advantage of sexual reproduction, offspring with the combination of the chosen male's and the female's genes have to be about twice as fit as if she had mated with a random male. At least two different (or more) specific MHC alleles have to exist at any time to provide an offspring with maximum resistance. We expect that especially healthy females choose healthy males. Of course, females that suffer from parasites will be less likely to reproduce and they might be avoided as mates. Furthermore, sampling different potential mates will be more costly for infected and thus weaker females. Such increased costs of sequential mate choice reduce selectivity (Milinski & Bakker 1992, Wong & Jennions 2003). Bakker et al. (1999) found that female mate choice in three-spined sticklebacks also depends on her health condition when there is no cost of choice; females in poor condition even preferred dull males. Resistant hermaphroditic snails (*Biomphalaria glabrata*) actively refused to copulate as females with an infected partner (Webster et al. 2003). Jennions & Petrie (1997) list further examples. A question that awaits an empirical answer is: Do offspring have a combination of currently advantageous MHC alleles more frequently than would be expected from random mating?

INDIVIDUALS SHOULD BE OPTIMALLY, NOT MAXIMALLY, MAJOR HISTOCOMPATIBILITY COMPLEX HETEROZYGOUS

If having three MHC class I loci is better than having one, why are there not far more MHC loci? This could, in principle, be easily achieved by duplication of loci (Parham & Ohta 1996). For improving resistance, sex would be unnecessary if everybody had all MHC variants that are present in the population. The probable explanation is that each time a distinct MHC molecule is added to the MHC repertoire, all T-cell clones that can recognize self-peptides bound to that molecule must be removed in order to maintain self tolerance. It seems that the low number of MHC loci present in, e.g., humans and mice is about optimal to balance out the advantages of presenting an increased range of foreign peptides and the disadvantages of an increased loss of T cells from the repertoire (Janeway et al. 2001, Lawlor et al. 1990).

Clearing Up the Mystery of T-Cell Selection

How does our immune system avoid attacking our own cells? In uninfected cells, peptides derived from self-proteins fill the peptide binding groove of the mature MHC class I and class II molecules and are presented to T cells. Because the central role of T cells is to stimulate immune responses, it is crucially important that mature T cells do not react with cells presenting self-peptides. T-cell receptors recognize features both of the peptide and of the MHC molecule to which it is bound. There are many T-cell lines each with a different T-cell receptor. Each T-cell receptor type is specific for a particular combination of MHC molecule and peptide antigen. The contact area of the peptide, which interacts with the T-cell receptor, is different from its anchor residues, which are decisive for fitting into the binding groove of the MHC molecule. Both self- and parasite-derived peptides with the same anchors bind to the same MHC molecule.

During complex selection processes, the initial highly variable repertoire of developing thymocytes is purged of T-cell clones that recognize self-peptides bound to MHC molecules. In an initial positive selection step, the T-cell receptor expressed on T cells is checked for its general capability to interact with peptide/MHC complexes; specificity does not matter here, as a diverse repertoire of T cells can be positively selected on a single-type peptide/MHC complex. In a subsequent step, self-tolerance of the repertoire is achieved by eliminating or otherwise inactivating the emerging self-reactive T-cell clones. It is in the negative selection process where specificity really matters. Increasing the number of individual MHC molecules expressed in an individual potentially increases both positively and negatively selected T-cell clones but has the most pronounced effect on negative selection. While in positive selection, T-cell receptors are, to a first approximation, selected for reactivity against self-MHC (with the peptide having primarily a structural, i.e., stabilizing role), whereas negative selection focuses on the peptide ligand of the MHC molecule. It is difficult to predict the effect on the eventual outcome of T-cell differentiation in terms of the diversity of the repertoire when the number of MHC molecules increases. Several modeling studies have attempted to provide a numerical estimation and explanation of the observed MHC optimum in natural populations. In light of the above it is conceivable (T. Boehm, personal communication) that models predicting an optimal number of MHC alleles per individual, which assign greater weight to negative selection (Nowak et al. 1992), reflect the natural situation better than those that attribute more weight to the positive selection step (Borghans et al. 2003).

During negative selection the vast majority of T-cell lines are eliminated (Janeway et al. 2001, Lawlor et al. 1990). It is reasonable that there will be quite severe limits on the number of different MHC molecules expressed in the thymus, above which the increase of novel, positively selected T-cell clones will not compensate for the deletion of T-cell clones necessary for establishing self-tolerance. So no one individual can be a high responder to all potential pathogens. An individual with all the MHC molecules required to present the necessary peptides would probably have no T cells left to respond to them (Borghans et al. 2003, De Boer & Perelson 1993, Lawlor et al. 1990, Nowak et al. 1992) and an optimal number of different MHC alleles per individual is predicted.

Both Parasites and Mate Choice Select for an Optimal Number of Major Histocompatibility Complex Alleles per Individual

A pragmatic approach to understanding why individuals have a low MHC diversity is to study the selective forces acting on individuals with fewer or more different MHC alleles. Reusch et al. (2001) determined the number of different MHC class IIB alleles of individual sticklebacks from an interconnected system of natural lakes. Three-spined sticklebacks have probably six loci of MHC class IIB (Sato et al. 1998), and thus up to 12 different alleles. There are many fish with fewer than six different alleles (**Figure 2a**), which is difficult to explain if there are six loci. Thus individual fish either differ in the number of their loci (e.g., Málaga-Trillo et al. 1998) or, because of recent gene duplications (Reusch & Langefors 2005, Reusch et al. 2004), very similar alleles that cannot be distinguished may be found at different loci so that a fish may have, e.g., four alleles of the same "kind." Contrary to the heterozygote advantage hypothesis most fish in natural populations had either five or six different alleles (mean of 5.8) (**Figure 2a**). Wegner et al. (2003b) not only confirmed these results but identified all macro parasites in wild-caught fish: The lowest diversity of parasites and the lowest number of parasites per species were found in fish with an intermediate number of MHC class IIB alleles (calculated minimum in fish with 5.17 alleles), which appear most resistant.

To test this observational finding in an experiment, offspring from different clutches were bred parasite free in the lab from wild-caught sticklebacks and exposed singly to three locally abundant parasites in a natural dosage (Wegner et al. 2003a). Again fish with an intermediate number of 5.82 MHC class IIB alleles were

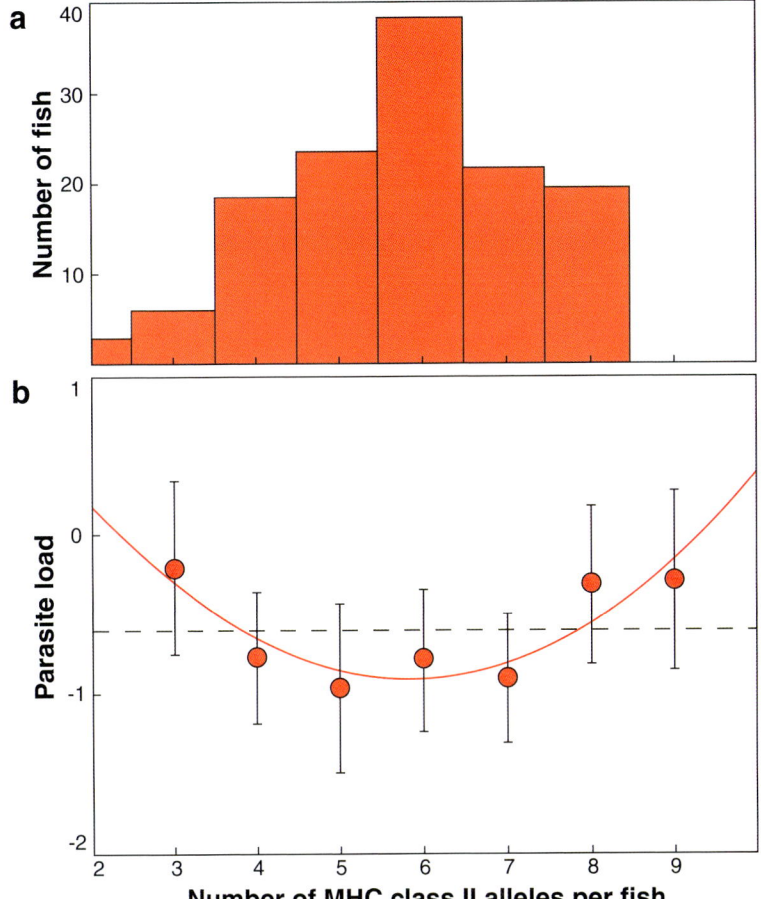

Figure 2

(*a*) Frequency distribution of the number of major histocompatibility complex (MHC) class II alleles detectable in 144 three-spined sticklebacks from a lake system. The mean number of MHC alleles was 5.8. (from Reusch et al. 2001). (*b*) Relationship between the number of expressed MHC class II molecules and mean parasite load, expressed as summed residuals from general linear model (GLM) analysis. The function matches a quadratic polynomial with a minimum of 5.82 alleles. (From Wegner et al. 2003a.)

least infected (**Figure 2b**). This experimental evidence shows that multiple parasites can select for an immunogenetic optimum of about six different MHC class II alleles per individual. Interestingly, fish with an intermediate MHC diversity had the lowest expression of an innate immune trait, the respiratory burst reaction, in which reactive oxygen molecules are generated to kill pathogens (Kurtz et al. 2004). Obviously fish with both sub- and superoptimal MHC allele numbers tried to compensate for their deficit with this costly innate immune response (Kurtz et al. 2006). Fish with fewer MHC alleles had higher expression of these genes, suggesting compensatory up-regulation (Wegner et al. 2006). Nevertheless they had higher parasite loads. When lab-bred sticklebacks (as in Wegner et al. 2003a) were exposed in cages to the natural parasite fauna in their parents' lake during the exceptionally warm summer of 2003, significantly more of those fish with an intermediate number of different MHC class IIB alleles survived this period (Wegner 2004). Mice that were heterozygous at all MHC loci were even slightly less resistant than mice that were homozygous at all loci

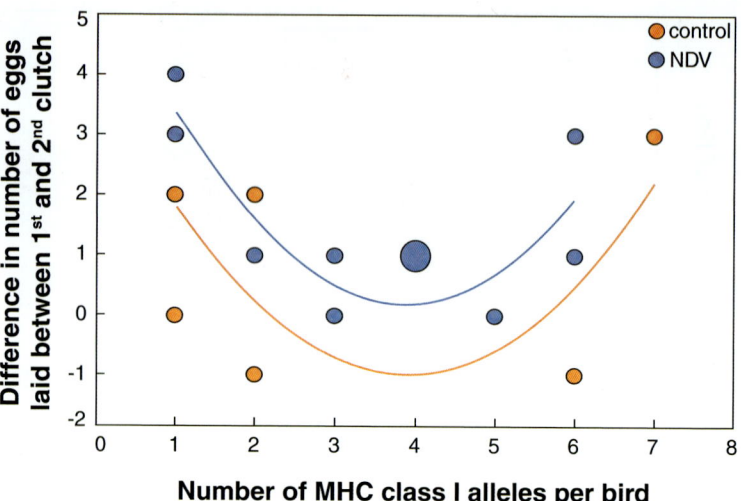

Figure 3

Variation in the number of eggs laid by house sparrows between the first and replacement clutch as a function of the number of MHC alleles for both immune treatment groups. The Newcastle disease virus (NDV) group is compared to the control (PBS) group. Circles vary in size according to the number of observations (1 or 4). (From Bonneaud et al. 2004.)

when challenged with different strains of *Salmonella*; no individuals with intermediate heterozygosity were tested (P. Ilmonen, personal communication). Neff (2004) found that overall heterozygosity seems optimized in a natural population of bluegill sunfish.

Further evidence for optimal individual MHC diversity comes from house sparrows, *Passer domesticus* (Bonneaud et al. 2004). Females with intermediate numbers of MHC class I alleles produced the largest first clutches. After this clutch had been removed their immune system was manipulated with a vaccine against the Newcastle disease virus. A "terminal investment hypothesis" predicts that "infected" individuals should invest more in a replacement clutch than do controls. Vaccination produced this effect. Also the number of different MHC class I alleles per individual influenced investment in the second clutch. Females with intermediate MHC allelic diversity maintained the large clutch sizes they had produced on their first attempt, whereas females with a small or large number of MHC alleles laid more eggs in the replacement clutch. This is precisely what was assumed if they expected to suffer more from the virus (**Figure 3**).

Are sticklebacks with the optimal number of different MHC class IIB alleles the most frequent class in nature (**Figure 2a**) as a result of early selection by parasites or already through female choice? When wild-caught females chose between the odors of two males in a flow channel, they did not prefer an MHC-dissimilar male to a similar one (Reusch et al. 2001). When they chose between males that had either few or many different MHC class IIB alleles, females with many alleles themselves preferred males with few alleles and those with few alleles preferred males with many alleles (Aeschlimann et al. 2003, Reusch et al. 2001). Using an algorithm that can detect an overall optimum in the number of MHC alleles that females try to achieve for their offspring, when choosing between any two males, researchers showed that the number significantly approached such an optimum and that this optimum is close to 5.2 different MHC class IIB alleles (Aeschlimann et al. 2003, Milinski 2003a). This "mate choice optimum" is close to both the immunogenetic optimum (Wegner et al. 2003a) and the distribution of genotypes in nature (Reusch et al. 2001).

Predictions from a Rule for Mate Choice to Achieve an Optimal Number of Major Histocompatibility Complex Alleles

Female sticklebacks use self-reference to complement their own set of alleles with a more or less diverse set of male alleles such that the combined diversity reaches an optimum for the offspring (Aeschlimann et al. 2003, Milinski 2003a). This rule automatically leads to preferences for dissimilar males in somewhat inbred populations and for similar males in outbred populations. This rule might be compatible with studies that found a (usually weak) preference for MHC dissimilar mates, no preference, or even a preference for similar conspecifics (e.g., Jacob et al. 2002). Wedekind et al. (1995) and Wedekind & Füri (1997) found a preference among Swiss students from Bern University for the odor of t-shirts that had been worn by MHC-dissimilar students, whereas Jacob et al. (2002) found a preference for MHC-similar men among women from one community tested with t-shirts worn by men of different ethnicity who had only very few MHC alleles that matched the women's. Despite the seemingly contradictory results (McClintock et al. 2002, Wedekind 2002) the studies might be perfectly compatible if the test persons in both studies had aimed at an optimal number of combined different MHC alleles. A rule to prefer just mates with intermediate levels of MHC dissimilarity to achieve the optimum (Penn 2002, Penn & Potts 1999) might sometimes miss the goal as the above example indicates.

Because females can probably perceive only MHC genotypes, not haplotypes, at least in the context of precopulatory mate choice (Ziegler et al. 2005), their mate choice will be imperfect. The mating between two optimal genotypes with six alleles each might result in a mixture of genotypes when the haplotypes of each parent consist of packages (gametes) of two and four different alleles. The offspring would have on average six alleles, but only 50% of them exactly, because of Mendelian mixing (Wegner et al. 2004). Oocytes "knowing" their MHC haplotype could choose perfectly among sperm signaling their MHC haplotype. Suggestive evidence comes from mice (Rülicke et al. 1998) and salmon (Yeates 2005).

Just preferring a mate with whom the female achieves the optimal MHC diversity for her offspring would again be only a best-of-bad-job rule. The best rule would include those alleles that provide resistance against the current parasites: From several males with the optimal number of complementary alleles, the female might select the healthiest and probably most resistant one by preferring the most brightly colored male to obtain his "good genes" (Barber et al. 2001; Milinski & Bakker 1990, 1992) to achieve what Hamilton & Zuk (1982) envisioned without the present knowledge of MHC-based mate choice. A two-step choice is predicted (Aeschlimann et al. 2003): From a distance the female selects by smell the male offering the right complementary number of MHC alleles, then she checks from his "revealing handicaps" whether he is healthy. Bonneaud et al. (2006) found that paired male house sparrows did not have very few MHC class I alleles, were not too dissimilar to their mate, and carried certain specific "coadapted" alleles. Howard & Lively (2004) modeled whether mate choice should take either good or complementary genes for resistance into account. The combination of these modes now awaits modeling.

Different Optima among Species, Populations, and Individuals?

The number of MHC loci varies among species (e.g., one locus each for MHC class I and II in salmon, six loci for class I and three for class II in humans) and sometimes among individuals within one species (e.g., Málaga-Trillo et al. 1998). If there is an optimal number of MHC variants per individual, why does this optimum differ among species? More loci seem easily achieved through duplication and interlocus gene conversion (e.g., Parham & Ohta 1996, Reusch & Langefors 2005, Reusch et al. 2004). With each additional locus the potential number of MHC alleles per individual increases and thus the risk of autoimmune diseases if T-cell selection is not perfect or cross-reactive immune responses occur (e.g., Borghans et al. 2003, Janeway et al. 2001, Ridgway & Fathman 1998). Only with usually high parasite diversity does the optimum number of loci pay off; otherwise fewer might be better. Salmon may afford to have only one locus for each MHC class because they return to their native river. Wegner et al. (2003a) found a much lower number of parasite species in sticklebacks from rivers than in lake fish. Accordingly about twice as many different MHC class II alleles existed in lake populations compared to river populations. The mean number of different MHC class II alleles per stickleback was lower in river (4.98 ± 0.15 SE) than in lake populations ($6.34 + 0.11$ SE; K.M. Wegner, unpublished data), suggesting a lower optimum in river populations. Do river female sticklebacks aim for a lower number of MHC alleles for their offspring than lake fish? This needs to be studied. Even perceived increased parasite pressure may shift the chosen MHC optimum upward. This might explain an "intriguing finding" (Penn 2002) by Rülicke et al. (1998) showing that mice produced more MHC-heterozygous offspring when challenged with mouse hepatitis virus.

Fragrant Genes Help Damenwahl

MHC genes are "fragrant"; that is, their gene products or specific ligands smell. These natural chemosignals must be structurally polymorphic enough to reveal the allelic composition of every potential MHC genotype. Because soluble MHC molecules were found in blood of mice, rats, and humans, Singh et al. (1987) proposed that the ability of MHC molecules to associate in a selective way with other small molecules could also be the mechanism by which a unique mixture of volatile, endogenous metabolites is transported by MHC glycoproteins from the blood into the urine. Supporting this "carrier hypothesis," Yamazaki et al. (1999) suggested that "soluble MHC gene products themselves bind circulating odorants selectively [presumably after they lost their bound peptide (Falk et al. 1991)] and then release them to a minimal degree in serum," suggesting carboxylic acids as likely odorants that Singer et al. (1997) had found in mixtures differing between mouse MHC types. Penn & Potts (1998a) call this "the peptide hypothesis." An alternative hypothesis assumed that MHC genes alter the relative abundance of indigenous microflora and thereby body odor (Singh et al. 1990). However, Yamazaki et al. (1990) failed to find a loss of MHC-determined odor in germ free mice.

Two recent studies (Leinders-Zufall et al. 2004, Milinski et al. 2005) tested what could be called "the MHC ligand peptide hypothesis." Because the range of peptides displayed by an individual's MHC molecules mirrors the diversity of its MHC alleles, the peptides themselves that are bound to MHC molecules may be the polymorphic signal. Leinders-Zufall et al. (2004) found that MHC ligand peptides function as stimuli for a subset of olfactory neurons in the vomeronasal organs of mice. Each peptide elicited a unique activation pattern. This specificity depended on the anchor residues of the peptide with which it binds to "its" MHC molecule. A neuron did not distinguish between peptides with the same anchors but different residues in other positions. It did not, however, respond when only the anchors were mutated to alanine (which is not known to serve as an anchor). These olfactory neurons responded exactly to those residues of the peptide that mirror the anchor specificity of the MHC molecule. Do MHC peptides also function during social recognition? Bruce (1959) found that pregnancy failed when a recently mated female mouse was, after removing the stud male, housed with a new male from a different strain (or presented only with its odor). Yamazaki et al. (1986) showed that females block pregnancy when the odors of new and stud males are from MHC congenic strains that differ only by one MHC class I allele. Pregnancy was not blocked when the new male came from the stud male's strain. Leinders-Zufall et al. (2004) induced the "Bruce effect" when, e.g., a BALB/c female that had mated with a C57BL/6 male was presented with a BALB/c specific MHC peptide, but pregnancy was not blocked when she was presented with a C57BL/6 specific MHC peptide. This result shows that MHC ligand peptides function as individuality signals in mice (see also Boehm & Zufall 2006), and it suggests that the function of the Bruce effect is not to exchange a less heterozygous with a more heterozygous embryo, but probably to avoid an undue investment by the female in a pregnancy that would likely result in infant killing (vom Saal & Howard 1982) by the new male.

Are MHC ligand peptides used in actual mate choice decisions? If peptides are the natural odor signal that reveals a male's MHC alleles, it should be possible to manipulate this information by supplementing further peptides. Because a female stickleback prefers the scent of a male that offers the optimal complementation of her own alleles (Aeschlimann et al. 2003, Milinski 2003a), supplementing the same mix of four different synthetic MHC peptides should increase the attractiveness of a suboptimal male but should render an optimal or superoptimal male unattractive. When gravid females chose between spiked and unspiked water from the tank of one male per female, they preferred the peptide side when the combined MHC diversity of the pair was below the optimum, but they avoided the spiked side when it was at or above the optimum (**Figure 4**). However, choosing females ignored the same mix of synthetic peptides when their anchor residues were changed to alanine (Milinski et al. 2005). Like mouse olfactory neurons, stickleback noses concentrate on the anchor residues of the peptide. Because MHC peptides interact with natural MHC-related signals in both mice and sticklebacks in a predictable way, peptides are probably part

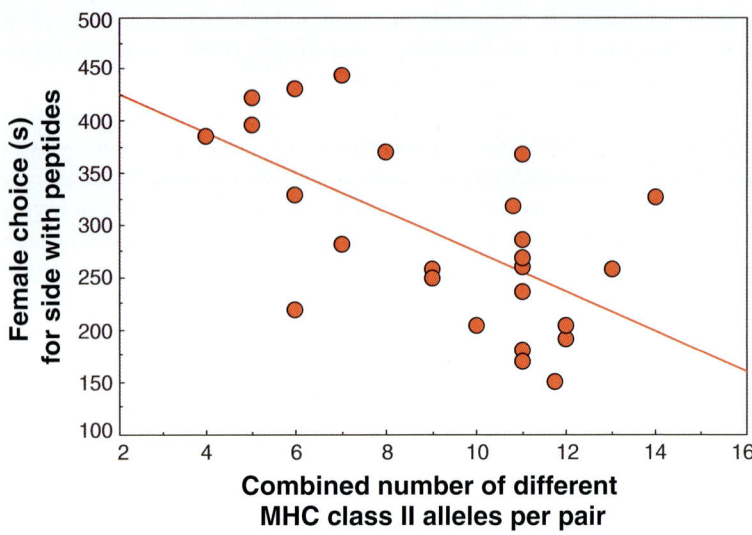

Figure 4

In a two-choice flow channel, single gravid female three-spined sticklebacks chose between the water from the tank of a single male supplemented with solvent only or four peptides in solvent. The time (a total of 600 s) females spent in the side of the flow channel to which peptides were supplemented is shown as a function of the combined number of different MHC class II alleles of the pair. Offspring of a pair with a combined diversity of 10 alleles would have about 5 alleles because each parent has haploid gametes. When females were used in two successive gravidities with different males (or males were tested twice with different females), respective data were entered as means per male or female to avoid pseudoreplication. (From Milinski et al. 2005.)

of the natural perfume also in other vertebrates. The genetics of MHC-related odor perception is not known. However, Ziegler et al. (2000) and Younger et al. (2001) found a cluster of highly polymorphic genes for olfactory receptors linked to the MHCs in both mice and humans. The function of these genes is still unknown.

Major Histocompatibility Complex and Perfume Use

Humans have used fragrances for at least 5000 years in all cultures. Only a small subsample of fragrances of sometimes-strange origin (e.g., amber, musk) of all potential natural ingredients is found in traditional perfumes. "This suggests there may be something very special about them" (Stoddart 1990). People have great individual differences in preference for their own perfume, which is a puzzle for perfumologists. Milinski & Wedekind (2001) found with 137 MHC-typed students, who scored 36 natural perfume ingredients for use on self or a potential partner, that persons who shared a common MHC allele had a strong preference for the same ingredients but only for use on self. Thus MHC alleles seem to determine individual perfume preference. This suggests that perfumes function as amplifiers of MHC-related body odors, because in former times men in various cultures carried their neckerchiefs in their armpits during a dance ready to fan their "sweethearts" (Vollrath & Milinski 1995), the natural MHC signal has obviously restricted volatility, as is the case with peptides. One of the many constituents of each natural perfume ingredient may mimic a specific component of the MHC-related signal (Milinski 2003b). Archunan & Dominic (1990) found in mice that stud males anointed with a commercial perfume failed to protect implantation in females: A perfume induced the Bruce effect. Because it is unlikely that this perfume happened to mimic the stud male's MHC odor, its signal was probably understood as a new male's MHC-related odor. Interestingly some olfactory neurons of the vomeronasal organ of mice reacted to odorants that are common in perfumes (Sam et al. 2001).

SEX MUST HAVE A SHORT-TERM ADVANTAGE IN INVERTEBRATES AND PLANTS LACKING A MAJOR HISTOCOMPATIBILITY COMPLEX

Sexual reproduction and mate choice must be older than the MHC system (see also Boehm 2006) because invertebrates and plants reproduce sexually but lack an MHC. Both invertebrates and vertebrates (the latter of which also have an adaptive MHC system) have an "innate immune system" that had been thought to be fast but unspecific, nonpolymorphic, and without a memory—a blow to the Red Queen hypothesis for the evolution of sex in invertebrates. However, evidence is accumulating that invertebrates [e.g., bumblebees (Schmid-Hempel et al. 1999), *Daphnia* (Carius et al. 2001), copepods (Kurtz & Franz 2003), and snails (Webster et al. 2003)] are polymorphic in their genetic resistance to specific parasites, indicating the potential for coevolution based on frequency-dependent selection. Kurtz & Franz (2003) provided evidence even for memory in invertebrate immunity. What about plants? Balanced polymorphism appears to be the inevitable result of selection for self/nonself discrimination, including diverse mating-recognition systems in plants (Richman 2000). Plants express a sophisticated molecular system for recognition of and responses to potentially pathogenic microorganisms, which resembles the innate immune system (Gómez-Gómez 2004). All the necessary ingredients for a short-term advantage for sexual reproduction seem to be present in both invertebrates and plants. The Red Queen may be of universal validity.

ACKNOWLEDGMENTS

I thank Thomas Boehm, Barbara Uchanska-Ziegler, and Andreas Ziegler for their insightful suggestions and comments on the manuscript, and Petteri Ilmonen for allowing me to cite his unpublished results.

LITERATURE CITED

Aeschlimann PB, Häberli MA, Reusch TBH, Boehm T, Milinski M. 2003. Female sticklebacks *Gasterosteus aculeatus* use self-reference to optimize MHC allele number during mate selection. *Behav. Ecol. Sociobiol.* 54:119–26

Apanius V, Penn D, Slev PR, Ruff LR, Potts WK. 1997. The nature of selection on the major histocompatibility complex. *Crit. Rev. Immunol.* 17:179–224

Archunan G, Dominic CJ. 1990. Stud male-induced protection of implantation in food-deprived mice: Masking effect of an artificial scent on pheromonal odour. *Indian J. Exp. Biol.* 28:371–72

Arkush KD, Giese AR, Mendonca HL, McBride AM, Marty GD, et al. 2002. Resistance to three pathogens in the endangered winter-run chinook salmon (*Oncorhynchus tshawytscha*): effects of inbreeding and major histocompatibility complex genotypes. *Can. J. Fish. Aquat. Sci.* 59:966–75

Bakker TCM, Künzler R, Mazzi D. 1999. Sexual selection—Condition-related mate choice in sticklebacks. *Nature* 401:234

Barber I, Arnott SA, Braithwaite VA, Andrew J, Huntingford FA. 2001. Indirect fitness consequences of mate choice in sticklebacks: offspring of brighter males grow slowly but resist parasitic infections. *Proc. R. Soc. London Ser. B* 268:71–76

Bell G. 1982. *The Masterpiece of Nature*. London: Croom Helm

Bernatchez L, Landry C. 2003. MHC studies in nonmodel vertebrates: what have we learned about natural selection in 15 years? *J. Evol. Biol.* 16:363–77

Bodmer WF. 1972. Evolutionary significance of Hl—a system. *Nature* 237:139–45

Boehm T. 2006. Co-evolution of a primordial peptide-presentation system and cellular immunity. *Nat. Rev. Immunol.* 6:79–84

Boehm T, Zufall F. 2006. MHC peptides and the sensory evaluation of genotype. *Trends Neurosci.* 29:100–7

Bonhoeffer S, Chappey C, Parkin NT, Whitcomb JM, Petropoulos CJ. 2004. Evidence for positive epistasis in HIV-1. *Science* 306:1547–50

Bonneaud C, Chastel O, Federici P, Westerdahl H, Sorci G. 2006. Complex *Mhc*-based mate choice in a wild passerine. *Proc. R. Soc. B* 273:1111–16

Bonneaud C, Mazuc J, Chastel O, Westerdahl H, Sorci G. 2004. Terminal investment induced by immune challenge and fitness traits associated with major histocompatibility complex in the house sparrow. *Evolution* 58:2823–30

Bonneaud C, Richard M, Faivre B, Westerdahl H, Sorci G. 2005. An Mhc class I allele associated to the expression of T-dependent immune response in the house sparrow. *Immunogenetics* 57:782–89

Borghans JAM, Noest AJ, De Boer RJ. 2003. Thymic selection does not limit the individual MHC diversity. *Eur. J. Immunol.* 33:3353–58

Briles WE, Briles RW, Taffs RE, Stone HA. 1983. Resistance to a malignant lymphoma in chickens is mapped to subregion of major histocompatibility (B) complex. *Science* 219:977–79

Brown JL, Eklund A. 1994. Kin recognition and the major histocompatibility complex—an integrative review. *Am. Nat.* 143:435–61

Brown RE, Singh PB, Roser B. 1987. The major histocompatibility complex and the chemosensory recognition of individuality in rats. *Physiol. Behav.* 40:65–73

Bruce HM. 1959. An exteroceptive block to pregnancy in the mouse. *Nature* 184:105

Carius HJ, Little TJ, Ebert D. 2001. Genetic variation in a host-parasite association: Potential for coevolution and frequency-dependent selection. *Evolution* 55:1136–45

Carrington M, Nelson GW, Martin MP, Kissner T, Vlahov D, et al. 1999. HLA and HIV-1: heterozygote advantage and B*35-Cw*04 disadvantage. *Science* 283:1748–52

Carroll L. 1872. *Through the Looking Glass*. London: Macmillan

Clarke B, Kirby DRS. 1966. Maintenance of histocompatibility polymorphisms. *Nature* 211:999–1000

Cooper MD, Alder MN. 2006. The evolution of adaptive immune systems. *Cell* 124:815–22

Crow JF, Kimura M. 1965. Evolution in sexual and asexual populations. *Am. Nat.* 909:439–50

Darwin C. 1871. *The Descent of Man and Selection in Relation to Sex*. London: Murray

De Boer RJ, Borghans JAM, van Boven M, Kesmir C, Weissing FJ. 2004. Heterozygote advantage fails to explain the high degree of polymorphism of the MHC. *Immunogenetics* 55:725–31

De Boer RJ, Perelson AS. 1993. How diverse should the immune system be? *Proc. R. Soc. London Ser. B* 252:171–75

de la Pena M, Elena SF, Moya A. 2000. Effect of deleterious mutation-accumulation on the fitness of RNA bacteriophage MS2. *Evolution* 54:686–91

De Tomaso AW, Nyholm SV, Palmeri KJ, Ishizuka KJ, Ludington WB, et al. 2005. Isolation and characterization of a protochordate histocompatibility locus. *Nature* 438:454–59

De Visser JAGM, Hoekstra RF, van den Ende H. 1997. An experimental test for synergistic epistasis and its application in *Chlamydomonas*. *Genetics* 145:815–19

Doherty PC, Zinkernagel RM. 1975. Enhanced immunological surveillance in mice heterozygous at the H-2 gene complex. *Nature* 256:50–52

Ebert D, Hamilton WD. 1996. Sex against virulence: the coevolution of parasitic diseases. *Trends Ecol. Evol.* 11:79–82

Edwards SV, Hedrick PW. 1998. Evolution and ecology of MHC molecules: from genomics to sexual selection. *Trends Ecol. Evol.* 13:305–11

Egid K, Brown JL. 1989. The major histocompatibility complex and female mating preferences in mice. *Anim. Behav.* 38:548–50

Ekblom R, Sæther SA, Grahn M, Fiske P, Kålås JA, et al. 2004. Major histocompatibility complex variation and mate choice in a lekking bird, the great snipe (*Gallinago media*). *Mol. Ecol.* 13:3821–28

Eklund A, Egid K, Brown JL. 1991. The major histocompatibility complex and mating preferences of male mice. *Anim. Behav.* 42:693–94

Elena SF. 1999. Little evidence for synergism among deleterious mutations in a nonsegmented RNA virus. *J. Mol. Evol.* 49:703–7

Elena SF, Lenski RE. 1997. Test of synergistic interactions among deleterious mutations in bacteria. *Nature* 390:395–98

Falk K, Rötzschke O, Stevanovic S, Jung G, Rammensee HG. 1991. Allele-specific motifs revealed by sequencing of self-peptides eluted from MHC molecules. *Nature* 351:290–96

Fisher RA. 1930. *The Genetical Theory of Natural Selection*. London: Univ. Press

Freeman-Gallant CR, Meguerdichian M, Wheelwright NT, Sollecito SV. 2003. Social pairing and female mating fidelity predicted by restriction fragment length polymorphism similarity at the major histocompatibility complex in a songbird. *Mol. Ecol.* 12:3077–83

Gómez-Gómez L. 2004. Plant perception systems for pathogen recognition and defense. *Mol. Immunol.* 41:1055–62

Grimholt U, Larsen S, Nordmo R, Midtlyng P, Kjoeglum S, et al. 2003. MHC polymorphism and disease resistance in Atlantic salmon (*Salmo salar*); facing pathogens with single expressed major histocompatibility class I and class II loci. *Immunogenetics* 55:210–19

Hamilton WD. 1980. Sex versus nonsex versus parasite. *Oikos* 35:282–90

Hamilton WD, Axelrod R, Tanese R. 1990. Sexual reproduction as an adaptation to resist parasites (a review). *Proc. Natl. Acad. Sci. USA* 87:3566–73

Hamilton WD, Zuk M. 1982. Heritable true fitness and bright birds: a role for parasites? *Science* 218:384–87

Harf R, Sommer S. 2005. Association between major histocompatibility complex class II DRB alleles and parasite load in the hairy-footed gerbil, *Gerbillurus paeba*, in the southern Kalahari. *Mol. Ecol.* 14:85–91

Hedrick PW. 2002. Pathogen resistance and genetic variation at MHC loci. *Evolution* 56:1902–8

Hedrick PW, Kim TJ. 1999. Genetics of complex polymorphisms: Parasites and maintenance of the major histocompatibility complex variation. In *Evolutionary Genetics: From Molecules to Morphology*, ed. RS Singh, CB Krimbas, pp. 204–34. Cambridge, UK: Cambridge Univ. Press

Hill AVS. 1999. Immunogenetics: Defense by diversity. *Nature* 398:668–69

Hill AVS, Allsopp CEM, Kwiatkowski D, Anstey NM, Twumasi P, et al. 1991. Common West African HLA antigens are associated with protection from severe malaria. *Nature* 352:595–600

Hill GE. 1991. Plumage coloration is a sexually selected indicator of male quality. *Nature* 350:337–39

Horton R, Wilming L, Rand V, Lovering RC, Bruford EA, et al. 2004. Gene map of the extended human MHC. *Nat. Rev. Genet.* 5:889–99

Houde AE, Torio AJ. 1992. Effect of parasitic infection on male color pattern and female choice in guppies. *Behav. Ecol.* 3:346–51

Howard RS, Lively CM. 1994. Parasitism, mutation accumulation and the maintenance of sex. *Nature* 367:554–57

Howard RS, Lively CM. 2004. Good vs complementary genes for parasite resistance and the evolution of mate choice. *BMC Evol. Biol.* 4:48

Hughes AL, Nei M. 1992. Models of host-parasite interaction and MHC polymorphism. *Genetics* 132:863–64

Jacob S, McClintock MK, Zelano B, Ober C. 2002. Paternally inherited HLA alleles are associated with women's choice of male odor. *Nat. Genet.* 30:175–79

Jaenike J. 1978. An hypothesis to account for the maintenance of sex within populations. *Evol. Theory* 3:191–94

Janeway C, Travers P, Walport M, Shlomchik M. 2001. *Immunobiology: The Immune System in Health and Disease*. New York: Garland. 732 pp.

Jeffery KJM, Bangham CRM. 2000. Do infectious diseases drive MHC diversity? *Microbes Infect.* 2:1335–41

Jeffery KJM, Usuku K, Hall SE, Matsumoto W, Taylor GP, et al. 1999. HLA alleles determine human T-lymphotropic virus-I (HTLV-I) proviral load and the risk of HTLV-I-associated myelopathy. *Proc. Natl. Acad. Sci. USA* 96:3848–53

Jennions MD, Petrie M. 1997. Variation in mate choice and mating preferences: A review of causes and consequences. *Biol. Rev. Philos. Soc. Cambridge* 72:283–327

Jordan WC, Bruford MW. 1998. New perspectives on mate choice and the MHC. *Heredity* 81:127–33

Kavaliers M, Choleris E, Pfaff DW. 2005. Recognition and avoidance of the odors of parasitized conspecifics and predators: Differential genomic correlates. *Neurosci. Biobehav. Rev.* 29:1347–59

Kavaliers M, Colwell DD. 1995. Discrimination by female mice between the odours of parasitized and nonparasitized males. *Proc. R. Soc. London Ser. B* 261:31–35

Keightley PD, Eyre-Walker A. 2000. Deleterious nutations and the evolution of sex. *Science* 290:331–33

Kelley J, Walter L, Trowsdale J. 2005. Comparative genomics of natural killer cell receptor gene clusters. *PLoS Genet.* 1:129–39

Kondrashov AS. 1988. Deleterious mutations and the evolution of sexual reproduction. *Nature* 336:435–40

Kurtz J, Franz K. 2003. Evidence for memory in invertebrate immunity. *Nature* 425:37–38

Kurtz J, Kalbe M, Aeschlimann PB, Häberli MA, Wegner KM, et al. 2004. Major histocompatibility complex diversity influences parasite resistance and innate immunity in sticklebacks. *Proc. R. Soc. London Ser. B* 271:197–204

Kurtz J, Wegner KM, Kalbe M, Reusch TBH, Schaschl H, et al. 2006. MHC genes and oxidative stress in sticklebacks: an immuno-ecological approach. *Proc. R. Soc. London Ser. B* 273:1407–14

Ladle RJ. 1992. Parasites and sex: catching the Red Queen. *Trends Ecol. Evol.* 7:405–8

Landry C, Garant D, Duchesne P, Bernatchez L. 2001. "Good genes as heterozygosity": the major histocompatibility complex and mate choice in Atlantic salmon (*Salmo salar*). *Proc. R. Soc. London Ser. B* 268:1279–85

Langefors Å, Lohm J, Grahn M, Andersen Ø, von Schantz T. 2001. Association between major histocompatibility complex class IIB alleles and resistance to *Aeromonas salmonicida* in Atlantic salmon. *Proc. R. Soc. London Ser. B* 268:479–85

Lawlor DA, Zemmour J, Ennis PD, Parham P. 1990. Evolution of class-I MHC genes and proteins—from natural-selection to thymic selection. *Annu. Rev. Immunol.* 8:23–63

Leinders-Zufall T, Brennan P, Widmayer P, Chandramani P, Maul-Pavicic A, et al. 2004. MHC class I peptides as chemosensory signals in the vomeronasal organ. *Science* 306:1033–37

Lively CM. 1987. Evidence from a New Zealand snail for the maintenance of sex by parasitism. *Nature* 328:519–21

Lohm J, Grahn M, Langefors Å, Andersen Ø, Storset A, von Schantz T. 2002. Experimental evidence for major histocompatibility complex-allele-specific resistance to a bacterial infection. *Proc. R. Soc. London Ser. B* 269:2029–33

Málaga-Trillo E, Zaleska-Rutczynska Z, McAndrew B, Vincek V, Figueroa F, et al. 1998. Linkage relationships and haplotype polymorphism among cichlid MHC class II B loci. *Genetics* 149:1527–37

Maynard Smith J. 1976. *The Evolution of Sex*. Cambridge, UK: Cambridge Univ. Press. 222 pp.

Mays JHL, Hill GE. 2004. Choosing mates: good genes versus genes that are a good fit. *Trends Ecol. Evol.* 19:554–59

McClintock MK, Schumm P, Jacob S, Zelano B, Ober C. 2002. In reply to: The MHC and body odors: arbitrary effects caused by shifts of mean pleasantness. *Nat. Genet.* 31:237–38

Meagher S, Penn DJ, Potts WK. 2000. Male-male competition magnifies inbreeding depression in wild house mice. *Proc. Natl. Acad. Sci. USA* 97:3324–29

Milinski M. 2003a. The function of mate choice in sticklebacks: optimizing MHC genetics. *J. Fish Biol.* 63(Suppl. A):1–16

Milinski M. 2003b. Perfumes. In *Evolutionary Aesthetics*, ed. E Voland, K Grammer, pp. 325–39. Berlin/Heidelberg/New York: Springer-Verlag

Milinski M, Bakker TCM. 1990. Female sticklebacks use male coloration in mate choice and hence avoid parasitized males. *Nature* 344:330–33

Milinski M, Bakker TCM. 1992. Costs influence sequential mate choice in sticklebacks, *Gasterosteus aculeatus*. *Proc. R. Soc. London Ser. B* 250:229–33

Milinski M, Griffiths S, Wegner KM, Reusch TBH, Haas-Assenbaum A, Boehm T. 2005. Mate choice decisions of stickleback females predictably modified by MHC peptide ligands. *Proc. Natl. Acad. Sci. USA* 102:4414–18

Milinski M, Wedekind C. 2001. Evidence for MHC-correlated perfume preferences in humans. *Behav. Ecol.* 12:140–49

Møller AP. 1990. Effects of a hematophagous mite on the barn swallow (*Hirundo rustica*): a test of the Hamilton and Zuk hypothesis. *Evolution* 44:771–84

Moritz C, McCallum H, Donnellan S, Roberts JD. 1991. Parasite loads in parthenogenetic and sexual lizards (*Heteronotia binoei*): support for the Red Queen hypothesis. *Proc. R. Soc. London Ser. B* 244:145–49

Muller HJ. 1932. Some genetic aspects of sex. *Am. Nat.* 66:118–32

Muller HJ. 1964. The relation of recombination to mutational advance. *Mutat. Res.* 1:2–9

Neff BD. 2004. Stabilizing selection on genomic divergence in a wild fish population. *Proc. Natl. Acad. Sci. USA* 101:2381–85

Nowak MA, Tarczy-Hornoch K, Austyn JM. 1992. The optimal number of major histocompatibility complex molecules in an individual. *Proc. Natl. Acad. Sci. USA* 89:10896–99

Ober C, Weitkamp LR, Cox N, Dytch H, Kostyu D, Elias S. 1997. HLA and mate choice in humans. *Am. J. Hum. Genet.* 61:497–504

Olsson M, Madsen T, Nordby J, Wapstra E, Ujvari B, et al. 2003. Major histocompatibility complex and mate choice in sand lizards. *Proc. R. Soc. London Ser. B* 270:254–56

Olsson M, Madsen T, Wapstra E, Silverin B, Ujvari B, Wittzell H. 2005. MHC, health, color, and reproductive success in sand lizards. *Behav. Ecol. Sociobiol.* 58:289–94

Parham P, Ohta T. 1996. Population biology of antigen presentation by MHC class I molecules. *Science* 272:67–74

Paterson S, Wilson K, Pemberton JM. 1998. Major histocompatibility complex variation associated with juvenile survival and parasite resistance in a large unmanaged ungulate population (*Ovis aries* L.). *Proc. Natl. Acad. Sci. USA* 95:3714–19

Penn D, Potts W. 1998a. How do major histocompatibility complex genes influence odor and mating preferences? *Adv. Immunol.* 69:411–36

Penn D, Potts WK. 1998b. Chemical signals and parasite-mediated sexual selection. *Trends Ecol. Evol.* 13:391–96

Penn DJ. 2002. The scent of genetic compatibility: sexual selection and the major histocompatibility complex. *Ethology* 108:1–21

Penn DJ, Potts WK. 1999. The evolution of mating preferences and major histocompatibility complex genes. *Am. Nat.* 153:145–64

Piertney SB, Oliver MK. 2006. Review. The evolutionary ecology of the major histocompatibility complex. *Heredity* 96:7–21

Potts WK, Manning CJ, Wakeland EK. 1991. Mating patterns in seminatural populations of mice influenced by MHC genotype. *Nature* 352:619–21

Potts WK, Manning CJ, Wakeland EK. 1994. The role of infectious disease, inbreeding and mating preferences in maintaining MHC genetic diversity: an experimental test. *Philos. Trans. R. Soc. London Ser. B* 346:369–78

Potts WK, Wakeland EK. 1990. The maintenance of MHC polymorphism. *Immunol. Today* 11:39–40

Reusch TBH, Häberli MA, Aeschlimann PB, Milinski M. 2001. Female sticklebacks count alleles in a strategy of sexual selection explaining MHC polymorphism. *Nature* 414:300–2

Reusch TBH, Langefors Å. 2005. Inter- and intralocus recombination drive MHC class IIB gene diversification in a teleost, the three-spined stickleback *Gasterosteus aculeatus*. *J. Mol. Evol.* 61:531–41

Reusch TBH, Schaschl H, Wegner KM. 2004. Recent duplication and interlocus gene conversion in major histocompatibility class II genes in a teleost, the three-spined stickleback. *Immunogenetics* 56:427–37

Richardson DS, Komdeur J, Burke T, von Schantz T. 2005. MHC-based patterns of social and extrapair mate choice in the Seychelles warbler. *Proc. R. Soc. B* 272:759–67

Richman A. 2000. Invited review. Evolution of balanced genetic polymorphism. *Mol. Ecol.* 9:1953–63

Ridgway WM, Fathman CG. 1998. The association of MHC with autoimmune diseases: Understanding the pathogenesis of autoimmune diabetes. *Clin. Immunol. Immunopathol.* 86:3–10

Ridley M. 2004. *Evolution*. Oxford, UK: Blackwell. 751 pp. 3rd ed.

Rülicke T, Chapuisat M, Homberger FR, Macas E, Wedekind C. 1998. MHC-genotype of progeny influenced by parental infection. *Proc. R. Soc. London Ser. B* 265:711–16

Sam M, Vora S, Malnic B, Ma WD, Novotny MV, Buck LB. 2001. Odorants may arouse instinctive behaviours. *Nature* 412:142

Sanjuan R, Moya A, Elena SF. 2004. The contribution of epistasis to the architecture of fitness in an RNA virus. *Proc. Natl. Acad. Sci. USA* 101:15376–79

Sato A, Figueroa F, Murray BW, Málaga-Trillo E, Zaleska-Rutczynska Z, et al. 2000. Nonlinkage of major histocompatibility complex class I and class II loci in bony fishes. *Immunogenetics* 51:108–16

Sato A, Figueroa F, O'hUigin C, Steck N, Klein J. 1998. Cloning of major histocompatibility complex (Mhc) genes from threespine stickleback, *Gasterosteus aculeatus*. *Mol. Mar. Biol. Biotechnol.* 7:221–31

Schmid-Hempel P, Puhr K, Krüger N, Reber C, Schmid-Hempel R. 1999. Dynamic and genetic consequences of variation in horizontal transmission for a microparasitic infection. *Evolution* 53:426–34

Singer AG, Beauchamp GK, Yamazaki K. 1997. Volatile signals of the major histocompatibility complex in male mouse urine. *Proc. Natl. Acad. Sci. USA* 94:2210–14

Singh PB, Brown RE, Roser B. 1987. MHC antigens in urine as olfactory recognition cues. *Nature* 327:161–64

Singh PB, Herbert J, Roser B, Arnott L, Tucker DK, Brown RE. 1990. Rearing rats in a germ-free environment eliminates their odors of individuality. *J. Chem. Ecol.* 16:1667–82

Slade RW, McCallum HI. 1992. Overdominant vs frequency-dependent selection at MHC loci. *Genetics* 132:861–62

Stet RJM, Kruiswijk CP, Dixon B. 2003. Major histocompatibility lineages and immune gene function in teleost fishes: The road not taken. *Crit. Rev. Immunol.* 23:441–71

Stoddart DM. 1990. *The Scented Ape: The Biology and Culture of Human Odour*. Cambridge, NY: Cambridge Univ. Press

Suri A, Walters JJ, Kanagawa O, Gross ML, Unanue ER. 2003. Specificity of peptide selection by antigen-presenting cells homozygous or heterozygous for expression of class II MHC molecules: The lack of competition. *Proc. Natl. Acad. Sci. USA* 100:5330–35

Takahata N, Nei M. 1990. Allelic genealogy under overdominant and frequency-dependent selection and polymorphism of major histocompatibility complex loci. *Genetics* 124:967–78

Thursz MR, Thomas HC, Greenwood BM, Hill AVS. 1997. Heterozygote advantage for HLA class-II type in hepatitis B virus infection. *Nat. Genet.* 17:11–12

Tiwari JL, Terasaki PI. 1985. *HLA and Disease Associations*. New York: Springer-Verlag

Tregenza T, Wedell N. 2000. Genetic compatibility, mate choice and patterns of parentage: invited review. *Mol. Ecol.* 9:1013–27

Trivers R. 1985. *Social Evolution*. Menlo Park, CA: Benjamin/Cummings. 462 pp.

VanValen L. 1973. A new evolutionary law. *Evol. Theory* 3:1–30

Vollrath F, Milinski M. 1995. Fragrant genes help Damenwahl. *Trends Ecol. Evol.* 10:307–8

vom Saal FS, Howard LS. 1982. The regulation of infanticide and parental behavior: implications for reproductive success in male mice. *Science* 215:1270–72

von Schantz T, Göransson G, Andersson G, Fröberg I, Grahn M, et al. 1989. Female choice selects for a viability-based male trait in pheasants. *Nature* 337:166–69 von Schantz T, Wittzell H, Göransson G, Grahn M, Persson K. 1996. MHC genotype and male ornamentation: genetic evidence for the Hamilton-Zuk model. *Proc. R. Soc. London Ser. B* 263:265–71

Webster JP, Hoffman JI, Berdoy M. 2003. Parasite infection, host resistance and mate choice: battle of the genders in a simultaneous hermaphrodite. *Proc. R. Soc. London Ser. B* 270:1481–85

Wedekind C. 2002. The MHC and body odors: arbitrary effects caused by shifts of mean pleasantness. *Nat. Genet.* 31:237

Wedekind C, Füri S. 1997. Body odour preferences in men and women: do they aim for specific MHC combinations or simply heterozygosity? *Proc. R. Soc. London Ser. B* 264:1471–79

Wedekind C, Seebeck T, Bettens F, Paepke AJ. 1995. MHC-dependent mate preferences in humans. *Proc. R. Soc. London Ser. B* 260:245–49

Wegner KM. 2004. *Major histocompatibility genes, polymorphism and balancing selection—the case of parasites and sticklebacks*. PhD thesis. Christian-Albrechts-Univ., Kiel. 118 pp.

Wegner KM, Kalbe M, Kurtz J, Reusch TBH, Milinski M. 2003a. Parasite selection for immunogenetic optimality. *Science* 301:1343

Wegner KM, Kalbe M, Kurtz J, Reusch TBH, Milinski M. 2004. Response to comment on "Parasite selection for immunogenetic optimality". *Science* 303:957

Wegner KM, Kalbe M, Rauch G, Kurtz J, Schaschl H, et al. 2006. Genetic variation in MHC class II expression and interactions with MHC sequence polymorphism in three-spined sticklebacks. *Mol. Ecol.* 15:1153–64

Wegner KM, Reusch TBH, Kalbe M. 2003b. Multiple parasites are driving major histocompatibility complex polymorphism in the wild. *J. Evol. Biol.* 16:224–32

West SA, Peters AD, Barton NH. 1998. Testing for epistasis between deleterious mutations. *Genetics* 149:435–44

Westerdahl H. 2003. *Avian MHC: variation and selection in the wild*. PhD thesis. Lund Univ., Lund. 145 pp.

Westerdahl H, Hansson B, Bensch S, Hasselquist D. 2004. Between-year variation of MHC allele frequencies in great reed warblers: selection or drift? *J. Evol. Biol.* 17:485–92

Wloch DM, Borts RH, Korona R. 2001. Epistatic interactions of spontaneous mutations in haploid strains of the yeast *Saccharomyces cerevisiae*. *J. Evol. Biol.* 14:310–16

Wong BBM, Jennions MD. 2003. Costs influence male mate choice in a freshwater fish. *Proc. Natl. Acad. Sci. USA* 270:S36–38

Yamazaki K, Beauchamp GK, Imai Y, Bard J, Phelan SP, et al. 1990. Odortypes determined by the major histocompatibility complex in germ-free mice. *Proc. Natl. Acad. Sci. USA* 87:8413–16

Yamazaki K, Beauchamp GK, Matsuzaki O, Kupniewski D, Bard J, et al. 1986. Influence of a genetic difference confined to mutation of H-2K on the incidence of pregnancy block in mice. *Proc. Natl. Acad. Sci. USA* 83:740–41

Yamazaki K, Beauchamp GK, Singer AG, Bard J, Boyse EA. 1999. Odortypes: Their origin and composition. *Proc. Natl. Acad. Sci. USA* 96:1522–25

Yamazaki K, Boyse EA, Mike V, Thaler HT, Mathieson BJ, et al. 1976. Control of mating preferences in mice by genes in the major histocompatibility complex. *J. Exp. Med.* 144:1324–35

Yamazaki K, Yamaguchi M, Andrews PW, Peake B, Boyse EA. 1978. Mating preferences of F_2 segregants of crosses between MHC-congenic mouse strains. *Immunogenetics* 6:253–59

Yeates SE. 2005. *Fertilisation dynamics in Atlantic salmon*. PhD thesis. Univ. East Anglia, Norwich. 149 pp.

Younger RM, Amadou C, Bethel G, Ehlers A, Lindahl KF, et al. 2001. Characterization of clustered MHC-linked olfactory receptor genes in human and mouse. *Genome Res.* 11:519–30

Zelano B, Edwards SV. 2002. An Mhc component to kin recognition and mate choice in birds: Predictions, progress, and prospects. *Am. Nat.* 160:S225–37

Ziegler A, Ehlers A, Forbes S, Trowsdale J, Uchanska-Ziegler B, et al. 2000. Polymorphic olfactory receptor genes and HLA loci constitute extended haplotypes. In *Major Histocompatibility Complex*, ed. M Kasahara, pp. 110–30. Tokyo: Springer-Verlag

Ziegler A, Kentenich H, Uchanska-Ziegler B. 2005. Female choice and the MHC. *Trends Immunol.* 26:496–502

Zuk M, Thornhill R, Ligon JD. 1990. Parasites and mate choice in red jungle fowl. *Am. Zool.* 30:235–44

Some Evolutionary Consequences of Being a Tree

Rémy J. Petit[1] and Arndt Hampe[1,2]

[1] Institut National de la Recherche Agronomique, UMR Biodiversity, Genes and Communities, F-33610 Cestas, France; email: petit@pierroton.inra.fr

[2] Consejo Superior de Investigaciones Científicas, Estación Biológica de Doñana, Integrative Ecology Group, E-41080 Sevilla, Spain; email: arndt@ebd.csic.es

Key Words

allometric scaling, evolutionary rate, gene flow, growth form, life history

Abstract

Trees do not form a natural group but share attributes such as great size, longevity, and high reproductive output that affect their mode and tempo of evolution. In particular, trees are unique in that they maintain high levels of diversity while accumulating new mutations only slowly. They are also capable of rapid local adaptation and can evolve quickly from nontree ancestors, but most existing tree lineages typically experience low speciation and extinction rates. We discuss why the tree growth habit should lead to these seemingly paradoxical features.

"Few there are [...] who seem to clearly realize how broad a lesson on the life-history of plants is written in the trees that make the great forest regions of the world." Clarke 1894

Substitution rate: number of nucleotide differences that accumulate in a sequence per unit of time, usually much more than one generation

1. INTRODUCTION

The importance of trees for sustaining life in general and biodiversity in particular can hardly be overstated. An estimated 27% of the terrestrial surface of Earth is (still) covered by forests (FAO World Resources 2000–2001), and trees make up around 90% of Earth's biomass (Whittaker 1975). Not surprisingly, forests also harbor the vast majority of the world's terrestrial biodiversity. Estimates of global tree species richness range from a low 60,000 (Grandtner 2005) to 100,000 taxa (Oldfield et al. 1998), that is, as much as 15% to 25% of the 350,000–450,000 vascular plants of the world (Scotland & Wortley 2004). Unfortunately, ongoing deforestation (estimated at 9.4 million hectares per year in the 1990s) and other human-induced changes have brought >10% of the world's tree species close to extinction (Oldfield et al. 1998). The impact of global change will depend to a great extent on the reaction of trees and the ecosystems they sustain (e.g., Ozanne et al. 2003; Petit et al. 2004a, 2005b). Mitigating these harmful consequences requires knowledge of tree biodiversity and evolution. However, trees are not only overexploited but also understudied in many respects, because their size and life span make them difficult subjects for experimental investigations (Linhart 1999).

The tree growth habit has evolved many times. This is probably the reason why few attempts have been made over the past several decades to consider trees collectively and discuss their mode of evolution. This apparent lack of interest contrasts with a strong tradition in earlier years (e.g., Arber 1928; Clarke 1894; Grant 1963, 1975; Sinnott 1916; Stebbins 1958). The current interest in comparative biology, thanks to the development of accurate phylogenies and powerful analytical methods, should help revive this tradition. Far from representing a problem, the multiple origins of trees will actually facilitate this work, as each distinct tree lineage can be viewed as an independent evolutionary experiment. Comparative analyses should help elucidate if typical tree features such as tallness, longevity, and fecundity affect their evolutionary dynamics.

From an evolutionary standpoint, trees have several intriguing and apparently paradoxical features. In particular, they often have high levels of genetic diversity but experience low nucleotide substitution rates and low speciation rates. They also combine high local differentiation for adaptive traits with extensive gene flow. Moreover, exceptional maintenance of species integrity in the face of abundant interspecific gene flow seems to be the rule in trees.

In this review, we first compare existing definitions of the tree habit and then identify and discuss trees' major ecological characteristics. Second, we examine why trees generally harbor such high levels of genetic diversity and can adapt rapidly to local conditions. Third, we ask why trees have such a low pace of evolution at longer timescales, both in terms of DNA sequence and character change within lineage and in terms of diversification rate. Finally, we discuss how to reconcile the observations of rapid microevolution and slow macroevolution. Much of the earlier work on the evolutionary consequences of the tree growth habit dates from the 1950s (Stebbins 1950, 1958, 1974; Grant 1958). This type of analysis, in which life history traits and reproductive characteristics of plants were viewed as an integrated set of attributes (the so-called genetic system) contributing to adjusting levels of genetic diversity to the ecological demand, has lost popularity. However, the need for broad synthetic approaches aimed at organizing and interpreting the growing body of knowledge on these topics is greater than ever.

Lignophytes: plants having an external layer of porous bark and an internal core of wood produced by the cambium

2. WHAT IS A TREE? MAJOR FEATURES AND CONSTRAINTS OF THE TREE GROWTH HABIT

2.1. What Is a Tree?

With one known exception—*Prototaxites*, a 9-m-high tree-like holobasidiomycete or lichen that dominated the land flora 350–400 Myr ago (Hueber 2001, Selosse 2002)—all organisms ever considered to be trees are vascular plants (tracheophytes). As such, they share features such as indefinite and flexible growth, modular structure, lack of clear separation between germline and soma, reversible cellular differentiation, great phenotypic plasticity and physiological tolerance, and presence of haploid and diploid multicellular generations (Bradshaw 1972). Evolution of trees cannot be understood without due consideration of these attributes.

The particular character of the tree growth form has always been recognized and, since Theophrastus (born c. 370 BC), botanists have generally distinguished between trees, shrubs, and herbs. From a functional point of view, trees share a number of features, such as large size, long life span, and a self-supporting woody perennial trunk, but not one is really exclusive. According to Van Valen (1975), a tree is, in the ecological sense, "any tall woody plant." However, trees are generally distinguished from shrubs and vines, so most researchers prefer to be more specific. For instance, for Thomas (2000), "a tree is any plant with a self-supporting, perennial woody stem"; for Donoghue (2005) the tree growth habit is characterized by "tall plants, with a thickened single trunk, branching well above ground level"; and for Niklas (1997) a tree is "any perennial plant with a permanent, woody, self-supporting main stem or trunk, ordinarily growing to a considerable height, and usually developing branches at some distance above the ground." The modulations introduced express the need to accommodate situations where plants generally considered to be trees adopt unusual habit or size in some environments. Finally, somewhat arbitrary definitions can be found in the forestry literature, for inventory purposes: "Trees are woody plants with one erect perennial stem, a definitely formed crown, a height of at least 4 m and a stem diameter at breast height of at least 5 cm" (Little 1979).

The presence of wood is sometimes taken as an argument to circumscribe trees to the lignophytes (see Niklas 1997). Interestingly, recent molecular genetic and genomic studies in *Populus* and *Arabidopsis* have shown that the genes responsible for cambium function and woody growth are not unique to woody plants: Genes involved in the vascular cambium of woody plants are also expressed in the regulation of the shoot apical meristem of *Arabidopsis* (Groover 2005). This might explain why woodiness can evolve so readily (as observed in many island radiations; e.g., Böhle et al. 1996, Carlquist 1974) and led Groover (2005) to conclude that forest trees "constitute a contrived group of plants that have more in common with herbaceous relatives than we foresters like to admit."

According to Arber (1928), one needs to go beyond textbook definitions and acknowledge that the difference between trees and other plants is mostly a question of scale. Below, we provide an account of the prominent features of the tree growth habit from an ecological standpoint. In so doing, we follow Arber (1928) and stress questions of scale and allometry.

Allometry: the study of size and its consequences

2.2. Prominent Tree Features

The woody habit involves a series of ecological benefits and constraints that have contributed to the dominance of trees across many ecosystems worldwide and to their scarcity or complete absence from others. According to Harper (1977, p. 599), the major advantage of a woody growth habit is that "it can give perenniality to height." These two components are tightly linked, as a high stature can obviously not be attained without the corresponding life span. Tallness and longevity are also the prerequisites for another central feature of trees: their large, sometimes huge lifetime reproductive output, despite a somewhat delayed maturity.

Although it is clear that these characteristics have been molded by selection pressures (Niklas 1997), they are subject to a diversity of anatomical, physiological, or ontogenetic constraints (e.g., Mencuccini et al. 2005, Niklas 1997, Rowe & Speck 2005, Silvertown et al. 2001). Major steps to understand the primary causes of the evolution of the tree growth habit have been made by simulating adaptive walks through the morphospace of early vascular land plants (Niklas 1997). These studies indicate that growing tall is indeed an adaptive process; in particular, "tree-like morphologies bearing lateral planated branching systems or foliage leaves occupy adaptive peaks" (Niklas 1997). Altogether, the tree growth form can be viewed as an integrated ecological strategy involving many trade-offs (**Table 1**). In the following we discuss implications of the tree habit and outline the major characteristic of trees' life cycle.

2.2.1. Tallness.
Trees grow tall where resources are abundant, stresses are minor, and competition for light takes place (e.g., Falster & Westoby 2003, King 1990, Loehle 2000). Large size enables them to create a physical and chemical environment that influences their own performance and that of interacting organisms (e.g., Boege & Marquis 2005, Herwitz et al. 2000, Ricklefs & Latham 1992). High stature helps mitigate the effects of disturbances that take place primarily at ground level, such as grazing and trampling by large herbivores or fires (Ordóñez et al. 2005), but it makes trees highly susceptible to other disturbances such as wind (Gutschick & BassiriRad 2003, Loehle 1988, Rowe & Speck 2005). Growing tall requires the development of resistant supporting and protective tissues. This generates high costs of maintenance, reduces growth rates, and limits the existence of trees to areas that provide a minimum long-term input of energy, water, and nutrients (Ward et al. 2005a, Wardle et al. 2004).

Table 1 Some advantages and drawbacks of the tree growth form

Advantage	Drawback
Great potential of biomass gain	High maintenance costs
High competition after successful establishment	Extremely high recruit mortality
Endurance to short-term resource depletion	Increased probability of suffering catastrophic events
Escape from disturbances at ground level (e.g., grazing, fire)	Exposure to physical disturbances above ground (in particular wind)
Life-long increase in storage capacity and fecundity	High investment in supporting tissues and defense mechanisms reduces overall allocation to reproduction
Great lifetime fecundity	Delayed maturity
Little dependence on particular reproductive events	Trade-off between present reproductive output and future growth, survival, and reproduction
Attraction of mutualists (e.g., pollinators, seed dispersers, herbivore predators)	Attraction of antagonists (e.g., herbivores, pathogens)
Satiation of enemies (e.g., mast-fruiting)	Satiation of mutualists (e.g., geitonogamy, disperser satiation)
Large pollen and seed production and release height facilitate gene dispersal	Low plant density complicates mating and increases pollen limitation
Relatively little seed limitation of recruitment	Strong limitation of suitable sites and time windows for recruitment
Effective population size close to adult population size	Large differences in life spans exacerbate inequality in individual lifetime fecundity
Local adaptation favored by strong selection during early life stages	Local adaptation hindered by high gene flow
Reduced accumulation of mutations per unit of time	Increased mutation rate per generation
Strong inbreeding depression increases outcrossing rate and maintains genetic diversity	Lifelong accumulation of somatic mutations results in susceptibility to inbreeding depression
Long life span reduces extinction risk	Long generation time reduces speciation rate
Extensive intra- and interspecific gene flow reduces extinction risk	Extensive intra- and interspecific gene flow reduces speciation rate
Slower evolution than mutualists results in greater share of resources for host trees (Red King effect)	Slower evolution than antagonists results in host trees lagging behind in arms races (Red Queen effect)

Tall stature also tends to increase gene flow. For instance, the height of release is often (yet not necessarily) related to the median transport distance of wind-dispersed pollen and seeds (Nathan et al. 2002, Okubo & Levin 1989, Portnoy & Willson 1993). Tall, conspicuous plants with large flower or fruit displays also tend to attract disproportionately many animal pollinators and seed dispersers (Ghazoul 2005). The latter holds likewise for antagonists such as herbivores or pathogens, however.

Height is probably the plant trait that has most often been included in comparative studies. Westoby et al. (2002) consider it a leading dimension of plant ecological strategies, as it conveys knowledge on many other aspects of species' ecology. The upper limit to plant height has been the object of many studies and debate that will not be reviewed in detail here. Two major hypotheses coexist, the respiration hypothesis and the hydraulic-limitation hypothesis (e.g., Mencuccini et al. 2005). However,

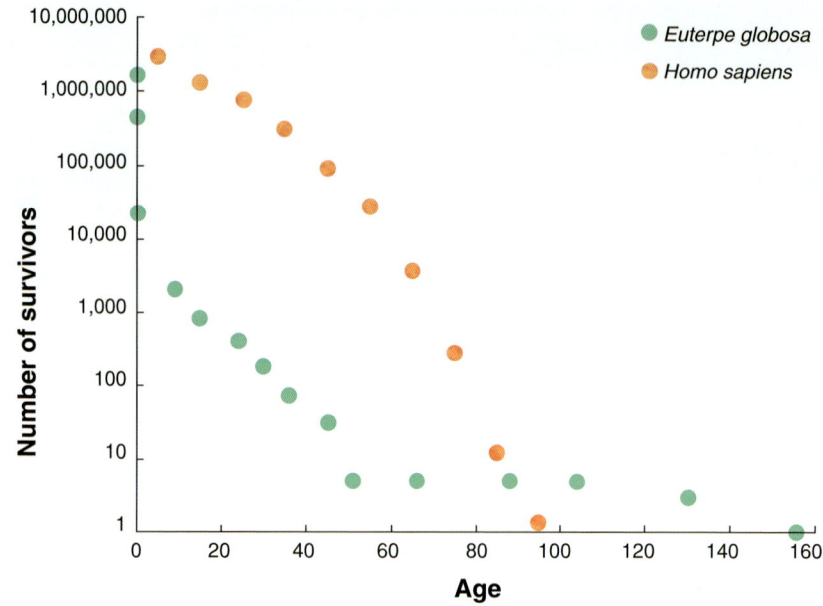

Figure 1
Survivorship curves for a large plant (the palm *Euterpe globosa*) and a large animal (*Homo sapiens*). Note the concave versus convex relationship first noticed by Szabó (1931). Data for the tree is from Van Valen (1975).

Generation time: the time from seed to seed

evolution should not lead to a single limiting factor, according to the principle of equalization of marginal returns on alternative expenditures (Westoby et al. 2002). Hence, species are not expected to grow as tall as physically possible because of various trade-offs, for example with reproduction, wood density (and hence longevity), or leaf mass (e.g., Loehle 1988).

2.2.2. Extended life cycle. Although great advances have been made in our understanding of initial recruitment processes in forests and in the modeling of vegetation dynamics (e.g., Clark et al. 1999, Loehle 2000), extremely few studies have considered the entire tree life cycle. A notable exception is the work of Van Valen (1975), who presented the first complete life table for a tree, the tropical palm *Euterpe globosa*. According to his computations, only one seed in one million produces a shoot that reaches the canopy in this species. The resulting demographic curve, when expressed in logarithmic terms, is highly convex, contrary to that of many large animals, as first pointed out by Szabó (1931) (**Figure 1**). In this palm species, generation time was estimated to be 101 years, a value intermediate between age at maturity (50 years) and maximum observed life span (156 years). These values underline the need to distinguish between age at maturity, generation time, and life span, which are often inappropriately used interchangeably in the literature.

2.2.3. Seed production. Estimates of lifetime reproductive output for trees are rare (Moles et al. 2004), but it is clear that many trees produce prodigious numbers of seeds. Reproduction is costly and trade-offs with vegetative growth are well-known (e.g., Obeso 2002). Niklas & Enquist (2003) proposed an allometric model for reproduction in seed plants that shows that the annual reproductive biomass scales with the two-thirds power of the standing shoot biomass; in other words, allocation to reproduction decreases with size. Hence, larger plants would produce comparatively fewer seeds if seed size scaled isometrically with plant size. Aarssen (2005) tested this latter relation for 600 North American species and found that seed length increases only at about half the rate of plant height, indicating that the prevalent evolutionary trend (i.e., the deviation from allometric scaling) is toward comparatively smaller seeds, thus maintaining fecundity at the expense of seed provisioning. Moreover, the variability in seed length grows disproportionately with plant height. Similar results were obtained by Moles et al. (2004) for plant and seed mass. Aarssen (2005) argued that the observed patterns might be a simple consequence of the fact that the spectrum of possible seed sizes broadens with plant size.

Contrary to animals, plant fecundity usually increases more or less continually through an individual's life (Franco & Silvertown 1996). Hence, the lifetime seed production of trees is typically orders-of-magnitude greater than that of herbs, even though decreasing allocation results in a lower annual output of seeds per unit of canopy (Moles et al. 2004). Unfortunately, as for many other relationships, it remains unclear if individual variation in lifetime fecundity is greater in trees than in herbs or if it scales isometrically.

Finally, much attention has been paid to the phenomenon of mast seeding in trees (i.e., the synchronous intermittent production of large seed crops). Overall, it appears that fruit crop size scales positively with its among-year variability (Kerkhoff & Ballantyne 2003). But it remains unclear whether this phenomenon results mostly from weather conditions or represents an evolved plant reproductive strategy to improve pollination efficiency and outcrossing levels (in wind-pollinated species), and/or to increase offspring survival through predator satiation. Recent meta-analyses of extensive data sets indicate that both components may be involved to varying extents (Kelly & Sork 2002).

2.2.4. Establishment. As trees tend to live in comparatively stable habitats and generation turnover is slow, only an extremely small fraction of the seeds produced during an individual's lifetime will eventually survive to maturity. This has important consequences for trees' evolution. First, the considerable selection potential during early life stages should favor local adaptation of recruits, particularly for traits that enhance competitive ability (such as early growth and delayed maturity). By contrast, selective culling during trees' establishment appears to have little influence on population demography (Franco & Silvertown 1996). Second, because much of the density-dependent mortality takes place before maturity in trees, their effective population size should be closer to the actual adult census size compared to herbs, contributing to preserve genetic diversity (Dodd & Silvertown 2000).

2.2.5. Age at maturity. One classical trade-off in population dynamics is that between early growth and age at maturity. Precocity of reproduction has a great influence on the potential growth rate of a population (Harper 1977). Only very stringent competition for resources (e.g., light) during the early life of trees can select for delayed maturity. Among trees, there is a great variation in age at maturity. Woody angiosperms tend to reproduce sooner than gymnosperms [modal class is 1–5 years compared to 6–20 years (Verdú 2002)]. Age at maturity has received some attention by molecular biologists. Genetic manipulations demonstrate that juvenile trees can be induced to flower by modifying the expression of a single gene, e.g., *LFY* in transgenic poplars (reviewed in Martín-Trillo & Martínez-Zapater 2002). Hence, as for secondary growth, the evolution of shortened maturity does not require profound genetic changes at the molecular level. (The converse is not necessarily true, however; the evolution of delayed maturity might be more complex.)

2.2.6. Longevity. A long life span is favored in stable habitats as long as it remains advantageous to allocate resources to future reproduction. Great longevity provides several obvious advantages. First, once successfully established, plants can endure periods of environmental stress while taking advantage of relatively short pulses of less harsh conditions. In particular, long-lived species can endure periodic reproductive failures without direct negative demographic consequences (Ashman et al. 2004, Calvo & Horvitz 1990). This flexibility might explain why woody plants generally display stronger pollen limitation than herbs (Knight et al. 2000). Second, spreading reproduction over many years boosts lifetime reproductive output. However, a long life span also means that individuals have to cope with variable environmental conditions including catastrophic events (Gutschick & BassiriRad 2003). Hence, allocations to growth, reproduction, and survival need to be adjusted throughout lifetime. Such plasticity would in turn contribute to enlarge trees' potential habitats (e.g., Hampe & Bairlein 2000, Jónsson 2002), resulting in considerable buffering against extinction (Hampe & Petit 2005).

2.2.7. Senescence. The extreme longevity observed in woody plants makes them useful models for senescence research and trees have actually started to attract the interest of gerontologists (e.g., Flanary & Kletetschka 2005, Lanner 2002, Larson 2001). As pointed out by Williams (1957), the degree of senescence is a function of the lifetime distribution of reproductive effort, so senescence should be far lower in organisms that increase reproduction with age, like trees.

Extreme conditions (e.g., low temperatures, drought or wind) are associated with the occurrence of particularly old and slow-growing trees (e.g., Laberge et al. 2000, Lanner 2002, Larson et al. 1999), suggesting that low metabolism contributes to their delayed senescence. Until recently, it was generally assumed that whole-organism metabolic rate scales with the three-fourths power of body mass in all organisms (Gillooly et al. 2001). Hence, trees would inherently experience reduced metabolic rates simply owing to their size. However, Reich et al. (2006) have shown that the metabolic rate of plants (including herbs, woody plant seedlings and young saplings) instead scales approximately isometrically with plant size, thereby discarding allometry as a possible source of reduced metabolic rate in trees. Nevertheless, the remarkable amount of resources that woody plants need to invest in supporting structures and defenses (such as a thick bark or defensive chemicals) is generally related to a reduction of growth rate and, hence, of metabolism (Loehle 1988).

Trees dispose of a suite of active and passive mechanisms to repair, isolate, or replace deteriorated tissues (Loehle 1988). These can greatly increase life span thanks to the modular structure of plant growth and to the fact that at least some cell lines inside meristems retain the juvenile ability to contribute to new growth (Lanner 2002). Low extrinsic mortality and efficient repair mechanisms would promote resource allocation to repair (especially early in life), resulting in delayed growth rate and maturity, large size, and a dramatic increase in survival and maximum life span (Cichón 1997). Empirical support for this notion comes from a demographic analysis of herbaceous and woody plants (Silvertown et al. 2001) that detected increasing age-specific mortality near the maximum life span (that is, signs of senescence) only in the longest-lived species. So far, however, little evidence exists for whole-tree senescence in terms of changes in gene expression that might indicate genetically controlled aging mechanisms (Diego et al. 2004), although there are preliminary data indicating that both telomere length and telomerase activity could be involved in tree longevity (Flanary & Kletetschka 2005).

Inbreeding depression: the reduction in performance of progeny derived from selfing

Self-incompatibility systems: methods preventing self-fertilization in hermaphrodites through recognition and rejection of pollen expressing the same allelic specificity as that expressed in the pistils

3. TREES HAVE HIGH LEVELS OF GENETIC DIVERSITY AND EXPERIENCE RAPID MICROEVOLUTION

Comparative surveys based on molecular markers have consistently indicated that trees have more genetic diversity within their populations than herbaceous plants and shrubs (e.g., Hamrick et al. 1979; Hamrick & Godt 1989, 1996; Nybom 2004). However, genome-wide estimates of nucleotide diversity in plants are still too few to see if this trend also holds at the sequence level (Neale & Savolainen 2004). Tree populations are also less genetically structured than herbaceous plants (Hamrick & Godt 1989, 1996; Nybom 2004). Finally, trees appear to be capable of rapid adaptation to new conditions (e.g., Petit et al. 2004a). Below, we discuss possible causes that might account for these observations.

3.1. Mating System

3.1.1. Trees are predominantly outcrossed.
Although many trees can self, not one is predominantly selfing (Hamrick & Godt 1996). Clarifying the causes of this marked association between life form and mating system is of utmost importance because mating system has major evolutionary consequences; in particular, it has been repeatedly shown to be one of the best predictors of the genetic structure of populations, both at presumably neutral markers (e.g., Hamrick & Godt 1989, 1996; Schoen & Brown 1991) and at quantitative traits (Charlesworth & Charlesworth 1995).

3.1.2. Proximate causes.
Trees are primarily outcrossing as a consequence of mechanisms that enforce allogamy, like inbreeding depression, self-incompatibility, or dioecy. First, strong early acting inbreeding depression is particularly frequent in trees (e.g., Husband & Schemske 1996, Sorensen 1999). It ensures that all adult plants eventually result from outcrossing. Given the formidable life-long reproductive capacity

of trees resulting in high juvenile mortality and hence in "convex" demographic curves (**Figure 1**), early acting inbreeding depression might represent a demographically acceptable strategy. Although some tree populations have been identified that are largely purged of their inbreeding depression (e.g., Sorensen 2001), they are very rare and appear to have experienced a bottleneck. Second, self-incompatible species are on average markedly more long-lived than self-compatible ones, even among perennials (Ehrlén & Lehtilä 2002). RNase-based self-incompatibility is currently considered the ancestral state in the majority of eudicots (Igic & Kohn 2001), so any difference between trees and herbs would imply a more rapid loss of self-incompatibility in herbs. Third, dioecy is consistently more frequent in woody plants than in herbs (Vamosi & Vamosi 2004). Dioecy has frequently evolved following colonization of oceanic islands, along with increased size and woodiness (e.g., Böhle et al. 1996). In small colonist populations, the accumulation of deleterious mutations could cause male sterility and precipitate the evolution of gynodioecy and ultimately dioecy.

3.1.3. Ultimate causes. One possible explanation for the relation between life span and mating system is that the reproductive assurance granted by selfing would be of less significance in long-lived perennials, because failures to reproduce one year do not compromise their life-long fitness (Ashman et al. 2004, Calvo & Horvitz 1990). In support of this, seed augmentation experiments indicate that seed limitation is most prevalent in early successional habitats (Turnbull et al. 2000), where selfing species are most common.

Morgan et al. (1997) have suggested that temporally fluctuating inbreeding depression could instead represent the major cause of the allogamous mating system of long-lived plants. Inbreeding depression is spread over many years in trees because of their greater longevity; this should exacerbate selection against inbred individuals because of the multiplicative effects of inbreeding depression. They also note that iteroparous perennial plants, if self-pollinated via modes of selfing that provide reproductive assurance, would potentially suffer from an additional fitness cost: that of between-season seed discounting, i.e., the loss of opportunities to produce outcrossed seed in a year with great availability of pollinators. In theory, this factor (as well as inbreeding depression over many seasons) could act as a further selective force preventing the evolution of selfing in trees.

Alternatively, the outcrossed mating system of trees could directly result from their large body size rather than from their longevity. Trees' stature necessarily leads to an elevated number of mitotic cell divisions per generation, which results in a higher incidence of deleterious recessive mutations in the gametes. Using models that allow inbreeding depression of populations to evolve and assuming that deleterious mutations accumulate on a per-generation basis, Morgan (2001) showed that perenniality should result in a reduction of inbreeding depression (by making selfing-induced purging more efficient), and in inbreeding depression being caused by increasingly recessive, rather than partially dominant, mutations. Although the latter prediction holds true [deleterious mutations are typically recessive in trees (cf. Williams & Savolainen 1996)], the first prediction is not met: Perennials experience higher not lower inbreeding depression compared to annuals. By comparing empirical data on selfing rate and inbreeding depression, Scofield & Schultz (2006) showed that for the same selfing rate, high-stature plants tend to have lower inbreeding coefficients. This implies that they have much higher inbreeding depression than low-stature plants, suggesting high deleterious mutation rates per generation, in line with experimental evidence (see Section 4.1). Scofield & Schultz (2006) further predicted that high-stature plants should have progeny with essentially zero fitness when selfed, which is well supported by experimental evidence (e.g., Sorensen 1999).

Mutation rate: the probability of genetic change per generation

3.2. Gene Flow

3.2.1. Intraspecific gene flow. Trees seem to experience remarkably high levels of gene flow. A growing body of research indicates that pollen flow over 5 or 10 km is not uncommon, both in the tropics and in temperate settings, and for both wind- and animal-pollinated trees (**Table 2**). These large field estimates are backed by modeling studies and by investigations of pollen viability (Katul et al. 2006, Schueler et al. 2005). Similarly, regular long-distance seed dispersal events spanning several

Table 2 Examples of long distance pollen dispersal in trees inferred with genetic markers

Species	Pollination system	Genetic marker	Location	Distance	Ref.
Fraxinus excelsior (Oleaceae)	Wind	Microsatellites	Relic woodlands in Scotland	53% of successful pollination by immigrant pollen in a catchment at >10 km from other populations	1
Dinizia excelsa (Fabaceae)	Bees	Microsatellites	Manaus, Brazil	Mean pollen dispersal distance of 1.5 km; pollen transport up to 3.2 km.	2
Populus trichocarpa (Salicaceae)	Wind	Microsatellites	Pacific Northwest, USA	27% of matings from individuals located beyond 2.7 km and up to 9.8 km	3
Cecropia obtusifolia (Moraceae)	Wind	Allozymes	Southern Mexico	A population at 6 km accounted for 27% of the offspring and another at 14 km accounted for 9%	4
Ficus spp. (Moraceae)	Wasps	Allozymes	Central Panama	Pollen dispersal estimated to occur routinely over 5.8–14.2 km between widely spaced trees	5
Pinus sylvestris (Pineaceae)	Wind	Microsatellites	Central Spain	4.3% of matings with pollen from >30 km	6
Ceiba pentandra (Malvaceae)	Bats	Microsatellites	Central Amazonia	Several matings >5 km; up to 18.6 km	7
Swietenia humilis (Meliaceae)	Small butterflies, bees etc.	Microsatellites	Coastal plain, Honduras	Direct distance of pollen flow >4.5 km	8

1. Bacles et al. (2005), 2. Dick et al. (2003b), 3. DiFazio et al. (2004), 4. Kaufman et al. (1998), 5. Nason et al. (1998), 6. Robledo-Arnuncio & Gil (2005), 7. Gribel, cited in Ward et al. (2005b), 8. White et al. (2002).

kilometers have been reported (Bacles et al. 2006, Gaiotto et al. 2003, Godoy & Jordano 2001), even if the bulk of gene flow is usually mediated by pollen (Petit et al. 2005a).

These observations suggest that trees could experience comparatively more gene flow than herbs with the same mating system. Their high stature makes the world smaller for them. Large individuals necessarily grow at lower density, which implies a greater absolute distance between potential mates and increases pollen dispersal distances (Ward et al. 2005b). Frequent long-distance pollen movements should buffer tree populations against diversity loss resulting from fragmentation (Hamrick et al. 1991, White et al. 2002).

3.2.2. Interspecific gene flow. According to many botanists (e.g., Grant 1958, 1963; Stebbins 1950, 1958), long-lived woody perennials engage more readily in interspecific matings than other plants. Such comparisons are difficult, however, and few studies have attempted to quantify this trend. An exception is the review of Ellstrand et al. (1996), which shows that hybrids are more frequently detected in outcrossing perennials. But further work on this topic is clearly needed, as few tree-rich floras have been examined for the frequency of hybrids. In principle, this apparent propensity of trees to hybridize could at least partly account for the high levels of genetic diversity observed. It might also represent a means to colonize new habitats (Petit et al. 2004b).

3.2.3. Large effective population size. Contrary to large animals, trees can have huge global census sizes. For instance, European beech forests cover some 17 million ha, which should represent 1.5–2 billion mature individuals, assuming that there are around 100 adult trees per hectare (disregarding seedlings and saplings, most of which will not make it to the reproductive stage). Even those tree species found typically at low density (as is typical in the tropics) can have, in fact, fairly large global population sizes, because they belong to "predictable oligarchies that dominate several thousand square kilometers of forest" (Pitman et al. 2001). A few narrow endemic tree species do exist, but these are found either on oceanic islands or represent relicts that were historically more widespread, as in the case of some Mediterranean trees (Petit et al. 2005b). Large effective population size implies that polymorphisms can persist during extended periods of time. The recent finding of trans-species shared polymorphisms in allopatric tree species that have diverged over 13 Myr ago has been interpreted in this light (Bouillé & Bousquet 2005). More studies are needed to determine if such ancient polymorphisms are frequent in trees.

3.3. Asexual Reproduction

Whereas no selfing tree species has been described so far, trees with predominantly asexual reproduction exist, although they seem to be rare (Thomas 1997). Contrary to selfing, asexual reproduction does not expose recessive deleterious mutations to selection. At low population size, asexual reproduction might better preserve heterozygosity than outcrossing, at least in the short term. For instance, the sole case of paternal apomixis ever described in plants is for a relict cypress species in the Tassili desert, which consists of fewer than 200 adult trees (Pichot et al. 2001). Further studies are needed to test whether asexual reproduction is actually less frequent in trees than in herbs.

3.4. Chromosome Number

A high basic number of chromosomes should promote diversity through its effect on recombination (Grant 1958, 1975, p. 448). Levin & Wilson (1976) have estimated that tree genera have a mean basic chromosome number of 13.1 compared to only 9.3 for herbaceous plants. The growing number of studies that estimate linkage disequilibrium from within-species sequence data should eventually allow for the testing of possible differences between trees and other plants in the recombination parameter. The first studies point to particularly rapid decay of linkage disequilibrium in trees, with polymorphic nucleotide sites a few hundred base pairs apart often being uncorrelated (Neale & Savolainen 2004).

In principle, polyploidy should also help preserve genetic diversity by increasing the number of copies of each gene. However, no study seems to have compared its prevalence in trees and in herbs, although other correlates of polyploidy have been identified (Ramsey & Schemske 1998).

3.5. Diversifying Selection

Trees are exposed to highly heterogeneous biotic and abiotic conditions within their individual lifetimes and across their ranges. Linhart & Grant (1996) estimate that short-lived plants harbor on average 10–30 taxa of parasites and herbivores, compared to over 200 for larger long-lived species, resulting in far greater complexity of selection in the latter. These parasites and herbivores can exert different selection pressures at different life stages, from seed to seedling, juveniles, and mature stages. Although this is not unique to trees, the heterogeneity of selection pressures is exacerbated by trees' longevity and by the diversity of organisms with whom they are interacting (Boege & Marquis 2005, Linhart & Grant 1996). Similarly, extreme climatic events are likely to occur within trees' lifetimes (Gutschick & BassiriRad 2003). This allows a complex interplay of frequency-dependent, balancing or episodic selection pressures that could contribute to the maintenance of genetic diversity.

Following foundation of a new population by a single individual, a loss of diversity is expected, even in self-incompatible species. However, genetic diversity can then be quickly re-established if seeds sired by immigrant pollen have greater fitness (Richards 2000), which is another form of frequency-dependent selection. Perhaps as a consequence of this preserved store of variation, invasive populations of trees can adapt within a few generations to new conditions (Petit et al. 2004a).

Finally, and most importantly, extensive gene flow does not seem to compromise local adaptation in trees. Trees commonly combine substantial genetic differentiation at quantitative traits (Q_{ST}) with little differentiation at molecular markers (F_{ST}) (McKay & Latta 2002). Computing average Q_{ST} and F_{ST} values for allogamous herbs and trees from table 1 of McKay & Latta (2002) shows that although trees have much lower differentiation at molecular markers (F_{ST} of 0.05 versus 0.17), indicating higher gene flow among tree populations, differentiation at quantitative traits was similar in the two groups (Q_{ST} of 0.34 versus 0.35). Trees' large fecundity and the resulting strong selection of recruits could account for this observation (Le Corre & Kremer 2003).

3.6. Age at maturity

Delayed maturity in trees could dramatically reduce founder events during invasions, thereby preserving genetic diversity (Austerlitz et al. 2000). At the time when the first individuals start to reproduce, a non-negligible part of the space available for establishment will already be occupied by juveniles from seeds that arrived years before. In contrast, an annual plant colonizing an empty site can reproduce the first year and quickly fill the available space with its offspring. Everything else being equal, this should result in a much sharper loss of diversity and much greater differentiation in annuals. Simulations show that the key factor in avoiding founder effects is indeed delayed reproduction and not overlapping generations (Austerlitz et al. 2000).

4. PACE OF EVOLUTION IN TREES

Sinnott (1916) first argued that generation time should affect the rapidity of evolutionary change in trees as compared to herbs. Here we consider whether trees are indeed characterized by different mutation rates, nucleotide substitution rates, and patterns of diversification in comparison with other plants.

4.1. Mutation Rates

The large genetic diversity identified during population genetic surveys of trees has led some researchers to infer that trees have higher mutation rates than herbaceous plants (e.g., Linhart 1999). Trees' high genetic load (Klekowski 1988) seems to support this prediction. However, although trees might be expected to have higher per-generation mutation rates than other plants (because of the "chemostat-like" postzygotic accumulation of somatic mutations in the apical initials during plant growth; Klekowski & Godfrey 1989), it does not follow that they accumulate more mutations per unit of time. The arguments are as follows: (*a*) Metabolic rates seem to be lower in trees than in other plants (see Section 2.2.2). (*b*) Trees experience less recombination events per unit of time because of their longer generation time. (*c*) Assuming that mutations occur predominantly at cell division, trees should accumulate less mutations per unit of time compared to short-lived plants, because cell divisions corresponding to germination and flowering occur on a per-generation, not on a per-growth-season, basis. (*d*) Ontogenetic patterns of cell divisions could promote genomic stasis by allowing mutant cells to be eliminated (thereby compensating for the absence of an immune surveillance system capable of eliminating cells with deviant phenotypes, as is found in some animals). In trees, such ontogenetic pathways include logarithmic cell divisions and highly branched phenotypes, as well as particular patterns of branch senescence (Klekowski et al. 1989). The extension in girth of the cambium, accomplished through the initiation of new radial cell files in excess of the number needed to achieve growth, has also been interpreted as a mechanism destined to facilitate the elimination of somatic mutations from the meristematic population (Mellerowicz et al. 2001).

We therefore expect that fewer somatic mutations should get fixed per unit of time in perennials than in annuals. In other words, the per-generation increase in mutation rate in perennials would be less than predicted from their difference in generation time. This seems supported by the work of Klekowski & Godfrey (1989) who estimated that mutation rate in mangrove trees is 25 times that of annual plants, although differences in generation times would predict a mutation rate >100 times as large. Future molecular studies might provide data on somatic mutation rates in annuals versus perennials, as indicated by a few promising attempts relying on microsatellites in long-lived trees (Cloutier et al. 2003, O'Connell & Ritland 2004). Similarly, if more studies confirm that trees have lower nucleotide diversity than expected from surveys based on genetic markers (Neale & Savolainen 2004), this would support the idea that trees accumulate fewer mutations per unit of time than do other plants (see Sidebar).

4.2. Substitution Rates

Although estimates of mutation rates remain rare in trees, evidence has now accumulated that shows perennials evolve more slowly at the DNA sequence level, for chloroplast, mitochondrial, and nuclear genes, particularly at silent sites (**Table 3**). Differences can be quite pronounced, but their causes are still under discussion. Some researchers consider the generation-time effect is an unlikely explanation for plants because cells continue to divide throughout their lives and do not rest like germline cells in animals (but see Section 3.1). Another (nonexclusive) hypothesis is that substitution rates would be driven by speciation events. Rates of substitution and diversification are correlated in angiosperms (Barraclough & Savolainen 2001, Jobson & Albert 2002, Xiang et al. 2004). This might be caused by differences in population

---

ARE MUTANT CELLS SUBJECT TO POSITIVE SELECTION WITHIN THE CROWN OF TREES?

Although the diploid (or polyploid) nature of plants and the presence of stratified meristems should lower the immediate phenotypic impact of somatic mutations, there have been repeated claims that somatic variation can play a role in generating immediately selectable variation among plant parts, especially in long-lived tree species. However, the evidence is not compelling and the topic remains highly controversial (reviewed in Gill et al. 1995; see also the discussion in the *Journal of Evolutionary Biology*, 2004, volume 17, issue 6).

Table 3 Comparison of nucleotide substitution rates in long-lived versus short lived plants

Dataset	Sample size	DNA sequences	Main conclusion	Ref.
Lupinus (Fabaceae)	44 taxa	ITS	Most taxa on long branches of the phylogenetic tree are annuals	1
Sidalcea (Malvaceae)	28 taxa	ITS and ETS	Annual species have up to 7 times higher molecular evolutionary rates than perennials	2
Seed plants	43 species	*rbc*L	Annuals evolve more rapidly than perennials, especially at nonsynonymous sites, owing to a recent acceleration of substitution rates	3
Seed plants	33–63 spp.	*atp*B – *rbc*L	Significant negative correlation between substitution rates and perenniality, especially at silent sites	4
Grasses versus palms	3 spp.	*rbc*L, *Adh*, *atp*A	Grasses evolve more rapidly than palms at silent sites at all three genes corresponding to the three plant genomes: ~3.7 times for *rbc*L, ~2.5 times for *Adh*, and ~6.7 times for *atp*A	5
Lentibulariaceae	69 spp.	7 loci from the three genomes	No relationship between substitution rate and generation time	6
Angiosperms	15 spp.	*rps*3 intron	Annual taxa evolve up to 10–15 times faster than perennials for substitution and indel rates; first demonstration of differences between annuals and perennials in noncoding DNA	7
Angiosperms	24 spp. pairs	ITS1 and ITS2	Annual species evolve faster in 60% of the cases but the trend is not significant	8

1. Ainouche & Bayer (1999), 2. Andreasen & Baldwin (2001), 3. Bousquet et al. (1992), 4. Duminil, Grivet, Ollier, Jeandroz & Petit, unpublished results, 5. Eyre-Walker & Gaut (1997), 6. Jobson & Albert (2002), 7. Laroche & Bousquet (1999), 8. Whittle & Johnston (2003).

size, as speciation represents a form of bottleneck. In fact, large population sizes and extensive gene flow have been suggested as the causes of the low rates of evolution in trees (e.g., Bousquet et al. 1992). Other explanations rely on body size, which would, along with temperature, affect the rate of DNA evolution through their relation with metabolic rate (e.g., Gillooly et al. 2005; but see Reich et al. 2006).

4.3. Diversification Rates

Slow sequence evolution is often associated with morphological stasis (Barraclough & Savolainen 2001, Soltis et al. 2002), so trees are predicted to have lower diversification rates than other plants. Reduced rates of diversification in trees have been suggested long ago by comparing species richness at similar taxonomic levels, and they were explained by the generation-time effect (Sinnott 1916). Subsequently, growth form has been included in most treatments of diversification of angiosperms, along with mode of pollen and seed dispersal. An early study that confirmed a reduced rate of diversification in trees while controlling for the age of lineage is that of Levin & Wilson (1976). However, the first study that used phylogenetically independent contrasts to investigate the effects of the growth habit on diversification was conducted by Dodd et al. (1999). They found that a majority of transitions in growth form (75–84%) were from woody to herbaceous modes and that diversification in these new herbaceous lineages was consistently more rapid. There was no major exception to the rule that the change from woodiness to herbaceousness results in increased species richness. Verdú (2002) extended these analyses by focusing on woody angiosperms only and by classifying the plants according to their age at maturity (as a surrogate for generation time), while controlling for pollination and seed dispersal mode. He could confirm the relationship at all taxonomic levels considered, implying that trees have lower diversification rates than shrubs. Trees have also low rates of karyotypic evolution (an order of magnitude lower than herbs according to Levin & Wilson 1976). Tree species are therefore much older than herbaceous species (e.g., Levin & Wilson 1976, Magallón & Sanderson 2001). These prolonged species life spans imply low rates of extinction, given the low rates of speciation.

Nevertheless, tree species can appear rapidly under some circumstances, such as in islands (e.g., Baldwin & Sanderson 1998, Böhle et al. 1996). In Hawaii, most evolutionary changes are from herbaceous to woody growth forms (Price & Wagner 2004). However, several of the woody taxa there have retained characteristics of herbaceous plants such as short generation time and specialization to ephemeral habitats. There are also a number of tree genera that are relatively species-rich (*Acacia*, *Eucalyptus*, *Prunus*, *Quercus*, *Salix*...), but they generally include shrubs or treelets that might have driven the radiation. Similarly, there has been a recent report of extraordinary rapid diversification in a neotropical tree genus (*Inga*), but the corresponding species are considered to have low generation times for trees (Richardson et al. 2001), so all these examples do not contradict the rule. In fact, many widely distributed rain forest tree species appear to be of great age and to have experienced morphological stasis, as is suggested by phylogeographic studies and by comparisons of the woody flora of the New World and Old World tropics (Dick et al. 2003a).

As for substitution rates, various explanations have been proposed to explain these differences in diversification rates. Trees shape the communities and buffer their own environment, which could reduce their evolutionary rates. The observation that forest herb species experience morphological stasis (Ricklefs & Latham 1992) suggests that the stability of forest environments could contribute to the reduced extinction rates. This hypothesis deserves further investigation. High intra- and interspecific gene flow and large population sizes could also reduce the likelihood of divergence and speciation, whereas the elevated individual life span should allow trees to persist under difficult conditions, thereby reducing extinction risks arising from demographic stochasticity (Hampe & Petit 2005). On the contrary, the suggestion that increased level of within-species genetic variability promotes speciation (because it is available for conversion to species differences) does not appear to be supported by the available evidence (Avise 1977).

4.4. Tree Evolution and Biotic Interactions

Arms races between host trees and their pathogens and herbivores are expected, thereby promoting fast rates of evolution, as illustrated by the Red Queen model (Van Valen 1973). However, antagonistic organisms generally have much shorter generation times than trees and may easily evolve new features within the lifetime of their host. This asymmetry could be compensated for in various ways: (*a*) There might be differences in the amount of segregating genetic variation. For instance, trees could rely on rapid adaptive changes thanks to their increased levels of genetic diversity compared to that of their pathogens. (*b*) Trees also maintain populations of enemies of their pathogens and herbivores; in particular, mutualist microorganisms with similarly short generation times could mediate defense against antagonists, as in the case of protective ant-plant interactions corresponding to antiherbivore defenses "worn on the outside" (Heil & McKey 2003) or in fungal endophytes that limit pathogen damages (Arnold et al. 2003). (*c*) Specialized antagonists generally exert a weaker selective pressure on the host tree than vice versa (see, e.g., Benkman et al. 2003 for an elegant case study involving crossbills, which rely strongly on conifer seeds, and lodgepole pine).

In contrast to antagonistic interactions, mutualistic interactions could favor low rates of evolution in trees. The Red King model of Bergstrom & Lachmann (2003) uses a game-theoretical approach to show that the "slowest runner" can dominate the coevolutionary process. On an evolutionary timescale, slow evolution effectively ties the hands of a species, allowing it to "commit" to threats and thus "bargain" more effectively with its mutualistic partner over the course of the coevolutionary process. Because mutualistic and antagonistic relationships are often not easily differentiated under natural conditions and transitions from mutualism to antagonism may be frequent (Thompson 2005), a component of each model might apply to many real-world interactions.

The longevity of trees relative to that of their associated microorganisms could also directly select for the formation of mutualisms: Favoring tree performance benefits the microorganisms by preserving a stable environment for its offspring, especially in taxa that experience limited dispersal. This idea is very similar to the notion that spatial structure tends to favor mutualism, a well-established principle (Yamamura et al. 2004).

5. CONCLUSIONS

Trees have dominated terrestrial ecosystems for over 370 million years (Niklas 1997), a testimony to their evolutionary success. We have shown that the rapid rate of microevolution often reported in tree populations is not incompatible with their slow rate of macroevolution: Trees possess features that allow them to preserve genetic diversity during extended periods of time. This, in combination with their large juvenile population sizes, enables strong and variable selection. Such a strategy, which results in great potentials for local adaptation despite low evolutionary rates, appears to be the key to their success from an evolutionary standpoint. Yet, explaining the origin of this seemingly paradoxical evolutionary strategy proves difficult, as the effects of size, generation time, longevity, age at maturity, fecundity, and other potential explanatory factors are often difficult to tease apart. For instance, the predominantly allogamous mating system of trees have been interpreted to be the consequence of either their great longevity or their large size; similarly, the lower nucleotide substitution rates of trees have been attributed to increased generation time but also to decreased metabolic rate. Further comparative studies are needed to disentangle these factors, allowing a better understanding of the way plant evolution scales with size and longevity.

Compared with studies of short-lived herbaceous plants, a change in timescale is needed to investigate the factors that shape tree evolution: Rare and extreme events become inevitable when life span increases. They should therefore shape trees' physiology and ecology and determine their resilience as populations and species. Because the lifetime of most trees by far exceeds the professional lifetime of biologists, innovative interdisciplinary approaches are required to better understand their evolution.

SUMMARY POINTS

1. Trees are an extremely polyphyletic assemblage, but they share key characters such as great size, height, and longevity, which explain their ecological success.

2. In demographic terms, trees (and other large plants) have little in common with large animals. Most importantly, they experience much less senescence effects and their often prodigious fecundity increases continually with increasing size.

3. Trees have high levels of genetic diversity within populations but little differentiation among populations, due to their outcrossed mating system, their aptitude for extensive gene flow and diversifying selection, and their large population sizes.

4. Trees experience markedly slower mutation, nucleotide substitution, and speciation rates than other plants.

5. As a consequence, the tree growth form combines a great potential for rapid microevolution with slow rates of macroevolution.

6. Identifying the major causes of tree evolution is difficult, because potential factors such as longevity, size, or fecundity are often tightly interconnected.

FUTURE ISSUES

1. More comparative studies (including phylogenetic corrections) are required to disentangle the various factors affecting the ecology and evolution of trees.

2. Further studies on allometric scaling should help distinguishing between true compensatory adaptations to the tree habit and mere consequences of trees' size and longevity.

3. Model species with their complete genomes sequenced will provide a powerful tool for identifying the genetic mechanisms that are involved in growth form changes.

4. Further comparisons with the evolutionary consequences of size and longevity in animals should be of great interest.

ACKNOWLEDGMENTS

We thank Jean Bousquet, Santiago González Martínez, Berthold Heinze, Pedro Jordano, Antoine Kremer, Andrew Lowe, Douglas Schemske, Daniel Schoen, and Marc-André Selosse for discussions or for comments that helped improve this manuscript. Support from the European Commission and the Spanish Ministry of Science and Technology (grants A025383-ACORNDISP, REN2003-00273 and HF2005-0257) is greatly acknowledged.

LITERATURE CITED

Aarssen LW. 2005. Why don't bigger plants have proportionately bigger seeds? *Oikos* 111:199–207

Ainouche AK, Bayer RJ. 1999. Phylogenetic relationships in *Lupinus* (Fabaceae: Papilionoideae) based on internal transcribed spacer sequences (ITS) of nuclear ribosomal DNA. *Am. J. Bot.* 86:590–607

Andreasen K, Baldwin BG. 2001. Unequal evolutionary rates between annual and perennial lineages of checker mallows (Sidalcea, Malvaceae): evidence from 18S-26S rDNA internal and external transcribed spacers. *Mol. Biol. Evol.* 18:936–44

Arber A. 1928. The tree habit in angiosperms: its origin and meaning. *New Phytol.* 27:69–84

Arnold EA, Mejía LC, Kyllo D, Rojas EI, Maynard Z, et al. 2003. Fungal endophytes limit pathogen damage in a tropical tree. *Proc. Natl. Acad. Sci. USA* 100:15649–54

Ashman TL, Knight TM, Steets JA, Amarasekare P, Burd M, et al. 2004. Pollen limitation of plant reproduction: ecological and evolutionary causes and consequences. *Ecology* 85:2408–21

Austerlitz F, Mariette A, Machon N, Gouyon PH, Godelle B. 2000. Effects of colonization processes on genetic diversity: differences between annual plants and tree species. *Genetics* 154:1309–21

Avise JC. 1977. Genic heterozygosity and rate of speciation. *Paleobiology* 3:422–32

Bacles CFE, Burczyk J, Lowe AJ, Ennos RA. 2005. Historical and contemporary mating patterns in remnant populations of the forest tree *Fraxinus excelsior* L. *Evolution* 59:979–90

Bacles CFE, Lowe AJ, Ennos RA. 2006. Effective seed dispersal across a fragmented landscape. *Science* 311:628

Baldwin BG, Sanderson MJ. 1998. Age and rate of diversification of the Hawaiian silversword alliance (Compositae). *Proc. Natl. Acad. Sci. USA* 95:9402–6

Barraclough TG, Savolainen V. 2001. Evolutionary rates and species diversity in flowering plants. *Evolution* 55:677–83

Benkman CW, Parchman TL, Favis A, Siepielski AM. 2003. Reciprocal selection causes a coevolutionary arms race between crossbills and lodgepole pine. *Am. Nat.* 162:182–94

Bergstrom CT, Lachmann M. 2003. The Red King effect: when the slowest runner wins the coevolutionary race. *Proc. Natl. Acad. Sci. USA* 100:593–98

Boege K, Marquis RJ. 2005. Facing herbivory as you grow up: the ontogeny of resistance in plants. *Trends Ecol. Evol.* 20:441–48

Margin note: Shows that in mutualistic interactions, the more slowly evolving species gains a greater benefit, a counter-intuitive (and controversial) result.

Böhle UR, Hilger HH, Martin WF. 1996. Island colonization and evolution of the insular woody habit in *Echium* L. (Boraginaceae). *Proc. Natl. Acad. Sci. USA* 93:11740–45

Bouillé M, Bousquet J. 2005. Trans-species shared polymorphisms at orthologous nuclear gene loci among distant species in the conifer *Picea* (Pinaceae): implications for the long-term maintenance of genetic diversity in trees. *Am. J. Bot.* 92:63–73

Bousquet J, Strauss SH, Doerksen AH, Price RA. 1992. Extensive variation in evolutionary rates of *rbc*L gene sequence. *Proc. Natl. Acad. Sci. USA* 89:7844–48

Bradshaw AD. 1972. Some of the evolutionary consequences of being a plant. *Evol. Biol.* 5:25–47

Calvo RN, Horvitz CC. 1990. Pollinator limitation, cost of reproduction, and fitness in plants: a transition-matrix demographic approach. *Am. Nat.* 136:499–516

Carlquist S. 1974. *Island Biology*. New York: Columbia Univ. Press

Charlesworth D, Charlesworth B. 1995. Quantitative genetics in plants: the effect of the breeding system on genetic variability. *Evolution* 49:911–20

Cichón M. 1997. Evolution of longevity through optimal resource allocation. *Proc. R. Soc. London Ser. B* 264:1383–88

Clark JS, Beckage B, Camill P, Cleveland B, HilleRisLambers J, et al. 1999. Interpreting recruitment limitation in forests. *Am. J. Bot.* 86:1–16

Clarke HL. 1894. The meaning of tree life. *Am. Nat.* 28:465–72

Cloutier D, Rioux D, Beaulieu J, Schoen DJ. 2003. Somatic stability of microsatellite loci in Eastern white pine, *Pinus strobus* L. *Heredity* 90:247–52

Dick CW, Abdul-Salim K, Bermingham E. 2003a. Molecular systematic analysis reveals cryptic Tertiary diversification of a widespread tropical rain forest tree. *Am. Nat.* 162:691–703

Dick CW, Etchelecu G, Austerlitz F. 2003b. Pollen dispersal of tropical trees (*Dinizia excelsa*: Fabaceae) by native insects and African honeybees in pristine and fragmented Amazonian rainforest. *Mol. Ecol.* 12:753–64

Diego LB, Berdasco M, Fraga MF, Cañal MJ, Rodríguez R, Castresana C. 2004. A *Pinus radiata* AAA-ATPase, the expression of which increases with tree ageing. *J. Exp. Bot.* 55:1597–99

DiFazio SP, Slavov GT, Burczyk J, Leonardi S, Strauss SH. 2004. Gene flow from tree plantations and implications for transgenic risk assessment. In *Forest Biotechnology for the 21st Century*, ed. C Walter, M Carson, pp. 405–22. Kerala, India: Res. Signpost

Dodd ME, Silvertown J. 2000. Size-specific fecundity and the influence of lifetime size variation upon effective population size in *Abies balsamea*. *Heredity* 85:604–9

Dodd ME, Silvertown J, Chase MW. 1999. Phylogenetic analysis of trait evolution and species diversity variation among angiosperm families. *Evolution* 53:732–44

The first study that used phylogenetic correction methods to examine the relationship between plant growth form and species richness.

Donoghue MJ. 2005. Key innovations, convergence, and success: macroevolutionary lessons from plant phylogeny. *Paleobiology* 31:77–93

Demonstrates how key innovations, including the tree growth habit, were actually the result of the accumulation of smaller advances.

Ehrlén J, Lehtilä K. 2002. How perennial are perennial plants? *Oikos* 98:308–22

Ellstrand NC, Whitkus R, Rieseberg LH. 1996. Distribution of spontaneous plant hybrids. *Proc. Natl. Acad. Sci. USA* 93:5090–93

Eyre-Walker A, Gaut BS. 1997. Correlated rates of synonymous site evolution across plant genomes. *Mol. Biol. Evol.* 14:455–60

Falster DS, Westoby M. 2003. Plant height and evolutionary games. *Trends Ecol. Evol.* 18:337–43

FAO World Resour. 2000–2001. *Food and agricultural organization of the United Nations and Forest Stewardship Council.* http://www.fao.org

Flanary BE, Kletetschka G. 2005. Analysis of telomere length and telomerase activity in tree species of various life-spans, and with age in the bristlecone pine *Pinus longaeva. Biogerontology* 6:101–11

Franco M, Silvertown J. 1996. Life history variation in plants: an exploration of the fast-slow continuum hypothesis. *Philos. Trans. R. Soc. London Ser. B* 351:1341–48

Gaiotto FA, Grattapaglia D, Vencovsky R. 2003. Genetic structure, mating system, and long-distance gene flow in heart of palm (*Euterpe edulis* Mart.). *J. Hered.* 94:399–406

Ghazoul J. 2005. Pollen and seed dispersal among dispersed plants. *Biol. Rev.* 80:413–33

Gill DE, Chao L, Perkins SL, Wolf JB. 1995. Genetic mosaicism in plants and clonal animals. *Annu. Rev. Ecol. Syst.* 26:423–44

Gillooly JF, Allen AP, West GB, Savage VM, Brown JH. 2005. The rate of DNA evolution: effects of body size and temperature on the molecular clock. *Proc. Natl. Acad. Sci. USA* 102:140–45

Gillooly JF, Brown JH, West GB, Savage VM, Charnov EL. 2001. Effects of size and temperature on metabolic rate. *Science* 293:2248–51

Godoy JA, Jordano P. 2001. Seed dispersal by animals: exact identification of source trees with endocarp DNA microsatellites. *Mol. Ecol.* 10:2275–83

Grandtner MM. 2005. *Elsevier's Dictionary of Trees.* Volume 1: *North America.* Amsterdam: Elsevier

Grant V. 1958. The regulation of recombination in plants. *Cold Spring Harbor Symp. Quant. Biol.* 23:337–63

Grant V. 1963. *The Origin of Adaptations.* New York: Columbia Univ. Press

Grant V. 1975. *Genetics of Flowering Plants.* New York: Columbia Univ. Press

Groover AT. 2005. What genes make a tree a tree? *Trends Plant Sci.* 10:210–14

Gutschick VP, BassiriRad H. 2003. Extreme events as shaping physiology, ecology, and evolution of plants: toward a unified definition and evaluation of their consequences. *New Phytol.* 160:21–42

Hampe A, Bairlein F. 2000. Modified dispersal-related traits in disjunct populations of bird-dispersed *Frangula alnus* (Rhamnaceae): a result of its Quaternary distribution shifts? *Ecography* 23:603–13

Hampe A, Petit RJ. 2005. Conserving biodiversity under climate change: the rear edge matters. *Ecol. Lett.* 8:461–67

Hamrick JL, Godt MJ. 1989. Allozyme diversity in plant species. In *Plant Population Genetics, Breeding, and Genetic Resources*, ed. AHD Brown, MT Clegg, AL Kahler, BS Weir, pp. 43–63. Sunderland, MA: Sinauer

Hamrick JL, Godt MJW. 1996. Effects of life history traits on genetic diversity in plant species. *Philos. Trans. R. Soc. London Ser. B* 351:1291–98

Following the sequencing of the poplar genome, trees are shown to differ little from herbs at the molecular level.

Hamrick JL, Linhart YB, Mitton JB. 1979. Relationships between life history characteristics and electrophoretically detectable genetic variation in plants. *Annu. Rev. Ecol. Syst.* 10:173–200

Hamrick JL, Murawski DA. 1991. Levels of allozyme diversity in populations of uncommon Neotropical tree species. *J. Trop. Ecol.* 7:395–99

Harper JL. 1977. *Population Biology of Plants*. London: Academic

Heil M, McKey D. 2003. Protective ant-plant interactions as model systems in ecological and evolutionary research. *Annu. Rev. Ecol. Evol. Syst.* 34:425–53

Herwitz SR, Slye RE, Turton SM. 2000. Long-term survivorship and crown area dynamics of tropical rain forest canopy trees. *Ecology* 81:585–97

Hueber FM. 2001. Rotted wood-alga-fungus: the history and life of *Prototaxites* Dawson 1859. *Rev. Palaeobot. Palynol.* 116:123–58

Husband BC, Schemske DW. 1996. Evolution of the magnitude and timing of inbreeding depression in plants. *Evolution* 50:54–74

Igic B, Kohn JR. 2001. Evolutionary relationships among self-incompatibility RNases. *Proc. Natl. Acad. Sci. USA* 98:13167–71

Jobson R, Albert VA. 2002. Molecular rates parallel diversification contrasts between carnivorous plant sister lineages. *Cladistics* 18:127–36

Jónsson TH. 2002. Stature of subarctic birch in relation to growth rate, lifespan and tree form. *Ann. Bot.* 94:753–62

Katul GG, Williams CG, Siqueira M, Poggi D, Porporato A, et al. 2006. Spatial modelling of transgenic conifer pollen. In *Landscapes, Genomics, and Transgenic Conifers*, ed. CG Williams. Dordrecht: Springer-Verlag. 265 pp.

Kaufman SR, Smouse PE, Alvarez-Buylla ER. 1998. Pollen-mediated gene flow and differential male reproductive success in a tropical pioneer tree, *Cecropia obtusifolia* Bertol. (Moraceae). *Heredity* 81:164–73

Kelly D, Sork VL. 2002. Mast seeding in perennial plants: Why, how, where? *Annu. Rev. Ecol. Syst.* 33:427–47

Kerkhoff AJ, Ballantyne F. 2003. The scaling of reproductive variability in trees. *Ecol. Lett.* 6:850–56

King DA. 1990. The adaptive significance of tree height. *Am. Nat.* 135:809–28

Klekowski EJ Jr. 1988. Genetic load and its causes in long-lived plants. *Trees* 2:195–203

Klekowski EJ Jr. 1998. Mutation rates in mangroves and other plants. *Genetica* 102/103:325–31

Klekowski EJ Jr, Godfrey PJ. 1989. Ageing and mutation in plants. *Nature* 340:389–91

Klekowski EJ Jr, Kazarinova-Fukshansky N, Fukshansky L. 1989. Patterns of plant ontogeny that may influence genomic stasis. *Am. J. Bot.* 76:185–95

Knight TM, Steets J, Mitchell RJ, Johnston M, Burd M, et al. 2000. Pollen limitation of plant reproduction: pattern and process. *Annu. Rev. Ecol. Evol. Syst.* 36:467–97

Laberge MJ, Payette S, Bousquet J. 2000. Life span and biomass allocation of stunted black spruce clones in the subarctic environment. *J. Ecol.* 88:584–93

Lanner RM. 2002. Why do trees live so long? *Ageing Res. Rev.* 1:653–71

Laroche J, Bousquet J. 1999. Evolution of the mitochondrial *rps3* intron in perennial and annual angiosperms and homology to *nad5* intron 1. *Mol. Biol. Evol.* 16:441–52

An introductory overview on tree ageing and on the intrinsic mechanisms that extend life span.

Larson DW. 2001. The paradox of great longevity in a short-lived tree species. *Exp. Geront.* 36:651–73

Larson DW, Matthes U, Gerrath JA, Gerrath JM, Nekola JC, et al. 1999. Ancient stunted trees on cliffs. *Nature* 398:382–83

Le Corre V, Kremer A. 2003. Genetic variability at neutral markers, quantitative trait loci and trait in a subdivided population under selection. *Genetics* 164:1205–19

Levin DA, Wilson AC. 1976. Rates of evolution in seed plants: Net increase in diversity of chromosome numbers and species numbers through time. *Proc. Natl. Acad. Sci. USA* 73:2086–90

Linhart YB. 1999. Variation in woody plants: molecular markers, evolutionary processes and conservation biology. In *Molecular Biology of Woody Plants. For. Sci.*, ed. SM Jain, SC Minocha, 64:341–74. Dordrecht: Kluwer Acad.

Linhart YB, Grant MC. 1996. Evolutionary significance of local genetic differentiation in plants. *Annu. Rev. Ecol. Syst.* 27:237–77

Little EL Jr. 1979. *Checklist of United States Trees (Native and Naturalized)*. U.S. Dept. Agric., Agric. Handb. 541, p. 375. Washington, DC: USDA

Loehle C. 1988. Tree life history strategies: the role of defenses. *Can. J. For. Res.* 18:209–22

> A synthetic analysis of tree defenses and of their trade-offs with growth and other features.

Loehle C. 2000. Strategy space and the disturbance spectrum: a life-history model for tree species coexistence. *Am. Nat.* 156:14–33

Magallón S, Sanderson MJ. 2001. Absolute diversification rates in angiosperm clades. *Evolution* 55:1762–80

Martín-Trillo M, Martínez-Zapater JM. 2002. Growing up fast: manipulating the generation time of trees. *Curr. Opin. Biotechnol.* 13:151–55

McKay JK, Latta RG. 2002. Adaptive population divergence: markers, QTL and traits. *Trends Ecol. Evol.* 17:285–91

Mellerowicz EJ, Baucher M, Sundberg B, Boerjan W. 2001. Unravelling cell wall formation in the woody dicot stem. *Plant Mol. Biol.* 47:239–74

Mencuccini M, Martínez-Vilalta J, Vanderklein D, Hamid HA, Korakaki E, et al. 2005. Size-mediated ageing reduces vigour in trees. *Ecol. Lett.* 8:1183–90

Moles AT, Falster DS, Leishman MR, Westoby M. 2004. Small-seeded species produce more seeds per square meter of canopy per year, but not per individual lifetime. *J. Ecol.* 92:384–96

Morgan MT. 2001. Consequences of life history for inbreeding depression and mating system evolution in plants. *Proc. R. Soc. London Ser. B* 268:1817–24

Morgan MT, Schoen DJ, Bataillon T. 1997. The evolution of self-fertilization in perennials. *Am. Nat.* 150:618–38

Nason JD, Herre EA, Hamrick JL. 1998. The breeding structure of a tropical keystone plant species. *Nature* 391:685–87

Nathan R, Katul GG, Horn HS, Tomas SM, Oren R, et al. 2002. Mechanisms of long-distance dispersal of seeds by wind. *Nature* 418:409–13

Neale DB, Savolainen O. 2004. Association genetics of complex traits in conifers. *Trends Plant Sci.* 9:325–30

Niklas KJ. 1997. *The Evolutionary Biology of Plants*. Chicago: Univ. Chicago Press

> A classic book on plant evolution that includes an extended discussion on selective advantages of the tree habit.

Niklas KJ, Enquist BJ. 2003. An allometric model for seed plant reproduction. *Evol. Ecol. Res.* 5:79–88

Nybom H. 2004. Comparison of different nuclear DNA markers for estimating intraspecific genetic diversity in plants. *Mol. Ecol.* 13:1143–55

Obeso JR. 2002. The costs of reproduction in plants. *New Phytol.* 155:321–48

O'Connell LM, Ritland K. 2004. Somatic mutations at microsatellite loci in western redcedar (*Thuja plicata*: Cupressaceae). *J. Hered.* 95:172–76

Okubo A, Levin SA. 1989. A theoretical framework for data analysis of wind dispersal of seeds and pollen. *Ecology* 70:329–38

Oldfield S, Lusty C, MacKinven A. 1998. *The World List of Threatened Trees*. Cambridge, UK: World Conserv. Press

Ordóñez JL, Retana J, Espelta JM. 2005. Effects of tree size, crown damage, and tree location on postfire survival and cone production of *Pinus nigra* trees. *For. Ecol. Manag.* 206:109–17

Ozanne CMP, Anhuf D, Boulter SL, Keller M, Kitching RL, et al. 2003. Biodiversity meets the atmosphere: global view of forest canopies. *Science* 301:183–87

Petit RJ, Bialozyt R, Garnier-Géré P, Hampe A. 2004a. Ecology and genetics of tree invasions: from recent introductions to Quaternary migrations. *For. Ecol. Manag.* 197:117–37

> An extensive review of the genetic consequences of trees' invasions providing examples of rapid evolution in new environments.

Petit RJ, Bodénès C, Ducousso A, Roussel G, Kremer A. 2004b. Hybridization as a mechanism of invasion in oaks. *New Phytol.* 161:151–64

Petit RJ, Duminil J, Fineschi S, Hampe A, Salvini D, Vendramin GG. 2005a. Comparative organisation of chloroplast, mitochondrial and nuclear diversity in plant populations. *Mol. Ecol.* 14:689–701

Petit RJ, Hampe A, Cheddadi R. 2005b. Climate changes and tree phylogeography in the Mediterranean. *Taxon* 54:877–85

Pichot CE, Maâtaoui M, Raddi S, Raddi P. 2001. Surrogate mother for endangered *Cupressus*. *Nature* 412:39

Pitman NCA, Terborgh JW, Silman MR, Núñez P, Neill DA, et al. 2001. Dominance and distribution of tree species in upper Amazonian terra firme forests. *Ecology* 82:2101–17

Portnoy S, Willson MF. 1993. Seed dispersal curves—behavior of the tails of the distribution. *Evol. Ecol.* 7:25–44

Price JP, Wagner WL. 2004. Speciation in Hawaiian angiosperm lineages: cause, consequence, and mode. *Evolution* 58:2185–200

Ramsey J, Schemske DW. 1998. Pathways, mechanisms and rates of polyploid formation in flowering plants. *Annu. Rev. Ecol. Syst.* 29:477–501

Reich PB, Tjelker MG, Machado JL, Oleksyn J. 2006. Universal scaling of respiratory metabolism, size and nitrogen in plants. *Nature* 439:457–61

Richards C. 2000. Inbreeding depression and genetic rescue in a plant metapopulation. *Am. Nat.* 155:383–94

Richardson JE, Pennington RT, Pennington TD, Hollingsworth PM. 2001. Rapid diversification of a species-rich genus of neotropical rain forest trees. *Science* 293:2242–45

Ricklefs RE, Latham RE. 1992. Intercontinental correlation of geographic ranges suggests stasis in ecological traits of relict genera of temperate perennial herbs. *Am. Nat.* 139:1305–21

Robledo-Arnuncio JJ, Gil L. 2005. Patterns of pollen dispersal in a small population of *Pinus sylvestris* L. revealed by total-exclusion paternity analysis. *Heredity* 94:13–22

Rowe N, Speck T. 2005. Plant growth forms: an ecological and evolutionary perspective. *New Phytol.* 166:61–72

Schoen DJ, Brown AHD. 1991. Intraspecific variation in population gene diversity and effective population size correlates with the mating system in plants. *Proc. Natl. Acad. Sci. USA* 88:4494–97

Schueler S, Schlünzen KH, Scholz F. 2005. Viability and sunlight sensitivity of oak pollen and its implications for pollen-mediated gene flow. *Trees* 19:154–61

Scofield DG, Schultz ST. 2006. Mitosis, stature and evolution of plant mating systems: low-Φ and high-Φ plants. *Proc. R. Soc. London Ser. B* 273:275–82

> A provocative paper arguing that plants' mating system is affected by their size, not by their generation time.

Scotland RW, Wortley AH. 2004. How many species of seed plants are there? *Taxon* 52:101–4

Selosse M-A. 2002. *Prototaxites*: a 400 Myr old giant fossil, a saprophytic holobasidiomycete, or a lichen? *Mycol. Res.* 106:642–44

Silvertown J, Franco M, Ishiwara RP. 2001. Evolution of senescence in iteroparous perennial plants. *Evol. Ecol. Res.* 3:393–412

Sinnott EW. 1916. Comparative rapidity of evolution in various plant types. *Am. Nat.* 50:466–78

> The first paper that explicitly links generation time and evolutionary rate, overlooked during the topic's rediscovery in the 1980s.

Soltis PS, Soltis DE, Savolainen V, Crane PR, Barraclough TG. 2002. Rate heterogeneity among lineages of tracheophytes: Integration of molecular and fossil data and evidence for molecular living fossils. *Proc. Natl. Acad. Sci. USA* 99:4430–35

Sorensen FC. 1999. Relationship between self-fertility, allocation of growth, and inbreeding depression in three coniferous species. *Evolution* 53:417–25

Sorensen FC. 2001. Effect of population outcrossing rate on inbreeding depression in *Pinus contorta* var. *murrayana* seedlings. *Scand. J. For. Res.* 16:391–403

Stebbins GL. 1950. *Variation and Evolution in Plants*. London: Oxford Univ. Press

Stebbins GL. 1958. Longevity, habitat, and release of genetic variability in the higher plants. *Cold Spring Harbor Symp. Quant. Biol.* 23:365–78

Stebbins GL. 1974. *Flowering Plants: Evolution Above the Species Level*. Cambridge, MA: Harvard Univ. Press

Szabó I. 1931. The three types of mortality curve. *Q. Rev. Biol.* 6:462–63

Thomas P. 2000. *Trees: Their Natural History*. Cambridge, UK: Cambridge Univ. Press

Thomas SC. 1997. Geographic parthenogenesis in a tropical forest tree. *Am. J. Bot.* 84:1012–15

Thompson JN. 2005. *The Geographic Mosaic of Coevolution*. Chicago: Univ. Chicago Press

Turnbull LA, Crawley MJ, Rees M. 2000. Are plant populations seed-limited? A review of seed sowing experiments. *Oikos* 88:225–38

Vamosi JC, Vamosi SM. 2004. The role of diversification in causing the correlates of dioecy. *Evolution* 58:723–31

Van Valen L. 1973. A new evolutionary law. *Evol. Theory* 1:1–30

Van Valen L. 1975. Life, death, and energy of a tree. *Biotropica* 7:260–69

Verdú M. 2002. Age at maturity and diversification in woody angiosperms. *Evolution* 56:1352–61

Ward JK, Harris JM, Cerling TE, Wiedenhoeft A, Lott MJ, et al. 2005a. Carbon starvation in glacial trees recovered from the La Brea tar pits, southern California. *Proc. Natl. Acad. Sci. USA* 102:690–94

Ward M, Dick CW, Gribel R, Lowe AJ. 2005b. To self, or not to self... A review of outcrossing and pollen-mediated gene flow in neotropical trees. *Heredity* 95:246–54

Wardle DA, Walker LR, Bardgett RD. 2004. Ecosystem properties and forest decline in contrasting long-term chronosequences. *Science* 305:509–13

Westoby M, Falster DS, Moles AT, Vesk PA, Wright IJ. 2002. Plant ecological strategies: some leading dimensions of variation between species. *Annu. Rev. Ecol. Syst.* 33:125–59

White GM, Boshier DH, Powell W. 2002. Increased pollen flow counteracts fragmentation in a tropical dry forest: An example from *Swietenia humilis* Zuccarini. *Proc. Natl. Acad. Sci. USA* 99:2038–42

Whittaker RH. 1975. *Communities and Ecosystems*. London: Macmillian. 2nd ed.

Whittle CA, Johnston MO. 2003. Broad-scale analysis contradicts the theory that generation time affects molecular evolutionary rates in plants. *J. Mol. Evol.* 56:223–33

Williams GC. 1957. Pleiotropy, natural selection and the evolution of senescence. *Evolution* 11:393–411

Williams CG, Savolainen O. 1996. Inbreeding depression in conifers: Implications for breeding strategy. *For. Sci.* 42:102–17

Xiang QY, Zhang WH, Ricklefs RE, Qian H, Chen ZD, et al. 2004. Regional differences in rates of plant speciation and molecular evolution: a comparison between eastern Asia and eastern North America. *Evolution* 58:2175–84

Yamamura N, Higashi M, Behera N, Wakano JY. 2004. Evolution of mutualism through spatial effects. *J. Theor. Biol.* 226:421–28

RELATED REVIEWS

Bateman RM, Crane PR, DiMichele WA, Kenrick PR, Rowe NP, et al. 1998. Early evolution of land plants: phylogeny, physiology, and ecology of the primary terrestrial radiation. *Annu. Rev. Ecol. Syst.* 29:263–92

Byers DL, Waller DM. 1999. Do plant populations purge their genetic load? Effects of population size and mating history on inbreeding depression. *Annu. Rev. Ecol. Syst.* 30:479–513

Charlesworth B, Charlesworth D. 1989. Inbreeding depression and its evolutionary consequences. *Annu. Rev. Ecol. Syst.* 18:237–68

Loveless MD, Hamrick JL. 1984. Ecological determinants of genetic structure in plant populations. *Annu. Rev. Ecol. Syst.* 15:65–95

Mindell DP, Thacker CE. 1996. Rates of molecular evolution: phylogenetic issues and applications. *Annu. Rev. Ecol. Syst.* 27:279–303

Tiffney BH. 2004. Vertebrate dispersal of seed plants through time. *Annu. Rev. Ecol. Evol. Syst.* 35:1–29

Late Quaternary Extinctions: State of the Debate

Paul L. Koch[1] and Anthony D. Barnosky[2]

[1]Department of Earth and Planetary Sciences, University of California, Santa Cruz, California 95064; email: pkoch@pmc.ucsc.edu

[2]Department of Integrative Biology and Museums of Paleontology and Vertebrate Zoology, University of California, Berkeley, California 74720; email: barnosky@berkeley.edu

Key Words

mammal, overkill, paleoclimate, paleoecology, Pleistocene

Abstract

Between fifty and ten thousand years ago, most large mammals became extinct everywhere except Africa. Slow-breeding animals also were hard hit, regardless of size. This unusual extinction of large and slow-breeding animals provides some of the strongest support for a human contribution to their extinction and is consistent with various human hunting models, but it is difficult to explain by models relying solely on environmental change. It is an oversimplification, however, to say that a wave of hunting-induced extinctions swept continents immediately after first human contact. Results from recent studies suggest that humans precipitated extinction in many parts of the globe through combined direct (hunting) and perhaps indirect (competition, habitat alteration) impacts, but that the timing and geography of extinction might have been different and the worldwide magnitude less, had not climatic change coincided with human impacts in many places.

INTRODUCTION

Fifty thousand years ago, ecosystems around the globe were populated with large animals that are now extinct. On continents worldwide, about 90 genera of mammals weighing ≥44 kg disappeared (see **Supplemental Table 1**; follow the Supplemental Material link from the Annual Reviews home page at **http://www.annualreviews.org/**). North America had proboscideans, giant ground sloths, camels, saber-tooth cats, and a giant beaver, among others. In Eurasia, woolly mammoth and rhinoceros, and giant deer with antlers spanning 3 m were common. South America hosted the car-sized glyptodont and the three-toed litoptern, which resembled a horse with a camel's neck and a short elephantine trunk. In Australia, the rhinoceros-like *Diprotodon*, the largest marsupial that ever lived, coexisted with hog-sized wombats. Timing varied across the globe, but by 10,000 years ago, these animals had vanished except at very high latitudes, on islands (where the extinction of large and small animals was more recent), and in Africa (where many large animals survive today) (Martin & Steadman 1999).

This late Quaternary extinction (LQE) was recognized by the nineteenth century, when explanations included climatic catastrophes, gradual climate change, and overkill by human hunters (Grayson 1984). The debate took on new life after the revolution in ^{14}C dating, as Martin and colleagues began to articulate and test more explicit overkill hypotheses (Martin 1966, 2005; Mosimann & Martin 1975). Growing knowledge of shifts in Quaternary climate, vegetation, and animal communities, as well as doubts about the capabilities and impacts of human hunters, led others to offer more focused environmental extinction hypotheses (Guilday 1967, Slaughter 1967). Some argued for a "one-two punch" combining human impacts and climatic changes (Barnosky 1989, Haynes 2002a). The debate has produced a vast literature spanning archaeology, (paleo)ecology, and climatology. **Table 1** summarizes the main hypotheses advanced and explored over the past 25 years to explain the LQE.

Here, we offer a continent-by-continent summary of the LQE, focusing mostly on mammals. We do not discuss Holocene island extinctions in any detail because most researchers accept that anthropogenic factors were pivotal. We evaluate the main hypotheses proposed to explain the LQE and focus on the impacts on herbivores, as their extinction by any means would precipitate extinctions among carnivores and scavengers. We define megafauna as animals weighing ≥44 kg, "large" as animals between 44 kg and 10 kg, and "small" as animals <10 kg. Counts of extinct versus extant genera are based on our vetting of the literature (Barnosky et al. 2004b). Species counts and body-size data follow Smith et al. (2003), with minor additions from Brook & Bowman (2004) and Johnson (2002). Ages of events are reported in units of 1000 calendar years before present (kyr BP). Ages reported in ^{14}C years were converted to calendar years using CALIB 4.4 for ages between 0 and 20,050 years BP (Stuiver et al. 1998) and the GISP2-tuned equation from Bard et al. (2004) for ages between 20,050 and 45,000 years BP.

LQE: late Quaternary extinction

Overkill hypothesis: hypothesis that extinction results because human hunting causes death rates to exceed birth rates in prey species

Megafauna: animals with a body mass greater than 44 kg

kyr BP: 1000 years before present

Table 1 Hypotheses to explain the late Quaternary extinctions

Type or name	Description
Environmental hypotheses	
Catastrophes	Megadrought, rapid cooling, bolide impact?
Habitat loss	Preferred habitat types lost or too fragmented
Mosaic-nutrient hypothesis	Loss of floras with high local diversity
Co-evolutionary disequilibrium	Disruption of coevolved plant-animal interactions due to flora rearrangement
Self-organized instability	Collapse of system due to intrinsic dynamics
Human impacts other than hunting	
Habitat alteration	Loss or fragmentation of viable habitat due to human impacts, including fires
Introduced predators	Direct predation by dogs, rats, cats, pigs, etc.
Hyper-disease	Introduction of virulent diseases
Overkill hypotheses	
Blitzkrieg	Rapid loss of prey due to overhunting
Protracted overkill	Loss of prey after prolonged interaction with predator
Combined hypotheses	
Keystone megaherbivores	Ecosystem collapse due to loss of landscape altering megaherbivores, perhaps with increase in fire
Prey-switching	Nonhuman carnivores switch prey as humans usurp preferred prey
Predator avoidance	Herbivores restricted to inviable refugia

Modified from Burney & Flannery (2005).

WHAT HAPPENED?

In North America south of Alaska, 34 Pleistocene genera of megafaunal mammals did not survive into the Holocene; two mammalian orders (Perissodactyla, Proboscidea) were eliminated completely (**Table 2**, **Supplemental Table 1**). Six of the extinct genera survived elsewhere, and within genera that survived, species were lost. Besides the megafauna, large and small mammal genera became extinct, including an antilocaprid (*Capromeryx*), a canid (*Cuon*, an Asian survivor), a skunk (*Brachyprotoma*), and a rabbit (*Aztlanolagus*). At the species level, the extinction was total for mammals larger than 1000 kg, greater than 50% for size classes between 1000 and 32 kg, and 20% for those between 32 and 10 kg (**Figure 1**, **Supplemental Table 2**). There were also cases of pseudoextinction at the species level. For example, genetic analyses suggest that Holocene bison from south of the Laurentide ice sheet (*Bison bison*) are descendents of larger, morphologically distinct Pleistocene bison from the region (*Bison antiquus*) (Shapiro et al. 2004). Robust temporal brackets place the last occurrences for 15 genera between 15.6 and 11.5 kyr BP (**Figure 2**, discussion of dates is based on review by Barnosky et al. 2004b unless noted).

The South American LQE was even more profound, with the loss of 50 megafaunal genera (~83%, **Table 2**). Three orders of mammals disappeared (Notoungulata,

Figure 1

Percent extinction by body size class for five continents. Data are the percent of Pleistocene species that became extinct in the late Quaternary. Data for the histogram and a description of data sources are provided in **Supplemental Table 2**.

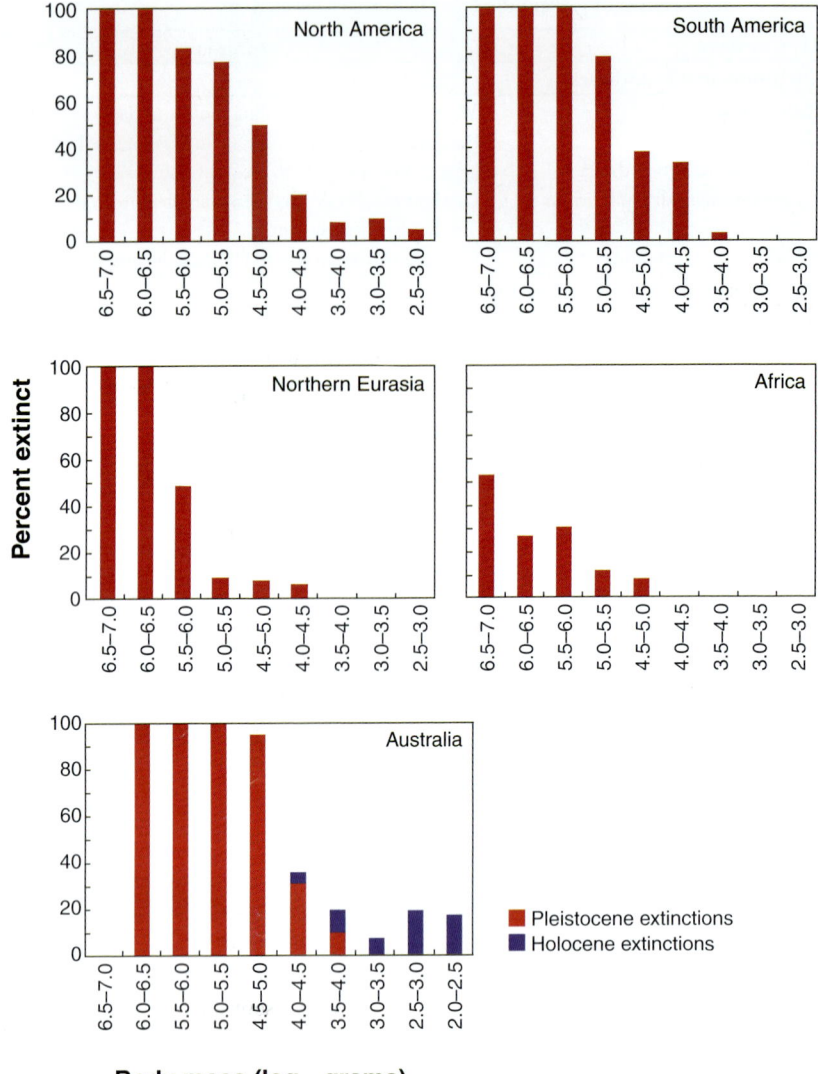

Proboscidea, Litopterna), as did all megafaunal xenarthrans. At the species level, the extinction was total for mammals larger than 320 kg, was high for the size class between 320 and 100 kg, and was moderate for size classes between 100 and 10 kg (**Figure 1**, **Supplemental Table 2**). The timing of extinction is poorly constrained. Most taxa are only dated biostratigraphically as members of the Lujanian South American Land Mammal Age (100 to 11.5 kyr BP) (**Supplemental Table 1**). Many reported ^{14}C dates do not pass rigorous criteria for accepting dates (Grayson 1991), including reports that suggest survival of South American megafauna into the Holocene, which

Table 2 Continental extinctions of mammalian megafauna at the generic level

	Number globally extinct genera	Number extinct genera surviving on other continents	Number Holocene survivors	% extinct
Africa	7	3	38	21
Australia[a]	14	—	2	88
Eurasia[b]	4	5	17	35
North America[c]	28	6	13	72
South America	48	2	10	83

[a]Australia also has seven extinct (and no surviving) genera of megafaunal reptiles and birds.
[b]Eurasia encompasses only northern Asia, because insufficient data exist to include southern Asia.
[c]Table does not account for the late survival of mammoths on Bering Sea islands.

we therefore consider equivocal. Robust dates indicate the survival of equids (*Hippidion*, *Equus*), ground sloths (*Eremotherium*, *Mylodon*, *Nothrotherium*) (Steadman et al. 2005), and *Cuvieronius* (a proboscidean) into the latest Pleistocene (15.6–11.5 kyr BP). It is likely that many other taxa were also lost near the Pleistocene-Holocene boundary, but we emphasize that this remains to be demonstrated.

Northern Eurasia (Europe and northern Asia) lost 9 genera (35%) (**Table 2**, **Supplemental Table 1**). It is unclear if Pleistocene forms of bison and camel became extinct or survived by evolution into Holocene species (Stuart 1991). Comprehensive dating campaigns in Eurasia and the extension of Eurasian steppe biomes in Alaska and the Yukon reveal that the extinction occurred in two pulses (Guthrie 2003, 2004; MacPhee et al. 2002; Stuart et al. 2002, 2004). In Eurasia, warm-adapted megafauna that were abundant during prior interglacials (straight-tusked elephants, hippos) became extinct between 48.5 and 23.5 kyr BP. In Alaska and the Yukon, hemionid horses and short-faced bears disappeared at 35.4 and 24.8 kyr BP, respectively. A coalescence analysis of DNA sequences suggests that bison populations began to decline in size in Beringia at roughly this time (e.g., 35–40 kyr BP) and attributed bison demise to climate change rather than human impacts (Shapiro et al. 2004). However, different treatment of the same data by some of the same researchers contradicts this conclusion (Drummond et al. 2005), estimating instead a severe bottleneck coincident with earliest evidence for abundant humans in Beringia around 10 kyr BP. The second pulse of extinctions began in the latest Pleistocene and hit cold-adapted animals. Mammoths dropped in abundance across Eurasia and Alaska after 14 kyr BP, but survived into the Holocene on the Taimyr Penisula and the Wrangell and Pribilof Islands. Giant Irish deer dropped in abundance and began to dwarf about 13 kyr BP before disappearing from Europe and Siberia in the Holocene. Muskox appear to have reduced genetic diversity sometime between 18 and 10 kyr BP, and became extinct in Eurasia in the Holocene (MacPhee et al. 2005). In Alaska, caballoid horses began to dwarf at 29.2 kyr BP and became extinct 15 kyr BP. While the largest megafaunal size classes were eliminated by the LQE, the impact on mammals in the size range from 1000 to 100 kg was less than in the Americas (**Figure 1**, **Supplemental Table 2**).

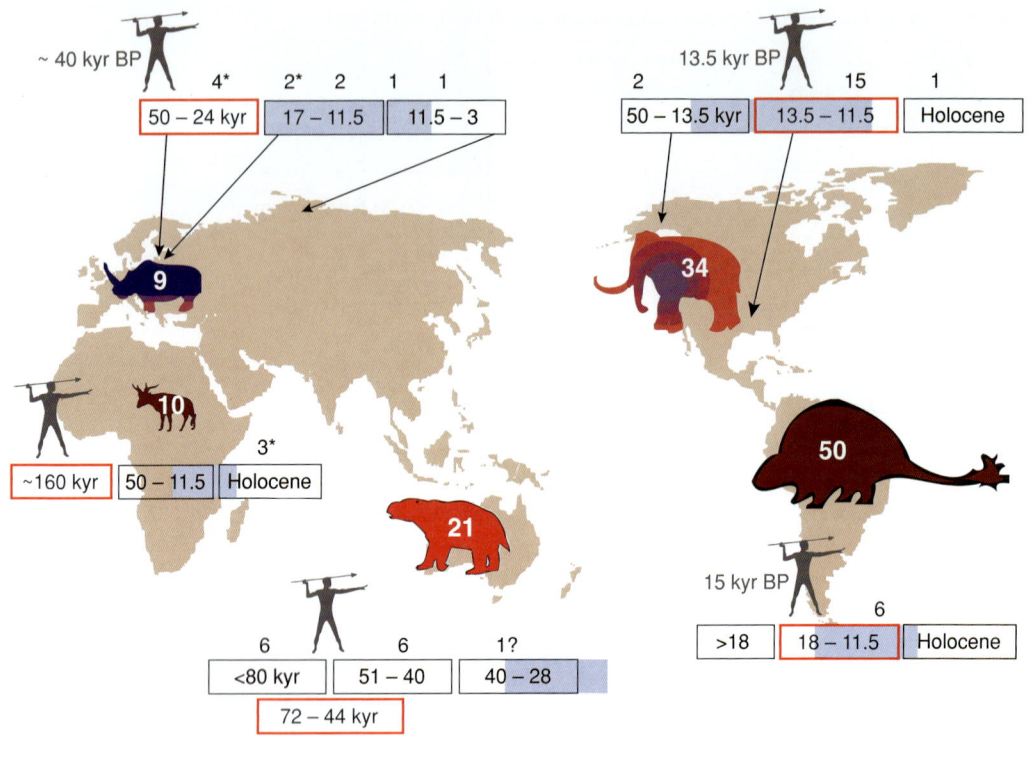

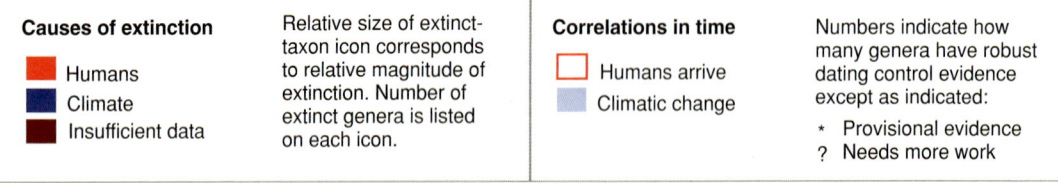

Figure 2

Chronology of the late Quaternary extinction, climate change, and human arrival on each continent. The timing of extinction for each genus was judged as robust or provisional based on previous publications that evaluated quality of dates (see Barnosky et al. 2004b). For humans, the earliest generally accepted arrival of *Homo sapiens sapiens* is indicated. Pre-*sapiens* hominins were present in Eurasia and Africa much earlier.

Australia lost 14 of its 16 genera of Pleistocene mammalian megafauna along with all megafaunal reptiles (6 genera) and *Genyornis*, a giant flightless bird (Flannery & Roberts 1999) (**Table 1**, **Supplemental Table 1**). Three marsupial families became extinct (diprotodontids, palorchestids, thylacoleonids), and the extinction removed entire guilds, such as megafaunal browsers and carnivores. The LQE was catastrophic for large body sizes, with the complete loss of all animals larger than 100 kg, almost total extinction for animals between 100 and 32 kg, and lesser impacts on smaller size classes. A recent analysis of dates of last occurrence included only articulated skeletons, assuming that such carcasses are unlikely to be mixed from other deposits (Roberts et al. 2001). Of the 21 extinct megafaunal genera, 12 lasted to 80 kyr BP, and at least 6 persisted to between 51 kyr BP and 40 kyr BP (**Figure 2**). Yet disarticulated skeletons of megafauna occur at many sites younger than 40 kyr BP, especially those with signs of human occupation (Johnson 2005). An especially contested site is Cuddie Springs, where lithic artifacts and extinct megafauna are in firm association in sediments dated between 40 kyr BP and 32 kyr BP (Trueman et al. 2005, Wroe et al. 2004). The scarcity of articulated carcasses in younger nonanthropogenic contexts may be evidence that megafaunal abundance declined at ∼40 kyr BP, with ultimate extinction occurring at a more recent date (Johnson 2005).

Africa is a fortunate anomaly in this story. Although 10 genera of Pleistocene megafauna (21%) disappeared, extinction of 7 of these cannot be bracketed tighter than in the last 100 kyr BP, and three went extinct in the Holocene (**Table 1**, **Supplemental Table 1**, **Figure 2**). Yet even in Africa, the species-level extinction was most intense for larger megafauna (40% to 50% for mammals between 10,000 and 1000 kg and moderate for those between 1000 and 320 kg) (**Figure 1**).

WAS THE LATE QUATERNARY EXTINCTION UNUSUAL?

To evaluate hypotheses for the LQE, the event must be viewed in context. Early work on extinction dynamics in North America, which has the best synthesized Cenozoic mammal record, offered little support for the idea that the LQE was exceptionally intense, though most studies noted that larger mammals were severely affected (Barnosky 1989, Gingerich 1984). However, recent analyses that account for temporal and sampling biases in the fossil record recognize a LQE spike that exceeds all but one other extinction in the past 55 million years (Alroy 1999). Even taking into account that larger animals are preferentially impacted by extinction events, and that smaller species are increasingly affected as extinction intensity rises, by Alroy's metrics the LQE appears more selective than any extinction event in the preceding 65 million years of mammalian history in North America.

Studies of Cenozoic extinction dynamics are not available for other continents, but the size selectivity of the LQE has been examined. Before the extinction, body size distributions were bimodal and similar in North America, South America, and Africa, whereas Australia had a right-skewed distribution (Lyons et al. 2004a). On all four continents, however, extinct species were significantly larger than surviving species, and for the Americas and Australia, almost the entire large mode or tail of the distribution was removed. The probability of extinction increases with body mass, without a strong threshold in mass separating extinct and extant mammals (**Figure 1**, **Supplemental Table 2**) (see also Brook & Bowman 2004).

ENVIRONMENTAL HYPOTHESES

Description of Environmental Hypotheses

The LQE roughly coincided with the most recent glacial-interglacial transition, leading some to conclude that extinction was due to environmental change. Early hypotheses calling for a sudden climatic catastrophe (deep freeze, drought) have been discredited (Grayson 1984). The more lasting hypotheses focus on the ecological effects of climate change that would lead to extinction: habitat loss hypotheses, the mosaic-nutrient hypothesis, coevolutionary disequilibrium, and self-organized instability. Habitat loss hypotheses (HLH) argue that as climate changed, areas with adequate conditions to maintain megafauna either disappeared entirely or became too small and/or fragmented to support viable populations (e.g., Barnosky 1986, Ficcarelli et al. 2003, King & Saunders 1984). The mosaic-nutrient hypothesis (MNH) is a special case of habitat loss. It argues that climate change reduced the growing season and local plant diversity, and also increased plant antiherbivore defenses, all of which reduced the carrying capacity for herbivores (Guthrie 1984). HLHs, including the MNH, tend to offer regionally and taxonomically specific explanations for extinction; they are difficult to generalize to explain the broad pattern of extinction across diverse habitats and many clades. Co-evolutionary disequilibrium (CED) is a more general hypothesis. It posits that the high herbivore diversity of Pleistocene ecosystems was maintained by extensive resource partitioning, analogous to the grazing succession of modern African savannas, and that an extremely rapid glacial-interglacial transition reorganized floras, disrupting this tightly coevolved system (Graham & Lundelius 1984). Self-organized instability (SOI) argues that extinction results from a slight perturbation that is amplified into a catastrophe by dynamics intrinsic to complex, multicomponent ecosystems with interacting subunits (Forster 2004). The key to extinction under HLH, MNH, CED, and to some extent SOI, is the claim that the last glacial-interglacial transition was unusually large and unusually rapid relative to earlier glacial-interglacial transitions, too fast for animal adaptation and/or redistribution in climate space.

Habitat loss hypotheses (HLH): as climate changed, areas with adequate conditions to maintain megafauna either disappeared or became too small/fragmented to support populations

Mosaic-nutrient hypothesis (MNH): climate change reduced growing season and plant diversity, and increased plant defenses, reducing herbivore carrying capacity and leading to extinction

Co-evolutionary disequilibrium (CED): disruption of extensive Pleistocene networks of resource partitioning by rapid glacial-interglacial transition led to extinction

Self-organized instability (SOI): extinction results from a slight perturbation amplified into a catastrophe by dynamics intrinsic to complex, multicomponent ecosystems with interacting subunits

Plausibility of Environmental Hypotheses

The response of the modern biota to anthropogenic perturbations offers insights into the plausibility of environmental hypotheses. Anthropogenic warming has led to changes in seasonal activities, geographic range, local abundance, species composition, phenotype, and genotype (Root et al. 2003, Walther et al. 2002). Most cases involve impacts on taxa of lower trophic level and smaller body sizes. However, climate change has precipitated geographic range and relative abundance shifts in large mammals (e.g., Post & Forchhammer 2002). Global extinctions of some smaller animals are attributed to current warming (Pounds et al. 2006), and the experiment is still in progress, with simulations suggesting that substantial extinction may result (Thomas et al. 2004).

The effects of habitat fragmentation and loss have been studied to understand controls on the diversity and viability of species and populations. A loss of species diversity with habitat fragmentation is often observed in natural and simulated systems (Crooks 2002, Lindenmayer et al. 2000, Wahlberg et al. 1996). As diversity drops, some species irrupt following release from predation or competition, which can itself lead to ecological reorganization (Terborgh et al. 2001). Organisms at high trophic levels, or which occur at low abundance, and those with low rates of dispersal, are susceptible to local extinction with habitat fragmentation. Herbivorous megafauna possess one trait predisposing them to extinction by habitat fragmentation (low abundance), but often exhibit pronounced dispersal and migration.

The response of faunas to climate change on islands that lack humans offers a means to assess the plausibility of environmental hypotheses. New Zealand has a rich Pleistocene paleontological record, and there is ample evidence for strong climatic and environmental changes across the Pleistocene-Holocene boundary. Yet the diverse avifauna of New Zealand, which included megafauna such as moas and more abundant smaller birds, persisted through these climate shifts only to collapse in the late Holocene following first contact with humans and rats (Holdaway 1999). A similar pattern is emerging for extinct sloths in the West Indies (Steadman et al. 2005). On the other hand, in at least two cases, islands without humans record the demise of megafaunal populations. Giant deer on Ireland became extinct coincident with a major shift in vegetation, pointing to habitat loss as the cause of local extinction (Barnosky 1986). Guthrie (2004) argued that postglacial sea-level rise made Bering Sea islands too small to sustain mammoth populations.

In summary, climate and habitat change affect geographic ranges and population density, and so plausibly may affect the extinction vulnerability of animals. On some islands, environmental change probably caused extinction. It is not clear, however, which megafaunal species would be most susceptible to specific environmental changes, or how population dynamics in susceptible species would translate into extinction in the late Quaternary. Contradicting the plausibility of environmental causes for extinction are island cases in which megafaunal extinction clearly followed first human contact, but did not result from earlier environmental change (Burney et al. 2004, Holdaway 1999).

Are the Assumptions of Environmental Hypotheses Correct?

Dietary assumptions about Pleistocene herbivores under environmental hypotheses can be tested with stable isotope data (Koch et al. 1994). Carbon isotope data can discriminate between diets rich in C_3 plants (trees, shrubs, cool-climate grasses) and those rich in C_4 plants (warm-climate grasses). In addition, animals feeding deep under closed canopy forests may have especially low carbon isotope values. The strong resource partitioning among herbivores assumed under CED is supported by isotopic studies from the United States. (**Supplemental Table 3**). For example, the three North American proboscidean genera all had very different diets, as did the three camelids. In contrast, proboscideans in South America seem to have had less specialized diets containing a broad mix of both C_3 and C_4 plants (Prado et al. 2005), which is in line with the dietary assumptions of the MNH, but does not support the assumptions of CED.

Other key assumptions of the CED, MNH, and HLH are unverified. These hypotheses assume that local plant diversity was higher in Pleistocene biomes, and that Holocene biomes were more homogeneous with vast tracts of forest or grassland or tundra rather than more integrated biomes with a mix of vegetation types. Compelling pollen or macrofloral evidence that Pleistocene biomes had higher alpha diversity than Holocene biomes is lacking. In some regions, such as the eastern United States, Alaska, and northern Eurasia, there is evidence that Holocene forests replaced the more open vegetation that dominated under full and late glacial climates. However, robust paleoecological reconstructions of the structure of Pleistocene biomes are in their infancy (e.g., Williams et al. 2000). As this work proceeds, vertebrate paleoecologists should push for inferences about changes in plant alpha diversity, functional type, growing season, antiherbivore defenses, and net primary production.

LGM: last glacial maximum

Chronologic Tests of Environmental Hypotheses

Given the unusual nature of the LQE, most environmental hypotheses require that climate or ecosystem structure at the time of extinction was unusual relative to conditions earlier in the Pleistocene. Therefore it is important to look beyond the Last Glacial Maximum (LGM), recognizing that Pleistocene faunas survived many earlier events only to succumb during the most recent glacial-interglacial transition. Barnosky et al. (2004b) offered a more complete discussion of climatic and vegetation changes associated with the LQE; we summarize their conclusions here.

Benthic marine oxygen isotope records, which chiefly record the waxing and waning of continental ice sheets (Raymo 1994), show that the most recent deglaciation was neither more rapid nor of greater magnitude than other shifts in the past million years (**Figure 3**). The Holocene is not (yet) the most extreme deglaciation; the LGM, while extreme, was similar to several prior glacial advances. Climate shifts around the Americas and Eurasia at the time of extinction, as measured by continuous records of ocean surface temperature, were large but not unusual (**Figure 3**). Even if one accepts the younger dates for the LQE in Australia, it preceded the LGM, at a time when ocean records are almost invariant. There is evidence from Australian lake records that the LQE occurred during a prolonged arid interval (Magee et al. 2004), but it is not yet clear if this aridity was unusual relative to earlier time intervals. Finally, Johnson (2005) has noted that late, contested megafaunal sites occur in a range of habitats; they are not restricted to more mesic, coastal sites as expected if aridity in the Australian interior was the dominant factor driving extinction.

Multivariate analysis of pollen records from around the globe show that in Europe, northern Asia and the Americas, the LQE occurred at a time when vegetation was changing rapidly, but that the situation is less clear for Australia (Barnosky et al. 2004b). Tests of the uniqueness of floras at the LQE are possible with long, high-resolution pollen records spanning more than one glacial cycle. Few such records are publicly available on pollen databases for quantitative analysis, but from inspection of published records it appears that floras of prior interglacials were similar to those of

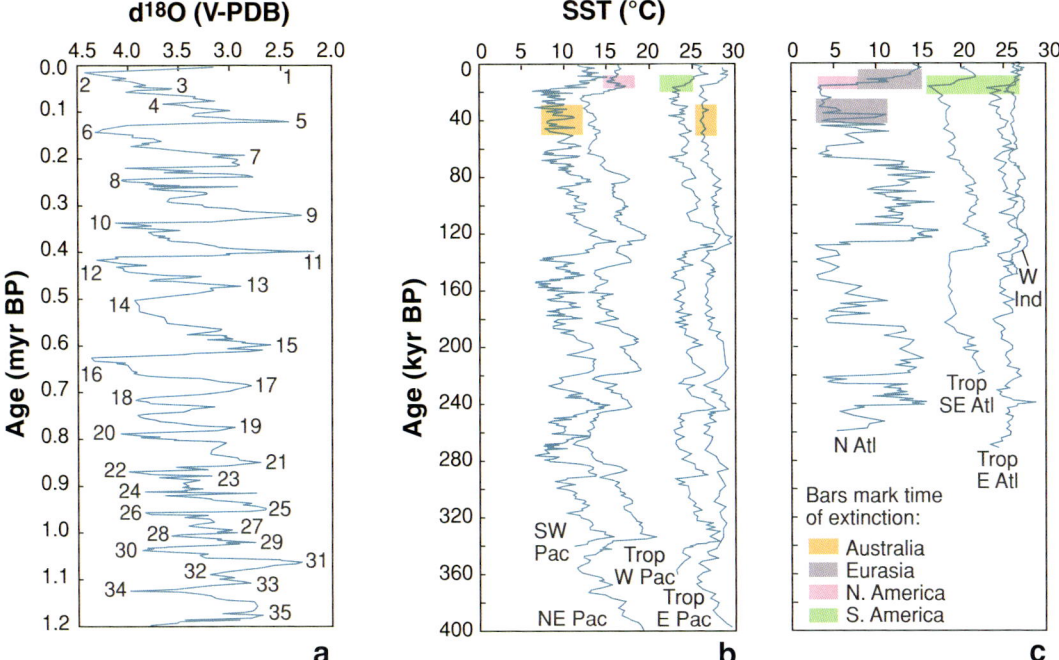

Figure 3

(*a*) Oxygen isotope data from benthic foraminifera at North Atlantic DSDP site 607. Numbers refer to marine isotope stages (MIS). The majority of well-constrained extinctions and drops in abundance occurred in MIS 2 and 3. (*b*) and (*c*) Sea surface temperature (SST) records from the Pacific and Atlantic/Indian Oceans, respectively. Colored bars indicate the time of extinction on nearby continents. Core location, core label, and type of SST estimate: northeast Pacific, ODP 1020, alkenone; tropical east Pacific, TR 163–9, Mg/Ca; tropical west Pacific, ODP 806B, Mg/Ca; southwest Pacific, MD 97–2120, Mg/Ca; west Indian, MD 85674, alkenone; north Atlantic, K 708–1, foraminiferal transfer function (average of August and February); tropical east Atlantic, GeoB 1112, Mg/Ca; tropical southeast Atlantic, GeoB 10,285, alkenone. For core K 708–1, ^{14}C ages were converted to calendar years. All other age models are as reported in primary publications. Modified from figure 2 in Barnosky et al. (2004), where primary citations for data are reported.

the Holocene, and that rapid transitions in floral composition are typical of glacial-interglacial transitions. So far, no evidence has shown that the states or rates of change in vegetation or climate at the time of the LQE were unprecedented relative to earlier in the Pleistocene. Finally, records of floral change in regions that did not experience strong extinction in the late Pleistocene (e.g., Thailand, New Zealand) vary as much as records from places that did (e.g., eastern North America).

HLHs assume that preferred habitats for extinct species shrank just before the extinction. Two recent studies cast doubt on this assumption for extinct North American megafauna. To explore the relationship between ecosystem change and herbivore abundance, Robinson et al. (2005) studied the concentration of pollen and spores of the dung fungus *Sporormiella* (a proxy for megafaunal biomass) at mastodon and stagmoose fossil sites in southeastern New York. Drops in *Sporormiella* (and presumably megafaunal) abundance spanned a millennium. At most sites, megafaunal abundance dropped as spruce was rising in abundance or dominant, falsifying the hypothesis of King & Saunders (1984) that linked the disappearance of mastodons to the loss of spruce forest habitats. The abundance-drop also preceded the homogenization of floras at the beginning of the Holocene, falsifying the MNH of Guthrie (1984) for this region. Martínez-Meyer et al. (2004) took a more theoretical approach. They modeled where the climate-space occupied by eight Pleistocene mammal species would occur on the present landscape and concluded that niche-space for six of the extinct species actually increased in the Holocene. For the other two, it did not decrease enough to explain their extinction by habitat loss alone.

Do Environmental Hypotheses Explain the Selectivity of the Late Quaternary Extinction?

The determinants underpinning extinction selectivity are coming into focus. In a study of mammalian victims and survivors in Australia, Eurasia, the Americas and Madagascar, Johnson (2002) used clade-specific allometric relationships to estimate reproductive rates. In the nine groups he studied, the species with low reproductive rates had high probabilities of extinction, regardless of their body size. The probability of extinction exceeds 0.5 when the reproductive rate is ≤ 1 offspring per female per year. For all surviving mammals in these regions that reproduce this slowly, most live in arctic, alpine, or deep forest habitats or are arboreal or nocturnal; only three surviving slow breeders lack all these traits. Because large body size is correlated with slow breeding, large animals would be more susceptible to extinction under any environmental or anthropogenic impact that targeted slow breeders. Environmental causes could explain the survival of alpine or deep-forest animals if those habitats increased at the last glacial-interglacial transition. However, it is difficult to explain why extinction susceptibility would be lower for nocturnal or arboreal animals under a model that invokes environmental change as a sole cause of the LQE.

Among herbivores, extinction is not related to diet in any simple way. In North America, where the most isotopic data on diet are available, victims include grazers (mammoths) and a wide range of browsers (antilocaprids, a musk-ox-like bovid, camelids, tapirs, peccaries, stagmoose, mastodons) (**Supplemental Table 3**). CED predicts that herbivores with very broad dietary tolerances would fare better than those with highly specialized diets, yet animals with little dietary specificity (e.g., *Camelops, Hemiauchenia, Equus*) were victims. The MNH gains support from the observation that the animals with the most extreme diets survived (e.g., *Bison*, the end-member for grazing, and *Odocoileus*, an end-member for browsing). If floras did become more homogeneous locally, with uninterrupted stretches of forest, grassland or tundra, we might expect these "hyper"-grazers and "hyper"-browsers to be favored. Some researchers have argued that environmental change (either natural or anthropogenic) led to a loss of woodland and shrubland in Australia, preferentially impacting browsers. Yet a recent study revealed that large browsers and grazers were victims of the LQE in proportion to their species abundance prior to the extinction (Johnson & Prideaux 2004); the extinction did not preferentially target browsers.

Body size decreased near and across the Pleistocene-Holocene boundary in victims and survivors of the LQE (Guthrie 1984, 2003; King & Saunders 1984). Size reduction has been viewed as evidence for nutritional limitation under "decaying" environmental conditions. For example, in the relatively small giant deer from Ireland, nutritional stress prior to extinction is indicated by the small antlers relative to body size (Barnosky 1985, 1986). However, populations that lingered into the early Holocene on the Isle of Man seem not to have been nutritionally stressed, judging by their antler proportions (Gonzalez et al. 2000). Body size decreased in many taxa in prior warm intervals (Kurtén 1968) and it is unclear if the magnitude of size reduction in the Holocene is unusual. Finally, size reduction is also a possible outcome of selection by human predators if they preferentially targeted large animals either because of added food payoff or as trophies (e.g., Olsen et al. 2004). The overall lesson is that body-size arguments are complex. Depending on the situation, they can support environmental hypotheses or anthropogenic impacts or simply indicate ecophenotypic variation of little consequence to extinction per se.

When the LQE debate heated up in the 1960s, little information existed on mammal communities of earlier Pleistocene times outside Europe. In Europe, cold-stage faunas, which include reindeer, arctic lemming, mammoth, and woolly rhinoceros, alternated with warm-stage faunas, which include wild boar, giant Irish deer, red and fallow deer, straight-tusked elephants, and different rhinoceroses (Stuart 1991, 1999). Holocene faunas differ from earlier warm interval faunas only in their lack of typical large community members. Records of mammalian community response to prior glacial-interglacial transitions outside of Europe reveal that diversity and the size, trophic, and taxonomic structure of communities changed more from the late Pleistocene to the late Holocene than they had in the previous million years (Moriarty et al. 2000, Barnosky et al. 2004a). The climatic shifts that precipitated faunal change at earlier glacial-interglacial transitions more strongly affected lower size and trophic categories. In contrast, at the time of the LQE, the range shifts and abundance changes at lower size and trophic levels, which were similar to those at earlier transitions, were accompanied by megafaunal extinctions.

On balance, it is not possible to explain many aspects of the LQE solely by existing environmental hypotheses. Some selectivity with respect to diet matches expectations of the mosaic-nutrient and habitat loss hypotheses, but key assumptions about floral change underpinning these models are unverified. At a continental or global scale, there is little evidence for the complete elimination of favored habitats. Indeed, existing models suggest expansion of favored climate-space. No environmental hypothesis has explained why the last glacial-interglacial transition should have spared slow-breeding arboreal or nocturnal mammals. Finally, there is no compelling evidence that the last glacial-interglacial transition was so different from previous ones that wholesale extinction of megafaunal species would result, in view of the fact that those species had survived many previous glacial-interglacial transitions.

The conclusion that existing environmental hypotheses are unable to explain many aspects of the LQE does not mean that environmental change did not contribute to the extinction. Climate has shifted dramatically over the past 50,000 years, and the geographic ranges, abundance, and morphology of both large and small mammals have changed in response. Some regions did become inhospitable to prior inhabitants. For example, the replacement of steppe grasslands with tundra and forest in parts of Beringia must have been detrimental to grazers. While environmental drivers seem inadequate as a total explanation for the LQE, they in fact seem necessary to explain some details of the event, particularly the timing and magnitude at a regional level.

Future Work on Environmental Impacts on Extinction

Progress on environmental contributions to the LQE will require a more quantitative approach. As biome reconstructions yield more detailed information on factors of import to herbivores, quantitative models of animal energetics and foraging must be used to test whether Pleistocene biomes were more capable of sustaining large mammals than early Holocene biomes (Matheus 1997, Moen et al. 1999). Methods to study population viability in parks and preserves can be applied as well. O'Regan et al. (2002) applied the VORTEX model, which inputs life history data and tracks the fate of animal populations, to assess the minimum range needed to sustain a large carnivore (*Panthera gombaszoegensis*) in southern Europe during glacial intervals. They concluded that Mediterranean islands were too small; persistence for >1 kyr required an area equivalent to the Italian peninsula. A similar approach could be used to ask whether range restriction to northern Siberia in the Holocene would be fatal for cold-stage mammals. Studies such as that by Martínez-Meyer et al. (2004) should illuminate the extent to which niches may or may not have been reduced for many species.

The extraction of mitochondrial DNA (mtDNA) from Quaternary fossils is becoming a reliable source of information on extinct populations and species (Hadly et al. 1998, Wayne et al. 1999). Phylogeographic studies of Pleistocene brown bears (*Ursus arctos*) from Alaska and Canada point to geographic shifts that precede the arrival of humans, suggesting a role for environmental change and/or interactions with competitors or diseases arriving from Asia (Barnes et al. 2002, Leonard et al. 2000). Paleogenetic data potentially can clarify if the typical response of Pleistocene populations to environmental change or human occupation is a regional demographic crash (Shapiro et al. 2004, Drummond et al. 2005), especially when treated in the context of newly developing phylochronologic techniques (Hadly et al. 2004). An understanding of the demographic consequences of climate change is central to any explanation for large mammal extinction.

An idea meriting further exploration is that ecological threshold events can cause dramatic restructuring of ecosystems (Scheffer et al. 2001). In concept, coevolutionary disequilibrium is such a model, but it has been presented in only qualitative terms. Forster (2004) has taken the threshold concept further by exploring the idea that the Australian extinction reflects the collapse of a system poised at a critical state and "pushed over the edge" by the arrival of humans with their catholic diets. To paraphrase Solé et al. (2002), highly diverse ecosystems may be poised at the cusp between stable and unstable states because of the tension between a tendency toward higher diversity (due to immigration and speciation) and the constraints on diversity imposed by the increasing numbers of ecological links, a process he called SOI. SOI makes predictions about the time series behavior of communities that could be tested. Furthermore, simulations suggest that SOI will be a property of highly diverse ecosystems (tropical forests, coral reefs); it is unclear if such dynamics would characterize more species-poor temperate and boreal systems, or if they can explain the size-selectivity of the LQE.

HUMAN IMPACTS OTHER THAN HUNTING

Three indirect anthropogenic impacts have received attention as potential agents of extinction: habitat alteration, introduced predators, and "hyperdisease." By indirect impacts we mean anthropogenic influences other than intentional killing of prey. We also note that habitat alteration might involve habitat loss, in which niches disappear, or habitat fragmentation, in which critical niches remain but are reduced in extent and connectedness.

Human habitat alteration by clearing land and the use of fire is a factor in Holocene extinctions on islands (Burney et al. 2004, Diamond 1984, Steadman & Martin 2003). Flannery (1994) and Miller et al. (1999, 2005) extended this idea to Australia, arguing for more frequent intense fires after the LQE as fuel loads rose with the loss of herbivores and as Aboriginal people began to use fire. The primary evidence for habitat alteration comes from abundance shifts in pollen records and dramatic increases in charcoal abundance (Edwards & Macdonald 1991). The coincidence of such shifts with faunal loss supports the role of habitat alteration in many island studies. The Australian case is more difficult because of chronologic problems, inconsistencies among charcoal records, and debate as to whether charcoal spikes are natural or anthropogenic (Bowman 1998). Robinson et al. (2005) extended these ideas to eastern North America, noting that the concentration of charcoal (indicating human-set fires) increased after the megafaunal dung spore *Sporormiella* dropped in abundance. It is not clear, however, that this increase in fire use led to major habitat alteration; to our knowledge, there is no evidence for extensive, anthropogenically driven habitat alteration in the Americas coincident with the LQE.

When humans arrived on the Pacific islands, they brought along pigs and dogs, and inadvertently introduced rats, all of which functioned as new predators. The impacts on island faunas were immense. Many islands lacked terrestrial predators and thus had faunas that were both behaviorally and evolutionarily naïve to being preyed upon (Wroe et al. 2004). The main islands of New Zealand lost 60% of their avifauna in the last 2000 years. Holdaway (1999) conducted an elegant analysis of this extinction, considering both the ecology of the prey and the time of arrival and ecology of potential predators. He argued that the avifauna was sequentially stripped of species that were most susceptible to newly arrived predators (e.g., rats, humans, dogs, etc.).

The role of introduced nonhuman predators in Australia and the Americas is less clear, but seems too small to explain the LQE. Those regions had native terrestrial

Hyper-disease hypothesis: extinction results from introduction of hypervirulent, hyperlethal diseases to immunologically naïve species on first contact with humans and associated fauna

carnivores, so prey would not have been highly naïve. Dingoes did not reach Australia until 3.5 kyr BP, long after the LQE (Johnson & Wroe 2003). Domestic dogs must have come to the New World with humans, but the time of their arrival is a question. The oldest archaeological evidence for domestic dogs in the New World dates from 9–10 kyr BP, after the first appearance of abundant humans (12–14 kyr BP) and after the LQE, though Fiedel (2005) makes cogent arguments that dogs were present earlier. In any case, coyotes and wolves preyed on North American megafauna for at least one million years prior to the LQE (Anderson 2004), so the introduction of domestic, pack-hunting canids is unlikely to have significantly affected the predator-prey balance, absent their use in hunts coordinated by humans.

MacPhee & Marx (1997) explored a third indirect-effect hypothesis, that extinction might result from the introduction of hypervirulent, hyperlethal diseases (hyperdiseases) to immunologically naïve species in the Americas and Australia on first contact with humans and their associated fauna (hyper-disease hypothesis; HDH). To cause the sudden collapse of so many species, such diseases would have to have reservoir species in which they are not fatal, be an entirely new pathogen on the affected landmasses, kill infected species rapidly, spread by vectors other than reproductive behavior, and "jump" species easily.

We do not favor the hyperdisease idea for several reasons. First, such extremely lethal, cross-species pathogens are not known today. Perhaps the closest analog is West Nile virus, which recently invaded North America and infects many bird species and humans. West Nile virus has not yet led to extinctions, and if they do occur, they are highly unlikely to show the phylogenetic diversity and strong size-bias of the LQE (Lyons et al. 2004b).

Second, at the time of the LQE, North and South America were not isolated continents. Faunal exchange among North America, Eurasia, and Africa occurred throughout the late Cenozoic (Woodburne & Swisher 1995), and the Americas have exchanged taxa for the past 4 million years. MacPhee & Marx (1997) recognized that regular exchange rules out typical animal-borne diseases. For hyperdisease to be a viable mechanism, they argued it would have to be hosted or transmitted by humans or their associates, have emerged just before human dispersal to Australia, and have then disappeared or lost virulence.

Third, rapid extinction immediately postcontact (within a few generations) is required under the hyperdisease model. Evidence for human arrival in the Americas thousands of years prior to the LQE (discussed below) would falsify this model in the Americas.

In summary, none of the indirect-effect models makes strong predictions regarding the size-selectivity of extinction. While anthropogenic habitat alteration and/or human-introduced predators were factors in many island extinctions, and may have contributed to Holocene extinctions in Australia, it is unlikely they were dominant causes of the LQE in the Americas and Australia. Finally, while diseases transmitted from humans and associated fauna might have contributed to population declines in some species, the hyperdisease model seems implausible as a general explanation for the LQE.

OVERKILL HYPOTHESES

Under overkill hypotheses, extinction occurs because hunting causes death rates to exceed birth rates in prey species. Period. Hunters need not be large game specialists, or gain most of their calories from hunting. Most deaths in the victims need not be due to hunting; hunting merely adds to the cumulative number of deaths by some critical amount. The extinction could be sudden or gradual. It could happen at first contact with humans or after millennia of interaction. This wide range of possibilities does not make overkill untestable. Rather, it requires the framing of specific overkill models with quantitative parameters and predictions about people and their prey that can be compared with data.

The blitzkrieg hypothesis has been the focus of much debate and testing (Martin 1973, Mosimann & Martin 1975). Under blitzkrieg, extinction at a continental scale occurs rapidly, within 500 to 1000 years, due to human geographic expansion and population growth fueled by intense predation on large game. Early versions called for extinctions to travel in a wave across the landscape from the point of first human arrival, with extinction at a local scale occurring in decades rather than centuries, but later models relaxed this constraint (Whittington & Dyke 1984). Blitzkrieg assumes high kill rates are possible because naïve prey lack behavioral and evolutionary adaptations to escape human predation. Martin (1973) viewed blitzkrieg as a solution to two perceived problems with the overkill hypothesis. First, prey naïveté was invoked to explain why the magnitude of extinction was high in the Americas and Australia, but low in the region of human origins, Africa and Eurasia. Second, Martin speculated that because the interaction between humans and extinct prey would have been brief, blitzkrieg explained why so few archaeological sites in the Americas and Australia contain extinct fauna.

Blitzkrieg hypothesis: human geographic expansion and population growth fueled by large game hunting leads to rapid extinction

Prey naïveté: failure of prey species to recognize the threat presented by new predator species or to respond by fight or flight

Plausibility of Overkill Hypotheses

Hunting by nonindustrial humans has been a factor in the extinction of island endemics and populations of marine birds and mammals (Diamond 1984). These societies may maximize their short-term harvesting rates without regard for the long-term sustainability of prey, and take highly ranked prey if they are encountered, even after they have become rare (Alvard 1993, Winterhalder 2001). Extrapolating the behaviors of extant hunter-gatherers to prehistoric humans is problematic, however, given that prehistoric environments differed from anything today, that modern hunter-gatherers have been limited to relatively unproductive landscapes that are not suitable for agriculture, and that human behavior and material culture have evolved.

Some researchers are skeptical that humans armed solely with stone and wood tools could hunt large animals to extinction. All acknowledge, however, that the weapons carried by late Pleistocene humans in Eurasia and the Americas could be used to hunt megafauna because stone points have been found in the body cavities of extinct species (Grayson & Meltzer 2002). The efficacy of the tool kit carried by early Australians is a greater question (Wroe et al. 2004). Stone spear points and dogs, which are used by modern hunter-gatherers that take large megafauna (>300 kg), were not present in Australia until the mid-Holocene. Proponents of overkill suggest the early Australians may have used wooden weapons, but these are less lethal than stone tipped spears. Finally, unlike the situation in Eurasia and the Americas, there are no sites with extinct megafauna that were clearly killed by humans, and only one site where extinct mammalian megafauna are associated with stone artifacts. These observations cast doubt on the plausibility of overkill in Australia, but are insufficient to falsify the hypothesis.

Simulations of Overkill: Single-Species and Related Models

Simple predator-prey models offer insights into the plausibility of overkill. Many simulations treat prey as a single species, with population parameters (initial biomass, replacement rate [r], and carrying capacity [K]) that are varied in sensitivity analyses (**Supplemental Table 4**). Human population dynamics are either modeled using parameters (initial biomass, r, K, migration rate) that respond to prey population size, or else population densities are set at fixed values that are varied in sensitivity tests. In some models, hunting efficiency is varied as well, and prey can lose naïveté.

Many early simulations with coupled human and prey population dynamics yielded overkill; however, stability analyses suggested that extinction was a mathematically inevitable outcome of these models (Belovsky 1988). Several simulations with uncoupled human and prey population dynamics counter-intuitively found that overkill was less likely for large, slow-breeding prey, perhaps because they fixed human population densities at a very low value or assumed that hunting efficiency dropped as this single prey became rare (Brook & Bowman 2002, Choquenot & Bowman 1998).

Single-prey simulations implicitly assume that humans are switching to alternate resources as prey become scarce, but they do not consider the properties of these resources that might make overkill more or less likely. Winterhalder & Lu (1997) developed multiprey simulations based on optimal foraging theory with coupled human and prey population dynamics. In these simulations, if large prey are present, they are always part of the diet; if they breed slowly, they go extinct when other, rapidly reproducing prey are available. Humans in these simulations take large prey when encountered, but are sustained by fast-breeding, "fall-back" prey such as tubers or small animals. Belovsky (1988) constructed an optimal foraging model to examine the demographic consequences of differences in nutrition when humans consumed hunted versus gathered resources. He found that prey extinction increased as primary production rose (i.e., when gathered foods become more abundant). Prior to prey extinction, people in high productivity environments had a smaller percentage of meat in their diets than people in low productivity environments, but humans were so abundant that prey extinction was likely.

These simple models offer a key insight that has not received enough consideration in the overkill debate. Extinction of prey due solely to overkill is highly unlikely if human diets are composed of a just few types of hunted megafauna. Such human populations will exhibit boom-bust dynamics that follow those of their prey. The most likely outcome in such situations is that human populations will go extinct, not prey populations. For large, slow-breeding animals, the most dangerous predators are omnivores who are sustained by small game and gathered food. A conclusion from these foraging models is that the traditional view of overkill, which attributes extinction to voracious predation by Pleistocene humans who specialized in hunting large game, is flawed, as are attempts to falsify overkill by noting that late Pleistocene humans had broad diets. Indeed, a recent study of the potential for megafaunal specialization by North American Paleoindians in a quantitative, optimal foraging framework suggested that foragers should have taken a wide array of taxa including not only proboscideans, but also ungulates and smaller game (Byers & Ugan 2005).

Models and Prey Naïveté

Wroe et al. (2004) critiqued the idea that prey naïveté in the Americas and Australia would have been great enough to permit blitzkrieg. Completely naïve prey that fail to protect themselves or flee from a predator are restricted to remote islands that lack all large terrestrial carnivores. Such island species can have deeply rooted behaviors (e.g., ground nesting) that place them in peril when terrestrial predators arrive. They may also have morphological adaptations (e.g., flightlessness) that leave them vulnerable. On continents with carnivores, such morphological and behavioral adaptations are rare, and potential prey soon learn to flee from new predators (Berger et al. 2001). Given the long history of predator-prey interactions in the Americas and Australia, we agree that it is unlikely that megafauna would have remained naïve very long to newly introduced predators, even ones as clever as humans. Prey naïveté might have had some impact if the response to predators did not involve flight or hiding, but rather grouping or fighting [as Martin (2005) suggests for ground sloths]; these tactics would not be effective when facing armed humans.

Brook & Bowman (2002) critiqued models that do not allow predators to develop antipredation strategies. This criticism is valid for single-prey simulations (as predator avoidance would make the single prey increasingly difficult to locate), but is of less concern for multiprey simulations, in which predators take rare, predator-shy prey encountered incidentally while hunting more abundant prey.

Tests of Overkill with Realistic Predator-Prey Models

Models show what is possible, not what actually happened. Ideally, they establish quantitative predictions constrained by input parameters that portray reasonable assumptions about starting points and "rules" that govern the process under examination. The most useful models are those that clearly specify the input parameters, how robust the predictions are in terms of statistical probabilities, and how predictions can be tested against archaeological or paleontological data. It is the iterative feedback between what models predict and what the data actually say that gives both models and data meaning. For example, arguments about the number of kill sites in the fossil record have been used to support (Mosimann & Martin 1975) or refute (Grayson & Meltzer 2003) blitzkrieg. None of these arguments is robust in the absence of a clear statement of how many kill sites would be expected given the total number of appropriately aged sites, probabilities of fossil preservation derived from firm taphonomic studies, and realistic quantitative models of human-faunal interactions that include multiple species.

Quantitative predator-prey models have proven useful in studying the extinction of particular species, such as Eurasian mammoths, moas in New Zealand, or megafauna in northern Australia (**Supplemental Table 4**). The most comprehensive model coupled human and prey population dynamics to simulate predation on 41 large species and an undifferentiated secondary resource (plants, small game) in North America (Alroy 2001b). Hunting efficiency, the geography of invasion, and competitive interactions were varied, and all simulations assumed that hunters nonselectively took prey as encountered. Overkill occurred for a range of parameter values, although an error in the parameterization of prey r values makes it difficult to assess these results (Slaughter & Skulan 2001). In a recalculation of the best-fit trial with appropriate r values, the model correctly predicted the fate of 34 out of 41 species, with a median time to extinction of 895 years (Alroy 2001a). In that trial hunters obtained 30% of their calories from large mammals and occurred at densities of 28 people per 100 km^2, both within the range of values for modern hunter-gatherers. As in more generalized optimal-foraging models, the key to overkill was a relatively high human population density subsidized by smaller, faster-breeding prey. Hunting ability matters too, with greater hunting success leading to greater extinction rates, but overkill occurred even when success rates were fairly low.

This realistic simulation is an elegant first step, but interpretation of its results is complicated by assumptions about prey dispersal, geographic range, and carrying capacity that should be subject to sensitivity tests. More realistic models would allow for variation in primary production across the modelled landscape and through time and for feedbacks between primary production and prey population dynamics and would explore selective hunting strategies. New models could make explicit predictions for the archaeological record by calculating the fractions of individuals that die at human hands and, given plausible estimates of preservation potential, how many of these bodies are likely to be found and recognized as cases of human predation.

Chronologic Tests of Overkill

A linchpin of the overkill argument is the purported synchrony of the first appearance of substantial numbers of technologically sophisticated humans with the last records of megafauna (Martin & Steadman 1999). For Holocene extinctions on islands, the pattern is crystal clear; island ecosystems collapse following contact with humans, though the timescale may vary from centuries to millennia, depending on differences in abiotic, biotic, and cultural factors (Rolett & Diamond 2004, Steadman & Martin 2003). Many continental extinctions are now dated with enough security to know that they too occurred postcontact.

In North America south of the Laurentide ice sheet, at least 15 species became extinct coincident with the appearance of Clovis-style tools (13.5 kyr BP to 11.5 kyr BP) near or in the Younger Dryas climate event (Grayson 2006, Grayson & Meltzer

2003) (**Figure 2**). Debate continues about pre-Clovis humans in the Americas, but there is no evidence for large numbers of people before Clovis times (Fiedel & Haynes 2004). In South America, humans were present in coastal Chile and Patagonia at 15.5 kyr BP to 14.8 kyr BP (Alberdi et al. 2001, Meltzer et al. 1997), and sites younger than 11.5 kyr BP are common across the continent (Dillehay 2000). It seems clear that the megafauna went extinct in the late Pleistocene, probably after humans arrived, but comprehensive analyses are needed.

Humans or their predecessors have been in Eurasia for 2 million years, with good evidence for predation on large game back to 400 kyr BP (Thieme 1997). This chronology falsifies blitzkrieg, but other human impacts cannot be rejected. The pre*sapiens* hominins who lived and hunted in Europe from 400 kyr BP to 50 kyr BP without inducing extinctions had simple tools, and isotopic data suggest they were highly carnivorous with diets focusing on large open country herbivores, such as mammoths and woolly rhinoceros (Bocherens et al. 2005); they may have waxed and waned in response to changes in their prey (Kuhn & Stiner 2001). In contrast, the first megafaunal extinctions beginning 45 kyr BP roughly coincide with the arrival and spread across Europe of anatomically modern humans (*Homo sapiens sapiens*), with their diverse suites of tools and broad diets (Mellars 2004). The extinctions and population crashes that began 14 kyr BP may be associated with a rise in population densities at the end of the glacial interval (Bar-Yosef 2002).

As discussed above, the timing of megafaunal extinction in Australia remains contentious. Humans arrived sometime between 72 and 44 kyr BP (Brook & Bowman 2002). If we accept the earlier dates for extinction (50–40 kyr BP), then human arrival and megafaunal extinction roughly coincide. If the younger dates for extinction are correct, then extinction is temporally decoupled from the first appearance of humans. Either way, the seemingly long period of overlap between megafauna and humans casts doubt on blitzkrieg.

Africa is the continent of origin of *Homo sapiens*, and many human ancestors. *Homo sapiens* appeared in Africa 195 kyr BP (McDougall et al. 2005), long before the few poorly dated Pleistocene or better-dated Holocene extinctions.

Archaeological Tests of Overkill

The harshest critiques of overkill have come from archaeologists, who argue there is no "empirical" evidence of overkill in the Americas and Australia. Thousands of diagnostic fluted points are known from the late Pleistocene of North America, but our understanding of human-megafaunal interaction and human diets are based on a few dozen sites (Haynes 2002b). There are large regions of the United States that had people for which we have no archaeofaunal evidence at all (Waguespack & Surovell 2003). This sparse record reflects a bias against bone preservation (e.g., southeastern United States) and a lack of study.

A recent rigorous vetting of putative human-megafaunal interaction required either unambiguous association of megafauna with Clovis artifacts or clear evidence of bone modification by humans (Grayson & Meltzer 2002). Using these criteria, four extinct genera were recognized as being hunted by humans in North America

(*Mammuthus*, *Mammut*, *Equus*, and *Camelops*) at 14 sites. The culling excluded ∼50 other putative associations of humans with *Mammuthus* or *Mammut*. Also excluded were sites where megafaunal bones or ivory were fashioned into tools (Haynes 2002b), because this material could have been scavenged. More lax definitions of association suggest that late Pleistocene humans in North America regularly hunted large game (Haynes 2002b, Waguespack & Surovell 2003).

Even with its holes, the North American record is much better known than those of South America, Australia, Africa, and much of northern Asia. Haynes (2002b) notes there are fewer than a dozen well-documented proboscidean kill sites in Africa older than mid-Holocene. In South America, evidence for use of extinct megafauna by humans exists at a few sites (Fiedel & Haynes 2004, Meltzer et al. 1997), and associations of artifacts and extinct megafauna are generally accepted, although the criteria for accepting associations is not consistently specified (Dillehay 2000, Scheinsohn 2003). In Australia, there is no evidence for human butchery of extinct megafauna (Mulvaney & Kamminga 1999) and only one site (Cuddie Springs) has lithic artifacts associated with extinct megafaunal mammals (Field & Dodson 1999).

Is it possible to assess the claim that the paucity of kill sites falsifies overkill? Even with a quantitative model of overkill, we do not know the probability of a kill site being preserved. Barnosky et al. (2004b) used the FAUNMAP database to explore this issue for North America. Humans became abundant there in the late glacial period. Roughly 25% of the mammoth sites from this interval have firm archaeological associations (as vetted by Grayson & Meltzer 2002). Barnosky et al. (2004b) tested whether other late Pleistocene mammals have rates of archaeological association significantly different from that of mammoths (**Table 3**). A chi-square analysis showed that for many taxa, even complete absence from archaeological sites does not refute the hypothesis of human association at the same frequency as accepted *Mammuthus*-human associations. Put another way, the apparent paucity of archeological associations with most taxa is not strong evidence that humans did not hunt them. Rather, it reflects that there are few well-dated late glacial sites—either with or without archaeological associations—that preserve extinct mammals other than mammoths. There were two exceptions. *Mammut* and *Platygonus* were significantly less frequently associated with humans than were mammoths, suggesting that humans probably hunted them less. For *Mammut*, this situation would be reversed if some of the many putative *Mammut* butcher sites were better documented. For *Platygonus*, there are few sites with even potential archaeological associations, so the result seems solid.

The frequency of late glacial North American mammoth sites accepted to be in archaeological association is quite high (25%). If these sites do represent kills, and if killing was a new mode of death, it would indicate a 30% rise in mortality. Megafauna can increase their intrinsic rate of increase or survival (which allows more offspring) when food is abundant, as might occur when hunters thin the herd (Fisher 1996). It is unlikely, however, that they could tolerate such a high level of new mortality unless it was strongly skewed toward males. That said, we emphasize that biases in site preservation, discovery, and description make it very unwise to assume anything about the fraction of animals taken by hunting from site or carcass counts at this point.

Table 3 Clovis-age kill-sites and total number of late-glacial occurrences of some extinct mammals documented in the FAUNMAP database

Taxon	# of kill sites[a]	Total # of late-glacial sites[b]	Ratio of kill sites to all sites	X^{2c}
Arctodus	0	8	0	0.15
Camelops	1	15	0.06	0.17
Capromeryx	0	1	0	0.61
Equus	2	17	0.11	0.30
Glossotherium	0	8	0	0.15
Hemiauchenia	0	5	0	0.25
Holmensina	0	3	0	0.37
Mammut	2	68	0.03	**0.001**
Mammuthus	12	45	0.25	
Megalonyx	0	4	0	0.31
Platygonus	0	16	0	**0.05**
Smilodon	0	3	0	0.37
Tapirus	0	6	0	0.21

[a]Sites with robust evidence for human predation (Grayson & Meltzer 2002).
[b]All late-glacial age (18.5–10.5 kyr BP) sites with the genus contained in the online FAUNMAP database (**http://museum.state.il.us/research/faunmap**), except *Tapirus*, which is from published literature (FAUNMAP Working Group 1994).
[c]Chi-square probability that the ratio for the genus is the same as the ratio for *Mammuthus*. Bold text indicates significantly fewer archaeological associations than expected relative to *Mammuthus*.

Archaeological data can sometimes reveal causes of extinction. For example, on some islands, large accumulations of bones from vulnerable prey occur in archaeological middens immediately after first contact (Holdaway & Jacomb 2000, Steadman & Martin 2003). This is compelling evidence of overkill. Such examples usually are found where faunas lacked bone-crushing scavengers and the extinction was mid- to late Holocene. The LQE in Australia and the Americas occurred much earlier, and in the Americas, bone crushers were present. It is unreasonable to expect catastrophic assemblages representing the last generations of extinct fauna in these regions.

The archaeological record has also revealed cases where prey species were depleted over an interval of time. Signs of prey depletion include a drop in ratio of high-yield prey relative to low-yield prey, increased human diet breadth, and a size decrease in prey species with indeterminate growth (Grayson 2001). Demonstration of prey depletion requires large samples spanning multiple time intervals, typically generated at stable residential or food processing sites. People in the Americas at the time of the LQE were highly mobile. Most late glacial archaeological sites are interpreted as briefly occupied hunting camps with relatively sparse faunal data (Kelly & Todd 1988), and thus are not amenable to study of resource depletion.

Testing Overkill with Data on Faunal Population Dynamics

Fisher (1987, 1996) has suggested an approach to test for the contributions of overkill and environmental change to proboscidean extinctions. Measurements of the thickness of annual growth increments in tusks may reveal key life history parameters, such as onset and shut down of reproduction and calving interval. When under nutritional stress, modern elephants mature later, cease reproduction earlier, and nurse longer (and thus increase the calving interval), all of which lead to reduced or negative population growth. In contrast, when resources are replete (often due to over hunting) elephants mature earlier, cease reproduction later, and wean faster (decreasing the calving interval). Thus overkill and environmental deterioration lead to mutually exclusive predictions regarding life history change. This approach requires further development and is labor intensive, but it may yield the "cleanest" answer about the fates of several key victims of the extinction.

Future Work on Overkill Hypotheses

More work is needed to refine the dates of first human contact and last records of megafauna, especially in South America and Australia. Consistent criteria for determining whether megafaunal remains were produced by human hunting need to be applied worldwide. Taphonomic work must focus on evaluating how fossil abundance relates to the relative abundances of animals, and on the probabilities of preservation of anthropogenically versus naturally generated carcasses. These are all daunting, long-term tasks.

Simulations may offer further breakthroughs if they are designed to address specific questions and yield predictions that can be tested with archaeological and paleontological data. Among questions that would be fruitful to explore are the following. Is it plausible that late Pleistocene North Americans, living at a density of 28 people per 100 km^2 and obtaining 30% of their calories from hunting large prey over 500 to 1500 years (Alroy 2001a,b), would leave such a sparse archaeological record for relatively abundant prey species other than proboscideans? Or is it possible that many prey species were not as abundant as overkill models assume, perhaps due to a climatically driven drop in carrying capacity? Can realistic models simulate the extinction of megafauna in Australia, where some researchers question whether human populations were ever large enough to drive overkill (Wroe et al. 2004), or the coexistence of humans and megafauna in Africa?

COMBINED HYPOTHESES

Various models have been proposed that require the confluence of anthropogenic, ecologic, and climatic phenomena.

Owen-Smith (1987, 1999) recognized the profound ability of megaherbivores to tranform landscapes, creating mosaic habitats that promote the diversity of other herbivores and carnivores. His Keystone Herbivore hypothesis posits that the loss of critical megaherbivores, either from overkill or climate change, would lead to shifts in vegetation and cascading negative impacts on other species. Zimov et al. (1995) offered a provocative version of this hypothesis, arguing that the Beringian climate permits two vegetational states, the soggy, moss-dominated, nutrient-poor tundra present today, and a drier, grassy, nutrient-rich steppe that characterized the

Pleistocene. They suggested that steppe vegetation was promoted by megaherbivore disturbance, and that the spread of tundra coincident with the Pleistocene-Holocene boundary was an effect of megafaunal extinction. Whatever its cause, widespread tundra was detrimental to grassland species such as bison and horses. One prediction of these models is that proboscideans, the most abundant megaherbivores, should disappear before major floral change and before the disappearance of other taxa. At present, data from Eurasia, Alaska, and probably central North America suggest proboscideans were among the last to go. However, drops in proboscidean abundance, not presence/absence, drive landscape change. Microstratigraphic analysis to estimate megafaunal herbivore abundance from dung spore abundance may offer a key test. In eastern North America, at least, it appears that megaherbivore abundance does drop well before final extinction and major floral reorganization (Robinson et al. 2005).

Late Pleistocene faunas had a diverse suite of extinct carnivores, many of which were larger than surviving carnivores. If humans began to reduce populations of preferred prey (e.g., proboscideans, sloths, etc.), or even scavenged the carcasses of megafauna, other carnivores might have intensified their predation on other herbivores (horses, peccaries, etc.). Terrestrial carnivores are certainly capable of prey switching on rapid timescales, though predator densities might drop if new prey were more difficult to hunt. It will be difficult to test this idea. Careful taphonomic analysis of predators from before and during the extinction event has revealed evidence for an increase in the intensity of carcass utilization (Van Valkenburgh 1993). This observation is consistent with the expectation of competitive interference with humans, but also with an environmentally driven drop in the abundance of prey.

A third possibility is that predator avoidance strategies limited Pleistocene mammals to unsustainably small geographic refuges that lacked humans. An observation that would support this idea would be the development of cryptic behaviors postcontact. The observation that most surviving, slow-breeding animals outside of Africa occur in habitats that humans have difficulty occupying at high density is consistent with this idea (Johnson 2002). An issue that has not been examined is whether late surviving populations of extinct species were restricted to small subsets of viable habitat (as demonstrated by their distribution in prior interglacials), or if the viable habitat from which they were absent had high human population densities in the late Pleistocene.

These combined models highlight that the extinction of any species is likely to involve a complex interaction of factors that vary in space and time. If we want to understand the proximate mechanisms of extinction, the most profitable approach will be a series of wholistic, species-by-species analyses that consider animal behavior and life history, environmental pressures, the impact of human predation—as well as indirect human impacts—and a realistic evaluation of taphonomic processes (Fisher 1987, Grayson et al. 2001).

SUMMARY

The Pleistocene extinction debate has progressed greatly in the past decade. It is now clear that species with low reproductive rates were hard hit, regardless of body size. Even so, North American evidence indicates that, in comparison to earlier extinctions, the LQE was exceptional in its impact on large-bodied animals and that it was one of the two biggest of the past 55 million years. In North America, then, the tenet that the LQE was unusual both in intensity and in affecting large mammals preferentially has withstood the test of time. We suspect that data from other continents will mimic the North American situation.

These unusual characteristics are consistent with variants of anthropogenic models, but are difficult to explain with models that rely solely on environmental causes. Thus they provide the strongest support for a human contribution to the LQE. However, it is an oversimplification to say that an abrupt wave of hunting-induced extinctions swept continents right after first human contact. It is this extreme view of overkill, which irks many critics, that has been subjected to most archaeological and chronologic tests. Blitzkrieg can now be firmly rejected in western Europe, Siberia, Alaska, and probably Australia and central and eastern North America. A more complicated portrait of the LQE is emerging from recent studies. Taken as a whole, recent studies suggest that humans precipitated the extinction in many parts of the globe through combined direct (hunting) and perhaps indirect (competition, habitat alteration and fragmentation) impacts, but that late Quaternary environmental change influenced the timing, geography, and perhaps magnitude of extinction. Put another way, absent the various impacts of *Homo sapiens sapiens*, it is highly unlikely global ecosystems would have experienced a mass extinction of large, slow-breeding animals in the late Quaternary. But, absent concurrent rapid climatic change evident in many parts of the globe, some species may have persisted longer.

Overall, evidence from Eurasia points to a role for climate change in pacing the decline of late Pleistocene megafaunal populations, but whether or not these declines would have led to extinction absent coincident human impacts is not yet clear. An idea that merits further testing is that the arrival and population expansion of modern humans began to fragment megafaunal ranges, restricting populations to inviable Arctic refugia. Further work on Neanderthal paleoecology should illuminate why these early humans did not precipitate megafaunal extinctions earlier in the Pleistocene. Megafaunal scavenging (rather than hunting) or extreme reliance on megafaunal prey are two plausible explanations.

Australian evidence suggests that megafaunal extinction followed human arrival. Unfortunately, neither the arrival of humans nor the timing of megafaunal extinction can be bracketed within less than 10 kyr, though the extinction probably lags behind human arrival by enough time to preclude blitzkrieg. There is no simple association between extinction and environmental change, and climates and floras at the LQE were not unusual relative to states earlier in the Pleistocene, when climatic transitions did not induce extinction. To complicate matters further, early Australians lacked efficient stone weapons, there are no uncontested butcher sites, and some doubt that human populations were large enough to have a substantial impact on prey. Thus while the timing and the uniqueness of the event point to an anthropogenic impact of some sort, it is not clear if it was through hunting or habitat alteration, or that environmental change did not contribute as well.

In South America, data on extinction chronology are accumulating but await critical analysis. It is likely that humans hunted megafauna, and that the LQE postdates human arrival, but not much more can be said at present. In contrast, robust dating verifies simultaneous climate change and the first contact with substantial human populations in the conterminous United States, where extinctions were rapid and pronounced. Support for human impacts includes (*a*) indisputable hunting of two extinct species, (*b*) clustering of extinctions within 1500 years (perhaps less) of first contact with Clovis hunters, (*c*) widespread distribution of Clovis hunters, (*d*) simulations, and (*e*) more pronounced extinction than in earlier glacial-interglacial transitions. On a broader North American scale, the demise of megafaunal species without significant human presence in Alaska is consistent with a stronger role for climate at the edges of species' geographic ranges.

Africa remains an anomaly. Perhaps coevolution between humans and their prey did give megafauna an advantage (Martin 1984). Yet many African animals do not use cryptic habitats, and many lack defensive behaviors that would deter armed, group-hunting humans. If we accept that similarly armed and technologically sophisticated humans precipitated an extinction of most large mammals elsewhere around the globe, the lack of extinction in Africa is troubling. We offer three speculations on this issue.

First, perhaps small game and gathered food were so abundant that they were more remunerative than larger prey (Wroe et al. 2004). Yet with abundant resources, human populations should have grown until larger game was added to human diets, though perhaps not enough to impact prey populations. Second, perhaps human population density across Africa was lower than in Europe or the Americas. One intriguing possibility is that as the site of human origins, Africa harbors the greatest density of coevolved pathogens and parasites (Martin 1984). Humans would have been released from some of these warm-adapted pathogens as they moved to and through cooler regions. Finally, perhaps between human disease hot-zones and the patchiness of its ecosystems, Africa offered more refugia from human predation than other continents.

We believe it is time to move beyond casting the Pleistocene extinction debate as a simple dichotomy of climate versus humans. Human impacts were essential to precipitate the event, just as climate shifts were critical in shaping the expression and impact of the extinction in space and time. The unanswered questions now revolve around why some species succumbed to this combination of expanding human populations and climate change as it played out globally over thousands of years, whereas other species emerged essentially unscathed. Such understanding is essential for informed predictions about the future of surviving biota.

BROADER ECOLOGICAL IMPLICATIONS

From the perspective of conservation biology, a significant implication is that the intersection of dramatic climate change with human impacts on fauna is especially pernicious. The late Quaternary event provides a sobering example of what to expect as contemporary exploding human population densities collide with higher-than-normal rates of climate change under a "business as usual" model (Houghton et al. 2001): accelerated extinction rates and continued wholesale restructuring of Earth's ecosystems.

In addition, the extinction affects our understanding of the adaptations and ecology of surviving species. Some species are relicts of disrupted coevolutionary partnerships (Barlow 2001, Janzen & Martin 1982). Examples include plants that have lost their primary agents of seed dispersal or that are replete with defenses for herbivores that no longer exist, herbivores that are "overdesigned" for all existing predators, and scavengers such as condors that have no naturally occurring carcasses to eat in continental settings.

The recent data continue to support Martin's (1970) contention that modern ecosystems are unique in having vast populations of one species (humans) while lacking the array of megafauna that had populated terrestrial ecosystems for at least the past 20 million years. The net effect has been simplification of and loss of redundancy in food webs, which may not portend well for the stability of communities (Bengtsson et al. 2003).

Finally, such observations raise vexing value judgments for restoration ecology. Should we attempt to "resurrect" ecosystem processes that were in place before the LQE by introducing phylogenetically related species with ecological preferences similar to those of extinct megafauna (Donlan et al. 2005, Martin 2005)? Or should the goal be restoration of species and processes that were in place a few hundred years ago in ecosystems that were "wild," but clearly highly modified relative to their condition prior to prehistoric human occupation? While the feasibility of resurrecting Pleistocene ecosystems in the Americas and Eurasia is open to debate, a clear benefit of such efforts would be increasing the geographic range of various existing megafauna such as elephants and rhinoceros, which today are under threat of extinction from a variety of human-induced pressures.

The last frontier for megafaunal extinctions is the oceans. Marine communities are in the throes of a modern megafaunal extinction event (Hughes et al. 2003, Jackson et al. 2001, Steneck et al. 2002). Populations have crashed, but with a few notable exceptions (Caribbean monk seal, Steller's sea cow), species have not yet been driven to extinction, so marine communities are not yet beyond saving. The Quaternary extinction has taught us that the unfortunate intersection of human activities and climate change wiped out whole sectors of Earth's terrestrial ecosystems. Knowing that, are we willing to let overexploitation, habitat alteration, and climate change (this time anthropogenically induced) exterminate the marine megafauna and forever change those communities as well?

SUMMARY POINTS

1. Between fifty and ten thousand years ago, most large mammals became extinct everywhere except Africa.

2. This extinction, with its extreme focus on large and slow-breeding animals, was unusual relative to extinctions earlier in the Cenozoic.

3. The unusual body-size selectivity of the extinction, and its rough synchrony with the global geographic expansion of modern humans are compelling evidence that the extinction was precipitated by human activities, especially hunting.

4. Climate change likely affected the timing, geography, and perhaps magnitude of this anthropogenically triggered extinction.

5. The intersection of rapid climate change with initial human contact seemed especially deadly for megafauna.

6. The extinction of so many species in near-time raises vexing questions for ecologists and conservation biologists.

ACKNOWLEDGMENTS

We thank R.S. Feranec and A.B. Shabel for helping to compile information and S.L. Wing for his careful review of the manuscript. Members of the extinctions discussion groups and our labs at UC Santa Cruz and UC Berkeley helped in crystallizing these ideas. Research on climate's role in ecosystem change was in part funded by National Science Foundation grants.

LITERATURE CITED

Alberdi MT, Miotti L, Prado JL. 2001. *Hippidion saldiasi* Roth, 1899. (Equidae, Perissodactyla), at the Piedra Museo Site (Santa Cruz, Argentina): its implications for the regional economy and environmental reconstruction. *J. Archaeol. Sci.* 28:411–19

Alroy J. 1999. Putting North America's end-Pleistocene megafaunal extinction in context: large-scale analyses of spatial patterns, extinction rates, and size distributions. See MacPhee 1999, pp. 105–43

Alroy J. 2001a. Did human hunting cause extinction? *Science* 294:1459–62

Alroy J. 2001b. A multispecies overkill simulation of the end-Pleistocene megafaunal mass extinction. *Science* 292:1893–96

Alvard MS. 1993. Testing the "ecologically noble savage" hypothesis: interspecific prey choice in Piro hunters of Amazonian Peru. *Hum. Ecol.* 21:355–87

Anderson E. 2004. The carnivora from Porcupine Cave. In *Biodiversity Response to Climate Change in the Middle Pleistocene, The Porcupine Cave Fauna from Colorado*, ed. AD Barnosky, pp. 141–54. Berkeley: Univ. Calif. Press

Bard E, Rostek F, Menot-Combes G. 2004. Radiocarbon calibration beyond 20,000 ^{14}C yr BP by means of planktonic foraminifera of the Iberian Margin. *Quat. Res.* 61:204–14

Barlow C. 2001. *The Ghosts of Evolution*. New York: Basic Books. 254 pp.

Barnes I, Matheus P, Shapiro B, Jensen D, Cooper A. 2002. Dynamics of Pleistocene population extinctions in Beringian brown bears. *Science* 295:2267–70

Barnosky AD. 1985. Taphonomy and herd structure of the extinct Irish elk, *Megaloceros giganteus*. *Science* 228:340–44

Barnosky AD. 1986. "Big Game" extinction caused by late Pleistocene climatic change: Irish Elk (*Megaloceros giganteus*) in Ireland. *Quat. Res.* 25:128–35

Barnosky AD. 1989. The late Pleistocene event as a paradigm for widespread mammal extinction. In *Mass Extinctions: Processes and Evidence*, ed. SK Donovan, pp. 235–54. London: Belhaven

Barnosky AD, Bell CJ, Emslie SD, Goodwin HT, Mead JI, et al. 2004a. Exceptional record of mid-Pleistocene vertebrates differentiates climatic from anthropogenic ecosystem perturbations. *Proc. Natl. Acad. Sci. USA* 101:9297–302

Barnosky AD, Koch PL, Feranec RS, Wing SL, Shabel AB. 2004b. Assessing the causes of Late Pleistocene extinctions on the continents. *Science* 306:70–75

Bar-Yosef O. 2002. The Upper Paleolithic revolution. *Annu. Rev. Anthropol.* 31:363–93

Belovsky GE. 1988. An optimal foraging-based model of hunter-gatherer population dynamics. *J. Anthropol. Archaeol.* 7:329–72

Bengtsson J, Angelstam P, Elmqvist T, Emanuelsson U, Folke C, et al. 2003. Reserves, resilience and dynamic landscapes. *Ambio* 32:389–96

Berger J, Swenson JE, Persson IL. 2001. Recolonizing carnivores and naive prey: Conservation lessons from Pleistocene extinctions. *Science* 291:1036–39

Bocherens H, Drucker DG, Billiou D, Patou-Mathis M, Vandermeersch B. 2005. Isotopic evidence for diet and subsistence pattern of the Saint-Cesaire I Neanderthal: review and use of a multi-source mixing model. *J. Hum. Evol.* 49:71–87

Bowman DMJS. 1998. Tansley Review No. 101: The impact of Aboriginal landscape burning on the Australian biota. *New Phytol.* 140:385–410

Brook BW, Bowman DMJS. 2002. Explaining the Pleistocene megafaunal extinctions: Models, chronologies, and assumptions. *Proc. Natl. Acad. Sci. USA* 99:14624–27

Brook BW, Bowman DMJS. 2004. The uncertain blitzkrieg of Pleistocene megafauna. *J. Biogeogr.* 31:517–23

Burney DA, Burney LP, Godfrey LR, Jungers WL, Goodman SM, et al. 2004. A chronology for late prehistoric Madagascar. *J. Hum. Evol.* 47:25–63

Burney DA, Flannery TF. 2005. Fifty millennia of catastrophic extinctions after human contact. *TREE* 20:395–401

Byers DA, Ugan A. 2005. Should we expect large game specialization in the late Pleistocene? An optimal foraging perspective on early Paleoindian prey choice. *J. Archaeol. Sci.* 32:1624–40

Choquenot DM, Bowman JS. 1998. Marsupial megafauna, Aborigines and the overkill hypothesis: Application of predator-prey models to the question of Pleistocene extinction in Australia. *Glob. Ecol. Biogeogr. Lett.* 7:167–80

Crooks KR. 2002. Relative sensitivities of mammalian carnivores to habitat fragmentation. *Conserv. Biol.* 16:488–502

Diamond JM. 1984. Historic extinctions: A Rosetta Stone for understanding prehistoric extinctions. In *Quaternary Extinctions: A Prehistoric Revolution*, ed. PS Martin, RG Klein, pp. 824–62. Tucson: Univ. Ariz. Press

Dillehay TD. 2000. *The Settlement of the Americas: A New Prehistory*. New York: Basic Books. 371 pp.

Donlan CJ, Greene HW, Berger J, Bock CE, Bock JH, et al. 2005. Re-wilding North America. *Nature* 436:913–14

Drummond AJ, Rambaut A, Shapiro B, Pybus OG. 2005. Bayesian coalescent inference of past population dynamics from molecular sequences. *Mol. Biol. Evol.* 22:1185–92

Edwards KJ, MacDonald GM. 1991. Holocene palynology. II. Human influence and vegetation change. *Prog. Phys. Geogr.* 15:364–91

FAUNMAP Work. Group. 1994. FAUNMAP: A database documenting Late Quaternary distributions of mammal species in the United States. *Ill. State Mus. Sci. Pap.* 25:1–690

Ficcarelli G, Coltorti M, Moreno-Espinosa M, Pieruccini PL, Rook L, Torre D. 2003. A model for the Holocene extinction of the mammal megafauna in Ecuador. *J. South Am. Earth Sci.* 15:835–45

Fiedel SJ. 2005. Man's best friend—mammoth's worst enemy? A speculative essay on the role of dogs in Paleoindian colonization and megafaunal extinction. *World Archaeol.* 37:11–25

Fiedel SJ, Haynes G. 2004. A premature burial: comments on Grayson and Meltzer's "Requiem for overkill". *J. Archaeol. Sci.* 31:121–31

Field J, Dodson JR. 1999. Late Pleistocene megafauna and archaeology from Cuddie Springs, south-eastern Australia. *Proc. Prehist. Soc.* 65:275–301

Fisher DC. 1987. Mastodont procurement by Paleoindians of the Great Lakes Region: Hunting or scavenging? In *The Evolution of Human Hunting*, ed. MH Nitecki, DV Nitecki, pp. 309–421. Chicago, IL: Plenum

Fisher DC. 1996. Extinction of proboscideans in North America. In *The Proboscidea: Evolution and Palaeoecology of Elephants and Their Relatives*, ed. J Shoshani, P Tassy, pp. 296–315. New York: Oxford Univ. Press

Flannery TF. 1994. *The Future Eaters: An Ecological History of the Australasian Lands and People*. Chatswood: Reed Books. 432 pp.

Flannery TF, Roberts RG. 1999. Late Quaternary extinctions in Australasia: an overview. See MacPhee 1999, pp. A239–55

Forster MA. 2004. Self-organised instability and megafaunal extinctions in Australia. *Oikos* 103:235–39

Gingerich PD. 1984. Pleistocene extinctions in the context of origination-extinction equilibria in Cenozoic mammals. See Martin & Klein 1984, pp. 211–22

Gonzalez S, Kitchener AC, Lister AM. 2000. Survival of the Irish elk into the Holocene. *Nature* 405:753–54

Graham RW, Lundelius ELJ. 1984. Coevolutionary disequilibrium and Pleistocene extinction. See Martin & Klein 1984, pp. 223–49

Grayson DK. 1984. Nineteenth-century explanations of Pleistocene extinctions: a review and analysis. See Martin & Klein 1984, pp. 5–39

Grayson DK. 1991. Late Pleistocene mammalian extinctions in North America: taxonomy, chronology, and explanations. *J. World Prehist.* 5:193–231

Grayson DK. 2001. The archaeological record of human impacts on animal populations. *J. World Prehist.* 15:1–68

Grayson DK. 2006. Early Americans and Pleistocene Mammals in North America. In *Handbook of North American Indians: Environment, Origins, and Population*, Vol. 3, In press. Washington, DC: Smithson. Inst. Press

Grayson DK, Delpech F, Rigaud JP, Simek JF. 2001. Explaining the development of dietary dominance by a single ungulate taxon at Grotte XVI, Dordogne, France. *J. Archaeol. Sci.* 28:115–25

Grayson DK, Meltzer DJ. 2002. Clovis hunting and large mammal extinction: A critical review of the evidence. *J. World Prehist.* 16:313–59

Grayson DK, Meltzer DJ. 2003. A requiem for North American overkill. *J. Archaeol. Sci.* 30:585–93

Guilday JE. 1967. Differential extinction during late-Pleistocene and recent times. See Martin & Wright 1967, pp. 121–40

Guthrie RD. 1984. Mosaics, allelochemics, and nutrients: an ecological theory of Late Pleistocene megafaunal extinctions. See Martin & Klein 1984, pp. 259–98

Guthrie RD. 2003. Rapid body size decline in Alaskan Pleistocene horses before extinction. *Nature* 426:169–71

Guthrie RD. 2004. Radiocarbon evidence of mid-Holocene mammoths stranded on an Alaskan Bering Sea island. *Nature* 429:746–49

Hadly EA, Kohn MH, Leonard JA, Wayne RK. 1998. A genetic record of population isolation in pocket gophers (*Thomomys talpoides*) during Holocene climatic change. *Proc. Natl. Acad. Sci. USA* 95:6893–96

Hadly EA, Ramakrishnan U, Chan YL, van Tuinen M, O'Keefe K, et al. 2004. Genetic response to climatic change: Insights from ancient DNA and phylochronology. *PLos Biol.* 2:1600–9

Haynes G. 2002a. The catastrophic extinction of North American mammoths and mastodonts. *World Archaeol.* 33:391–416

Haynes G. 2002b. *The Early Settlement of North America: The Clovis Era*. Cambridge, UK: Cambridge Univ. Press. 345 pp.

Holdaway RN. 1999. Introduced predators and avifaunal extinction in New Zealand. See MacPhee 1999, pp. 189–238

Holdaway RN, Jacomb C. 2000. Rapid extinction of the moas (Aves: Dinorninthiformes): model, test, and implications. *Science* 287:2250–54

Houghton JT, Ding Y, Griggs DJ, Noguer M, van der Linden PJ, Xiaosu D, eds. 2001. *Climate Change 2001: The Scientific Basis*. New York: Cambridge Univ. Press

Hughes TP, Baird AH, Bellwood DR, Card M, Connolly SR, et al. 2003. Climate change, human impacts, and the resilience of coral reefs. *Science* 301:929–33

Jackson JBC, Kirby MX, Berger WH, Bjorndal KA, Botsford LW, et al. 2001. Historical overfishing and the recent collapse of coastal ecosystems. *Science* 293:629–38

Janzen DH, Martin PS. 1982. Neotropical anachronisms—the fruits the gomphotheres ate. *Science* 215:19–27

Johnson CN. 2002. Determinants of loss of mammal species during the Late Quaternary 'megafauna' extinctions: life history and ecology, but not body size. *Proc. R. Soc. London Ser. B* 269:2221–27

Johnson CN. 2005. What can data on the late survival of Australian megafauna tell us about the cause of their extinction? *Quat. Sci. Rev.* 24:2167–72

Johnson CN, Prideaux GJ. 2004. Extinctions of herbivorous mammals in the late Pleistocene of Australia in relation to their feeding ecology: No evidence for environmental change as cause of extinction. *Aust. Ecol.* 29:553–57

Johnson CN, Wroe S. 2003. Causes of extinction of vertebrates during the Holocene of mainland Australia: arrival of the dingo, or human impact? *Holocene* 13:941–48

Kelly RL, Todd LC. 1988. Coming into the country: Early Paleoindian hunting and mobility. *Am. Antiq.* 53:231–44

King JE, Saunders JJ. 1984. Environmental insularity and the extinction of the American mastodont. See Martin & Klein 1984, pp. 315–39

Koch PL, Fogel ML, Tuross N. 1994. Tracing the diets of fossil animals using stable isotopes. In *Stable Isotopes in Ecology and Environmental Science*, ed. K Lajtha, RH Michener, pp. 63–92. Oxford: Blackwell Sci.

Kuhn SL, Stiner MC. 2001. The antiquity of hunter-gatherers. See Panter-Brick et al. 2001, pp. 99–142

Kurtén B. 1968. *Pleistocene Mammals of Europe*. Chicago: Aldine. 317 pp.

Leonard JA, Wayne RK, Cooper A. 2000. Population genetics of ice age brown bears. *Proc. Natl. Acad. Sci. USA* 97:1651–54

Lindenmayer DB, McCarthy MA, Parris KM, Pope ML. 2000. Habitat fragmentation, landscape context, and mammalian assemblages in southeastern Australia. *J. Mamm.* 81:787–97

Lyons SK, Smith FA, Brown JH. 2004a. Of mice, mastodons, and men: human mediated extinctions on four continents. *Evol. Ecol. Res.* 6:339–58

Lyons SK, Smith FA, Wagner PJ, White EP, Brown JH. 2004b. Was a 'hyperdisease' responsible for the late Pleistocene megafaunal extinction? *Ecol. Lett.* 7:859–68

MacPhee RDE, ed. 1999. *Extinctions in Near Time: Causes, Contexts, and Consequences*. New York: Kluwer Acad./Plenum. 394 pp.

MacPhee RDE, Marx PA. 1997. The 40,000 year plague: humans, hyperdiseases, and first-contact extinctions. In *Natural Change and Human Impact in Madagascar*, ed. SM Goodman, BR Patterson, pp. 169–217. Washington, DC: Smithson. Inst. Press

MacPhee RDE, Tikhonov AN, Mol D, de Marliave C, Van der Plicht H, et al. 2002. Radiocarbon chronologies and extinction dynamics of the late Quaternary mammalian megafauna of the Taimyr Peninsula, Russian Federation. *J. Archaeol. Sci.* 29:1017–42

MacPhee RDE, Tikhonov AN, Mol D, Greenwood AD. 2005. Late Quaternary loss of genetic diversity in muskox (*Ovibos*). *BMC Evol. Biol.* 5:49

Magee JW, Miller GH, Spooner NA, Questiaux D. 2004. Continuous 150 k.y. monsoon record from Lake Eyre, Australia: Insolation-forcing implications and unexpected Holocene failure. *Geology* 32:885–88

Martin PS. 1966. African and Pleistocene overkill. *Nature* 212:339–42

Martin PS. 1970. Pleistocene niches for alien animals. *BioScience* 20:218–21

Martin PS. 1973. The discovery of America. *Science* 179:969–74

Martin PS. 1984. Prehistoric overkill: the global model. See Martin & Klein 1984, pp. 354–403

Martin PS. 2005. *Twilight of the Mammoths: Ice Age Extinctions and the Rewilding of America*. Berkeley: Univ. Calif. Press. 240 pp.

Martin PS, Klein RD, eds. 1984. *Quaternary Extinctions: A Prehistoric Revolution.* Tucson: Univ. Ariz. Press. 892 pp.

Martin PS, Steadman DW. 1999. Prehistoric extinctions on islands and continents. See MacPhee 1999, pp. 17–55

Martin PS, Wright HEJ, eds. 1967. *Pleistocene Extinctions: The Search for a Cause.* New Haven: Yale Univ. Press. 453 pp.

Martínez-Meyer E, Peterson AT, Hargrove WW. 2004. Ecological niches as stable distributional constraints on mammal species, with implications for Pleistocene extinctions and climate change projections for biodiversity. *Glob. Ecol. Biogeogr.* 13:305–14

Matheus PE. 1997. *Paleoecology and ecomorphology of the giant short-faced bear in eastern Beringia.* PhD thesis. Univ. Alaska, Fairbanks. 282 pp.

McDougall I, Brown FH, Fleagle JG. 2005. Stratigraphic placement and age of modern humans from Kibish, Ethiopia. *Nature* 433:733–36

Mellars P. 2004. Neanderthals and the modern human colonization of Europe. *Nature* 432:461–65

Meltzer DJ, Grayson DK, Ardila G, Barker AW, Dincauze DF, et al. 1997. On the Pleistocene antiquity of Monte Verde, southern Chile. *Am. Antiq.* 62:659–63

Miller GH, Fogel ML, Magee JW, Gagan MK, Clarke SJ, Johnson BJ. 2005. Ecosystem collapse in Pleistocene Australia and a human role in megafaunal extinction. *Science* 309:287–90

Miller GH, Magee JW, Johnson BJ, Fogel ML, Spooner NA, et al. 1999. Pleistocene extinction of *Genyornis newtoni*: Human impact on Australian megafauna. *Science* 283:205–8

Moen RA, Pastor J, Cohen Y. 1999. Antler growth and extinction of Irish Elk. *Evol. Ecol. Res.* 1:235–49

Moriarty KC, McCulloch MT, Wells RT, McDowell MC. 2000. Mid-Pleistocene cave fills, megafaunal remains and climate change at Naracoorte, South Australia: towards a predictive model using U-Th dating of speleothems. *Palaeogeogr. Palaeoclim. Palaeoecol.* 159:113–43

Mosimann JE, Martin PS. 1975. Simulating overkill by paleoindians. *Am. Sci.* 63:304–13

Mulvaney DJ, Kamminga J. 1999. *Prehistory of Australia.* Washington, DC: Smithson. Inst. Press. 480 pp.

Olsen EM, Heino M, Lilly GR, Morgan MJ, Brattey J, et al. 2004. Maturation trends indicative of rapid evolution preceded the collapse of northern cod. *Nature* 428:932–35

O'Regan HJ, Turner A, Wilkinson DM. 2002. European Quaternary refugia: a factor in large carnivore extinction? *J. Quat. Sci.* 17:789–95

Owen-Smith N. 1987. Pleistocene extinctions: The pivotal role of megaherbivores. *Paleobiology* 13:351–62

Owen-Smith N. 1999. The interaction of humans, megaherbivores, and habitats in the late Pleistocene extinction event. See MacPhee 1999, pp. 57–70

Panter-Brick C, Layton RH, Rowley-Conwy P, eds. 2001. *Hunter-Gatherers: An Interdisciplinary Perspective.* Cambridge, UK: Cambridge Univ. Press. 341 pp.

Post E, Forchhammer MC. 2002. Synchronization of animal population dynamics by large-scale climate. *Nature* 420:168–71

Pounds JA, Bustamante MR, Coloma LA, Consuegra JA, Fogden MPL. 2006. Widespread amphibian extinctions from epidemic disease driven by global warming. *Nature* 439:161–67

Prado JL, Alberdi MT, Azanza B, Sanchez B, Frassinetti D. 2005. The Pleistocene Gomphotheriidae (Proboscidea) from South America. *Quat. Int.* 126/128:21–30

Raymo ME. 1994. The initiation of northern hemisphere glaciation. *Annu. Rev. Earth Planet. Sci.* 22:353–83

Roberts RG, Flannery TF, Ayliffe LK, Yoshida H, Olley JM, et al. 2001. New ages for the last Australian megafauna: Continent-wide extinction about 46,000 years ago. *Science* 292:1888–92

Robinson GS, Burney LP, Burney DA. 2005. Landscape paleoecology and megafaunal extinction in southeastern New York state. *Ecol. Monogr.* 75:295–315

Rolett B, Diamond J. 2004. Environmental predictors of pre-European deforestation on Pacific islands. *Nature* 431:443–36

Root TL, Price JT, Hall KR, Schneider SH, Rosenzweig C, et al. 2003. Fingerprints of global warming on wild animals and plants. *Nature* 421:57–60

Scheffer M, Carpenter S, Foley JA, Folke C, Walker B. 2001. Catastrophic shifts in ecosystems. *Nature* 413:591–96

Scheinsohn V. 2003. Hunter-gatherer archaeology in South America. *Annu. Rev. Anthropol.* 32:339–61

Shapiro B, Drummond AJ, Rambaut A, Wilson MC, Matheus PE, et al. 2004. Rise and fall of the Beringian steppe bison. *Science* 306:1561–65

Slaughter BH. 1967. Animal ranges as a clue to late-Pleistocene extinction. See Martin & Wright 1967, pp. 155–67

Slaughter R, Skulan J. 2001. Did human hunting cause extinction? *Science* 294:1459–62

Smith FA, Lyons SK, Ernest SKM, Jones KE, Kaufman DM, et al. 2003. Body mass of late Quaternary mammals. *Ecology* 84:3403

Solé RV, Alonso D, McKane A. 2002. Self-organized instability in complex ecosystems. *Philos. Trans. R. Soc. London Ser. B* 357:667–81

Steadman DW, Martin PS. 2003. The late Quaternary extinction and future resurrection of birds on Pacific islands. *Earth Sci. Rev.* 61:133–47

Steadman DW, Martin PS, MacPhee RDE, Jull AJT, McDonald HG, et al. 2005. Asynchronous extinction of late Quaternary sloths on continents and islands. *Proc. Natl. Acad. Sci. USA* 102:11763–68

Steneck RS, Graham MH, Bourque BJ, Corbett D, Erlandson JM, et al. 2002. Kelp forest ecosystems: biodiversity, stability, resilience and future. *Environ. Conserv.* 29:436–59

Stuart AJ. 1991. Mammalian extinctions in the Late Pleistocene of northern Eurasia and North America. *Biol. Rev.* 66:453–562

Stuart AJ. 1999. Late Pleistocene megafaunal extinctions: a European perspective. See MacPhee 1999, pp. 257–70

Stuart AJ, Kosintsev PA, Higham TFG, Lister AM. 2004. Pleistocene to Holocene extinction dynamics in giant deer and woolly mammoth. *Nature* 431:684–89

Stuart AJ, Sulerzhitsky LD, Orlova LA, Kuzmin YV, Lister AM. 2002. The latest woolly mammoths (*Mammuthus primigenius* Blumenbach) in Europe and Asia: a review of the current evidence. *Quat. Sci. Rev.* 21:1559–69

Stuiver M, Reimer PJ, Bard E, Beck JW, Burr GS, et al. 1998. INTCAL98 radiocarbon age calibration, 24,000–0 cal BP. *Radiocarbon* 40:1041–83

Terborgh J, Lopez L, Nuñez VP, Rao M, Shahabuddin G, et al. 2001. Ecological meltdown in predator-free forest fragments. *Science* 294:1923–26

Thieme H. 1997. Lower Palaeolithic hunting spears from Germany. *Nature* 385:807–10

Thomas CD, Cameron A, Green RE, Bakkenes M, Beaumont LJ, et al. 2004. Extinction risk from climate change. *Nature* 427:145–48

Trueman CNG, Field JH, Dortch J, Charles B, Wroe S. 2005. Prolonged coexistence of humans and megafauna in Pleistocene Australia. *Proc. Natl. Acad. Sci. USA* 102:8381–85

Van Valkenburgh B. 1993. Tough times at La Brea: tooth breakage in large carnivores of the late Pleistocene. *Science* 261:456–59

Waguespack NM, Surovell TA. 2003. Clovis hunting strategies, or how to make out on plentiful resources. *Am. Antiq.* 68:333–52

Wahlberg N, Moilanen A, Hanski I. 1996. Predicting the occurrence of endangered species in fragmented landscapes. *Science* 273:1536–38

Walther GR, Post E, Convey P, Menzel A, Parmesan C, et al. 2002. Ecological responses to recent climate change. *Nature* 416:389–95

Wayne RK, Leonard JA, Cooper A. 1999. Full of sound and fury: The recent history of ancient DNA. *Annu. Rev. Ecol. Syst.* 30:457–77

Whittington SL, Dyke B. 1984. Simulating overkill—experiments with the Mosimann and Martin model. See Martin & Klein 1984, pp. 451–65

Williams JW, Webb T III, Richard PJH, Newby P. 2000. Late Quaternary biomes of Canada and the eastern United States. *J. Biogeogr.* 27:585–607

Winterhalder B. 2001. The behavioural ecology of hunter-gatherers. See Panter-Brick et al. 2001, pp. 12–38

Winterhalder B, Lu F. 1997. A forager-resource population ecology model and implications for indigenous conservation. *Conserv. Biol.* 11:1354–64

Woodburne MO, Swisher CC III. 1995. Land mammal high resolution geochronology, intercontinental overland dispersals, sea-level, climate, and vicariance. In *Geochronology, Time Scales and Global Stratigraphic Correlation*, ed. WA Berggren, DV Kent, M-P Aubry, J Hardenbol. SEPM Spec. Publ. 54, pp. 336–64

Wroe S, Field J, Fullagar R, Jermin LS. 2004. Megafaunal extinction in the Late Quaternary and the global overkill hypothesis. *Alcheringa* 28:291–31

Zimov SA, Chuprynin VI, Oreshko AP, Chapin FS III, Reynolds JF, et al. 1995. Steppe-tundra transition: A herbivore-driven biome shift at the end of the Pleistocene. *Am. Nat.* 146:765–94

Innate Immunity, Environmental Drivers, and Disease Ecology of Marine and Freshwater Invertebrates

Laura D. Mydlarz, Laura E. Jones, and C. Drew Harvell

Department of Ecology and Evolutionary Biology, Cornell University, Ithaca, New York 14853; email: ldm42@cornell.edu, lej4@cornell.edu, cdh5@cornell.edu

Key Words

cellular immunity, ecological immunity, outbreak, prophenoloxidase pathway

Abstract

Despite progress in the past decade, researchers struggle to evaluate the hypothesis that environmental conditions compromise immunity and facilitate new disease outbreaks. In this chapter, we review known immunological mechanisms for selected phyla and find that there are critical response pathways common to all invertebrates. These include the prophenoloxidase pathway, wandering phagocytic cells, cytotoxic effector responses, and antimicrobial compounds. To demonstrate the links between immunity and the environment, we summarize mechanisms by which immunity is compromised by environmental conditions. New environmental challenges may promote emergent disease both through compromised host immunity and introduction of new pathogens. Such challenges include changing climate, polluted environment, anthropogenically facilitated pathogen invasion, and an increase in aquaculture. The consequences of these environmental issues already manifest themselves as increased mortality on coral reefs, pathogen range expansion, and transmission of disease from aquaculture to natural populations, as we summarize in a final section on recent marine epizootics.

INTRODUCTION

Host-pathogen interactions are strongly governed by the environment (Harvell et al. 2004). Recent increases in disease outbreaks among marine organisms (Ward & Lafferty 2004) may be caused either by introduction of new pathogens or a change in the environment (Harvell et al. 2004, Lafferty et al. 2004, McCallum et al. 2004). Host resistance is a first line of defense and its success is a major determinant of whether a new pathogen or a changed environment will result in a disease outbreak. Thus the hypothesis that changing environmental conditions compromise host immunity and lead to disease outbreaks should be carefully considered. This review critically evaluates this hypothesis by examining recent advances in immune mechanisms of marine invertebrates, the known interactions of immunity and environment, and the evidence for compromised immunity in recent outbreaks of disease in the ocean.

The invertebrate immune system is based on self/nonself recognition and cellular and humoral processes. To date, no true adaptive components have been identified in these innate systems (Soderhall & Cerenius 1998), though elements suggesting memory and specificity are seen in invertebrates as basal as sponges (Bigger et al. 1982, Hildemann et al. 1979). Exploration of immunity in invertebrates is dominated by mechanistic studies of model organisms, with little attention to natural populations (Little et al. 2005). Thus, identifying what is known about key components of the innate immune system in different invertebrate phyla, and how these are affected by changing environmental conditions, remains a critical priority. Invertebrates are especially attractive study systems owing to the nonadaptive nature of their innate immune systems, the propensity they have for straightforward experimental manipulation, and the diversity of environments they inhabit.

From invertebrate model systems we have learned that the primary components of innate immunity fall into three categories that define the effectiveness of an immune response. First, the organism distinguishes between self and nonself; second, the organism mounts a defensive response that can kill or disable the invader; and finally, the organism recognizes and can eliminate its own damaged or diseased cells. These requirements lead to the three essential components of innate immunity: phagocytosis (cell-mediated); activation of humoral responses leading to opsonization, melanization, and coagulation (cell-free); and the production of humoral antimicrobial compounds (cell-free).

Below we review immunity in phylogenetically widespread invertebrates, including basal phyla such as Porifera and Cnidaria and representative phyla from Lophotrochozoa, Ecdysozoa, and Deuterostomia. We focus on marine phyla that are ecologically, environmentally, and commercially relevant, that are not model organisms, and that have suffered substantial losses because of disease outbreaks. Most of the groups we select are affected by changing climate, water quality, and pollutants. We then examine how environmental factors, namely climate change and environmental pollution, modulate these immune responses. Finally, we explore the ecological effects of diseases in invertebrates by documenting disease outbreaks in natural populations.

PRIMARY ELEMENTS OF THE INNATE IMMUNE SYSTEM IN MARINE AND FRESHWATER INVERTEBRATES

In the following and in **Table 1**, we describe by phylum the known elements of the innate immune system in marine invertebrates, focusing on the following where known: (*a*) recognition responses such as lectins, pattern recognition proteins, cell adhesion factors; (*b*) cell-free or humoral responses characterized by antibacterial peptides, small molecule antifungal and antibacterial compounds; (*c*) cellular responses, such as physio-chemical barriers (melanin), phagocytosis, granulated vesicles, antioxidant enzymes, and adhesive proteins; (*d*) communication and integration of immune function pathways including prophenyloxidase, complement system, cytokines, eicosanoids, and Toll-like receptors (TLRs), for example; and (*e*) evidence for primitive immune memory and specificity.

TLR: Toll-like receptor

Phylum Porifera

Sponges are the most basal of the Metazoa. Not only are true tissues absent in Porifera, but most body cells are totipotent—capable of changing function and form as needed. Thus despite the fact that they are large, multicellular animals, in many ways they function like organisms that are unicellular in complexity.

Recognition. As early as the latter part of the nineteenth century, Wilson (1891) demonstrated the remarkable capacity of sponge cells to reaggregate after being dissociated. This surprisingly complex behavior was later found to involve a number of molecules and signaling pathways. Adhesion of sponge cells involves the molecules galactin, integrin, fibronectin or fibronectin-like proteins, and collagen (Brower et al. 1997, Labat-Robert et al. 1981, Muller et al. 1999). In the sponge *Microciona prolifera*, calcium ions serve as an intracellular messenger in stimulus-response coupling during cell-cell aggregation (Dunham et al. 1983, Weissman et al. 1986), and carbohydrate self-recognition mediates cellular adhesion (Haseley et al. 2001). Allogenic rejection has a cellular component: Interactions between incompatible sponges activate exopinacocytes (cells forming the external surface of the sponge) and various mesohyl cells, some of which contain the secondary metabolites so common in sponges (Gaino et al. 1999).

Responses: cell-free. A large number of bioactive compounds, many of which have potential pharmacological applications (e.g., antimicrobial, antifungal, antiinflammatory, antitumor, and cytoxic bioactivities) have been discovered in sponges. Some of these compounds are products of the sponge cells themselves, and some are secreted by symbionts (e.g., Becero & Paul 2004, Rifai et al. 2004).

Responses: cell-based. Archeocytes are wandering, totipotent, amoeboid cells capable of differentiating into virtually every other cell type present in a sponge. They are large, phagocytic cells that play a major role in digestion, defense, and food transport. As such, they possess digestive enzymes, including protease, lipase, and amylase, and can accept phagocytized material from other cells, for example, the choanocytes

Table 1 Invertebrate immune parameters

Organism/Phyla	Pathogen/elicitor	Immune parameter characterized	Reference[2]
Porifera			
Sponge *Ephydatia mulleri*	N/S[1]	Self-recognition, cell-cell adhesion, fibronectin-like protein	1
Sponge *Microciona prolifera*	N/S	Carbohydrate self-recognition, cellular adhesion and aggregation, activation of protein kinase C, calcium signaling	2
Demosponge *Geodia cydonium*	N/S	Adhesion molecules: galactin, integrin, fibronectin, collagen; allogenic rejection involves upregulation of phenylalanine hydroxylase, an enzyme initiating the pathway to melanin synthesis	3
Sponges (four spp.)	N/S	Alloimmune memory revealed by consistently accelerating second-set graft rejections	4
Sponge	N/S	Enhanced self/nonself recognition upon second exposure to foreign sponge tissue	5
Sponge	N/S	Activation of exopinacocytes and mesohyl (granule-bearing) cell types	6
Sponge *Suberites domuncula*	Gram-positive bacteria	Activation of endocytosis, lysozyme release	7
Cnidaria			
Sea anemone	N/S	Self/nonself recognition	8
Sea anemone *Anemonia viridis*, Coral *Goniopora stokesi*	N/S	Antioxidant enzymes, superoxide dismutase, catalase, glutathione peroxidase associated with granulated vesicles	9
Sea anemone *Aiptasia pallida*	N/S	Nitric oxide synthase activity	10
Gorgonian coral *Swiftia exserta*	Foreign particles (inert)	Activation of phagocytic cells	11
Gorgonian coral *Gorgonia ventalina*	*Aspergillus sydowii* (fungal pathogen)	Melanin, lipid antifungal and antibacterials	12
Scleractinian coral	N/S	Immune specificity, primitive memory	13
Scleractinian coral	Cyanobacteria, other microbes	Mucus accumulation and shedding, secretion of antimicrobial compounds, nitric oxide synthase activity	14
Gorgonian and scleractinian corals	Bacterial	Wound healing by motile phagocytic cells, tissue reorganization	15
Gorgonian coral *Plexaurella fusifera*	N/S	Wound healing by amoebocytes	16
Hydroid *Hydractinia*	Fungal, nematode pathogens	Chitinase expression	17
Sea anemone	Gram-negative bacteria	Amoebocytes engage in phagocytosis, reactive oxygen species production, soluble microbials	18
Mollusca			
Marine bivalves	Bacterial pathogens	Phagocytosis, humoral components (lectins, lysozymes), reactive oxygen species, small antimicrobial peptides	19

(Continued)

Table 1 (*Continued*)

Organism/Phyla	Pathogen/elicitor	Immune parameter characterized	Reference[2]
Marine bivalves	Protozoan *Perkinsus*	Phagocytosis and encapsulation by hemocytes, secretion of toxic polypeptides, opsonization	20
Clam *Ruditapes philippinarum*	Self-nonself recognition, protozoan *Perkinsus* sp.	Self/nonself recognition by lectins, lectin binds specifically to the surface of purified hypnospores from *Perkinsus* spp.	21
Marine bivalves	Bacterial	Hemocytes activity; mobile, highly clumping, phagocytic, chemotactic, and bactericidal	22
Oyster *Crassostrea virginica*	Bacteria: *Vibrio* spp., (normal gut flora)	Cell-free hemolymph agglutinizes *Vibrio cholerae*, but not 79 other bacterial strains	23
Oyster *Crassostrea gigas*	Viruses	Oyster hemolymph has detectable antiviral activity against HSV, IPNV	24
Oyster *Saccostrea glomerata*	Protozoan *Marteilia sydneyi*	Phenoloxidase activity induced and suppressed in infected areas	25
Bivalves, *Mytilus edulis*, *Chlamys islandica*	Bacterial	Lysozymes and small antibacterial proteins	26
Crustacea			
Crustaceans, arthropods	Self-nonself	Induction of prophenoloxidase triggered by lippopolysaccarides, peptidoglycans	27
Crayfish	N/S	Cell adhesion, encapsulation, nodule formation, degranulation, opsonization, and phagocytosis	28
Horseshoe crab *Tachypleus tridentatus*	Wound-healing	Induced coagulation cascade, prophenoloxidase, Toll-like receptor	29
Horseshoe crab *Tachypleus tridentatus*	Bacterial	Pathogen recognition: lectins	30
Crab *Carcinus maenas*	Bacterial/microbial	Hemocytes produce antibacterial protein	31
Spiny lobster	Gram-negative bacteria	Induced bactericidal molecules. Memory: secondary bactericidal responses enhanced over primary responses	32
Cladoceran *Daphnia magna*	Bacterial	Specificity: strain-specific immunity conferred to offspring	33
Copepod *Macrocyclops albidus*	Parasitic tapeworm	Memory and specificity	34
Shrimp, blue crab, crayfish	Variety of viruses	Antiviral substances bind specifically to a variety of RNA and DNA viruses	35
Shrimp *Penaeus monodon*	Bacterial, variety of viruses	Lysozyme activity and antimicrobial peptides. Protein with C-lectin-like domain with strong antiviral activity	36
Echinodermata			
Sea star	Bacterial	Lysozyme activity, reactive oxygen species production	37
Sea star *Asterias rubens*	Nonspecific	Primitive cytokines and cytokine receptors	38
Sea urchin *Lytechinus pictus*	Nonspecific; evidence of memory	Immune response to grafted tissue similar to vertebrates, accelerated rejection rates with second-set allografts	39

(*Continued*)

Table 1 (*Continued*)

Organism/Phyla	Pathogen/elicitor	Immune parameter characterized	Reference[2]
Sea urchin	Microbial	Humoral agents: lectins, agglutinins, and lysins. Complement proteins analoguous to vertebrate complement C3	40
Sea urchin *Strongylocentrotus purpuratus*	Bacterial	Complement homologue SpC3 functions as Opsonin	41
Urochordata			
Colonial ascidians	Self/nonself	Allorecognition, same colonies fuse while nonrelated colonies form lesions. C-type lectins	42
Tunicates *Halocynthia aurantium, Styela plicata*	Bacterial	Antibacterial peptides	43
Tunicate	Lippopolysaccharides	Cell migration, chemotaxis	44
Tunicates *Ciona intestinalis, Boltenia villosa*	N/S	Toll-like receptors	45

[1]N/S, not specific.
[2]1) Labat-Robert et al. 1981; 2) Dunham et al. 1983, Weissman et al. 1986, Haseley et al. 2001; 3) Muller et al. 1999; 4) Bigger et al. 1982; 5) Hildemann et al. 1979; 6) Gaino et al. 1999; 7) Thakur et al. 2005; 8) Lubbock 1980; 9) Hawkridge et al. 2000; 10) Morall et al. 2000; 11) Olano & Bigger 2000; 12) Petes et al. 2003, Kim et al. 2000a,b; 13) Hildemann et al. 1977; 14) Koh 1997; 15) Bigger & Hildemann 1982, Mullen et al. 2004; 16) Meszaros & Bigger 1999; 17) Mali et al. 2004; 18) Hutton & Smith 1996; 19) Canesi et al. 2002; 20) Villalba et al. 2004; 21) Bulgakov et al. 2004; 22) Pruzzo et al. 2005; 23) Tamplin & Fisher 1989; 24) Olicard et al. 2005; 25) Peters & Raftos 2003; 26) Olsen et al. 2003, Cerenius & Soderhall 2004; 27) Nilsen et al. 1991; 28) Johanssen 1999, Holmblad & Soderhall 1999; 29) Nagai & Kawabata 2000, Nagai et al. 2001, Inamori et al. 2004; 30) Saito et al. 1997, Inamori et al. 1999; 31) Relf et al. 1999; 32) Weinheimer et al. 1969, Evans et al. 1968, Evans et al. 1969; 33) Little et al. 2003; 34) Kurz & Franz 2003; 35) Pan et al. 2000; 36) Sotelo-Mundo et al. 2003, Bachere et al. 2000; 37) Bachali et al. 2004, Coteur et al. 1999; 38) Legac et al. 1996; 39) Coffaro & Hinegardner 1977; 40) Gross et al. 1999, Al-Sharif et al. 1998; 41) Clow et al. 2004; 42) Khalturin et al. 2004; 43) Jang et al. 2002, Tincu et al. 2003; 44) Raftos et al. 1991; 45) Azumi et al. 2003.

PPG: peptidoglycan

(specialized collar cells). As the primary macrophage of the sponge, however, they are nonselective in their phagocytosis, thus not specifically immune cells. Nonetheless, intracellular digestion functions as a powerful component of immunity in sponges. Sponge lysozyme is released by mesohyl-type cells after exposure to peptidoglycan (PPG) from the cell walls of Gram-positive bacteria (Thakur et al. 2005).

Primitive immune memory. Consistent with the early discovery of self-recognition in sponges, isografts of sponges fuse compatibly, but allografts are invariably rejected. In addition, reaction times of sponges to allografts on second and third exposure are reduced, suggesting short-term immune memory. Indeed, although sponges lack an "organized" circulatory system, the immunologic memory spreads rapidly through the body of the sponge and lingers for several weeks (Bigger et al. 1982, Hildemann et al. 1979). Finally, in the desmosponge *Geodia cydonium*, allogenic rejection involves the upregulation of phenylalanine hydroxylase, which is an enzyme initiating the pathway to melanin formation (Muller et al. 1999).

Phylum Cnidaria

The Cnidaria comprise a diverse phylum that includes jellyfish, soft and stony corals, sea anemones, and the laboratory model system *Hydra*.

Recognition. Sessile colonial invertebrates, like the Cnidaria, have the ability and basic need to distinguish between their own tissues and those of unrelated conspecifics (Hidaka 1985). Typically, genetically identical or closely related colonies will fuse, whereas unrelated conspecifics reject. Cnidarians employ a diverse set of responses against allogenic tissues; these include extrusion of mesenterial filaments, and growth of sweeper tentacles and hyperplastic stolons. Allorecognition responses within the hydractinians are by far the best understood, and thus may serve as an example (Grosberg et al. 1997). There are three possible outcomes after contact between two *Hydractinia* colonies: fusion, transitory fusion, and rejection (Cadavid 2005). In fusion, compatible colonies adhere to one another and can eventually share a gastrovascular system. In rejection, fusion does not occur and interacting tissues swell owing to migration of nematocysts to the contact area (Cadavid 2005, Cadavid et al. 2004).

SOD: superoxide dismutase
CAT: catalase
GPX: glutathione peroxidase
ROS: reactive oxygen species

Responses: cell-free. The antioxidant enzymes, superoxide dismutase (SOD), catalase (CAT), and glutathione peroxidase (GPX) are associated with granulated vesicles, the endosymbiotic algae, and with cnidae (nematocysts). SOD and CAT are localized in all forms of cnidae. The presence of both SOD and CAT in the accumulation bodies of endosymbiotic algae is consistent with a hypothesized role of these bodies in digestion and cell-aging (Hawkridge et al. 2000). Nitrate oxide synthase activity has been observed in the sea anemone *Aiptasia pallida* (Morall et al. 2000). In the hydroid *Hydractinia* sp., chitinase is expressed in epidermal tissues following metamorphosis in response to fungal pathogens and nematode parasites; it is not detectable in larval stages, however (Mali et al. 2004). Gorgonian corals produce antifungal and antibacterial compounds (Jensen et al. 1996; Kim et al. 2000a,b). Crude extracts from the sea fan *Gorgonia ventalina* inhibit germination of fungal spores and have antibacterial activity against the bacterium *Listonella anguillarum* (Kim et al. 2000a). Antifungal and antibacterial activity is highest at the colony edges (youngest tissue) and lowest in the medial and central portions of the fan (Kim et al. 2000b). Lipid extracts from scleractinian corals are active against six species of marine bacteria (Koh 1997).

Responses: cellular. Like the sponges, cnidarians also possess motile phagocytic cells. In gorgonian and scleractinian corals, amoebocytes aid in wound healing and tissue reorganization (Bigger & Hildemann 1982). In the Scleractinia, these amoebocytes are few, scattered within mesoglea. In gorgonians they are abundantly scattered throughout the thicker mesoglea (Mullen et al. 2004). In gorgonians, trauma induces activation of phagocytic cells, including granular amoebocytes; it furthermore induces phagocytosis in cells not usually phagocytic. Phagocytic cells in this case include granular amoebocytes, epidermal cells, sclerocytes, mesogleal cells, and gastrodermal cells (Olano & Bigger 2000), all of which respond to the presence of bacteria. In sea anemones challenged with Gram-negative bacteria, amoebocytes engage in phagocytosis, reactive oxygen species (ROS) production, and production of soluble microbials (Hutton & Smith 1996). In *Plekaurella fusifera*, motile amoebocytes aid in wound healing (Meszaros & Bigger 1999). In gorgonian corals, infection by *Aspergillus sydowii* induces a band of melanization (a physical barrier formed from polymerized quinones) locally around the fungal lesion as well as nodule formation (Mullen et al. 2004, Petes et al. 2003).

Primitive immune memory. Tissue transplantation immunity in scleractinian and gorgonian corals has a specific (short-term) memory component, as shown by significantly accelerated rejection rates for second-set grafts (Hildemann et al. 1977, Salter-Cid & Bigger 1991).

Phylum Mollusca (Bivalvia)

The molluscs are a diverse phylum of coelomates that includes the familiar snails, slugs, clams, mussels, and octopi. We limit discussion here to the Bivalvia (clams, oysters, mussels, etc.). Owing to their filter-feeding habits, bivalves accumulate large numbers of bacteria, which are both a source of nourishment and an immune challenge.

Recognition. As in other invertebrates, bivalves possess a suite of adhesion molecules that aid in self/nonself recognition and bind reversibly to carbohydrate-containing molecules of foreign cells. The properties of lectins and their recognition roles in bivalve host defense are well documented in several commercial species, including mussels and oysters (Bulgakov et al. 2004, Fisher 1992). Among these lectins is the recently isolated manila clam lectin, which binds to the surface of hypnospores from *Perkinsu*s sp., a protozoan parasite of manila clams and other bivalves (Bulgakov et al. 2004).

Responses: cell-free. Among the many weapons in the chemical arsenal of bivalves are humoral defense factors such as agglutinins (lectins), ROS, antimicrobial peptides, and lysozymal enzymes (Canesi et al. 2002). Bivalve lysozymes have both immune and digestive function (abundant in both the stylus and digestive gland); immune-defense and digestive functions may thus act contemporaneously, as bacteria filtered from the water represent both nourishment and a threat. In bivalves, multiple lysosymes are involved in self-defense against pathogenic bacteria (Olsen et al. 2003). For example, in the bivalve *Chlamys islandica*, an antibacterial lysosyme-like protein (chlamysn) isolated from the viscera inhibits all growth of Gram-positive and -negative bacteria. Interestingly, this protein is active at cold temperatures but remains stable and active when heated (Nilsen et al. 1999). In addition, bivalve hemocytes secrete antimicrobial polypeptides; hemocytes from the clam *Tapes decussatus* secrete a polypeptide that is toxic to *Perkinsus* sp. (Villalba et al. 2004). There is relatively little literature on the antiviral defenses of bivalves; protease-inhibiting peptide isolated from the Pacific oyster *Crassostrea gigas* was effective against HIV-1 (Lee & Maruyama 1998), and oyster hemolymph was found to have broad antiviral activity against HSV, IPNV, and other viruses (Olicard et al. 2005).

Responses: cellular. Circulating cells or hemocytes are primarily responsible for immune defense. Hemocytes are mobile, clumping, phagocytic, and chemotactic

(Pruzzo et al. 2005). Bacterial susceptibility to hemocyte attack in different bivalve species may be due to variation in phagocytic hemocyte affinity to different bacterial products, the presence (or absence) of appropriate opsonizing molecules, and the abilities of hemocytes to bind to the surface of the bacteria (Canesi et al. 2002). Surface interactions between bacteria and hemocytes lead to rapid bacterial clearance. Bacterial cells with fimbria (fringes) are cleared more rapidly owing to enhanced cell-cell adhesion with hemocytes (Canesi et al. 2001). Hemocytes in the eastern oyster, *Crassostrea virginica*, recognize and phagocytose *Perkinsus marinus* cells. However, in contrast to filtered bacteria, hemocytes have only limited ability to kill *P. marinus* merozoites, the reproductive life-stage of the parasite (La Peyre 1993).

LPS: lipopolysaccharides
proPO: prophenoloxidase
PO: phenoloxidase

Phylum Arthropoda (Class Crustacea)

The Arthopoda comprises coelomate organisms with a diversity of form, size, and lifestyle. Crustaceans (including crabs, lobster, and shrimp) are the most abundant and ecologically relevant in the context of innate immunity of the marine arthropods.

Recognition. Fibrogen-related molecules in hemolymph function as self/nonself recognition lectins in the horseshoe crab (Gokudan et al. 1999, Kairies et al. 2001). Horseshoe crab lectins recognize specific antigens of bacterial lipopolysaccharides (LPS) (Inamori et al. 1999, Saito et al. 1997). Conversion of prophenoloxidase (proPO) to active phenoloxidase (PO), and thus initiation of the melanization pathway, is thought to be part of the self/nonself recognition system in arthropods, and is triggered by very small amounts of molecules such as LPS, peptidoglycan, and β-1,3 glucans (Cerenius & Soderhall 2004).

Responses: cell-free. Crustacean hemocytes produce an array of antimicrobial compounds as an efficient means of protection against systemic coelomic infection. Granular hemocytes in the shore crab (*Carcinus maenas*) produce an antibacterial protein active against Gram-positive marine bacteria (Relf et al. 1999). In the spiny lobster and rock lobster, induced bactericidal molecules have broad activity, covering a wide range of pathogens (Tzvetnenko et al. 2001, Weinheimer et al. 1969). In shrimp, blue crab, and crayfish, antiviral substances bind specifically to a variety of RNA and DNA viruses (Pan et al. 2000). In the white shrimp *Penaeus monodon* lysozyme activity is reported, and penaedins, antimicrobial peptides, isolated in the white shrimp are active against viruses, including Parvo-like viridae, Picornaviridae, and Baculoviridae (Bachere 2000, Bachere et al. 2000, Sotelo-Mundo et al. 2003).

Responses: cellular. Cell adhesion is necessary for immune responses, including encapsulation, nodule formation, opsonization, and phagocytosis [observed in crayfish (Johanssen 1999) and cladoceran (Little et al. 2004)]. Hemolymph proteins in crustaceans may facilitate the cell adhesion, which leads to initiation of phagocytosis and encapsulation. For example, peroxinectin, a hemolymph protein in crayfish (stored in granular hemocytes) is released concomitant with proPO activation, triggering cell adhesion, encapsulation, enhanced phagocytosis, and opsonization of foreign cells.

Peroxinectin and extracellular SOD may cooperate during respiratory burst to destroy ingested parasites (Holmblad & Soderhall 1999). Upon activation of the proPO pathway, enhanced cell adhesion activity is observed in black tiger shrimp hemolymph (Sritunyalucksana et al. 2001).

Communication and integration of immune responses. A primary means of defense in arthropods, including crustaceans, is the melanization of pathogen and damaged tissue, which is orchestrated by PO. The activation of the proPO pathway is triggered by the presence of potential pathogens, which in turn signals the induction of many other immune responses, including generation of cytotoxic and antimicrobial factors, and compounds that facilitate opsonization and encapsulation (Cerenius & Soderhall 2004). The TLR tToll plays a role in cell-signaling after wounding in horseshoe crab (Inamori et al. 1999). Horseshoe crab coagulation cascade promotes proPO activation, leading to conversion of hemocyanin to PO, and PO participates in wound healing, that is, repair of damaged exoskeleton (Nagai & Kawabata 2000, Nagai et al. 2001).

Primitive immune memory. There is evidence from a number of Crustacea that is suggestive of immune memory. Secondary bactericidal responses to Gram-negative bacteria are enhanced over primary responses in the spiny lobster (Evans et al. 1968, 1969). In addition, there are signs of immune specificity associated with adaptive immunity in higher metazoa: strain-specific immunity is conferred to offspring in the cladoceran *Daphnia magna* (Little et al. 2003) by mothers infected by a bacterial pathogen. Evidence of specificity and memory is found in copepods infected with a tapeworm parasite, with reduced success of infection on consecutive exposures to antigenically similar parasites (Kurz & Franz 2003).

Phylum Echinodermata

This phylum comprises some 7000 identified living species, among these the familiar sea star and sea urchin. They are coelomate Deuterosomes, are radially symmetric (fivefold symmetry) as adults, possess a calcareous endoskeleton arising from mesodermal tissue, and a distinguishing coelomic water-vascular system composed of fluid-filled canals.

Recognition. Empirical evidence implying the existence of mechanisms for determining self/nonself is seen in the rejection of unrelated conspecific allografts in sea urchins. In the sea urchin *Lytechinus pictus*, allografts of unrelated conspecifics were rejected within a month; second and third set grafts were rejected within 12 days, even after two-months delay following the first graft (Coffaro & Hinegardner 1977).

Responses: cell-free. Echinoderm coelomocytes produce a battery of humoral factors, including lectins, agglutinins, lysins, and soluble microbials (Gross et al. 1999). A sea star (bacterial) lysozyme has been identified, with the inferred function of the digestion of bacteria (Bachali et al. 2004). In the sea star, ROS produced by sea star immunocytes is stimulated by direct contact with bacteria or bacterial wall proteins (Coteur et al. 1999).

Responses: cellular. Echinoderms, like all other metazoans, possess motile coelomocytes, some of them phagocytic. The coelomocytes are produced in the the axial gland (axial organ), which is thought to be an ancestral lymphoid organ (Legac et al. 1996). Coelomocytes can be divided into five categories by morphology and physiology: spherulocytes, crystal cells, progenitor cells, vibratile cells, and amoebocytes (phagocytes) (Nair et al. 2005). As they are differentiated cells, coelomocytes are not specifically immune cells; in asteroids (as amoebocytes) and in urchins they phagocytose nonselectively, scavenging waste products from the body fluids. Yet in purple urchins, coelomocytes mount the cellular response to immune challenge via opsonization and phagocytosis, encapsulation, and production of antimicrobial agents (Gross et al. 1999). The cellular immune response of the sea star *Asterias rubens* consists of two differentiated amoebocyte lines: phagocytic macrophage-like cells and lymphocyte-like cells. Experiments show that the lymphocyte-like cells, upon stimulation, further subdivide into analogues of the vertebrate B and T cell lines (Legac et al. 1996). B-like and T-like cells have specific, nonoverlapping roles in cellular immunity (Leclerc & Brillouet 1988). Further empirical evidence for specialized function in sea star immune cell types is found in more recent experiments employing bacterial challenge. Sea star coelomocytes (amoebocytes) were observed to differentiate into two types, one of which responds very rapidly to bacteria; by contrast, the second is immunomodulated, responding in a density-dependent fashion to the pathogen (Coteur et al. 2002).

Communication and integration of immune response. There is evidence both of primitive complement and cytokine-like systems in the echinoderms. A simple complement system has been identified in purple sea urchin that is homologous to the vertebrate alternative pathway, is inducible by a challenge, and further activates coelomocytes (Gross et al. 1999). Sea urchin coelomocytes secrete complement proteins analogous to vertebrate complement component C3 (Al-Sharif et al. 1998). Coelomocyte gene expression in response to LPS results in a diverse complement system (Nair et al. 2005). Sea urchin complement homologue SpC3 functions as an opsonin (Clow et al. 2004). In the sea star *Asterias rubens*, molecules with interleukin-like function (IL-1 and IL-2-like) and receptors for these molecules have been identified (Legac et al. 1996).

Primitive immune memory. Evidence exists of immune memory in echinoderms. The sea urchin rejects allografts of unrelated conspecifics within 30 days. Acceptance of allografts is only observed in inbred animals, with the likelihood of acceptance increasing with the degree of inbreeding. In addition, there is accelerated rejection of second-set allografts, suggesting both primitive memory and specificity (Coffaro & Hinegardner 1977).

Phylum Chordata, Subphylum Urochordata (Tunicata)

As members of the phylum Chordata, Tunicates or Urochordates are bilaterally symmetric, possessing a dorsal notochord and pharyngeal slits at some point in their development. As chordates, the Tunicata have been unusually well studied because they are the closest of the invertebrate phyla in body plan and development to vertebrates, and thus deserve mention here.

Recognition. The tunicates employ a cell-based allogenic response, and natural killer (NK)-type cells expressing a C-type lectin binding domain are deployed in the task of recognizing and eliminating allogenic tissue (Khalturin et al. 2004). Pairs of *Botryllus schlosseri* colonies placed together either fuse, or develop lesions and reject within 24 hours. Participating in the rejection of incompatible colonies are the morula cells (see below), which gather at the site of tissue interaction and cause the formation of points of rejection (Khalturin et al. 2004). Elements of the lectin-mediated activation pathway such as galectins, C3-like component and M*asp*-like protease have been isolated from the tunicate *Clavelina picta* (Vasta et al. 1999).

Responses: cell-free. Many antimicrobial peptides have been isolated from hemocytes of the Urochordata; the following are select examples. Halocidin was isolated from the solitary tunicate *Halocynthia aurantium*. Halocidin and its synthetic congeners demonstrated antibacterial activity against the antibiotic-resistant bacteria *Staphylococcus aureus* and *Pseudomonas aeruginosa* (Jang et al. 2002). Plicatamide is a small octapeptide found in the hemocytes of *Styela plicata*, which exhibited fast-acting, broad antimicrobial activity (Tincu et al. 2003). Genes for the lytic pathway have been identified in the *Ciona intestinalis* genome. This pathway consists of soluble hemolymph proteins with similar domains and organization to terminal vertebrate complement components C6–C9 (Azumi et al. 2003).

Responses: cellular. The tunicates are endowed with a suite of differentiated hemocytes, all with specialized function. In the solitary tunicate *Styela clava* there are two types of granulocytes, phagocytic hyaline cells, and lymphocyte-like hemoblasts (Sawada et al. 1993). Although hyaline cells are most aggressively phagocytic, granulocytes also actively phagocytize, and are more mobile and abundant. Hemoblasts are lymphocyte-like cells, and are also thought to function as proliferative stem cells. These observations are recently confirmed and expanded in the colonial ascidian *Botrylloides leachi*, which was found to have hemoblasts as a circulating stem cell pool and at least five specialized hemocyte differentiation pathways. These pathways lead variously to a phagocytic cell line (hyaline cells and macrophage-like cells), a cytotoxic line (granulocyte-like cells and morula cells), a vacuolated cell line involved in storage of catabolites, compartment amoebocytes, and a second granulated cell line including highly migratory trophic cells (Cima et al. 2001).

Communication and integration of immune response. An interleukin 1-like molecule (cytokine-like) stimulates proliferation of tunicate immune cells, both granulocyte and lymphocyte-like (Raftos et al. 1991). In addition, bacterial LPS significantly enhances the cell migration in tunicate hemocytes, and directional chemotaxis is stimulated by two tunicate hemolymph proteins. Hemocyte migration under chemotactic stimulation is both directional and two to three times faster than untreated controls (Raftos et al. 1998). The tunicates *Ciona intestinalis* and *Boltenia villosa* utilize TLRs and the corresponding signaling pathways (Azumi et al. 2003, Davidson & Swalla 2002).

Synopsis

Ability to distinguish self is a basic property of all metazoans. It is well developed in sponges, the most basal of the Metazoa, and is a cell-based system involving a number of molecules and signaling pathways, including a calcium-dependent signaling system and the molecules galactin, integrin, fibronectin-like proteins, and collagen (Brower et al. 1997, Labat-Robert et al. 1981, Muller et al. 1999). Sessile colonial invertebrates, including most Cnidaria, distinguish between their own tissues and those of unrelated conspecifics. Closely related (sharing fusibility alleles) or genotypically exact colonies fuse, whereas less-related conspecifics reject (Grosberg et al. 1997). In the Crustacea, self/nonself recognition arises both through fibrogen-related molecules in hemolymph that function as self/nonself recognition lectins (Gokudan et al. 1999, Kairies et al. 2001) and TLR-mediated activation of the proPO pathway leading to melanization. Note that the cnidarians possess TLRs, but lack key transcriptional factors belonging to the NF-kB family, which are required in Toll-signaling pathways (Zheng et al. 2005). These animals also activate the melanization pathway, but it is likely mediated in a slightly different fashion given the incomplete nature of the Toll-signaling pathway in the Cnidaria.

Phagocytosis is perhaps the most primitive mode of defense, and wandering phagocytic cells are found in all metazoans. As observed in the Porifera, phagocytosis is primarily a means of acquiring nutrients with a secondary defense function. It thus seems likely that in more complex animals, nutritional phagocytosis evolved to be (or was co-opted as) a means of immune defense. Note that in the cnidarians, trauma induces phagocytosis in cells that normally have other functions. Cnidarian amoebocytes respond to presence of local bacterial invaders by becoming phagocytic and engaging in production of soluble microbials (Hutton & Smith 1996). Coelomate invertebrates have specialized cell lines. The most complex cellular differentiation described here is observed in the deuterosome tunicates and echinoderms, as both possess proliferating lymphocyte-like immune cells as well as nonproliferating phagocytic cell lines. In the crustaceans and other coelomates, enhanced phagocytosis appears to be tied to the PO pathway; upon activation of this pathway, heightened phagocytosis along with cell adhesion and encapsulation are triggered (Zheng et al. 2005).

Chemical warfare is another important defense tool for invertebrates. Sponge and cnidarian cells, or the symbionts they host, secrete a large number of antioxidants and bioactive compounds, many of which have potential pharmacological applications (Becero & Paul 2004, Rifai et al. 2004). Animals possessing a true body cavity (here, the molluscs, the crustaceans, the echinoderms, and the tunicates) likely developed the ability to secrete antimicrobial peptides, mediated by a signaling cascade employing TLRs, as an efficient means of protection against systemic coelomic infection (Zheng et al. 2005). Note that in coelomate invertebrates, for example the sea urchin, coelomic cavity fluid is rapidly cleared of microbes by phagocytic activity, suggesting an efficient signaling mechanism.

ENVIRONMENTAL FACTORS AND IMMUNOCOMPETENCE

Environmental stressors have long been thought to negatively impact invertebrate immune systems, although information linking environmental pressures, reduced immunity, and disease outbreaks are few (LeMoullac & Haffner 2000, Malham et al. 2003). Altered environmental conditions can affect immunity directly, by changing components of the immune responses, or indirectly, by inducing general stress responses. In invertebrate studies, deciphering the effects of environmental stressors as direct or indirect responses on immunity is complex and remains relatively mechanistic. These types of studies, however, represent the first step toward integrating specific physiological information with immune function, and thus permit prediction and modeling of the potential effects of environmental factors on immune response. In **Table 2** and this section, we review documented links between environmental stress and impaired immunity in experimental situations and in natural populations of marine and aquatic invertebrates.

Chemical Pollutants

The field of ecotoxicology aids in the understanding of how chemical pollutants affect immunocompetence of invertebrates (see review by Galloway & Depledge 2001). Many studies link exposure to heavy metals, polycyclic aromatic hydrocarbons, and pesticides to altered immune function in laboratory studies, but establishing a direct link between changes in immune function and increased susceptibility to disease has remained elusive. In addition, the studies of the effects of anthropogenic stress and pollutant stress in natural populations in situ remain scarce (Galloway & Depledge 2001).

Studies using bivalves such as the mussel *Mytilus edulis* (Coles et al. 1995, Pipe et al. 1999), the oyster *Crassostrea virginica* (Anderson 1994, Baier-Anderson & Anderson 2000), crustaceans including the prawn *Macrobrachium rosenbergii* (Cheng & Wang 2001) and the lobster *Nephrops norvegicus* (Hernroth et al. 2004), and echinoderms such as sea star *Asterias rubens* (Coteur et al. 2005), have demonstrated that cellular responses are affected by the metals copper, cadmium, and manganese. These studies are laboratory-based and employ controlled dosages of the pollutant, yielding straightforward results. Hemocyte counts and phagocytic activity are observed to decrease and the production of ROS to increase with heavy metal stress. When natural populations of invertebrates are exposed to environmental pollution, the data are more ambiguous. In three French rivers, Vasseur & Leguille (2004) transported

Table 2 Environmental factors and immunocompetence

Stressor	Organism	Immune function measured	Reference*
Pollutants			
Biocides	Oyster *Crassostrea virginica*	Hemocyte activation, phagocytosis, reduced pyridine nucleotides and ROS production	1
Copper	Mussel *Mytilus edulis*	Hemocyte counts, pathogenic activity	2
Copper	Oyster *Crassostrea virginica*	Hemocyte function, ROS production	3
Copper sulfate	Prawn *Macrobrachium rosenbergii*	Hemocyte density, PO activity, respiratory burst	4
Cadmium	Mussel *Mytilus edulis*	Immune response alteration	5
Manganese	Lobster *Nephrops norvegicus*	Hemocyte counts and degranulation, PO activity, expression of the Runt-domain protein	6
Cadmium	Sea star *Asterias rubens*	Phagocytosis, production of ROS	7
Natural pollution salinity gradient	Sea star *Asterias rubens*	Coelomic amoebocyte concentration (CAC) and production of ROS by amoebocytes	8
Harbor dredging	Shrimp *Crangon crangon*	Total hemocyte count, PO activity	9
Sites contaminated with heavy metals, sewage and industrial outfalls	Mussel *Mytilus edulis*	Molecular characteristics of the hemocytes, phagocytosis, superoxide generation	10
Natural river pollution	Bivalve *Unio* sp.	Glutathione, glutathione reductase, selenium dependent glutathione peroxidase, nonselenium-dependent glutathione peroxidase, lipid peroxidation	11
Immunostimulants	Lobster *Homarus gammarus*	Granulocyte activity in vitro	12
Physical and Chemical Changes			
Temperature	Bivalve *Mya arenaria*	Mitochondrial respiration, energetic coupling to phosphorylation, production of ROS in mitochondria	13
Temperature	Bivalve *Laternula elliptica*	ROS production by mitochondria	14
Temperature	Sponge *Axinella polypoides*	Electrophysiological methods, ADP-ribosyl cyclase, abscisic acid production	15
Temperature	Coral *Acropora grandis*	Intracellular calcium and a series of stress-proteins, including heme oxygenase	16
Seasonal Temperature	Crab *Carcinus maenus*	Hemocyte antibacterial activity	17
Temperature	Shrimp *Penaeus californiensis*	Hemolymph proPO and plasma protein	18
Temperature and copper	Mussel *Mytilus edulis*	Total and differential hemocyte counts, production of intracellular superoxide, phagocytosis	19

(Continued)

Table 2 *(Continued)*

Stressor	Organism	Immune function measured	Reference*
Temperature and copper	Coral *Porites cylindrica*	Stress measured as respiration and primary production	20
Temperature, pollutants, injury	Sea Urchin	Coelomocytes expression of heat shock protein (hsp70)	21
Elevated pO_2 and solar irradiance	Sponge *Petrosia ficiformis*	Oxygen radical scavengers; superoxide dismutase, catalase, glutathione peroxidases and total oxyradical scavenging capacity (small molecules)	22
Hypoxia	Oyster *Crassostrea virginica*	Production of ROS in hemocytes to produce	23
Hypoxia	Shrimp *Penaeus stylirostris*	Hemocyte counts, hemocyte differentiation, respiratory burst, PO activity, mortality	24
Hypercapnic hypoxia	Crab *Callinectes sapidus*	Hemolymph clearance of bacteria	25
Hypercapnic hypoxia	Shrimp *Litopenaeus vannamei*, *Palaemonetes pugio*	Total hemocyte count, mortality	26
Hypoxia and TBT	Oyster *Crassostrea virginica*	Total hemocyte counts, mortality, ROS production, lysozyme activity	27
Other Disturbances			
Mechanical disturbance	Abalone *Haliotis tuberculata*	Number of circulating hemocytes and migratory activity, phagocytosis, respiratory burst, noradrenaline and dopamine levels	28
Wounding	Amphipod *Gammarus pulex*	PO activity	29
Wounding and feeding conditions	Crustacean *Daphnia magna*	PO activity	30

*1) Baier-Anderson & Anderson 2000; 2) Pipe et al. 1999; 3) Anderson 1994; 4) Cheng & Wang 2001; 5) Coles et al. 1995; 6) Hernroth et al. 2004; 7) Coteur et al. 2005; 8) Coteur et al. 2003; 9) Smith et al. 1995; 10) Dyrynda et al. 1998; 11) Vasseur & Leguille 2004; 12) Hauton & Smith 2004; 13) Abele et al. 2002; 14) Heise et al. 2003; 15) Zocchi et al. 2001; 16) Fang et al. 1997; 17) Chisholm & Smith 1994; 18) Vargas-Albores et al. 1998; 19) Parry & Pipe 2004; 20) Nyström et al. 2001; 21) Matranga et al. 2000; 22) Regoli et al. 2000; 23) Boyd & Burnett 1999; 24) Le Moullac et al. 1998; 25) Holman et al. 2004; 26) Mikulski et al. 2000; 27) Anderson et al. 1998; 28) Malham et al. 2003; 29) Plaistow et al. 2003; 30) Mucklow & Ebert 2003.

the bivalve *Unio* sp. to control sites and to sites up- and downstream from sources of pollution. They measured antioxidant and lipid peroxide levels. Only in one of three rivers did bivalves downstream from a pollutant source exhibit decreased antioxidant activity and higher lipid peroxidation indicative of impaired immune response. The other two rivers were equally if not more contaminated, but the bivalves did not show any significant changes in antioxidant levels, illustrating the difficulty in examining one environmental parameter, such as pollution, in a continually changing environment.

In an English study, Dyrynda et al. (1998) examined mussels from known contaminated sites and comparable uncontaminated sites, and measured immune components including total hemocyte count, hemocyte phagocytic activity, generation of superoxide anion, N-acetyl-glucosamidase actitvity, chymotrypsin activity, lysosomal volume, peroxidase, and PO activity. Of the immune parameters, only hemocyte production of superoxide anion and glucosaminidase were different in mussels from all three contaminated sites over uncontaminated sites, while other immune parameters were different at only one or two sites. Again, this study of natural populations in contaminated sites exemplifies the variability of immune defenses between different environmental locations and different levels of contamination. Although results from these studies are ambiguous relative to controlled experiments, they are nonetheless extremely important in characterizing impaired immune response in invertebrates associated with contamination, and in establishing appropriate invertebrate indicator species and relevant immune measurements.

Temperature

Temperature stress, both warming and cooling, can compromise physiological and immune function in freshwater and marine organisms. In the bivalves *Mya arenia* (Abele et al. 2002) and *Laternula elliptica* (Heise et al. 2003), temperature stress caused mitochondrial decoupling owing to ROS membrane damage. This type of physiological stress can impair defense responses against pathogens or lead to direct oxidative damage to tissues. In the sponge *Axinella polypoides*, temperature stress induced a signaling cascade that involved heat-gated cation channels and the stimulation of abscisic acid, a scavenger of reactive oxygen (Zocchi et al. 2001). In the coral *Acropora grandis*, heat stress induced the production of heat shock proteins and heme-oxygenases (Fang et al. 1997). A more direct influence of temperature on disease was characterized in the mussel *Mytilus edulis* (Parry & Pipe 2004), the freshwater prawn *Macrobrachium* (Cheng & Chen 1998), and the brown shrimp *Penaeus californiensis* (Vargas-Albores et al. 1998). In the mussel, temperature and copper stress caused significant decrease in hemocyte function to clear the bacterial pathogen *Vibrio tubiashii*. Increasing temperature and pH between 8.8 and 9.5 caused increased mortality in the freshwater prawn infected by bacteria relative to controls. Increasing water temperature by approximately 12°C above ambient for the brown shrimp caused decreased activity of the prophenyloxidase system (Vargas-Albores et al. 1998). Also important when considering temperature stress is that decreases as well as increases in temperature may compromise immunity, and seasonal temperature change leaves certain invertebrates vulnerable to disease at specific times of the year. For example, the antimicrobial properties of hemocytes from the crab *Carcinus maenus* were the least active when water temperatures dropped in February and reached their highest activity in August (Chisholm & Smith 1994).

For organisms with obligate algal symbionts, temperature stress uniquely affects immune function and general physiology. Invertebrates including anemones, sponges, and corals harbor photosynthetic dinoflagellates that can cause daily changes in the oxidative state of the host tissues (Regoli et al. 2000; Richier et al. 2001, 2003). This has led to symbiotic organisms having an enhanced or more efficient antioxidant system and increased tolerance to oxidative stress. Although these studies have not examined the relationship of oxidative stress tolerance to disease, the role of ROS and the ability to regulate the oxidative state of tissues during pathogen invasion and utilize reactive oxygen as a respiratory burst is an important aspect of defense. For example, in amoebocytes isolated from sea anemones, phagocytosis and production of reactive oxygen conferred antibacterial activity and reduced growth of the Gram-negative bacterium *Psychrobacter immobilis* (Hutton & Smith 1996).

Although symbiosis may provide enhanced tolerance to changes in the oxidative state of tissues, which in turn can provide some immune protection, symbiosis gone awry can cause a state of hyperoxia and harm the tissue of the host. The disruption of the fine balance between symbiont and host can have dire physiological consequences. The coral bleaching phenomenon is an example of the complete breakdown between symbiosis, causing mass mortality and potentially leaving corals vulnerable to disease. Cellular events concomitant with bleaching include the release of intact endoderm host cells and the symbionts they contain, suggesting that thermal stress causes dysfunction in cnidarian host cell adhesion (Gates et al. 1992). The direct physiological and mechanistic link between water temperature warming, bleaching, and impaired immunity in corals has not yet been examined and is further discussed in this article.

Environmental Hypoxia

High summer temperatures and continued climate warming can cause more frequent seasonal hypoxia, which can be accompanied by elevated levels of CO_2 and lowered pH. Hypercapnia, hypoxia, and low pH can affect benthic, sessile organisms, especially those living in estuarine habitats, and these conditions have been associated with mortality and disease outbreaks (Boyd & Burnett 1999). Several groups have conducted studies to understand the association of disease-related mortality and hypoxic conditions. In the oyster *Crassostrea virginica* (Boyd & Burnett 1999), hypoxia caused a reduction in the ability of oyster hemocytes to produce ROS, thereby reducing their ability to kill pathogen invaders. In another study, *Crassostrea virginica* (Anderson et al. 1998), naturally infected with the parasite *Perkinsus marinus*, were subjected to the pollutant tributyltin and hypoxia. Although tributyltin alone did not cause mortality above control levels, hypoxia exacerbated this effect and increased oyster mortality synergistically with tributyltin. The effects of hypercapnia and hypoxia on bacterial clearance was tested in the crab *Callinectes sapidus* (Holman et al. 2004) and the shrimps *Litopenaeus vannamei* and *Palaemonetes pugio* (Mikulski et al. 2000). In these experiments organisms kept in hypercapnic hypoxia were exposed to live pathogenic bacteria. In the shrimp *Litopenaeus vannamei*, and in the crab *Callinectes sapidus*, total hemocyte counts following bacterial infection were reduced in the hypoxia treatments. In both shrimp species, mortality was increased relative to controls in the hypercapnic hypoxia treatments.

The relationship between hypoxia and ability to fight pathogens is very important when considering the long-term effects of increasing ocean acidity and water temperature. Benthic sediment habitats play crucial roles in ecosystem processes (by providing the water column with key nutrients, for example) and the bacterial and large animal communities they support are important to maintaining this function. Bioturbation of the sediments by shrimp, crabs and urchins is crucial to nutrient exchange to support phytoplankton growth and subsequently other members of the food chain. If disease occurrences aggravated by more frequent hypoxia cause a shift in the sediment-dwelling invertebrate communities, the impact on the benthic ecosystem water column and even characteristics of the topography and sediment of a tidal flat may be large (Mouritson et al. 1998).

Physical Injury

Fluctuations in environmental conditions are not the only extrinsic factors that can affect immunity. Physical injury, wounds, and food supply can alter immune parameters. Henry & Hart (2005, and references therein) discuss the effects on immunocompetence of wound regeneration in corals, sponges, and other invertebrates. They suggest that the resources allocated for allorecognition and cell-mediated immune responses may be depleted during regeneration of wounds. In addition, cells involved in both wound healing and immunity, such as amoebocytes (Olano & Bigger 2000) and archeocytes, may be limited in their ability to fight disease during active regeneration of wounds. This constitutes double jeopardy, as wound sites provide entry for pathogenic microorganisms.

Malham et al. (2003) studied the mollusc *Haliotis turberculata* using a mechanical disturbance similar to handling in aquaculture to evaluate the effects of physical pressure on the immune system. The effects of this stress were transient, but marked decrease in hemocyte numbers, migratory ability, and phagocytic ability of *Haliotis* hemocytes were observed. Their experiments demonstrate a direct link between physical stress and impaired immunity. Whether the physical stress is natural, such as in the case of predation or hurricanes, or anthropogenic injury, such as bottom trawling or handling in aquacultural settings, these factors cannot be ruled out when considering susceptibility to disease and may contribute not only to impaired immunity but may provide opportunistic pathogens entryway into the organism.

Very few studies provide a direct link between any meaningful component of immune function and susceptibility to disease. In work with the crustacean *Gammarus pulex* (Plaistow et al. 2003), animals with more naturally occurring wounds (as measured by counting melanized spots) exhibited less proPO activity, indicating a decreased ability to produce melanin as a cellular response to infection. In the copepod *Daphnia magna* (Mucklow & Ebert 2003, Mucklow et al. 2004), experimentally wounding individuals induced higher levels of proPO 24 hours postwounding. Higher levels of proPO were also observed in well-fed *Daphnia*. Furthermore, those animals with higher proPO activity were more resistant to infection by the bacterial pathogen *Pasteuria ramosa*. Mucklow & Ebert's experiment exemplifies the link between health of the organism and resistance to infection as measured by an important immune parameter, melanin production.

Synopsis

The use of field populations to study the links between anthropogenic effects, impaired immunity, and disease is complicated by natural variability in phenotype and genotype, seasonal variation in level of defense, and local environmental variation in study sites. For the field of ecotoxicology this is an ongoing issue; laboratory experiments clearly demonstrate causal links between exposure to pollutants and impaired immune function, but verification of these relationships in the field remains problematic (Galloway & Depledge 2001). As demonstrated in this review, field surveys of natural populations exposed to pollutants can be ambiguous and often raise questions about the nutritional and reproductive status of the organism studied. In addition there are the confounding effects of local environment and season. Future challenges are to resolve these issues so relationships between anthropogenic effects and disease are more conclusive and thus may inform conservation and environmental management efforts.

Also important to conservation and environmental management are the effects of a changing climate on immune parameters. The main outcomes of a warming ocean are decreased oxygen levels in coastal waters and increased prolonged temperature anomalies. Although several studies show that hypoxia can directly affect immunocompetence of marine invertebrates, the same cannot be said for temperature effects. Based on terrestrial models, temperature can affect pathogen growth and range, which in turn can increase the severity of disease (Harvell et al. 2002), but the mechanistic links between temperature and immune function are not yet resolved.

Many studies that do incorporate temperature, pH, and other environmental conditions into the study of invertebrate disease are driven by the aquaculture industry. In shrimp and oyster aquaculture disease limits production and therefore economic returns (Bachere et al. 2004). Problems that have risen from the aquaculture industry have exposed the lack of existing knowledge on basic immune defenses of invertebrates. In culture, invertebrates are densely populated, subject to changing environmental conditions, such as in oxygen and pH, and increased concentration of potential pathogens (Bachere et al. 2004). To counteract these problems and prevent disease, some industries have begun adding immunostimulants to the diets of the farm-raised invertebrates. In one study of this practice (Hauton & Smith 2004), the immunostimulants (glucans, bacterial cell walls) intended to boost immune response in the lobster *Homarus gammarus* turned out to be mildly cytotoxic to lobster hemocytes.

DISEASE OUTBREAKS INVOLVING MARINE INVERTEBRATES

The major disease outbreaks recorded in marine and freshwater invertebrates over the past 30 years are summarized in **Table 3**. Diseases of marine invertebrates are the focus of this review, as there are fewer recorded events in freshwater invertebrates, and include outbreaks in invertebrates from most phyla, from sponges to echinoderms. The groups best represented are those of commercial or ecological importance, such as mollusks, crustaceans, corals, and echinoderms. This unequal distribution of disease outbreaks amoung phylogenetic groups and emphasis on commercially important groups suggests that diseases of marine invertebrates are seriously understudied, especially if there are no cataloged cases of disease outbreaks in major groups like polychaetes, urochordates, or nematodes. Similar imbalance occurs with freshwater invertebrates. Although records of disease or parasite outbreaks exist for snails (Lively & Jokela 1996) and the crustacean *Daphnia* (Ebert 2005, Mitchell et al. 2004, Mucklow & Ebert 2003), there are scant records of diseases for the majority of freshwater invertebrates. The work of Ebert and Little's groups on epizootics of *Daphnia* represents some of the most comprehensive work in a natural, aquatic invertebrate population. *Daphnia magna* is infected with three microsporidian parasites, and a bacterial pathogen (*Pasteuria ramose*) (Ebert 2005). Several of these infections reach epizootic proportions under different environmental conditions. Overall, the greatest epizootics correlated with warm temperatures and fishless conditions, both situations that trigger high host densities (Ebert 2005). Capaul & Ebert (2003) detected microevolutionary change in laboratory microcosms of *D. magna* populations exposed to microsporidians. Although the frequencies of the different clones shifted, it is not known if these populations became more resistant after selection. Mucklow & Ebert (2003) showed that wounded *D. magna* have upregulated PO. This group has done pioneering work on evolutionary ecology of disease in natural populations of freshwater *Daphnia*, and yet important questions remain about the links between immunity and disease susceptibility.

Observations of diverse invertebrates in the ocean suffering epizootics prompt the question, do other freshwater molluscs, such as unionid clams, or even zebra mussels, escape disease outbreak? Or is it merely that such events remain undetected and underreported in freshwater? Understanding the true distribution of disease among invertebrate groups is important because it yields clues to the interplay between immunity and disease risk. For example, documenting what we suspect is a disproportionate level of disease outbreaks in corals might suggest either a particular vulnerability to disease in cnidarians, or an unusual change in environmental conditions, such as temperature stress, which allows opportunistic pathogens to invade.

Another pattern observable in **Table 3** is the paucity of data on viral diseases. Typically pathogens reported to cause outbreaks in marine invertebrates are bacterial or protozoan, with fungi also accounting for significant mortality, even though viruses are present in seawater at concentrations between 10^6–10^8/ml seawater (Wilson et al. 2005). Most of the data available on viral diseases and immunity comes from commercially important invertebrates. Viral diseases are known in farmed shrimp and in some bivalves (Lightner & Redman 1998, Renault & Novoa 2004, Shields & Behringer 2004, Stentiford et al. 2004), and viral-like particles have been reported in the coral *Pavona danai* (Wilson et al. 2005) and the sea anemones *Metridium senile* and *Anemonia viridis* (Wilson & Chapman 2001, Wilson et al. 2005). To date no viral diseases are reported to cause cnidarian die offs and no antiviral defenses in this group have been identified. Antiviral compounds have been detected in oysters (Olicard et al. 2005), and in the shrimp *Fenneropenaeus indicus*, where oral administration of inactivated white spot syndrome virus protects the animal from subsequent infection

Table 3 Disease outbreaks among natural populations of selected marine invertebrates adapted and updated from Kim et al. 2005. For each reported outbreak, location, pathogen identity, if known, and environmental correlates are shown: T = temperature, ND = no data, sal = salinity, turb = turbidity, hur = hurricane, ? = unknown.

Start date	Host species	Outbreak location	Pathogen identity	Estimated mortality (%)	Env. correlates	Reference[*]
1938	Sponges	N Caribbean	fungus?	70–95	ND	1
1946	*Crassostrea* (oyster)	Gulf Coast, USA	*Perkinsus marinus*	extensive	high T, sal	2
1967	*Crassostrea angulata* (oyster)	France	iridovirus	extensive	ND	3
1975	*Heliaster* (starfish)	W USA	?	>90	high T	4
1980	*Strongylocentrotus* (urchin)	NW Atlantic	amoeba?	>50	ND	5
1980	*Ostrea* (oyster)	Netherlands	*Bonamia ostreae*	extensive	ND	6
1981	*Acropora* (coral)	Caribbean-wide	bacterium?	>90	ND	7
1982	*Gorgonia* (coral)	Central America	?	extensive	high T	8
1982	*Haliotis* (abalone)	Australia	*Perkinsus* sp.	extensive	high T	9
1983	Scleractinian Corals	Caribbean-wide	microbial consortium	seasonal	?	10
1983	*Patinopecten* (scallop)	W Canada	*Perkinsus qugwadi*	extensive	ND	11
1983	*Diadema* (urchin)	Caribbean-wide	bacterium?	>95	high T	12
1985	*Paralithodes platypus* (crab)	?	herpes virus	?	?	13
1985	*Haliotis* (abalone)	NE Pacific	?	>95	high T	14
1986	*Ostrea chilensis* (oyster)	New Zealand	*Bonamia exitiosa*	>90	dredging	15
1987	*Ruditapes philippinarum* (clam)	France	*Vibrio tapetis*	?	T	16
1988	*Argopecten* (scallop)	N Caribbean	protozoan	extensive	ND	1

Year	Host	Location	Pathogen	Mortality	Conditions	Ref.
1989	*Argopecten* (scallop)	E Canada	*Perkinsus karlssoni*	extensive	ND	17
1991	*Macrobrachium rosenbergii*	Taiwan	*Metschnikowia bicuspidata*	>50	ND	18
1995	*Strongylocentrotus* (urchin)	Norway	nematode?	~90	ND	19
1995	*Gorgonia* (coral)	Caribbean-wide	fungus	extensive	ND	20
1995	*Dichocoenia* and others (coral)	Florida, USA	bacterium	<38	seasonal	21
1996	*Diploria* and others (coral)	Puerto Rico	bacterium	extensive	seasonal, hur	22
1996	*Tapes decussates* (clam)	NW Spain	*Perkinsus olseni*	?	high T	23
1997	*Litopenaeus setiferus* (shrimp)	SE USA	WSSV Virus	<10	ND	24
1999	Gorgonian corals	NW Mediterranean	protozoan/fungi?	>20	high T	25
2000	*Crangon crangon* (shrimp)	Scotland	bacilliform virus (CcBV)	100	?	26
2002	*Montipora* (coral)	Australia	bacterial consortium	extensive	high T	27
2002	*Panulirus argus* (lobster)	SE USA	virus	≤37	?	28
2002	*Homarus americanus* (lobster)	NE USA	not isolated	extensive	high T	29
2002	*Echinopora/Montipora* (coral)	E Africa	fungi?	extensive	ND	30
2003	krill	NW USA	*Collinia* sp (ciliate)	ND	ND	31
2004	*Mercenaria* (clams)	New York, USA	QPX	<10	ND	32

*1) Galtsoff et al. 1939, Smith 1939; 2) Soniat 1996; 3) Renault & Novoa 2004; 4) Dungan et al. 1982; 5) Miller & Colodey 1983, Scheibling & Stephenson 1984, Jones & Scheibling 1985; 6) van Banning 1991; 7) Antonius 1981, Gladfelter 1982; 8) Guzmán & Cortéz 1984; 9) Lester et al. 1990; 10) Rützler et al. 1983, Carlton & Richardson 1995; 11) Bower et al. 1998; 12) Lessios et al. 1984; 13) Sparks & Morado 1986; 14) Lafferty & Kuris 1993; 15) Cranfield et al. 2005; 16) Paillard 2004; 17) McGladdery et al. 1991; 18) Su et al. 2005; 19) Skadsheim et al. 1995; 20) Nagelkerken et al.1997; 21) Richardson et al. 1998; 22) Bruckner & Bruckner 1997; 23) Villalba et al. 2005; 24) Chapman et al. 2004; 25) Cerrano et al. 2000; 26) Stentiford et al. 2004; 27) Jones et al. 2004; 28) Shields & Behringer 2004; 29) Dove et al. 2004a; 30) Dove et al. 2004b; 31) Gómez-Gutiérrez et al. 2003; 32) Dove et al. 2004b.

by white spot virus (Bright et al. 2005). In addition, the physiological effects of viruses on infected organisms of commercial value are being documented; for example oxidative stress and depressed levels of antioxidant defenses were detected in shrimp infected with white spot virus (Mohankumar & Ramasamy 2006). From the work on commercial invertebrate species, inferences may be drawn about the viral immunity of related, ecologically important species.

Also evident (**Tables 2** and **3**) is the observation that increasing temperature may be an important environmental factor and a possible facilitator of disease outbreaks in the ocean, even though understanding of the mechanistic effects of temperature on immunity is in its infancy. Below we further examine hypotheses behind this linkage, using two disease systems as examples, molluscan infection by *Perkinsus* and coral bleaching and disease.

- *Perkinsus*. Well-documented outbreaks involving temperature as a facilitator involve the protozoan *Perkinsus* spp. in various commercially important bivalve species. *Perkinsus* suppresses host immune responses, such as lysozyme activity, as well as causes hemocyte mortality (Romestand et al. 2002). *Perkinsus* outbreaks represent a clear example of disease range expansion due to climate warming. In both the Southeast and in Northern Europe the geographic range of the *Perkinsus* spp. that infects oysters expands in warmer months and is geographically limited by cold winters (Cook et al. 1998). *Perkinsus* spread in eastern oyster (*Crassostrea virginica*) populations from New York to Maine during warming winters in the 1990s (Cook et al. 1998). These range expansions combined with the effects of warming waters and increased hypoxia on bivalve immunity (Anderson et al. 1998, Parry & Pipe 2004) make the *Perkinsus* epizootic a compelling example of the complex interactions between host immunocompetence, parasite resilience, and environment. Gaffney & Bushak (1996) showed variation in resistance among strains of oyster exposed to *Perkinsus*, although the mechanism of immunity is unknown.

- Coral bleaching. The best example of climate warming specifically facilitating disease outbreaks in compromised hosts is likely to be corals. Corals have undergone massive bleaching events that are directly driven by warm sea surface temperatures (Hoegh-Guldberg & Salvat 1995). Coral bleaching occurs when the symbiotic algae are expelled from coral tissue during thermal stress, causing the coral to appear white or bleached. This is a nonlethal event unless the temperature stress is of protracted duration. However, bleaching is widely hypothesized to trigger susceptibility to opportunistic microorganisms, because stressed hosts are known to be more susceptible to disease. In situ observations (Bruckner & Bruckner 1997, Harvell et al. 2001, Jones et al. 2004; see also Selig et al. 2006) show disease outbreaks occurring concurrently with bleaching events when warmer than usual water temperatures are present. For example, in the 1998 El Niño warming event in the Caribbean, Harvell et al. (2001) recorded a mass mortality of the gorgonian, *Briareum asbestinum*, caused by a cyanobacterial pathogen attacking bleached colonies. Jones et al. (2004) recorded a similar event in scleractinain corals on the Great Barrier Reef in

2002. Selig et al. (2006) documented a link between warm temperature anomalies and disease outbreaks of white syndrome in Australia in 2002.

Missing from these studies, however, is an understanding of why heat-stressed corals are more susceptible to disease. Whether warm water temperatures directly and detrimentally interfere with immune components or indirectly affect the organism through the bleaching phenomenon remains unclear. What is apparent, however, is that the sensitivity of symbiotic organisms to temperature variation may leave corals unusually susceptible to changes in physiological state and immune function. Efforts to unveil this link between temperature stress and coral disease require careful experimentation with host immune responses and with pathogen virulence and infectivity. Pathogen response to increased temperatures may be a key element in the dynamics of coral diseases. For example, *Aspergillus sydowii*, the fungal pathogen of the sea fan disease aspergillosis, grows at a faster rate at higher temperatures (Alker et al. 2001), and the bacterial pathogen of hard corals, *Vibrio corallilyticus* (Ben-Haim et al. 2003), produces more lytic proteins when grown at the elevated temperatures that increase its virulence. Adhesion ability, a critical virulence factor in the causative agent in coral bleaching (*Vibrio shiloi*) is also temperature sensitive (Toren et al. 1998). In addition to adhesion, production of antialgal toxins and SOD (which detoxifies superoxide anion) are also temperature-dependent virulence factors, which seem to be induced in *V. shiloi* by elevated seawater temperatures (Banin et al. 2003).

Linking Immunity and Disease Outbreaks in Nature

Although there are considerable data on invertebrate immunity (**Table 1**) and many documented disease outbreaks in nature (**Table 3**), these two areas have not been appropriately linked to address two important questions:

- What are the major ecologically important immunological measures?
- How is the probability of infection related to immunity of an organism in a changing environment?

Ecologists studying disease in natural populations continue to seek biologically relevant, quantifiable measures of immunocompetence that are appropriate for field deployment. However, the search for a common measure of immunity in invertebrates is complicated by the sheer diversity of the invertebrate fauna in body plan, function, life history, and the option of measuring either cellular or humoral components of immunity. In the case of oysters, hemocyte function measured as hemocyte counts, phagocytic activity, and production of ROS are proven measures of immunity (Anderson 1994, Bachere et al. 2004, Baier-Anderson & Anderson 2000, Boyd & Burnett 1999). In crabs or lobsters, hemocytes phagocytize bacteria and secrete antibacterial compounds (Bachere et al. 2004, Holman et al. 2004). In corals, it is still unclear which measures of immunity are most predictive of resistance to a given pathogen, as both cellular and humoral components have an influence on bacterial and fungal pathogens, and critical immune components may vary with pathogen. Thus in the study of invertebrate immunity, the tools used by ecologists will undoubtedly differ depending on the organism, pathogen, and ecological setting. Some good examples are outlined in **Table 2**, where immunity in natural populations of crustaceans (Le Moullac & Haffner 2000; Smith et al. 1995) and bivalves (Dyrynda et al. 1998; Vasseur & Leguille 2004) was measured using a suite of cellular immune responses. In *Drosophila melanogaster*, multiple components of antibacterial immunity vary in natural populations and affect successful infection by the bacterial pathogen, *Serratia marcescens* (Lazzaro et al. 2004).

Another ongoing problem with the study of immunity in natural populations is the lack of diagnostics for many microbially based diseases. Improvements in these diagnostic methods will put tools in the hands of ecologists to evaluate and study disease outbreaks as they are occurring. This in turn can lead scientists to discover the environmental cofactors implicated in an outbreak, as well as the levels of immunocompetence in the host invertebrates. Hence, using field-deployable immunity assays, we can measure immune parameters in situ in natural populations. We can then begin to estimate the effects of natural selection on invertebrate immunity, estimations that will lead to studies of the evolution of immunity in natural populations (Altizer et al. 2003).

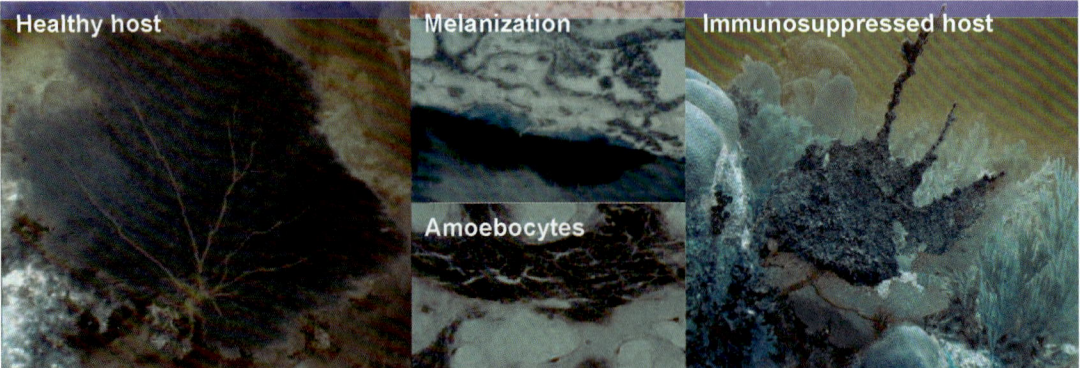

Figure 1

A hypothetical model of the effects of changing environmental conditions on the immune responses and resistance of the sea fan *Gorgonia ventalina* to pathogens. Outcome of this interaction is either a healthy immunocompetent host or an immunosuppressed host with disease. Photographs by L. Mydlarz and J. Bruno.

CONCLUSIONS

Invertebrates are phylogenetically diverse and have evolved a multiplicity of efficient defense strategies to defend against microbial attack. Marine invertebrates are especially challenged by their environment with high bacterial and viral loads, pollutants that can spread from their point source, and through changing states of oxygenation and temperature. Data on the effects of environmental factors on immunity are scarce and mechanistic, and are typically focused on economically valuable species. However, the study of marine invertebrate ecological immunity is advancing rapidly, and critical signaling pathways and cytotoxic responses are being elucidated. From the data presented in this review it is clear that several defense and signaling mechanisms are important in the innate immune responses to changing environment and pathogens. Examples are the proPO pathway and the ability of immune and phagocytic cells to clear or encapsulate pathogen invaders (**Figure 1**). However, there are almost no field studies estimating how these components vary in nature, or how variation in immunity affects susceptibility to infection in natural populations.

Thus, via collaborative efforts between ecologists, immunologists, cell biologists, and physiologists, we may expand the current understanding of innate immunity in naturally occurring, ecologically important species. With greater understanding of the connections between environment and organismal immunity, we can make predictions about the effects of changing climate and environment on immunocompetence and disease outbreaks.

ACKNOWLEDGMENTS

We would like to thank Cindy Liefer, Kurt A. Mckean, Dieter Ebert, Bette Willis, Steve Palumbi, and Jessica R. Ward for helpful comments and reviews. We would like to acknowledge funding from NSF/NIH Ecology of Infectious Disease Program (NSF OCE-0326705).

LITERATURE CITED

Abele D, Heise K, Pörtner HO, Puntarulo S. 2002. Temperature-dependence of mitochondrial function and production of reactive oxygen species in the intertidal mud clam *Mya arenaria*. *J. Exp. Biol.* 205:1831–41

Alker AP, Smith GW, Kim K. 2001. Characterization of *Aspergillus sydowii* (Thom and Church) a fungal pathogen of Carribean sea fan corals. *Hydrobiologia* 460:105–11

Al-Sharif WZ, Sunyer JO, Lambris JD, Smith LC. 1998. Sea urchin coelomocytes specifically express a homologue of the complement component C3. *J. Immunol.* 160:2983–97

Altizer S, Harvell CD, Friedle E. 2003. Rapid evolutionary dynamics and disease threats to biodiversity. *Trends Ecol. Evol.* 18:589

Anderson RS. 1994. Haemocyte derived reactive oxygen intermediate production in four bivalve molluscs. *Dev. Comp. Immunol.* 18:89–96

Anderson RS, Brubacher LL, Ragone Clavo LM, Unger MA, Burreson EM. 1998.

Effects of tributyltin and hypoxia on the progression of *Perkinsus marinus* infections and host defense mechanisms in oyster, *Crassostrea virginica* (Gmelin). *J. Fish Dis.* 21:371–80

Antonius A. 1981. The "band" diseases in coral reefs. *Proc. 4th Int. Coral Reef Symp.* 2:7–14

Azumi K, De Santis R, De Tomaso A, Rigoutsos I, Yoshizaki F, et al. 2003. Genomic analysis of immunity in a Urochordate and the emergence of the vertebrate immune system: "waiting for Godot." *Immunogenetics* 55:570–81

Bachali S, Bailly X, Jolles J, Jolles P, Deutsch JS. 2004. The lysozyme of the starfish *Asterias rubens*–A paradigmatic type i lysozyme. *Eur. J. Biochem.* 271:237–42

Bachere E. 2000. Shrimp immunity and disease control. *Aquaculture* 191:3–11

Bachere E, Destomieux D, Bulet P. 2000. Peneidins, antimicrobial peptides of shrimp: a comparison with other effectors of innate immunity. *Aquaculture* 191:71–88

Bachere E, Gueguen Y, Gonzalez M, de Lorgeril J, Garnier J, Romestand B. 2004. Insights into the antimicrobial defense of marine invertebrates: the penaeid shrimps and the oyster *Crassostrea gigas*. *Immunol. Rev.* 198:149–68

Baier-Anderson C, Anderson RS. 2000. The effects of chlorothalonil on oyster hemocyte activation: phagocytosis, reduced pyridine nucleotides, and reactive oxygen species production. *Environ. Res.* 83:72–78

Banin E, Vassilakos D, Orr E, Martinez RJ, Rosenberg E. 2003. Superoxide dismutase is a virulence factor produced by the coral bleaching pathogen *Vibrio shiloi*. *Curr. Microbiol.* 46:418–22

Becero MA, Paul VJ. 2004. Effects of light and depth on secondary metabolites and cyanobacterial symbionts of the sponge *Dysidea granulosa*. *Mar. Ecol. Prog. Ser.* 280:115–28

Ben-Haim Y, Zicherman-Keren M, Rosenberg E. 2003. Temperature-regulated bleaching and lysis of the coral Pocillopora damicornis by the novel pathogen *Vibrio coralliilyticus*. *Appl. Environ. Microbiol.* 69:4236–42

Bigger CH, Hildemann WH. 1982. Cellular defense systems of the Coelenterata. In *The Reticuloendothelial System*, ed. N Cohen, MM Sigel, pp. 59–87. New York: Plenum

Bigger CH, Jokiel PL, Hildemann WH, Johnston IS. 1982. Characterization of alloimmune memory in a sponge. *J. Immunol.* 129:1570–72

Bower SM, Blackbourn J, Meyer GR. 1998. Distribution, prevalence, and pathogenicity of the protozoan *Perkinsus qugwadi* in Japanese scallops, *Patinopecten yessoensis*, cultured in British Columbia, Canada. *Can. J. Zool.* 76:954–59

Boyd JN, Burnett LE. 1999. Reactive oxygen intermediate production by oyster hemocytes exposed to hypoxia. *J. Exp. Biol.* 202:3135–43

Bright SIS, Manjusha M, Pai SS, Rosamma P. 2005. *Fenneropenaeus indicus* is protected from white spot disease by oral administration of inactivated white spot syndrome virus. *Dis. Aquat. Org.* 66:265–70

Brower DL, Brower SM, Hayward DC, Ball EE. 1997. Molecular evolution of integrins: Genes encoding integrin b subunits from a coral and a sponge. *Proc. Natl. Acad. Sci. USA* 94:9182–87

Bruckner AW, Bruckner RJ. 1997. Outbreak of coral disease in Puerto Rico. *Coral Reefs* 16:260

Bulgakov AA, Park KI, Choi KS, Lim HK, Cho M. 2004. Purification and characterization of a lectin isolated from the Manila clam *Ruditapes philippinarum* in Korea. *Fish Shellfish Immunol.* 16:487–99

Cadavid LF. 2005. Self/nonself discrimination in Basal Metazoa: Genetics of allorecognition in the hydroid *Hydractinia*. *Integr. Comp. Biol.* 45:623–30

Cadavid LF, Powell AE, Nicotra ML, Moreno M, Buss LW. 2004. An invertebrate histocompatibility complex. *Genetics* 167:357–65

Canesi L, Gallo G, Gavioli M, Pruzzo C. 2002. Bacteria-hemocyte interactions and phagocytosis in marine bivalves. *Micros. Res. Tech.* 57:469–76

Canesi L, Pruzzo C, Tarsi R, Gallo G. 2001. Surface interactions between *Escherichia coli* and hemocytes of the Mediterranean mussel *Mytilus galloprovincialis* Lam. leading to efficient bacterial clearance. *Appl. Environ. Microbiol.* 67:464–68

Capaul M, Ebert D. 2003. Parasite mediated selection in experimental populations of *Daphnia magna*. *Evolution* 57:249–60

Carlton R, Richardson L. 1995. Oxygen and sulfide dynamics in a horizontally migrating cyanobacterial mat: black band disease of corals. *FEMS Microbiol. Ecol.* 18:155–62

Cerenius L, Soderhall K. 2004. The prophenoloxidase-activating system in invertebrates. *Immunol. Rev.* 198:116–26

Cerrano C, Bavestrello G, Bianchi CN, Cattaneo-Vetti R, Bava S, et al. 2000. A catastrophic mass-mortality episode of gorgonians and other organisms in the Ligurian Sea (Northwestern Mediterranean), summer 1999. *Ecol. Lett.* 3:284–93

Chapman RW, Browdy CL, Savin S, Prior S, Wenner E. 2004. Sampling and evaluation of white spot syndrome virus in commercially important Atlantic penaeid shrimp stocks. *Dis. Aquat. Org.* 59:179–85

Cheng W, Chen JC. 1998. *Enterococcus*-like infections in *Macrobrachium rosenbergii* are exacerbated by high pH and temperature but reduced by low salinity. *Dis. Aquat. Org.* 34:103–8

Cheng W, Wang CH. 2001. The susceptibility of the giant freshwater prawn *Macrobrachium rosenbergii* to *Lactococcus garvieae* and its resistance under copper sulfate stress. *Dis. Aquat. Org.* 47:137–44

Chisholm JRS, Smith VJ. 1994. Variation of antibacterial activity in the hemocytes of the shore crab, *Carcinus maenas*, with temperature. *J. Mar. Biol. Assoc. UK* 74:979–82

Cima F, Perin A, Burighel P, Ballarin L. 2001. Morpho-functional characterization of hemocytes of the compound ascidian *Botrylloides leachi* (Tunicata, Ascidiacea). *Acta Zool.* 82:261–74

Clow LA, Raftos DA, Gross PS, Smith LC. 2004. The sea urchin complement homologue, SpC3, functions as an opsonin. *J. Exp. Biol.* 207:2147–55

Coffaro KA, Hinegardner RT. 1977. Immune response in the sea urchin *Lytechinus pictus*. *Science* 197:1389–90

Coles JA, Farley SR, Pipe RK. 1995. Alteration of the immune response of the common marine mussel *Mytilus edulis* resulting from exposure to cadmium. *Dis. Aquat. Org.* 22:59–65

Cook T, Folli M, Klinck J, Ford S, Miller J. 1998. The relationship between increasing sea surface temperature and the northward spread of *Perkinsus marinus* (Dermo) disease epizootics in oysters. *Estuar. Coast. Shelf Sci.* 46:587–97

Coteur G, DeBecker G, Warnau M, Jangoux M, Dubois P. 2002. Differentiation of immune cells challenged by bacteria in the common European starfish, *Asterias rubens* (Echinodermata). *Fish Shellfish Immunol.* 12:187–200

Coteur G, Gillan D, Joly G, Pernet P, Dubois P. 2003. Field contamination of the starfish *Asterias rubens* by metals. Part 2: Effects on cellular immunity. *Environ. Toxicol. Chem.* 22:2145–51

Coteur G, Gillan D, Pernet P, Dubois P. 2005. Alteration of cellular immune responses in the seastar *Asterias rubens* following dietary exposure to cadmium. *Aquat. Toxicol.* 73:418–21

Coteur G, Warnau M, Jangoux M, Dubois P. 1999. Reactive oxygen species (ROS) production by amoebocytes of *Asterias rubens* (echinodermata). *Fish Shellfish Immunol.* 12:187–200

Cranfield HJ, Dunn A, Doonan IJ, Michael KP. 2005. *Bonamia exitiosa* epizootic in *Ostrea chilensis* from Foveaux Strait, southern New Zealand between 1986 and 1992. *ICES J. Mar. Sci.* 62:3–13

Davidson B, Swalla BJ. 2002. A molecular analysis of ascidian metamorphosis reveals activation of an innate immune response. *Development* 129:4739–51

Dove ADM, Bowser PR, Cerrato RM. 2004a. Histological analysis of an outbreak of QPX disease in wild hard clams *Mercenaria mercenaria* in New York. *J. Aquat. Anim. Health* 16:246–50

Dove ADM, LoBue C, Bowser P, Powell M. 2004b. Excretory calcinosis: a new fatal disease of wild American lobsters *Homarus americanus*. *Dis. Aquat. Org.* 58:215–21

Dungan M, Miller T, Thomson D. 1982. Catastrophic decline of a top carnivore in the Gulf of California rocky intertidal zone. *Science* 216:989–91

Dunham PC, Anderson C, Rich AM, Weissmann G. 1983. Stimulus-response coupling in sponge cell aggregation: Evidence for calcium as an intracellular messenger. *Proc. Natl. Acad. Sci. USA* 80:4756–60

Dyrynda EA, Pipe RK, Burt GR, Ratcliffe NA. 1998. Modulations in the immune defenses of mussels (*Mytilus edulis*) from contaminated sites in the UK. *Aquat. Toxicol.* 42:169–85

Ebert D. 2005. *Ecology, Epidemiology, and Evolution of Parasitism in Daphnia* (Internet). Bethesda, MD: Natl. Libr. Med., Natl. Cent. Biotechnol. Inf.

Evans EE, Painter B, Evans ML, Weinheimer P, Acton RT. 1968. An induced bactericidin in the spiny lobster, *Panulirus argus*. *Proc. Soc. Exp. Biol. Med.* 128:394–98

Evans EE, Weinheimer PF, Painter B, Acton RT, Evans ML. 1969. Secondary and tertiary responses of the induced bactericidin from the West Indian spiny lobster, *Panulirus argus*. *J. Bacteriol.* 98:943–46

Fang L, Huang SP, Lin K. 1997. High temperature induces the synthesis of heat-shock proteins and the elevation of intracellular calcium in the coral *Acropora grandis*. *Coral Reefs* 16:127–31

Fisher WS. 1992. Occurrence of agglutinin in the pallial cavity mucus of oysters. *J. Exp. Mar. Biol.* 162:1–13

Gaffney P, Bushak D. 1996. Genetic aspects of disease resistance in oysters. *J. Shellfish Res.* 15:135–40

Gaino EG, Bacestrello G, Magnino G. 1999. Self/nonself recognition in sponges. *Ital. J. Zool.* 66:229–316

Galloway TS, Depledge MH. 2001. Immunotoxicology in invertebrates: Measurement and ecotoxicological relevance. *Ecotoxicology* 10:5–23

Galtsoff PS, Brown HH, Smith CL, Smith FGW. 1939. Sponge mortality in the Bahamas. *Nature* 143:807–8

Gates RD, Baghdasarian G, Muscatine L. 1992. Temperature stress causes host cell detachment in symbiotic cnidarians: Implications for coral bleaching. *Biol. Bull.* 182:324–32

Gladfelter WB. 1982. White-band disease in *Acropoa palmata* implications for the structure and growth of shallow reefs. *Bull. Mar. Sci.* 32:639–43

Gokudan S, Muta T, Tsuda R, Koori K, Kawahara T, Seki N. 1999. Horseshoe crab acetyl group-recognizing lectins involved in innate immunity are structurally related to fibrinogen. *Proc. Natl. Acad. Sci. USA* 96:10086–91

Gómez-Gutiérrez J, Peterson WT, De Robertis A, Brodeur RD. 2003. Mass mortality of krill caused by parasitoid ciliates. *Science* 301:339

Grosberg RK, Hartt MW, Levitan DR. 1997. Is allorecognition specificity in *Hydractinia symbiolongicarpus* controlled by a single gene? *Genetics* 145:857–60

Gross P, Al-Sharif WZ, Clow LA, Smith LC. 1999. Echinoderm immunity and the evolution of the complement system. *Dev. Comp. Immunol.* 23:429–44

Guzmán H, Cortéz J. 1984. Mortand de *Gorgonia flabellum* Linnaeus (Octocorallia: Gorgoniidae) en la Costa Cribe de Costa Rica. *Rev. Biol. Trop.* 32:305–8

Harvell CD, Aronson R, Baron N, Connell J, Dobson A, et al. 2004. The rising tide of ocean diseases: unsolved problems and research priorities. *Front. Ecol.* 2:375–82

Harvell CD, Kim K, Quirolo C, Weir J, Smith G. 2001. Coral bleaching and disease: contributors to 1998 mass mortality in *Briareum asbestinum* (Octocorallia, Gorgonacea). *Hydrobiologia* 460:97–104

Harvell CD, Mitchell CE, Ward JR, Altizer S, Dobson AP, et al. 2002. Climate warming and disease risks for terrestrial and marine biota. *Science* 296:2158–62

Haseley SR, Vermeer HJ, Kamerling JP, Vliegenthart JFG. 2001. Carbohydrate self-recognition mediates marine sponge cellular adhesion. *Proc. Natl. Acad. Sci. USA* 98:9419–24

Hauton C, Smith VJ. 2004. In vitro cytotoxicity of crustacean immunostimulants for lobster (*Homarus gammarus*) granulocytes demonstrated using the neutral red uptake assay. *Fish Shellfish Immunol.* 17:65–73

Hawkridge JM, Pipe RK, Brown BE. 2000. Localization of antioxidant enzymes in the cnidarians *Anemonia viridis* and *Goniopora stokesi*. *Mar. Biol.* 137:1–9

Heise K, Puntarulo S, Pörtner HO, Abele D. 2003. Production of reactive oxygen species by isolated mitochondria of the Antarctic bivalve *Laternula elliptica* (King and roderip) under heat stress. *Comp. Biochem. Physiol. C* 134:79–90

Henry L, Hart M. 2005. Regeneration from injury and resource allocation in sponges and corals: a review. *Int. Rev. Hydrobiol.* 90:125–58

Hernroth B, Baden SP, Holm K, André T, Söderhäll I. 2004. Manganese induced immune suppression of the lobster, *Nephrops norvegicus*. *Aquat. Toxicol.* 70:223–31

Hidaka M. 1985. Tissue compatibility between colonies and between newly settled larvae of *Pocillopora damicornis*. *Coral Reefs* 4:111–16

Hildemann WH, Johnson IS, Jokiel PL. 1979. Immunocompetence in the lowest metazoan phylum. *Science* 204:420–22

Hildemann WH, Raison RL, Cheung G, Hull CJ, Akaka L, Okamoto J. 1977. Immunological specificity and memory in a scleractinian coral. *Nature* 270:219–23

Hoegh-Guldberg O, Salvat B. 1995. Periodic mass-bleaching and elevated sea temperatures: bleaching of outer reef slope communities in Moorea, French Polynesia. *Mar. Ecol. Prog. Ser.* 121:181–90

Holman JD, Burnett KG, Burnett LE. 2004. Effects of hypercapnic hypoxia on the clearance of *Vibrio campbellii* in the Atlantic Blue Crab, *Callinectes sapidus* Rathbun. *Biol. Bull.* 206:188–96

Holmblad T, Soderhall K. 1999. Cell adhesion molecules and antioxidative enzymes in a crustacean, possible role in immunity. *Aquaculture* 172:111–23

Hutton DMC, Smith VJ. 1996. Antibacterial properties of isolated amoebocytes from the sea anemone *Actinia equine*. *Biol. Bull.* 191:441–51

Inamori K, Ariki S, Kawabata S. 2004. A Toll-like receptor in horseshoe crabs. *Immunol. Rev.* 198:106–15

Inamori K, Saito T, Iwaki D, Nagira T, Iwanaga S, et al. 1999. A newly identified horseshoe crab lectin with specificity for blood group A recognizes specific O-antigens of bacterial lipopolysaccharides. *J. Biol. Chem.* 274:3272–78

Jang WS, Kim KN, Lee YS, Nam MH, Lee IH. 2002. Halocidin: a new antimicrobial peptide from hemocytes of the solitary tunicate, *Halocynthia aurantium*. *FEBS Lett.* 521:81–86

Jensen PR, Harvell CD, Wirtz K, Fenical W. 1996. The incidence of antimicrobial activity among Caribbean gorgonians. *Mar. Biol.* 125:411–20

Johanssen MW. 1999. Cell adhesion molecules in invertebrate immunity. *Dev. Comp. Immunol.* 23:303–15

Jones GM, Scheibling RE. 1985. *Paramoeba* sp. (Amoebida, Paramoebidae) as the possible causative agent of sea urchin mass mortality in Nova Scotia. *J. Parasitol.* 71:559–65

Jones RJ, Bowyer J, Hoegh-Guldberg O, Blackall LL. 2004. Dynamics of a temperature-related coral disease outbreak. *Mar. Ecol. Prog. Ser.* 281:63–77

Kairies N, Beisel HG, Fuentes-Prior P, Tsuda R, Muta T, et al. 2001. The 2.0 Angstrom crystal structure of tachylectin 5A provides evidence for the common origin of innate immunity and the blood coagulation systems. *Proc. Natl. Acad. Sci. USA* 98:13519–24

Khalturin K, Panzer Z, Cooper MD, Bosch TCG. 2004. Recognition strategies in the innate immune system of ancestral chordates. *Mol. Immunol.* 41:1077–87

Kim K, Dobson AP, Gulland FMD, Harvell CD. 2005. Diseases and the conservation of marine biodiversity. In *Marine Conservation Biology*, ed. EA Norse, LB Crowder, pp. 149–68. Washington, DC: Island Press

Kim K, Harvell CD, Kim PD, Smith GW, Merkel SM. 2000a. Fungal disease resistance of Caribbean sea fan corals (Gorgonia spp.). *Mar. Biol.* 136:259–67

Kim K, Kim PD, Alker AP, Harvell CD. 2000b. Antifungal properties of gorgonian corals. *Mar. Biol.* 137:393–401

Koh EGL. 1997. Do scleractinians corals engage in chemical warfare against microbes? *J. Chem. Ecol.* 23:379–98

Kurz J, Franz K. 2003. Evidence for memory in invertebrate immunity. *Nature* 425:37–38

Labat-Robert J, Robert L, Auger C, Lethias C, Garrone R. 1981. Fibronectin-like protein in Porifera: Its role in cell aggregation. *Proc. Natl. Acad. Sci. USA* 78:6261–65

Lafferty K, Kuris A. 1993. Mass mortality of abalone *Haliotis cracherodii* on the California Channel Islands: test of epidemiological hypotheses. *Mar. Ecol. Prog. Ser.* 96:239–48

Lafferty KD, Porter J, Ford SE. 2004. Are diseases increasing in the ocean? *Annu. Rev. Ecol. Evol. Syst.* 35:31–54

La Peyre JF. 1993. *Studies on the oyster pathogen Perkinsus marinus (ampicomplexa): and in vitro propagation.* PhD diss., Va. Inst. Mar. Sci., Gloucester Point

Lazzaro BP, Sceurman BK, Clark AG. 2004. Genetic basis of natural variation in *D. melanogaster* antibacterial immunity. *Science* 303:1873–76

Leclerc M, Brillouet C. 1988. The heterogeneity of antibody-like membrane receptors in subpopulations of sea-star axial organ cells. *Cell Biol. Int. Rep.* 12:465–70

Lee TG, Maruyama S. 1998. Isolation of HIV-1 protease inhibiting peptides from the thermolysin hydrolysate of oyster proteins. *Biochem. Biophys. Res. Commun.* 253:604–8

Legac E, Vaugier GL, Bousqet F, Bajelan M, Leclerc M. 1996. Primitive cytokines and cytokine receptors in invertebrates: the sea star *Asterias rubens* as a model of study. *Scand. J. Immunol.* 44:375–80

Le Moullac G, Haffner P. 2000. Environmental factors affecting immune responses in Crustaceae. *Aquaculture* 191:121–31

Le Moullac G, Soyez C, Saulnier D, Ansquer D, Avarre JC, Levy P. 1998. Effect of hypoxic stress on the immune response and the resistance to vibriosis of the shrimp, *Penaeus stylirostris*. *Fish Shellfish Immunol.* 8:621–29

Lessios HA, Roberston DR, Cubit JD. 1984. Spread of *Diadema* mass mortality through the Caribbean. *Science* 226:335–37

Lester RJG, Goggin CL, Sewell KB. 1990. *Perkinsus* in Australia. In *Pathology in Marine Science*, ed. TC Cheng, pp. 189–99. San Diego, CA: Academic

Lightner DV, Redman RM. 1998. Strategies for control of viral diseases of shrimp in the Americas *Fish Pathol.* 33:165–80

Little TJ, Colbourne JK, Crease TJ. 2004. Molecular evolution of Daphnia immunity genes: Polymorphism in a Gram negative binding protein and an alpha-2-macroglobulin. *J. Mol. Evol.* 59:498–506

Little TJ, Hultmark D, Read AF. 2005. Invertebrate immunity and the limits of mechanistic immunology. *Nat. Immunol.* 6:651–54

Little TJ, O'Connor B, Colegrave N, Watt K, Read AF. 2003. Maternal transfer of strain-specific immunity in an invertebrate. *Curr. Biol.* 13:489–92

Lively CM, Jokela J. 1996. Clonal variation for local adaptation in a host-parasite interaction. *Proc. R. Soc. London Ser. B* 262:891–97

Lubbock R. 1980. Clone-specific cellular recognition in a sea anemone. *Proc. Natl. Acad. Sci. USA* 77:6667–69

Malham SK, Lacoste A, Gélébart F, Cueff A, Poulet SA. 2003. Evidence for a direct link between stress and immunity in the Mollusc *Haliotis tuberculata*. *J. Exp. Zool. A* 295:136–44

Mali B, Möhrlen F, Frohme M, Frank U. 2004. A putative double role of a chitinase in a cnidarian: pattern formation and immunity. *Dev. Comp. Immunol.* 28:973–81

Matranga V, Toia G, Bonaventura R, Muller EG. 2000. Cellular and biochemical responses to environmental and experimentally induced stress in sea urchin coelomocytes. *Cell Stress Chaperones* 5:113–20

McCallum H, Kuris A, Harvell CD, Lafferty K, Smith G, Porter J. 2004. Does terrestrial epidemiology apply to marine systems? *Trends Ecol. Evol.* 19:585–91

McGladdery SE, Cawthorn RJ, Bradford BC. 1991. *Perkinsus karssoni* n. sp. (Apicomplexa) in bay scallops *Argopecten irradians*. *Dis. Aquat. Org.* 10:127–37

Meszaros A, Bigger CH. 1999. Qualitative and quantitative study of wound healing processes in the coelenterate, *Plexaurella fusifera*: spatial, temporal, and environmental (light attenuation) influences. *J. Invert. Pathol.* 73:321–31

Mikulski CM, Burnett LE, Burnett KG. 2000. The effects of hypercapnic hypoxia on the survival of shrimp challenged with *Vibrio parahaemolyticus*. *J. Shellfish Res.* 19:301–11

Miller R, Colodey A. 1983. Widespread mass mortalities of the Green sea urchin in Nova Scotia, Canada. *Mar. Biol.* 73:263–67

Mitchell SE, Read AF, Little TJ. 2004. The effect of a pathogen epidemic on the genetic structure and reproductive strategy of the crustacean *Daphnia magna*. *Ecol. Lett.* 7:848–58

Mohankumar K, Ramasamy P. 2006. White spot syndrome virus infection decreases the activity of antioxidant enzymes in *Fenneropenaeus indicus*. *Virus Res.* 115:69–75

Morall CE, Galloway TS, Trapido-Rosenthal HG, Depledge MH. 2000. Characterization of nitric oxide synthase activity in the tropical sea anemone *Aiptasia pallida*. *Comp. Biochem. Physiol. B* 125:483–91

Mouritson KN, Mouritson LT, Jensen KT. 1998. Change of topography and sediment characteristics of an intertidal mud-flat following mass mortality of the amphipod *Corophium volutator*. *J. Mar. Biol. Assoc. UK* 78:1167–80

Mucklow PT, Ebert D. 2003. Physiology of immunity in the water flea *Daphnia magna*: environmental and genetic aspects of phenoloxidase activity. *Physiol. Biochem. Zool.* 76:836–42

Mucklow PT, Vizoso DB, Jensen KH, Refardt D, Ebert D. 2004. Variation for phenoloxidase activity and its relation to parasite resistance within and between populations of *Daphnia magna*. *Proc. R. Soc. London Ser. B* 271:1175–83

Mullen K, Peters EC, Harvell CD. 2004. Coral resistance to disease. In *Coral Health and Disease*, ed. E Rosenberg, Y Loya, pp. 377–99. New York: Springer-Verlag

Muller WG, Koziol C, Muller IM, Wiens M. 1999. Towards an understanding of the molecular basis of immune responses in sponges: The marine demosponge *Geodia cydonium* as a model. *Micros. Res. Tech.* 44:219–36

Nagai T, Kawabata S. 2000. A link between blood coagulation and prophenoloxidase activation in arthropod host defense. *J. Biol. Chem.* 275:29264–67

Nagai T, Osaki T, Kawabata S. 2001. Functional conversion of hemocyanin to prophenoloxidase by horseshoe crab antimicrobial peptides. *J. Biol. Chem.* 276:27166–70

Nagelkerken I, Buchan K, Smith GW, Bonair K, Bush P, et al. 1997. Widespread disease in Caribbean sea fans. II. Patterns of infection and tissue loss. *Mar. Ecol. Prog. Ser.* 160:255–63

Nair SV, Del Valle H, Gross PS, Terwilliger DP, Smith LC. 2005. Macroarray analysis of coelomocyte gene expression in response to LPS in the sea urchin. Identification of unexpected immune diversity in an invertebrate. *Physiol. Genomics* 22:33–47

Nilsen IW, Øverbø K, Sandsdalen E, Sandaker E, Sletten K, Myrnes B. 1999. Protein purification and gene isolation of chlamysin, a cold-active lysozyme-like enzyme with antibacterial activity. *FEBS Lett.* 464:153–58

Nyström M, Nordemar I, Tedengren M. 2001. Simultaneous and sequential stress from increased temperature and copper on the metabolism of the hermatypic coral *Porites cylindrica*. *Mar. Biol.* 138:1225–31

Olano CT, Bigger CH. 2000. Phagocytic activities of the gorgonian coral *Swiftia exserta*. *J. Invertebr. Pathol.* 76:176–84

Olicard C, Renault T, Torhy C, Benmansour A, Bourgougnon N. 2005. Putative antiviral activity in hemolymph from adult Pacific oysters, *Crassostrea gigas*. *Antivir. Res.* 66:147–52

Olsen OM, Nilsen IW, Sletten K, Myrnes B. 2003. Multiple invertebrate lysozymes in blue mussel (*Mytilus edulis*). *Comp. Biochem. Physiol. B* 136:107–15

Paillard C. 2004. A short-review of brown ring disease, a vibriosis affecting clams, *Ruditapes philippinarum* and *Ruditapes decussatus*. *Aquat. Living Res.* 17:467–75

Pan J, Kurosky A, Xu B, Chopra AK, Coppenhaver DH, et al. 2000. Broad antiviral activity in tissues of crustaceans. *Antivir. Res.* 48:39–47

Parry HE, Pipe RK. 2004. Interactive effects of temperature and copper on immunocompetence and disease susceptibility in mussels (*Mytilus edulis*). *Aquat. Toxicol.* 69:311–25

Peters R, Raftos DA. 2003. The role of phenoloxidase suppression in QX disease outbreaks among Sydney rock oysters (*Saccostrea glomerata*). *Aquaculture* 223:29–39

Petes LE, Harvell CD, Peters EC, Webb MAH, Mullen KM. 2003. Pathogens compromise reproduction and induce melanization in Caribbean sea fans *Mar. Ecol. Prog. Ser.* 264:161–71

Pipe RK, Coles JA, Carissan FM, Ramanathan K. 1999. Copper induced immunomodulation in the marine mussel *Mytilus edulis*. *Aquat. Toxicol.* 46:43–54

Plaistow SJ, Outreman Y, Moret Y, Rigaud T. 2003. Variation in the risk of being wounded: an overlooked factor in studies of invertebrate immune function? *Ecol. Lett.* 6:489–94

Pruzzo C, Gallo G, Canesi L. 2005. Persistence of vibrios in marine bivalves: the role of interactions with hemolymph components. *Environ. Microbiol.* 7:761–72

Raftos DA, Cooper EL, Habicht GS, Beck G. 1991. Invertebrate cytokines: Tunicate cell proliferation stimulated by an interleukin-1 type molecule. *Proc. Natl. Acad. USA* 88:9518–22

Raftos DA, Stillman DL, Cooper EL. 1998. Chemotactic responses of tunicate (Urochordata, Ascidiacea) hemocytes in vitro. *J. Invertebr. Pathol.* 72:44–49

Regoli F, Cerrano C, Chierici E, Bompadre S, Bavestrello G. 2000. Susceptibility to oxidative stress of the mediterreanean demosponge *Petrosia ficiformis*: role of endosymbionts and solar irradiance. *Mar. Biol.* 137:453–61

Relf JM, Chisholm JRF, Kemp GD, Smith VJ. 1999. Purification and characterization of a cysteine-rich 11.5-kDa antibacterial protein from the granular hemocytes of the shore crab, *Carcinus maenas Eur. J. Biochem.* 264:350–57

Renault T, Novoa B. 2004. Viruses infecting bivalve molluscs. *Aquat. Living Res.* 17:397–409

Richardson L, Goldberg W, Kuta K, Aronson R, Smith G, et al. 1998. Florida's mystery coral-killer identified. *Nature* 392:557–58

Richier S, Furla P, Plantivaux A, Merle PL, Allemand D. 2001. Symbiosis-induced adaptation to oxidative stress. *J. Exp. Biol.* 208:277–85

Richier S, Merle PL, Furla P, Pigozzi D, Sola F, Allemand D. 2003. Characterization of superoxide dismutases in anoxia- and hyperoxia tolerant symbiotic cnidarians. *Biochim. Biophys. Acta* 1621:84–91

Rifai S, Fassouane A, Kijjoa A, Van Soest R. 2004. Antimicrobia activity of Utenospongin B, a metabolite from the marine sponge *Hippospongia communis* collected from the atlantic coast of Morocco. *Mar. Drugs* 2:147–53

Romestand B, Corbier F, Roch P. 2002. Protease inhibitors and hemagglutinins associated with resistance to the protozoan parasite, *Perkinsus marinus*, in the Pacific oyster, *Crassostrea gigas*. *Parasitology* 125:323–29

Rützler K, Santavy D, Antonius A. 1983. The black band diseases of Atlantic reef corals. III. Distribution, ecology, and development. *PSZNI Mar. Ecol.* 4:329–58

Saito T, Hatada M, Iwanaga S, Kawabata S. 1997. A newly identified horseshoe crab lectin with binding specificity to O-antigen of bacterial lipopolysaccharides. *J. Biol. Chem.* 272:30703–8

Salter-Cid L, Bigger CH. 1991. Alloimmunity in the gorgonian coral *Swiftia exserta*. *Biol. Bull* 181:127–34

Sawada T, Zhang J, Cooper EL. 1993. Classification and characterization of the hemocytes in *Styela clava*. *Biol. Bull.* 184:87–96

Scheibling RE, Stephenson RL. 1984. Mass mortality of *Strongylocentrotus droebachiensis* (Echinodermata: Echinoidea) off Nova Scotia, Canada. *Mar. Biol.* 78:153–64

Selig ER, Harvell CD, Bruno JF, Willis BL, Page CA, et al. 2006. Analyzing the relationship between ocean temperature anomalies and coral disease outbreaks at broad spatial scales. In *Coral Reefs and Climate Change: Science and Management*, ed. JT Phinney, WJ Skirving, J Kleypas, O Hoegh-Guldberg, AE Strong. Washington, DC: Am. Geophys. Union. In press

Shields JD, Behringer DC. 2004. A new pathogenic virus in the Caribbean spiny lobster *Panulirus argus* from the Florida Keys. *Dis. Aquat. Org.* 59:109–18

Skadsheim A, Christie H, Leinaas HP. 1995. Population reduction of *Strongylocentrotus droebachiensis* (Echinodermata) in Norway and the distribution of its endoparasite *Echinomermella matsi* (Nematoda). *Mar. Ecol. Prog. Ser.* 119:199–209

Smith FGW. 1939. Sponge mortality at British Honduras. *Nature* 143:785

Smith VJ, Swindlehurst RJ, Johnston PA, Vethaak AD. 1995. Disturbance of host defense capability in the common shrimp *Crangon crangon* by exposure to harbor dredge spoils. *Aquat. Toxicol.* 32:43–58

Soderhall K, Cerenius L. 1998. Role of the prophenoloxidase-activating system in invertebrate immunity. *Curr. Opin. Immunol.* 10:23–28

Soniat TM. 1996. Epizootiology of *Perkinsus marinus* disease of eastern oysters in the Gulf of Mexico. *J. Shellfish Res.* 15:35–43

Sotelo-Mundo RR, Islas-Osuna MA, de-la-Re-Vega E, Hernandez-Lopez J, Vargas-Albores F, Yepiz-Plascencia G. 2003. cDNA cloning of the lysozyme of the white shrimp *Penaeus vannamei. Fish Shellfish Immunol.* 15:325–31

Sparks AK, Morado JF. 1986. A herpes-like virus disease in the blue king crab *Paralithodes platypus. Dis. Aquat. Org.* 1:115–18

Sritunyalucksana K, Wongsuebsantati K, Johansson MW, Söderhäll K. 2001. Peroxinectin, a cell adhesive protein associated with the proPO system from the black tiger shrimp, *Penaeus monodon. Dev. Comp. Immunol.* 25:353–63

Stentiford GD, Bateman K, Feist SW. 2004. Pathology and ultrastructure of an intranuclear bacilliform virus (IBV) infecting brown shrimp *Crangon crangon* (Decapoda: Crangonidae). *Dis. Aquat. Org.* 58:89–97

Su HY, Wang PC, Lien YY, Tsai MA, Liu SS, et al. 2005. Upregulation of actin-like gene expression in giant freshwater prawns *Macrobrachium rosenbergii* infected with *Metschnikowia bicuspidate. Dis. Aquat. Org.* 66:175–80

Tamplin ML, Fisher WS. 1989. Occurrence and characteristics of agglutination of *Vibrio cholerae* by serum from the eastern oyster, *Crassostrea virginica. Appl. Environ. Microbiol.* 55:2882–87

Thakur NL, Perovic-Ottstadt S, Batel R, Korzhev M, Diehl-Seifert B, et al. 2005. Innate immune defense of the sponge *Suberites domuncula* against Gram-positive bacteria: induction of lysozyme and AdaPTin. *Mar. Biol.* 146:271–82

Tincu JA, Menzel LP, Azimov R, Sands J, Hong T, et al. 2003. Plicatamide, an antimicrobial octapeptide from *Styela plicata* hemocytes. *J. Biol. Chem.* 278:13546–53

Toren A, Landau L, Kushmaro A, Loya Y, Rosenberg E. 1998. Effect of temperature on adhesion of *Vibrio* strain AK-1 to *Oculina patagonica* and on coral bleaching. *Appl. Environ. Microbiol.* 64:1379–84

Tzvetnenko E, Fotedar S, Evans L. 2001. Antibacterial activity in the hemolymph of the western rock lobster, *Panulirus cygnus. Mar. Freshw. Res.* 52:1407–12

Van Banning P. 1991. Observations on bonamiasis in the stock of the European flat oyster, *Ostrea edulis*, in the Netherlands, with special reference to the recent developments in Lake Grevelingen. *Aquaculture* 93:205–11

Vargas-Albores F, Baltazar PH, Clark GP, Barajas FM. 1998. Influence of temperature and salinity on the yellowlegs shrimp, *Penaeus californiensis* Holmes, prophenoloxidase system. *Aquacult. Res.* 29:549–53

Vasseur P, Leguille C. 2004. Defense systems of benthic invertebrates in response to environmental stressors. *Environ. Toxicol.* 19:433–36

Vasta GR, Quesenberr M, Ahmed H, O'Leary N. 1999. C-type lectins and galectins mediate innate and adaptive immune functions: their roles in the complement activation pathway. *Dev. Comp. Immunol.* 23:401–20

Villalba A, Casas SM, Lopez C, Carballal MJ. 2005. Study of perkinsosis in the carpet shell clam *Tapes decussatus* in Galicia (NW Spain). II. Temporal pattern of disease dynamics and association with clam mortality. *Dis. Aquat. Org.* 65:257–67

Villalba A, Reece KS, Ordas MC, Casas SM, Figueras A. 2004. Perkinsosis in molluscs. A review. *Aquat. Living Resour.* 17:411–32

Ward JR, Lafferty KD. 2004. The elusive baseline of marine disease: Are diseases in ocean ecosystems increasing? *PLoS Biol.* 2:542–47

Weinheimer PF, Acton RT, Sawyer S, Evans EE. 1969. Specificity of the induced Bactericidin of the West Indian Spiny Lobster, *Panulirus argus*. *J. Bacteriol.* 98:947–48

Weissman G, Azaroff L, Davidsons S, Dunham P. 1986. Synergy between phorbol esters, 1-oleyl-2-acetylglycerol, urushiol, and calcium ionophore in eliciting aggregation of marine sponge cells. *Proc. Natl. Acad. Sci. USA* 83:2914–18

Wilson HV. 1891. Notes on the development of some sponges. *J. Morphol.* 5:511–19

Wilson WH, Chapman DM. 2001. Observation of virus-like particles in thin sections of the plumose anemone, *Metridium senile*. *J. Mar. Biol. Assoc. UK* 81:879–80

Wilson WH, Dale AL, Davy JE, Davy SK. 2005. An enemy within? Observations of virus-like particles in reef corals. *Coral Reefs* 24:145–48

Zheng L, Zhang L, Lin H, McIntosh MT, Malacrida AR. 2005. Toll-like receptors in invertebrate innate immunity (minireview). *Invert. Surviv. J. (ISJ)* 2:105–13

Zocchi E, Carpaneto A, Cerrano C, Bavestrello G, Giovine M, et al. 2001. The temperature-signaling cascade in sponges involves a heat-gated cation channel, abscisic acid, and cyclic ADP-ribose. *Proc. Natl. Acad. Sci. USA* 98:14859–64

Experimental Methods for Measuring Gene Interactions

Jeffery P. Demuth and Michael J. Wade

Department of Biology, Indiana University, Bloomington, Indiana 47405-3700;
email: jpdemuth@indiana.edu, mjwade@indiana.edu

Key Words

epistasis, quantitative genetics, QTL, microarray, analysis of variance, line cross analysis

Abstract

The role of epistasis in evolution has long been contentious. Resolving the issue requires empirical measurements that are statistically adequate and evolutionarily relevant. We review experimental methods for measuring epistasis, some that are commonly used but weak and others that are less frequently used but stronger. We review statistical genetic methods based on analyses of variances and means as well as molecular genetic methods for detecting gene interactions. We also highlight relevant empirical studies that illustrate the implementation of particular methods. In spite of the inherent weaknesses of most methods, epistasis is surprisingly common. We conclude with a discussion of how technologies for investigating genome-wide epistasis are bridging the gap between physiological and statistical epistasis for model organisms.

INTRODUCTION

ANOVA: analysis of variance

Epistasis is the physiological and/or statistical interaction of two or more genes in producing phenotypes and the variation among them. Epistasis is important in many areas of evolutionary theory and human genetics. Epistasis affects evolution in peripheral isolates, reproductive isolation, Muller's ratchet, speciation by introgressive hybridization, developmental homeostasis and canalization, the evolution of sex and recombination, inbreeding and outbreeding depression, and heterosis. The differing interpretations of epistasis by Fisher and Wright initiated a longstanding controversy over its importance in adaptive evolution (Coyne et al. 1997, 2000; Goodnight & Wade 2000; Schlichting & Pigliucci 1998; Wade & Goodnight 1998), a controversy whose resolution requires relevant empirical measurements.

In human genetics, the functional annotation of the human genome and our understanding of the role of genes in complex diseases depend on elucidating the actions of genes both alone and in combination. Complex diseases such as asthma, schizophrenia, autism, hypertension, diabetes, many cancers, and multiple sclerosis are more prevalent in the human population than single-gene diseases such as cystic fibrosis in large part because gene interactions limit the effectiveness of natural selection. Whenever the interaction of two genes causes a disease, an allele can be a good gene in one genetic background but a bad gene in another, with the disease manifest only in that subset of individuals bearing the specific multilocus allelic combination. Thus epistasis is one cause of the low penetrance of complex diseases, and the power of some standard methods of gene detection is reduced because the frequency of such a gene may differ slightly or not at all between collections of diseased and healthy individuals (Cardon & Bell 2001). Furthermore, a disease gene discovered in one pedigree with one genetic background may not be replicable in another pedigree with a different background (Hirschhorn et al. 2002, Ioannidis et al. 2001). With replicability as the gold standard of gene discovery, variations in sign and magnitude of gene effects caused by epistasis are problematic (Goodnight 1988; Wade 1992, 2001, 2002).

Epistasis is equally important to the microbes that cause disease and the emergence of pathogens. Influenza pathogenicity in humans and its resistance to antiviral drugs are associated with specific gene combinations of the surface glycoproteins hemaglutinin and neuraminidase (Peiris et al. 2004). Of the 126 possible two-gene combinations of these influenza surface-coat proteins, only two (H5N1 and H7N7) are highly pathogenic in poultry and only three (H1N1 in 1918, H2N2 in 1957, and H3N2 in 1968) have resulted in human pandemics (Kuiken et al. 2003, Peiris et al. 2004).

Detecting and quantifying epistasis require demonstrating that the effects of a gene change in sign and/or magnitude with changes in genetic context. Thus, the experimental designs for measuring epistasis tend to include methods for varying both the frequencies of focal genes as well as the genetic contexts in which they are expressed. In classic breeding designs, epistasis is detected as failure of an additive model to account for some of the variation in phenotype observed in a series of controlled crosses in the same way that interactions are detected in a factorial design using analysis of variance (ANOVA) (Wade 1992, 1996). Such designs can be biased against detecting epistasis or against detecting specific kinds of epistasis (see below).

In this review, we present methods for detecting and quantifying the phenotypic effects of epistasis. Because different experimental methods measure epistasis in different ways or measure some kinds of epistasis more effectively than others, we distinguish among them in terms of the applicability of their results to evolutionary genetic theory. Classical methods are based on the statistical signature of epistasis on the phenotypic means of complex crosses, whereas newer, molecular methods—e.g., quantitative trait loci (QTLs)—can be directly gene-based. Detecting and quantifying epistasis between QTLs are central to many areas in population genetics and evolutionary biology. It is also conceptually desirable to quantify metaphorical references to coadapted gene complexes and the release of cryptic gene-gene interactions to better evaluate whether they can bear the weight they are accorded in evolutionary explanation.

QTL: quantitative trait locus

Statistical Genetics and Epistasis

In **Table 1**, the average effect of a gene substitution at locus A is defined using the linear regression of genotypic value on dosage of the A allele weighted by the genotypic frequencies (Falconer 1985, Fisher 1941, Templeton 1987), assuming for simplification that the genotypic frequencies are in multilocus Hardy-Weinberg equilibrium, although deviations in natural or artificial populations necessarily change the regression. The breeding value of a genotype falls on the regression line and equals the portion of its phenotype transmissible to its offspring. Reciprocally, an individual's breeding value equals the average breeding value of its parents. The variance in breeding values is proportional to the additive genetic variance, the numerator of narrow-sense heritability. One can predict the response of a population to selection

Table 1 Genotypic values for nine unordered genotypes used to calculate the average effect of the A allele as the regression of genotypic mean value on dosage of A alleles

	AA	Aa	aa
BB	$a + a_2$	d_A	$-a - a_2$
Bb	a	d_A	$-a$
bb	$a - a_2$	d_A	$-a + a_2$
Mean	$a + a_2 (p_B - q_b)$	d_A	$-a - a_2(p_B - q_b)$
Dosage of A	2	1	0

Population mean: $M = a(p_A - q_a) + 2p_A q_a d_A + a_2(p_A - q_a)(p_B - q_b)$.
Average effect: linear regression coefficient of genotypic value on the number of A
  alleles = covariance(genotypic value, dosage of A)/variance(dosage of A).
Covariance(genotypic value, dosage of A) = $\{2p_A 2(a + a_2[p_B - q_b]) + 2p_A q_a d_A\}$
  $- \{2p_A\}\{a(p_A - q_a) + 2p_A q_a d_A + a_2(p_A - q_a)(p_B - q_b)\}$.
Covariance(genotypic value, dosage of A) = $2p_A q_a(a + d_A[q_a - p_A] + a_2[p_B - q_b])$.
Variance(dosage of A) = $4p_A 2 + 2p_A q_a - (2p_A)2 = 2p_A q_a$.
Average effect of A = $2p_A q_a(a + d_A[q_a - p_A] + a_2[p_B - q_b])/2p_A q_a$.
Average effect of A = $(a + d_A[q_a - p_A] + a_2[p_B - q_b])$.
No epistasis, $a_2 = 0$: average effect of A = $(a + d_A[q_a - p_A])$.
With epistasis, $a_2 \neq 0$: average effect of A = $(a + d_A[q_a - p_A] + a_2[p_B - q_b])$.

Metapopulation: population of populations

given the selection differential (the deviation of the parent mean from the mean of the population) and the heritability, which tends to be less than one owing to the contribution of the environment to total phenotypic variation. Fisher (1930) included epistatic variance in the nonheritable environmental variance, indicating it was not relevant to predicting the response to selection in large, randomly mating populations. However, epistasis affects the definition of the average effect of a gene substitution.

By the regression definition, the average effect of a gene substitution equals the quantity $a + d_A(q_a - p_A) + a_2(p_B - q_b)$, where a is the additive contribution to genotypic value, d_A is the dominance deviation, and a_2 is the additive-by-additive interaction between the loci A and B (**Table 1**). The term interaction in this context is the same as that used for the interaction of factors in an ANOVA but different from the concept of nonindependence of gene action used to define epistasis in population genetics. In the absence of dominance and epistasis (i.e., $d_A = a_2 = 0$), the average effect is constant and unchanging, independent of its own frequency and that of other genes. However, with dominance (i.e., $d_A \neq 0$, $a_2 = 0$), the average effect of an A allele depends on the difference in frequency of alleles at this locus ($q_a - p_A$) and is greater when allele A is rare than when it is common. Similarly, with epistasis (i.e., $d_A = 0$, $a_2 \neq 0$), the average effect of the A allele depends on the difference in frequency of alleles at the B locus ($p_B - q_b$) and is greater when the B allele is common and smaller when B is rare. If the genetic background at the B locus is fixed for the B allele (i.e., $p_B = 1$), then the average effect of the A allele is $a + a_2$. If the b allele is fixed (i.e., $q_b = 1$), then A's average effect is $a - a_2$. With dominance or epistasis, the average effect is not an invariant property of an allele. For example, setting the true additive effect, a, equal to zero does not necessarily make A's average effect equal zero. One implication of this is that on a single genetic background (similar to that of the large populations postulated by Fisher, where allele frequencies at all loci change slowly or not at all), the average effect appears to be a constant despite dominance and epistasis. However, a single genetic background is empirically insufficient for determining whether or to what degree the average effect depends on other genes for detecting additive and epistatic effects as separate contributions. To partition the additive effect, a, away from the epistatic effect, a_2, one must measure the effect in different backgrounds. A central empirical issue is the degree to which the kinds of variations in genetic background that occur from deme to deme by random genetic drift and local natural selection in natural metapopulations cause biologically significant variations in the average effects of a gene substitution. Some experimental designs better address these questions than others.

Analysis of Variance

ANOVA is a branch of statistics originally formulated by Fisher (1918a,b, 1922). The types of experimental designs used in ANOVA approaches in genetic analysis can be broken down into general categories depending on which types of relatives are observed—e.g., full sibs, half-sibs, both full and half-sibs, cousins, or more-distantly related individuals (Lynch & Walsh 1998). Because the original goal of ANOVA approaches was to estimate the heritability or additive genetic variance of traits, designs often do not incorporate the relationships necessary for measuring epistasis. Furthermore, some designs may unreliably estimate the additive genetic variance when epistasis is present.

There are at least three difficulties fundamental to ANOVA-based approaches regarding epistasis: (*a*) There is always less power to detect interactions (epistasis) than main (additive) effects because interactions are measured by smaller sample sizes, and their detection depends on the magnitude of the main effects; (*b*) estimates of the main effects are unreliable when interactions are present (Neyman 1935), and low replication or unbalanced design increases the unreliability; and (*c*) genetic experiments often involve the estimation of effects with variable, and sometimes unknown, underlying treatments (gene frequencies). The weaknesses of the ANOVA approach limit our ability to extrapolate findings from one population to the next. The following thought experiment, suggested originally by Hayman (1958), illustrates the difficulties of ANOVA-based estimates of genetic effects in the presence of gene interactions.

In a typical two-factor design with three treatments per factor (A_1, A_2, and A_3 and B_1, B_2, and B_3), the observation for A_1 is averaged over all factor B treatments. If there is an interaction between A_1 and B_1, it is measured by the deviation of the $A_1 \times B_1$ combination from the average effects of A_1 and B_1 across all treatments of the other factor. Thus, even in a balanced design where all factor-treatment combinations contain the same number of observations, the interaction terms have fewer degrees of freedom relative to the main effects. Furthermore, when the marginal effect of A_1 is much smaller than that of B_1, interaction effects are more likely to be interpreted as a result of a main effect of factor B.

Let us assume this first experiment is used to estimate the main effects of factors A and B. Now we conduct a second experiment in which we retain the treatments of factor A but change those of B to B_4, B_5, and B_6. The estimated main effects of each factor in the second experiment may be different from those in the first if interactions are present. In other words, the effects of factor A cannot be compared between experiments because factor B is not constant between experiments. This may be acceptable when an experiment is sufficient for a given time and place. However, if the three treatments of factors A and B are three genotypes at two diallelic loci, our estimates of genetic effects vary from one experiment (i.e., population, line, or pedigree) to another on the basis of the respective allele frequencies. The peculiarly genetic difficulty is that the differences in allele frequency from one experiment to another are often unknown.

Among-Deme Variation in Sire-Breeding Values (Sire-by-Deme Interaction)

Within the limitations of the ANOVA approach, some measures of epistasis are more evolutionarily relevant than others. For metapopulations, the central evolutionary question is to what degree does genetic subdivision cause interdemic differences in mean phenotype and interdemic variation in average allelic effects. In practice, average effects are measured as sire-breeding values, and the standard half-sib design can be extended to metapopulations using methods from animal science developed for

EPD: expected progeny difference

estimating nonadditive genetic effects within and between breeds (Elzo 1990, 1994). The intrabreed expected progeny difference (EPD) equals one half the estimated sire-breeding value and is used to rank sires with respect to the difference in progeny traits when bred to a reference population. However, as Elzo (1994, p. 3055) states, "[u]nfortunately, intrabreed EPD cannot be used to compare bulls of the same or different breeds for crossbreeding purposes because they consider only additive genetic effects... and they ignore nonadditive genetic effects." With epistasis, the ranking of sires within breeds may change their ranking for progeny traits across breeds. The genetic differences among breeds as well as those among demes in metapopulations are the result of both random drift and selection. Similar to the intrabreed EPD, the ranking of sires by progeny fitness within one deme may differ in another deme (Wade 1985, 2000).

Sires randomly chosen from a breed, a source population, or a deme within a metapopulation can be crossed to groups of dams from other breeds or demes. The breeding value of a sire is now the mean value of its offspring taken across all dams from all demes or breeds. It can be partitioned into three additional components: (a) the average value of offspring of the i-th sire mated to dams from all demes, (b) the average value of offspring of dams from the j-th deme mated to all sires, and (c) the average value of offspring of the i-th sire with dams from the j-th deme, the sire-by-deme interaction term. In a metapopulation, variation among sires in the first term may not be as predictive of the response to selection as it is in a large randomly mating population without genetic subdivision. Without gene interactions, the third term is zero, and the variance of the second term equals the among-deme variance in mean phenotype and should be proportional to F_{ST}. In the absence of epistasis, among-deme variance causes sires to vary in breeding value from deme to deme, but it does not affect the relative magnitude or ranking of the sire-breeding values (Goodnight 2004, Wade & Goodnight 1998). The sire-by-deme interaction term measures the extent to which gene effects differ from one deme to another and are manifest as changes in the relative breeding values of the tested sires.

The sire-by-deme interaction variance can be nonzero even when there is no among-deme variance in mean phenotype (figures 9.2 and 9.4 in Goodnight 2004). The among-deme variance in sire-breeding values can be expressed by estimating the intraclass correlation among sire-breeding values across demes. In the absence of interactions, the correlation should be one so that sires maintain their relative ranking. However, in the presence of gene interactions, the intraclass correlation is less than one because all components of interaction reduce the correlation, and additive-by-additive epistatic interactions can result in changes in the rank of sire-breeding values across demes (Goodnight 2000, 2004). Such changes in average allelic effects on fitness, from positive in a conspecific background to negative in a heterospecific background, are an essential aspect of the genetics of reproductive isolation and speciation (Coyne & Orr 2004, Johnson & Wade 1996, Moreno 1994) (see below).

The variance in breeding values among sires from the j-th deme with respect to the mean of the j-th deme (as opposed to the mean of the metapopulation) measures the local additive variance, useful for describing the response to local selection and adaptation occurring within the j-th deme. The average of this variance across demes is the average rate of local adaptation across the entire metapopulation. The ratio of the sire-by-deme interaction variance to this average within-deme variance in breeding values is a measure of the degree to which uniform selection in all demes acting on the same genes nevertheless results in a genetic diversification among the demes (Goodnight 2004, Wade & Goodnight 1998). By estimating the components of variance in breeding value and the intraclass correlation across demes of sire-breeding values for different values of F_{ST}, we can determine which levels of genetic population structure are as evolutionarily important as the claims of Wright and which are insignificant as claimed by Fisher.

In laboratory studies of experimental metapopulations using flour beetles, researchers have observed large, heritable differences in mean deme fitness for a variety of different combinations of effective population size (N_e) and migration rate, even for values of F_{ST} less than 0.10 (McCauley & Wade 1980, 1981; Wade 1978, 1979, 1980, 1982, 1984, 1985, 1988, 1990; Wade & Goodnight 1991; Wade & McCauley 1980, 1984). Artificial among-deme selection by both differential extinction and differential migration demonstrated that there was biologically significant genetic variation among demes, but the variance among demes in sire-breeding values or allelic effects was not measured. Only one study has measured the among-deme variance in sire-breeding values for fitness, using demes differentiated by random drift, F_{ST} value of ~0.15 (Wade 1985, 2000). In that experiment, the among-deme component of variance in breeding value for relative fitness was 18 times greater than the among-sire component, a value 54 times greater than that predicted by a strictly additive model. The average correlation of sire-breeding values across demes was 0.24 (s.d = 0.28), much less than the value of 1.0 predicted for a strictly additive model.

Effective population size (N_e): the number of individuals in an idealized population with the same sampling variance as the actual number of breeding individuals

LCA: line cross analysis

Line Cross Analysis

Line cross analysis (LCA) uses least-squares regression to evaluate the genetic architecture underlying divergence in mean phenotypes of two groups on the basis of the expected segregation of parental genes in hybrid lines (Cavalli 1952; Cockerham 1954, 1980; Hayman 1958, 1960; Hill 1982; Kempthorne 1954; Lynch 1991; Lynch & Walsh 1998; Mather & Jinks 1982). Detection of epistasis via LCA depends on the estimation of means rather than variances. Because for a given sample size means are measured with greater accuracy than variances, LCA can be a more-powerful approach (Fenster et al. 1997). Despite this fundamental difference, the underlying statistical machinery for LCA- and the ANOVA-based approaches is similar in that both are variations of general linear models. The ANOVA approach has traditionally been used to estimate the contribution of epistasis to genetic architectures within populations and experimental lines, whereas LCA has primarily been used to measure epistasis between populations, lines, or species (Demuth & Wade 2005). **Figure 1** illustrates the principle underlying LCA in which only additive and dominance effects influence the difference in a phenotype between two parental populations, P_1 and P_2. By adding regression parameters that are products of the single locus coefficients, we can extend LCA to account for deviations from the simple pattern in **Figure 1** that arise as a consequence of epistasis.

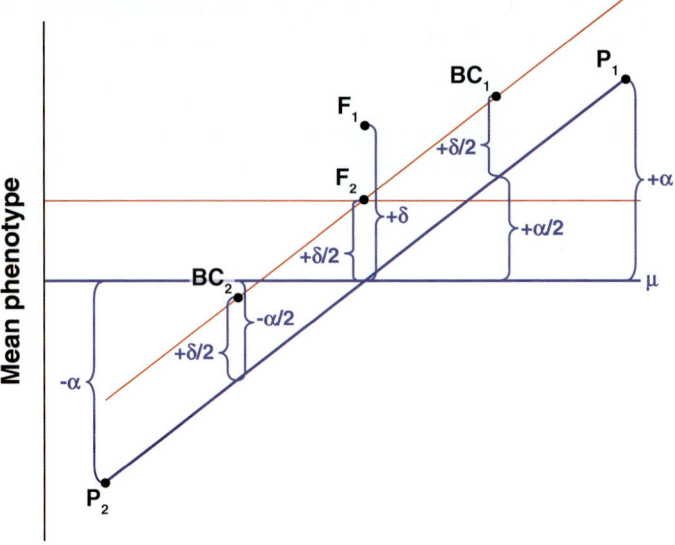

Figure 1
Example of line cross analysis where only additive and dominance effects differentiate two populations (P_1 and P_2). Brackets demarcate the contributions of composite additive (α) and dominance (δ) effects to hybrid F1, F2, and backcrosses toward P_1 and P_2 (BC_1 and BC_2, respectively). Blue lines and text indicate the assignment of effects using the mean of P_1 and P_2 (μ) as a reference. The red lines illustrate how using the F_2 mean as a reference requires rescaling the additive and dominance contributions.

There are a few important points to consider regarding LCA interpretation. First, the mean phenotypes of hybrid lines are manifestations of the composite genetic effects across all loci; therefore, if individual genes differ in sign, their net effects on the phenotype may be zero. Thus failure to detect a composite dominance effect, for example, does not necessarily imply there is no dominance for that phenotype. Rather it suggests that on average, genes from one population are not dominant to those in the other population. This potential for cancellation means that LCA provides a conservative estimate for whether a given genetic effect contributes to differences between parental populations. If desired, one can further investigate the cancellation of effects by analyzing within-line variances (Mather & Jinks 1982).

Second, LCA detects the genetic basis for differences between P_1 and P_2 but not the architecture underlying the phenotype within either population. Hence, it informs us about the product of evolution but does not allow us to project the future direction of evolution. A good illustration of this point is the conversion of additive-by-additive variance to additive variance (Goodnight 1987, 1988). For two loci fixed differently in alternate populations, the evolutionary fate of alleles at a segregating partner locus in each population is determined entirely by their additive effects. When both loci segregate in derived hybrid lines, LCA may detect the additive-by-additive epistasis between them.

There is one caveat regarding the genetic architecture of traits revealed by LCA. When differing lines derived from a single species or population are used as the P_1 and P_2 lines in an LCA, we can assess the contributions of various genetic effects within the original population (e.g., Kelly 2005). This use of LCA has long been a staple of plant and animal breeders and is philosophically and mathematically foundational to standard methodologies for mapping QTLs. However, as discussed above, methods that homogenize the genetic background when creating P_1 and P_2 may yield results that are not generalizable to other genetic backgrounds.

Because LCA measures the magnitude of genetic effects as deviations from the mean of a reference population, a final consideration is the choice of reference. **Figure 1** shows how a change in scale from the midparent to deviations from the F_2 mean necessitates redefining the individual expectations for line means. Models based on either reference are interconvertable (Van Der Veen 1959), so in most cases the choice of scale may ultimately be inconsequential to the detection of epistasis. However, the choice of reference becomes more important for QTL studies because it can complicate the interpretation of the magnitude of QTL effects (Yang 2004, Zeng et al. 2005).

Hypothesis Testing

LCA provides a basis for tests evaluating hypotheses about the genetic underpinnings of population divergence. Simple scaling tests for the presence of epistasis are of the general form wherein the expected sum of a particular set of line means incorporating only additive and dominance parameters is zero. For example, in the absence of epistasis, the scaling test utilizing the parent populations, F_1 and F_2 lines, has the expectation $S = 4\bar{z}(F_2) - 2\bar{z}(F_1) - \bar{z}(P_1) - \bar{z}(P_2) = 0$, where $\bar{z}(i)$ is the observed mean of line i.

Although there are numerous variants of simple scaling tests that utilize different line combinations (e.g., Anderson & Kempthorne 1954, Hayman 1957, Jinks 1956, Mather & Jinks 1982), the joint-scaling test provides a more-general method for testing hypotheses about genetic architecture (Cavalli 1952, Hayman 1960). The weighted error sum of squares

$$\chi^2 = \sum_{i=1}^{k} \frac{(\bar{z}_i - \hat{z}_i)}{Var(\bar{z}_i)} \qquad 1.$$

provides an χ^2 distributed statistic for testing goodness-of-fit between the observed means, $\bar{z}_i$, and the expected means on the basis of LCA, $\hat{z}_i$. Because adding parameters to account for different genetic effects always improves the overall model fit (i.e., reduces χ^2), standard protocol is to test increasingly complex models while using the difference in χ^2 values as a likelihood ratio for assessing the significance of the improvement. In principle, joint scaling can be used to test LCA models with any number of specified effect parameters. In practice, analyses are limited by the number of different types of crosses available that must be one more than the number of parameters being simultaneously tested to allow at least one degree of freedom. The

G × E:
genotype-by-environment interaction

Statistical epistasis:
deviation from the additive expectation that depends on the genotypic combination at two or more loci

ability to distinguish between LCA models is also determined by the replication of each line.

Experimental Results

Several recent studies provide examples of the utility of LCA and joint scaling for evolutionary analyses by crossing populations of a single species that span a broad range of geographic and genetic distances. For example, analysis of marine copepod (*Tigriopus californicus*) populations along the western coast of North America show that both the beneficial effects of dominance and the detrimental effects of breaking up coadapted gene complexes are magnified by increasing evolutionary distance. Maternal effects, diagnosed by incorporating means of reciprocal crosses, and digenic epistasis are common components of divergence in *T. californicus*. Subsequent work in this system demonstrates that much of the observed F_2 breakdown in divergent populations is a result of cytonuclear coadaptation (Edmands & Burton 1999, Willett & Burton 2001), and this cytonuclear epistasis has a large genotype-by-environment interaction (G × E) component (Willett & Burton 2003).

Similarly, experiments by Fenster & Galloway (2000; Galloway & Fenster 2001) demonstrate that cytonuclear interactions, autosomal epistasis, and G × E are involved in the divergence of fitness-related traits in populations of partridge pea (*Chamaecrista fasciculata*) from Maryland, Illinois, and Kansas, some of which exist in close proximity (100 m). Bradshaw & Holzapfel (2000) found that the degree of epistasis underlying divergence in photoperiodism in the mosquito *Wyeomyia smithii* depended more on time of divergence than distance between populations. The mosquito studies also demonstrate that there can be substantial changes in the composite genetic effects underlying fitness, even when the population means do not differ. Recently, LCA of selection lines derived from within wild mosquito populations showed there is also significant epistasis underlying photoperiodic response within populations (Bradshaw et al. 2005).

Finally, Demuth & Wade (2006) used LCA on an extensive set of crosses reared in different environments to demonstrate that maternal effects, digenic epistasis, cytoplasmic effects, and G × E are common components of divergence among closely related populations of the red flour beetle, *Tribolium castaneum*. Furthermore, many of the phenotypes present in crosses between *T. castaneum* populations are similar to those found in the interspecific cross between *T. castaneum* and *T. freemani*. In a few population crosses, a weak form of Haldane's rule is also present, suggesting the propensity of the heterogametic sex to suffer the penalty of hybridization may accompany the speciation process rather than occur strictly postspeciation.

Molecular Dissection of Interacting Genes

Above we review the framework for elucidating how unseen, underlying genetic factors can be inferred from measured phenotypes. Provided their weaknesses are taken into account, these methods provide measures of statistical epistasis that are often sufficient to make evolutionary predictions. However, there is no necessary link between the statistical characterization of epistasis and the physiological interaction of genetic factors. Although statistical epistasis likely implies physiological epistasis, the reverse is not true, as physiological interactions may produce additive outputs. A bridge between statistical and physiological epistasis lies in the molecular dissection of complex traits (Carlborg & Haley 2004, Moore & Williams 2005). The more-integrated understanding of statistical and physiological epistasis is increasingly possible, even in nonmodel systems, as technology makes the use of molecular markers more accessible and efficient.

> **Physiological epistasis:** physical interaction of two genes where the degree or pattern of expression is altered by expression of at least one other gene

There are two related approaches to mapping the molecular basis of phenotypes. Linkage mapping depends on known pedigrees or controlled crosses and measures the degree of recombination between an observed marker locus and the unobserved locus controlling variation in the trait of interest. In contrast, association mapping assesses differences in genotypes at marker loci between two populations that differ in the trait of interest (e.g., cases and controls in disease studies). Typically, linkage analysis is more useful for coarse-scale mapping because its resolution depends on recombination between the marker locus and trait locus, whereas association studies are more useful for fine mapping but require a marker in tight linkage disequilibrium with the locus affecting the trait. Association studies are the more-powerful approach to identifying loci with small individual effects (Gordon & Finch 2005, Long & Langley 1999, Risch & Merikangas 1996), and perhaps epistatic loci (Marchini et al. 2005), thus providing motivation for the human HapMap (the International HapMap Consortium 2005). However, the cost involved in attaining the requisite marker density for association studies is prohibitive for most evolutionary studies. Because linkage and association approaches share many of the same difficulties detecting and quantifying effects of epistatic loci, and because several recent reviews have focused on association studies (e.g., Freimer & Sabatti 2004, Gordon & Finch 2005, Marchini et al. 2005), our discussion below focuses on the most-commonly used linkage approach in evolutionary studies, QTL analysis.

Quantitative Trait Loci

A QTL is a position in the genome that explains variation in a quantitative phenotype. In a typical QTL mapping study, individuals of known pedigree are measured for the relevant phenotype and genotyped for polymorphic marker loci. The relationships between phenotypic values and specific marker polymorphisms depends on recombination between the marker(s) and QTLs (Lander & Botstein 1989, Soller et al. 1976). Since the late 1990s, QTL analyses have become the most-commonly utilized approach to understanding complex traits. The December 2005 Science Citation Index lists more than 3600 published theoretical, empirical, and review articles concerning QTL analysis since January 2000. Because the technical literature on QTL mapping is too extensive to cover in the present context, we refer the reader to the following reviews, and citations therein, as an introduction to QTL methodologies (Carlborg et al. 2003, Doerge 2002, Doerge et al. 1997, Lynch & Walsh 1998). Additionally, Yandell (2005) provides an extensive list of QTL references, including analysis software, and Erickson et al. (2004) summarize the utility of the QTL approach for studies of evolution.

The ability of QTL analysis to detect epistatic loci is generally poor because many of the weaknesses inherent in detecting statistical epistasis not only carry over but are compounded in QTL studies (Mackay 2001). The increased difficulty arises in part because the detection of QTLs and the assignment of their effects can be difficult to disentangle. There are at least three issues that make epistasis problematic for QTL experiments: (*a*) Estimating interaction effects requires larger samples sizes; (*b*) searching for an interacting QTL involves a large number of comparisons; and (*c*) epistasis limits our ability to interpret the effect of individual loci within and among studies.

Sample Size

The power to detect a QTL is proportional to the amount of phenotypic variance it explains; however, as the number of QTLs contributing to a trait increases, each QTL's average contribution must decrease (Lander & Botstein 1989). Therefore, as the number of QTLs expected to contribute to the phenotype increases, the number of individuals measured must increase. Several factors contribute to the exact number of individuals necessary for adequate power to detect QTLs, including marker type, number, and distribution; type of mapping population; and accuracy of phenotypic measures.

All else equal, epistasis exacerbates the need for large sample sizes because as the number of interacting loci increases, the number of individuals in any particular genotypic class decreases. This is the same difficulty as estimating interaction effects in the ANOVA approaches (see above). If interacting loci are linked, the problem is compounded. The result of low, and often unequal, sampling is that measures of phenotypes controlled by interacting loci have larger sampling variance. The problem of utilizing too few individuals in QTL studies is compounded because not only may interacting QTLs be missed, but the effects assigned to detected QTLs are also exaggerated, an artifact known as the Beavis effect (Beavis 1994, 1998).

The importance of statistical artifacts associated with low power depends on the type of question the study is meant to address. They are less relevant for studies aimed at identifying the primary causative factors underlying a trait; however, they are an important bias if the aim is to ascertain the relative importance of epistasis in evolution or the distribution of effects fixed during adaptation (Erickson et al. 2004). The concern is particularly acute in the presence of epistasis where the labels, major and minor effect, may have little general meaning because their sign and magnitude depend on allele frequencies at other loci [i.e., they are specific to the background of the particular study (Templeton 2000)].

Multiple Tests

A second set of issues that makes epistasis difficult to detect in QTL studies is a simple consequence of combinatorics. Even if we consider only epistasis between L QTLs that have significant single-locus effects, the number of pairwise combinations is $L^*(L-1)/2$. Each of these combinations is subject to tests of differing genetic architecture (as in LCA). This number may be manageable when the number of QTLs detected is small. However, given that an epistatic QTL may have no single locus effect, L becomes the number of markers, rather than the number of QTLs. The problem is multiplied when one considers additional traits in a single study.

Several approaches have been developed that are theoretically capable of detecting epistasis among QTLs (with or without single locus effects) in experimental populations (Carlborg et al. 2000, Chase et al. 1997, Kao et al. 1999, Sen & Churchill 2001) and in humans or other organisms in which investigators utilize natural pedigrees (Almasy & Blangero 1998, Liu et al. 2002, Lunetta et al. 2000, Pratt et al. 2000, Rabinowitz 1997, Schaid 1996, Umbach & Weinberg 1997, Wu 2000). However, computational and multiple-testing problems are an obstacle to widespread implementation. In other words, from a methodological perspective, the difficulty is not identifying models that incorporate epistasis; the problem is identifying the best model from the ocean of possibilities (Carlborg & Haley 2004, Doerge 2002).

Efforts to address the computational problem of multiple testing limit the number of comparisons by considering only combinations that are conditional on one or both loci having significant single locus effects (e.g., Cheverud & Routman 1995, Lark et al. 1995). In principle, this strategy is undesirable for the biological reason that it may overlook epistatic loci with no individual effects (Frankel & Schork 1996). Association-mapping simulations also suggest that full searches over all pairwise interactions have greater statistical power, even in the face of a large increase in the number of tests (Marchini et al. 2005). However, Storey et al. (2005) showed empirically that a procedure where searches are conditioned on the single QTL explaining the most variance in the trait was more powerful than searching all possible two-way interactions. This has the appealing stepwise property that, in principle, multilocus interactions can be searched by conditioning on interacting QTLs that explain the most variance in the previous search.

Development of more-efficient algorithms and the incorporation of parallel processing have drastically reduced computational limitations (Foster 2001, Ljungberg et al. 2004). However, the large number of potential tests also poses the statistical problem of inflated type I error rate (false positives). Recent large-scale analyses have adopted the false discovery rate (FDR) as a means of determining the appropriate threshold for QTL significance (e.g., Brem & Kruglyak 2005, Brem et al. 2005). The FDR is the ratio of false positives to the total number of significant loci (Benjamini & Hochberg 1995). The utility of the FDR to control for false QTL discovery is an active area of research (Benjamini & Yekutieli 2005, Storey & Tibshirani 2003, Storey et al. 2004), and thus far a stepwise-FDR procedure seems to strike the best balance between being too conservative and accepting too many false positives (Benjamini & Yekutieli 2005).

An alternative to the approaches above is the development of Bayesian methods for QTL analysis (Sen & Churchill 2001). Baysian methods address many of the aforementioned weaknesses of epistatic QTL detection by breaking the detection of QTLs and assignment of effects into distinct parts. This allows the use of information gained during the initial estimation of QTL genotypes in the search for the best model of QTL effects. Using this method, Montooth et al. (2003) found significant epistasis

FDR: false discovery rate

for loci involved in flight performance and energy metabolism in *Drosophila* among loci that showed no significant single locus effects.

Biology of Interactions

The prevalence of unreplicable mapping studies is not solely a result of inadequate statistical methodology. The last difficulty in detecting an epistatic QTL is one of biological complexity. It seems widely appreciated, yet often ignored, that when there is epistasis, the effects assigned to component loci are unreliable, whether all interacting loci are detected. **Figure 2** illustrates how a phenotype determined solely by the interaction between homozygotes at two loci (additive-by-additive epistasis) can result in the missassignment of QTL effects and conflicting results among experiments. Wade (2001) also demonstrated in a relatively simple three-locus model, in which two loci regulate a third candidate locus, that an allele at the candidate locus can appear additive, dominant, recessive, overdominant, underdominant, or neutral, depending on the allele frequencies at the regulatory loci. The manifestation of these apparent effects is independent of whether any of the loci have individual effects (Wade 2001, 2002).

Experimental Design

Given the high costs of time and money involved in conducting experiments of the necessary scope, it is worth considering whether some experimental designs are more efficient than others with respect to mapping epistatic loci. There is relatively little discussion in the QTL literature that specifically addresses the statistical power of different crossing designs for detecting interacting loci. For example, Darvasi (1998) provides an excellent comparison of the power of several crossing designs for dissecting quantitative traits; however, only additive and dominance effects are considered in the calculations. It seems intuitive that methods with more power to detect an individual QTL should also be good at detecting epistasis, but this need not be the case. Indeed, an example from body weight in mice (see below) shows how a relatively powerful design, coupled with insufficient analytical tools, can result in misleading results.

Demant (2003) provides a thorough review of the wide variety of breeding designs that have been developed to maximize the potential to detect QTL underlying

Figure 2

A consequence of epistasis on assigning quantitative trait locus (QTL) effects and repeatability. Two separate experiments derive high- and low-phenotype lines from a population that is segregating at two loci (A and B). In the process of selection, the two experiments inadvertently fix alternate loci. The resulting F_2 mapping populations suggest there is a single, additive QTL that explains all the phenotypic variance but is not repeatable between experiments. In reality, neither locus has an additive effect. Blue bars indicate phenotypic value for genotypes present in the respective populations. Gray bars indicate genotypes that are absent in the QTL analysis owing to experimental design.

Founder population

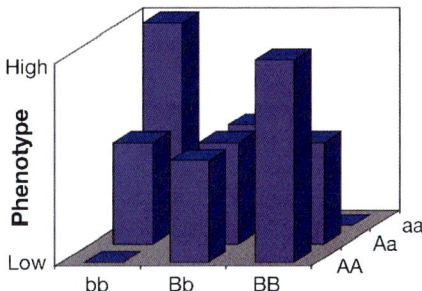

	AA	Aa	aa
BB	+a	0	-a
Bb	0	0	0
bb	-a	0	+a

Underlying genetic model: No marginal (additive) effect for either locus. All phenotypic variation due to interaction between homozygous loci (i.e., additive-by-additive epistasis)

Two different experiments create high and low lines for QTL analysis via selection and inbreeding

Experiment 1

Locus A unknowingly fixed for A in both lines

High line = AABB

Low line = AAbb

Experiment 2

Locus B unknowingly fixed for B in both lines

High line = AABB

Low line = aaBB

F_2 mapping population 1

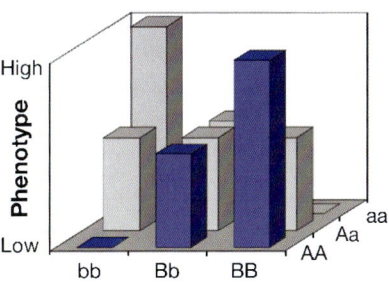

F_2 mapping population 2

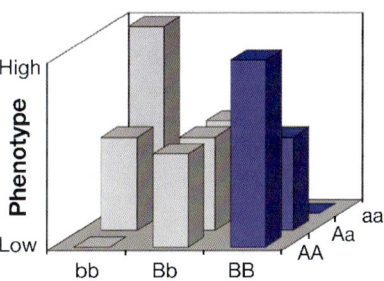

Conclusion 1

Locus B is an additive effect QTL with no interactions

Conclusion 2

Locus A is an additive effect QTL with no interactions

RCS: recombinant congenic strain

complex traits. Importantly, in the context of evolutionary studies, all artificial-breeding designs depend to some extent on the choice of parents and therefore lose their immediate connection with nature. Furthermore, experimental approaches that homogenize the genetic background (e.g., **Figure 2**) may create a QTL with large individual effects that are not transferable to other genetic backgrounds (Wade 2001). One suggested alternative to this trade-off between the ability to detect a QTL and evolutionary relevance is to map QTLs in natural hybrid zones; however, this presents several additional problems, such as an uneven distribution of genotypic classes (Gardner et al. 2000). A second alternative includes producing hybrids among several genetically distinct populations and then intercrossing the hybrids in a way that maintains maximal heterogeneity. This strategy underlies the most-elaborate experimental resource for QTL analysis of complex traits, the collaborative cross, which consists of ∼1000 recombinant inbred–mouse lines derived from eight parental lines (Complex Trait Consortium 2004).

As a compromise between capturing biological reality, the power to detect an epistatic QTL, and logistical efficiency, we propose the synthetic approach incorporating recombinant congenic strains (RCS) (Demant & Hart 1986) and recombinant progeny testing depicted in **Figure 3**. This strategy provides an efficient method to detect epistatic partners. If 40 RCS are developed (20 D1-RCS and 20 D2-RCS), it is possible to test ∼93% of each donor genome for interacting loci in the alternate donor genome. Some advantages to this design are as follows: (*a*) RCS lines need only be genotyped for all markers once, and (*b*) after the RCS lines containing epistatic loci are identified, additional backcrossing can further partition the genome, so identifying position is limited only by recombination. Additional backcross generations can be tested in the same ways but need only be genotyped for the markers in the original RCS.

Experimental Results

The studies of Cheverud and Routman on body weight in mice provide an example of initial efforts to comprehensively account for an epistatic QTL. Their mapping population consisted of 535 F_2 mice genotyped for 76 microsatellite markers. The initial search for epistasis between all unlinked pairs of QTLs with significant single locus effects (2443 pairwise combinations × 4 potential components of epistasis = 9772 tests) found that each QTL interacted with at least one other QTL (Routman & Cheverud 1997). A later whole-genome scan for two-locus interactions included 331,478 tests and took 20 days to compute (Cheverud 2000). In both cases, they used

Figure 3

Breeding design utilizing recombinant congenic strains (RCS) to discover interacting partners in different genomes. For simplicity, only one pair of homologous autosomes is shown. Crosses between donor 1 (D1, *red*) and donor 2 (D2, *yellow*) produce the phenotype of interest (*blue ovals*), but crosses between either D1 and host (H, *gray*) or D2 and H do not produce the phenotype. Therefore, pieces of each donor genome can be introgressed into the H background and subsequently tested against the alternative donor.

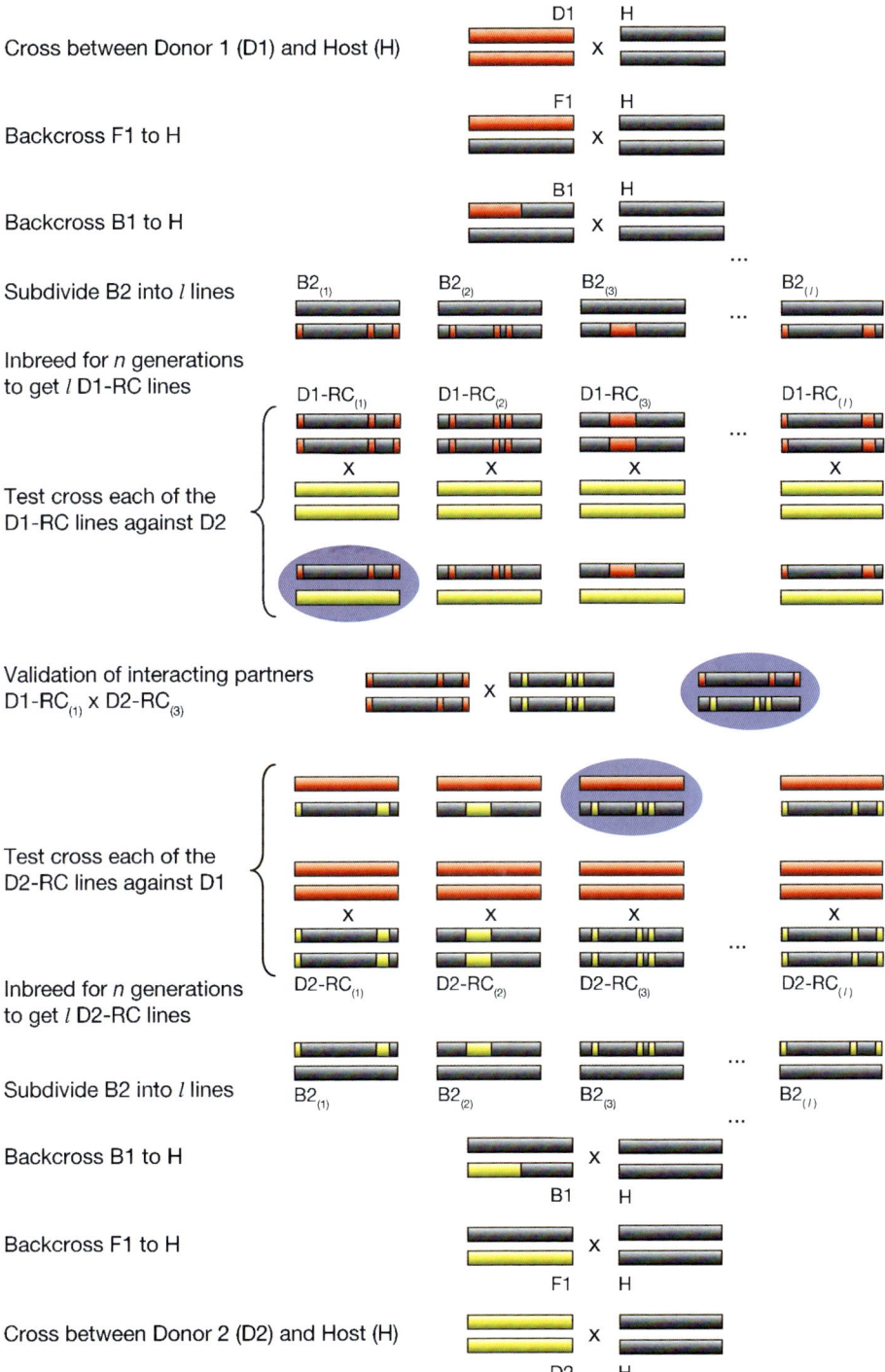

Bonferonni-based corrections for significance because computing threshold values by randomization or bootstrapping was not computationally feasible at the time, and FDR had not been incorporated into QTL methodologies.

The two searches of the same data produced different estimates of the relative frequencies of different types of epistasis. The earlier QTL-by-QTL survey found dominance-by-dominance to be most frequent type of epistasis, whereas the whole-genome scan found roughly equal frequencies of each type of epistasis. The difference between analyses was attributed to a bias in the QTL-by-QTL methodology because in populations with intermediate allele frequencies, as in the F_2 mapping population, loci with strong additive-by-additive effects tend to cancel each other, resulting in one or both loci going undetected. This can be visualized by considering the case where QTLs are mapped in the founding population in **Figure 2**. If allele frequencies are balanced, neither the A nor B locus shows a significant QTL effect, and searches for epistasis that rely on individually significant QTL find no QTLs for the trait.

Other empirical investigations bear out the necessity of large sample sizes for accurate estimation of epistatic effects. Studies using fewer individuals typically find less epistasis than large studies (reviewed in Carlborg & Haley 2004), although a few large studies have detected little epistasis (Flint et al. 2004, Richards et al. 2002, Zeng et al. 2000). There are also numerous cases in which epistatic effects are large enough that even intermediate sample sizes, in the range of 100–200 individuals, were sufficient (Cockerham & Zeng 1996, Cox et al. 1999, Damerval et al. 1994, Doebley et al. 1995, Eshed & Zamir 1996, Gallais & Rives 1993, Lark et al. 1995a, Leips & Mackay 2000, Li et al. 1997, Long et al. 1995, Shook & Johnson 1999). Considering the general weaknesses of QTL analysis for detecting interactions, it is premature to make general conclusions about the role of epistasis when most efforts to detect interactions have required the detection of significant single locus effects. Recent efforts have demonstrated an ability to detect QTLs with no single locus effects (Carlborg et al. 2003; Culverhouse et al. 2002, 2004; Montooth et al. 2003); however, few studies have yet to employ the necessary methods or sample sizes. The daunting realization of these studies is that experimental populations of 852 F_2 individuals (Carlborg et al. 2003) or groups of 95 recombinant inbred lines (Montooth et al. 2003) were needed to provide sufficient statistical power to detect epistatic loci with no individual effects.

The Genomic Frontier

Individual studies seeking to dissect the molecular basis of complex phenotypes have necessarily concentrated on one or a few organismal traits of evolutionary interest. In contrast, DNA microarrays can make the expression level of every gene in an organism's genome a measurable trait. Microarrays occur in several formats but analyses are all based on differential hybridization of sample DNA or RNA to thousands of reference sequences fixed on the array (e.g., The Chipping Forecast III 2005). **Figure 4** illustrates how combining QTL analysis and microarray expression profiles can highlight the interactions underlying gene regulation. Knowing the correspondence of QTL and gene-specific expression has the potential to yield

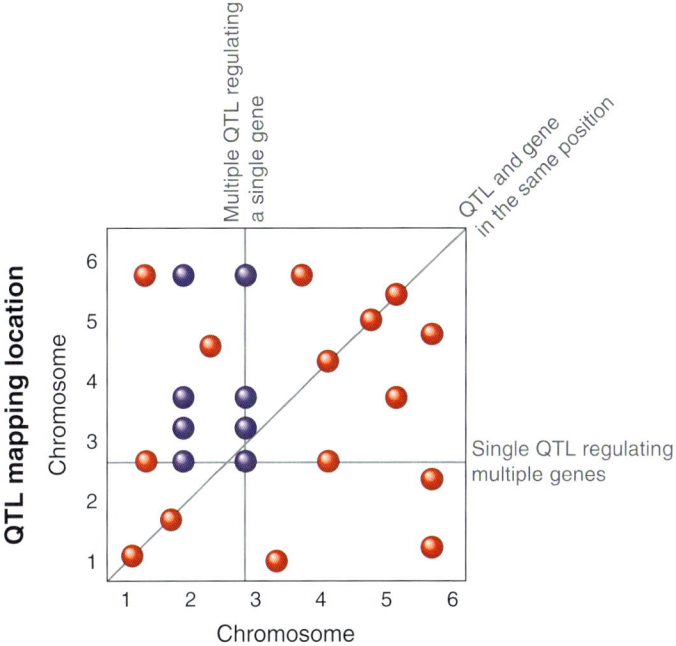

Figure 4

Combining microarray expression profiling with quantitative trait locus (QTL) analysis offers a new perspective on how genetic variation translates into variation in gene expression. Transcript-level variation is subject to QTL analysis. Transcripts with significant QTLs are then plotted according to QTL position (Y-axis) and their physical location in the genome (X-axis, known from prior genome sequencing). Points on the diagonal are *cis*-regulated (i.e., variation in the gene sequence affects expression level). Points lying along the gray horizontal line represent a case where genetic variation at one locus explains expression variation in multiple genes (*trans*-regulation). Points lying along the gray vertical line represent a case where genetic variation at multiple loci is required to explain expression of a single gene (multigenic regulation). Expression variation in multiple genes that is explained by the same set of QTLs (*blue circles*) may also indicate a shared molecular pathway. Modified from Jansen (2003).

important evolutionary insights, such as the relative roles of *cis*- and *trans*-regulation of expression-level variation (Jansen 2003, Jansen & Nap 2001).

The marriage of QTL and microarray methods also provides a unified view of statistical and physiological measures of epistasis. Kruglyak and colleagues (Brem & Kruglyak 2005; Brem et al. 2002, 2005; Storey et al. 2005; Yvert et al. 2003) utilized the above approach to assess genetic architectures underlying transcript-level variation between wild and laboratory yeast strains. Some pertinent findings include the following: (*a*) The vast majority of transcript variation has a polygenic basis; (*b*) twenty-one percent of transcripts with transgressive segregation (i.e., expression outside the range of either parent) are a result of epistasis; (*c*) sixteen percent of highly heritable transcripts and 14% of all traits show statistical epistasis for trait variation; (*d*) forty percent of highly heritable transcripts have no detectable additive QTLs, suggesting an even greater contribution of epistasis; and (*e*) many statistically significant epistasis are more common between loci where only one locus is significant by its individual effect than between loci where both loci are identified as significant QTLs.

Although it provides more data, the use of genomic tools does not overcome the traditional shortcomings inherent in measuring interactions. In fact, the volume of data obtained in single studies exacerbates statistical problems such as multiple testing. Even so, genome-wide analyses capable of examining both physiological and statistical epistasis promise an ability to address the evolutionary relevance of epistasis in ways that Fisher and Wright may not have dreamed possible. The extent of both physiological and statistical epistasis in even a simple eukaryote, such as yeast, supports the view of Wright who argued that epistasis is ubiquitous and should not be ignored in evolutionary studies (Wright 1968).

SUMMARY POINTS

1. There is still no direct and necessary connection from physiological to statistical epistasis. Although statistical epistasis implies physiological epistasis, the reverse is not true.

2. Variations in the sign and magnitude of gene effects caused by epistasis are problematic for the gene's eye view of evolution, particularly in small, subdivided populations.

3. Detecting and quantifying epistasis require demonstrating that the effects of a gene change in sign and/or magnitude with changes in genetic context, and this makes epistasis hard to measure.

4. Different experimental methods measure epistasis in different ways or measure some kinds of epistasis more effectively than others.

5. Given the high costs of time and money involved in conducting experiments of the necessary scope, some experimental designs are more efficient than others with respect to mapping epistatic loci.

6. Although it provides more data, the use of genomic tools does not overcome the traditional shortcomings inherent in measuring interactions. The volume of data obtained in single studies exacerbates statistical problems such as multiple testing.

FUTURE ISSUES

1. Understanding the complex interactome that lies between the genotype and phenotype remains a major challenge for evolutionary studies and for

functional annotation of genomes. Meeting this challenge requires further development of methods for measuring epistasis.

2. The marriage of traditional methods and novel genomic tools offers promise for a unified view of statistical and physiological epistasis not previously available.

ACKNOWLEDGMENTS

This paper was supported by NSF DEB 0206628 and GM065414–01A to M.J.W. J.P.D. was also supported during the course of this work by NSF IGERT in Evolution Development and Genomics 9972830 and NSF DDIG 0206628.

LITERATURE CITED

Almasy L, Blangero J. 1998. Multipoint quantitative-trait linkage analysis in general pedigrees. *Am. J. Hum. Genet.* 62:1198–211

Anderson VL, Kempthorne O. 1954. A model for the study of quantitative inheritance. *Genetics* 39:883–98

Beavis WD. 1994. *The power and deceit of QTL experiments: lessons from comparative QTL studies.* Presented at Annu. Corn Sorghum Res. Conf., 49th, Washington, DC

Beavis WD. 1998. QTL analyses: power, precision and accuracy. In *Molecular Dissection of Complex Traits*, ed. AH Paterson, pp. 145–62. Boca Raton, FL: CRC Press

Benjamini Y, Hochberg Y. 1995. Controlling the false discovery rate—a practical and powerful approach to multiple testing. *J. R. Stat. Soc. Ser. B* 57:289–300

Benjamini Y, Yekutieli D. 2005. Quantitative trait loci analysis using the false discovery rate. ***Genetics*** **171:783–89**

> Details the use of FDR statistics for QTL analysis.

Bradshaw WE, Haggerty BP, Holzapfel CM. 2005. Epistasis underlying a fitness trait within a natural population of the pitcher-plant mosquito, *Wyeomyia smithii*. *Genetics* 169:485–88

Bradshaw WE, Holzapfel CM. 2000. The evolution of genetic architectures and the divergence of natural populations. See Wolf et al. 2000, pp. 245–63

Brem RB, Kruglyak L. 2005. The landscape of genetic complexity across 5,700 gene expression traits in yeast. *Proc. Natl. Acad. Sci. USA* 102:1572–77

Brem RB, Storey JD, Whittle J, Kruglyak L. 2005. Genetic interactions between polymorphisms that affect gene expression in yeast. *Nature* 436:701–3

Brem RB, Yvert G, Clinton R, Kruglyak L. 2002. Genetic dissection of transcriptional regulation in budding yeast. *Science* 296:752–55

Cardon LR, Bell JI. 2001. Association study designs for complex diseases. *Nat. Rev. Genet.* 2:91–99

Carlborg O, Andersson L, Kinghorn B. 2000. The use of a genetic algorithm for simultaneous mapping of multiple interacting quantitative trait loci. *Genetics* 155:2003–10

Carlborg O, Haley CS. 2004. Epistasis: too often neglected in complex trait studies? *Nat. Rev. Genet.* 5:618–25

Carlborg O, Kerje S, Schutz K, Jacobsson L, Jensen P, Andersson L. 2003. A global search reveals epistatic interaction between QTL for early growth in the chicken. *Genome Res.* 13:413–21

Cavalli LL. 1952. An analysis of linkage in quantitative inheritance. In *Quantitative Inheritance*, ed. ECR Reeve, CH Waddington, pp. 135–44. London: HMSO

Chase K, Adler FR, Lark KG. 1997. Epistat: a computer program for identifying and testing interactions between pairs of quantitative trait loci. *Theor. Appl. Genet.* 94:724–30

Cheverud JM. 2000. Detecting epistasis among quantitative trait loci. See Wolf et al. 2000, pp. 58–81

Cheverud JM, Routman EJ. 1995. Epistasis and its contribution to genetic variance components. *Genetics* 139:1455–61

Cockerham CC. 1954. An extension of the concept of partitioning hereditary variance for analysis of covariances among relatives when epistasis is present. *Genetics* 39:859–82

Cockerham CC. 1980. Random and fixed effects in plant genetics. *Theor. Appl. Genet.* 56:119–31

Cockerham CC, Zeng ZB. 1996. Design III with marker loci. *Genetics* 143:1437–56

Complex Trait Consortium. 2004. The Collaborative Cross, a community resource for the genetic analysis of complex traits. *Nat. Genet.* 36:1133–37

Cox NJ, Frigge M, Nicolae DL, Concannon P, Hanis CL, et al. 1999. Loci on chromosomes 2 (NIDDM1) and 15 interact to increase susceptibility to diabetes in Mexican Americans. *Nat. Genet.* 21:213–15

Coyne JA, Barton NH, Turelli M. 1997. Perspective: a critique of Sewall Wright's shifting balance theory of evolution. *Evolution* 51:643–71

Coyne JA, Barton NH, Turelli M. 2000. Is Wright's shifting balance process important in evolution? *Evolution* 54:306–17

Coyne JA, Orr HA. 2004. *Speciation*. Sunderland, MA: Sinaur. 480 pp.

Culverhouse R, Klein T, Shannon W. 2004. Detecting epistatic interactions contributing to quantitative traits. *Genet. Epidemiol.* 27:141–52

Culverhouse R, Suarez BK, Lin J, Reich T. 2002. A perspective on epistasis: limits of models displaying no main effect. *Am. J. Hum. Genet.* 70:461–71

Damerval C, Maurice A, Josse JM, Devienne D. 1994. Quantitative trait loci underlying gene-product variation—a novel perspective for analyzing regulation of genome expression. *Genetics* 137:289–301

Darvasi A. 1998. Experimental strategies for the genetic dissection of complex traits in animal models. *Nat. Genet.* 18:19–24

Demant P. 2003. Cancer susceptibility in the mouse: genetics, biology and implications for human cancer. *Nat. Rev. Genet.* 4:721–34

Reviews breeding designs developed to maximize the chances of detecting genes underlying complex traits.

Demant P, Hart AAM. 1986. Recombinant congenic strains—a new tool for analyzing genetic traits determined by more than one gene. *Immunogenetics* 24:416–22

Demuth JP, Wade MJ. 2005. On the theoretical and empirical framework for studying genetic interactions within and among species. *Am. Nat.* 165:524–36

Demuth JP, Wade MJ. 2006. Population differentiation in the beethe *Tribolium castaneum* I: genetic architechture. *Evolution*. In press

Doebley J, Stec A, Gustus C. 1995. Teosinte Branched1 and the origin of maize—evidence for epistasis and the evolution of dominance. *Genetics* 141:333–46

Doerge RW. 2002. Mapping and analysis of quantitative trait loci in experimental populations. *Nat. Rev. Genet.* 3:43–52

Doerge RW, Zeng ZB, Weir BS. 1997. Statistical issues in the search for genes affecting quantitative traits in experimental populations. *Stat. Sci.* 12:195–219

Edmands S, Burton RS. 1999. Cytochrome c oxidase activity in interpopulation hybrids of a marine copepod: a test for nuclear-nuclear or nuclear-cytoplasmic coadaptation. *Evolution* 53:1972–78

Elzo MA. 1990. Covariances among sire by breed group of dam interaction effects in multibreed sire evaluation procedures. *J. Anim. Sci.* 68:4079–99

Elzo MA. 1994. Restricted maximum-likelihood procedures for the estimation of additive and nonadditive genetic variances and covariances in multibreed populations. *J. Anim. Sci.* 72:3055–65

Erickson DL, Fenster CB, Stenoien HK, Price D. 2004. Quantitative trait locus analyses and the study of evolutionary process. *Mol. Ecol.* 13:2505–22

Eshed Y, Zamir D. 1996. Less-than-additive epistatic interactions of quantitative trait loci in tomato. *Genetics* 143:1807–17

Falconer DS. 1985. A note on Fisher's average effect and average excess. *Genet. Res.* 46:337–47

Fenster CB, Galloway LF. 2000. The contribution of epistasis to the evolution of natural populations. See Wolf et al. 2000, pp. 232–44

Fenster CB, Galloway LF, Chao L. 1997. Epistasis and its consequences for the evolution of natural populations. *Trends Ecol. Evol.* 12:282–86

Fisher RA. 1918a. The causes of human variability. *Eugen. Rev.* 10:213–20

Fisher RA. 1918b. The correlations between relatives on the supposition of Mendelian inheritance. *Trans. R. Soc. Edinb.* 52:399–433

Fisher RA. 1922. On the mathematical foundations of theoretical statistics. *Philos. Trans. R. Soc. London Ser. A* 222:309–68

Fisher RA. 1930. *The Genetical Theory of Natural Selection*. Oxford, UK: Clarendon

Fisher RA. 1941. Average excess and average effect of a gene substitution. *Ann. Eugen.* 11:53–63

Flint J, DeFries JC, Henderson ND. 2004. Little epistasis for anxiety-related measures in the DeFries strains of laboratory mice. *Mamm. Genome* 15:77–82

Foster JA. 2001. Evolutionary computation. *Nat. Rev. Genet.* 2:428–36

Frankel WN, Schork NJ. 1996. Who's afraid of epistasis? *Nat. Genet.* 14:371–73

Freimer N, Sabatti C. 2004. The use of pedigree, sib-pair and association studies of common diseases for genetic mapping and epidemiology. *Nat. Genet.* 36:1045–51

Gallais A, Rives M. 1993. Detection, number and effects of QTLs for a complex character. *Agronomie* 13:723–38

Galloway LF, Fenster CB. 2001. Nuclear and cytoplasmic contributions to intraspecific divergence in an annual legume. *Evolution* 55:488–97

Gardner K, Buerkle A, Whitton J, Rieseberg L. 2000. Inferring epistasis in wild sunflower populations. See Wolf et al. 2000, pp. 264–79

Goodnight CJ. 1987. On the effect of founder events on epistatic genetic variance. *Evolution* 41:80–91

Goodnight CJ. 1988. Epistasis and the effect of founder events on the additive genetic variance. *Evolution* 42:441–54

Goodnight CJ. 2000. Quantitative trait loci and gene interaction: the quantitative genetics of metapopulations. *Heredity* 84:587–98

Goodnight CJ. 2004. Gene interaction and selection. In *Plant Breeding Reviews*, ed. J Jannick, pp. 269–90. New York: John Wiley & Sons

Goodnight CJ, Wade MJ. 2000. The ongoing synthesis: a reply to Coyne, Barton, and Turelli. *Evolution* 54:317–24

Gordon D, Finch SJ. 2005. Factors affecting statistical power in the detection of genetic association. *J. Clin. Invest.* 115:1408–18

Hayman BI. 1957. Interaction, heterosis and diallel crosses. *Genetics* 42:336–55

Hayman BI. 1958. The separation of epistatic from additive and dominance variation in generation means. *Heredity* 12:371–90

Hayman BI. 1960. The separation of epistatic from additive and dominance variation in generation means. *Genetica* 31:371–90

Hill WG. 1982. Dominance and epistasis as componenets of heterosis. *Z. Tierz. Zuechtungsbiol.* 99:161–68

Hirschhorn JN, Lohmueller K, Byrne E, Hirschhorn K. 2002. A comprehensive review of genetic association studies. *Genet. Med.* 4:45–61

International HapMap Consortium. 2005. A haplotype map of the human genome. *Nature* 437:1299–20

Ioannidis JPA, Ntzani EE, Trikalinos TA, Contopoulos-Ioannidis DG. 2001. Replication validity of genetic association studies. *Nat. Genet.* 29:306–9

Jansen RC. 2003. Studying complex biological systems using multifactorial perturbation. *Nat. Rev. Genet.* 4:145–51

Jansen RC, Nap JP. 2001. Genetical genomics: the added value from segregation. *Trends Genet.* 17:388–91

Jinks JL. 1956. The F2 and backcross generations from a set of diallel crosses. *Heredity* 10:1–30

Johnson NA, Wade MJ. 1996. Genetic covariances within and between species: indirect selection for hybrid inviability. *J. Evol. Biol.* 9:205–14

Kao CH, Zeng ZB, Teasdale RD. 1999. Multiple interval mapping for quantitative trait loci. *Genetics* 152:1203–16

Kelly JK. 2005. Epistasis in monkeyflowers. *Genetics* 171:1917–31

Kempthorne O. 1954. The correlation between relatives in a random mating population. *Proc. R. Soc. London Ser. B* 143:103–13

Kuiken T, Rimmelzwaan GF, Van Amerongen G, Osterhaus A. 2003. Pathology of human influenza A (H5N1) virus infection in cynomolgus macaques (*Macaca fascicularis*). *Vet. Pathol.* 40:304–10

Lander ES, Botstein D. 1989. Mapping Mendelian factors underlying quantitative traits using Rflp linkage maps. *Genetics* 121:185–99

Lark KG, Chase K, Adler F, Mansur LM, Orf JH. 1995. Interactions between quantitative trait loci in soybean in which trait variation at one locus is conditional upon a specific allele at another. *Proc. Natl. Acad. Sci. USA* 92:4656–60

Leips J, Mackay TFC. 2000. Quantitative trait loci for life span in *Drosophila melanogaster*: interactions with genetic background and larval density. *Genetics* 155:1773–88

Li ZK, Pinson SRM, Park WD, Paterson AH, Stansel JW. 1997. Epistasis for three grain yield components in rice (*Oryza sativa* L). *Genetics* 145:453–65

Liu Y, Tritchler D, Bull SB. 2002. A unified framework for transmission-disequilibrium test analysis of discrete and continuous traits. *Genet. Epidemiol.* 22:26–40

Ljungberg K, Holmgren S, Carlborg O. 2004. Simultaneous search for multiple QTL using the global optimization algorithm DIRECT. *Bioinformatics* 20:1887–95

Long AD, Langley CH. 1999. The power of association studies to detect the contribution of candidate genetic loci to variation in complex traits. *Genome Res.* 9:720–31

Long AD, Mullaney SL, Reid LA, Fry JD, Langley CH, Mackay TFC. 1995. High resolution mapping of genetic factors affecting abdominal bristle number in *Drosophila melanogaster*. *Genetics* 139:1273–91

Lunetta KL, Faraone SV, Biederman J, Laird NM. 2000. Family-based tests of association and linkage that use unaffected sibs, covariates, and interactions. *Am. J. Hum. Genet.* 66:605–14

Lynch M. 1991. The genetic interpretation of inbreeding depression and outbreeding depression. *Evolution* 45:622–29

Lynch M, Walsh B. 1998. *Genetics and Analysis of Quantitative Traits*. Sunderland, MA: Sinaur *Contains detailed statistical background pertaining to ANOVA, LCA, and QTL analysis.*

Mackay TFC. 2001. The genetic architecture of quantitative traits. *Annu. Rev. Genet.* 35:303–39

Marchini J, Donnelly P, Cardon LR. 2005. Genome-wide strategies for detecting multiple loci that influence complex diseases. *Nat. Genet.* 37:413–17

Mather K, Jinks JL. 1982. *Biometrical Genetics*. New York: Chapman & Hall

McCauley DE, Wade MJ. 1980. Group selection—the genetic and demographic basis for the phenotypic differentiation of small populations of *Tribolium castaneum*. *Evolution* 34:813–21

McCauley DE, Wade MJ. 1981. The populational effects of inbreeding in *Tribolium*. *Heredity* 46:59–67

Montooth KL, Marden JH, Clark AG. 2003. Mapping determinants of variation in energy metabolism, respiration and flight in *Drosophila*. *Genetics* 165:623–35

Moore JH, Williams SM. 2005. Traversing the conceptual divide between biological and statistical epistasis: systems biology and a more modern synthesis. *BioEssays* 27:637–46

Moreno G. 1994. Genetic architecture, genetic behavior, and character evolution. *Annu. Rev. Ecol. Syst.* 25:31–44

Neyman J. 1935. Statistical problems in agricultural experimentation. *Suppl. J. R. Stat. Soc.* 2:107–80

Peiris JSM, Yu WC, Leung CW, Cheung CY, Ng WF, et al. 2004. Re-emergence of fatal human influenza A subtype H5N1 disease. *Lancet* 363:617–19

Pratt SC, Daly MJ, Kruglyak L. 2000. Exact multipoint quantitative-trait linkage analysis in pedigrees by variance components. *Am. J. Hum. Genet.* 66:1153–57

Rabinowitz D. 1997. A transmission disequilibrium test for quantitative trait loci. *Hum. Hered.* 47:342–50

Richards BKS, Belton BN, Poole AC, Mancuso JJ, Churchill GA, et al. 2002. QTL analysis of self-selected macronutrient diet intake: fat, carbohydrate, and total kilocalories. *Physiol. Genomics* 11:205–17

Risch N, Merikangas K. 1996. The future of genetic studies of complex human diseases. *Science* 273:1516–17

Routman EJ, Cheverud JM. 1997. Gene effects on a quantitative trait: two-locus epistatic effects measured at microsatellite markers and at estimated QTL. *Evolution* 51:1654–62

Schaid DJ. 1996. General score tests for associations of genetic markers with disease using cases and their parents. *Genet. Epidemiol.* 13:423–49

Schlichting CD, Pigliucci M. 1998. *Phenotypic Evolution: A Reaction Norm Perspective*. Sunderland, MA: Sinauer

Sen S, Churchill GA. 2001. A statistical framework for quantitative trait mapping. *Genetics* 159:371–87

Shook DR, Johnson TE. 1999. Quantitative trait loci affecting survival and fertility-related traits in *Caenorhabditis elegans* show genotype-environment interactions, pleiotropy and epistasis. *Genetics* 153:1233–43

Soller M, Brody T, Genizi A. 1976. On the power of experimental designs for the detection of linkage between marker loci and quantitative loci in crosses between inbred lines. *Theor. Appl. Genet.* 47:35–39

Storey JD, Akey JM, Kruglyak L. 2005. Multiple locus linkage analysis of genomewide expression in yeast. *PLoS Biol.* 3:e267

Storey JD, Taylor JE, Siegmund D. 2004. Strong control, conservative point estimation and simultaneous conservative consistency of false discovery rates: a unified approach. *J. R. Stat. Soc. Ser. B* 66:187–205

Storey JD, Tibshirani R. 2003. Statistical significance for genomewide studies. *Proc. Natl. Acad. Sci. USA* 100:9440–45

Templeton AR. 1987. The general relationship between average effect and average excess. *Genet. Res.* 49:69–70

Templeton AR. 2000. Epistasis and complex traits. See Wolf et al. 2000, pp. 41–57

The Chipping Forecast III. 2005. *Nat. Genet.* 37 (Suppl.)

Umbach DM, Weinberg CR. 1997. Designing and analyzing case-control studies to exploit independence of genotype and exposure. *Stat. Med.* 16:1731–43

Van Der Veen JH. 1959. Tests of nonallelic interaction and linkage for quantitative characters in generations derived from two diploid pure lines. *Genetica* 30:201–32

Wade MJ. 1978. Critical-review of models of group selection. *Q. Rev. Biol.* 53:101–14

Wade MJ. 1979. Primary characteristics of *Tribolium* populations group selected for increased and decreased population-size. *Evolution* 33:749–64

Wade MJ. 1980. Group selection, population-growth rate, and competitive ability in the flour beetles, *Tribolium* spp. *Ecology* 61:1056–64

A full issue of *Nature Genetics* devoted to developments in microarray technology and projected utility.

Wade MJ. 1982. Group selection—migration and the differentiation of small populations. *Evolution* 36:949–61

Wade MJ. 1984. Changes in group-selected traits that occur when group selection is relaxed. *Evolution* 38:1039–46

Wade MJ. 1985. The effects of genotypic interactions on evolution in structured populations. In *Genetics: New Frontiers, Proc. XV Int. Congr. Genet.*, pp. 283–90. Oxford, UK: Oxford Univ. Press

Wade MJ. 1988. Group selection for increased and decreased competitive ability in subdivided populations of *Tribolium castaneum*. *Bull. Soc. Popul. Ecol.* 44:3–11

Wade MJ. 1990. Genotype-environment interaction for climate and competition in a natural-population of flour beetles, *Tribolium castaneum*. *Evolution* 44:2004–11

Wade MJ. 1992. Sewall Wright: gene interactions and the shifting balance theory. In *Oxford Series on Evolutionary Biology*, ed. J Antonovics, DJ Futuyma, pp. 35–62. Oxford, UK: Oxford Univ. Press

Wade MJ. 1996. Adaptation in subdivided populations: kin selection and interdemic selection. In *Adaptation*, ed. MR Rose, GV Lauder, pp. 381–405. San Diego: Academic

Wade MJ. 2000. Epistasis as a genetic constraint within popualtions and an accelerant of adaptive divergence among them. See Wolf et al. 2000, pp. 213–31

Wade MJ. 2001. Epistasis, complex traits, and mapping genes. *Genetica* 112:59–69

Wade MJ. 2002. A gene's eye view of epistasis, selection and speciation. *J. Evol. Biol.* 15:337–46

Demonstrates how, in the presence of epistasis, Mendelian labels such as dominant or recessive may change with genetic background.

Wade MJ, Goodnight CJ. 1991. Wright shifting balance theory—an experimental study. *Science* 253:1015–18

Wade MJ, Goodnight CJ. 1998. Perspective. The theories of Fisher and Wright in the context of metapopulations: when nature does many small experiments. *Evolution* 52:1537–53

Wade MJ, McCauley DE. 1980. Group selection—the phenotypic and genotypic differentiation of small populations. *Evolution* 34:799–812

Wade MJ, McCauley DE. 1984. Group selection—the interaction of local deme size and migration in the differentiation of small populations. *Evolution* 38:1047–58

Willett CS, Burton RS. 2001. Viability of cytochrome *c* genotypes depends on cytoplasmic backgrounds in *Tigriopus californicus*. *Evolution* 55:1592–99

Willett CS, Burton RS. 2003. Environmental influences on epistatic interactions: viabilities of cytochrome *c* genotypes in interpopulation crosses. *Evolution* 57:2286–92

Wolf JB, Brodie ED III, Wade MJ, eds. 2000. *Epistasis and the Evolutionary Process*. New York: Oxford Univ. Press

Wright S. 1968. *Evolution and the Genetics of Populations. Vol. 1: Genetic and Biometric Foundations*. Chicago: Univ. Chicago Press

Wu RL. 2000. Partitioning of population genetic variance under multiplicative-epistatic gene action. *Theor. Appl. Genet.* 100:743–49

Yandell B. 2005. *Quantitative trait loci (QTL) references*. **http://www.stat.wisc.edu/~yandell/statgen/reference/qtl.html**

Yang RC. 2004. Epistasis of quantitative trait loci under different gene action models. *Genetics* 167:1493–505

Yvert G, Brem RB, Whittle J, Akey JM, Foss E, et al. 2003. Trans-acting regulatory variation in *Saccharomyces cerevisiae* and the role of transcription factors. *Nat. Genet.* 35:57–64

Zeng ZB, Liu JJ, Stam LF, Kao CH, Mercer JM, Laurie CC. 2000. Genetic architecture of a morphological shape difference between two drosophila species. *Genetics* 154:299–310

Zeng ZB, Wang T, Zou W. 2005. Modeling quantitative trait loci and interpretation of models. *Genetics* 169:1711–25

Corridors for Conservation: Integrating Pattern and Process

Cheryl-Lesley B. Chetkiewicz, Colleen Cassady St. Clair, and Mark S. Boyce

Department of Biological Sciences, University of Alberta, Edmonton, Alberta, T6G 2E9, Canada; email: cbc2@ualberta.ca, cstclair@ualberta.ca, boyce@ualberta.ca

Key Words

behavior, habitat selection, movement, resource selection functions, scale

Abstract

Corridors are commonly used to connect fragments of wildlife habitat, yet the identification of conservation corridors typically neglects processes of habitat selection and movement for target organisms. Instead, corridor designs often are based on binary patterns of habitat suitability. New technologies and analytical tools make it possible to better integrate landscape patterns with behavioral processes. We show how resource selection functions can be used to describe habitat suitability with continuous and multivariable metrics and review methods by which animal movement can be quantified, analyzed, and modeled. We then show how the processes of habitat selection and movement can be integrated with landscape features using least-cost paths, graph theory, and step selection functions. These tools offer new ways to design, implement, and study corridors as landscape linkages more objectively and holistically.

INTRODUCTION

Corridor: regions of the landscape that facilitate the flow or movement of individuals, genes, and ecological processes

Matrix: component of the landscape that is neither patch nor corridor

Corridors are cornerstones of modern conservation. Traditionally, corridors have been viewed as linear strips of habitat that facilitate the movement of organisms through landscapes (Puth & Wilson 2001). Corridors, often in association with the charismatic megafauna whose populations they are designed to conserve, are a fundamental component of wildland conservation, particularly in North America where many regional and several continental-scale corridor initiatives are underway (Nelson et al. 2003, Noss 2003). International corridors foster new levels of transboundary conservation, elevating corridors from an ecological to a political and socioeconomic tool (Zimmerer et al. 2004). Despite the widespread application of corridors, much current practice causes them to fall far short of their conservation promise. On-the-ground applications of corridors usually are based on simplistic depictions of habitats that are assumed to provide the associated ecological processes. Typically, corridor applications proceed with little species-specific information and limited evaluation, and they are rarely published or reviewed in scientific journals (Vos et al. 2002; but see Beier et al. 2006). In some cases, corridors, selected for their political appeal, are being plunked down willy-nilly on landscapes that already have been carved up for other purposes. This makes the provision of practical corridor guidelines for managers as big a challenge today as it was over a decade ago (Hobbs 1992).

A grizzly bear (*Ursus arctos*) tagged as "99" and his victim provide a compelling study in the failings of this approach. This young male bear wandered into the fringes of the burgeoning town of Canmore, Alberta, in late May, 2005. After showing indifference to human encounters, it was captured on a local golf course and relocated by government conservation officers. A week later, "99" was detected in a designated wildlife corridor above the town of Canmore, one that was a scant 1000 m wide, perforated with human-use trails, and sandwiched between a recently built golf course and steep slopes above the townsite. By day's end, both the bear and a young woman were dead, and the world tuned in to Alberta's first grizzly-caused human fatality in seven years. Critics were quick to blame the wildlife policy that relocated the bear. But the bigger failing occurred years previously with the designation of the corridor. Corridors based on scant biological data supported Canmore's rapid development during the 1990s, obliterating much of the wildlife habitat in this montane valley. Too little fertile and connected habitat remains in the valley that contains Canmore to support grizzly bear movement to adjacent protected areas in the Canadian Rocky Mountains (Herrero 2005). Indeed, examining movements of three other grizzly bears in this area suggests that the designated corridors actually are avoided, and the oft-assumed distinction between corridor and matrix is not apparent (**Figure 1a**). Despite various planning guidelines supporting corridor designations (BCEAG 1999), the corridor designs in Canmore require important modifications, at least for grizzly bears. We suggest that more sophisticated approaches to corridor designs not only are possible but essential if corridors are to realize their potential for conserving biodiversity.

Although they have limitations (reviewed by Hilty et al. 2006), corridors have been promoted widely as a conservation strategy. Since their introduction as a tool for game management in the 1940s (reviewed by Harris & Scheck 1991), over 700

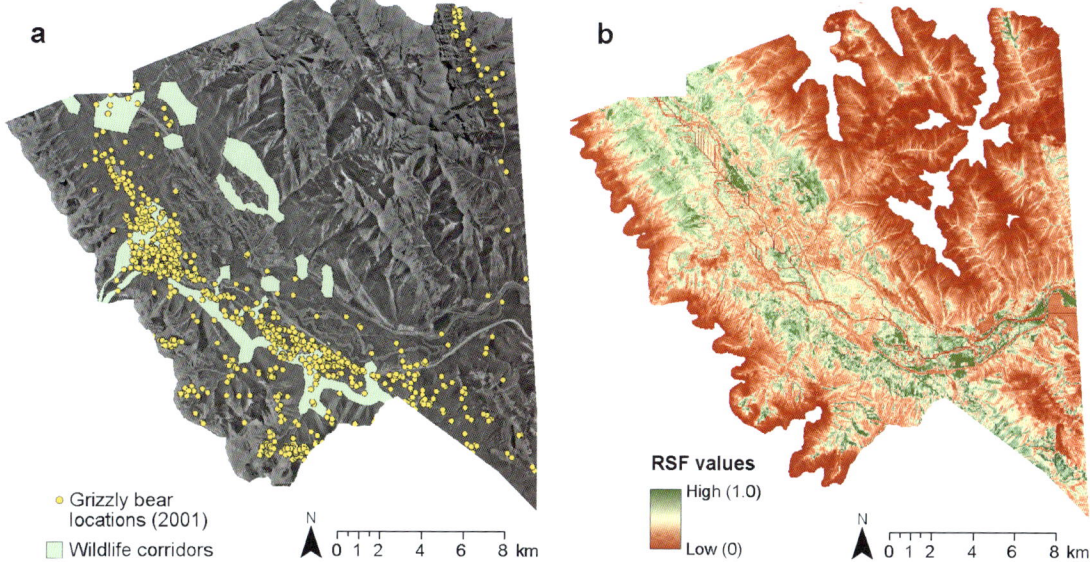

Figure 1

Telemetry locations for three grizzly bears during 2001 and designated wildlife corridors in the Canmore region of the Bow Valley, Alberta, Canada (*a*) were used to generate a resource selection function (RSF; *b*) (BCEAG 1999; C.-L. Chetkiewicz, unpublished data). An RSF was created using logistic regression to compare topographic and vegetation variables at grizzly bear telemetry locations obtained during 2001 with those at random points within the combined home ranges of the three bears. Applying the RSF in a geographic information system (GIS) identifies areas likely to support grizzly bear occupancy. Areas of high relative probability of occurrence (*green*) could be used to evaluate or amend corridor designations or guide recommendations to amend current corridor designations.

scientific papers concerning corridors have been published. Most acknowledge that the purpose of corridors is to counter the effects of habitat loss and fragmentation, which are important causes of biodiversity loss worldwide (Dirzo & Raven 2003, Sih et al. 2000). Corridors are expected to slow these effects by increasing the movement of individuals among otherwise-isolated populations (e.g., Gilbert et al. 1998, Gonzalez et al. 1998), thereby rescuing populations from stochastic local extinctions (e.g., Brown & Kodric-Brown 1977, Reed 2004), maintaining genetic diversity (e.g., Hale et al. 2001, Mech & Hallett 2001), and retaining ecological processes (Bennett 1999, Haddad & Tewskbury 2006, Hilty et al. 2006, Levey et al. 2005, Soulé & Terborgh 1999). Additionally, corridors might serve to provide routes and habitats for movement of organisms responding to climate change (Channell & Lomolino 2000). Other approaches to conserving biodiversity might be more effective than corridors (Hannon & Schmiegelow 2002, Schultz 1998) or offer better return on the investment of limited conservation dollars (Hobbs 1992, Simberloff & Cox 1987, Simberloff et al. 1992). We do not address these issues here. Rather, we assume that corridors will continue to occupy the conservation toolbox and ask how that tool can be used most effectively.

Pattern: spatial relationship between components within a landscape or ecosystem

Process: interaction between spatial components and the flow of energy, individuals, or genes

Functional connectivity: the degree to which a landscape facilitates or impedes movement among resource patches

Scale: the spatial or temporal dimension of an object or process, characterized by both grain (resolution) and extent

One important impediment to the effective use of corridors is the gap between their intended purpose and actual application, which generates a dichotomy between pattern and process. By pattern, we mean the composition and spatial configuration of habitats (Turner et al. 2001, Wiens 1995) and snapshots of organism distribution derived from censuses. By process, we mean the ways animals actually move within landscapes to cause patterns of distribution and drive related ecological processes. Probability of movement then determines the functional connectivity of landscapes (Taylor et al. 1993, Tischendorf & Fahrig 2000a,b). Despite the fact that the process of animal movement provides the impetus for corridor design and application, it is the pattern of landscape structure that dictates most of the research, planning, and application of corridors (Beier & Noss 1998, Vos et al. 2002). Extensive review (Beier & Noss 1998) found corroboration between corridor patterns and process-based metrics such as immigration and colonization rates in fewer than half of the studies. Since that time, dozens more observational and experimental studies have focused on corridors. A few emphasize processes (e.g., Berggren et al. 2002, Levey et al. 2005, Sieving et al. 2000). More often corridor designations are based—as they were in Canmore—on patterns of remaining habitat that appear (to human observers) to be connected in a simplified and binary depiction of the landscape.

The enduring bias of binary landscapes in corridor plans and studies stems partly from the ecological theory supporting corridor designs. Island biogeography (MacArthur & Wilson 1967) offered the stepping stones that others generalized to corridors (Diamond et al. 1976, Wilson & Willis 1975). Metapopulation theory (Hanski & Gilpin 1997) inferred the processes of dispersal, colonization, and local extinction in binary habitat patches with different spatial configurations (Dunning et al. 1992, Fahrig & Merriam 1994). Landscape ecology (Turner 2005) reinforced the patch-corridor-matrix paradigm by quantifying habitat configuration and composition patterns mainly with tools that juxtapose habitat and nonhabitat (e.g., McGarigal et al. 2002, Turner & Gardner 1991). Together, these theories have vastly increased appreciation of the relationships between habitat patterns and populations, but they have done so in a way that promotes corridors as archetypically linear and static features (Beier & Noss 1998, Hobbs 1992, Saunders & Hobbs 1991) in binary landscapes.

This simplistic, pattern-based view of corridors as habitats has resonated with ecologists because of its tractability (Goodwin 2003, Goodwin & Fahrig 2002) and scale versatility (Calabrese & Fagan 2004), but it has important limitations. First, it assumes that movement is categorically facilitated by corridors and impeded by the matrix (Baum et al. 2004, Rosenberg et al. 1997, Simberloff et al. 1992), whereas real landscapes create a continuum of influences on movement (Puth & Wilson 2001). Second, this simplified, categorical view of corridors homogenizes species and spatial scales for corridor planning, whereas functional connectivity is inevitably species-specific (Goodwin 2003, Lidicker 1999, Puth & Wilson 2001). In fact, corridors may not be beneficial to some species (Boswell et al. 1998, Collinge 2000, Hannon & Schmiegelow 2002, Schmiegelow et al. 1997, Schultz 1998), but even the potential disadvantages of corridors—e.g., disease transmission (Hess 1994)—may be outweighed by their benefits (McCallum & Dobson 2002). Thus, pattern-based approaches to corridor planning may not make appropriate provisions for all or even most of the species for which a corridor is designed, and corridor structure may be both insufficient and unnecessary to promote movement. Better integration of pattern and process is critically important to corridor design.

Several researchers have distinguished the pattern and process components of corridors (Bennett 1999, Rosenberg et al. 1997) and landscape connectivity more generally (Bélisle 2005, Fahrig 2003, Tischendorf & Fahrig 2000b). Others have acknowledged that corridors are more than linear structures in binary landscapes (Beier & Noss 1998, Hobbs 1992) and instead are places on the landscape that facilitate the movement of individuals, promote genetic exchange, and support ecological processes (Forman 2002, Puth & Wilson 2001). Broadening the concept of corridors to "linkages" allows them to support these processes without being linear, continuous, or even structurally distinct from the surrounding landscape (Bennett 1999). We amplify these views by suggesting that a greater emphasis on the processes of habitat selection and movement could address several fundamental questions that pattern-based approaches tend to neglect. We do not attempt to answer these questions but review new approaches and tools that can be used to identify, design, and test corridors for conservation more effectively.

First, should corridors promote certain types of movement? Corridors often are assumed to facilitate dispersal but this might not be the only movement type relevant to corridor designs. Moreover, it is frequently difficult to know the motivation of moving organisms (Lima & Zollner 1996). Instead of assuming this motivation, we could identify habitats that are associated statistically with short-range foraging movements versus longer-distance movements (e.g., Johnson et al. 2002). This approach makes it possible to separate movement into types, some of which might be targeted by corridor designs, even without identifying their underlying motivation.

Second, should corridors increase movement rates relative to movement in other habitats (Haddad & Tewksbury 2005, Puth & Wilson 2001)? Individuals have more tortuous pathways in good quality habitat and move further and faster over unfavorable terrain (Crist et al. 1992, Johnson et al. 1992, With 1994). However, individuals that move more sometimes suffer higher mortality (Biro et al. 2003; J.L. Frair, E.H. Merrill, J.R. Allen & M.S. Boyce, submitted). Moreover, high movement rates in corridors may not correlate with the functional connectivity of a landscape (Bélisle 2005).

Third, is habitat quality as important as movement characteristics in designing corridors? Even if animals use corridors only to travel between suitable patches, they are unlikely to do so if they perceive that habitats within the corridor are unsuitable. Organisms use a wide variety of mechanisms to select suitable habitats (Danchin et al. 2001, Stamps 2001) and knowing the details of habitat selection might be as important to corridor design as it is to identifying suitable habitats for other purposes.

Fourth, if corridors result in ecological traps or sinks (e.g., Weldon & Haddad 2005; but see Little et al. 2002), is their corridor function necessarily compromised? Only occasional movement is necessary to maintain gene flow (Mills & Allendorf 1996), and infrequent dispersal may be sufficient to sustain demographic rescue

Model: a way of describing the behavior of a process in order to predict its future or understand its past

GIS: geographic information system

(Hanski 2001). Corridors might provide these benefits to adjacent populations over large timescales, even if they lessen the survival and reproductive success of most of the individuals that use them.

Despite more than 20 years of research on corridors, few corridor studies lend insight into these questions. Rather than review the latest empirical studies that focus on corridors, we explore recent advances in technology and quantitative methods that make it easier to answer these questions by integrating pattern and process. These tools could revolutionize our ability to design and manage corridors to ensure that they are accomplishing conservation objectives. This review is intended to identify those opportunities by showing how we can develop gradient-based habitat selection models and probabilistic movement models to identify corridors in complex, real-world landscapes.

HABITAT SELECTION PROCESSES

Habitat selection is the behavioral process used by individuals when choosing resources (Johnson 1980) and habitats. These choices occur at a variety of spatial and temporal scales that range from finding food resources within a season, to defining home ranges during a lifetime, to expansion of ranges across generations (Johnson 1980; **Table 1**). The motivation for habitat selection is presumably to maximize individual fitness (Garshelis 2000) with consequences for distribution and density across different habitats (Morris 2003). The behavioral mechanisms that play a role in habitat selection for residency, such as conspecific attraction, habitat imprinting (reviewed by Stamps 2001), natal home range cues (Cooper et al. 2002), and public information (Danchin et al. 2001), logically apply to the selection of habitats for movement (i.e., corridors) as well. Even during dispersal movements, animals must forage, sleep, avoid predators, and either seek out or avoid conspecifics. They do not have the omniscience that a geographic information system (GIS) provides us for visualizing corridors and must instead continuously assess habitat for its suitability.

Table 1 Movement and habitat selection processes in relation to spatial scales and structures (adapted from Ims 1995, Johnson 1980)

Spatial scale	Habitat selection (after Johnson 1980)	Movement type (after Ims 1995)	Spatial structure
Resource Patch	Food items within the patch (fourth order)	Food items search (foraging)	Food item distribution Food patch shape and size Small-scale obstructions
Habitat Patch	Patches within home range (third order)	Patch searching, traplining, territory patrolling	Food patch configuration Shelter Abiotic factors and topography
Patch Mosaic	Selection of home range (second order)	Dispersal	Patch distribution Landscape features
Region	Geographical Range (first order)	Migration	Large scale topography barriers

The assumed dichotomy between patch and corridor is likely perceived by animals as a continuum.

A second false dichotomy applies to the way corridors are typically viewed as connecting areas of habitat, in a "sea" of inhospitable matrix. We know that organisms often use the so-called matrix as habitat (Berry et al. 2005, Haila 2002, Rosenzweig 2003) and that it can increase the viability of adjacent populations (e.g., Vandermeer & Carvajal 2001). Moreover, the matrix can affect interpatch movements, particularly for species that respond to boundaries between habitat types (Bender & Fahrig 2005), and determine the use of embedded corridors and stepping stones (Baum et al. 2004). Thus, organisms actually occupy a spectrum of habitats in nearly every landscape type. The artificial dichotomy of patch and matrix creates fundamental difficulties for understanding species responses to fragmented habitats (Fischer et al. 2004, McIntyre & Hobbs 1999). Fortunately, habitats can be described instead as probabilistic functions of multiple landscape attributes.

> **Resource selection function (RSF):** statistical models defined to be proportional to the probability of use of a resource unit
>
> **ZIP:** zero-inflated Poisson

Resource Selection Functions

Habitats can be characterized using resource selection functions (RSFs), defined to be any function that is proportional to the probability of use of a resource unit (Manly et al. 2002). A resource unit is a sampling unit of the landscape, e.g., a pixel or grid cell. Predictor variables (covariates) are habitat attributes that can be used to predict the relative probability of use for a resource unit (Manly et al. 2002).

A number of sampling designs can be used to estimate an RSF, e.g., a random sample of resource units could be drawn and examined for the presence or absence of an organism (Boyce & McDonald 1999). Model coefficients can be estimated using logistic regression if occurrence is recorded as absence-presence (0, 1), or an alternative link function might be used for count data, such as Poisson regression or zero-inflated Poisson (ZIP) regression (Nielsen et al. 2005). Alternatively a sample of occupied resource units could be contrasted with a random sample of landscape locations using a logistic discriminant function (Johnson et al. 2006). Predictive ability of an RSF can be assessed using k-fold cross-validation methods outlined by Johnson et al. (2006).

Such an RSF can be applied in a GIS to map the relative probability of use across the landscape, in contrast with binary maps of habitat versus nonhabitat. For most organisms, patterns of use of a landscape are much more complex than simple binary characterizations of habitat. These models can be used to identify habitat associations for animals at multiple scales (e.g., Boyce 2006, Carroll et al. 2001).

Using Resource Selection Functions to Delineate Corridors

By depicting landscapes as probabilistic functions, RSF models offer an important departure from categorical representations of corridors, patch, and matrix habitat. Although RSF models tell us nothing about the movement of animals per se, they allow us to identify habitats that are likely to support occupancy. For example, we used the telemetry locations for three grizzly bears in the Canmore region of the Bow Valley, Alberta, Canada (**Figure 1a**), to generate an RSF that compared topographic and vegetation variables at telemetry locations with those at random locations in their combined home ranges (**Figure 1b**). Applying the RSF in a GIS illustrates areas of high probability of occupancy (*green*) and their proximity to one another as well as to areas of lower probability of occupancy (*red*). This approach provides a powerful framework for locating potential corridors or evaluating current corridor designations (**Figure 1a**).

Although characterizing habitats used by organisms would appear to be a fundamental first step in identifying corridors, caveats are appropriate. Use of habitats does not necessarily mean that the habitats are productive ones, and in the worst case used habitats might be sinks or traps (Kristan 2003, Pulliam 1988). Yet, 85% of avian studies have found that habitats used more intensively by a bird species were also those in which reproductive success was highest (Bock & Jones 2004). Nonetheless, corridors may sometimes represent poor-quality habitats that still facilitate movement (Haddad & Tewksbury 2005).

MOVEMENT PROCESSES

Organisms are motivated to move to forage, avoid predators, find breeding opportunities, access seasonal or ephemeral resources, and expand ranges (Bennett 1999, Ims 1995), generating movements scaled within foraging patches of a few square centimeters to transcontinental migrations. Ims (1995) offered four categories of movement—foraging, searching, dispersal, and migration—that are strikingly similar to a hierarchy of habitat selection described earlier by Johnson (1980; **Table 1**). All of these categories are relevant to corridors (Bennett 1999), but dispersal tends to be emphasized as most pertinent (reviewed by Vos et al. 2002), particularly for spatially structured populations (reviewed by Clobert et al. 2001). Yet corridors also may be critical for maintaining seasonal migrations (e.g., Powell & Bjork 1995) or for access to resources within a home range (e.g., Nielsen et al. 2004a). With so many contexts for movement and such a fundamental role in population dynamics, it is surprising that movement as a process is seldom explicit in corridor planning. This lack of emphasis has been caused, in part, by the difficulty of quantifying movement.

Techniques for Measuring Movement

Turchin (1998) identified two empirical approaches for measuring movement: Eulerian and Lagrangian. Eulerian approaches measure population metrics by recording the redistribution of large numbers of marked or unmarked individuals at specific locations. Individuals have been marked using leg-bands in birds, radioisotope labels and dyes in insects, or otolith dyes in fish (reviewed by Southwood & Henderson 2000). Subsequent recaptures, resightings, or recovery provide an estimation of movement rates (reviewed by Bennetts et al. 2001). In contrast, Lagrangian approaches characterize the magnitude, speed, and directionality of individual movements with a variety of techniques. For insects, movement paths have been recorded using numbered flags (e.g., Schultz 1998) or harmonic radar systems (e.g., Cant et al. 2005), whereas movement pathways for vertebrates can be recorded using snow tracking

(Whittington et al. 2005) or radiotelemetry (Millspaugh & Marzluff 2001). Movement paths are quantified by velocity, step lengths, degree of directionality, and measures of tortuosity (Turchin 1998). Eulerian and Lagrangian approaches provide different but complementary methods for understanding animal movements across a landscape.

GPS: global positioning system

In general, Eulerian approaches do not provide the same detail of movement information as Lagrangian approaches, but they make it possible to describe movement over much larger spatial and temporal scales. Eulerian approaches employing genetic techniques (Nathan 2005, Nathan et al. 2003, Webster et al. 2002) or stable isotopes (reviewed by Hobson 2005, Rubenstein & Hobson 2004) are rapidly evolving and offer particular promise to reveal landscape connectivity for populations. Because individuals are "marked" with a unique genotype or isotopic signature, the frequency of various markers from different sources can be identified. Genetic techniques offer enough precision to provide an estimate of dispersal movements within one or more generations (Waser & Strobeck 1998). For example, Proctor et al. (2004) measured genetic similarity to estimate dispersal distances for grizzly bears and showed that animals moved with a series of short stepping stone–like movements rather than a few long-distance dispersal movements. Genetic approaches also can be used to measure the effect of corridor patterns on gene flow (e.g., Aars & Ims 1999, Mech & Hallett 2001) or to document that some organisms moved through corridors (e.g., Coffman et al. 2001). These methods may be complemented with Lagrangian approaches to show how individual movements influence gene flow (e.g., Keyghobadi et al. 2005).

Many applications of Lagrangian approaches have involved small organisms (e.g., Schultz 1998) and experimental systems (e.g., Haddad 2000), but global positioning systems (GPSs) radiotelemetry can provide detailed movement information over much broader spatial and temporal scales (reviewed by Millspaugh & Marzluff 2001). Obviously, GPS radiocollars increase the practicality of collecting movement information for wide-ranging organisms, but handheld GPS also can be combined with field observations or conventional telemetry to support equivalent spatial grain and extent for animals that are too small to wear GPS collars or to offset the relatively high costs of GPS radiotelemetry. GPS technology provides exciting new potential to use Lagrangian data to design and evaluate corridors. The ideal approach might engage both Eulerian and Lagrangian methods.

Quantifying Movement Processes

Kernohan et al. (2001) described three nonexclusive categories of quantitative approaches for characterizing movement: (*a*) summarizing movement pathways with turning angles, fractal dimensions, and step lengths; (*b*) modeling movement with random walks or their variations (Turchin 1998); and, (*c*) identifying patterns in movement data retrospectively to distinguish different movement types (e.g., Morales et al. 2004). The first approach, quantifying movement pathways as turning angles, step lengths (**Figure 2**), and fractal dimensions offers several advantages. First, these metrics can be used to associate movement types with landscape features. For example, cougars (*Puma concolor*) moving more than 100 m at any one time tended to

Figure 2

Example of how a movement pathway (*a*) can be quantified into step lengths (*b*) and turning angles (*c*) for a cougar, CACO1, during 2000–2001 in the Canmore region of the Bow Valley, Alberta, Canada (C.-L. Chetkiewicz, unpublished data).

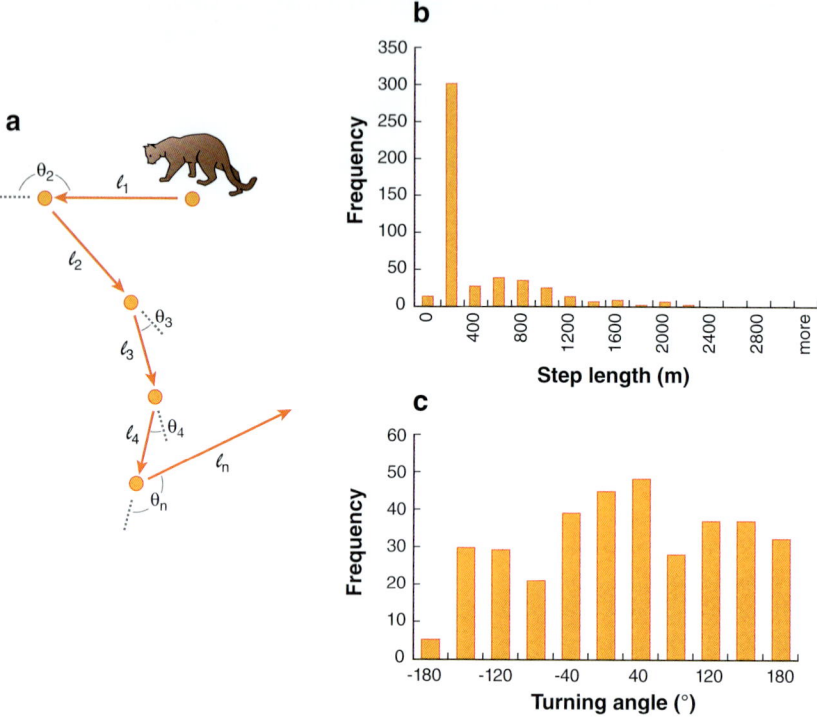

CRW: correlated random walk

have straighter movements and moved faster through urbanized areas (Dickson et al. 2005). Second, these metrics can be used to parameterize movement rules for spatially explicit models. Such a model was created from movement data for beetles to evaluate the effect of hedgerow width on movement rates (Tischendorf et al. 1998). A final advantage of quantifying movement pathways is they can be used to examine responses to edges or habitat boundaries. For example, eastern bluebirds (*Sialia sialis*) typically flew parallel to edges in an experimentally fragmented field system emphasizing the role of edges in directing and channelling flight pathways (Levey et al. 2005).

The second approach characterizes movements according to a mechanistic model, typically derived from diffusion theory and approximations of random walks (Turchin 1998). For example, Gustafson & Gardner (1996) simulated self-avoiding random walkers to explore the effects of landscape heterogeneity on movement patterns and identify frequently traversed portions of the landscapes that might denote corridors. In another application, a correlated random walk (CRW) diffusion model was used to simulate movements by grizzly bears to illustrate how land ownership and habitat information could reveal potential dispersal routes (Boone & Hunter 1996). Even if real organisms usually violate some of the assumptions of general movement models (Bergman et al. 2000), CRWs can be useful null models for distinguishing different movement types (Austin et al. 2004) and identify opportunities for corridors.

The third approach for quantifying movement is to identify types of movement retrospectively. An early method for achieving this was fractal dimension (fractal **D**),

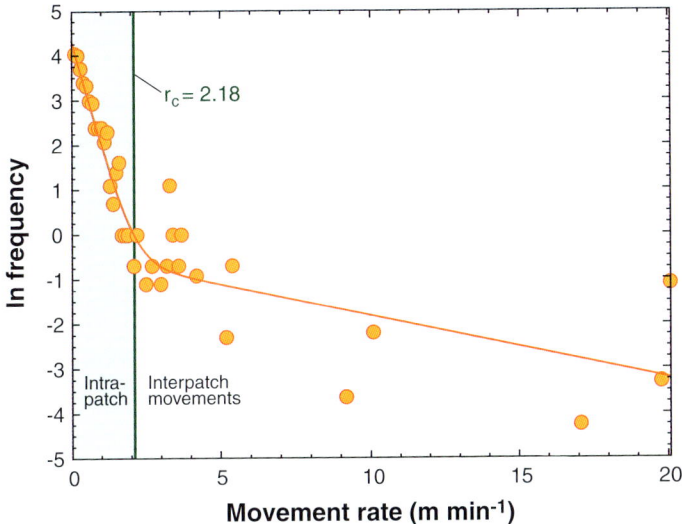

Figure 3

The $\log_e$ frequency distribution of movement rates for a caribou is assessed using a broken-stick method to calculate a scale criterion (r_c). Movement rates less than r_c represent intrapatch movement behaviors, whereas movement rates greater than r_c represent interpatch movement behaviors (C. Johnson, unpublished data).

but this technique was typically applied to small organisms at limited spatial scales (reviewed by Nams 2005). GPS technology makes it possible to apply similar approaches at much broader spatial and temporal scales. For example, Johnson et al. (2002) used a nonlinear ("broken stick") curve-fitting procedure to define two types of movement behavior for caribou (*Rangifer tarandus*) in British Columbia. This approach used variation in the frequency of movement rates to define a threshold value that could differentiate between intrapatch movements (short, high-frequency moves below the threshold) and interpatch movements (larger, less-frequent moves greater than the threshold) (**Figure 3**). We might expect that longer-step, interpatch movements would better characterize habitats used as corridors.

Once different movement states are identified, they can be combined with RSF-based habitat characterizations to align behavioral states with landscape features. Morales et al. (2004) used a latent model structure based on turn angles and step lengths to identify two behaviors: "encamped" (step lengths were small, turning angles were high) or "exploratory" (step lengths were several kilometers long, turning angles were low) for wapiti (*Cervus elaphus*) in Ontario. They then identified landscape features correlated with these states. Frair et al. (2005) used a similar approach to identify three types of movement behavior in wapiti and then related these behaviors to landscape conditions including wolf (*Canis lupus*) predation risk and cover.

The three approaches to quantifying movement we have described here have two important attributes. First, all are readily applied to a variety of temporal and spatial scales. Previous use of different approaches for small and large organisms has polarized the corridor literature (Haddad et al. 2000, Noss & Beier 2000). Although generalizations that transcend spatial scales for management are challenging (Boyce 2006), it is sometimes possible to derive movement mechanisms at one scale and apply them to other scales (e.g., Ims et al. 1993). In other cases, movement processes may not generalize across scales (e.g., Fortin et al. 2005b). For example, highway-crossing structures designed as corridors for grizzly bears are frequently used by large animals (Clevenger & Waltho 2005), but almost completely avoided by microtine rodents (McDonald & St. Clair 2004).

A second useful attribute of quantifying movement is that it provides a means of identifying important differences among individuals. For example, female grizzly bears appear much less willing to cross barriers than males (Gibeau et al. 2002). Individual variation generally has been viewed as an inconvenience in wildlife studies but might be profitably examined and incorporated in studies of both habitat selection and movement with random effects (Gillies et al. 2006). Similarly, latent class models (McCulloch et al. 2002) can be used to identify how individual motivation affects both habitat selection and movement. Understanding individual variation in movement and habitat selection may be an important aspect of corridor planning, particularly if the individuals targeted for conservation (e.g., adult females) exhibit specific preferences or behaviors.

MARRYING PATTERN AND PROCESS TO CORRIDOR DESIGNS

A main impediment to advancing corridor study and planning is the missing integration between patterns of landscape composition and configuration, and the processes of habitat selection and movement. In this section we review what we consider to be the most promising approaches for advancing that integration. One of the earliest applications of this sort is percolation theory (With 1997, 2002), which examines movement within spatially structured systems representing neutral landscapes. In these landscapes, a lattice grid of "habitat" cells can be connected structurally (lattice percolation) or via movement rules (bond percolation) (With 2002). Species-specific responses to real landscapes, such as gap-crossing abilities (e.g., Desrochers 2003, St. Clair et al. 1998) and responses to edges (e.g., Haddad 1999, Schultz 1998), can be used to define movement rules for percolation models (With 2002). For example, Williams & Snyder (2005) used common "neighbor rules" from percolation theory to evaluate how habitat corridors could be restored to maintain percolating clusters, an assemblage of connected habitat cells, across the extent of simulated neutral landscapes. This application showed how landscape connectivity could be optimized to maintain percolating clusters while minimizing both corridor length and the number of nonhabitat cells that needed to be restored. Surprisingly, a meandering corridor sometimes generated lower costs (measured with both the number of restored cells and corridor length) than the shortest straight-line corridor between habitat cells. In this case, percolation theory based on movement rules identified a nonintuitive approach to corridor design.

Least-cost path analysis is a GIS-based approach similar to percolation theory except it involves estimating movement costs between two points from the suitability of intervening habitat. Parameters are based on descriptions for suitable habitats derived either from the literature or expert opinion (e.g., wolves are unlikely to occur above 1500 m; Singleton et al. 2002), and a raster grid based on accumulated distance weighted such that suitable habitats have lower movement costs than unsuitable habitats. The least-cost path analysis evaluates the costs of moving between two habitat nodes by comparing the cumulative weighted distance between the cell and the two nodes. This approach has been used to map and visualize corridors (e.g., Rouget et al. 2006, Singleton et al. 2004) but is typically based on assumptions about movement and habitat suitability that are rarely validated (Clevenger et al. 2002). Tools like RSF offer new ways to quantify landscapes for least-cost path models. For example, inverse values generated from RSF models based on sighting data for three carnivores in the Rocky Mountains (Carroll et al. 2001) could be used to generate a cost surface to explore regional corridors between protected areas. Similarly, landscape features that characterized the risk of mortality for grizzly bears in Alberta (Nielsen et al. 2004b) could be used to generate a cost surface to explore local corridor designations. If these multivariable characterizations of habitats could be combined with movement processes, a better measure of functional landscape connectivity (sensu Taylor et al. 1993) would result.

Graph theory offers particular promise for measuring landscape connectivity holistically by combining the movement emphasis of percolation theory and the habitat modeling potential of least-cost path modeling. Graph theory evolved for transportation and computer networks (Cantwell & Forman 1993) and only recently has been applied to assessments of landscape connectivity (Urban & Keitt 2001). Graph-theoretic approaches combine landscape data, typically derived from a GIS, with movement data measured as either a dispersal distance (D'Eon et al. 2002) or a random draw from a dispersal kernel generated as a function of dispersal probability with distance (Havel & Medley 2006). A lattice describes the connections between pairwise combinations of resource patches (nodes), which can be quantified as dispersal distances (edges) or weighted by other movement metrics such as tortuosity. If the distance between a pair of nodes is less than or equal to the movement threshold used, the nodes are connected. The sum of these connections can be scaled up to assess the connectivity of the entire network using a variety of metrics such as correlation length and distance to cluster edge (Calabrese & Fagan 2004). Greater correlation lengths, for example, result from an increase in the sizes of clusters suggesting greater landscape connectivity. Best of all, these process-based metrics of connectivity are readily visualized on maps to explore the effects of adding or removing connections between nodes (e.g., corridors) or resource patches (Bunn et al. 2000, Urban & Keitt 2001). For example, Urban (2005) created a graph for the wood thrush (*Hylocichla mustelina*) in North Carolina using habitat patches as defined in a GIS as nodes and movement thresholds of 2500 m to define graph edges (**Figure 4**). The resulting graph effectively identified functional corridor locations by showing how the loss of two small patches would break the single connected graph into three separate components. Importantly, these locations did not fit a conventional corridor description of linear and connected habitat and their identification was driven by information about bird movement. A pattern-based approach to corridor designation would have been less likely to have identified these corridor locations.

Although graph theory typically relies on a binary depiction of habitat (nodes), it is possible to identify these nodes probabilistically with an RSF (B.L. Schwab, C. Woudsma, S.E. Nielsen, G.B. Stenhouse, S.E. Franklin & M.S. Boyce, submitted).

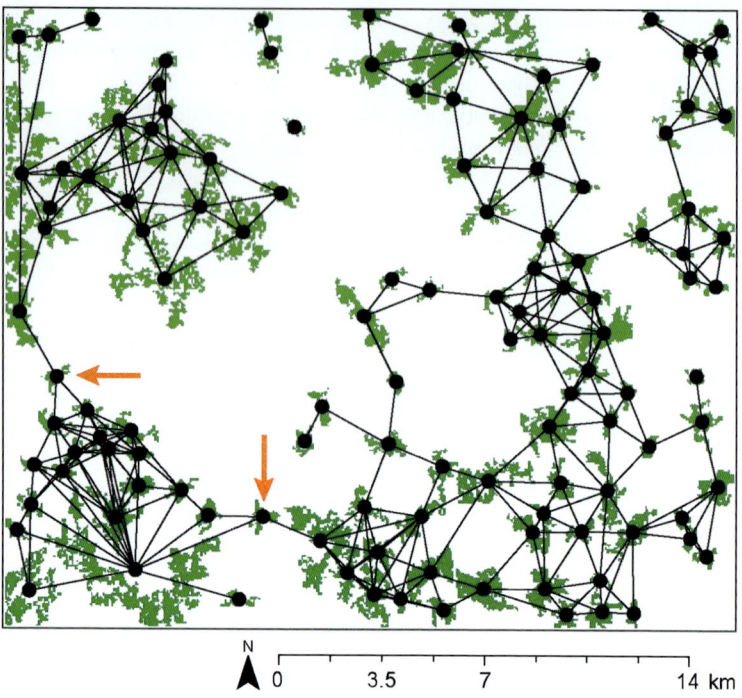

Figure 4

Graph depicting connectivity for wood thrush in a North Carolina landscape. The graph was generated using nodes generated from forest patches in a geographic information system and edges based on a dispersal distance of 2500 m. Corridor locations can be visualized between nodes where the loss of a single forest patch (*arrows*) would alter connectivity across the landscape by breaking the graph into separate components. Figure adapted from Urban (2005).

LZP: linkage zone prediction

Schwab and colleagues developed an RSF for grizzly bears in Alberta to locate areas where bears were more likely to occur (high RSF). These areas were then used to generate nodes (habitat patches) and the inverse of the RSF (i.e., 1/RSF) was used to generate a cost surface as a surrogate for movement. Least-cost path modeling was then applied to this 1/RSF cost surface and the resulting paths were compared to paths created with out-of-sample GPS location data. These data aligned with the cost surface estimated from 1/RSF showing that it performed well as a predictor of movement. This approach provides an exciting advance over previous least-cost methods such as linkage zone prediction (LZP) models. LZP models typically predict the relative probability of movement through an area by integrating qualitative scores for a number of GIS layers. For example, an LZP model for grizzly bears integrated human features, linear disturbance elements, visual cover, and riparian habitat (Singleton et al. 2002). However, an LZP model does not incorporate quantitative information about habitat or movement and generally is not validated with empirical data (Carroll et al. 2001).

Combining graph theory with RSF models offers a technique for quantifying connectivity in general and corridors in particular because it explicitly combines spatial topology with resource selection (Wagner & Fortin 2005). Because graph theory summarizes the spatial relationships between landscape elements (configuration and composition) in a concise way (Calabrese & Fagan 2004, D'Eon et al. 2002, Urban & Keitt 2001), it is especially helpful in anticipating the effects of adding or deleting particular landscape elements. Graphs also may be used to model effects of landscape on movements in two ways. First, if one uses qualitative measures or values derived from movement data in different habitats (Manseau et al. 2002), nodes can be assigned with different weights or resistance to movement (Cantwell & Forman 1993). Second, directionality can be applied to the graph edges in the form of vectors (Urban & Keitt 2001), overcoming the enduring problem of ignoring anisotropy in landscape connectivity (Bélisle 2005). And finally, graphs can be constructed with fairly modest data (Urban 2005) to provide a useful visual tool for considering corridor placement for several species simultaneously or evaluating their associated land costs (Williams 1998).

Step selection function (SSF): statistical models of landscape effects on movement probability

A second new approach for integrating landscape pattern and movement processes uses conditional logistic regression to quantify movement probabilities across landscapes using step selection functions (SSF), a technique similar to RSF. Instead of characterizing telemetry locations in an RSF, Fortin et al. (2005a) compared each step (i.e., a segment between locations on the landscape) made by wapiti with random steps having the same starting point to model the effects of landscape heterogeneity on movement. They found wapiti movements were influenced by distance to roads, cover, and wolf predation risk. With this approach, areas of high movement probability quantified by the SSF could be used to predict movement distance and direction in the context of a specific landscape, which is the essence of corridor design (sensu Haddad & Tewksbury 2006). SSF also could be used in combination with information on movement behavior at boundaries or edges to provide stronger support for corridor designations, without reliance on categorical landscape depictions.

Graph theory and SSFs are two ways that pattern and process can be integrated better in corridor designs and studies, but many other approaches are likely possible. For example, the currency of travel cost, so extensively employed in analytical models of optimal foraging behavior (Stephens & Krebs 1986), has barely been investigated in the context of landscape connectivity (Bélisle 2005). More generally, we advocate using behaviorally informed or process-driven methods to model habitat use and movement to identify landscape locations with high need or potential for corridor functions, rather than assuming these functions based on perceptions of habitat structural connectivity. We suggest that this approach offers several important advantages for designing and assessing corridors. First, movement processes reflect an organism's perception of landscape (Lima & Zollner 1996, Olden et al. 2004), which undoubtedly varies among individuals as well as species. Second, a focus on movement behavior lets one identify whether or not corridors alter movement rates, a critical dimension of corridor efficacy (Simberloff & Cox 1987, Simberloff et al. 1992). Finally, a better understanding of movement processes can be used to evaluate the effect of corridors on related key processes for individuals (dispersal, reproduction, and survival, e.g., Dzialak et al. 2005), populations (rates of immigration, emigration, persistence, and recolonization, e.g., Berggren et al. 2002, Coffman et al. 2001), and communities (biodiversity, predator-prey interactions, trophic cascades, e.g., Haddad & Tewksbury 2006).

CONCLUSIONS

"Corridors are not *the* answer to our conservation problems" (Noss 1987), but they could be used better to fulfil the promise they offer to conservation. We believe that the limitations to identifying and designing effective corridors can be traced to insufficient understanding of the processes that govern use of corridors by species of conservation interest. Behavioral processes of habitat selection and movement determine how animals use landscapes and thereby are fundamental to the identification and evaluation of corridors. We have reviewed a new generation of technological and analytical tools that allow us to quantify both habitat selection and movements with the expectation that these will allow us to approach corridors more holistically and objectively.

The Canmore example given in the introduction provides an illustration of the approach we advocate and, indeed, are attempting (C.-L. Chetkiewicz, unpublished data). There, we could conduct an RSF analysis for grizzly bears using sightings, mortality locations, and data from telemetered animals (e.g., **Figure 1*a***) to identify habitats with high probabilities of use. Then we could use SSF to identify factors that promote movement across the landscape. RSFs would identify landscape characteristics supporting grizzly bear occurrence outside designated corridors (e.g., **Figure 1*a*,*b***) and an SSF could be used to identify habitat characteristics that promote different movement behaviors. We could also use RSF and SSF models to explore important variation among individuals (e.g., habituated versus nonhabituated animals) in habitat selection and movement processes. Together, this information could be used to identify locations for mitigation (e.g., enhancing habitat, removing attractants, limiting human use or infrastructure) both inside and outside currently designated corridors. For example, the removal of human infrastructure and associated human use was highly successful in restoring connectivity for wolves on the outskirts of the town of Banff, Alberta (Duke et al. 2001). These approaches might also make it possible to combine humans and wildlife more safely in areas that appeal to both groups because of the wild areas they still contain.

We believe that more attention to the processes of habitat selection and movement will greatly strengthen our ability to identify and design effective corridors for conservation, and we suggest that this attention will bear importantly on the four fundamental questions we posed in the introduction. There we asked (*a*) if certain types of movement were more pertinent to corridors, (*b*) if corridor designs should promote faster movement, (*c*) if habitat selection is as important as movement parameters in identifying corridors, and (*d*) if corridors can promote gene flow and rescue effects even if they function as ecological traps and sinks? Answers to these questions are just beginning to emerge.

Unfortunately, even with these answers, we are unlikely to have general prescriptions for corridor designs for multiple species (e.g., Beier & Loe 1992). When Bunn et al. (2000) used a graph-theoretic approach to show that American mink (*Mustela vison*) perceived the landscape as connected, they could not generalize this result to prothonotary warblers (*Protonotaria citrea*) in the same landscape. By contrast, Haddad et al. (2003) found that corridors created in their experimental field system facilitated movement for a number of species. Thus, the best features for corridors are unknown and, even when they can be identified, may not translate well to other species, locations, and scales. That corridors have no universal rules should not really surprise us; it is a fact of most of ecology (Lawton 1999). Habitat needs for charismatic umbrella species (Simberloff 1998) like grizzly bears might encompass the needs of some species within the ecosystem (Carroll et al. 2001) and can be helpful in lobbying public support needed to meet those needs. A reasonable approach might be to identify the species and their source habitats that likely matter most in a given system (Beier et al. 2006), learn something about their actual processes of habitat selection and movement, and then use this information to restore, retain, or manage habitat in a way that will promote functional connectivity. This general approach appears to work well, but it could work better with more information about the critical processes with which animals use and move through habitat. In-depth study in the countries that can afford to support this level of investigation may well produce some guidelines, if not prescriptions, for the many countries in the world where biodiversity is being lost very rapidly and where there is neither time nor resources to spare.

In sum, we hope we have provided some new ideas and tools for sagacious input into the design and evaluation of corridors for conservation. Although they may seem daunting, many of the analytical techniques we describe are becoming quite tractable and could be used by land managers and planners now. We hope that the more process-based examination of habitat and movement we espouse can function to integrate humans better with other animals, particularly in the interface between urban and rural, and semirural and wilderness areas where many of these problems occur (McKinney 2002). Anticipating future landscapes by acknowledging how humans will directly and indirectly (e.g., climate change) affect them is critical if we are to retain our biological heritage. Conservation corridors could play an important role in ameliorating these effects and bring us closer to integrating the needs of humans and other organisms so that, at least sometimes, both parties win (sensu Rosenzweig 2003).

SUMMARY POINTS

1. Corridors are commonly applied in conservation and land-use planning to structurally connect otherwise noncontiguous patches, typically within a binary landscape.

2. Ecological processes such as habitat selection and movement are often assumed to occur via these connections, yet process-based metrics are rarely used to evaluate designated corridors. Process-based approaches to designing and planning corridors are lacking.

3. Resource selection functions describe habitat selection patterns in a continuous and multivariate way, and movement patterns can be quantified to examine movement behavior across the landscape.

4. Processes of habitat selection and movement can be integrated with landscape features using a variety of approaches. These tools offer new ways to design, implement, and study corridors as landscape linkages.

ACKNOWLEDGMENTS

We thank Paul Beier, Andrew Bennett, Jacqui Frair, Nick Haddad, Rolf Ims, and Dan Simberloff for comments on earlier versions of the manuscript and the Statistical Ecology Discussion Group at the University of Alberta for ideas, especially E. Bayne, M. Hebblewhite, M. Krkosek, M. Tremblay, and T. Swift. We especially thank Chris Johnson, Barb Schwab, and Dean Urban for allowing us to use their unpublished data and original figures. This work was supported by the Alberta Ingenuity Fund, the Alberta Conservation Association, the Wilburforce Foundation, the Wildlife Conservation Society, Wildlife Habitat Canada, and the Natural Science and Engineering Research Council of Canada.

LITERATURE CITED

Aars J, Ims RA. 1999. The effect of habitat corridors on rates of transfer and interbreeding between vole demes. *Ecology* 80:1648–55

Austin D, Bowen WD, McMillan JI. 2004. Intraspecific variation in movement patterns: modeling individual behaviour in a large marine predator. *Oikos* 105:15–30

Baum KA, Haynes KJ, Dillemuth FP, Cronin JT. 2004. The matrix enhances the effectiveness of corridors and stepping stones. *Ecology* 85:2671–76

Beier P, Loe S. 1992. A checklist for evaluating impacts to wildlife movement corridors. *Wildl. Soc. Bull.* 20:434–40

Beier P, Noss RF. 1998. Do habitat corridors provide connectivity? ***Conserv. Biol.*** **12:1241–52**

Beier P, Penrod KL, Luke C, Spencer WD, Cabañero C. 2006. South coast missing linkages: restoring connectivity to wildlands in the largest metropolitan area in the United States. See Crooks & Sanjayan 2006, pp. 555–86

Bélisle M. 2005. Measuring landscape connectivity: the challenge of behavioral landscape ecology. *Ecology* 86:1988–95

Bender DJ, Fahrig L. 2005. Matrix structure obscures the relationship between interpatch movement and patch size and isolation. *Ecology* 86:1023–33

Bennett AF. 1999. *Linkages in the Landscape: The Role of Corridors and Connectivity in Wildlife Conservation.* Gland, Switzerland: IUCN. 254 pp.

Bennetts RE, Nichols J-D, Lebreton JD, Pradel R, Hines JE, Kitchens WM. 2001. Methods for estimating dispersal probabilities and related parameters using marked animals. See Clobert et al. 2001, pp. 3–17

Marginal note: This review on corridor effects found that experiments and observations of moving individuals supported the efficacy of corridor designs for connectivity.

Berggren A, Birath B, Kindvall O. 2002. Effect of corridors and habitat edges on dispersal behavior, movement rates, and movement angles in Roesel's bush-cricket (*Metrioptera roeseli*). *Conserv. Biol.* 16:1562–69

Bergman CM, Schaefer JM, Luttich SN. 2000. Caribou movement as a correlated random walk. *Oecologia* 123:364–74

Berry O, Tocher MD, Gleeson DM, Sarre SD. 2005. Effect of vegetation matrix on animal dispersal: genetic evidence from a study of endangered skinks. *Conserv. Biol.* 19:855–64

Biro PA, Post JR, Parkinson EA. 2003. From individuals to populations: prey fish risk-taking mediates mortality in whole-system experiments. *Ecology* 84:2419–31

Bock CE, Jones ZF. 2004. Avian habitat evaluation: should counting birds count? *Front. Ecol. Environ.* 2:403–10

Boone RB, Hunter ML Jr. 1996. Using diffusion models to simulate the effects of land use on grizzly bear dispersal in the Rocky Mountains. *Landscape Ecol.* 11:51–64

Boswell GP, Britton NF, Franks NR. 1998. Habitat fragmentation, percolation theory and the conservation of a keystone species. *Proc. R. Soc. London Ser. B* 265:1921–25

Bow Corridor Ecosystem Advisory Group (BCEAG). 1999. *Wildlife Corridor and Habitat Patch Guidelines for the Bow Valley, Rep. T/411.* Calgary: BCEAG. 33 pp.

Boyce MS. 2006. Scale of resource selection functions. *Diversity Distrib.* 12:269–76

Boyce MS, McDonald LL. 1999. Relating populations to habitats using resource selection functions. *Trends Ecol. Evol.* 14:268–72

Brown JH, Kodric-Brown A. 1977. Turnover rates in insular biogeography: effect of immigration on extinction. *Ecology* 58:445–49

Bunn AG, Urban DL, Keitt TH. 2000. Landscape connectivity: a conservation application of graph theory. *J. Environ. Manage.* 59:265–78

Calabrese JM, Fagan WF. 2004. A comparison-shopper's guide to connectivity metrics. *Front. Ecol. Environ.* 2:529–36

Cant ET, Smith AD, Reynolds DR, Osborne JL. 2005. Tracking butterfly flight paths across the landscape with harmonic radar. *Proc. R. Soc. London Ser. B* 272:785–90

Cantwell MD, Forman RTT. 1993. Landscape graphs: ecological modeling with graph theory to detect configurations common to diverse landscapes. *Landscape Ecol.* 8:239–55

Carroll C, Noss RF, Paquet PC. 2001. Carnivores as focal species for conservation planning in the Rocky Mountain region. *Ecol. Appl.* 11:961–80

Channell R, Lomolino MV. 2000. Dynamic biogeography and conservation of endangered species. *Nature* 403:84–86

Clevenger AP, Waltho N. 2005. Performance indices to identify attributes of highway crossing structures facilitating movement of large mammals. *Biol. Conserv.* 121:453–64

Clevenger AP, Wierzchowski J, Chruszcz B, Gunson K. 2002. GIS-generated, expert-based models for identifying wildlife habitat linkages and planning mitigation passages. *Conserv. Biol.* 16:503–14

Clobert J, Danchin E, Dhondt AA, Nichols JD, eds. 2001. *Dispersal.* London: Oxford Univ. Press. 452 pp.

Coffman CJ, Nichols JD, Pollock KH. 2001. Population dynamics of *Microtus pennsylvanicus* in corridor-linked patches. *Oikos* 93:3–21

Collinge SK. 2000. Effects of grassland fragmentation on insect species loss, colonization, and movement patterns. *Ecology* 81:2211–26

Cooper CB, Walters JR, Priddy J. 2002. Landscape patterns and dispersal success: simulated population dynamics in the brown treecreeper. *Ecol. Appl.* 12:1576–87

Crist TO, Guertin DS, Wiens JA, Milne BT. 1992. Animal movement in heterogeneous landscapes: an experiment with *Eleodes* beetles in shortgrass prairie. *Funct. Ecol.* 6:536–44

Crooks KR, Sanjayan MA, eds. 2006. *Connectivity Conservation*. Cambridge: Cambridge Univ. Press. 736 pp.

Danchin E, Heg D, Doligez B. 2001. Public information and breeding habitat selection. See Clobert et al. 2001, pp. 243–58

D'Eon RG, Glenn SM, Parfitt I, Fortin M-J. 2002. Landscape connectivity as a function of scale and organism vagility in a real forested landscape. *Conserv. Ecol.* Vol. 6. http://www.consecol.org/vol6/iss2/art10

Desrochers A. 2003. Bridging the gap: linking individual bird movement and territory establishment rules with their patterns of distribution in fragmented forests. In *Animal Behavior and Wildlife Conservation*, ed. M Festa-Bianchet, M Apollonio, pp. 63–76. Washington, DC: Island Press

Diamond JM, Terborgh J, Whitcomb RF, Lynch JF, Opler PA, et al. 1976. Island biogeography and conservation: Strategy and limitations. *Science* 193:1027–32

Dickson BG, Jenness JS, Beier P. 2005. Influence of vegetation, topography, and roads on cougar movement in southern California. *J. Wildl. Manage.* 69:264–76

Dirzo R, Raven PH. 2003. Global state of biodiversity and loss. *Annu. Rev. Environ. Resour.* 28:137–67

Duke DL, Hebblewhite M, Paquet PC, Callaghan C, Percy M. 2001. Restorating a large-carnivore corridor in Banff National Park. In *Large Mammal Restoration: Ecological and Sociological Challenges in the 21st Century*, ed. DS Maehr, RF Noss, JL Larkin, pp. 261–75. Washington, DC: Island Press

Dunning JB, Danielson BJ, Pulliam HR. 1992. Ecological processes that affect populations in complex landscapes. *Oikos* 65:169–75

Dzialak MR, Lacki MJ, Larkin JL, Carter KM, Vorisek S. 2005. Corridors affect dispersal initiation in reintroduced peregrine falcons. *Anim. Conserv.* 8:421–30

Fahrig L. 2003. Effects of habitat fragmentation on biodiversity. *Annu. Rev. Ecol. Evol. Syst.* 34:487–515

Fahrig L, Merriam G. 1994. Conservation of fragmented populations. *Conserv. Biol.* 8:50–59

Fischer J, Lindenmayer DB, Fazey I. 2004. Appreciating ecological complexity: Habitat contours as a conceptual landscape model. *Conserv. Biol.* 18:1245–53

Forman RTT. 2002. Foreward. See Gutzwiller 2002, pp. vii–x

Fortin D, Beyer HL, Boyce MS, Smith DW, Duchesne T, Mao JS. 2005a. Wolves influence elk movements: Behavior shapes a trophic cascade in Yellowstone National Park. *Ecology* 86:1320–30

Fortin D, Morales JM, Boyce MS. 2005b. Elk winter foraging at fine scale in Yellowstone National Park. *Oecologia* 145:335–43

Frair JL, Merrill EH, Visscher DR, Fortin D, Beyer HL, Morales JM. 2005. Scales of movement by elk (*Cervus elaphus*) in response to heterogeneity in forage resources and predation risk. *Landscape Ecol.* 20:273–87

Garshelis DL. 2000. Delusions in habitat evaluation: measuring use, selection, and importance. In *Research Techniques in Animal Ecology: Controversies and Consequences*, ed. L Boitani, TK Fuller, pp. 111–64. New York: Columbia Univ. Press

Gibeau ML, Clevenger AP, Herrero S, Wierzchowski J. 2002. Grizzly bear response to human development and activities in the Bow River Watershed, Alberta, Canada. *Biol. Conserv.* 103:227–36

Gilbert F, Gonzalez A, Evans-Freke I. 1998. Corridors maintain species richness in the fragmented landscapes of a microecosystem. *Proc. R. Soc. London Ser. B* 265:577–82

Gillies CG, Hebblewhite M, Nielsen SE, Krawchuk MA, Aldridge CL, et al. 2006. Application of random effects to the study of resource selection by animals. *J. Anim. Ecol.* 75:887–98

Gonzalez A, Lawton JH, Gilbert FS, Blackburn TM, Evans-Freke I. 1998. Metapopulation dynamics, abundance, and distribution in a microecosystem. *Science* 281:2045–47

Goodwin BJ. 2003. Is landscape connectivity a dependent or independent variable? *Landscape Ecol.* 18:687–99

Goodwin BJ, Fahrig L. 2002. How does landscape structure influence landscape connectivity? *Oikos* 99:552–70

Gustafson EJ, Gardner RH. 1996. The effect of landscape heterogeneity on the probability of patch colonization. *Ecology* 77:94–107

Gutzwiller KJ, ed. 2002. *Applying Landscape Ecology in Biological Conservation*. New York: Springer-Verlag. 518 pp.

Haddad N. 2000. Corridor length and patch colonization by a butterfly, *Junonia coenia*. *Conserv. Biol.* 14:738–45

Haddad NM. 1999. Corridor use predicted from behaviors at habitat boundaries. *Am. Nat.* 153:215–27

Haddad NM, Bowne DR, Cunningham A, Danielson BJ, Levey DJ, et al. 2003. Corridor use by diverse taxa. *Ecology* 84:609–15

Haddad NM, Rosenberg DK, Noon BR. 2000. On experimentation and the study of corridors: Response to Beier and Noss. *Conserv. Biol.* 14:1543–45

Haddad NM, Tewksbury JJ. 2005. Low-quality habitat corridors as movement conduits for two butterfly species. *Ecol. Appl.* 15:250–57

Haddad NM, Tewksbury JJ. 2006. Impacts of corridors on populations and communities. See Crooks & Sanjayan 2006, pp. 390–415

Haila Y. 2002. A conceptual genealogy of fragmentation research: From island biogeography to landscape ecology. *Ecol. Appl.* 12:321–34

Hale ML, Lurz PWW, Shirley MDF, Rushton S, Fuller RM, Wolff K. 2001. Impact of landscape management on the genetic structure of red squirrel populations. *Science* 293:2246–48

Hannon SJ, Schmiegelow FKA. 2002. Corridors may not improve the conservation value of small reserves for most boreal birds. *Ecol. Appl.* 12:1457–68

> Movement behavior at habitat boundaries incorporated into simulation models confirmed corridors directed movements of habitat-restricted butterflies.

Hanski I. 2001. Population dynamic consequences of dispersal in local populations and in metapopulations. See Clobert et al. 2001, pp. 283–98

Hanski I, Gilpin ME, eds. 1997. *Metapopulation Biology: Ecology, Genetics, and Evolution*. San Diego: Academic. 512 pp.

Hansson L, Fahrig L, Merriam G, eds. 1995. *Mosaic Landscapes and Ecological Processes*. London: Chapman & Hall. 380 pp.

Harris LD, Scheck J. 1991. From implications to applications: the dispersal corridor principle applied to the conservation of biological diversity. See Saunders & Hobbs 1991, pp. 189–220

Havel JE, Medley KA. 2006. Biological invasions across spatial scales: Intercontinental, regional, and local dispersal of cladoceran zooplankton. *Biol. Invasions* 8:459–73

Herrero S, ed. 2005. *Biology, demography, ecology and management of grizzly bears in and around Banff National Park and Kananaskis Country*. ESGBP Final Rep., Univ. Calgary, Alberta, Canada. 248 pp.

Hess GR. 1994. Conservation corridors and contagious disease: a cautionary note. *Conserv. Biol.* 8:256–62

Hilty JA, Lidicker WZ Jr, Merenlender AM, eds. 2006. *Corridor Ecology: the Science and Practice of Linking Landscapes for Biodiversity Conservation*. Washington, DC: Island Press. 323 pp.

Hobbs RJ. 1992. The role of corridors in conservation: solution or bandwagon. *Trends Ecol. Evol.* 7:389–92

Hobson KA. 2005. Using stable isotopes to trace long-distance dispersal in birds and other taxa. *Divers. Distrib.* 11:157–64

Ims RA. 1995. Movement patterns related to spatial structures. See Hansson et al. 1995, pp. 85–109

Ims RA, Rolstad J, Wegge P. 1993. Predicting space use responses to habitat fragmentation: Can voles *Microtus oeconomus* serve as an experimental model system (EMS) for capercaillie grouse *Tetrao urogallus* in boreal forest? *Biol. Conserv.* 63:261–68

Johnson AR, Wiens JA, Milne BT, Crist TO. 1992. Animal movements and population dynamics in heterogeneous landscapes. *Landscape Ecol.* 7:63–75

Johnson CJ, Nielsen SE, Merrill EH, McDonald TL, Boyce MS. 2006. Resource selection functions based on use-availability data: Theoretical motivation and evaluation methods. *J. Wildl. Manag.* 70:347–57

Johnson CJ, Parker KL, Heard DC, Gillingham MP. 2002. Movement parameters of ungulates and scale-specific responses to the environment. *J. Anim. Ecol.* 71:225–35

Johnson DH. 1980. The comparison of usage and availability measurements for evaluating resource preference. *Ecology* 61:65–71

Kernohan BJ, Gitzen RA, Millspaugh JJ. 2001. Analysis of animal space use and movements. See Millspaugh & Marzluff 2001, pp. 126–66

Keyghobadi N, Roland J, Strobeck C. 2005. Genetic differentiation and gene flow among populations of the alpine butterfly, *Parnassius smintheus*, vary with landscape connectivity. *Mol. Ecol.* 14:1897–909

Kristan WBI. 2003. The role of habitat selection behavior in population dynamics: Source-sink systems dynamics and ecological traps. *Oikos* 103:457–68

Lawton JH. 1999. Are there general laws in ecology? *Oikos* 84:177–92

Levey DJ, Bolker BM, Tewksbury JJ, Sargent S, Haddad NM. 2005. Effects of landscape corridors on seed dispersal by birds. Science 309:146–48

Lidicker WZ Jr. 1999. Responses of mammals to habitat edges: An overview. *Landscape Ecol.* 14:333–43

Lima SL, Zollner PA. 1996. Towards a behavioral ecology of ecological landscapes. *Trends Ecol. Evol.* 11:131–35

Little SJ, Harcourt RG, Clevenger AP. 2002. Do wildlife passages act as prey-traps? *Biol. Conserv.* 107:135–45

MacArthur RH, Wilson EO. 1967. *The Theory of Island Biogeography*. Princeton: Princeton Univ. Press. 203 pp.

Manly BFJ, McDonald LL, Thomas DL, McDonald TL, Erikson WP. 2002. *Resource Selection by Animals: Statistical Design and Analysis for Field Studies*. New York: Kluwer. 221 pp.

Manseau M, Fall A, O'Brien D, Fortin M-J. 2002. National parks and the protection of woodland caribou: a multi-scale landscape analysis. *Res. Links* 10:24–28

McCallum H, Dobson A. 2002. Disease, habitat fragmentation and conservation. *Proc. R. Soc. London Ser. B* 269:2041–49

McCulloch CE, Lin H, Slate EH, Turnbull BW. 2002. Discovering subpopulation structure with latent class mixed models. *Stat. Med.* 21:417–29

McDonald W, St. Clair CC. 2004. Elements that promote highway crossing structure use by small mammals in Banff National Park. *J. Appl. Ecol.* 41:82–93

McGarigal K, Cushman SA, Neel MC, Ene E. 2002. *FRAGSTATS: Spatial pattern analysis program for categorical maps*. Comp. software prog. Univ. Mass., Amherst. **http://www.umass.edu/landeco/research/fragstats/fragstats.html**

McIntyre S, Hobbs R. 1999. A framework for conceptualizing human effects on landscapes and its relevance to management and research models. *Conserv. Biol.* 13:1282–92

McKinney ML. 2002. Urbanization, biodiversity, and conservation. *BioScience* 52:883–90

Mech SG, Hallett JG. 2001. Evaluating the effectiveness of corridors: a genetic approach. *Conserv. Biol.* 15:467–74

Mills LS, Allendorf FW. 1996. The one-migrant-per-generation rule in conservation and management. *Conserv. Biol.* 10:1509–18

Millspaugh JJ, Marzluff JM, eds. 2001. *Radio Tracking and Animal Populations*. San Diego: Academic. 474 pp.

Morales JM, Haydon DT, Frair J, Holsinger KE, Fryxell JM. 2004. Extracting more out of relocation data: Building movement models as mixtures of random walks. *Ecology* 85:2436–45

Morris DW. 2003. Toward an ecological synthesis: a case for habitat selection. *Oecologia* 136:1–13

Nams VO. 2005. Using animal movement paths to measure response to spatial scale. *Oecologia* 143:179–88

Small-scale movement data of birds in corridors were used to predict the effects of corridors on seed dispersal at the landscape scale.

Nathan R. 2005. Long-distance dispersal research: Building a network of yellow brick roads. *Divers. Distrib.* 11:125–30

Nathan R, Perry G, Cronin JT, Strand AE, Cain ML. 2003. Methods for estimating long-distance dispersal. *Oikos* 103:261–73

Nelson JG, Day JC, Sportza LM, eds. 2003. *Protected Areas and the Regional Planning Imperative in North America: Integrating Nature Conservation and Sustainable Development*. Calgary: Univ. Calgary Press. 429 pp.

Nielsen SE, Boyce MS, Stenhouse GB. 2004a. Grizzly bears and forestry I. Selection of clearcuts by grizzly bears in west-central Alberta, Canada. *For. Ecol. Manag.* 199:51–65

Nielsen SE, Herrero S, Boyce MS, Mace RD, Benn B, et al. 2004b. Modelling the spatial distribution of human-caused grizzly bear mortalities in the Central Rockies ecosystem of Canada. *Biol. Conserv.* 120:101–13

Nielsen SE, Johnson CJ, Heard DC, Boyce MS. 2005. Can models of presence-absence be used to scale abundance? Two case studies considering extremes in life history. *Ecography* 28:197–208

Noss RF. 1987. Corridors in real landscapes: a reply to Simberloff and Cox. *Conserv. Biol.* 1:159–64

Noss RF. 2003. A checklist for wildlands network designs. *Conserv. Biol.* 17:1270–75

Noss RF, Beier P. 2000. Arguing over little things: response to Haddad et al. *Conserv. Biol.* 14:1546–48

Olden JD, Schooley RL, Monroe JB, Poff NL. 2004. Context-dependent perceptual ranges and their relevance to animal movements in landscapes. *J. Anim. Ecol.* 73:1190–94

Powell GVN, Bjork R. 1995. Implications of intratropical migration on reserve design: a case study using *Pharomachrus mocinno*. *Conserv. Biol.* 9:354–62

Proctor MF, McLellan BN, Strobeck C, Barclay RMR. 2004. Gender-specific dispersal distances of grizzly bears estimated by genetic analysis. *Can. J. Zool.* 82:1108–18

Pulliam HR. 1988. Sources, sinks, and population regulation. *Am. Nat.* 132:652–61

Puth LM, Wilson KA. 2001. Boundaries and corridors as a continuum of ecological flow control: lessons from rivers and streams. *Conserv. Biol.* 15:21–30

Reed DH. 2004. Extinction risk in fragmented habitats. *Anim. Conserv.* 7:181–91

Rosenberg DK, Noon BR, Meslow EC. 1997. Biological corridors: Form, function, and efficacy. *BioScience* 47:677–87

Rosenzweig ML. 2003. Reconciliation ecology and the future of species diversity. *Oryx* 37:194–205

Rouget M, Cowling RM, Lombard AT, Knight AT, Kerley GIH. 2006. Designing large-scale conservation corridors for pattern and process. *Conserv. Biol.* 20:549–61

Rubenstein DR, Hobson KA. 2004. From birds to butterflies: animal movement patterns and stable isotopes. *Trends Ecol. Evol.* 19:256–63

Saunders DA, Hobbs RJ, eds. 1991. *Nature Conservation. 2: The Role of Corridors*. Chipping Norton, Aust.: Surrey Beatty & Sons. 442 pp.

Schmiegelow FKA, Machtans CS, Hannon SJ. 1997. Are boreal birds resilient to forest fragmentation? An experimental study of short-term community responses. *Ecology* 78:1914–32

Schultz CB. 1998. Dispersal behavior and its implications for reserve design in a rare Oregon butterfly. *Conserv. Biol.* 12:284–92

Sieving KE, Willson MF, De Santo TL. 2000. Defining corridor functions for endemic birds in fragmented south-temperate rainforest. *Conserv. Biol.* 14:1120–32

Sih A, Jonsson BG, Luikart G. 2000. Habitat loss: ecological, evolutionary and genetic consequences. *Trends Ecol. Evol.* 15:132–34

Simberloff D. 1998. Flagships, umbrellas, and keystones: Is single-species management passé in the landscape era? *Biol. Conserv.* 83:247–57

Simberloff D, Cox J. 1987. Consequences and costs of conservation corridors. *Conserv. Biol.* 1:63–71

Simberloff D, Farr JA, Cox J, Mehlman DW. 1992. Movement corridors: conservation bargains or poor investments? *Conserv. Biol.* 6:493–504

Singleton PH, Gaines WL, Lehmkuhl JF. 2002. *Landscape permeability for large carnivores in Washington: A geographic information system weighted-distance and least-cost corridor assessment. Res. Pap. PNW-RP-549*. Portland, OR: U.S. Dept. Agric., For. Serv., Pac. Northwest Res. Stn. 89 pp.

Singleton PH, Gaines WL, Lehmkuhl JF. 2004. Landscape permeability for grizzly bear movements in Washington and southwestern British Columbia. *Ursus* 15:90–103

Soulé ME, Terborgh J, eds. 1999. *Continental Conservation. Scientific Foundations of Regional Reserve Networks*. Washington, DC: Island Press. 228 pp.

Southwood TRE, Henderson PA. 2000. *Ecological Methods*. Oxford: Blackwell Sci. 576 pp.

Stamps JA. 2001. Habitat selection by dispersers: integrating proximate and ultimate approaches. See Clobert et al. 2001, pp. 230–42

St. Clair CC, Bélisle M, Desrochers A, Hannon S. 1998. Winter responses of forest birds to habitat corridors and gaps. *Conserv. Ecol.* Vol. 2. **http://www.ecologyandsociety.org/vol2/iss2/art13**

Stephens DW, Krebs JR. 1986. *Foraging Theory*. Princeton, NJ: Princeton Univ. Press. 247 pp.

Taylor PD, Fahrig L, Henein K, Merriam G. 1993. Connectivity is a vital element in landscape structure. *Oikos* 68:571–73

Tischendorf L, Fahrig L. 2000a. How should we measure landscape connectivity? *Landscape Ecol.* 15:633–41

Tischendorf L, Fahrig L. 2000b. On the usage and measurement of landscape connectivity. *Oikos* 90:7–19

Tischendorf L, Irmler U, Hingst R. 1998. A simulation experiment on the potential of hedgerows as movement corridors for forest carabids. *Ecol. Model.* 106:107–18

Turchin P. 1998. *Quantitative Analysis of Movement*. Sunderland, MA: Sinauer. 383 pp.

Turner MG. 2005. Landscape ecology: what is the state of the science? *Annu. Rev. Ecol. Evol. Syst.* 36:319–44

Turner MG, Gardner RH, eds. 1991. *Quantitative Methods in Landscape Ecology*. New York: Springer-Verlag. 536 pp.

Turner MG, Gardner RH, O'Neill RV. 2001. *Landscape Ecology in Theory and Practice. Pattern and Process*. New York: Springer-Verlag. 401 pp.

Urban D, Keitt T. 2001. Landscape connectivity: a graph-theoretic perspective. *Ecology* 82:1205–18

Urban DL. 2005. Modeling ecological processes across scales. *Ecology* 86:1996–2006

Vandermeer J, Carvajal R. 2001. Metapopulation dynamics and the quality of the matrix. *Am. Nat.* 158:211–20

Vos CC, Baveco H, Grashof-Bokdam CJ. 2002. Corridors and species dispersal. See Gutzwiller 2002, pp. 84–104

Wagner HH, Fortin M-J. 2005. Spatial analysis of landscapes: Concepts and statistics. *Ecology* 86:1975–87

Waser PM, Strobeck C. 1998. Genetic signatures of interpopulation dispersal. *Trends Ecol. Evol.* 13:43–44

Webster MS, Marra PP, Haig SM, Bensch S, Holmes RT. 2002. Links between worlds: Unraveling migratory connectivity. *Trends Ecol. Evol.* 17:76–83

Weldon AJ, Haddad NM. 2005. The effects of patch shape on indigo buntings: Evidence for an ecological trap. *Ecology* 86:1422–31

Whittington J, St. Clair CC, Mercer G. 2005. Spatial responses of wolves to roads and trails in mountain valleys. *Ecol. Appl.* 15:543–53

Wiens JA. 1995. Landscape mosaics and ecological theory. See Hansson et al. 1995, pp. 1–26

Williams JC. 1998. Delineating protected wildlife corridors with multi-objective programming. *Environ. Model. Assess.* 3:77–86

Williams JC, Snyder SA. 2005. Restoring habitat corridors in fragmented landscapes using optimization and percolation models. *Environ. Model. Assess.* 10:239–50

Wilson EO, Willis EO. 1975. Applied biogeography. In *Ecology and Evolution of Communities*, ed. ML Cody, JM Diamond, pp. 522–34. Cambridge: Harvard Univ. Press

With KA. 1994. Ontogenetic shifts in how grasshoppers interact with landscape structure: An analysis of movement patterns. *Funct. Ecol.* 8:477–85

With KA. 1997. The application of neutral landscape models in conservation biology. *Conserv. Biol.* 11:1069–80

With KA. 2002. Using percolation theory to assess landscape connectivity and effects of habitat fragmentation. See Gutzwiller 2002, pp. 105–30

Zimmerer KS, Galt RE, Buck MV. 2004. Global conservation and multi-spatial trends in the coverage of protected-area conservation (1980–2000). *Ambio* 33:520–29

The Population Biology of Large Brown Seaweeds: Ecological Consequences of Multiphase Life Histories in Dynamic Coastal Environments

David R. Schiel[1] and Michael S. Foster[2]

[1]Marine Ecology Research Group, School of Biological Sciences, University of Canterbury, Christchurch, New Zealand; email: david.schiel@canterbury.ac.nz

[2]Moss Landing Marine Laboratories, Moss Landing, California 95039; email: foster@mlml.calstate.edu

Key Words

demography, Fucales, kelp, Laminariales, life history, microscopic phases, population biology

Abstract

Seaweed population biology has received far less attention than trophic dynamics, yet is critically important in establishing and maintaining algal communities. Complex life histories of habitat-forming kelps and fucoids, including spores, gametophytes, gametes, and microscopic and macroscopic benthic stages, must be considered within the context of their highly dynamic nearshore environments. We evaluate differences within and between kelps and fucoids in life histories as they affect population biology; dispersal and potential limitations in population establishment; macroscopic stages and variations in survival and longevity affecting stand structure; and microscopic stage responses to disturbance and variation in the physical environment. We suggest that the commonly made comparisons of seaweeds with terrestrial seed plants are misleading because of large differences in morphology, environments, and the ephemeral nature of propagule banks in the sea. We conclude that progress in understanding algal populations depends on better knowledge of microscopic stages and on feedback through density-dependent reproductive processes, dispersal, and settlement.

INTRODUCTION

Kelp: large brown seaweeds of the Order Laminariales

Fucoid: large brown seaweeds of the Order Fucales

Population biology: incorporation of ecology, demography, and life history characteristics and requirements of seaweeds

Seaweeds, particularly large brown algae in the Orders Laminariales (true kelps or laminarians) and Fucales (fucoids), commonly dominate intertidal and subtidal rocky reefs and, like terrestrial plants, provide much of the three-dimensional structure, biomass, habitat, and food for associated species. They are influenced by and also affect environmental factors such as light, turbulence, nutrients, and, in the case of intertidal species, moisture and temperature, thereby affecting myriad species in their associated communities. Understanding the population dynamics of marine algae, however, is enormously complicated because their life histories are multiphasic and occur in complex and variable coastal environments. Although seaweed populations have long been considered analogues of terrestrial plants (Darwin 1860), important differences between algae and seed plants tend to be obscured by a relatively poor functional understanding of algal life history requirements, particularly of early life phases, and a continuing focus on trophic dynamics in the environments they dominate (Jackson et al. 2001) rather than the population dynamics of the seaweeds themselves. It is our premise that understanding seaweed population biology critically underpins an ecological understanding of the communities they dominate.

Parallel research traditions, especially in ecology and phycology, have influenced perspectives on structuring process of algal stands, often reflecting different historical interests, questions, and approaches (cf., Fagerstrom & Westoby 1997). Ecologists most commonly have studied associated animals, especially barnacles, mussels, sea urchins, and fish, and top-down interactions that can affect stand structure (e.g., Estes & Duggins 1995, reviewed in Schiel 2004, Schiel & Foster 1986). In contrast, phycologists traditionally have been concerned with the taxonomy, biology, and physiology of the seaweeds themselves. Given the differences in focus between disciplines, it can be difficult to make ecological sense of complex macroalgal life histories and their environmental interactions, including intraspecific, density-dependent effects. This review attempts to resolve this by evaluating the population biology of large seaweeds, incorporating their entire life histories in an ecological context.

We build on past perspectives on algal community structure (Chapman 1995, Dayton 1985, Schiel & Foster 1986) by evaluating the population biology of kelps and fucoids in the broad sense through the incorporation of ecology, demography, and life history characteristics and requirements of the seaweeds themselves in their dynamic nearshore environments. We begin with a background to algal population biology, contrast critical features of life histories, and evaluate the differences in demographics of micro- and macroscopic stages of kelps and fucoids across four themes.

- Comparisons of seaweeds and seed plants: are they misleading with respect to seaweed population biology?
- Early life stages: is knowledge of their biology and density-dependent processes the key to understanding the population biology of kelps and fucoids?
- Dispersal: does this limit seaweed recruitment and population replenishment?
- Disturbance and variation in the physical environment: how do these factors influence seaweed populations through settlement and recruitment processes?

We conclude by highlighting approaches that may provide a better understanding of algal-dominated communities by incorporating the unique aspects of their life history attributes.

BACKGROUND

There are at least 100 species of kelps and 500 species of fucoids worldwide (Guiry et al. 2006), and good overviews of their biology and ecology can be found in recent texts (e.g., Graham & Wilcox 2000). Kelps and fucoids occur almost entirely on hard reefs from the intertidal zone to depths of over 100 m but usually shallower than 30 m. They are particularly abundant where the water temperature is generally lower than 20°C, often forming dense stands over large areas, with individuals of some species reaching 30 m in length. They clearly have considerable ecological importance in their distinctive communities as foundation (Dayton 1972) or autogenic (Lawton 1994) species. Both kelps and fucoids have free-living microscopic and macroscopic stages (**Figure 1**) that have different resource requirements and environmental needs. Their population biology and ecology, therefore, can be understood only in the context of their entire life histories, including all of the morphological, cytological, and reproductive phases, and the environments in which they live.

The characteristics that have traditionally described the Order Laminariales are being re-examined (Druehl et al. 1997), but the basic features that define their life histories remain unaltered and we refer to the traditional genera (Guiry et al. 2006) as "kelps." The great majority of kelps occur subtidally, whereas most fucoids occupy the intertidal and very shallow subtidal zones, usually having high light requirements that exclude them from deeper waters (Chapman 1995). Consequently, there are considerable differences between these algal groups in their responses to environmental variables, including desiccation, waves, sediments, temperature, often high grazer abundances (Vadas et al. 1992) and, increasingly, human disturbances (Kautsky et al. 1986, Keough & Quinn 1998, Malm et al. 2001), which vary considerably in scale and intensity along depth gradients. As a result, survival of all life stages can be highly variable.

Comparisons with Seed Plants

Seed plant population models, such as in Harper's (1977) classic book, *Population Biology of Plants*, often provide a conceptual context for seaweed studies. However, because of the considerable differences in the structure, life histories, and environments of the two groups, terrestrial plant models may not be well-suited to seaweeds, many of which have free-living microscopic stages (versus seeds and related seed banks), produce enormous quantities of spores and gametes with little documented "cost" (0.1–4% of total adult plant biomass; Joska & Bolton 1987, Neushul 1963, Vernet & Harper 1980), have rapid growth rates and maturity, relatively small dispersal distances, short life spans, and no known animal fertilization ("pollinators") or dispersal vectors. Although many ecologists and phycologists recognize these distinctive features, their relative importance for the maintenance of seaweed populations is not well established in the ecological literature.

a Kelps

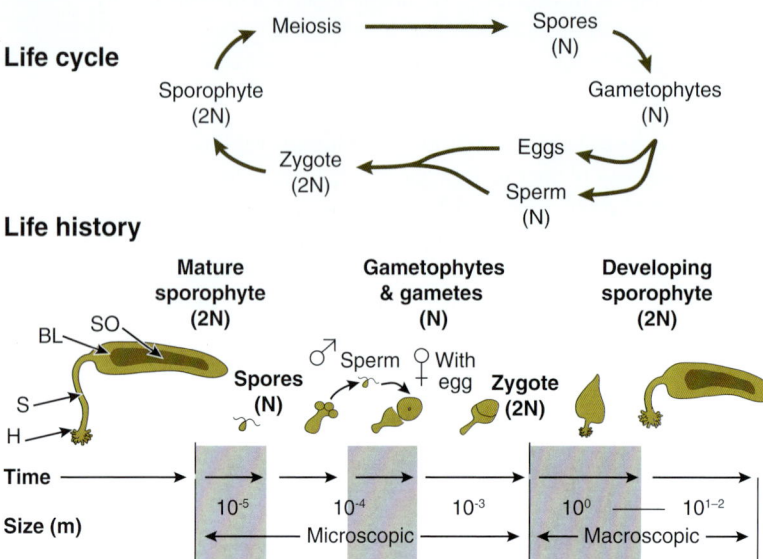

b Fucoids

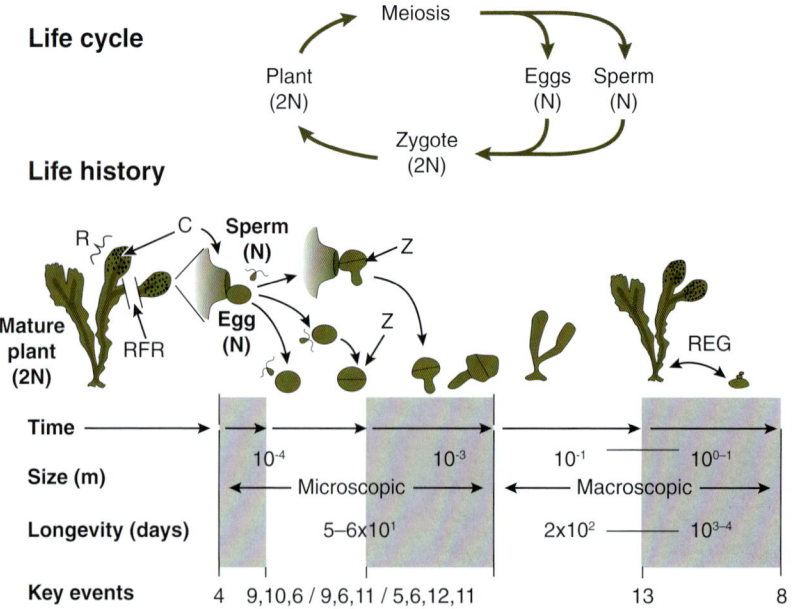

Harper (1977) emphasized the necessity of considering the entire life history or cycle of seed plants and the demographics of all stages to understand population dynamics. He considered the diploid stages of seed plants (i.e., seeds and plants as genets or genetically distinct individuals made up of modules or ramets) as the primary stages in population dynamics because their recruitment, survival, and fecundity are affected directly by the environment. In contrast, kelps and fucoids expose both their diploid phases (from microscopic zygotes to large, mature plants) and haploid stages (microscopic spores, gametophytes, gametes) to the environment. None of the microscopic stages is protected or nourished within the individuals that produce them or protected by specialized coverings analogous to seeds or seed coats. Generally all stages are photosynthetic, free-living individuals, and fertilization is external to adult plants (Graham & Wilcox 2000). Both 1N and 2N stages, which occur over four to five orders of magnitude in size (**Figure 1**), can therefore be of primary importance in different habitats and at different times. Although there are interesting similarities between portions of kelp and fucoid life histories and those of some seed plants (e.g., sperm dispersal by water and wind pollination), the differences reviewed here are much more ecologically informative.

Gametophyte: the microscopic haploid (1N) life stage of kelps that produces gametes

LIFE HISTORIES OF KELPS AND FUCOIDS

To understand the role of life histories in population biology it is necessary to recognize the distinctive features of large brown algae. Kelps have an extremely heteromorphic life history that is usually considered an adaptation to large, seasonal changes in the environment, including grazing (review in Clayton 1988). They often grow side-by-side with fucoids and other large algae that have very different life histories, so the kelp life history must also reflect phylogenetic constraints and perhaps unique characteristics of the environments in which it arose. Kelps have macroscopic diploid sporophytes that even in nonfloat-bearing species can be 15 m long (e.g., *Laminaria angustata*, Kawashima 1972), tiny haploid stages (spores and sperm) that can be planktonic, and microscopic benthic stages that are either haploid (gametophytes and eggs) or diploid (young sporophytes) (**Figure 1a**). Adult sporophytes produce flagellated spores (~7 μm in length) with a 1:1 sex ratio (e.g., Kain 1979). These settle and develop into free-living, dioecious gametophytes that are self-fertile. Eggs are extruded

Figure 1

Kelp (*a*) and fucoid (*b*) life cycles and life histories with emphasis on stages, sizes (plant length), longevities, and events of particular relevance to population dynamics. The fucoid life history illustrates three possible sequences leading from gamete production to attached zygotes or embryos. C, conceptacle (cavity in which gametangia develop); BL, blade; H, holdfast; R, receptacle (terminal portion of branch bearing conceptacles); REG, regeneration; RFR, reproductive frond release (in some fucoids); S, stipe; SO, sorus; Z, zygote. 1, spore production; 2, spore dispersal; 3, spore settlement; 4, gamete production; 5, sperm dispersal; 6, fertilization; 7, sporangia development; 8, sporophyte/adult plant mortality; 9, sperm and egg dispersal; 10, egg settlement; 11, zygote settlement; 12, zygote dispersal; 13, development of gametangia.

Sporophyte: the diploid (2N) stage of kelps

from the oogonium but remain attached to it, and are fertilized by sperm. In all species examined, eggs release a pheromone that induces sperm release and attracts sperm to eggs over distances of about 1 mm (Maier & Muller 1986, Muller 1981). The microscopic zygote grows on the female gametophyte and develops to become the mature, macroscopic sporophyte.

In contrast to kelps, meiosis in fucoids results in gametes instead of spores; fucoids do not have a free-living gametophyte phase. Fertilization is external (Graham & Wilcox 2000) and plants can be either monoecious or dioecious, depending on species. Gametangia are produced within sunken chambers (conceptacles) that develop as the diploid plants mature (**Figure 1b**). Eggs and biflagellated sperm are extruded in mucilage through a pore in the conceptacle, often still contained within their respective gametangia. Eggs of different species are 64–250 μm in diameter (Clayton 1992), considerably larger than the spores or eggs of kelp, which has consequences for the survival rates of early life stages. As in kelps, there is no obligate period of development from release to fertilization, but this process is more variable in fucoids than in kelps. Fertilization has been studied in a few fucoids, in which it occurs rapidly on or very near parent plants (Serrão et al. 1996). In the common intertidal *Fucus distichus*, for example, eggs can be fertilized while still loosely attached by mucilage to the parent, or in the water column or on the benthos after they are fully released (**Figure 1b**; Pearson & Brawley 1996). Low intertidal and subtidal populations of *Sargassum* and *Cystoseira* develop deciduous receptacles or fronds with receptacles that bear reproductive conceptacles (**Figure 1b**). Fertilization appears to occur primarily while eggs are still attached to the parent plants by mucilage envelopes or mucilaginous stalks (Nanba 1995), and the fertilized eggs begin development on the outside of receptacles (Chapman 1995).

Microscopic Stages

Most of what is known about the development and potential longevity of the microscopic stages of kelp comes from laboratory studies. In culture, spores develop into gametophytes, gametophytes produce gametangia, and gametangia produce gametes with little or no cell division if light (including blue light in most species), temperature, and nutrients are sufficient, each female producing one or two eggs (Lewis 1995, Luning & Neushul 1978). Annual giant kelp populations in Chile are an exception, as female gametophytes become filamentous even in good growth conditions, and can produce more than two eggs (Muñoz et al. 2004). These developments and fertilization can occur in as little as 10 days (**Figure 1a**; Lewis 1995). Under suboptimal conditions gametophytes can divide and grow to become multicellular filaments, with some capable of surviving 16–20 months in complete darkness, and reproducing if returned to optimal growth conditions (tom Dieck 1993, Yoneshigue-Valentin 1990). The growth rate of microscopic sporophytes can also vary depending on culture conditions (Kinlan et al. 2003, Norton & Burrows 1969). In fucoids, the microscopic zygotes and young juveniles take up to several months to develop to macroscopic size under optimal conditions, and growth may be delayed under less favorable conditions (Ang 1991a, Gunnill 1980). The potential for delayed growth in early life stages has important implications for population replenishment, especially following disturbances (discussed below).

Macroscopic Stages

Although some kelps are annuals (e.g., *Alaria marginata*, McConnico & Foster 2005; *Postelsia palmaeformis*, Dayton 1973; *Nereocystis luetkeana*, Amsler & Neushul 1989), most by far are perennial (e.g., most *Laminaria* species, Kain 1963; *Macrocystis pyrifera* in the northeast Pacific, Neushul 1963). At least one species is biennial (*Laminaria angustata* var. *longissima*, Kawashima 1972). Some perennial species with annual growth rings in the stipe have maximum ages around 15 years (*Pterygophora californica*, Hymanson et al. 1990, *Laminaria hyperborea*, Kain 1979). Growth and maturation of perennial sporophytes vary considerably within and among species, but all species appear to produce spores during their first 1–2 years and every year thereafter under suitable conditions. Most kelp sporophytes occur as distinct individuals. If ramets occur (e.g., the fronds of giant kelp), they can be vegetative or spore-producing (sporophyll) blade-stipe modules that arise from a common holdfast (illustrations and phylogenetic relationships in Druehl & Saunders 1992). Sporophytes can grow very fast (5–40 cm day^{-1}; Kawashima 1972, McConnico & Foster 2005, North 1972).

The morphology of mature sporophytes varies considerably among species but Neushul (1972) pointed out that the morphology of 1- to 5-cm tall juveniles is very similar, all resembling immature *Laminaria* spp. (the macroscopic sporophyte in **Figure 1a** but without a sorus). This has a bearing on field studies because juveniles of different species can be difficult to distinguish in mixed stands. Cell divisions in the meristematic region between the stipes and blades lead to the production of thallus parts; with few exceptions (e.g., *Laminaria sinclairii*, Markham 1968), blades and stipes are not initiated from other parts of the thallus and cannot regenerate if the meristematic region is lost. In stipe-bearing kelps, severe grazing in this region can result in plant death (Andrew & Jones 1990).

Fucoids are almost all perennials but individual biomass may vary greatly because of losses from grazing and turbulence, or the annual loss of most of the thallus (e.g., *Sargassum*, Kendrick & Walker 1994) or of large reproductive fronds (e.g., *Cystoseira osmundacea*, Schiel 1985b). Individuals of many species live 5–10 years (Chapman 1995; Gunnill 1980, 1986), and some like *Ascophyllum nodosum* may live over 100 years (Åberg 1992b). However, longevity is unknown for most species. Growth in fucoids is initiated by one or a few meristematic cells at the ends of fronds, and vegetative cells can become meristematic (Graham & Wilcox 2000). Unlike most kelps, fucoids have the ability to regenerate from holdfasts or even holdfast fragments (Kendrick & Walker 1994, McCook & Chapman 1992), which no doubt reduces their mortality from grazing and other disturbances. In some species, therefore, a single individual may alternate between macro- and microscopic forms, a variation in size within one generation similar to that between the two generations in kelps (Regeneration in **Figure 1b**).

DEMOGRAPHICS

Microscopic Stages

There has been a recent surge of interest in population inputs and the factors responsible for settlement, recruitment, and population maintenance (e.g., Kinlan & Gaines 2003, Schiel 2004). In large brown seaweeds, this process involves the supply, survival, and growth of microscopic stages from the release of spores, eggs, or sperm through to visible recruits in field conditions. Despite the life history of kelps being known for over 90 years (Sauvageau 1915), in situ studies of the microscopic stages are relatively few. Such studies are inhibited by the small size of these stages, morphological similarity among kelps and between kelps and other taxa, and the myriad other microscopic organisms, sediment, organic particles, and low-lying algae that occur in the same microenvironments. Therefore, these microscopic stages have commonly been lumped together in a "black box," with inputs of spore abundance and outputs of visible "recruits." This approach can be informative (e.g., McConnico & Foster 2005) but cannot determine where key demographic changes occur. If, for example, abundant spore release is followed by poor sporophyte recruitment it could be due to poor spore survival in the water column; low spore settlement; poor gametophyte development, survival, or maturation; reduced fertilization success; or high mortality of microscopic sporophytes. Here we evaluate what is known about these stages of development and survival through them.

Kelps are extremely fecund, producing up to 10^8–10^{12} spores individual^{-1} year^{-1} (DeWreede & Klinger 1988, Joska & Bolton 1987, McConnico & Foster 2005). In a dense stand of subtidal giant kelp, Graham (2003) found up to 54,000 spores L^{-1} at 3 cm above the benthos. Stands of *Alaria marginata* synchronously release up to 10^9 spores individual^{-1} hr^{-1} and probably saturate the area beneath adults with spores (McConnico & Foster 2005). Spore release in annual populations generally occurs in late autumn before sporophytes die (Buschmann et al. 2004, Dayton 1973, McConnico & Foster 2005). Of the few perennial species studied, some (e.g., *Pterygophora californica*) have discrete periods of spore production, some produce spores continuously (e.g., *Laminaria digitata*), or continuously but with seasonal peaks (e.g., *Macrocystis pyrifera*, *Ecklonia maxima*; Chapman 1984, Joska & Bolton 1987, Reed et al. 1997). It is unknown whether the spores are equally viable in all seasons.

Survival from spore to juvenile sporophyte is very difficult to determine, but a few studies provide some general estimates. McConnico & Foster (2005) found that densities of about 10^{10} spores m^{-2} produced around 400 visible (>1 cm) sporophytes m^{-2} of *Alaria marginata* on an intertidal reef in central California. Assuming that half the spores were female and one egg is produced per female gametophyte, proportional survival would be 8×10^{-6}. The only estimate of survival from spore release to settlement is from Chapman (1984), who used sporangial densities in two perennial *Laminaria* species in Nova Scotia to estimate spore release over a year; he estimated spore settlement from the number of visible sporophytes on ceramic tiles placed monthly into kelp stands and then cultured in the laboratory. Each species released about 10^{10} spores m^{-2}, and proportional survival from release to settlement was

2×10^{-3} for *L. longicruris* and 10^{-4} for *L. digitata* (**Figure 2a**). Deysher & Dean (1986) seeded small ropes with giant kelp spores at a density of $\sim 3.3 \times 10^8$ spores m^{-2}, outplanted the ropes, and then counted the microscopic sporophytes produced after 42 days in the field. In repeated experiments, the highest microscopic sporophyte density was 5×10^6 m^{-2}. Reed (1990a) seeded rocks with varying densities of spores, outplanted the rocks, and counted the number of visible sporophytes of *Macrocystis pyrifera* and *Pterygophora californica* produced. His results were also highly variable, but the greatest survival at spore densities of 10^6 m^{-2} (1 spore mm^{-2}) and 10^8 m^{-2} resulted in sporophyte densities of 10^2 and 10^4 m^{-2}, respectively, for the species (**Figure 2a**). At spore densities >10 mm^{-2}, proportionately fewer female gametophytes produced sporophytes but there were far more recruits overall.

Fucoid propagules (zygotes and embryos) begin at a considerably larger size than those of kelps; an egg is 10–40 times greater in diameter than a typical kelp spore. Newly settled embryos and germlings can be identified readily in the field under low-power magnification (Brawley & Johnson 1991), and are visible to the unaided eye from when they are weeks (Gunnill 1980) to a few months old (Ang 1991a). Consequently, there is more information on the survival of microscopic stages of fucoids than kelps under a variety of environmental conditions.

Like kelps, by far the greatest mortality in fucoids is in the microscopic stages, as illustrated by a few species for which detailed data could be derived (**Figure 2b**). Schiel (1988) found that 2×10^6 ova of adult *Sargassum sinclairii* produced around 3×10^4 benthic settlers and 1.5×10^3 visible recruits (**Figure 2b**). Therefore, about 5% of settlers made it through to recruitment. Ang (1991a,b) estimated that *Fucus distichus* produced around 1.4×10^7 eggs m^{-2} years^{-1}, 30% of which were released from conceptacles; about 1–10% of eggs in receptacles became settlers and of these 0.4–12% became visible recruits (Chapman 1995). In one of the only published accounts of detailed survival of fucoid embryos to juvenile stages in the field, Dudgeon & Petraitis (2005) found that across numerous treatments, survival ranged from about 0.3–3.8% over 400 days (**Figure 2b**). Mortality was greatest in the first few days before attachment was secure, and average life expectancies of initial germlings were only 4–25 days. Even though *Ascophyllum nodosum* stands may produce 10^9 eggs m^{-2} years^{-1} (Åberg & Pavia 1997), Dudgeon & Petraitis (2005) estimated a minimum period of 13 years for an individual to replace itself. *Fucus gardneri* had similarly poor but highly variable survival of recently settled embryos, with only 2 of 5395 embryos surviving to visible recruits (Wright et al. 2004). These survival rates, although small, are still several orders-of-magnitude greater than for the microscopic stages of kelps (**Figure 2a** versus **2b**). Most deleterious interactions occur in these and the early postrecruitment juvenile stages and are discussed below.

Macroscopic Stages

There are few studies that assess the survival of macroscopic stages of kelps by following cohorts, and most were done at a single site for one cohort. Furthermore, studies are difficult to compare because the starting size (commonly referred to as "visible") has varied from 1–30 cm and mortality is usually highest in smaller plants. As pointed out by De Wreede (1986) and Chapman (1993), the lack of standardization, combined with sparse information on the demography of microscopic stages, density-dependent mortality, and size variation within age groups diminishes the generality and usefulness of some demographic analyses, especially matrix models.

Despite these problems, some informative general trends have emerged (**Figure 2a**). Most kelp cohorts had greater mortality when plants were small,

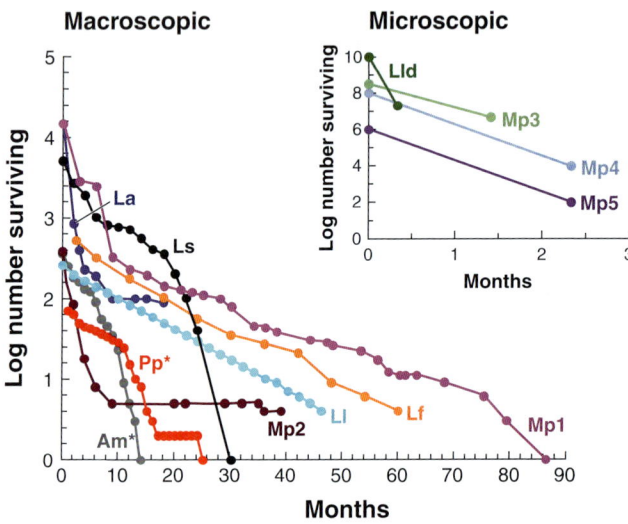

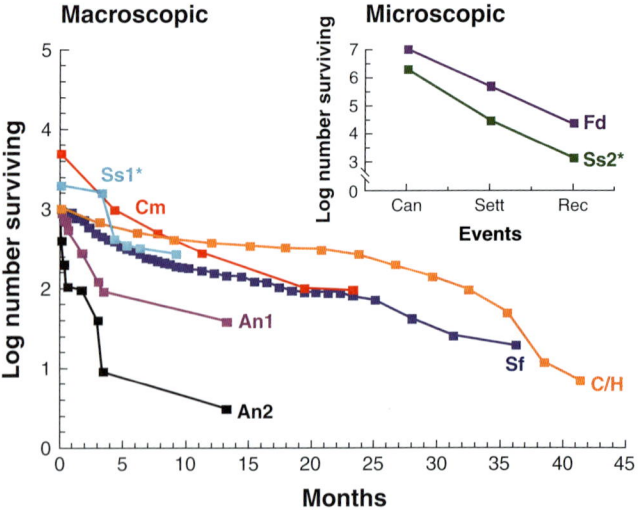

especially when initial densities were high. Black (1974), for example, found in an annual population of *Egregia laeviegata* that, at a natural density of 2000 juveniles m^{-2}, density-dependent mortality occurred as clusters of plants with intertwined holdfasts were removed in clumps by water motion. Mortality of giant kelp juveniles can also be density-dependent (Dean et al. 1989). Many of the survival curves indicate a roughly constant rate of mortality after juveniles begin to reach adult size, and some maintain this rate for all or a great portion of their lives. This may be followed by increased mortality of the oldest plants in some perennial populations (e.g., Mp1 in **Figure 2a**). In other perennial populations, mortality rates may decline as plants mature and dominate a cohort [**Figure 2a**; *M. pyrifera* (Mp2), *L. angustata*]. For giant kelp, there is considerable variation in the different segments of survivorship curves between sites (**Figure 2a**) and for larger plants within sites and among years (Dayton et al. 1992) and regions (Buschmann et al. 2004). Although variable, these data suggest that macroscopic sporophyte demography might best be understood if this portion of the life history is divided into three stages: small juveniles, large juveniles to young adults, and old adults. It is not surprising that very high sporophyte fecundity is needed to maintain dense stands of mature plants given the low survivorship of microscopic stages and young juveniles. The demographic patterns described above for various kelps are more similar to those of terrestrial ferns (Cousens 1988) than seed plants.

Studies of the dynamics of adult kelp stands composed of multiaged plants make up the bulk of the ecological literature on marine algae and have been reviewed thoroughly (Dayton 1985, Schiel & Foster 1986). However, their demographics have received far less attention. Surprisingly, the spatial distribution and density of stands within local sites can be relatively constant from year to year in annual populations, most likely the result of when spores are produced, their microhabitat requirements, and the facilitative effects of high adult densities on their microscopic stages, including

Figure 2

Survivorship curves for cohorts of kelps (*a*) and fucoids (*b*). Macroscopic cohorts began with juveniles 1–10 cm long for kelps and 1–3 cm for fucoids. Microscopic kelp cohorts began and ended with microscopic stages described below for each curve. Microscopic fucoid cohort abundance plotted against events rather than time: zygotes/embryos released at canopy height (Can), number settled (Sett), and number of recruits (Rec, 1–3 cm long). *denotes annual population. Kelps: Am, *Alaria marginata* (McConnico & Foster 2005); La, *Laminaria angustata* (Kawashima 1972); Lf, *L. farlowii* (Dayton et al. 1984); Ll, *L. longicruris* (Chapman 1986); Lld, *L. longicruris* and *digitata* (average, spores in water to spores settled; Chapman 1984); Ls, *L. saccharina* (Parke 1948); Mp1, *Macrocystis pyrifera* (Dayton et al. 1984); Mp2, *M. pyrifera* (Rosenthal et al. 1974); Mp3, *M. pyrifera* (settled spores to microscopic sporophytes; Deysher & Dean 1986); Mp4, *M. pyrifera* (high density of settled spores to visible sporophytes; Reed 1990a); Mp5, *M. pyrifera* (low density of settled spores to visible sporophytes; Reed 1990a); Pp, *Pelagophycus porra* (Coyer & Zaugg-Haglund 1982). Fucoids: An1, *Ascophyllum nodosum* (best survivorship; Dudgeon & Petraitis 2005); An2, *A. nodosum* (worst survivorship; Dudgeon & Petraitis 2005); C/H, *Cystoseira osmundacea/Halidrys dioica* (Gunnill 1986); Cm, *Carpophyllum maschalocarpum* (Schiel 1985a); Fd, *Fucus distichus* (Ang 1991b, Chapman 1995); Sf, *Silvetia fastigiata* (Gunnill 1980), Ss1 and Ss2, *Sargassum sinclairii* (Schiel 1985a).

reduction of potential competitors (Dayton 1973, McConnico & Foster 2005). It would be informative to know if similar consistency is maintained in low-density stands of these species. In a very thorough depletion study (survivorship of a "cohort" composed of all plants in a stand) of kelps, Chapman (1984) found replacement rates of 1–2 plants m^{-2} years^{-1} in stands of two perennial *Laminaria* spp. with densities of 1–3 plants m^{-2}. These densities are at the low end of the range of densities for mature, perennial, stipitate kelps (e.g., 0.3–7 m^{-2} for *Laminaria farlowii* (Dayton et al. 1984), 2–45 for *Ecklonia radiata* (Schiel & Choat 1980), 1.5–58 for *Pterygophora californica* (De Wreede 1984, Hymanson et al. 1990). This gradual loss and continual replacement was similar to that described for giant kelp at protected sites where densities of adult plants were about 0.1 m^{-2} (Graham et al. 1997). Adult giant kelp densities can, however, be highly variable within and among stands (Schiel & Foster 1986). Population age frequency distributions have been obtained for some long-lived kelps and used to intuit population dynamics (De Wreede 1984, Kain 1963, Klinger & De Wreede 1988). These show that age distributions can be unimodal (dominated by middle-aged or young plants), multimodal or stable, indicating that, as in giant kelp, different population dynamics occur within and among species.

As for kelps, there are few complete life tables for any fucoid species. Compiling these can be difficult because of the very high densities at which many populations occur, the entwinement of holdfasts and branching of plants that can make ramets difficult to distinguish from genets, and their often turbulent environment. Survival curves for several species (**Figure 2b**), therefore, are indicative only because there is such wide variation in survival depending on conditions (Chapman & Johnson 1990, Wright et al. 2004).

Fucoid life histories can be considered in four stages: (*a*) production of gametes, fertilization, and presettlement, (*b*) settlement and development through microscopic benthic phases, (*c*) the juvenile period when many density-dependent processes are manifested, and (*d*) mature adults. From the visible stage of recruitment (∼1 cm), plants generally appear to have an order-of-magnitude-greater survival than kelp recruits in their first year or two. For example, Gunnill (1980) generated survivorship curves for numerous cohorts of intertidal *Silvetia* (=*Pelvetia*) *fastigiata*. Over 700–1200 days, survival from visible recruitment ranged from about 1–24%, with 50% mortality typically occurring in the first 120 days and relatively little mortality after one year (**Figure 2b**). Similarly, mortality of a *Cystoseira/Halidrys* complex was initially high, then stabilized for a year before increasing again (Gunnill 1986). Subtidal, primarily annual *Sargassum sinclairii* that recruited at around 2000 m^{-2} had 13% survival after one year, when plants died back following reproduction (**Figure 2b**, Schiel 1985a). A perennial fucoid, *Carpophyllum maschalocarpum*, that recruited at 5000 m^{-2} had only 2% survival after two years (but still 95 plants m^{-2}) when plants became reproductive. For 15 cohorts of *Fucus distichus*, Ang (1991a) found that survival over one year ranged from about 5–10%, and after three years from 0.2–5%, with wide variation in the slopes of survival curves. Mortality was generally greater (40%) in the first two months than at later times. For the intertidal *Fucus gardneri*, monthly survival of large plants over one year averaged >90%, yielding about 28% survival after 1 year, but during periods of long daytime exposure the average monthly survival was <65%

(Wright et al. 2004). Some fucoids, such as the intertidal *Ascophyllum nodosum*, can be extremely long-lived as adults, and have little population turnover. Åberg (1992a,b) estimated that individuals can live 50–60 years where ice scour affects habitats and >120 years without ice effects. The greatest annual mortality of *A. nodosum* of about 60–80% occurred in the smallest recruits (Åberg 1992a).

Studies on both kelps and fucoids show that by far the greatest mortality occurs in the early life stages and that, except where grazing by macroinvertebrates such as sea urchins is intense (Chapman & Johnson 1990, Schiel & Foster 1986), the patterns of algal stand structure are largely determined in the microscopic early life stages. We evaluate some of the factors affecting these stages below.

LONG-LIVED MICROSCOPIC STAGES OR SEED BANKS?

Some "storage" of microscopic stages, often referred to as "banks" (Chapman 1986, Edwards 2000, Hoffmann & Santelices 1991), undoubtedly occurs in large brown algae but it is far different from the seed banks found in the terrestrial environment. In large brown algae, prolonged development of microscopic stages is critical to the population dynamics of annual species, because macroscopic plants may be absent for several months (Dayton 1973, McConnico & Foster 2005), and of perennial species, because of the often catastrophic loss of plants from grazing, severe water motion, and other abiotic stresses (Dayton et al. 1992, Underwood 1999). Unlike seeds, however, kelp and fucoid propagules have no protective structures [the exception is Saito's (1975) observation of *Undaria pinnatifida* gametophytes forming thick-walled "resting stages" at high temperatures] or known innate dormancy, are far more transitory, and the surrounding medium offers none of the protection afforded seeds in terrestrial soils. Rather than true dormancy, algal propagules simply appear able to delay or prolong development when conditions are unfavorable.

In culture, kelp gametophytes can be long-lived and reproduce if returned to optimal conditions, and the growth of microscopic sporophytes can be delayed (discussed earlier). However, there is little evidence that this occurs in the field (for a possible exception, see Ladah et al. 1999). Delayed development can occur in the embryo or germling stages of fucoids in very cold water (Worm et al. 1999) or when other seasonal conditions are unfavorable (Clayton 1988, DeWreede 1986). Most species, however, probably go through their microscopic stages within a few months in the field. Deysher & Dean (1986), for example, found that densities of microscopic sporophytes on outplanted substrates seeded with giant kelp spores peaked at about 40 days and then declined. Given the numerous intrinsic and extrinsic factors that affect early life stages, including grazing, scour, sediment, temperature stress, shading, and overgrowth (Vadas et al. 1992), the ability to persist in adverse benthic conditions is probably not great.

At present, it would seem best to understand better how long these microscopic stages may persist in nature, the relative contributions of older versus younger stages to juvenile plant abundance and, after disturbances of various degrees, how these contribute to population replenishment and persistence in such variable environments. A stretched and perhaps inappropriate analogy to terrestrial seed banks could inhibit more scientifically fruitful investigations that simply consider these algal stages as what they are, microscopic plants.

DENSITY-DEPENDENT EFFECTS

Kelps and fucoids have complex density-dependent relationships and, as for terrestrial plants, most of the literature in this area involves thinning and growth processes of postrecruitment stages across densities. From an early paper showing positive effects of density on growth and reproduction of postrecruits in a kelp and a fucoid, and comparisons with terrestrial plants (Schiel & Choat 1980), the role of density-dependence in seaweeds has been controversial, with varying degrees of positive and negative effects being reported (Ang & Dewreede 1992, Cousens & Hutchings 1983, Creed et al. 1998, Kendrick 1994, Neushul & Harger 1985, Reed 1990b, Scrosati 2005). However, the demographic and population consequences of different densities are far from clear in most species, particularly when entire life histories are considered. The dominance and suppression effects seen in terrestrial plants occur at high densities in algae (Ang & DeWreede 1992, Reed 1990b), but there are positive effects of high densities that are often ignored. Elevated local densities may increase dispersal distances and gametophyte densities (discussed below), enhance fertilization (Pearson & Brawley 1996), reduce physical stress in intertidal habitats by retaining moisture (Johnson et al. 1998), dampen water motion (Schiel 1985b), reduce per capita loss from fish (Dayton & Tegner 1984) and sea urchin grazing (Harrold & Reed 1985, Konar 2000, Mattison et al. 1977), reduce juvenile sea urchin densities (Andrew & Choat 1985), and suppress recruitment and growth of competitors (Choi & Norton 2005, Reed 1990b).

A major distinction between terrestrial plants and large brown algae is that high densities of the latter are generally required to achieve fertilization. In kelps, high adult densities lead to increased spore densities, and peaks in the latter occur when adult densities are high and spore release is synchronous (McConnico & Foster 2005, Reed et al. 1997). Such high densities may be critical because fertilization requires male and female gametophytes to be in very close proximity. High spore densities, in combination with increased water motion, also increase the probability of long-range dispersal (Gaylord et al. 2002, Reed et al. 2006). Conversely, when water motion is severe and plants sparse, dilution of spores may greatly compromise recruitment processes (Gaylord et al. 2002). Fucoids require critical sperm concentrations for high fertilization rates (Berndt et al. 2002). These may be achieved in some species through retention of gametes by mucilage on parent conceptacles, thereby enhancing chances of successful fertilization in dense stands of dioecious species (Pearson & Brawley 1996) and perhaps outbreeding in monoecious species. Fertilization is also enhanced in some fucoids by the release of gametes during calm seas, a process that can be turned on and off over remarkably short time periods (Pearson & Brawley 1996). If generally true, such mechanisms may be important to the turnover and persistence of populations in a wide range of turbulent conditions.

As for the macroscopic stages, numerous negative effects of high density have been documented for germlings (Choi & Norton 2005, Steen & Scrosati 2004), but considerably more work is needed to understand density-dependent processes from gamete release to settlement and recruitment, and how these vary among species. The retention of moisture and amelioration of desiccation in dense aggregations (Johnson et al. 1998), elongation similar to the etiolation of terrestrial plants (Reed 1990b), and the quick formation of closed canopies may allow some species to compete more effectively with others (Choi & Norton 2005). It may well be the case that given their life histories, kelps and fucoids may require high adult densities to persist, and that negative effects such as inbreeding from self-fertilization (Raimondi et al. 2004) and thinning as populations develop are consequences of requirements of their early life stages.

Dispersal

The ability of species to disperse and replenish marine populations is an important area of research receiving much attention by ecologists (e.g., Kinlan & Gaines 2003). It was long thought from observations of recruitment around isolated kelp stands that spore dispersal was limited to a few meters, and this seems to be true for *Postelsia palmaeformis* based on recruit distribution (Dayton 1973) and genetic relatedness (Coyer et al. 1997). Reed et al. (1988) and subsequent studies by Gaylord et al. (2002), however, showed that propagules of the kelp *Macrocystis pyrifera* and the fucoid *Sargassum muticum* can disperse >1 km, and genetic data indicate long-distance dispersal in the kelp *Alaria marginata* (Kusumo & Druehl 2000). Such dispersal depends on the intensity of turbulence of waves, the sinking speed of propagules and their release height above the substratum. Gaylord et al. (2002) noted that the dominant feature distinguishing passive dispersal on land from that in the sea is the density of the surrounding medium. Air is far less dense than seeds, but the densities of seawater and macroalgal propagules may differ by only a few percent (Denny 1993). Therefore, algal propagules tend to sink slowly and can be carried great distances before reaching the sea floor.

Spores likely behave as passive particles at scales greater than a few millimeters in moving water (Graham 2003). They lack cell walls, are photosynthetic, and have nutritional reserves that enable them to swim for up to five days in the light (Clayton 1992, Reed et al. 1992). However, they do not have an obligate planktonic period and may settle, attach, and begin development immediately after release (reviewed in Santelices 1990). Spores have behaviors that aid in settlement; they can be positively chemotactic to nutrients, are stimulated to settle by them (Amsler & Neushul 1990), and may exhibit thigmo- and rheotaxes (Fletcher & Callow 1992, Santelices 1990). These abilities are probably effective only at scales of a few millimeters or less under typical water motion conditions because spores are tiny and water is viscous to them (Gaylord et al. 2002). They may serve mainly to concentrate spore settlement in favorable microsites and increase the success rate of gametophyte development and fertilization. Spore attachment processes, especially how they vary among species, substrates, and water motion conditions, are largely unknown, but knowledge of them should provide valuable insights into the nature of microenvironmental settlement sites.

Gaylord et al. (2002) distinguished between colonization potential and dispersal potential. Where a minimum density of kelp spores is required for fertilization, the decline in density with distance may obviate effective colonization even if spores disperse far. A model by Kinlan et al. (2005) shows that algal propagules have far less ability for long-distance dispersal than other taxa such as invertebrates and fish, mainly because of their very short planktonic period. They comment that this is at odds with the known rapid spread of many algal species, to which other mechanisms must contribute. Reed et al. (2006) argued that dense colonization at moderate distances (kilometers to tens of kilometers) is more likely to result from the dispersal of spores from large aggregations of attached plants (i.e., kelp beds) than from a few drifting plants. This was shown by colonization patterns on an artificial reef (Reed et al. 2004). Although kelp drifting at the sea surface may produce viable spores (Macaya et al. 2005), the spores are so small and sink so slowly that the probability they will reach the sea bed in sufficient densities to colonize is likely to be low. Furthermore, severe inbreeding depression (Raimondi et al. 2004) may compromise survival and fecundity of plants resulting from drift dispersal of isolated individuals. Most kelps may rely primarily on spore dispersal via water motion to establish populations.

In all fucoids studied, zygotes or germlings appear to be the primary dispersal phase, but can disperse while still attached to detached, drifting, fertile parts of parental plants (Deysher & Norton 1982, Schiel 1985a). Fucoids, therefore, may be generally more effective than kelps at dispersing long distances and colonizing through rafting of reproductive material (**Figure 1b**, reproductive frond release). Unlike many kelps, most fucoids float or are neutrally buoyant (e.g., Norton & Mathieson 1983), and their propagules need only to attach and grow after release. Zygotes and germlings produce adhesives that aid in attachment to the substrate (Fletcher & Callow 1992). Some species, such as *Sargassum muticum*, *S. sinclairii*, and *Cystoseira osmundacea*, have fronds bearing reproductive structures that dehisce seasonally and float along the coast (Chapman 1995, Norton 1992). Some propagules may be released before reproductive structures become planktonic, enabling colonization of both local and distant sites (Deysher & Norton 1982, Schiel 1985a). Given the much larger mass of fucoid propagules relative to kelp spores, however, and that many fucoids release them near the substratum (e.g., *Fucus* spp.), it seems likely that propagules produced by attached fucoids do not travel far from parent plants (e.g., Johnson & Brawley 1998). Where large tracts of fucoids have been removed, for example, it can take many years for populations to recover (Jenkins et al. 1999, 2004; Underwood 1999).

DISTURBANCE AND ENVIRONMENTAL EFFECTS

Coastal marine environments are characterized by disturbances of different intensities and frequencies, from waves over a matter of seconds, to seasonal storm events and temperature fluctuations, episodic grazing, El Niño periods over years, and decadal oscillations (e.g., Dayton & Tegner 1984). These have numerous effects on communities, but most importantly on the demographic characteristics of growth, reproduction, and survival of species. Furthermore, the marine environment varies by tidal height, depth, and boundary-layer effects near the substratum. Here we consider some of the intrinsic and extrinsic factors highlighted by Vadas et al. (1992) that act on different life stages and still require work. The emphasis is on the early life stages before visible recruits appear. These stages are where most mortality occurs, and the combination of favorable circumstances define the "windows" of recruitment that maintain populations.

Water Motion

Water motion has long been recognized as having a major influence on the distribution and abundance of seaweeds (Lewis 1968). The removal of large stands during storms and resulting piles of wrack on the shore are the most spectacular manifestation of this, but there is a broad correlation between species distributions and wave exposure along coastlines (e.g., Bustamante & Branch 1996). Water motion has numerous documented effects on adults (Schiel & Foster 1986) and, no doubt, early life stages of kelps and fucoids. The latter are in the early stages of investigation. For many fucoids there is an apparent paradox of living in highly turbulent shallow environments and yet having relatively poor ability to fertilize eggs and attach in such conditions. This is solved by at least some fucoids by releasing gametes only in calm conditions. Serrão et al. (1996) found, for example, that *Fucus vesiculosus* had a high release of gametes and about 100% fertilization when water velocities were below 0.2 m s^{-1} and that greater water motion inhibited release by *F. distichus* and *Silvetia* (= *Pelvetia*) *fastigiata*. These results are supported by other studies on fucoids (Pearson & Brawley 1996), but it is still unclear the extent to which some species release gametes at low tide or if other mechanisms such as buoyancy by mucilage (Clayton 1992) allows effective release in turbulent conditions. Furthermore, the timing required for fertilization, only a matter of a few hours, may be inadequate for secure attachment of zygotes. Vadas et al. (1990) showed that even a low-velocity wave removed 99% of 15-minute old zygotes of *Ascophyllum nodosum* from experimental tiles. Even with refuges and postsettlement times up to 4 hours, 75–90% of zygotes were removed by a single wave. Beyond current speeds of 20 cm s^{-1}, attachment of this species was poor, whereas another fucoid, *Fucus evanescens*, attached more securely (Vadas et al. 1992). Similarly, Taylor & Schiel (2003) showed that the ability to attach across wave gradients was broadly correlated with the distribution of adult populations. The sheltered coast fucoid *Hormosira banksii* could attach effectively only in the calmest conditions and when zygotes remained undisturbed on the substratum for 12 hours. Only 5–8% of zygotes remained on exposed shores after a single tidal cycle. In contrast, the exposed coast species *Durvillaea antarctica* had >90% successful attachment in all conditions, even when zygotes were just an hour old. A fucoid characteristic of intermediate conditions, *Cystophora torulosa*, had intermediate survival in all conditions.

Fucoid zygotes attach initially by mucilage before rhizoids are formed to anchor them securely to substrata, a process that can take a couple of days (Clayton 1992, Pearson & Brawley 1996). Because of the time necessary for effective attachment, therefore, there is a very narrow window of opportunity during the calmest of conditions for some species to become established. This probably applies mostly in intertidal and very shallow subtidal environments where constant wave impacts are greatest. It may also apply to kelps as well as fucoids; Gaylord et al. (2002) found that spore attachment of giant kelp declined 30-fold between velocities of 15–25 cm s^{-1}. The processes involved in secure attachment and their role in the demography of species across environmental gradients is a fruitful area for further investigation.

Canopies

The suppressive effects of algal canopies on intra- and interspecific recruitment of large brown algae has been shown in many studies; removing algal canopies usually is followed by a burst of recruitment of many species owing to an increase in light (Reed & Foster 1984). Light reduction is exacerbated for microscopic stages on the sea bed where many low-lying species occur. In intertidal areas, however, direct exposure to sunlight and associated temperature stress can affect seaweeds adversely (e.g., Dethier et al. 2005), so the presence of a canopy can also be positive.

Tall algal canopies and dense patches of much shorter, highly branched species (turfs) can modify relationships between settlement of propagules and recruitment. Johnson & Brawley (1998) found that orders-of-magnitude-more settlement occurred beneath rather than just outside canopies of *Pelvetia compressa*, but subsequent recruitment was three times more common outside. Most recruits occurred in red algal turf, especially when compared to bare rock where no recruits were found. McConnico & Foster (2005) found greatest recruitment of the kelp *Alaria marginata* in patches of coralline algal turf and on residual *A. marginata* holdfasts even though spores probably saturated numerous other available benthic habitats. In the absence of tall canopies, juvenile kelp survivorship may be facilitated by subtidal algal turfs that protect the kelps from fish grazing (Harris et al. 1984). A disconnection between settlement and recruitment is probably common, not only because of suppression or facilitation by canopies but through interactions with grazers and other features of the benthic environment (e.g., Chapman & Johnson 1990). Given the generally high levels of disturbance to algal communities and the highly localized dispersal noted in most large algae, it seems that saturating the parental environment with propagules is an essential life history attribute.

Grazing, Sediments, and Nutrients

The effect of grazers on algal assemblages has been one of the most studied aspects of early postsettlement mortality (Vadas et al. 1992) and provides the basis for "top-down" models of trophic control of algal assemblages. Most knowledge about the interactions of consumers and resources in controlling algal communities comes from work on adult stages. Numerous studies show that grazing can exert strong localized control of algal community structure through selective or complete removal of large algae, but it is also highly variable in space and time (Chapman & Johnson 1990, Schiel & Foster 1986). Furthermore, grazing may interact with water motion, sedimentation, and many other variables, perhaps especially acting on the highly vulnerable early life stages on the substratum (Bertness et al. 2002), which are prey to a wide range of grazers (e.g., small crustaceans, snails, sea urchins), and potentially unfavorable nutrient and sediment conditions near the boundary layer. The recent discovery of kelp gametophytes living in situ as endophytes within the cell walls of other algae (Garbary et al. 1999) is significant as it suggests a possible means of protection from some adverse microenvironmental effects.

Sedimentation is potentially one of the most pervasive factors affecting colonization and survival of early life stages of marine algae (Airoldi 2003). Devinny & Volse (1978), in one of the few studies of sediment effects on spores and gametophytes, found that sediment concentrations of only 10 mg cm^{-2} could prevent giant kelp spores from settling and developing successfully, probably because they settled on unstable sediment grains. At a much larger scale, it is likely that the massive loss of giant kelp forests along the southern California coast in the late 1950s and 1960s was caused in part by increased sedimentation associated with sewer outfalls (Grigg & Kiwala 1970). Sediment chemistry and smothering can greatly affect the growth and survival of fucoid germlings, especially in organically rich deposits (Chapman & Fletcher 2002). However, the greatest effect of sediment is probably to prevent attachment, as shown by studies on kelp spores. Even fine sediments can reduce attachment of fucoids by >90% relative to controls, and there can be species-specific differences in the ability to attach to primary substrata in the presence of fine sediments (Schiel et al. 2006). On a large scale, changes in sedimentation have resulted in long-term changes in the distribution and abundance of fucoid species in the Baltic Sea (Kautsky et al. 1986). Seasonal influxes of sand can alter communities through differential tolerances of species to abrasion and burial (D'Antonio 1986), but the effects of sedimentation may be ameliorated by grazers in some situations (Bertness 1984).

Nutrients can have large effects on the dynamics of adult kelp populations, particularly at range margins where nutrients are variable (reviewed in Schiel & Foster 1986). At smaller spatial scales, a nutrient enrichment (N-P-K) and grazing experiment by Lotze et al. (2001) showed a grazing effect but no nutrient effect on *Fucus* germlings, although grazers caused a shift in dominance of benthic species that was enhanced by an increase in nutrients. Steen (2004) found in lab experiments that there was a significant interaction between temperature, nutrients, and competitor density on the survival of *F. serratus* and *F. evanescens* germlings. At a high temperature and a high level of nutrients (nitrate and ammonia) the green alga *Entermorpha* outgrew the fucoids, potentially displacing them in eutrophic conditions. However, in another study, grazers had a positive effect on the establishment of *Fucus* species by reducing benthic algae such as *Enteromorpha* (Worm et al. 1999). In other experiments on *F. serratus* and *F. evanescens*, Steen & Scrosati (2004) concluded that among even-aged germlings, nutrient competition was more important than competition for light in early life stages, especially at higher temperatures when fast growth increases nutrient demands. At larger scales, light attenuation from coastal eutrophication can potentially eliminate deep water kelp populations (Spalding et al. 2003).

In one of the earliest examples of interactive effects of environmental factors on recruitment windows for kelp in the field, Deysher & Dean (1986) showed that high recruitment of giant kelp in southern California occurred only when temperatures were lower than 16.3°C and light levels greater than 0.4 Einsteins m^{-2} day^{-1}, a coincidence that happens only occasionally in this environment. Artificially adding nutrients increased recruitment only in the late summer when natural temperatures were higher than 16°C and nutrients were low.

CONCLUSIONS

The recent ecological literature shows a welcome increase in studies that focus on early life stages and inputs to algal populations. These studies suggest that although there are interesting analogies between large brown algae and seed plants, differences between their life histories and environments require different conceptual approaches to understanding their population ecology. In both fucoids and kelps, the microscopic stages are morphologically distinct from diploid adults, exist in environments in or near the boundary layer that are very different from those of adults, can be tremendously abundant near dense stands of adults, and are hugely variable in their survival to recruit and juvenile stages. The "bottlenecks" of these stages are only beginning to be understood, and for relatively few of the dominant species on rocky reefs worldwide. These microscopic stages are directly dependent on dispersal and settlement of propagules, processes that also require much further investigation. Current knowledge also suggests interesting and important links between the density of macroscopic plants and the distribution, abundance, and survivorship of microscopic plants. Further progress in all these areas requires innovative techniques that cross ecological, phycological, physical, and chemical disciplines to be applied in field situations.

Despite this emerging literature, most models involving the dynamics of stands of large brown algae are based on trophic control through top-down processes. Although the generality of these models has been questioned (Elner & Vadas 1990, Foster 1992, Foster & Schiel 1988), they continue to be applied with little consideration of alternatives (e.g., Jackson et al. 2001). There is considerable evidence that even with great reductions in major predators, climatic and related physical and nutrient effects still have the greatest overall impacts on many kelp forests (Dayton et al. 1998). At this stage, it is difficult to resolve the parity of the scales of these different processes (Dayton & Tegner 1984). All act locally on algal stands but their pervasiveness along coastlines and with depth can be very different. These influences are not mutually exclusive but operate in different frames of time and space, most dramatically focussing on large losses rather than microscopic inputs. Understanding the feedbacks between these highly variable processes and reproductive output, dispersal, and survival through recruitment still requires considerable work.

Kelp and fucoid stands are affected globally by degradation of coastal habitats, especially through increased sedimentation and eutrophication, and potentially through seawater warming (Schiel et al. 2004). There is an increasing social mandate for science to help solve these problems (Hughes et al. 2005). Theory provides a context within which research can focus questions, but it is comparatively barren if it ignores the individual characteristics of important foundation species, their demographies, and how they make their living in complex and dynamic coastal environments.

SUMMARY POINTS

1. The population biology of large brown seaweeds is usually absent from models of algal community structure, yet the differing life histories, ecology, and demographics of kelps and fucoids from spores/gametes, through microscopic stages to settlement, recruitment, and maturity are critical to the establishment and persistence of algal stands.

2. Terrestrial plant models are not well suited to kelps and fucoids, which have life histories with free-living microscopic stages, produce enormous quantities of spores and gametes with little documented cost, have rapid growth rates and maturity, relatively small dispersal distances, short life spans, no known animal fertilization or dispersal vectors, and no true "seed banks" but rather delayed development for usually no more than a few months.

3. In comparison to kelps, fucoids often occur in highly turbulent nearshore environments, may release gametes during calm conditions when they are more likely to fertilize and attach, have no obligate period of development to fertilization after release from parent plants, produce far fewer, much larger propagules than kelps, and often have orders-of-magnitude-better survival through early life stages.

4. Density-dependent factors are critical for the establishment and maintenance of both kelps and fucoid stands, especially in the close proximity of gametes required for fertilization, dispersal processes that may require high densities of spores/gametes to establish populations, and feedbacks between adult densities, reproductive output, settlement, and recruitment.

5. Recruitment windows in both kelps and fucoids require the coincidence of critical levels of several conditions, especially of nutrients, light, temperature, and water motion, but are often limited by sedimentation, grazing, degree of canopy development, and microsite conditions.

FUTURE ISSUES

1. Dispersal: What are the feedbacks between generations and over what scales? To what degree do metapopulations exist? What barriers exist and over what scales do they operate? How do these relate to propagule movement, attachment, and water motion? Under what conditions can microstages be long-lived in nature? What can population genetics reveal about population isolation and connectivity?

2. Density-dependent effects on early life stages: What are the critical densities for effective settlement, fertilization, and recruitment and how do they vary among species? How do these relate to adult stand densities, reproductive timing and output, and dispersal distances?

3. Boundary layer effects: What critical interactions (physical, chemical, biological) occur at the level of microsites, and how do these scale up to wider processes to produce recruitment?

4. Species invasions: What critical features of life history stages, propagule distribution, and environments enable kelps and fucoids to get traction and spread in new habitats or, conversely, resist invasion by other species?

5. Multiple stressors: How do changing coastal environments, including nutrients, temperature, light, and water motion affect population dynamics through stage-specific processes?

ACKNOWLEDGMENTS

Our sincere thanks to R. DeWreede, M. Graham, D. Harvell, L. McConnico, D. Reed, and M. Winterbourn for many helpful comments on the manuscript, and to L. McMasters for preparing the figures. D.R.S. thanks the Andrew Mellon Foundation of New York and the New Zealand Foundation for Research, Science and Technology for supportive funding.

LITERATURE CITED

Abbott IA, Kurogi M, eds. 1972. *Contributions to the Systematics of Benthic Marine Algae of the North Pacific*. Kobe: Jpn. Soc. Phycol.

Åberg P. 1992a. A demographic study of two populations of the seaweed *Ascophyllum nodosum*. *Ecology* 73:1473–87

Åberg P. 1992b. Size-based demography of the seaweed *Ascophyllum nodosum* in stochastic environments. *Ecology* 73:1488–501

Åberg P, Pavia H. 1997. Temporal and multiple spatial scale variation in juvenile and adult abundance of the brown alga *Ascophyllum nodosum*. *Mar. Ecol. Prog. Ser.* 158:111–19

Airoldi L. 2003. The effects of sedimentation on rocky coast assemblages. *Oceanogr. Mar. Biol. Ann. Rev.* 41:161–236

Amsler CD, Neushul M. 1989. Diel periodicity of spore release from the kelp *Nereocystis luetkeana* (Mertens) Postels *et* Rupreecht. *J. Exp. Mar. Biol. Ecol.* 134:117–27

Amsler CD, Neushul M. 1990. Nutrient stimulation of spore settlement in the kelps *Pterygophora californica* and *Macrocystis pyrifera*. *Mar. Biol.* 107:297–304

Andrew NL, Choat JH. 1985. Habitat related differences in the survivorship and growth of juvenile sea urchins. *Mar. Ecol. Prog. Ser.* 27:155–61

Andrew NL, Jones GP. 1990. Patch formation by herbivorous fish in a temperate Australian kelp forest. *Oecologia* 85:57–68

Ang PO. 1991a. Age- and size-dependent growth and mortality in a population of *Fucus distichus*. *Mar. Ecol. Prog. Ser.* 78:173–87

Ang PO. 1991b. Natural dynamics of a *Fucus distichus* (Phaeophyceae, Fucales) population: reproduction and recruitment. *Mar. Ecol. Prog. Ser.* 78:71–85

Ang PO, De Wreede RE. 1992. Density-dependence in a population of *Fucus distichus*. *Mar. Ecol. Prog. Ser.* 90:169–81

Berndt M, Callow JA, Brawley SH. 2002. Gamete concentrations and timing and success of fertilization in a rocky shore seaweed. *Mar. Ecol. Prog. Ser.* 226:273–85

Bertness MD. 1984. Habitat and community modification by an introduced herbivorous snail. *Ecology* 65:370–81

Bertness MD, Trussell GC, Ewanchuk PJ, Silliman BR. 2002. Do alternate stable states exist in the Gulf of Maine rocky intertidal zone? *Ecology* 83:3434–48

Black R. 1974. Some biological interactions affecting intertidal populations of the kelp *Egregia laevigata*. *Mar. Biol.* 28:189–98

Brawley SH, Johnson LE. 1991. Survival of fucoid embryos in the intertidal zone depends upon developmental stage and microhabitat. *J. Phycol.* 27:179–86

Buschmann AH, Vásquez JA, Osorio P, Reyes E, Filún L, et al. 2004. The effect of water movement, temperature and salinity on abundance and reproductive patterns of *Macrocystis* spp. (Phaeophyta) at different latitudes in Chile. *Mar. Biol.* 145:849–62

Bustamante RH, Branch GM. 1996. Large scale patterns and trophic structure of southern African rocky shores: the role of geographic variation and wave-exposure. *J. Biogeogr.* 23:339–51

Chapman ARO. 1984. Reproduction, recruitment and mortality in two species of *Laminaria* in southwest Nova Scotia. *J. Exp. Mar. Biol. Ecol.* 78:99–109

Chapman ARO. 1986. Age versus stage: an analysis of age- and size-specific mortality and reproduction in a population of *Laminaria longicruris* Pyl. *J. Exp. Mar. Biol. Ecol.* 97:113–22

Chapman ARO. 1993. 'Hard' data for matrix modelling of *Laminaria digitata* (Laminariales, Phaeophyta) populations. *Hydrobiologia* 260/261:263–67

Chapman ARO. 1995. Functional ecology of fucoid algae: twenty-three years of progress. *Phycologia* 34:1–32

Good review of the biology and ecology of fucoid algae.

Chapman ARO, Johnson CR. 1990. Disturbance and organization of macroalgal assemblages in the Northwest Atlantic. *Hydrobiologia* 192:71–121

Chapman AS, Fletcher RL. 2002. Differential effects of sediments on survival and growth of *Fucus serratus* embryos (Fucales, Phaeophyceae). *J. Phycol.* 38:894–903

Choi HG, Norton TA. 2005. Competitive interactions between two fucoid algae with different growth forms, *Fucus serratus* and *Himanthalia elongata*. *Mar. Biol.* 146:283–91

Clayton MN. 1988. Evolution and life histories of brown algae. *Bot. Mar.* 31:379–87

Clayton MN. 1992. Propagules of marine macroalgae: structure and development. *Br. Phycol. J.* 27:219–32

Cousens MI. 1988. Reproductive strategies of pteridophytes. In *Plant Reproductive Ecology*, ed. JL Doust, LL Doust, pp. 307–28. Oxford, UK: Oxford Univ. Press

Cousens R, Hutchings MJ. 1983. The relationship between density and mean frond weight in monospecific seaweed stands. *Nature* 301:240–41

Coyer JA, Olsen JL, Stam WT. 1997. Genetic variability and spatial separation in the sea palm kelp *Postelsia palmaeformis* (Phaeophyceae) as assessed with M13 fingerprints and RAPDS. *J. Phycol.* 33:561–68

Coyer JA, Zaugg-Haglund AC. 1982. A demographic study of the elk kelp, *Pelagophycus porra* (Laminariales, Lessoniaceae), with notes on *Pelagophycus* x *Macrocystis* hybrids. *Phycologia* 21:399–407

Creed JC, Kain JM, Norton TA. 1998. An experimental evaluation of density and plant size in two large brown seaweeds. *J. Phycol.* 34:39–52

D'Antonio CM. 1986. Role of sand in the domination of hard substrata by the intertidal alga *Rhodomela larix*. *Mar. Ecol. Prog. Ser.* 27:263–75

Darwin C. 1860. *The Voyage of the Beagle*. Garden City, NY: Doubleday. (Reprinted in 1962)

Dayton PK. 1972. Toward an understanding of community resilience and the potential effects of enrichments to the benthos at McMurdo Sound, Antarctica. In *Proceedings of the Colloquium on Conservation Problems in Antarctica*, ed. BC Parker, pp. 81–94. Lawrence, KS: Allen

Dayton PK. 1973. Dispersion, dispersal and persistence of the annual intertidal alga *Postelsia palmaeformis*. *Ecology* 54:433–38

Dayton PK. 1985. *Ecology of kelp communities*. *Annu. Rev. Ecol. Syst.* 16:215–45

Dayton PK, Currie V, Gerrodette T, Keller BD, Rosenthal R, Ven Tresca D. 1984. Patch dynamics and stability of some California kelp communities. *Ecol. Monogr.* 54:253–89

Dayton PK, Tegner MJ. 1984. The importance of scale in community ecology: a kelp forest example with terrestrial analogs. In *A New Ecology: Novel Approaches To Interactive Systems*, ed. PW Price, CN Slobodchikoff, WS Gaud, pp. 457–81. New York: Wiley

Dayton PK, Tegner MJ, Edwards PB, Riser KL. 1998. Sliding baselines, ghosts, and reduced expectations in kelp forest communities. *Ecol. Appl.* 8:309–22

> Thoughtful perspective on kelp forest ecology and perceptions of structuring processes.

Dayton PK, Tegner MJ, Parnell PE, Edwards PB. 1992. Temporal and spatial patterns of disturbance and recovery in a kelp forest community. *Ecol. Monogr.* 62:421–45

Dean TA, Thies K, Lagos SL. 1989. Survival of juvenile giant kelp: the effects of demographic factors, competitors, and grazers. *Ecology* 70:483–95

Denny MW. 1993. *Air and Water: The Biology and Physics of Life's Media*. Princeton: Princeton Univ. Press

Dethier MN, Williams SL, Freeman A. 2005. Seaweeds under stress: manipulated stress and herbivory affect critical life-history functions. *Ecol. Monogr.* 75:403–18

Devinny JS, Volse LA. 1978. Effects of sediments on the development of *Macrocystis pyrifera* gametophytes. *Mar. Biol.* 48:343–48

De Wreede RE. 1984. Growth and age class distribution of *Pterygophora californica* (Phaeophyta). *Mar. Ecol. Prog. Ser.* 19:93–100

De Wreede RE. 1986. Demographic characteristics of *Pterygophora californica* (Laminariales, Phaeophyta). *Phycologia* 25:11–17

De Wreede RE, Klinger T. 1988. Reproductive strategies in algae. See Doust & Doust 1988, pp. 267–84

Deysher L, Norton TA. 1982. Dispersal and colonization in *Sargassum muticum* (Yendo) Fensholt. *J. Exp. Mar. Biol. Ecol.* 56:179–95

Deysher LE, Dean TA. 1986. In situ recruitment of sporophytes of the giant kelp, *Macrocystis pyrifera* (L.) C.A. Agardh: effects of physical factors. *J. Exp. Mar. Biol. Ecol.* 103:41–63 — Thorough, innovative field study that defined recruitment windows in giant kelp.

Doust JL, Doust LL, eds. 1988. *Plant Reproductive Ecology*. Oxford: Oxford Univ. Press

Druehl LD, Mayes C, Tan IH, Saunders GW. 1997. Molecular and morphological phylogenies of kelp and associated brown algae. *Plant Syst. Evol.* 11:221–35

Druehl LD, Saunders GW. 1992. Molecular explorations in kelp evolution. *Prog. Phycol. Res.* 8:47–83

Dudgeon SR, Petraitis PS. 2005. First year demography of the foundation species, *Ascophyllum nodosum*, and its community implications. *Oikos* 109:405–15 — Good experimental demographic study of early life stages of a fucoid.

Edwards MS. 2000. The role of alternate life-history stages of a marine macroalga: a seed bank analog? *Ecology* 81:2404–25

Elner RW, Vadas RL. 1990. Inference in ecology: the sea urchin phenomenon in the northwestern Atlantic. *Am. Nat.* 136:108–25

Estes JA, Duggins DO. 1995. Sea otters and kelp forests in Alaska: generality and variation in a community ecological paradigm. *Ecol. Monogr.* 65:75–100

Fagerstrom T, Westoby M. 1997. Population dynamics in sessile organisms: some general results from three seemingly different theory-lineages. *Oikos* 80:588–94

Fletcher RL, Callow ME. 1992. The settlement, attachment and establishment of marine algal spores. *Br. Phycol. J.* 27:303–29

Foster MS. 1992. How important is grazing to seaweed evolution and assemblage structure in the north-east Pacific? In *Plant-Animal Interaction in the Marine Benthos*, ed. DM John, SJ Hawkins, JH Price, pp. 61–85. Oxford, UK: Clarendon

Foster MS, Schiel DR. 1988. Kelp communities and sea otters: keystone species or just another brick in the wall? In *The Community Ecology of Sea Otters*, ed. GR VanBlaricom, JA Estes, pp. 92–115. Berlin: Springer-Verlag

Garbary DJ, Kim KY, Klinger T, Duggins D. 1999. Red algae as hosts for endophytic kelp gametophytes. *Mar. Biol.* 135:35–40

Gaylord B, Reed DC, Raimondi PT, Washburn L, McLean SR. 2002. A physically based model of macroalgal spore dispersal in the wave and current-dominated nearshore. *Ecology* 83:1239–51 — Nice interdisciplinary study of algal spore dispersal and its relationship to coastal hydrodynamics.

Graham LE, Wilcox LW. 2000. *Algae*. Upper Saddle River, NJ: Prentice Hall

Graham MH. 2003. Coupling propagule output to supply at the edge and interior of a giant kelp forest. *Ecology* 84:1250–64

Graham MH, Harrold C, Lisin S, Light K, Watanabe JM, Foster MS. 1997. Population dynamics of giant kelp *Macrocystis pyrifera* along a wave exposure gradient. *Mar. Ecol. Prog. Ser.* 148:269–79

Grigg RW, Kiwala RS. 1970. Some ecological effects of discharged wastes on marine life. *Calif. Fish Game* 56:145–55

Guiry MD, Rindi F, Guiry GM. 2006. *AlgaeBase version 4.0*. **http://www.algaebase.org**

Gunnill FC. 1980. Demography of the intertidal brown alga *Pelvetia fastigiata* in southern California, USA. *Mar. Biol.* 59:169–79

Gunnill FC. 1986. Demography of *Cystoseira osmundacea* and *Halidrys dioica* (Phaeophyta, Cystoseiraceae) in La Jolla, California, USA. *Bot. Mar.* 229:137–46

Harper JL. 1977. *Population Biology of Plants*. London: Academic

Harris LG, Ebeling AW, Laur DR, Rowley RJ. 1984. Community recovery after storm damage: a case of facilitation in primary succession. *Science* 224:1136–38

Harrold C, Reed DC. 1985. Food availability, sea urchin grazing, and kelp forest community structure. *Ecology* 66:1160–69

Hoffmann AJ, Santelices B. 1991. Banks of algal microscopic forms: hypotheses on their functioning and comparisons with seed banks. *Mar. Ecol. Prog. Ser.* 79:185–94

Hughes T, Bellwood D, Folke C, Steneck R, Wilson J. 2005. New paradigms for supporting the resilience of marine ecosystems. *Trends Ecol. Evol.* 20:380–86

Hymanson ZP, Reed DC, Foster MS, Carter JW. 1990. The validity of using morphological characteristics as predictors of age in the kelp *Pterygophora californica* (Laminariales, Phaeophyta). *Mar. Ecol. Prog. Ser.* 59:295–304

Jackson JBC, Kirby MX, Berger WH, Bjorndal KA, Botsford LW, et al. 2001. Historical overfishing and the recent collapse of coastal ecosystems. *Science* 293:629–38

Jenkins SR, Hawkins SJ, Norton TA. 1999. Direct and indirect effects of a macroalgal canopy and limpet grazing in structuring a sheltered intertidal community. *Mar. Ecol. Prog. Ser.* 188:81–92

Jenkins SR, Hawkins SJ, Norton TA. 2004. Long term effects of *Ascophyllum nodosum* canopy removal on mid shore community structure. *J. Mar. Biol. Assoc. UK* 84:327–29

> Long-term perspective on effects of fucoid canopy removal.

Johnson LE, Brawley SH. 1998. Dispersal and recruitment of a canopy-forming intertidal alga: the relative roles of propagules availability and postsettlement processes. *Oecologia* 117:517–26

> One of very few in situ studies on early life processes in fucoids, showing role of algal turfs.

Johnson MP, Hawkins SJ, Hartnoll RG, Norton TA. 1998. The establishment of fucoid zonation on algal-dominated rocky shores: hypotheses derived from a simulation model. *Funct. Ecol.* 12:259–69

Joska MA, Bolton JJ. 1987. in situ measurement of zoospore release and seasonality of reproduction in *Ecklonia maxima* (Alariaceae, Laminariales). *Br. Phycol. J.* 22:209–14

Kain JM. 1963. Aspects of the biology of *Laminaria hyperborea*. II. age, weight and length. *J. Mar. Biol. Assoc. UK* 43:129–51

Kain JM. 1979. A view of the genus *Laminaria*. *Oceanogr. Mar. Biol. Ann. Rev.* 17:101–61

Kautsky N, Kautsky H, Kautsky U, Waern M. 1986. Decreased depth penetration of *Fucus vesiculosus* (L.) since the 1940's indicates eutrophication of the Baltic Sea. *Mar. Ecol. Prog. Ser.* 28:1–8

Kawashima S. 1972. A study of life history of *Laminaria angustata* Kjellm. var. *longissima* Miyabe by means of concrete block. See Abbott & Kurogi 1972, pp. 93–109

Kendrick GA. 1994. Effects of propagules settlement density and adult canopy on survival of recruits of *Sargassum* spp. (Sargassaceae: Phaeophyta). *Mar. Ecol. Prog. Ser.* 103:129–40

Kendrick GA, Walker DI. 1994. Role of recruitment in structuring beds of *Sargassum* spp. (Phaeophyta) at Rottnest Island, Western Australia. *J. Phycol.* 30:200–8

Keough MJ, Quinn GP. 1998. Effects of periodic disturbances from trampling on rocky intertidal algal beds. *Ecol. Appl.* 8:141–61

Kinlan BP, Gaines SD. 2003. Propagule dispersal in marine and terrestrial environments: a community perspective. *Ecology* 84:2007–20

Kinlan BP, Gaines SD, Lester SE. 2005. Propagule dispersal and the scales of marine community processes. *Divers. Distrib.* 11:139–48

Kinlan BP, Graham MH, Sala E, Dayton PK. 2003. Arrested development of giant kelp (*Macrocystis pyrifera*, Phaeophyceae) embryonic sporophytes: a mechanism for delayed recruitment in perennial kelps? *J. Phycol.* 39:47–57

Klinger T, De Wreede RE. 1988. Stipe rings, age, and size in populations of *Laminaria setchellii* Silva (Laminariales, Phaeophyta) in British Columbia, Canada. *Phycologia* 27:234–40

Konar B. 2000. Seasonal inhibitory effects of marine plants on sea urchins: structuring communities the algal way. *Oecologia* 125:208–17

Kusumo HT, Druehl LD. 2000. Variability over space and time in the genetic structure of the winged kelp *Alaria marginata*. *Mar. Biol.* 136:397–409

Ladah LB, Zertuche-González JA, Hernández-Carmona G. 1999. Giant kelp (*Macrocystis pyrifera*, Phaeophyceae) recruitment near its southern limit in Baja California after mass disappearance during ENSO 1997–1998. *J. Phycol.* 35:1106–12

Lawton JH. 1994. What do species do in ecosystems? *Oikos* 71:367–74

Lewis JR. 1968. Water movements and their role in rocky shore ecology. *Sarsia* 34:13–36

Lewis RJ. 1995. Gametogenesis and chromosome number in *Postelsia palmaeformis* (Laminariales, Phaeophyceae). *Phycol. Res.* 43:61–64

Lotze HK, Worm B, Sommer U. 2001. Strong bottom-up and top-down control of early life stages of macroalgae. *Limn. Oceanogr.* 46:749–57

Luning K, Neushul M. 1978. Light and temperature demands for growth and reproduction of laminarian gametophytes in southern and central California. *Mar. Biol.* 45:297–309

Macaya EC, Boltaña S, Hinojosa IA, Macchiavello JE, Valdivia NA, et al. 2005. Presence of sporophylls in floating kelp rafts of *Macrocystis* spp. (Phaeophyceae) along the Chilean Pacific coast. *J. Phycol.* 41:913–22

Maier I, Muller DG. 1986. Sexual pheromones in algae. *Biol. Bull.* 170:145–75

Malm T, Kautsky L, Engkvist R. 2001. Reproduction, recruitment and geographical distribution of *Fucus serratus* L. in the Baltic Sea. *Bot. Mar.* 44:101–8

Markham JW. 1968. Studies on the haptera of *Laminaria sinclairii* (Harvey) Farlow, Anderson et Eaton. *Syesis* 1:125–31

Mattison JE, Trent JD, Shanks AL, Akin TB, Pearse JS. 1977. Movement and feeding activity of red sea urchins (*Strongylocentrotus franciscanus*) adjacent to a kelp forest. *Mar. Biol.* 39:25–30

McConnico LA, Foster MS. 2005. Population biology of the intertidal kelp, *Alaria marginata* Postels and Ruprecht: a nonfugitive annual. *J. Exp. Mar. Biol. Ecol.* 324:61–75

McCook LJ, Chapman ARO. 1992. Vegetative regeneration of *Fucus* rockweed canopy as a mechanism of secondary succession on an exposed rocky shore. *Bot. Mar.* 35:35–46

Muller DG. 1981. Sexuality and sex attraction. In *The Biology of Seaweeds*, ed. CS Lobban, MJ Wynne, pp. 661–74. Berkeley: Univ. Calif. Press

Muñoz V, Hernández-González MC, Buschmann AH, Graham MH, Vásquez JA. 2004. Variability in per capita oogonia and sporophyte production from giant kelp gametophytes (*Macrocystis pyrifera*, Phaeophyceae). *Rev. Chil. Hist. Nat.* 77:639–47

Nanba N. 1995. Egg release and germling development in *Sargassum horneri* (Fucales, Phaeophyceae). *Phycol. Res.* 43:121–25

Neushul M. 1963. Studies on the giant kelp, *Macrocystis*. II. Reproduction. *Am. J. Bot.* 50:354–59

Neushul M. 1972. Functional interpretation of benthic marine algal morphology. See Abbott & Kurogi 1972, pp. 47–73

Neushul M, Harger BWW. 1985. Studies of biomass yield from a near-shore macroalgal test farm. *J. Solar Energy Eng.* 107:93–96

North WJ. 1972. Observations on populations of *Macrocystis*. See Abbott & Kurogi 1972, pp. 75–92

Norton TA. 1992. Dispersal by macroalgae. *Br. Phycol. J.* 27:293–301

Norton TA, Burrows EM. 1969. Studies on marine algae of the British Isles. 7. *Saccorhiza polyschides* (Lightf.) Batt. *Br. Phycol. J.* 4:19–53

Norton TA, Mathieson AC. 1983. The biology of unattached seaweeds. *Prog. Phycol. Res.* 2:333–86

Parke M. 1948. Studies on British Laminariaceae. I. Growth in *Laminaria saccharina* (L.) Lamour. *J. Mar. Biol. Assoc. UK* 27:651–709

Pearson GA, Brawley SH. 1996. Reproductive ecology of *Fucus distichus* (Phaeophyceae): an intertidal alga with successful external fertilization. *Mar. Ecol. Prog. Ser.* 143:211–23

Raimondi PT, Reed DC, Gaylord B, Washburn L. 2004. Effects of self-fertilization in giant kelp, *Macrocystis pyrifera*. *Ecology* 85:3267–76

Reed DC. 1990a. The effects of variable settlement and early competition on patterns of kelp recruitment. *Ecology* 71:776–87

Reed DC. 1990b. An experimental evaluation of density dependence in a subtidal algal population. *Ecology* 71:2286–96

Reed DC, Amsler CD, Ebeling AW. 1992. Dispersal in kelps: factors affecting spore swimming competency. *Ecology* 73:1577–85

Reed DC, Anderson TW, Ebeling AW, Anghera M. 1997. The role of reproductive synchrony in the colonization potential of kelp. *Ecology* 78:2443–57

Reed DC, Foster MS. 1984. The effects of canopy shading on algal recruitment and growth in a giant kelp forest. *Ecology* 65:937–48

Reed DC, Kinlan BP, Raimondi PT, Washburn L, Gaylord B, Drake PT. 2006. A metapopulation perspective on patch dynamics and connectivity of giant kelp. In *Marine Metapopulations*, ed. JP Kritzer, PF Sale, pp. 353–86. San Diego: Academic

Reed DC, Laur DR, Ebeling AW. 1988. Variation in algal dispersal and recruitment: the importance of episodic events. *Ecol. Monogr.* **58:321–35**

> First in a series of important papers by Reed and colleagues on the population ecology of giant kelp. Demonstrates long-distance dispersal.

Reed DC, Schroeter SC, Raimondi P. 2004. Spore supply and habitat availability as sources of recruitment limitation in the giant kelp *Macrocystis pyrifera* (Phaeophyceae). *J. Phycol.* 40:275–84

Rosenthal RJ, Clarke WD, Dayton PK. 1974. Ecology and natural history of a stand of giant kelp, *Macrocystis pyrifera*, off Del Mar, California. *Fish. Bull.* 72:670–84

Saito Y. 1975. *Undaria*. In *Advances in Phycology in Japan*, ed. J Tokida, H Hirose, pp. 304–20. The Hague: Dr W Junk

Santelices B. 1990. Patterns of reproduction, dispersal and recruitment in seaweeds. *Oceanogr. Mar. Biol. Ann. Rev.* 28:177–76

Sauvageau C. 1915. Sur la sexualité heterogamique d'une Laminaire (*Saccorhiza bulbosa*). *C.R. Acad. Sci. Paris* 161:796–99

Schiel DR. 1985a. Growth, survival and reproduction of two species of marine algae at different densities in natural stands. *J. Ecol.* 73:199–217

Schiel DR. 1985b. A short-term demographic study of *Cystoseira osmundacea* (Fucales: Cystoseiraceae) in central California. *J. Phycol.* 21:99–106

Schiel DR. 1988. Algal interactions on subtidal reefs in northern New Zealand: a review. *NZ J. Mar. Freshw. Res.* 22:481–89

Schiel DR. 2004. The structure and replenishment of rocky shore intertidal communities and biogeographic comparisons. *J. Exp. Mar. Biol. Ecol.* 300:309–42

Schiel DR, Choat JH. 1980. Effects of density on monospecific stands of marine algae. *Nature* 285:324–26

Schiel DR, Foster MS. 1986. The structure of subtidal algal stands in temperate waters. *Oceanogr. Mar. Biol. Ann. Rev.* 24:265–307

Schiel DR, Steinbeck JR, Foster MS. 2004. Ten years of induced ocean warming causes comprehensive changes in marine benthic communities. *Ecology* 85:1833–39

Schiel DR, Wood SA, Dunmore RA, Taylor DI. 2006. Sediment on rocky intertidal reefs: effects on early postsettlement stages of habitat-forming seaweeds. *J. Exp. Mar. Biol. Ecol.* 331:158–72

Scrosati R. 2005. Review of studies on biomass-density relationships (including self-thinning lines) in seaweeds: Main contributions and persisting misconceptions. *Phycol. Res.* 53:224–33

Serrão EA, Pearson G, Kautsky L, Brawley SH. 1996. Successful external fertilization in turbulent environments. *Proc. Natl. Acad. Sci. USA* **93:5286–90**

> Highlights mechanisms for release of fucoid gametes in calm conditions.

Spalding H, Foster MS, Heine JN. 2003. Composition, distribution, and abundance of deep water (>30 m) macroalgae in central California. *J. Phycol.* 39:273–84

Steen H. 2004. Interspecific competition between *Enteromorpha* (Ulvales: Chlorophyceae) and *Fucus* (Fucales: Phaeophyceae) germlings: effects of nutrient concentration, temperature, and settlement density. *Mar. Ecol. Prog. Ser.* 278:89–101

Steen H, Scrosati R. 2004. Intraspecific competition in *Fucus serratus* and *F. evanescens* (Phaeophyceae: Fucales) germlings: effects of settlement density, nutrient concentration, and temperature. *Mar. Biol.* 144:61–70

Taylor DI, Schiel DR. 2003. Wave-related mortality in zygotes of habitat-forming algae from different experiments in southern New Zealand: the importance of "stickability". *J. Exp. Mar. Biol. Ecol.* 290:229–45

tom Dieck I. 1993. Temperature tolerance and survival in darkness of kelp gametophytes (Laminariales, Phaeophyta): ecological and biogeographical implications. *Mar. Ecol. Prog. Ser.* 100:253–64

Underwood AJ. 1999. Physical disturbances and their direct effect on an indirect effect: responses of an intertidal assemblage to a severe storm. *J. Exp. Mar. Biol. Ecol.* 232:125–40

Vadas RL, Johnson S, Norton TA. 1992. Recruitment and mortality of early postsettlement stages of benthic algae. *Br. Phycol. J.* 27:331–51

Vadas RL, Wright WA, Miller SL. 1990. Recruitment of *Ascophyllum nodosum*: wave action as a source of mortality. *Mar. Ecol. Prog. Ser.* 61:263–72

Vernet P, Harper JL. 1980. The costs of sex in seaweeds. *Biol. J. Linn. Soc.* 13:129–38

Worm B, Lotze HK, Boström C, Engkvist R, Labanauskas V, Sommer U. 1999. Marine diversity shift linked to interactions among grazers, nutrients and propagule banks. *Mar. Ecol. Prog. Ser.* 185:309–14

Wright JT, Williams SL, Dethier MN. 2004. No zone is always greener: variation in the performance of *Fucus gardneri* embryos, juveniles and adults across tidal zone and season. *Mar. Biol.* 145:1061–73

Yoneshigue-Valentin Y. 1990. The life cycle of *Laminaria abyssalis* (Laminariales, Phaeophyta) in culture. *Hydrobiologia* 204/205:461–66

Experimental study of several factors affecting early life stages of fucoids.

Living on the Edge of Two Changing Worlds: Forecasting the Responses of Rocky Intertidal Ecosystems to Climate Change

Brian Helmuth,[1] Nova Mieszkowska,[1] Pippa Moore,[2] and Stephen J. Hawkins[2,3]

[1]Department of Biological Sciences, University of South Carolina, Columbia, South Carolina 29208; email: helmuth@biol.sc.edu, nova@biol.sc.edu

[2]Marine Biological Association of the United Kingdom, Citadel Hill, PL1 2PB Plymouth, United Kingdom; e-mail: ppm@mba.ac.uk, sjha@mba.ac.uk

[3]School of Biological Sciences, University of Plymouth, PL4 8AA Plymouth, United Kingdom

Key Words

biogeography, ecophysiology, invertebrate, marine ecology, modeling

Abstract

Long-term monitoring shows that the poleward range edges of intertidal biota have shifted by as much as 50 km per decade, faster than most recorded shifts of terrestrial species. Although most studies have concentrated on species-range edges, recent work emphasizes how modifying factors such as regional differences in the timing of low tide can overwhelm large-scale climatic gradients, leading to a mosaic of environmental stress. We discuss how changes in the mean and variability in climatic regimes, as modified by local and regional factors, can lead to complex patterns of species distribution rather than simple range shifts. We describe how ecological forecasting may be used to generate explicit hypotheses regarding the likely impacts of different climatic change scenarios on the distribution of intertidal species and how related hindcasting methods can be used to evaluate changes that have already been detected. These hypotheses can then be tested over a hierarchy of temporal and spatial scales using coupled field and laboratory-based approaches.

INTRODUCTION

Anthropogenic increases in greenhouse-gas production and changes in land use have led to significant changes in global climate, and these trends are expected to continue far into the foreseeable future (Intergovernmental Panel on Climate Change 2001, Stainforth et al. 2005). Subsequently, understanding and forecasting the impacts of climatic change on the abundance and distribution of native and invasive species have become important endeavors in the fields of ecology and conservation biology (McCarty 2001, Parmesan & Galbraith 2004, Root & Schneider 1995, Wiens & Graham 2005).

This review summarizes research on the effects of climate change on rocky intertidal habitats, which are among the most-intensively experimentally studied ecosystems globally (Paine 1994), and sets forth an agenda for how and where the ecological effects of future changes in climate may be quantified. Rocky intertidal habitats (the regions between the high- and low-tide lines of coastlines) exist at the margins of both the terrestrial and marine realms; thus animals and algae in this ecosystem are subject to environmental challenges posed by both aquatic and aerial climatic regimes. As a result, rocky-shore organisms and assemblages may serve as early warning systems for the impacts of climate change (Barry et al. 1995; Harley et al. 2006; Hawkins et al. 2003; Mieszkowska et al. 2006; Sagarin et al. 1999; Southward et al. 1995; Thompson et al. 2002, 2004).

We draw upon insights from long-term monitoring studies (Hawkins et al. 2003, Southward 1991, Southward et al. 1995) and resurveys (Barry et al. 1995, Mieszkowska et al. 2005, Sagarin et al. 1999, Zacherl et al. 2003) carried out in the intertidal as an initial attempt to answer some of the key questions posed by ecologists faced with the dilemma of current global warming:

1. Which intertidal taxa show evidence of climatic impacts?
2. What is the rate and nature of change observed in intertidal systems?
3. Which elements of biological change must we investigate to detect and forecast past and future responses to climate change?

We highlight the importance of studies encompassing multiple spatial and temporal scales for the detection and quantification of biological responses to climatic shifts (Denny et al. 2004, Helmuth et al. 2002, Holtmeier & Broll 2005), emphasizing the necessity of considering physiological performance within the context of environmental regime (Helmuth et al. 2005, Hofmann et al. 2005). We then discuss the complementary application of ecological forecasting (Clark et al. 2001, Gilman et al. 2006) as a means of mechanistically predicting the physiological and ecological effects of future climate change scenarios on this sensitive ecosystem. Specifically, we address how forecasting (and related hindcasting) approaches can be used to generate hypotheses regarding the potential impacts of climate, and climate change, on the distribution of intertidal species, which can then be tested using coupled field and laboratory-based approaches. We argue that by generating explicit predictions of the direct physiological and ecological impacts of climate change and its indirect influences on biotic interactions, we will be better prepared to face future challenges. We emphasize the necessity of generating explicit hypotheses that can then be tested not just at poleward- and equatorial-range boundaries, but also at multiple locations within the ranges of species.

THE ROCKY INTERTIDAL ZONE

Intertidal invertebrates and algae are ectotherms that evolutionarily are of marine origin but must regularly contend with the terrestrial environment during each low tide. As such they provide a unique perspective on the relationships between both aquatic and terrestrial climatic regimes and organismal physiology and ecology. Indeed, largely because of the steep gradient in thermal and desiccation stresses that occurs during low tide, the rocky intertidal zone has long served as a natural laboratory for examining relationships between abiotic stresses, biotic interactions, and ecological patterns in nature (Bertness et al. 1999, Connell 1972, Hatton 1938, Orton 1929, Paine 1994, Somero 2002, Southward 1958, Wethey 1984). The upper limits of zonation of many intertidal organisms (i.e., their upward extent from the low-tide level) can be set directly by thermal and desiccation stresses (Connell 1972, Davenport & Davenport 2005, Somero 2002). Biotic interactions and behavior can also play an important role (Southward 1958, Underwood & Jernakoff 1981, Wolcott 1973), but such processes ultimately are dictated by the underlying environmental gradient (Raffaelli & Hawkins 1996; but see Wolcott 1973). Many intertidal organisms are therefore expected to display strong responses to changes in terrestrial climatic conditions (Fields et al. 1993, Lubchenco et al. 1993, Somero 2002). Specifically, if and when the zonational limit of a species is set by stresses related to terrestrial climate (such as increases in aerial body temperature), then we should observe a downward shift in upper limit of zonation as these stresses are magnified (Harley & Helmuth 2003, Wethey 1983). We should, however, only observe mortality events in locations where the upper limit is set by some factor related to terrestrial climate (as opposed to, for example, feeding or immersion time) (Davenport & Davenport 2005, Harley & Helmuth 2003, Southward 1958, Wolcott 1973).

Species can therefore be eliminated altogether from an intertidal region by changes in water temperature, upwelling regimes (Leslie et al. 2005) or oxygen levels (Service 2004), or when the upper limit of a prey species is squeezed down to the upper limit of its predator (Harley 2003). In these cases, we expect to observe contractions in species geographic distributions as climatic conditions exceed the species physiological threshold of tolerance. Conversely, when environmental conditions at a site become physiologically tolerable for the first time, we expect to see range expansions as new individuals are able to colonize. Understanding the impacts of climate change on the intertidal ecosystem thus requires an integrated approach that links changes in multiple environmental parameters to the physiological and ecological responses of organisms over a range of temporal and spatial scales and within a hierarchy of biological organization.

WEATHER, CLIMATE, AND CLIMATE CHANGE

Ecologists have long recognized the impacts of both weather and climate on natural ecosystems. Short-term localized atmospheric conditions (weather) fluctuate from hour to hour and day to day and encompass local air temperature, solar radiation, cloud cover, precipitation, and wind (Stenseth et al. 2003). Intertidal and subtidal organisms are also exposed to short-term fluctuations when the tide is in as a result

ENSO: El Niño Southern Oscillation

NAO: North Atlantic Oscillation

SST: sea surface temperature

of internal wave formation, upwelling, and changes in wave action, salinity, sedimentation, and oxygen availability (Dahlhoff & Menge 1996, Leichter et al. 1996, Leslie et al. 2005, Service 2004, Witman & Smith 2003). Although short-term fluctuations in the marine environment are often less variable in magnitude compared with those in the terrestrial environment, they have been shown to have significant physiological impacts (Dahlhoff 2004, Dahlhoff & Menge 1996, Leichter et al. 1996, Leslie et al. 2005). For example, rapid fluctuations in phytoplankton associated with upwelling can have significant influences on the growth of intertidal invertebrates (Dahlhoff & Menge 1996), and even moderately protracted periods of low oxygen can result in significant levels of mortality (Service 2004).

Climate is typically defined in the meteorological community as the mean of weather over a large temporal scale (30 years or more). The important distinction is therefore that weather is a much more temporally variable measure of what organisms experience at specific times. Moreover, climatic variables are often grouped over large spatial scales into climatic indices, such as the El Niño Southern Oscillation (ENSO) or the North Atlantic Oscillation (NAO), which tend not only to lump multiple environmental variables, but also tend to smooth small-scale and short-term changes in climate into large-scale patterns (Hallett et al. 2004, Stenseth et al. 2003). When considering environmental change there is therefore a trade-off (Stenseth et al. 2003) between manageability in terms of understanding, identifying, and predicting trends (climatic indices) and documenting the variability actually experienced by an organism or population at any given point in space and time (weather). In the next section, we examine the evidence for the impacts of climatic change on intertidal organisms, focusing on the utility of long-term monitoring programs conducted over nested spatial scales. We then discuss the need for an understanding of local patterns of environmental variables (Hallett et al. 2004) coupled with ecological and physiological responses of organisms to those environmental conditions (Hofmann 2005, Hofmann et al. 2005, Somero 2002).

HISTORY OF MARINE BIOGEOGRAPHICAL STUDIES

Many of the early biogeographic studies correlated species distributions with temperature (Andrewartha & Birch 1954, Hutchins 1947, Orton 1920) and still form the basis of ecological principles today. Hutchins (1947) provided a comprehensive synthesis of pioneering work and explicitly separated out the likely causes of poleward and equatorial limits. He emphasized the important role of reproduction and recruitment (termed repopulation) in setting distributions, building on work by Orton (1920). Hutchins (1947) argued that isotherms of sea surface temperature (SST) were correlated with the worldwide distribution of mussels (*Mytilus edulis*) and barnacles (*Semibalanus balanoides*), and he used these correlations to develop a general model to explain the role of thermal tolerance in setting range boundaries in coastal species.

Species distribution limits were grouped into four main zonal types: (*a*) the setting of poleward limits by survival during winter (cold stress) and equatorial limits by survival during summer (heat stress) (Southward 1958); (*b*) poleward limits determined by the inability to repopulate in summer and equatorial limits set by the inability to repopulate during winter (Bowman & Lewis 1986, Moore & Kitching 1939); (*c*) summer temperatures setting limits both by reducing survival at equatorial margins and repopulation at poleward margins (Ritchie 1927); and (*d*) winter temperatures setting limits by reducing survival at poleward margins and repopulation at equatorial margins (Hutchins 1947). Transient populations could occur beyond these margins, but these populations are often seasonal, only surviving until the extreme temperature event (cold winter or hot summer) occurs. Relic populations are occasionally found beyond range limits when a year class has survived a period of less-extreme climate, but these populations quickly become extinct when exposed to conditions that are too extreme for repopulation or survival. A more-recent extension of this theory is that population abundances are expected to be highest near the center of the species' distribution, where levels of physiological stress are assumed to be lowest, and then to decrease toward the range margins (Brown 1984).

The records collected during these early biogeographic studies provide baseline and time-series data for intertidal species prior to the onset of the current period of rapid warming. Surveys of the distribution and abundance of intertidal organisms made by various researchers in the region of Point Conception (northeast Pacific) were assimilated into a data set stretching back 18 decades (Smith & Gordon 1948). Hewatt (1937) made small-scale abundance surveys in the 1930s on the California coastline, but work was not as prolific as that in the northeast Atlantic, where Orton (1920) inspired a generation of researchers in postwar Britain. Detailed mapping of individual species followed in Europe (e.g., Crisp & Southward 1958, 1959; Fischer-Piette 1933, 1955; Fischer-Piette & Prenant 1956; Lewis 1964; Moore & Kitching 1939; Southward 1980, Southward & Crisp 1954b).

These pioneering surveys and somewhat ethically dubious transplantation experiments charted the distributions of intertidal species and investigated the factors setting species distributions. This work on the intertidal zone became integrated into a broader approach to understanding fluctuations in ecosystems of the western English Channel (Crisp & Southward 1958; Mieszkowska et al. 2006; Russell et al. 1971; Southward 1963, 1991; Southward et al. 1995, 2005). Southward & Crisp (1954a,b) resurveyed earlier work (Moore 1936, Moore & Kitching 1939) and showed that between the 1930s and 1950s, the northern barnacle *S. balanoides* had become rarer in the United Kingdom while the southern barnacle *Chthamalus stellatus* (later split into *C. stellatus* and *C. montagui*) had become more common (Southward & Crisp 1956). By the early 1960s, the role of climatic fluctuations in driving marine ecosystem responses, including those of intertidal zones, was well appreciated (Southward 1967). The extremely cold winter of 1962–1963 (Crisp 1964) and a shift to colder climatic conditions for the next two decades gave researchers further impetus.

At the assemblage level, Ballantine (1961) showed how the balance between algal domination in shelter and sessile invertebrates (particularly barnacles and mussels) on exposed shores shifted geographically on the European coastline, mediated by limpet grazing (Hawkins & Hartnoll 1983, Hawkins et al. 1992). Broitman et al. (2001) also observed this latitudinal pattern of changes in functional group dominance in Chile.

Research shifted focus from the ultimate and underlying factor of temperature and its influence on ectothermic species to the proximate causes of species-range edges. Processes proposed included a lack of consistent recruitment (Lewis 1986), barriers to dispersal (Kendall 1987), a lack of suitable habitat (Crisp & Southward 1958), and extreme events causing adult mortality (Crisp 1964). There was also a growing appreciation of the correlation between the fluctuations in climate (Southward & Crisp 1954b, 1956) and in the abundance of animals within ranges and at boundaries (Crisp & Southward 1958). As early as 1954, Southward & Crisp (1954b) proposed *S. balanoides* and *Chthamalus* spp. as indicator organisms to investigate species responses to changes in the general environment, as both taxa were likely sensitive to even small changes in temperature.

By the late 1980s, it became apparent that patterns driven by solar flux (using the sun-spot index) were beginning to break down, and anthropogenic influences were suggested (Southward 1991). From the late 1980s onward, conditions were obviously becoming much warmer, with consequent responses shown by intertidal species in Europe (Southward 1991, Southward et al. 1995).

EVIDENCE FOR THE IMPACTS OF CLIMATE CHANGE ON INTERTIDAL COMMUNITIES

An increasing number of recent studies have documented the impacts of climate change on intertidal invertebrates and algae, and these studies are summarized in **Table 1**. Studies carried out at the northern range edge of the southern neogastropod *Kelletia kelletii* in the California region of the northeast Pacific, for example, demonstrated that the boundary had shifted north from the late 1970s/early 1980s to the 2000s (Herrlinger 1981, Zacherl et al. 2003) (**Table 1**). Data from fossil records (Arnold 1903, Lohnhart & Tupen 2001), and surveys ranging from 1830s until modern times (Zacherl et al. 2003), showed that this was the first-recorded extension north of Point Conception. The range extension coincided with the steady increase in regional coastal SST that occurred during the second half of the twentieth century (Zacherl et al. 2003).

The blue mussel *M. edulis* has recently jumped a historic oceanographical barrier from the Norwegian mainland to the Svalbard archipelago, extending its distribution north by 500 km to Bear Island between 1977 and 1994 (Weslawski et al. 1997). By 2002, *M. edulis* had extended its range a further 500 km north to Isfjorden on Svalbard island, where it had not been recorded for 1000 years (Berge et al. 2005, Salvigsen et al. 1992). This 1000-km poleward extension of northern distributional limits over the course of the past three decades is thought to have arisen from the dispersal of planktonic larvae from a source population on the Norwegian mainland by an anomalously large northward transport of warm Atlantic water into the Greenland Sea region in recent years (Berge et al. 2005).

Northern and eastern range extensions have been identified around the British coastline for the southern trochid gastropods *Gibbula umbilicalis* and *Osilinus lineatus* (Mieszkowska et al. 2005, 2006) (**Table 1**) between the mid-1980s and the 2000s, in contrast to static range limits observed during the 1970s and early 1980s (Kendall

1987, Kendall & Lewis 1986). The ranges of the southern barnacles (*Chthamalus* spp. and *Balanus perforatus*) have also extended in Britain since the previous surveys in the 1950s (Herbert et al. 2003; Mieszkowska et al. 2005, 2006) (**Table 1**), and have also increased in abundance at survey sites along the English Channel coastline (Herbert et al. 2003). The eastern limit of the southern alga *Bifurcaria bifurcata* has re-extended along the English Channel in the past four decades (N. Mieszkowska, unpublished data) (**Table 1**). The previous record of *B. bifurcata* at its current range edge at Portland was in 1902. This was immediately prior to a sudden change in the marine climate in the English Channel from a warm phase into one of the coolest periods in modern record, and *B. bifurcata* was restricted to southwest England during this time (E. Burrows, unpublished data).

In general, the southern boundaries of cold-water species have shown less movement (e.g., range contractions of two species have been recorded) in Britain compared with northern boundaries of warm-water species (e.g., eight species have recorded range extensions), which may reflect that many species with northern ranges reach their absolute range limits further south in Europe, particularly in Portugal (Southward et al. 1995). The northern alga *Alaria esculenta* has, however, disappeared from the coastline of northeast Britain since the 1990s (R. Bowman, personal communication), and although present at some locations in southwest Britain, no survival occured in transplant experiments to locations where it was previously recorded before the 1950s (Vance 2004). Water temperatures greater than $16°C$ are fatal to mature sporophytes of *A. esculenta* (Widdowson 1970). Increases in sea and air temperatures in both the North Sea and English Channel have occurred between the original surveys and resurveys, and the poleward-range contraction of *A. esculenta* at its southern limits in Britain during this time has mirrored the poleward shift of the summer $16°C$ SST isotherm.

Abundances of many southern intertidal species (e.g., Chthamalid barnacles, trochids, *Patella depressa* and *B. perforatus*) have increased in the region of the biogeographic boundary in southwest England since the 1980s (Crisp & Southward 1958, Herbert et al. 2003, Mieszkowska et al. 2005). In contrast, many of the northern species have shown decreases in abundance (e.g., *Semibalanus balanoides* and *Patella vulgata*), although not to the same degree (Mieszkowska et al. 2006, Southward 1991, Southward et al. 1995). In general, where northern and southern pairs of species with similar ecological niches co-occur, a general shift in the abundance ratios in favor of southern species has been observed in Britain. Similar patterns have been observed in Monterey Bay, California, where researchers compared abundance data for over 130 species between the 1930s (Hewatt 1937) and the 1990s (Barry et al. 1995, Sagarin et al. 1999). This comparison showed a significant increase in the abundance of southern species of gastropods, anthozoans, and barnacles; a significant decrease in the abundance of northern anthozoan and limpet species; and no apparent trend for species classified as cosmopolitan in their distribution (Barry et al. 1995, Sagarin et al. 1999). These changes occurred in parallel to increases in summer coastal SST of $\sim 2°C$ and led to large alterations in local community structure.

Although factors including the overfishing of urchins and groundfish have affected coastal-community composition in the Gulf of Maine since the 1980s (Harris &

Table 1 Summary of major shifts in the biogeographic distributions of rocky intertidal species, and population-level changes observed within species ranges in response to the current period of climate warming. Southern species spread northward and eastward around the British Isles, including paradoxically southward around the cooler North Sea coastlines

Species	Geographic region	Impact	Survey	Resurvey
Mytilus edulis	Bear Island, Svalbard	Range extension 500 km N;[a] range extension 500 km N[b]	1979 (first record in 1000 years)	1994, 2004
Gibbula umbilicalis	North Scotland, south England, north Wales, south England	Range extension 55 km N, range extension 125 km E; increased recruitment success near range edges[c,d]	1985, 1985	2001–2004 2001–2004
Osilinus lineatus	North Wales, south England	Re-extension 40 km N, range extension 55 km E; increased recruitment success near range edges[d,e]	1964, 1980s, 1986	2001–2004 2001–2004
Chthamalus montagui	North Scotland	Range extension 75 km along cooler North Sea coast[d,f]	1980s	2001–2004
Chthamalus stellatus	North Scotland	Range extension 40 km along cooler North Sea coast[d,f]	1980s	2001–2004
Patella depressa	South England	Range extension 30 km E[d]	1964, 1980s	2001–2004
Patella ulyssiponensis	South England	Range extension 120 km E[d]	1964, 1980s	2004
Balanus perforatus	South England	Range extension 170 km E[g]	1960s	1993–2001
Bifurcaria bifurcata	South England	Re-extension 150 km E[d]	1902, 1936	2002
Patella rustica	North Portugal	Infilling of gap 39.3° N–43.2° N at southern end of range[h]	1950s	2003
Kelletia kelletii	California	Range extension 400 km N[i]	1970s	1980s
Codium fragile	Gulf of Maine	Range extension N; warmer summers facilitating reproduction[j]	1975–present	1990

			1950s	1990s, 2004
Alaria esculenta	North England, south England	Range contraction >120 km N[d]; range contraction 120 km E[f]		
Tectura tessulata	Irish Sea	Loss of species from south of Isle of Man[d]	1980s	2001–2004
Fissurella crassa	Chile	Range contraction 3° latitude N[k]	1962	1998–2000
Enoplachiton niger	Chile	Range contraction 3.5° latitude N[k]	1962	1998–2000
Echinoittorina peruviana	Chile	Range contraction 5° latitude N[k]	1962	1998–2000
Thais haematoma	Chile	Range contraction 8° latitude N[k]	1962	1998–2000
Gastropods, anthozoans, barnacles	California	Increases in abundance of warm-water species[l]	1930s	1998–2000
Anthozoans and limpets	California	Decreases in abundance of cold-water species[l]	1930s	1998–2000
Asterias forbesi	Gulf of Maine	Increase in abundance[m]	1979	1998
Asterias vulgaris	Gulf of Maine	Decrease in abundance[m]	1979	1998

[a] Weslawski et al. 1997.
[b] Berge et al. 2005.
[c] Kendall & Lewis 1986.
[d] Mieszkowska et al. 2005.
[e] Kendall 1987.
[f] Crisp & Southward 1958.
[g] Herbert et al. 2003.
[h] Lima et al. 2006.
[i] Zacherl et al. 2003.
[j] Harris & Tyrrell 2001.
[k] Rivadeneira & Fernandez 2005.
[l] Sagarin et al. 1999.
[m] Harris et al. 1998.

Tyrrell 2001), increasing sea temperatures are the driving factor behind the increases in abundance of the warm-water starfish *Asterias forbesi* and the concurrent decline in the cold-water congener *A. vulgaris* (Harris et al. 1998). Increasing sea temperatures have also facilitated the range extension of the warm-water alga *Codium fragile* into the Gulf owing to summer temperatures sufficient to allow successful reproduction since the mid-1970s (Harris & Tyrrell 2001).

Natural climatic phenomena such as fluctuations in the strength and direction of the ENSO index or the NAO index—where temperature, currents, sea level, and storm frequency may all be altered (Allison et al. 1998, Strub & James 2002)—can also be used as a proxy for the likely effects of anthropogenic climate change. Rising temperatures and altered currents as a result of ENSO events have been linked with changes in species distributions (Fields et al. 1993), including introduced species such as *Carcinus maenas* on the west coast of the United States (Yamada et al. 2005). Increased sea temperatures that result from ENSO-mediated relaxation of upwelling events have been used to document changes in the interaction strength between a keystone predator and its prey (Sanford 1999).

In summary, the locations of range edges of many intertidal species have shifted by as much as 50 km per decade (**Table 1**), much faster than most recorded shifts of terrestrial species (e.g., Parmesan & Galbraith 2004). Contractions in the equatorial-range limits have also been seen, but the rate and extent are less than the changes observed at the poleward limits of distribution. These biogeographic shifts have been accompanied by increases in the abundance of many species close to their poleward-range limits and changes in the relative abundances of warm- and cold-water species at long-term monitoring sites.

However, these records of fluctuating densities of warm- and cold-water species, and alterations in assemblage composition, highlight two important points. First, not all evidence of change has occurred at poleward and equatorial margins of species-range boundaries. Recently the range of the southern limpet *Patella rustica* has expanded in northern Portugal, bridging a historic distributional gap within the species distributional range during a period when the regional cold coastal upwelling was seen to weaken (Lima et al. 2006).

Second, although the evidence is clear that climate has had, and will continue to have, large impacts on at least some species in rocky intertidal ecosystems (**Table 1**), we should not expect to see impacts of climate change everywhere. For example, the distributional limits of many intertidal species are not all exhibiting the same trend or extent of poleward expansions, with some species showing little or no change despite exposure to warming sea and air temperatures (Rivadeneira & Fernandez 2005). Importantly, only through continuous monitoring can many of these potential changes in abundance and distribution be detected, as these responses can occur rapidly. Moreover, the challenge before us is not simply to monitor these responses, but to build on this existing knowledge in an attempt to forecast future changes and to understand how and when such responses are linked to climatic variability. Importantly, we must first understand the mechanisms by which organisms are impacted by their ambient physical environment, both directly through physiological performance and indirectly through mediation of biotic interactions.

PHYSIOLOGICAL RESPONSES TO CLIMATE FACTORS BY INTERTIDAL ORGANISMS

Intertidal invertebrates and algae live at the interface of the marine and terrestrial worlds. Many studies, for example, have shown that the production of heat shock proteins occurs after exposure to temperatures experienced during low tide and have suggested that the production of these proteins incurs some metabolic cost (Dahlhoff 2004, Hofmann 2005, Sanders et al. 1991, Somero 2002, Tomanek & Sanford 2003). The upper limits of zonation of many species of invertebrates have been correlated with maximum aerial temperatures (Davenport & Davenport 2005, Kennedy 1976, Somero 2002, Williams & Morritt 1995), and experimental manipulations of aerial body temperature by shading have resulted in changes in mortality, rates of predation, relative competitive ability, and species zonation patterns (Harley & Lopez 2003, Hatton 1938, Wethey 1984). Other studies have documented natural mortality events resulting from both high- and low-temperature extremes experienced during aerial exposure (Carroll & Highsmith 1996, Tsuchiya 1983).

When these upper limits of an organism are squeezed down to the upper limit of a predator or dominant competitor (Harley 2003), the subordinate or prey species is eliminated from the intertidal zone, and species are geographically limited by physiological stresses related to aerial exposure (Harley & Helmuth 2003). This observation is key because organisms are differentially affected by terrestrial climate and by the characteristics of the organism itself. As a result, two organisms exposed to the same climatic conditions can exhibit markedly different body temperatures and hence different levels of physiological stress (Helmuth 2002, Helmuth et al. 2005, Gilman et al. 2006; but see Wethey 1983). Such interspecific variation may be particularly important when considering the impacts of climate on predator/prey pairs, competitors, or native versus invasive species. For example, Helmuth (2002) has shown that at low tide, the predatory seastar *Pisaster ochraceus* may be up to 10° cooler than the prey (*Mytilus californianus*) that it is feeding on. Differences in the body temperature of species pairs may have implications for biogeographic boundaries. For example, if the upper limit of a prey species is shifted downward by climate change, but the upper limit of its predator remains unaffected, the prey species may be eliminated from the ecosystem (Harley 2003).

Numerous studies have also focused on the important role of water temperature, usually measured as SST, in setting species distributional limits (Frank 1975, Hutchins 1947). Water temperature has been shown to have strong effects on the metabolism and growth (Bayne et al. 1973, Phillips 2005), feeding behavior (Petraitis 1992; Sanford 1999, 2002), and reproduction (Hutchins 1947, Philippart et al. 2003) of intertidal and subtidal organisms. Water temperature also can have significant impacts on rates of larval development (Anil et al. 2001, Hoegh-Guldberg & Pearse 1995, Luppi et al. 2003).

Importantly, however, although various aspects of both terrestrial and aquatic climate drive the physiological performance and survival of coastal organisms, these factors often alternate with one another, and with nonclimate-related factors, in setting distributional limits. Many factors, such as temperature and food, and temperature and desiccation, are often interactive in their effects (Gilman 2006, Li & Brawley 2004). Harley & Helmuth (2003) correlated the upper zonation limits of mussels and barnacles with measurements of maximum body temperature at multiple sites in northwestern Washington State (United States) and found that local distributional limits were tightly correlated with temperature at some sites, but at others the correlation was weak. Their results suggest maximum body temperature alternated with some other environmental factor, such as feeding time, in setting upper zonation heights of these species in this region and again argue that limiting factors must be considered on a spatially explicit basis. In contrast, Sanford (1999, 2002) showed that the foraging behavior of the keystone predator *P. ochraceus* was driven predominantly by water temperature experienced during submersion at high tide and appeared to be uncorrelated with air temperature.

Taken collectively, these studies show that although climate is likely to have significant impacts on the local and geographic distributions of intertidal species, the specific climatic drivers that contribute to these patterns are likely complex and must be tested experimentally. In the above example, a predator (*Pisaster*) appears most susceptible to changes in water temperature, whereas its prey (*Mytilus* spp.) may be affected by both the aerial and aquatic climatic regimes. Moreover, these climatic drivers are likely to alternate in their importance as limiting factors from location to location, and simple statistical correlations between environmental factors and species-range limits are likely to yield poor results, especially when made over large geographic scales (Holtmeier & Broll 2005). In other words, it may be naïve to expect single environmental variables such as water temperature or air temperature to be correlated with population abundances and range limits at all sites and at all scales examined. Similarly, we should not expect environmental variables to be physiologically limiting at all range edges, as is commonly assumed by many environmental niche models; i.e., range edges are not always indicative of an organism's fundamental niche space.

In addition, researchers must also investigate the potential variation in the physiological and ecological responses of organisms to climatic and nonclimatic factors at different stages of their life history (**Figure 1**). For example, the onset of gametogenesis has been linked to increases or decreases in environmental temperature in many intertidal species, with duration and frequency of gonad activity also regulated by local thermal regimes (Bode et al. 1986, Kendall & Lewis 1986). Owing to the variation in reproductive mode between species, spawning, fertilization and larval release may respond to either thermal, physical, or biological cues including temperature changes (Bowman & Lewis 1986, Hawkins 1982, Lewis 1986), wave action, storminess (Carrington 2002, Lewis 1986), photoperiod (Brockington & Clarke 2001), and the timing of phytoplankton blooms (Clarke 1988, Connell 1961), all of which vary on local as well as regional scales (**Figure 1**). Many intertidal invertebrate species have a planktonic larval phase, the duration of which can be altered by water temperature (Hoegh-Guldberg & Pearse 1995). Dispersal of larvae is controlled by both local and regional hydrographic regimes, wave action, the length of time spent as a pelagic veliger, larval feeding mode, and larval size and shape (Pechenik 1999).

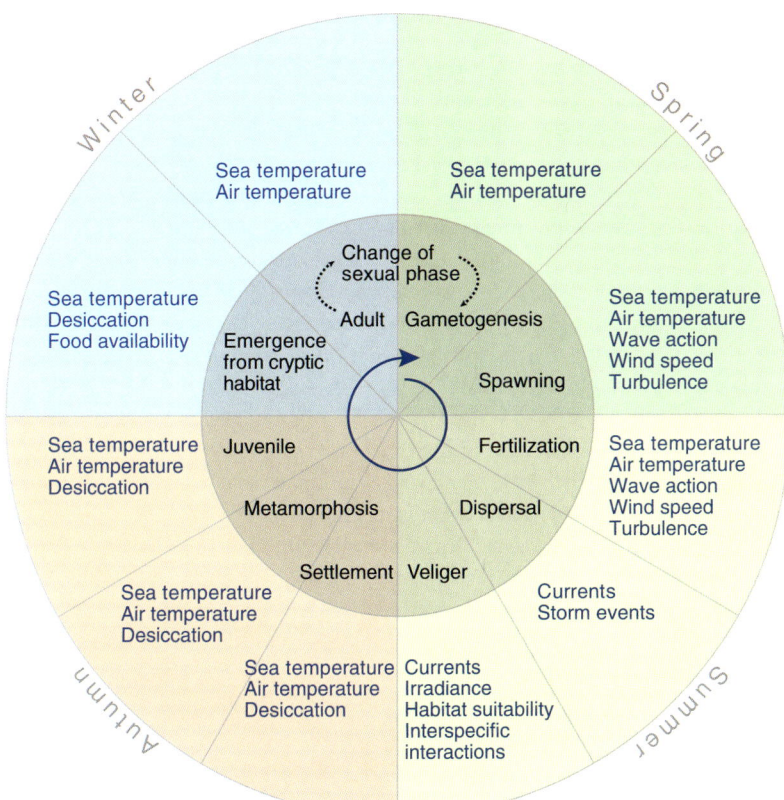

Figure 1

Variable importance of multiple climatic and nonclimatic factors on physiological performance and survival of *Patella depressa* during different stages of the limpet's life history.

Settlement may also occur in response to temperature changes (Pineda et al. 2002), but nonthermal cues including habitat suitability, irradiance, and intra- and interspecific interactions can also mediate or determine when and where larvae commence the sessile phase of the life cycle (Bertness et al. 1999, Wethey 1986). Postsettlement survival has also been shown to be affected by large-scale terrestrial climatic changes, as experienced during the cold winter of 1962–1963 (Crisp 1964, Kendall & Lewis 1986) and 1979 (Wethey 1985) and by the marine climate at the regional scale (Southward & Crisp 1956; N. Mieszkowska, unpublished data), but idiosyncratic species responses can modulate survival rates at the local scale and affect the extent of range contractions on a regional scale.

The interactive effects of climate must therefore be considered with respect to both the life history of each species and the variation in life-stage sensitivity to both climatic and nonclimatic environmental factors. This must be done within a spatially explicit context that includes aspects of local environmental variability (Schoch & Dethier 1996), as well as variability in physiological and genetic responses to those environmental signals (Hilbish 1981, Stillman 2003).

THE BASIS OF WEATHER AND CLIMATE IN DRIVING SPECIES DISTRIBUTIONS

Hutchins' (1947) general argument that large-scale climatic regimes determine the latitudinal distributions of marine species remains a baseline assumption in many ecological studies related to climatic change. It is obvious that polar communities experience colder conditions than tropical communities and that large-scale gradients in climate exist when looking over very large segments of latitude (Stenseth et al. 2003). Moreover, large-scale, integrative indices of climate can in some cases serve as a far more synthetic tool for looking at broad-scale patterns than individual weather records (Stenseth et al. 2003). To this end, most meta-analyses of the impacts of climate change on natural ecosystems have looked for evidence of latitudinal shifts in species-range boundaries at the northern and southern edges of their distributions (Parmesan & Yohe 2003).

Indeed, in many cases there is strong evidence of multispecies faunal breaks that appear to coincide with changes in climatic variables (Briggs 1974, Morris et al. 1980, Zacherl et al. 2003). Two factors in this assumption, however, tend to be ignored when considering the likely impacts of climate change, particularly when employing climate envelope–type approaches. First, not all latitudinal boundaries are set by climate. Discontinuities in larval dispersal may act to set distributional limits (Crisp & Southward 1958, Gaylord & Gaines 2000), which may be caused by localized hydrographic features such as the strong offshore currents that occur at Portland in the English Channel (Maddock & Pingree 1977), or by larger, basin-scale features such as increases in offshore Eckman transport along sections of the eastern Pacific seaboard (Connolly & Roughgarden 1998, Broitman et al. 2001).

Alternatively, range edges may be set by a lack of unsuitable habitat. For example, the eastern range edge of southern species of barnacles, limpets, topshells, and winkles in Europe has been limited to the western English Channel from the 1930s to the 1980s owing to the lack of rocky intertidal habitat as well as strong offshore currents at various headlands (Crisp & Southward 1958, Kendall 1987). Thus we should not expect to observe species-range shifts at all latitudinal boundaries because not all may be directly set by some aspect of climate. Key areas where climate change likely has large impacts are thus the transitional zones bounding those biogeographic provinces primarily delimited by climate. Such sharp faunal breakpoints corresponding to changes in thermal provinces occur on the western Atlantic at Cape Hatteras (Johnson 1934); the eastern Atlantic in the Bay of Biscay (Sauvageau 1897, Fischer-Piette 1955); southwest England, Wales, and Ireland (Forbes 1853); the east Pacific at Point Conception (Briggs 1974); the west Pacific at the East Cape (Schiel 2004); and the Cook Straight (Knox 1960) of New Zealand. Range shifts of cold- and warm-adapted species have already been documented in many of these regions (Herbert et al. 2003; Lima et al. 2006; Mieszkowska et al. 2005, 2006; Zacherl et al. 2003).

Second, spatial patterns of climatic conditions within species ranges are often far more complex than generally appreciated, so organisms may be presented with a mosaic pattern of environmental stress, rather than a simple latitudinal gradient. In other words, although latitudinal trends in climate may exist over very large spatial scales, considerable spatial variability may occur at scales that are biogeographically meaningful. Organisms respond physiologically not to some generic aspect of climate measured over large scales, but to the localized environmental conditions in which they live (Hallett et al. 2004). Especially in the case of sessile organisms, this sphere of influence may be exceedingly restricted and may be significantly affected by factors ranging in scale from the angle of the substratum on which the organism sits to the character of the local weather system (e.g., coastal fog) and patterns of local tides and waves (Helmuth & Hofmann 2001, Schoch & Dethier 1996). Although some degree of heterogeneity along broad-scale climatic gradients is certainly not surprising, several recent studies suggest that species and ecosystem responses to climate are far more complex than previously appreciated. Specifically, in some cases, the effects of climatic gradients may be subsumed by those of local modifying factors, even over large areas. In other words, patterns of weather may be more important than climate, and we need to pay close attention to how localized patterns of weather are averaged into large-scale climatic indices (Hallett et al. 2004).

For example, Holtmeier & Broll (2005) argue that large-scale descriptors of climate are a poor predictor of the sensitivity of treeline distribution to climate change and that the influence of local modifying factors such as recent land use and substratum aspect can override those of climate. Likewise, Helmuth et al. (2002) have shown that patterns of local weather, wave height, and tidal regime interact to create a thermal mosaic along the west coast of the United States, so maximum aerial body temperatures of intertidal mussels at sites in northern Washington and Oregon can exceed those of animals living 100–1000s of kilometers farther to the south, in southern California. Their results suggest that because of the influence of the timing of low tide, coupled with the impacts of wave splash and local weather, patterns of mussel body temperature do not become increasingly colder moving northward from California to Washington. As a result, mortality events at hot spots where periods of summertime low tides coincide with hot terrestrial climatic conditions should be expected rather than simple latitudinal range shifts. These results fit well with the observation that many intertidal species along the west coast of the United States do not display an abundant center distribution, as would be predicted by Brown's (1984) principle if levels of physiological stress were lowest in the center of their distribution and highest near their range edges (Gilman 2005, Sagarin & Gaines 2002, Sagarin & Somero 2006). Instead these species show a pattern where population numbers (Sagarin & Gaines 2002) and physiological indicators of stress (Sagarin & Somero 2006) wax and wane along the length of the species distribution. Such patterns strongly suggest that we require local patterns of intertidal organism temperature, and temperature anomalies, to decipher changes in assemblages over time.

Similarly, Hallett et al. (2004) correlated patterns of mortality in Soay sheep against records of local weather, and against broad-scale patterns of climate (the NAO index). They found that when they used a mechanistic relationship between local weather and mortality, patterns of local weather proved to be a much better predictor of population trends than did the large-scale NAO index. In other words, once the relationship between local weather variables such as rainfall and air temperature was considered within the context of the physiological response of the organism, local descriptors of weather served as a far better predictor of population response than did the NAO index (Hallett et al. 2004).

Taken cumulatively, these results indicate a need for clearer understanding of the relative importance of climatic indices versus localized factors in driving the responses of intertidal populations to climate change, and we should not expect to see evidence of climate change only at poleward- and equatorial-range margins (Gilman et al. 2006; Helmuth et al. 2002, 2005; Holtmeier & Broll 2005; Sagarin & Somero 2006). Thus, when local and regional factors are sufficiently large, they may create holes in species distributions, especially when they occur adjacent to habitat that is already unsuitable (for example, sections of muddy substratum uninhabitable by rocky intertidal organisms).

The question, therefore, is do micro- and mesoscale heterogeneity in the physical environment as driven by local modifying factors matter in terms of biogeographic responses to climate change? Microscale variability introduced by factors such as substratum angle may serve as an effective means of detecting early signs of the impacts of climate change, and this scale of heterogeneity may have important consequences for maintaining genetic polymorphism (Schmidt & Rand 2001). At larger scales, factors such as tidal regime may be sufficiently large in spatial extent to have significant biogeographic consequences (Helmuth et al. 2002).

Overall, the importance of heterogeneity in the physical environment is determined largely by the ability of populations to recover from disturbance events, which in turn is driven by the interaction between the life history of the larval stage and local and regional dispersal patterns (Broitman et al. 2005, Sotka et al. 2004). For example, if a mortality event occurs on a section of coastline, but larvae or gametes are able to bypass or repopulate that dead zone, then the presence of this small-scale hot spot has few or no biogeographic consequences. If, in contrast, a persistent dead zone is sufficiently large, or occurs adjacent to an area of unsuitable habitat such that propagules are no longer able to disperse across this gap, then barriers to genetic dispersal may be created. Similar arguments may be made in the case of range expansions. For example, the topshell *G. umbilicalis*, which has a short larval stage, has extended and consolidated its northern range limits to a greater extent than the barnacles *C. montagui* and *C. stellatus*, which have much longer planktonic dispersal times. This appears to be a result of high densities of recruits settling locally, facilitating colonization of neighboring shores, and rapidly increasing the range by small increments (N. Mieszkowska, unpublished data). The relative importance that larval dispersal and reproduction play in the regulation of heterogeneity within species distributions therefore depends on the spatiotemporal magnitude of the disturbance and the location of disturbance within the species range.

Predators and competitors may also play a key role in these patterns by restricting subordinate intertidal species from subtidal environments; if an intertidal population is eliminated owing to a heat-stress event, for example, but a healthy subtidal population remains, then the subtidal population may serve as a reservoir. If there are no subtidal populations, then the elimination of a species from the intertidal could have significant biogeographic consequences, but only if its absence creates gaps that cannot be spanned by larvae over time. The processes of reproduction, dispersal, larval mortality, and postrecruitment mortality, which can drive range expansions and contractions, must be therefore examined on a spatially explicit basis, encompassing areas within the range as well as the range edges. Specifically, whereas physiological responses to climate drive patterns of reproductive success, mortality, and species interactions, oceanographic regimes and coastal geomorphology determine levels of connectivity between habitats that are made suitable or unsuitable by climatic change (Broitman et al. 2005, Lima et al. 2006, Sotka et al. 2004). To this end, the application of ecological-forecasting techniques may potentially serve as an effective means of predicting patterns in environmental variables.

ECOLOGICAL FORECASTING

Ecological forecasting (and related hindcasting) uses first principles to relate patterns in environmental variables to the physiological responses of organisms. For example, weather data can be used to generate heat-budget models, which in turn can generate spatially and temporally explicit maps of aquatic and aerial body temperature (Helmuth 1999, Porter et al. 2002, Wethey 2002). From these patterns, and given explicit measurements of physiological responses of organisms to relevant environmental variables and of the indirect impacts of these responses on species interactions, we can then quantitatively anticipate effects on natural assemblages given various scenarios of global climate change. Specifically, forecasting the impact of climate change on the distribution and abundance of organisms requires that the following questions be addressed:

1. How are relevant environmental factors such as air and water temperature, food supply, and oxygen availability likely to change over appropriate temporal and spatial scales under future climate-change scenarios?
2. How are environmental factors such as air temperature and water temperature translated into physiologically relevant factors such as body temperature (Gilman et al. 2006, Helmuth 2002, Helmuth et al. 2002, Wethey 2002)? In many cases, organisms perceive and respond to environmental signals such as water velocities, air temperatures, and oxygen availabilities in different ways depending on factors such as organism size, morphology, and behavior. It may thus be impossible to decipher environmental patterns without considering the role of the organism in integrating and responding to these environmental signals (Helmuth et al. 2005).
3. How close are individuals living to the limits of their physiological tolerance within each portion of each species range (Chown et al. 2004, Hofmann 2005, Somero 2002, Southward 1958), and in cases where distributional limits (both local and geographic) are set directly by physiological limitation, what environmental factor(s) drives these limits? Namely, it is necessary to understand when and where organisms are living at the limits of their fundamental niche (i.e., are limited by some aspect of physiological tolerance) in contrast to when their patterns of distribution are affected by the presence of predators, competitors, and facilitators (i.e., species boundaries are representative of realized niche space) (Davis et al. 1998, Hampe 2004, Hutchins 1947). How do

multiple environmental factors (e.g., food availability, temperature, and pH) interact to drive susceptibility to climatic change (Dahlhoff 2004, Harley et al. 2006, Helmuth et al. 2005)?
4. In cases where species distributions are determined by biotic interactions, how are such interactions indirectly affected by climate (Burnaford 2004; Sanford 1999, 2002; Southward 1991; Wethey 1983, 2002)? Specifically, can mechanistic forecasts be made of the indirect impacts of climatic change on species interactions?
5. How rapidly can populations evolve in response to environmental changes (and interactions among them) (Clarke 2003)?
6. When cases of mortality and reproductive failure do occur, how do patterns of fertilization success and propagule dispersal affect the probability of successful recolonization (Broitman et al. 2005, Gaylord & Gaines 2000, Palumbi et al. 2003, Sotka et al. 2004)? Similarly, when sites become newly habitable, how does propagule dispersal affect the colonization of these sites or affect the invasibility by nonnative species? Finally, when large dead zones are created in the centers of species distributions, how do larval and gamete dispersal and organism life history interact with the size of the dead zone to affect the creation of disjunct populations (Sotka et al. 2004)? What impact does this have on the capacity of populations to evolve in response to rapid climate change (Clarke 2003)?

Satisfactorily answering these questions is clearly not a straightforward or easy task for any ecosystem, although significant progress has been made in most of these arenas. Answering them all for any one ecosystem is an incredibly ambitious challenge and likely impossible to enact for every species within an assemblage. However, by applying such techniques to keystone and habitat-forming species, considerable insight can be gained into the likely impacts of climatic change on population and community levels of organization. Moreover, a failure to include these aspects into a forecasting approach does not allow us to mechanistically predict where and when damage to ecosystems is likely to occur.

For example, as pointed out by several authors (Davis et al. 1998, Hampe 2004), most niche modeling approaches have assumed that all observed range boundaries are representative of a species' fundamental niche space. These studies have also assumed that physiological tolerance remains constant across a species range, a pattern shown to be inaccurate (Stillman 2003). Similarly, many studies assume that indicators of physiological stress such as aerial body temperature become increasingly stressful with decreasing latitude and often neglect both the importance of cold stress incurred at higher latitudes (Gilman 2005) and the importance of local and mesoscale processes on geographic patterns of intertidal stress in the middle of species ranges (Helmuth et al. 2002, 2005; Sagarin & Somero 2006). In many cases, studies have failed to provide a mechanistic link between environmental parameters such as air temperature and physiologically relevant metrics such as body temperature (Gilman et al. 2006), making it difficult to explicitly forecast how future changes in climate will impact the distribution and abundance of marine organisms.

As a result, the potential for committing both Type I errors (observing change where none truly exists) or, conversely, failing to see expected change in response to climatic variation can be large and is particularly great when physiological mechanisms are not taken into account (Helmuth et al. 2005). Many studies, for example, attempt to correlate changes in the abundance or distribution of a species with a single environmental parameter (for example, air or water temperature) along a broad-scale latitudinal gradient. There are several reasons why this approach is likely to fail. First, there is no a priori reason to expect a single environmental parameter to always act as the limiting factor at all locations along a species range. For example, for intertidal organisms, water temperature may alternate with aerial temperature stress or feeding time in setting species distributions along the species range (Harley & Helmuth 2003), and factors such as body temperature and food supply are often interactive in their effects (Gilman 2006). Second, the parameter in question may not always serve as an effective proxy for physiological stress; for example, air temperature is often a poor proxy for body temperature (Helmuth 2002, Gilman et al. 2006). It is therefore not surprising that attempts to correlate changes in water temperature with changes in intertidal species abundances have sometimes yielded mixed results (e.g., Schiel et al. 2004) because water temperature may in some cases be less important than aerial body temperature in driving physiological stress. Third, environmental measurements only have relevance within the context of an organism's tolerance to those environmental factors (Southward 1958, Stillman 2003).

Importantly, by explicitly predicting where lethal and sublethal stress should occur, researchers can avoid looking for evidence of climate change in the wrong place and at the wrong time. Ecological-forecasting approaches (the prediction of patterns of environmental stress using climatic databases, remote sensing data, global climate change models, and techniques such as heat-budget modeling) can be used to generate explicit hypotheses regarding where climate change is most likely to impact organisms and populations and where it is not. Similarly, hindcasting techniques can be used to generate maps of environmental stress that can be compared against historical records of species-range shifts to explicitly test hypotheses regarding the impacts of past climatic changes on marine communities. However, these techniques are only useful when they are combined with information on the indirect and direct effects of climate on physiological performance and population dynamics.

These approaches can be used to predict patterns of physiological stress in a spatially and temporally explicit manner and to generate explicit hypotheses that can then be tested under field conditions. The important distinction between this type of forecasting and larger-scale correlative approaches is that there is no a priori expectation of a simple relationship between a single environmental variable and the response of individuals, populations, or communities to climatic change. Instead, predictions can be tested on a spatially and temporally explicit basis, and environmental and biotic factors can alternate with one another in setting distributional limits across a cascade of scales.

An important potential weakness in the climate envelope approach that has been highlighted is that as currently used, it tends to ignore the indirect effects of climate on species interactions. Davis et al. (1998) rekindled the debate regarding the role of biotic versus abiotic factors in determining the geographic range of species. Most climate envelope (environmental niche modeling) approaches estimate species distribution patterns on the basis of environmental variables and then map suitable climate (niche) space, often ignoring the effects of species interactions and dispersal capabilities (Davis et al. 1998). Importantly, however, although these factors have not been addressed previously using such approaches, the inclusion of indirect effects in ecological-forecasting approaches is not precluded.

Instead of selection for either abiotic or biotic parameters, we argue for the incorporation of both types of data at local as well as regional spatial scales. Specifically, we suggest physiological and behavioral responses of biota to microclimate and local weather (Sanford 1999, 2002; Southward 1958) should be integrated into existing models of climate-driven biogeographic distribution (Helmuth et al. 2005, Hodkinson 1999, Porter et al. 2002). Together, this approach can be used to explicitly generate hypotheses as to when and where climate-mediated interactions, or climate alone, may set geographical distributions (Helmuth et al. 2005, Wethey 2002). For example, forecasting techniques can be used to generate spatially explicit maps of both subtidal and intertidal body temperatures of keystone species over a cascade of spatial and temporal scales. These maps can then be compared against lethal limits generated by physiological studies (Somero 2002) and studies exploring the indirect biotic influence of these impacts (Sanford 1999, Wethey 1983). These overlain maps can then be used to test the potential role of extreme temperatures and cumulative measures such as degree hours, as is currently used in subtidal environments such as coral reefs (Gleeson & Strong 1995), in setting present, past, and future range boundaries.

Such a mechanistic framework therefore recognizes that species interactions can both modify the effects of and be modified themselves by climate change (Burnaford 2004, Pincebourde & Casas 2006). Where species of cold- and warm-water origins coexist, changes in climate can alter the relative dominance of species and force competitively inferior species into refugia (Connell 1961; Southward 1967, 1991; Southward & Crisp 1954b; Wethey 1983, 1984) or cause localized extinctions to occur. The strength of predator-prey and facilitation interactions may also be affected by shifts in climate (Bertness et al. 1999; Sanford 1999, 2002; P. Moore, unpublished data), resulting in trophic mismatches and ecosystem destabilization (Power et al. 1999, Richardson & Schoeman 2004).

Importantly, such a framework can be used to predict the potential movement of invasive species. The introduction of invasive species has become more common throughout the twentieth century with the intensification of aquaculture and increased global maritime transport (Carlton & Geller 1993). The proliferation of invasive species may be aided by climatic change as local environmental conditions become suitable for initial survival and promote the earlier settlement and increased reproductive success of invasive species (Stachowicz et al. 2002). This competitive advantage may lead to reductions in the abundance or total extinction of native species from sites.

SUMMARY

A large body of evidence has conclusively demonstrated that Earth's climate is changing and that these changes in some cases have devastating impacts on natural ecosystems. Importantly, the results of long-term monitoring have shown that these changes are evident across a wide range of marine taxa and can be rapid, with major shifts in biogeographic distributions occurring on a subdecadal scale. Unless we continuously track changes in ecological communities alongside concomitant changes in key environmental variables, measured at scales relevant to organisms, we may miss important signature events such as extreme freezes that can leave their marks on assemblages for decades.

The challenge before biologists is to understand the mechanisms underlying the role of climate in driving physiological and ecological performance and the subsequent impacts on species distributions. An important consideration, however, is that climatic factors do not affect all ecological interactions, and not all impacts are negative. Moreover, when mortality events occur, they may be overwhelmed by patterns of larval dispersal. We should therefore not expect to see evidence of the impacts of climatic change everywhere we look. Conversely, we should be prepared to look beyond simple range shifts for impacts of climate change on intertidal organisms.

Subsequently, it is important to shift our emphasis away from the search for the impacts of climate change and toward a hypothesis-driven approach that recognizes the interactions between climatic and nonclimatic factors in determining patterns of species abundance and distribution (Parmesan et al. 2005). Clearly, climate change is having and will continue to have a large impact on many natural assemblages; the challenge is to forecast where and when such effects are most likely to occur, given our best estimates of future climatic change. For example, by comparing physiological tolerances against patterns of environmental variables in the field, we can determine where species are living at their realized versus fundamental niche spaces and therefore determine how close to the edges of their tolerance limits these organisms are likely to be (Somero 2002). Similarly, by looking for rapid changes in past climatic conditions—for example, inflection points where sudden cooling or heating events occurred (e.g., Crisp 1964)—we can compare our predictions of changes in faunal communities against the results of monitoring to test our understanding of community responses to climate change using hindcasting approaches. Importantly, such an approach dictates that we look not only at species-range boundaries, but also at multiple sites within species distributions.

Such a program is ambitious and difficult to attain for even a few species, much less all species over a large geographic region. Beginning with keystone and habitat-forming species may serve to explain much of the ecological impacts of climate change where they occur. To this end, the use of large-scale meta-analysis on the impacts of climate on natural ecosystems must continue, for it allows us to detect general patterns over large scales and is vital for informing policy. However, if we are to move forward in predicting the impacts of climate change, we must move beyond the search for more proof and toward methods that permit the explicit forecasting of where damage is likely to occur. Ecological forecasting is only one piece in an exceedingly large and complex puzzle, but it may serve as an effective means of initial triage, helping to determine which sites are most likely to sustain damage in coming years. Although the pioneers envisaged such an approach (Crisp & Southward 1958, Fischer-Piette 1933, Hutchins 1947, Lewis 1964, Orton 1920, Southward & Crisp 1954a), it is only with recent technological advances that we can expand on this knowledge to move from qualitative prediction to quantitative forecast.

SUMMARY POINTS

1. Long-term monitoring has shown that changes in the distribution, abundance, and range limits of intertidal species can occur rapidly, and that poleward-range edges have shifted by as much as 50 km per decade.
2. Climate change impacts intertidal species at multiple life history stages. Factors such as water temperature, aerial body temperature, and food supply are likely to alternate in their importance in space and time in setting species-range boundaries.
3. Recent evidence suggests that environmental heterogeneity along species distributions may be higher than expected. As a result, species may display responses to climate change at multiple sites within their ranges, and not just at poleward- and equatorial-range boundaries.
4. Many studies rely on environmental proxies for physiological stress (e.g., air or water temperature) and fail to consider the environment at the scale of the organism; often these patterns in air or water temperature can be misleading.
5. Although large-scale climate envelope approaches provide an initial approximation of the potential effects of climate change, they often ignore the indirect effects of warming and the influence of nonclimate factors in setting range boundaries.
6. Ecological-forecasting approaches provide a means of generating explicit hypotheses regarding the effects of climate change on species distribution patterns.

FUTURE ISSUES

1. Our understanding of the future effects of climate change mandates that we validate our hypotheses by the comparison of historical changes in climate with the resulting responses of ecological communities. The continuation of long-term monitoring projects is thus an essential part of any future forecasting program.
2. Our understanding of the indirect effects of climate change, for example, on rates of foraging and sublethal levels of physiological stress, is still limited.
3. The interaction between multiple physiological stressors related to climate change (e.g., aerial and aquatic body temperatures, pH) is still poorly understood for most intertidal organisms.

ACKNOWLEDGMENTS

B.H. and N.M. were supported by funding from NASA NNG04GE43G and by NOAA NA04NOS4780264. S.J.H., N.M., and P.M. were funded by the MarClim project, a consortium-funded project supported by English Nature; Scottish Natural Heritage; Countryside Council for Wales; Environment Agency; The Scottish Executive; States of Jersey; Department for Environment, Food, and Rural Affairs; The Crown Estate; Joint Nature Conservation Committee; and World Wildlife Fund. S.J.H. and P.M. were funded by NERC grants GR/02390 and NE/E000029/1, and S.J.H. by a NERC grant-in-aid fellowship. Professor Alan Southward is thanked for inspiring this work and for starting and continuing many of the most-valuable long-term data sets. The authors wish to thank David Wethey, Jerry Hilbish, and Sally Woodin for helpful comments on the manuscript and for many hours of thoughtful discussion. Many of the ideas included herein were prompted by the NSF-funded CORONA working group, and for this we are particularly grateful to Cliff Cunningham. This is publication number one of the Robin Hood Club for International Cultural Exchange.

LITERATURE CITED

Allison GW, Lubchenco J, Carr MH. 1998. Marine reserves are necessary but not sufficient for marine conservation. *Ecol. Appl.* 8:S79–92

Andrewartha HG, Birch LC. 1954. *The Distribution and Abundance of Animals*. Chicago: Univ. Chicago Press

Anil AC, Desai D, Khandeparker L. 2001. Larval development and metamorphosis in *Balanus amphitrite* Darwin (Cirripedia; Thoracica): significance of food concentration, temperature and nucleic acids. *J. Exp. Marine Biol. Ecol.* 263:125–41

Arnold R. 1903. *The Paleontology and Stratigraphy of the Marine Pliocene and Pleistocene of San Pedro*. San Francisco: Calif. Acad. Sci.

Ballantine WJ. 1961. A biologically-defined exposure scale for the comparative description of rocky shores. *Field Stud.* 1:73–84

Barry JP, Baxter CH, Sagarin RD, Gilman SE. 1995. Climate-related, long-term faunal changes in a California rocky intertidal community. *Science* 267:672–75

Bayne BL, Thompson RJ, Widdows J. 1973. Some effects of temperature and food on the rate of oxygen consumption by *Mytilus edulis* L. In *Effects of Temperature on Ectothermic Organisms*, ed. Weiser W. pp. 181–93. Berlin: Springer-Verlag

Berge J, Johnsen G, Nilsen F, Gulliksen B, Slagstad D. 2005. Ocean temperature oscillations enable reappearance of blue mussels *Mytilus edulis* in Svalbard after a 1000 year absence *Mar. Ecol. Prog. Ser.* 303:167–75

Bertness MD, Leonard GH, Levine JM, Bruno JF. 1999. Climate-driven interactions among rocky intertidal organisms caught between a rock and a hot place. *Oecologia* 120:446–50

Bode A, Lombas I, Anadon N. 1986. Preliminary studies on the reproduction and population dynamics of *Monodonta lineata* and *Gibbula umbilicalis* (Mollusca, Gastropoda) on the central coast of Asturias (N. Spain). *Hydrobiologia* 142:31–39

Bowman RS, Lewis JR. 1986. Geographical variation in the breeding cycles and recruitment of *Patella* spp. *Hydrobiologia* 142:41–56

Briggs JC. 1974. *Marine Biogeography*. New York: McMillan

Brockington S, Clarke A. 2001. The relative influence of temperature and food on the metabolism of a marine invertebrate. *J. Exp. Mar. Biol. Ecol.* 258:87–99

Broitman BR, Blanchette CA, Gaines SD. 2005. Recruitment of intertidal invertebrates and oceanographic variability at Santa Cruz Island, California. *Limnol. Oceanogr.* 50:1473–79

Broitman BR, Navarrete S, Smith F, Gaines SD. 2001. Geographic variation of southeastern Pacific intertidal communities. *Mar. Ecol. Prog. Ser.* 224:21–34

Brown JH. 1984. On the relationship between abundance and distribution of species. *Am. Nat.* 124:255–79

Burnaford JL. 2004. Habitat modification and refuge from sublethal stress drive a marine plant-herbivore association. *Ecology* 85:2837–49

Carlton JT, Geller JB. 1993. Ecological roulette: the global transport of nonindigenous marine organisms. *Science* 261:78–82

Carrington E. 2002. Seasonal variation in the attachment strength of blue mussels: causes and consequences. *Limnol. Oceanogr.* 47:1723–33

Carroll ML, Highsmith RC. 1996. Role of catastrophic disturbance in mediating *Nucella-Mytilus* interactions in the Alaskan rocky intertidal. *Mar. Ecol. Prog. Ser.* 138:125–33

Chown SL, Gaston KJ, Robinson D. 2004. Macrophysiology: large-scale patterns in physiological traits and their ecological implications. *Funct. Ecol.* 18:159–67

Clarke A. 1988. Seasonality in the Antarctic marine environment. *Comp. Biochem. Physiol. B Biochem. Mol. Biol.* 90B:461–73

Clarke A. 2003. Costs and consequences of evolutionary temperature adaptation. *Trends Ecol. Evol.* 18:573–81

Clark JS, Carpenter SR, Barber M, Collins S, Dobson A, et al. 2001. Ecological forecasts: an emerging imperative. *Science* 293:657–60

Connell JH. 1961. The influence of interspecific competition and other factors on the distribution of the barnacle *Chthamalus stellatus*. *Ecology* 42:710–23

Connell JH. 1972. Community interactions on marine rocky intertidal shores. *Annu. Rev. Ecol. Syst.* 3:169–92

Connolly SR. Roughgarden J. 1998. A latitudinal gradient in northeast Pacific intertidal community structure: evidence for an oceanographically based synthesis of marine community theory *Am. Nat.* 151 4:311–26

Crisp DJ. 1964. The effects of the severe winter of 1962–63 on marine life in Britain. *J. Anim. Ecol.* 33:165–210

Crisp DJ, Southward AJ. 1958. The distribution of intertidal organisms along the coasts of the English Channel. *J. Mar. Biol. Assoc. UK* 37:157–208

Crisp DJ, Southward AJ. 1959. The further spread of *Elminius modestus* in the British Isles to 1959. *J. Mar. Biol. Assoc. U.K.* 38:429–37

Dahlhoff EP. 2004. Biochemical indicators of stress and metabolism: applications for marine ecological studies. *Annu. Rev. Physiol.* 66:183–207

Dahlhoff EP, Menge BA. 1996. Influence of phytoplankton concentration and wave exposure on the ecophysiology of *Mytilus californianus*. *Mar. Ecol. Prog. Ser.* 144:97–107

Davenport J, Davenport JL. 2005. Effects of shore height, wave exposure and geographical distance on thermal niche width of intertidal fauna. *Mar. Ecol. Prog. Ser.* 292:41–50

Davis AJ, Lawton JH, Shorrocks B, Jenkinson LS. 1998. Individualistic species responses invalidate simple physiological models of community dynamics under global environmental change. *J. Anim. Ecol.* 67:600–12

Denny MW, Helmuth B, Leonard GH, Harley CDG, Hunt LJH, Nelson EK. 2004. Quantifying scale in ecology: lessons from a wave-swept shore. *Ecol. Monogr.* 74:513–32

Fields PA, Graham JB, Rosenblatt RH, Somero GN. 1993. Effects of expected global climate change on marine faunas. *Trends Ecol. Evol.* 8:361–67

Fischer-Piette E. 1933. Le contour géographique des côtes françaises et anglaises de la Manche et la répartition de *Mytilus edulis* et *Balanus perforatus*. *C. R. Soc. Biogeogr.* 86:70–72

Fischer-Piette E. 1955. Repartition, le long des cotes septentrionales de l'Espagne, des principales especes peuplant les rochers intercotidaux. *Ann. Inst. Oceanogr.* 31:37–124

Fischer-Piette E, Prenant M. 1956. Distribution des Cirripèdes intercôtidaux d'Espagne septentrionale. *Bull. Cent. Etud. Rech. Sci. Biarritz* 1:7–19

Forbes E. 1853. *The Natural History of Europe's Seas*. London: John van Vorst. 240 pp.

Frank PW. 1975. Latitudinal variation in the life history features of the black turban snail *Tegula funebralis* (Prosobranchis: Trochidae). *Mar. Biol.* 31:181–92

Gaylord B, Gaines SD. 2000. Temperature or transport? Range limits in marine species mediated solely by flow. *Am. Nat.* 155:769–89

Gilman S. 2005. A test of Brown's principle in the intertidal limpet *Collisella scabra* (Gould, 1846). *J. Biogeogr.* 32:1583–89

Gilman S. 2006. Life at the edge: an experimental study of a poleward range boundary. *Oecologia*. 148:270–79

Gilman SE, Wethey DS, Helmuth B. 2006. Variation in the sensitivity of organismal body temperature to climate change over local and geographic scales. *Proc. Natl. Acad. Sci. USA* 103(25):9560–65

Gleeson MW, Strong AE. 1995. Applying MCSST to coral reef bleaching. *Adv. Space Res.* 16:151–54

Hallett TB, Coulson T, Pilkington JG, Clutton-Brock TH, Pemberton JM, Grenfell BT. 2004. Why large-scale climate indices seem to predict ecological processes better than local weather. *Nature* 430:71–75

Hampe A. 2004. Bioclimate envelope models: what they detect and what they hide. *Glob. Ecol. Biogeogr.* 13:469–76

Harley CDG. 2003. Abiotic stress and herbivory interact to set range limits across a two-dimensional stress gradient. *Ecology* 84:1477–88

Harley CDG, Helmuth BST. 2003. Local and regional scale effects of wave exposure, thermal stress, and absolute vs effective shore level on patterns of intertidal zonation. *Limnol. Oceanogr.* 48:1498–508

Harley CDG, Hughes AR, Hultgren K, Miner BG, Sorte CJB, et al. 2006. The impacts of climate change in coastal marine systems. *Ecol. Lett.* 9:228–41

Harley CDG, Lopez JP. 2003. The natural history, thermal physiology, and ecological impacts of intertidal mesopredators, *Oedoparena* spp. (Diptera: Dryomyzidae). *Invertebr. Biol.* 122:61–73

Harris LG, Tyrrell MC. 2001. Changing community states in the Gulf of Maine: synergism between invaders, overfishing and climate change. *Biol. Invasions* 3:9–21

Harris LG, Tyrrell MC, Chester CM. 1998. Changing ecological patterns for two *Asterias* species in the southwestern Gulf of Maine over a 20 year period. In *Proceedings of Ninth International Echinoderm Conference*, ed. R Mooi, pp. 243–48. Rotterdam: AA Balkema

Hatton H. 1938. Essais de bionomie explicative sur quelques espèces intercotidales d'algues et d'animaux. *Ann. Inst. Oceanogr. Monaco* 17:241–48

Hawkins SJ, Hartnoll RG. 1982. Settlement patterns of *Semibalanus balanoides* (L.) in the Isle of Man (1977–1981). *J. Exp. Mar. Biol. Ecol.* 62:271–83

Hawkins SJ, Hartnoll RG. 1983. Grazing of intertidal algae by marine invertebrates. *Oceanogr. Mar. Biol. Annu. Rev. London* 21:195–282

Hawkins SJ, Hartnoll RG, Kain JM, Norton TA. 1992. Plant animal interactions on hard substrata in the north-east Atlantic. In *Plant Animal Interactions in the Marine Benthos*, ed. DM John, SJ Hawkins, JH Price. Oxford, UK: Clarendon Press

Hawkins SJ, Southward AJ, Genner MJ. 2003. Detection of environmental change in a marine ecosystem—evidence from the western English Channel. *Sci. Total Environ.* 310:245–56

Helmuth B. 2002. How do we measure the environment? Linking intertidal thermal physiology and ecology through biophysics. *Integr. Comp. Biol.* 42:837–45

Helmuth B, Carrington E, Kingsolver JG. 2005. Biophysics, physiological ecology, and climate change: Does mechanism matter? *Annu. Rev. Physiol.* 67:177–201

Helmuth BS, Harley CDG, Halpin P, O'Donnell M, Hofmann GE, Blanchette C. 2002. Climate change and latitudinal patterns of intertidal thermal stress. *Science* 298:1015–17

Helmuth BST. 1999. Thermal biology of rocky intertidal mussels: quantifying body temperatures using climatological data. *Ecology* 80:15–34

Herbert RJH, Hawkins SJ, Sheader M, Southward AJ. 2003. Range extension and reproduction of the barnacle *Balanus perforatus* in the eastern English Channel. *J. Mar. Biol. Assoc. U.K.* 83:73–82

Herrlinger TJ. 1981. Range extension of *Kelletia kelletii*. *Veliger* 24:78

Hewatt WG. 1937. Ecological studies on selected marine intertidal communities of Monterey Bay, California. *Am. Midl. Nat.* 18:161–206

Hilbish TJ. 1981. Latitudinal variation in freezing tolerance of *Malampus bidentatus* (Say) (Gastropoda: Pulmonata). *J. Exp. Mar. Biol. Ecol.* 52:283–97

Hodkinson ID. 1999. Species response to global environmental change or why eco-physiological models are important: a reply to Davis et al. *J. Anim. Ecol.* 68:1259–62

Hoegh-Guldberg O, Pearse JS. 1995. Temperature, food availability, and the development of marine invertebrate larvae. *Am. Zool.* 35:415–25

Hofmann GE. 2005. Patterns of Hsp gene expression in ectothermic marine organisms on small to large biogeographic scales. *Integr. Comp. Biol.* 45:247–55

Hofmann GE, Burnaford JL, Fielman KT. 2005. Genomics-fueled approaches to current challenges in marine ecology. *Trends Ecol. Evol.* 20:305–11

Holtmeier FK, Broll G. 2005. Sensitivity and response of northern hemisphere altitudinal and polar treelines to environmental change at landscape and local scales. *Global Ecol. Biogeogr.* 14:395–410

Hutchins LW. 1947. The bases for temperature zonation in geographical distribution. *Ecol. Monogr.* 17:325–35

Intergovermental Panel on Climate Change. 2001. Climate change 2001: the scientific basis. **http://www.grida.no/climate/ipcc_tar/wg1/figts-22.htm**

Johnson CW. 1934. List of marine mollusca of the Atlantic Coast from Labrador to Texas. Boston: Boston Soc. Nat. Hist. 203 pp.

Kendall MA. 1987. The age and size structure of some northern populations of the trochid gastropod *Monodonta lineata*. *J. Molluscan Stud.* 53:213–22

Kendall MA, Lewis JR. 1986. Temporal and spatial patterns in the recruitment of *Gibbula umbilicalis*. *Hydrobiologia* 142:15–22

Kennedy VS. 1976. Desiccation, higher temperatures and upper intertidal limits of three species of sea mussels (Mollusca: Bivalvia) in New Zealand. *Mar. Biol.* 35:127–37

Knox GA. 1960. Littoral ecology and biogeography of the southern oceans. *Proc. R. Soc. London Ser. B* 152:577–624

Leichter JJ, Wing SR, Miller SL, Denny MW. 1996. Pulsed delivery of subthermocline water to Conch Reef (Florida Keys) by internal tidal bores. *Limnol. Oceanogr.* 41:1490–501

Leslie HM, Breck EM, Chan F, Lubchenco J, Menge BA. 2005. Barnacle reproductive hotspots linked to nearshore ocean conditions. *Proc. Natl. Acad. Sci. USA* 102:10534–39

Lewis JR. 1964. *The Ecology of Rocky Shores*. London: English Univ. Press

Lewis JR. 1986. Latitudinal trends in reproduction, recruitment and population characteristics of some rocky littoral molluscs and cirripedes. *Hydrobiologia* 142:1–13

Li R, Brawley SH. 2004. Improved survival under heat stress in intertidal embryos (*Fucus* spp.) simultaneously exposed to hypersalinity and the effect of parental thermal history. *Mar. Biol.* 144:205–13

Lima FP, Queiroz N, Ribeiro PA, Hawkins SJ, Santos AM. 2006. Geographic expansion of a marine gastropod, *Patella rustica* Linnaeus, 1758, and its relation with unusual climatic events. *J. Biogeogr.* 33:812–22

Lohnhart SA, Tupen JW. 2001. New range records of 12 marine invertebrates: the role of El Nino and other mechanisms in southern and central California. *Bull. South. Calif. Acad. Sci.* 100:238–48

Lubchenco J, Navarrete SA, Tissot BN, Castilla JC. 1993. Possible ecological consequences to global climate change: nearshore benthic biota of Northeastern Pacific coastal ecosystems. In *Earth System Responses to Global Change*, ed. HA Mooney, ER Fuentes, BI Kronberg, pp. 147–66. San Diego: Academic

Luppi T, Spivak E, Bas C. 2003. The effects of temperature and salinity on larval development of *Armases rubripes* Rathbun, 1897 (Brachyura, Grapsoidea, Sesarmidae), and the southern limit of its geographical distribution. *Estuarine Coast. Shelf Sci.* 58:575–85

Maddock L, Pingree R. 1977. Numerical simulations of Portland Bill tidal eddies *Estuarine Coast. Mar. Sci.* 6:353–63

McCarty JP. 2001. Ecological consequences of recent climate change. *Conserv. Biol.* 15:320–31

Mieszkowska N, Kendall MA, Hawkins SJ, Leaper R, Williamson P, et al. 2006. Changes in the range of some common rocky shore species in Britain—a response to climate change? *Hydrobiologia* 555:241–51

Mieszkowska N, Leaper R, Moore P, Kendall MA, Burrows MT, et al. 2005. Assessing and predicting the influence of climatic change using intertidal rocky shore biota. *Occas. Publ. J. Mar. Biol. Assoc. U.K.* 20:1–53

Moore HB. 1936. The biology of *Balanus balanoides*. V. distribution in the Plymouth area. *J. Mar. Biol. Assoc. U.K.* 20:701–16

Moore HB, Kitching JA. 1939. The biology of *Chthamalus stellatus* (Poli). *J. Mar. Biol. Assoc. U.K.* 23:521–41

Morris RH, Abbott DP, Haderlie EC. 1980. *Intertidal invertebrates of California*. Palo Alto, CA: Stanford Univ. Press

Orton JH. 1920. Sea temperature, breeding and distribution of marine animals. *J. Mar. Biol. Assoc. U.K.* 2:299–366

Orton JH. 1929. On the occurrence of *Echinus esculentus* on the foreshore in the British Isles. *J. Mar. Biol. Assoc. U.K.* 16:289–96

Paine RT. 1994. *Marine Rocky Shores and Community Ecology: An Experimentalist's Perspective*. Oldendorf/Luhe, Germany: Ecology Inst.

Palumbi SR, Gaines SD, Leslie H, Warner RR. 2003. New wave: high-tech tools to help marine reserve research. *Front. Ecol. Environ.* 1:73–79

Parmesan C, Gaines S, Gonzalez L, Kaufman DM, Kingsolver J, et al. 2005. Empirical perspectives on species borders: from traditional biogeography to global change. *Oikos* 108:58–75

Parmesan C, Galbraith H. 2004. *Observed Impacts of Global Climate Change in the U.S.* Arlington, VA: Pew Cent. Glob. Clim. Change

Parmesan C, Yohe G. 2003. A globally coherent fingerprint of climate change impacts across natural systems. *Nature* 421:37–42

Pechenik JA. 1999. On the advantages and disadvantages of larval stages in benthic marine invertebrate life cycles. *Mar. Ecol. Prog. Ser.* 177:269–97

Petraitis PS. 1992. Effects of body size and water temperature on grazing rates of four intertidal gastropods. *Austr. J. Ecol.* 17:409–14

Philippart C, van Aken H, Beukema J, Bos O, Cadee G, Dekker R. 2003. Climate-related changes in recruitment of the bivalve *Macoma balthica*. *Limnol. Oceanogr.* 48:2171–85

Phillips NE. 2005. Growth of filter-feeding benthic invertebrates from a region with variable upwelling intensity. *Mar. Ecol. Prog. Ser.* 295:79–85

Pincebourde S, Casas J. 2006. Multitrophic biophysical budgets: thermal ecology of an intimate herbivore insect-plant interaction. *Ecol. Monogr.* 76(2):175–94

Pineda J, Riebensahm D, Medeiros-Bergen D. 2002. *Semibalanus balanoides* in winter and spring: larval concentration, settlement, and substrate occupancy. *Mar. Biol.* 140:789–800

Porter WP, Sabo JL, Tracy CR, Reichman OJ, Ramankutty N. 2002. Physiology on a landscape scale: plant-animal interactions. *Integr. Comp. Biol.* 42:431–53

Power AM, Delany J, Myers AA, O'Riordan RM, McGrath D. 1999. Prolonged settlement and prediction of recruitment in two sympatric Chthamalus species from south-west Ireland. *J. Mar. Biol. Assoc. U.K.* 79:941–43

Raffaelli DG, Hawkins SJ. 1996. *Intertidal Ecology*. London: Chapman & Hall. 250 pp.

Richardson AJ, Schoeman DS. 2004. Climate impact on plankton ecosystems in the northeast Atlantic. *Science* 305:1609–12

Ritchie J. 1927. Reports on the prevention of the growth of mussels in submarine shafts and tunnels at Westbank Electric Station, Portobello. *Trans. R. Scot. Soc. Arts* 19:1–20

Rivadeneira MM, Fernandez M. 2005. Shifts in southern endpoints of distribution in rocky intertidal species along the south-eastern Pacific coast. *J. Biogeogr.* 32:203–9

Root TL, Schneider SH. 1995. Ecology and climate: research strategies and implications. *Science* 269:334–41

Russell FS, Southward AJ, Boalch GT, Butler EI. 1971. Changes in biological conditions in the English Channel off Plymouth during the last half century. *Nature* 243:468–70

Sagarin RD, Barry JP, Gilman SE, Baxter CH. 1999. Climate related changes in an intertidal community over short and long time scales. *Ecol. Monogr.* 69:465–90

Sagarin RD, Gaines SD. 2002. Geographical abundance distributions of coastal invertebrates: using one-dimensional ranges to test biogeographic hypotheses. *J. Biogeogr.* 29:985–97

Salvigsen O, Forman SL, Miller GH. 1992. Thermophilous molluscs on Svalbard during the Holocene and their implications. *Polar Res.* 11:1–10

Sanders BM, Hope C, Pascoe VM, Martin LS. 1991. Characterization of the stress protein response in two species of *Collisela* limpets with different temperature tolerances. *Physiol. Zool.* 64:1471–89

Sanford E. 1999. Regulation of keystone predation by small changes in ocean temperature. *Science* 283:2095–97

Sanford E. 2002. Water temperature, predation, and the neglected role of physiological rate effects in rocky intertidal communities. *Integr. Comp. Biol.* 42:881–91

Sauvageau C. 1897. Note prèliminare sur les Algues marines du golfe de Gascogne. *J. Bot.* 11:166–307

Schiel DR. 2004. The structure and replenishment of rocky shore intertidal communities and biogeographic comparisons. *J. Exp. Mar. Biol. Ecol.* 300:309–42

Schiel DR, Steinbeck JR, Foster MS. 2004. Ten years of induced ocean warming causes comprehensive changes in marine benthic communities. *Ecology* 85:1833–39

Schmidt PS, Rand DM. 2001. Adaptive maintenance of genetic polymorphism in an intertidal barnacle: habitat- and life-stage-specific survivorship of *Mpi* genotypes. *Evolution* 55:1336–44

Schoch GC, Dethier MN. 1996. Scaling up: the statistical linkage between organismal abundance and geomorphology on rocky intertidal shorelines. *J. Exp. Mar. Biol. Ecol.* 201:37–72

Service RF. 2004. New dead zone off Oregon coast hints at sea change in currents. *Science* 305:1099

Smith AGG, Gordon M Jr. 1948. The marine mollusks and brachiopods of Monterey Bay, California, and vicinity. *Proc. Calif. Acad. Sci. Fourth Ser.* 26:147–45

Somero GN. 2002. Thermal physiology and vertical zonation of intertidal animals: optima, limits, and costs of living. *Integr. Comp. Biol.* 42:780–89

Sotka EE, Wares JP, Barth JA, Grosberg RK, Palumbi SR. 2004. Strong genetic clines and geographical variation in gene flow in the rocky intertidal barnacle *Balanus glandula*. *Mol. Ecol.* 13:2143–56

Southward AJ. 1958. Note on the temperature tolerances of some intertidal animals in relation to environmental temperatures and geographical distribution. *J. Mar. Biol. Assoc. U.K.* 37:49–66

Southward AJ. 1963. The distribution of some plankton animals in the English Channel and Western approaches. III. Theories about long-term biological changes, including fish. *J. Mar. Biol. Assoc. U.K.* 43:1–29

Southward AJ. 1967. Recent changes in abundance of intertidal barnacles in southwest England: a possible effect of climatic deterioration. *J. Mar. Biol. Assoc. U.K.* 47:81–95

Southward AJ. 1980 The western English Channel—an inconsistent ecosystem? *Nature* 285:361–66

Southward AJ. 1991. Forty years of changes in species composition and population density of barnacles on a rocky shore near Plymouth. *J. Mar. Biol. Assoc. U.K.* 71:495–513

Southward AJ, Crisp DJ. 1954a. The distribution of certain intertidal animals around the Irish coast. *Proc. R. Irish Acad.* 57:1–29

Southward AJ, Crisp DJ. 1954b. Recent changes in the distribution of the intertidal barnacles *Chthamalus stellatus* (Poli) and *Balanus balanoides* L. in the British Isles. *J. Anim. Ecol.* 23:163–77

Southward AJ, Crisp DJ. 1956. Fluctuations in the distribution and abundance of intertidal barnacles. *J. Mar. Biol. Assoc. U.K.* 35:211–29

Southward AJ, Hawkins SJ, Burrows MT. 1995. Seventy years' observations of changes in distribution and abundance of zooplankton and intertidal organisms in the western English Channel in relation to rising sea temperature. *J. Therm. Biol.* 20:127–55

Southward AJ, Langmead O, Hardman-Mountford NJ, Aitken J, Boalch GT, et al. 2005. Long-term oceanographic and ecological research in the western English Channel. *Adv. Mar. Biol.* 47:1–105

Stachowicz JJ, Terwin JR, Whitlatch RB, Osman RW. 2002. Linking climate change and biological invasion: ocean warming facilitates nonindigenous species invasions. *Proc. Natl. Acad. Sci. USA* 99:15497–500

Stainforth DA, Aina T, Christensen C, Collins M, Faull N, et al. 2005. Uncertainty in predictions of the climate response to rising levels of greenhouse gases. *Nature* 433:403–6

Stenseth NC, Ottersen G, Hurrell JW, Mysterud A, Lima M, et al. 2003. Studying climate effects on ecology through the use of climate indices: the North Atlantic Oscillation, El Niño Southern Oscillation and beyond. *Proc. R. Soc. Lond. Ser. B* 270:2087–96

Stillman J. 2003. Acclimation capacity underlies susceptibility to climate change. *Science* 301:65

Strub PT, James C. 2002. Altimeter-derived surface circulation in the large-scale NE Pacific Gyres. Part 2. 1997–1998 El Niño anomalies. *Prog. Oceanogr.* 53:185–214

Thompson RC, Crowe TP, Hawkins SJ. 2002. Rocky intertidal communities: past environmental changes, present status and predictions for the next 25 years. *Environ. Conserv.* 29:168–91

Thompson RC, Norton TA, Hawkins SJ. 2004. Physical stress and biological control regulate the producer-consumer balance in intertidal biofilms. *Ecology* 85:1372–82

Tomanek L, Sanford E. 2003. Heat-shock protein 70 (Hsp70) as a biochemical stress indicator: an experimental field test in two congeneric intertidal gastropods (Genus: *Tegula*). *Biol. Bull.* 205:276–84

Tsuchiya M. 1983. Mass mortality in a population of the mussel *Mytilus edulis* L. caused by high temperature on rocky shores. *J. Exp. Mar. Biol. Ecol.* 66:101–11

Underwood AJ, Jernakoff P. 1981. The effects of tidal height, wave-exposure, seasonality and rock-pools on grazing and the distribution of intertidal macroalgae in New South Wales. *J. Exp. Mar. Biol. Ecol.* 75:71–96

Vance T. 2004. *Loss of the northern species* Alaria esculenta *from Southwest Britain and the implications for macroalgal succession*. Masters thesis, Univ. of Plymouth, Plymouth, UK

Weslawski JW, Zajaczkowski M, Wiktor J, Szymelfenig M. 1997. Intertidal zone of Svalbard. 3. Littoral of a subarctic, oceanic island: Bjornoeya. *Polar Biol.* 18:45–52

Wethey DS. 1983. Geographic limits and local zonation: the barnacles *Semibalanus* (*Balanus*) and *Chthamalus* in New England. *Biol. Bull.* 165:330–41

Wethey DS. 1984. Sun and shade mediate competition in the barnacles *Chthamalus* and *Semibalanus*: a field experiment. *Biol. Bull.* 167:176–85

Wethey DS. 1985. Catastrophe, extinction, and species diversity: a rocky intertidal example. *Ecology* 66:445–56

Wethey DS. 1986. Ranking of settlement cues by barnacle larvae: influence of surface contour. *Bull. Mar. Sci.* 39:393–400

Wethey DS. 2002. Biogeography, competition, and microclimate: the barnacle *Chthamalus fragilis* in New England. *Integr. Comp. Biol.* 42:872–80

Widdowson TB. 1970. A taxonomic revision of the genus *Alaria* Greville. *Syesis* 4:11–49

Wiens JJ, Graham CH. 2005. Niche conservatism: integrating evolution, ecology, and conservation biology. *Annu. Rev. Ecol. Evol. Syst.* 26:519–39

Williams GA, Morritt D. 1995. Habitat partitioning and thermal tolerance in a tropical limpet, *Cellana grata*. *Mar. Ecol. Prog. Ser.* 124:89–103

Witman JD, Smith F. 2003. Rapid community change at a tropical upwelling site in the Galápagos Marine Reserve. *Biodivers. Conserv.* 12:25–45

Wolcott TG. 1973. Physiological ecology and intertidal zonation in limpets (*Acmaea*): a critical look at "limiting factors." *Biol. Bull.* 145:389–422

Yamada SB, Dumbauld BR, Kalin A, Hunt CE, Figlar-Barnes R, Randall A. 2005. Growth and persistance of a recent invader *Carcinus maenas* in estuaries of the northeastern Pacific. *Biol. Invasion* 7:309–21

Zacherl D, Gaines SD, Lonhart SI. 2003. The limits to biogeographical distributions: insights from the northward range extension of the marine snail, *Kelletia kelletii* (Forbes, 1852). *J. Biogeogr.* 30:913–24

Has Vicariance or Dispersal Been the Predominant Biogeographic Force in Madagascar? Only Time Will Tell

Anne D. Yoder and Michael D. Nowak

Department of Biology, Duke University, Durham, North Carolina 27708; email: anne.yoder@duke.edu, mdn3@duke.edu

Key Words

divergence time estimation, Gondwana, historical biogeography

Abstract

Madagascar is one of the world's hottest biodiversity hot spots due to its diverse, endemic, and highly threatened biota. This biota shows a distinct signature of evolution in isolation, both in the high levels of diversity within lineages and in the imbalance of lineages that are represented. For example, chameleon diversity is the highest of any place on Earth, yet there are no salamanders. These biotic enigmas have inspired centuries of speculation relating to the mechanisms by which Madagascar's biota came to reside there. The two most probable causal factors are Gondwanan vicariance and/or Cenozoic dispersal. By reviewing a comprehensive sample of phylogenetic studies of Malagasy biota, we find that the predominant pattern is one of sister group relationships to African taxa. For those studies that include divergence time analysis, we find an overwhelming indication of Cenozoic origins for most Malagasy clades. We conclude that most of the present-day biota of Madagascar is comprised of the descendents of Cenozoic dispersers, predominantly with African origins.

Ignoring temporal information obscures the connection between biogeographic patterns and their underlying causes.
Donoghue & Moore, 2003
You're either on the bus, or off the bus.
Ken Kesey, as quoted by Tom Wolfe, 1968

THE KEY TO GONDWANA

To have an interest in the historical biogeography of Madagascar necessitates a thorough understanding of the origins and gradual sundering of Gondwana. When the supercontinent Pangea began to divide into the southern continental landmasses (Gondwana) and northern continental landmasses (Laurasia), approximately 175 Ma, Madagascar was tucked away, deep within Gondwana (Rakotosolofo et al. 1999, Reeves et al. 2002). Very shortly after the division of the southern and northern components of Pangea, Madagascar began the long journey to its current state of remote isolation in the Indian Ocean. Presently, Madagascar is surrounded by a vast oceanic barrier on all sides. It is closest to continental Africa, approximately 400 km to the west, but lies 4000 km from India, 5000 km from Antarctica, and 6000 km from Australia. It is therefore remarkable to consider that each of these landmasses was at one time contiguous with Madagascar. As a consequence of these deep geological relationships, the island's foundations are comprised of ancient continental crust, some of which can be traced back to more than 3200 million Ma (de Wit 2003). Thus, although Madagascar's roughly three-quarter-million km^2 (about the size of Texas) comprises less than 0.4% of Earth's land surface area, it has existed as an essential constituent of the reconfiguration of the Southern Hemisphere for the past 165 million years.

Initially, Gondwana was a single contiguous supercontinent comprised of what would become Africa, South America, Antarctica, Australia, India, and Madagascar (**Figure 1a**). Shortly after separating from Laurasia (**Figure 1b**), rifting between western Gondwana (Africa plus South America) and eastern Gondwana (Madagascar plus India, Antarctica, and Australia) commenced (Briggs 2003) as evidenced by the vast outpourings of the Karoo volcanics (182 ± 1 Ma) (de Wit 2003). From that point, Madagascar and the rest of eastern Gondwana began to drift southward relative to Africa, sliding along the strike-slip fault known as the Davie Ridge (Bassias 1992, Reeves & de Wit 2000). There is a general consensus among geologists that this occurred sometime between 165 and 155 Ma (Agrawal et al. 1992, Briggs 2003, Rabinowitz et al. 1983, Reeves & de Wit 2000, Reeves et al. 2002, Scotese 2000). By 140 Ma (Seward et al. 2004), marine conditions are clearly prevalent along the entirety of Madagascar's west coast. Thus, although separation from Africa began as early as 165 Ma, there was a subsequent period of perhaps 20 million years wherein biotic exchange would have been likely between western and eastern Gondwana. Madagascar reached its current position with respect to Africa by 130–118 Ma (Harland et al. 1990, Rabinowitz et al. 1983, Seward et al. 2004). Although shifts in latitude and relationships to other landmasses remained dynamic for many millions of years subsequent to this positioning, it is certain that Madagascar's present geographic isolation relative to

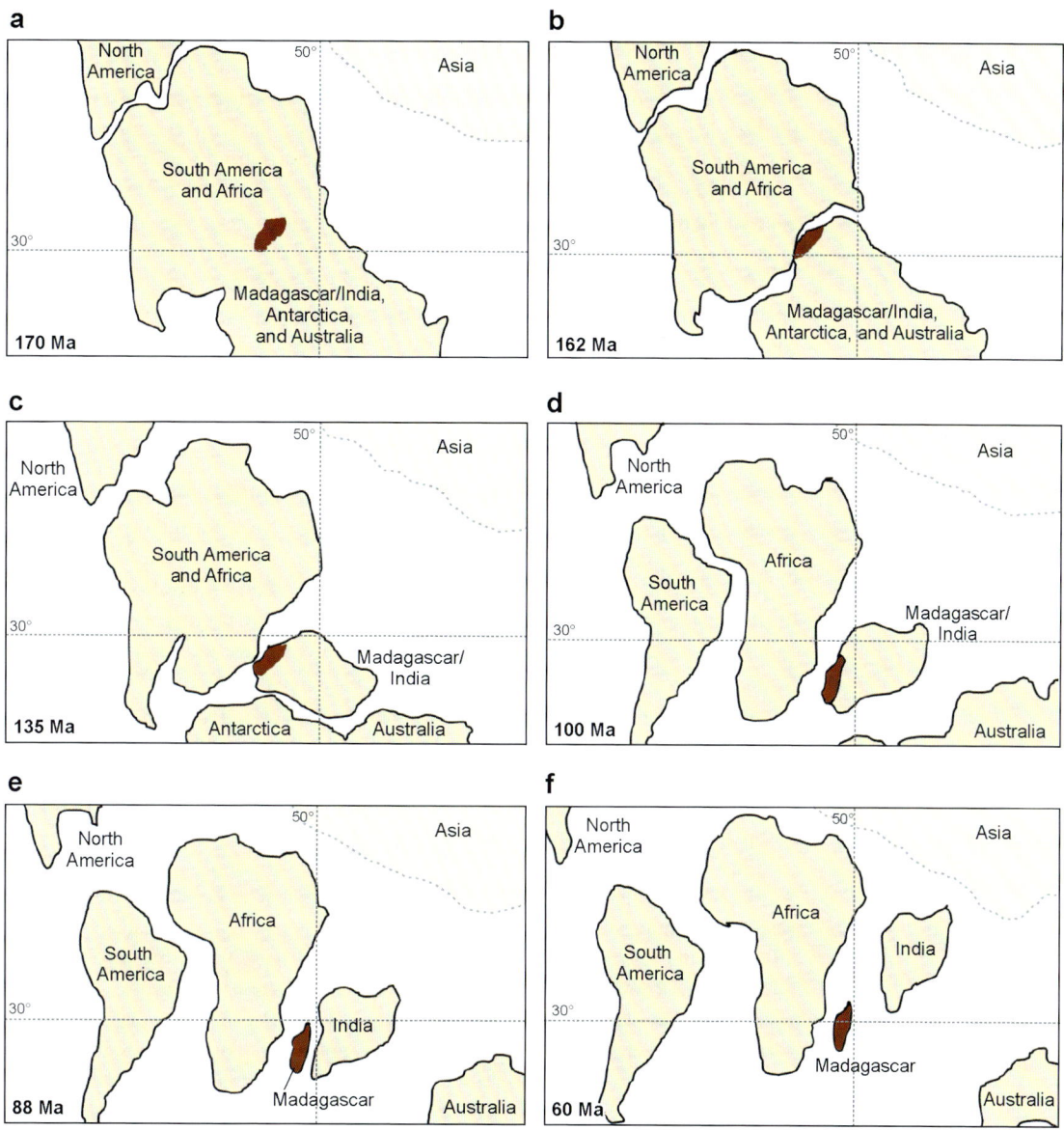

Figure 1

Paleoreconstructions of the breakup of Pangea, and Madagascar's subsequent geographic isolation. Redrawn from (Scotese 2000).

Africa has been stable for at least 118 million years, and probably, considerably longer. Moreover, the Mozambique Channel, separating Madagascar and Africa, is quite deep (3000 m, at the deepest point) and would not have been notably affected by changing sea levels after Madagascar's and Africa's final separation (Krause et al. 1997a).

From this point onward, Madagascar and India remained in close contact, forming the IndoMadagascar subcontinent (**Figures 1c** & **1d**), until the commencement of the northeasterly drift of the Seychelles-Indian block away from Madagascar's eastern margin (**Figure 1e**). The precise timing of India's separation from Madagascar has been variably dated from 100–95 Ma (Plummer 1996), to 97.6–80.3 Ma (Valsangkar et al. 1981), to 91.2 ± 0.2 Ma (Torsvik et al. 2000), to $87.6 + 0.6$ Ma (Storey et al. 1995). Consensus is emerging, however, that the time of the breakup was correlated with the position of Madagascar over the Marion hot spot. Storey et al. (1995) estimate that the southern tip of Madagascar's east coast was directly over the hot spot by 88 Ma, and it is this date that is most frequently cited in the literature, referencing the final separation of Madagascar and India. This appears to be supported by paleomagnetic data that indicate that the hot spot was beneath Madagascar's central east coast at 118 Ma, and at the south of Madagascar at 88 Ma (Torsvik et al. 1998). Thus, if the timing and position of the hot spot relates to the timing and direction of drifting, then the separation between Madagascar and India may well have occurred over a 30-million-year period, perhaps beginning in the north of Madagascar and progressing to the south (Seward et al. 2004). This is analogous to the model of Madagascar's separation from Africa in that it should be viewed as a prolonged and progressive process, rather than as a sudden event. Clearly, this distinction could potentially have enormous implications for the timing and capacity for biotic exchange between Madagascar and the other key parts of Gondwana. As added complexity, whereas most authorities reconstruct India's position as quite isolated, as it moved northward toward its ultimate collision with Asia in the earliest Eocene (Beck et al. 1995, Zhu et al. 2005) (**Figure 1f**), others argue that India's lack of a significantly endemic biota indicate that it must have remained within close proximity to other landmasses during its journey (Briggs 1989, 2003).

The discussion above has ignored the relative positions and movements of Antarctica and Australia. For the most part, geologists have concluded that Antarctica and Australia began their southward movement away from IndoMadagascar soon after the final separation of the latter from Africa (Briggs 2003), thus implying that terrestrial biotic exchange between these two continents and the remainder of Gondwana would have been impossible after 130 to 125 Ma. Recently, however, numerous investigators are modifying this view. Discovery and analysis of vertebrate fossils from the latest Cretaceous of Madagascar, India, and South America indicate a significant degree of cosmopolitanism among southern Gondwanan biota that can best be explained via terrestrial connections among these landmasses well into the Late Cretaceous. This suggests a subaerial contact between Antarctica and South America in the west, and between Antarctica and IndoMadagascar in the east, apparently existing until circa 80 Ma (Buckley & Brochu 1999; Buckley et al. 2000; Hay et al. 1999; Krause et al. 1997a,b; Sampson et al. 1998, 2001). Two such contacts have been suggested as potential routes for biotic exchange between Madagascar and South America, via Antarctica: the Kerguelen Plateau (KP) and the Gunnerus Ridge (GR) (**Figure 2**).

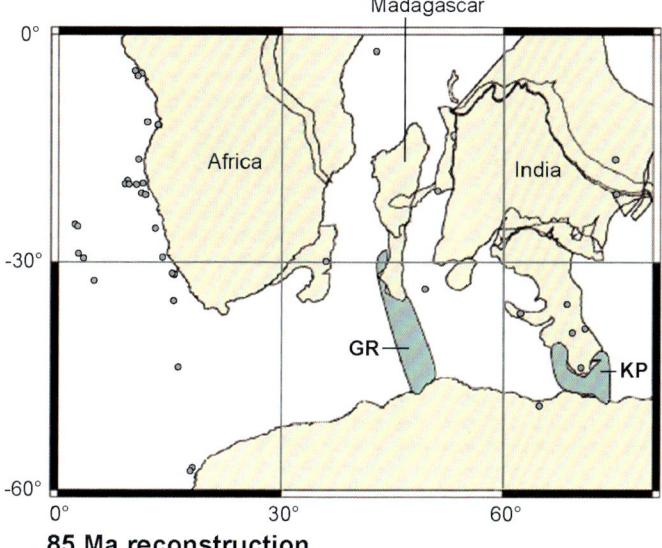

Figure 2

Reconstruction of potential dispersal routes in southern Gondwana at 85 Ma. Map from **http://www.odsn.de**; dots represent Ocean Drilling Program and Deep Sea Drilling Project sites; shorelines from (Hay et al. 1999) and (Case 2002). Figure is modeled after **Figure 1** in (Noonan & Chippendale 2006). KP, Kerguelen Plateau; GR, Gunnerus Ridge.

The implication of this biogeographic scenario is surprising because it suggests that the most recent vicariant events between Madagascar and other parts of Gondwana would be with South America, rather than with India, as more established views would have it. Finally, a minority view has argued that Madagascar actually served as a "cul-desac" within southern Gondwana for certain groups such as lemuriform primates, boine snakes, and iguanid lizards that would have arrived there from Laurasia via the Indian subcontinent (Rage 1996, Rage & Jaeger 1995). In this scenario, a terrestrial route existed from Asia to Madagascar via India and the Seychelles Plateau. According to its authors, the "northern hypothesis" need not be incompatible with the aforementioned "southern hypothesis," and indeed, may have been established after the disintegration of the KP and GR (Rage 2003).

Unraveling the historical biogeography of the Malagasy biota necessitates an analysis of many complex factors. The geographic/tectonic position of Madagascar through time must have had a dramatic influence on the past climate and biota on the island. Madagascar's current heterogeneous landscape and climatic profile contain a wealth of information on the historical influences of past climate and the evolution of the island's principle biomes. Although previous vegetation classifications have resulted in much more complex systems, reviewing these in detail is outside of the scope of this review. For this review three biome classifications that emphasize vegetation trends, topography, and climate (Du Puy & Moat 1996, Faramalala 1995, Lowry et al. 1997) are integrated in the classification of Madagascar's six principle biomes (**Figure 3**).

The current climatological landscape of Madagascar exhibits a remarkable east-west trend in precipitation, which is primarily the result of the orographic effect of the island's eastern mountain range. An evergreen lowland rainforest biome covers the east coast of the island and extends about 100 km inland up the eastern mountains. At

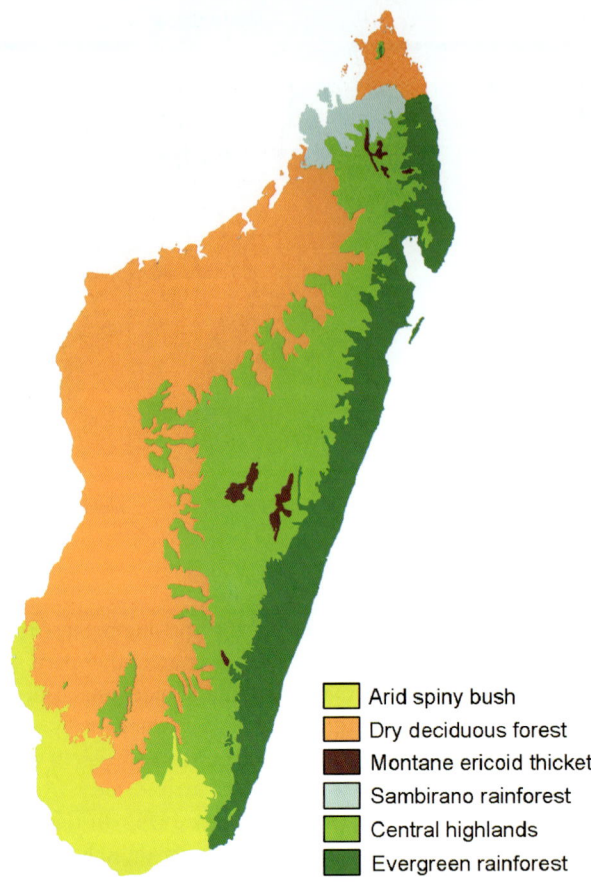

Figure 3
Illustration of biome distributions in Madagascar.

elevations above 800 m and extending well into the island's interior, the rainforest gives way to central highlands, which are dominated by moist montane forest. At higher elevations (i.e., above 2000 m) the central highlands give way to a high elevation thicket dominated by ericoid (Ericaceae) shrubs. On the western half of the island, below 800 m elevation, montane forest shifts to dry deciduous forest dominated by sclerophyllous trees and shrubs. Deforestation and agricultural grazing have heavily impacted both the western dry deciduous forest and the central highlands, and thus very little primary vegetation remains. These regions are now covered in secondary grasslands and are subject to frequent burning. The southwest extent of the dry deciduous forest becomes extremely arid as the influence of the 30-degree latitude subtropical arid belt becomes more pronounced. This region is called the "spiny bush" because of the floral dominance of the near-endemic angiosperm family Didieriaceae. In the extreme northwest, a region of the dry deciduous forest receives heavy seasonal rainfall from Indian monsoonal circulation. This seasonal influx of precipitation has produced a rainforest biome known as the Sambirano, which is geographically and biologically distinct from the eastern rainforest.

Such ecological heterogeneity within a limited land area begs questions about the sequence of geological and biological events that led to the assembly of these biomes. Most biologists agree that these are important questions, but few have ventured hypotheses. Wells (2003) provides the most detailed model for biome evolution to date. His approach relies heavily on paleogeographic reconstructions and paleoclimatic models, and has broad implications for the biogeographic history of Madagascar and the Southern Hemisphere as a whole. His contribution is of particular importance because it provides distinct hypotheses that can be tested with both geological and biological data. The hypothesis maintains that after Madagascar and India separated at 88 Ma, Madagascar lay within the 30-degree latitude subtropical arid belt and likely experienced a generally dry climate. India probably blocked any easterly influence of moisture from the Proto-Indian Ocean until about the Middle Eocene. In the Late Paleocene/Early Eocene, Madagascar (and Africa) drifted north toward the equator leaving the influence of the 30-degree latitude subtropical arid belt. This led to a significant increase in moisture from the north to the south and was amplified by the clearing of the Indian Ocean such that southeasterly trade winds were hitting the mountains of the Madagascar east coast, thus initiating the orographic precipitation that the eastern rainforest of Madagascar currently experiences. Through the remainder of the Cenozoic, Madagascar was tectonically and climatically rather stable. The last major climatic event to impact the Malagasy biota was the initiation of the Indian monsoons in the Late Miocene (∼8 Ma), perhaps coincident with the rise of the Tibetan Plateau (Rowley & Currie 2006). This then would have been the onset of heavy seasonal rains to the northwestern Sambirano region.

This explicit hypothesis of paleoclimatic history has important implications for the ages of Madagascar's principle biomes. It implies that the arid to seasonally dry biomes are much older than the humid and subhumid biomes. Wells (2003) interpretation is that the dominant biota in Madagascar in the Late Cretaceous and through the Paleocene was similar to the southern arid spiny bush. He envisions the effect of the island's northern drift in the Eocene as constricting the range of the arid spiny bush biome to the absolute south and southwest. Wells (2003) further suggests that current levels of endemism reflect the relative ages of biomes. Assuming that biome endemism is correlated with time, the biomes do show a striking concordance with this model of paleoenvironmental history. It remains for divergence time estimation of the endemic radiations contained within these discrete biomes to serve as an initial test of this complex model of biome evolution.

THE NATURALIST'S PROMISED LAND

It is not from mere curiosity that geologists, biogeographers, and biologists have developed a fascination for Madagascar. Rather, the island presents itself to even the most casual observer as a place of biological wonder. This impression is probably best captured in a quote from the eighteenth-century French naturalist, Philibert Commerson (Teissier 1859):

> May I announce to you that Madagascar is the naturalist's promised land? Nature seems to have retreated there into a private sanctuary, where she could work on different models from any she has used elsewhere. There you meet bizarre and marvelous forms at every step....

Recent biogeographers have focused more precisely on tabulating the species numbers, phylogenetic diversity, and levels of endemism of Madagascar's biota, finding that the patterns are truly astonishing. For example, 95% of the reptile species, 99% of amphibian species, and 100% of the island's land mammal species (excluding bats) occur nowhere else on Earth. Certain faunas are either poorly represented (e.g., only four orders of terrestrial eutherians are currently represented) or are completely absent (e.g., there are no salamanders, vipers, or varanid lizards), whereas other groups show unrivaled diversity (e.g., chameleons). Of the estimated 12,000 plant species, nearly 10,000 are unique to Madagascar. Moreover, within these endemic groups, the numbers of species can be spectacular. For example, there are at least 50 species of extant primates (suborder Lemuriformes), and when one also includes the more than 16 species of recently extinct subfossil lemurs, the numbers are even more impressive for an island fauna. For perspective, this means that >15% of living primate species are endemic to Madagascar, an island that comprises less than 0.4% of Earth's land surface area. Lizards show even greater levels of diversity, with 346 known species, 314 of which are endemic. As with many of the resident biotic groups, the herpetofauna are as remarkable for what is absent compared to Africa, and to a lesser extent, regions of the western Indian Ocean, as it is for the diversity that is present (Raxworthy 2003). For example, caecilians (Gymnophiones) occur on virtually every southern continent including South America, Africa, India, and even the Seychelles, but not on Madagascar. Worldwide, there are approximately 28 families of frogs, but apparently only 3 of these occur in Madagascar. Within these three families, however, there are more than 300 endemic species, representing nearly 4% of the world's amphibian fauna (Glaw & Vences 2003). This high diversity is even more remarkable given that of the 3 living amphibian clades, only the Anura are represented. Krause (1997a) has succinctly summarized this pattern of spectacular diversity within sporadic phylogenetic lineages as Madagascar's unique signature of "imbalance and endemism."

This pattern of imbalance, endemism, and diversity begs an obvious question: What force or forces have created the pattern? To many, the fact that diversity levels are so high suggest that many groups must have had very long histories in Madagascar, thereby allowing many millions of years for speciation and diversification. Gillespie & Roderick (2002), in an analysis of terrestrial arthropod distributions, distinguish two types of island systems: "Darwinian" islands and "fragment" islands. In the former case, the island will never have been in contact with other biotic reservoirs and is presumed to have an abundant supply of empty ecological niches. On these islands, species numbers will increase at first via colonization, which, over time, will be followed by the formation of neoendemics. According to this model, if isolation persists for vast periods of geological time, the ecological space will eventually be filled via speciation rather than immigration. In the case of fragment islands, Gillespie & Roderick posit that the ecological space will initially be filled due to contact with other landmasses, but that niches will open up as the fragment island becomes more isolated in a process that they call "relaxation." From this point, speciation among the initial residents will then create paleoendemics that over protracted time periods can result in high levels of endemism. Gillespie & Roderick (2002) place Madagascar squarely within the latter category. In looking at the data, and the patterns of vicariance and dispersal contained therein, we will argue that Madagascar does not comfortably fit into either category but instead shows patterns of neo-endemism that are as strong, if not considerably stronger, than patterns of paleoendemism.

HYPOTHESIS TESTING

Vicariance Versus Dispersal

The sundering of Gondwana over the past 165 million years or so has been a dynamical process. Madagascar's geographic position has been marked by temporal windows wherein biotic exchange would have been facilitated by contact with other landmasses, followed by a long period of progressive geographic isolation. Moreover, the timing and sequence of fragmentation and isolation may at times have been slowly progressive (e.g., Madagascar's long slide south, relative to Africa), or relatively abrupt (e.g., the hypothesized severing of southern connections via Antarctica). Finally, even the most recent conceivable connection to other landmasses via the Kerguelen Plateau at 80 Ma was at best contemporaneous with the temporal origins of major extant clades (e.g., placental mammals), though considerably more ancient than many of the subclades presently extant in Madagascar (e.g., feliform carnivorans). Thus, the present day biota of Madagascar must of necessity be comprised of groups whose presence is best explained by ancient vicariant events, as well as other groups that reached Madagascar via transoceanic dispersal. The challenge for biogeographers, therefore, is to distinguish between these two possibilities.

Prior to modern understanding of continental movements, the default explanation for the presence of biota in remote localities was "oceanic dispersal" (Nelson 1979). The validation of plate tectonics, however, yielded a sea change in the perception of dispersal hypotheses (Wiley 1988). Many researchers began to label dispersal hypotheses as bordering on the "miraculous" (Platnick 1981) or as "pseudoexplanations" (Croizat et al. 1974). Indeed, as de Queiroz (2005) so aptly states the case, biogeographers became "positively contemptuous" of dispersal scenarios. The tide has turned again, however. Numerous studies published within the past few years reveal case after case wherein biotic distributions can only be explained via hypotheses of dispersal (Cook & Crisp 2005, de Queiroz 2005, McDowall 2004, McGlone 2005). Thus, we have reached a state of the art wherein the majority of biogeographers are equally receptive to hypotheses of vicariance and dispersal. The essential question remains, however, of how to distinguish between the two, and with satisfactory degrees of confidence. Area cladograms have typically been employed for demonstrating vicariant distributions. If the phylogenetic history of an organismal group is congruent with the known sequence of vicariant events, then a vicariance scenario is supported (e.g., see Sparks & Smith 2005, **Figure 3**). Conversely, when phylogenies reveal lineages from one geographic area deeply nested within clades from another area, dispersal is typically inferred. More recently, temporal information has played a critical evidential role in supporting dispersal hypotheses (e.g., Poux et al. 2005, Yoder et al. 2003).

The assertion that temporal data are necessary for discriminating between vicariance and dispersal hypotheses makes an implicit assumption that divergence ages can be estimated with high degrees of confidence. Unfortunately, the fossil record of Madagascar places a severe handicap on temporal analysis of the Malagasy biota. Most notably, for the Tertiary of Madagascar, terrestrial fossil exposures are virtually nonexistent (Flynn & Wyss 2003, Krause 2003). Although it is true that phylogenetic and coalescent methods for estimating clade ages are growing increasingly sophisticated, they are by no means infallible (Drummond et al. 2006, Kishino et al. 2001, Thorne & Kishino 2002, Thorne et al. 1998, Yang & Rannala 2006), and indeed, should be viewed as rather gross approximations. Nonetheless, if we make the assumption that divergence times can be calculated with confidence, hypothesis discrimination becomes rather tractable. One "only" needs a complete geographic sampling for the clade of interest, a handsome amount of DNA sequence data for both the ingroup and outgroup taxa, and reliable temporal calibrations (e.g., fossils) on several nodes within the phylogeny. But what if an investigator is more ambitious, and actually wants to test for patterns of vicariance and dispersal across an entire biota, such as that of Madagascar? The task becomes daunting.

Not only does one need all of the information described above, one wants it for a global sample encompassing the multitude of taxa that compose a biota, from plants, to invertebrates, to vertebrates. And finally, how does one visualize the summation of so much information and in such a way as to discriminate among the alternate biogeographic hypotheses? **Figure 4** makes a first pass at visualizing a very simple null distribution of clade ages for both vicariant and dispersed biota in Madagascar. Here, we model the plants, invertebrates, and vertebrates separately, but visually compare their respective clade ages with respect to their fit to a hypothesized vicariant event. The plants are illustrated in green and are modeled such that 40% of clade ages are compatible with an 80 Ma vicariant event, leaving 60% as hypothesized transoceanic dispersers. The invertebrates are shown in blue, with 60% of clades compatible with an 80 Ma vicariance, and 40% as transoceanic dispersers. Finally, 20% of the vertebrates, illustrated in red, are modeled as vicariants, with 80% as dispersers. We modeled the vicariants to fit a gamma distribution with a shape parameter of two and a scale parameter of seven, and with a zero offset of 75 Ma. This accommodates the assumption that the steep, though not abrupt, left tail of the distribution (corresponding to more recent ages in the distribution) best models the progressive drift of two landmasses over time. Clearly, even in the most abrupt geological timescales, there is considerable time for "filter" dispersal across an ever-widening oceanic barrier. We make no claim that this model actually fits the distribution of clade ages for the biota of Madagascar. Rather, we present this figure as a means for potentially visualizing the distribution of clade ages in such a way as to test for patterns of vicariance versus dispersal in different biotic groups. It will require actual dated phylogenies of Malagasy biota to determine the fit of the data to this potential model.

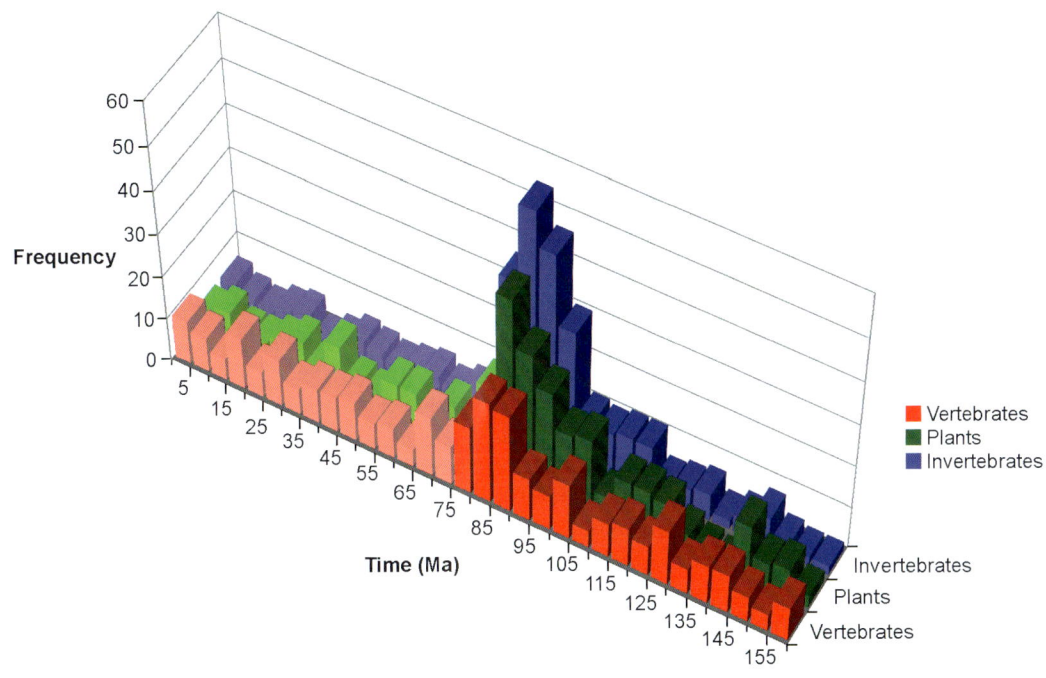

Figure 4

Hypothetical model of divergence dates for plants (*green*), invertebrates (*blue*), and vertebrates (*red*). The 80 Ma mark indicates the most recent plausible vicariance event for Malagasy biota.

THE DATA

The stage is now set for a careful examination of the data. Will the examination of the fossil data (such as they are) and the existing phylogenetic data allow us to distinguish the relative importance that vicariance and transoceanic dispersal have played in generating the biotic communities that presently exist in Madagascar? We believe so, despite the paucity of both data sources. As the sections below on plants, invertebrates, and vertebrates will illustrate, the number of phylogenetic studies devoted largely or strictly to Malagasy biota are few relative to the enormous number of taxa and evolutionary lineages present on the island. Even so, a clear indication emerges.

Phylogenetic Studies—Plants

The Malagasy flora is one of the most distinctive in the world. The most recent estimates of floral endemism are well above 90% at the species level (Goodman & Benstead 2005, Schatz 2000). It is widely agreed that both the vicariance resulting from the breakup of Gondwana and the action of long-distance dispersal were important mechanisms in forming the current levels of Malagasy floral endemism, but there is considerable contention regarding the relative importance of these two mechanisms. The dominant trend among botanists has been to emphasize the influence of Gondwanan vicariance (Grubb 2003, Leroy 1978).

Schatz (1996) was the first to put an emphasis on long-distance dispersal as a mechanism of origin for many elements of the Malagasy flora. He considered the Malagasy flora to be strongly influenced by dispersal from the Australian, Indian, and Southeast Asian floras. Schatz suggests that prevailing easterly winds across the Indian Ocean have fostered a condition in which continuous long-distance dispersal from these regions has been occurring since at least the Early Tertiary. If vicariance is the predominant mechanism underlying the biogeographic history of the Malagasy flora, the evolutionary signature of this vicariance should be significant in our review of phylogenetic studies. Instead, we could find no single phylogenetic study of plants with sufficiently old divergence time estimates (i.e., >80 Ma) to validate hypotheses of Gondwanan vicariance. This result is not surprising in light of the several divergence time analyses suggesting that the majority of angiosperm families with Malagasy elements are relatively young (Anderson et al. 2005, Bremer et al. 2004, Davis et al. 2005, Janssen & Bremer 2004). This is not to say that these vicariant patterns do not exist, but rather they may be relatively rare or expressed at deep taxonomic scales (e.g., familial or ordinal levels).

The most promising groups suggesting a signature of vicariance are those that molecular data place on long branches sister to large, disparate clades. Although none of these groups have been the focus of a detailed analysis of divergence times, they have been largely hypothesized to be relicts that have avoided extinction in the Malagasy flora (Schatz 1996). A few examples include *Takhtajania* (sister to all Winteraceae), *Didymeles* (sister to all Buxales), and *Humbertia* (sister to all Convolvulaceae). Many of the endemic Malagasy families (Didieriaceae, Sarcolaenaceae, Spherosepalaceae, Asteropeiaceae, and Physenaceae) have considerable diversity in Madagascar, and are likely the products either of vicariance or a single, ancient dispersal event. There is evidence for larger fossil distributions for some of these hypothesized relicts. For example, *Didymeles* pollen has been found in Paleocene sediments of Australia and New Zealand, and Sarcolaenaceae pollen has been found in Miocene sediments of South Africa (Schatz 2001). It is unknown whether these fossils represent extinct sister clades or the original home ranges of these groups.

Our review of the literature (**Supplemental Table 1**, follow the Supplemental Material link from the Annual Reviews home page at **http://www.annualreviews.org/**) suggests that despite the promise of detecting the signature of vicariance from a very few elements, the importance of dispersal to the assembly of the Malagasy flora cannot be denied. Furthermore, Africa appears by far to be the most important source of floral dispersal to Madagascar. Future divergence time estimates will clarify both the reality of a vicariant signature and the relative antiquity of the Malagasy flora.

Phylogenetic Studies: Invertebrates

Relative to the extraordinary diversity of invertebrates in Madagascar, there is a dearth of phylogenetic studies of these organisms, and for those that have been conducted, virtually all have focused on the insects. No doubt, this deficiency relates to the fact that invertebrate biologists are only beginning to scratch the surface in simply documenting and describing the biota. There are numerous higher-level taxa that are assumed to be relicts of ancient, pre-Jurassic Gondwanan vicariance, such as crayfish, notonemourine stoneflies, scarabaeine beetles, dipterans, aphids, and araneids (Paulian & Viette 2003), though virtually none of these groups have been subjected to phylogenetic analysis, much less divergence age analysis.

For relatively well-studied groups, such as the ants, hypotheses regarding biogeographic origins can be posed based simply on the distribution of related taxa. In the case of the ants, studies show a much greater affinity of Malagasy taxa to Afrotropical taxa, rather than to Indoasian groups (Fisher 1997, 2003). But with poorly studied groups, even modest phylogenetic analysis can reveal remarkable insights. Based on a study that included a single undescribed Malagasy species in the spider genus *Anelosimus*, investigators found that this species could be grouped within a clade of New World social species. This result provoked an expedition to Madagascar, which then revealed at least five additional periodic species from the initial species locality. When analyzed phylogenetically, the investigators found the Malagasy species to constitute a clade with a sister group relationship to a combined East African plus Americas clade (Agnarsson & Kuntner 2005).

Some studies have posited an explicit hypothesis of vicariant origins, such as a global distribution analysis of the Forficulina (Dermaptera) (Popham 2000). Though not phylogenetic in methodology, the Popham examines their distribution patterns and concludes that their center of origin was Gondwana, in the area that is now northeast Brazil and northwest Africa. From there, these insects are hypothesized to have spread both westward (into the Americas) and eastward to IndoMadagascar. The study also concludes that with the impact of India and Asia, a second center of origin was established for Asia. Presumably, therefore, the presence of the Forficulina in Madagascar represents a classic radiation of paleoendemics. Certainly, this appears reasonable based on distribution analysis but requires phylogenetic study and divergence age estimation for verification.

The most complex biogeographic histories among the invertebrates relate to the butterflies, given our current understanding of Malagasy invertebrate history. In the case of both the swallowtail butterflies (Zakharov et al. 2004) and the satyrine butterflies (Torres et al. 2001), a complex mix of vicariance history and dispersal seems necessary to explain their distribution in Madagascar and other parts of the Old World. In both cases, the researchers have evoked scenarios of Cretaceous vicariance, followed by endemic species radiation within Madagascar and subsequent dispersal from Madagascar to other areas. Certainly, the phylogenies show a strong tendency toward speciation within Madagascar with dispersal radiation from within. Unfortunately, neither study has explicitly addressed the geological timing for these events. The alternatives are that the ancestral Malagasy butterflies were present owing to vicariance, though as we know, this would imply that they existed in Madagascar at least 80 Ma, if not considerably earlier. Either way, it is clear that the subsequent dispersals out of Madagascar occurred sporadically over time, and almost certainly during the Cenozoic, at which time overwater dispersal would have been necessary.

Other invertebrate studies invoking dispersal include a study of the small minnow mayflies (Monaghan et al. 2005) and of fig wasps (Kerdelhue et al. 1999). As with the butterflies, the study of the mayflies also concludes that transoceanic dispersal between Madagascar and other landmasses occurred multiple times, though the timing of these events is presently obscure. Similarly, the fig wasp study posits long-distance dispersal, though in this case, it appears that dispersal occurred only once, and from Africa to Madagascar. All of these studies inferred dispersal from phylogeny and biogeography alone, given that divergence age estimation has not been an integral part of the investigation. A refreshing example of biogeographic, phylogenetic, and divergence age analysis, however, can be found in a recent study of the allodapine bee genus *Braunsapis* (Fuller et al. 2005). The results are striking and decisive in inferring two transoceanic dispersals, both of which occurred from Africa to Madagascar and within the relatively recent geological past. One is inferred to have occurred 13 Ma, and another much more recently at approximately 3 Ma. Here, it is worth noting that the divergence age estimates are sufficiently distinct that even though estimation errors are possible (if not probable), the relative difference between the two age estimates obviates the possibility that they could have occurred via a single event.

Phylogenetic Studies: Vertebrates

Fishes. Of all the vertebrate taxa to be considered herein, the cichlid fishes of Madagascar represent the most challenging puzzle for differentiating vicariance from dispersal. Despite previous claims that Madagascar's ichthyofauna is notably species poor (Kiener & Richard-Vindard 1972), recent intensive surveys show it to support the expected number of species given the island's land surface area (Sparks & Stiassny 2003). But, despite the abundance of species that potentially offer clues relating to biogeographic history, very few species groups have been the subject of phylogenetic investigation (Sparks & Stiassny 2003). Moreover, of the 14 living groups of freshwater fishes in Madagascar, only one (Ariidae, catfish) is potentially represented in the Cretaceous fossil record (Gottfried et al. 1998). This seems to mirror the pattern for other vertebrates, wherein it appears that there has been a near complete faunal turnover for Malagasy vertebrates, potentially demarcated by the K/T boundary (Krause 2003; Krause et al. 1997a, 1999) and possibly related to a hypothetical bollide impact near Bombay, India, that might have triggered the Deccan volcanism (Negi et al. 1993).

Among the evidence to be marshaled in support of ancient vicariance of the Malagasy freshwater fishes are (*a*) the near-perfect match of area cladograms with the temporal sequence of Gondwanan breakup for rainbow fishes (Bedotiidae), killifishes (Pachypanchax), and cichlids, (*b*) the observation that none of these clades shows a sister group relationship with African lineages, and (*c*) that for 70% of the primary lineages, sister groups are found on former Gondwanan fragments that are now separated from Madagascar by thousands of kilometers of open ocean (Sparks 2004; Sparks & Smith 2004a,b, 2005). Moreover, many of the Malagasy lineages are basal with respect to their relatives found on other landmasses (Sparks & Stiassny 2003). Finally, recent analysis of biogoegraphic patterns and fossil distributions indicates that the Cichlidae are indeed an ancient radiation with origins deep in the Mesozoic (T.J. Near, submitted).

Alternatively, paleontologists and others have argued that the island's freshwater fishes are descended from Cenozoic marine colonizers that are secondarily adapted to freshwater. Myers (1938) classified freshwater fishes into "primary" and "secondary" groups, depending upon the latter's demonstrated tolerance for at least brief periods of saltwater exposure. He placed cichlids in this category. Sparks & Smith (2005) marshal a combination of physiological, life history, and phylogenetic data to reject Myers' division, however, arguing that it is irrelevant to the question of Malagasy fish biogeography. Nonetheless, an explicitly molecular phylogenetic approach using divergence time estimation (Vences et al. 2001) posits a scenario in which the ancestral cichlids were widespread in Africa during the Early Cenozoic. From there, they dispersed to Madagascar and India, with modern cichlids in Africa displacing their ancestral lineages, thereby causing extinction with replacement, but with Malagasy and Indian basal lineages remaining. Though this extinction with replacement hypothesis may appear overly complex and ad hoc, Vences et al. (2001) point out that in Madagascar, cichlids recently introduced from Africa by human agency are replacing the native cichlids with devastating speed and efficiency. As potentially decisive support for the dispersal hypotheses, divergence ages for the Malagasy lineages are shown to be far too recent to be compatible with ancient vicariance (Murray 2001, Vences et al. 2001).

Herpetofauna. Contrary to the situation with the freshwater fishes, there have been quite a few molecular phylogenetic studies conducted on the amphibians and reptiles of Madagascar. In the case of the herpetofauna, the distinction between Cenozoic dispersers and ancient vicariants is fairly clear, and the predominant trend of thought is one of dispersal. Virtually all studies of Madagascar's frogs conclude that their presence in Madagascar is best explained by dispersal, typically from Africa (Vences et al. 2003a, 2004). This runs counter to all traditional hypotheses that argue that amphibian life history and physiology negate the possibility of transoceanic dispersal. Vences et al. (2003b) offer evidence from two endemic species of the genus *Mantellidae* that occur on the island of Mayotte, which is entirely volcanic and surrounded by sea depths of more than 3500 m. Although both species were assumed to have been human introductions from Madagascar, a divergence age analysis reveals that they are far too old to have been the products of human agency. The notable exception to this apparent pattern of Cenozoic transoceanic dispersal in Madagascar's frogs can be found in the analysis of ranid frogs conducted by Bossuyt & Milinkovitch (2001). Their study concludes that ranid origins began on the IndoMadagascar subcontinent, with the retention of ancestral lineages in Madagascar, their extinction in India relating to the formation of the Deccan Traps, and subsequent release into Eurasia upon the collision of India with Asia. Divergence age analysis appears to support the hypothesis, though it should be noted that the primary calibration point applied in the analysis is the 88 Ma separation of Madagascar and India. Thus, one can potentially argue a case of circularity in the derivation of their divergence ages in support of vicariance.

The situation with Madagascar's reptiles is similar in that studies of plated lizards (A. Raselimanana, personal communication; Odierna et al. 2002), colubrid snakes (Nagy et al. 2003), geckos (Austin et al. 2004), tortoises (Caccone et al. 1999), scincid lizards (Mausfeld et al. 2000), and chameleons (Raxworthy et al. 2002) all show distributions and ages (where calculated) that are most consistent with Cenozoic transoceanic dispersal. Perhaps the most remarkable study in this regard is that of the chameleons, which posits a minimum of five dispersal events among Madagascar, Africa, India, and the Seychelles (Raxworthy et al. 2002). A recent study of the boine snakes, podocnemid turtles, and iguanid lizards comes to quite a different conclusion, however. Using multiple gene loci and Bayesian divergence age analysis (Kishino et al. 2001, Thorne & Kishino 2002, Thorne et al. 1998), along with area cladogram analysis, this study concludes that these three taxa reside in Madagascar owing to a vicariant history (Noonan & Chippendale 2006). The calculation of crown node ages ranging from 76–90 Ma for the Malagasy clades, as well as the geographic distribution of the endemic taxa and their outgroups, is taken as compelling evidence for a protracted connection between South America and Madagascar via Antarctica, as suggested by Hay et al. (1999). The authors decidedly conclude that the data are neither compatible with ancient African/Madagascar vicariance, nor with Cenozoic dispersal.

Birds. Given that birds have the capacity for flight, and are thus assumed to be ready dispersers, it comes as no surprise that all molecular phylogenetic studies of Malagasy birds have concluded that they arrived on Madagascar via dispersal. Detailed studies of vangids (Yamagishi et al. 2001), ground rollers (Kirchman et al. 2001), warblers (Beresford et al. 2005), kingfishers (Marks & Willard 2005), sunbirds (Warren et al. 2003), bulbuls (Warren et al. 2005), kestrels (Groombridge et al. 2002), and asities (Prum 1993) have posited dispersal, typically from Africa. Moreover, for those studies in which divergence ages were determined (Groombridge et al. 2002; Warren et al. 2003, 2005), it appears that several groups and/or species are very recent arrivals (<4 Ma). Again, given flight abilities in birds, it is not terribly surprising that the avifauna of Madagascar might be strictly comprised of Cenozoic transoceanic dispersers. What is surprising, however, has been the discovery that the diversity of Madagascar's avifauna is explained by considerably fewer dispersal events than might be supposed.

This finding is best characterized by the studies of Yamagishi et al. (2001) and Cibois et al. (1999, 2001). In the former case, the researchers found that two genera, *Tylas* and *Newtonia*, both of which had previously been placed in different families, are actually members of the Malagasy Vangidae. Even more remarkably, the latter of the two studies found that 9 of 13 species of Malagasy songbirds belong to a single endemic clade. Prior to these studies, it had been assumed that these species were the product of at least three independent colonizations, rather than the single colonization and subsequent radiation posited by the Cibois et al. (2001) study. Moreover, the diversity of morphologies and ecologies typical of the birds examined by both studies is persuasive evidence of Madagascar's power to generate radiations of neoendemics (sensu, Gillespie & Roderick 2002).

Seemingly, the single exception to scenarios of dispersal for Madagascar's avifauna relates to the recently extinct elephant bird (genus *Aepyornis*). This giant ratite (the largest bird to have ever lived) was one of the many victims of the Holocene megafaunal extinction in Madagascar. Given the consistent agreement between fossil-based (Cracraft 2001) and molecular age estimates (Cooper et al. 2001) for ratite origins in the Late Cretaceous, many ornithologists have assumed that the elephant bird came to reside in Madagascar via Gondwanan vicariance. Indeed, the molecular phylogenetic age estimates for basal ratites suggest that elephant birds and ostriches may have entered IndoMadagascar via the Kerguelen Plateau, with ostriches eventually arriving in Eurasia via the collision of India and Asia (Cooper et al. 2001).

Mammals. Only four major lineages (classified as Orders) of extant, strictly terrestrial mammals are endemic to Madagascar: the tenrecs (Afrotheria, formerly classified as Insectivora), the lemurs (Order Primates), the nesomyine rodents (Family Muridae), and the carnivorans (Order Carnivora). Madagascar also supports a notable bat fauna, comprised of both micro- and megachiropterans. Of these, only one family, Myzopodidae (the sucker-footed bats), is endemic. With the exception of the bats, all of the terrestrial mammals of Madagascar have been subjected to rigorous molecular phylogenetic analysis, and all groups have had divergence age estimates generated by at least one study. The results of these analyses are strikingly uniform. In all four groups, phylogenetic analysis demonstrated each group to be monophyletic with its sister group found in Africa. These patterns lead to the conclusion that each clade is the product of a single colonization and subsequent radiation of neoendemics within the island. This is true of the lemurs (Goodman et al. 1994; Yoder 1994, 1997; Yoder et al. 1996), rodents (Jansa & Weksler 2004, Poux et al. 2005), tenrecs (Douady et al. 2002, Olson 1999, Olson & Goodman 2003, Poux et al. 2005), and carnivorans (Yoder & Flynn 2003, Yoder et al. 2003). Also, there is a consistent finding among divergence age studies that all crown group ages are post-Mesozoic (Poux et al. 2005; Yang & Yoder 2003; Yoder & Yang 2004; Yoder et al. 1996, 2003), with the oldest estimated age being for the lemurs in the earliest part of the Cenozoic (Yang & Yoder 2003, Yoder & Yang 2004, Yoder et al. 2003). Moreover, a recently published ancient DNA study confirms that the giant subfossil lemurs were members of the same lemuriform clade and are thus descended from the same Cenozoic colonist that produced the extant lemurs (Karanth et al. 2005). The sole molecular phylogenetic study thus far conducted on the bats of Madagascar reveals a slightly different story in finding that the genus *Triaenops* (Family Hipposideridae) is the product of two colonizations of Madagascar, both of which occurred very recently (<1 Ma) (A.L. Russell, J. Ranivo, E.P. Palkovacs, S.M. Goodman, A.D. Yoder, et al., submitted).

Summary of Emergent Patterns in the Data

The data presented above point to two generalities: (*a*) that there are numerous endemic clades of Malagasy taxa whose closest sister group relationships are to African taxa, and (*b*) that there appears to be an overwhelming indication of Cenozoic dispersal. **Figures 5** and **6** summarize these data and confirm the impression. **Figure 5** illustrates the fact that nearly half of Malagasy plants, invertebrates, and vertebrates studied show sister group relationships to African taxa. Moreover, the consideration of outgroup nodes (data available in **Supplemental Table 1**) suggests that the majority of lineages ancestral to Malagasy endemics had their origins in Africa. Finally, **Figure 6** shows that for those groups in which divergence ages have been estimated,

Figure 5

Pie charts representing geographic distributions of Madagascar sister taxa. Phylogenies that show Madagascar taxa sister to taxa distributed in more than one geographic region were scored such that each region was counted once, thus inflating these values with respect to regions not represented, but not with respect to other equally represented regions.

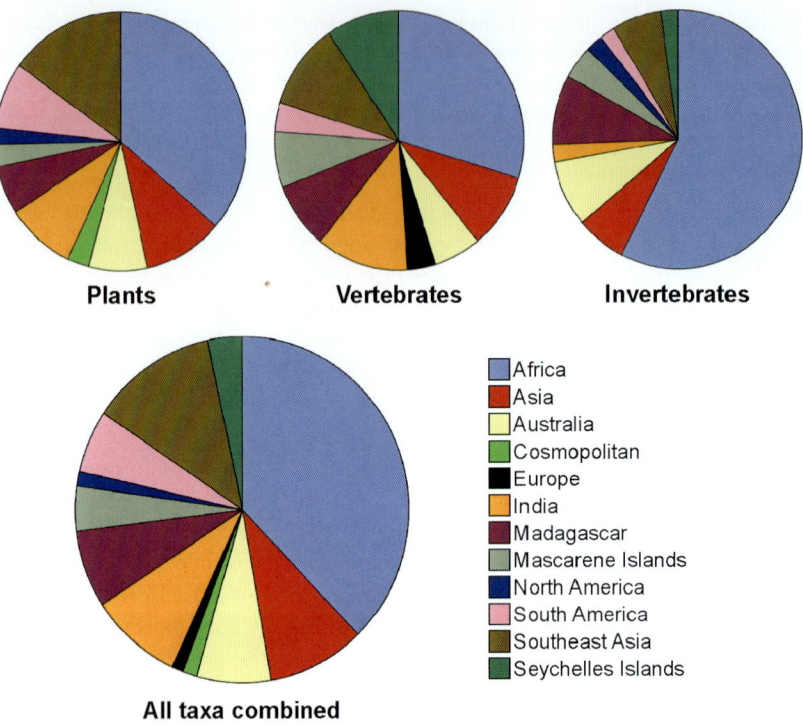

the vast majority of both crown and stem ages fall within the Cenozoic. The inescapable inference therefore is that the living biota is predominantly comprised of neoendemics that have evolved from transoceanic dispersers. Certainly, this strains credulity for some (e.g., Stankiewicz et al. 2006), especially given our knowledge of Madagascar's extreme and long-standing isolation. Moreover, the number of studies, relative to the number of endemic taxa, is paltry and we must therefore proceed with caution in drawing our conclusions.

It is not a novel observation to view the living Malagasy biota as "immigrants" (Krause 2003, Krause et al. 1997a, Simpson 1952). Indeed, Simpson (1952), in disputing hypotheses of landbridge connections between Madagascar and other landmasses, pointed out that the very limited representation of mammalian taxa in Madagascar is a definitive indication that the probability of colonization must have been "exceedingly low." It is here that one must revisit the impression that transoceanic dispersal hypotheses border on the "miraculous" (Platnick 1981). Since Simpson's time, landbridge hypotheses have continued to be invoked (e.g., McCall 1997), though they have not received empirical support (e.g., Poux et al. 2005, Yoder et al. 2003). Thus, rafting remains as the predominant explanation, at least for the vertebrates extant terrestrial mammals. Though there is compelling evidence that there may once have been two small subaerial exposures in the Mozambique Channel during the Miocene that could have broken the trip from Africa into three north to south stages of 295,

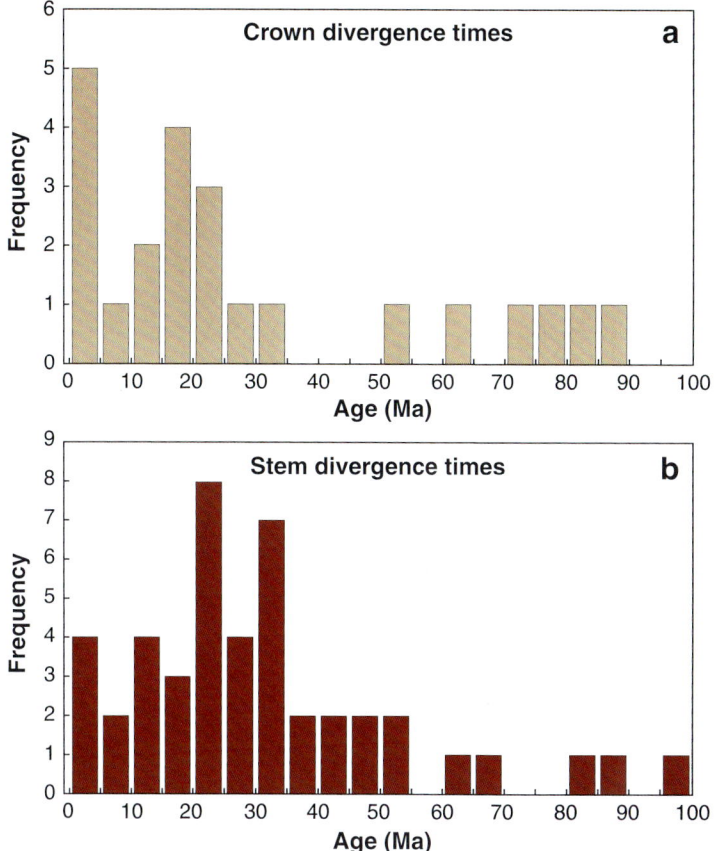

Figure 6
Divergence time estimates for Malagasy clades (see Supplemental **Table 1** for references). (*a*) Estimates of crown divergence times. (*b*) Estimates of stem divergence times. The scale on the x-axis is in millions of years before present (Ma).

210, and 125 km, and in the direction of the prevailing currents (Bassias 1992, Krause et al. 1997a, Malod et al. 1991), this would have been for a limited geological period, and would still have required long sea voyages as well as multiple and improbable colonizations.

Perhaps a more useful model can be drawn from present-day phenomena. Krause (1997a) reviews contemporary reports of floating "islands" of vegetation, often with standing trees and mammalian inhabitants, observed in remote oceanic locations, tens and hundreds of kilometers from land (Carlquist 1965, King 1962, Millot 1953). Certainly, there have been numerous empirical reports of transoceanic dispersal in lizards during recent history (Calsbeek & Smith 2003, Carranza et al. 2000, Censky et al. 1998). Add to this evidence the likelihood that the ancestors of today's Malagasy mammals were either hibernators, or had other physiological capacities for reduced metabolic demands (Kappeler 2000), and the "miraculous" becomes slightly more routine. Finally, given their less stringent life history attributes, credulity is far less strained by hypotheses of long-distance dispersal in plants and invertebrates (Barnes et al. 2006, Cowie & Holland 2006, Monaghan et al. 2005, Thiel & Gutow 2005). Given the data on hand, and the remarkable consistency of the trend across evolutionary lineages, we must consider the following to be the most credible hypothesis for the time being: Madagascar is an island primarily comprised of neoendemics that are the descendents of Cenozoic waif dispersers.

LITERATURE CITED

Agnarsson I, Kuntner M. 2005. Madagascar: an unexpected hot spot of social *Anelosimus* spider diversity (Araneae: Theidiidae). *Syst. Entomol.* 30:575–92

Agrawal PK, Pandey OP, Negi JG. 1992. Madagascar: a continental fragment of the paleo-super Dharwar craton of India. *Geology* 20:543–46

Anderson CL, Bremer K, Friis EM. 2005. Dating phylogenetically basal eudicots using *rbcL* sequences and multiple fossil reference points. *Am. J. Bot.* 92:1737–48

Austin JJ, Arnold EN, Jones CG. 2004. Reconstructing an island radiation using ancient and recent DNA: the extinct and living day geckos (Phelsuma) of the Mascarene islands. *Mol. Phylogenet. Evol.* 31:109–22

Barnes DKA, Hodgson DA, Convey P, Allen CS, Clarke A. 2006. Incursion and excursion of Antarctic biota: past, present and future. *Glob. Ecol. Biogeogr.* 15:121–42

Bassias Y. 1992. Petrological and geochemical investigation of rocks from the Davie Fracture Zone (Mozambique Channel) and some tectonic implications. *J. Afr. Earth Sci.* 15:321–39

Beck RA, Burbank DW, Sercombe WJ, Riley GW, Barndt JK, et al. 1995. Strategraphic evidence for an early collision between northwest India and Asia. *Nature* 373:55–58

Beresford P, Barker FK, Ryan PG, Crowe TM. 2005. African endemics span the tree of songbirds (Passeri): molecular systematics of several evolutionary 'enigmas.' *Proc. R. Soc. London Ser. B.* 272:849–58

Bossuyt F, Milinkovitch MC. 2001. Amphibians as indicators of early tertiary "out-of-India" dispersal of vertebrates. *Science* 292:93–95

Bremer K, Friis EM, Bremer B. 2004. Molecular phylogenetic dating of asterid flowering plants shows Early Cretaceous diversification. *Syst. Biol.* 53:469–505

Briggs JC. 1989. The historic biogeography of India isolation or contact? *Syst. Zool.* 38:322–32

Briggs JC. 2003. The biogeographic and tectonic history of India. *J. Biogeogr.* 30:381–88

Buckley GA, Brochu CA. 1999. An enigmatic new crocodile from the Upper Cretaceous of Madagascar. In *Cretaceous Fossil Vertebrates*, ed. D Unwin, pp. 147–55. London: Paleontol. Assoc.

Buckley GA, Brochu CA, Krause DW, Pol D. 2000. A pug-nosed crocodyliform from the Late Cretaceous of Madagascar. *Nature* 405:941–44

Caccone A, Amato G, Gratry OC, Behler J, Powell JR. 1999. A molecular phylogeny of four endangered Madagascar tortoises based on MtDNA sequences. *Mol. Phylogenet. Evol.* 12:1–9

Calsbeek R, Smith TB. 2003. Ocean currents mediate evolution in island lizards. *Nature* 426:552–55

Carlquist S. 1965. *Island Life*. Garden City, NY: Nat. Hist. Press

Carranza S, Arnold EN, Mateo JA, López-Jurado LF. 2000. Long-distance colonization and radiation in gekkonid lizards, *Tarentola* (Reptilia: Gekkonidae), revealed by mitochondrial DNA sequences. *Proc. R. Soc. London Ser. B* 267:637–49

Case JA. 2002. A new biogeographic model for dispersal of Late Cretaceous vertebrates into Madagascar and India. *J. Vertebr. Paleontol.* 22(Suppl. 3):42A

Censky EJ, Hodge K, Dudley J. 1998. Over-water dispersal of lizards due to hurricanes. *Nature* 395:556

Cibois A, Pasquet E, Schulenberg TS. 1999. Molecular systematics of the Malagasy babblers (Passeriformes: timaliidae) and warblers (Passeriformes: sylviidae), based on cytochrome *b* and 16S rRNA sequences. *Mol. Phylogenet. Evol.* 13:581–95

Cibois A, Slikas B, Schulenberg TS, Pasquet E. 2001. An endemic radiation of Malagasy songbirds is revealed by mitochondrial DNA sequence data. *Evol. Int. J. Org. Evol.* 55:1198–1206

Cook LG, Crisp MD. 2005. Directional asymmetry of long-distance dispersal and colonization could mislead reconstructions of biogeography. *J. Biogeogr.* 32:741–54

Cooper A, Lalueza-Fox C, Anderson S, Rambaut A, Austin J, et al. 2001. Complete mitochondrial genome sequences of two extinct moas clarify ratite evolution. *Nature* 409:704–7

Cowie RH, Holland BS. 2006. Dispersal is fundamental to biogeography and the evolution of biodiversity on oceanic islands. *J. Biogeogr.* 33:193–98

Cracraft J. 2001. Avian evolution, Gondwana biogeography and the Cretaceous-Tertiary mass extinction event. *Proc. R. Soc. London Ser. B.* 268:459–69

Croizat L, Nelson G, Rosen DE. 1974. Centers of origin and related concepts. *Syst. Zool.* 23:265–87

Davis CC, Webb CO, Wurdack KJ, Jaramillo CA, Donoghue MJ. 2005. Explosive radiation of Malpighiales supports a mid-Cretaceous origin of modern tropical rain forests. *Am. Nat.* 165:E36–65

de Queiroz A. 2005. The resurrection of oceanic dispersal in historical biogeography. *Trends Ecol. Evol.* 20:68–73

de Wit MJ. 2003. Madagascar: Heads it's a continent, tails its an island. *Annu. Rev. Earth Planet. Sci.* 31:213–48

Douady CJ, Catzeflis F, Kao DJ, Springer MS, Stanhope MJ. 2002. Molecular evidence for the monophyly of tenrecidae (mammalia) and the timing of the colonization of Madagascar by Malagasy Tenrecs. *Mol. Phylogenet. Evol.* 22:357–63

Drummond AJ, Ho SYW, Phillips MJ, Rambaut A. 2006. Relaxed phylogenetics and dating with confidence. *PLOS Biol.* 4:e88

Du Puy DJ, Moat J. 1996. A refined classification of the vegetation types of Madagascar, and their current distribution. See Lourenço 1996, pp. 205–18

Faramalala MH. 1995. Formations végétales et domaine forestier de Madagascar. Washington, DC: Conserv. Int., map

Fisher BL. 1997. Biogeography and ecology of the ant fauna of Madagascar (Hymenoptera: Formicidae). *J. Nat. Hist.* 31:269–302

Fisher BL. 2003. Formicidae, ants. See Goodman & Benstead 2003, pp. 811–19

Flynn JJ, Wyss AR. 2003. Mesozoic terrestrial vertebrate faunas: the early history of Madagascar's vertebrate diversity. See Goodman & Benstead 2003, pp. 34–40

Fuller S, Schwarz M, Tierney S. 2005. Phylogenetics of the allopadine bee genus Braunsapis: historical biogeography and long-range dispersal over water. *J. Biogeogr.* 32:2135–44

Gillespie RG, Roderick GK. 2002. Arthropods on islands: colonization, speciation, and conservation. *Annu. Rev. Entomol.* 47:595–632

Glaw F, Vences M. 2003. Introduction to the amphibians. See Goodman & Benstead 2003, pp. 883–98

Goodman M, Bailey WJ, Hayasaka K, Stanhope MJ, Slightom J, Czelusniak J. 1994. Molecular evidence on primate phylogeny from DNA sequences. *Am. J. Phys. Anthropol.* 94:3–24

Goodman SM, Benstead JP, eds. 2003. *The Natural History of Madagascar*. Chicago: Univ. Chicago Press, 1709 pp.

Goodman SM, Benstead JP. 2005. Updated estimates of biotic diversity and endemism for Madagascar. *Oryx* 39:1–5

Goodman SM, Patterson BD. 1997. *Natural Change and Human Impact in Madagascar*. Washington, DC: Smithson. Inst. Press, 432 pp.

Gottfried MD, Randriamiarimana LL, Rabarison JA, Krause DW. 1998. Late Cretaceous fish from Madagascar: implications for Gondwanan biogeography. *J. Afr. Earth Sci.* 27:91–92

Groombridge JJ, Jones CG, Bayes MK, van Zyl AJ, Carrillo J, et al. 2002. A molecular phylogeny of African kestrels with reference to divergence across the Indian Ocean. *Mol. Phylogenet. Evol.* 25:267–77

Grubb PJ. 2003. Interpreting some outstanding features of the flora and vegetation of Madagascar. *Perspect. Plant Ecol. Evol. Syst.* 6:125–46

Harland WB, Armstrong RL, Cox AV, Craig LE, Smith AG, Smith DG. 1990. *A Geological Time Scale*. Cambridge: Cambridge Univ. Press. 263 pp.

Hay WW, DeConto RM, Wold CN, Wilson KM, Voigt S, et al. 1999. Alternative global Cretaceous paleogeography. In *Evolution of the Cretaceous Ocean Climate System*, ed. E Barrera, CC Johnson, pp. 1–47. Geol. Soc. Am. Spec. Pap., Boulder, CO

Jansa SA, Weksler M. 2004. Phylogeny of muroid rodents: relationships within and among major lineages as determined by IRBP gene sequences. *Mol. Phylogenet. Evol.* 31:256–76

Janssen T, Bremer K. 2004. The age of major monocot groups inferred from 800+ *rbcL* sequences. *Botan. J. Linn. Soc.* 146:385–98

Kappeler PM. 2000. Lemur origins: rafting by groups of hibernators? *Folia Primatol.* 71:422–25

Karanth KP, Delefosse T, Rakotosamimanana B, Parsons TJ, Yoder AD. 2005. Ancient DNA from giant extinct lemurs confirms single origin of Malagasy primates. *Proc. Natl. Acad. Sci. USA* 102:5090–95

Kerdelhue C, Le Clainche I, Rasplus JY. 1999. Molecular phylogeny of the Ceratosolen species pollinating Ficus of the subgenus Sycomorus sensu stricto: biogeographical history and origins of the species-specificity breakdown cases. *Mol. Phylogenet. Evol.* 11:401–14

Kiener A, Richard-Vindard G. 1972. Fishes of the continental waters of Madagascar. In *Biogeography and Ecology in Madagascar*, ed. R Battistini, G Richard-Vindard, pp. 477–99. The Hague: W. Junk

King W. 1962. The occurence of rafts for dispersal of land animals into the West Indies. *Q. J. Fla. Acad. Sci.* 45–52

Kirchman JJ, Hackett SJ, Goodman SM, Bates JM. 2001. Phylogeny and systematics of ground rollers (Brachypteraciidae) of Madagascar. *Auk* 118:849–63

Kishino H, Thorne JL, Bruno WJ. 2001. Performance of a divergence time estimation method under a probabilistic model of rate evolution. *Mol. Biol. Evol.* 18:352–61

Krause DW. 2003. Late Cretaceous vertebrates of Madagascar: a window into Gondwanan biogeography at the end of the age of dinosaurs. See Goodman & Benstead 2003, pp. 40–47

Krause DW, Hartman JH, Wells NA. 1997a. Late Cretaceous vertebrates from Madagascar: implications for biotic change in deep time. See Goodman & Patterson 1997, pp. 3–43

Krause DW, Prasad GVR, von Koenigswald W, Sahni A, Grine FE. 1997b. Cosmopolitanism among Gondwanan Late Cretaceous mammals. *Nature* 390:504–7

Krause DW, Rogers RR, Forster CA, Hartman JH, Buckley GA, Sampson SD. 1999. The Late Cretaceous vertebrate fauna of Madagascar: implications for Gondwanan paleobiogeography. *GSA Today* 9:1–7

Leroy JF. 1978. Composition, origin, and affinities of the Madagascan vascular flora. *Ann. Mo. Bot. Gard.* 65:535–89

Lourenço WR. 1996. *Biogéographie de Madagascar*. Paris: Ed. ORSTOM, 588 pp.

Lowry PPI, Schatz GE, Phillipson PB. 1997. The classification of natural and anthropogenic vegetation in Madagascar. See Goodman & Patterson 1997, pp. 93–123

Malod JA, Mougenot S, Raillard S, Maillard A. 1991. Novelles constraintes sur la cinematique de Madagascar: les structures de la chaine Davie. *C. R. Acad. Sci. (Paris), Ser. 2* 312:1639–46

Marks BD, Willard DE. 2005. Phylogenetic relationships of the Madagascar pygmy kingfisher (Ispidina madagascariensis). *Auk* 122:1271–80

Mausfeld P, Vences M, Schmitz A, Veith M. 2000. First data on the molecular phylogeography of scincid lizards of the genus Mabuya. *Mol. Phylogenet. Evol.* 17:11–14

McCall R. 1997. Implications of recent geological investigations of the Mozambique Channel for the mammalian colonization of Madagascar. *Proc. R. Soc. London Ser. B* 264:663–65

McDowall RM. 2004. What biogeography is: a place for process. *J. Biogeogr.* 31:345–51

McGlone MS. 2005. Goodbye Gondwana. *J. Biogeogr.* 32:739–40

Millot J. 1953. Le continent de Gondwana et les méthodes de raisonnement de la biogéographie classique. *Ann. Sci. Nat. Zool. (Sér. 11)* 15:185–219

Monaghan MT, Gattolliat JL, Sartori M, Elouard JM, James H, et al. 2005. Trans-oceanic and endemic origins of the small minnow mayflies (Ephemeroptera, Baetidae) of Madagascar. *Proc. R. Soc. B* 272:1829–36

Murray AM. 2001. The fossil record and biogeography of the Cichlidae (Actinopterygii: Labroidei). *Biol. J. Linn. Soc.* 74:517–32

Myers GS. 1938. Fresh-water fishes and West Indian zoogeography. *Annu. Rep. Smithson. Inst.* 1937:339–64

Nagy ZT, Joger U, Wink M, Glaw F, Vences M. 2003. Multiple colonization of Madagascar and Socotra by colubrid snakes: evidence from nuclear and mitochondrial gene phylogenies. *Proc. R. Soc. B* 270:2613–21

Negi JG, Agrawal PK, Pandey OP, Singh AP. 1993. A possible K-T boundary bolide impact site offshore near Bombay and triggering of rapid Deccan volcanism. *Phys. Earth Planet. Inter.* 76:189–97

Nelson G. 1979. From Candolle to Croizat: comments on the history of biogeography. *J. Hist. Biogeogr.* 11:269–305

Noonan B, Chippendale PT. 2006. Vicariant origin of Malagasy reptiles supports Late Cretaceous Antarctic landbridge. *Am. Nat.* In press

Odierna G, Canapa A, Andreone F, Aprea G, Barucca M, et al. 2002. A phylogenetic analysis of cordyliformes (Reptilia: Squamata): comparison of molecular and karyological data. *Mol. Phylogenet. Evol.* 23:37–42

Olson LE. 1999. *Systematics, evolution, and biogeography of Madagascar's Tenrecs (Mammalia: Tenrecidae)* PhD thesis. Univ. Chicago

Olson LE, Goodman SM. 2003. Phylogeny and biogeography of Madagascar's tenrecs (Lipotyphla, Tenrecidae). See Goodman & Benstead 2003, pp. 1235–42

Paulian R, Viette P. 2003. An introduction to the terrestrial and freshwater invertebrates. See Goodman & Benstead 2003, pp. 503–11

Platnick NI. 1981. The progression rule or progress beyond the rules in biogeography. In *Vicariance Biogeography: a Critique*, ed. G Nelson, DE Rosen, pp. 144–50. New York: Columbia Univ. Press

Plummer PS. 1996. The Amirante Ridge/trough complex: response to rotational transform rift/drift between Seychelles and Madagascar. *Terra Nova* 8:34–47

Popham EJ. 2000. The geographical distribution of the Dermaptera (Insecta) with reference to continental drift. *J. Nat. Hist.* 34:2007–27

Poux C, Madsen O, Marquard E, Vieites DR, de Jong WW, Vences M. 2005. Asynchronous colonization of Madagascar by the four endemic clades of primates, tenrecs, carnivores, and rodents as inferred from nuclear genes. *Syst. Biol.* 54:719–30

Prum RO. 1993. Phylogeny, biogeography, and evolution of the broadbills (Eurylaimidae) and asities (Philepittidae) based on morphology. *Auk* 110:304–24

Rabinowitz PD, Coffin MF, Falvey D. 1983. The separation of Madagascar and Africa. *Science* 220:67–69

Rage JC. 1996. Le peuplement animal de Madagascar: une comosante venue de Laurasie est-elle envisageable? See Lourenço 1996, pp. 27–35

Rage JC. 2003. Relationships of the Malagasy fauna during the Late Cretaceous: northern or southern routes? *Acta Palaeontol. Pol.* 48:661–62

Rage JC, Jaeger JJ. 1995. The sinking Indian raft: a response to Thewissen and McKenna. *Syst. Biol.* 44:260–64

Rakotosolofo NA, Torsvik TH, Ashwal LD, Eide EA, de Wit MJ. 1999. The Karoo Supergroup revisited and Madagascar-Africa fits. *J. Afr. Earth Sci.* 29:135–51

Raxworthy CJ. 2003. Introduction to the reptiles. See Goodman & Benstead 2003, pp. 934–49

Raxworthy CJ, Forstner MR, Nussbaum RA. 2002. Chameleon radiation by oceanic dispersal. *Nature* 415:784–87

Reeves CV, de Wit M. 2000. Making ends meet in Gondwana: retracing the transforms of the Indian Ocean and reconnecting shear zones. *Terra Nova* 12:272–80

Reeves CV, Sahu BK, de Wit M. 2002. A re-examination of the paleo-position of Africa's eastern neighbours in Gondwana. *J. Afr. Earth Sci.* 34:101–8

Rowley DB, Currie BS. 2006. Palaeo-altimetry of the late Eocene to Miocene Lunpola basin, central Tibet. *Nature* 439:677–81

Sampson SD, Carrano MT, Forster CA. 2001. A bizzare predatory dinosaur from the late Cretaceous of Madagascar. *Nature* 409:504–6

Sampson SD, Witmer LM, Forster CA, Krause DW, O'Connor PM, et al. 1998. Predatory dinosaur remains from Madagascar: Implications for the Cretaceous biogeography of Gondwana. *Science* 280:1048–51

Schatz GE. 1996. Malagasy/Indo-australo-malesian phytogeographic connections. See Lourenço 1996, pp. 73–84

Schatz GE. 2000. Endemism in the Malagasy tree flora. In *Diversity and Endemism in Madagascar*, ed. WR Lourenço, SM Goodman, pp. 1–9. Paris: Mém. Soc. Biogéogr.

Schatz GE. 2001. *Generic Tree Flora of Madagascar*. Cumbria, UK: Kew, 490 pp.

Scotese CR. 2000. *PALEOMAP Project: Earth History* (paleogeographic maps). Dept. Geol., Univ. Texas, Arlington

Seward D, Grujic D, Schreurs G. 2004. An insight into the breakup of Gondwana: identifying events through low-temperature thermochronology from the basement rocks of Madagascar. *Tectonics* 23:C3007

Simpson GG. 1952. Probabilities of dispersal in geologic time. *Bull. Am. Mus. Nat. Hist.* 99:163–76

Sparks JS. 2004. Molecular phylogeny and biogeography of the Malagasy and South Asian cichlids (Teleostei: Perciformes: Cichlidae). *Mol. Phylogenet. Evol.* 30:599–614

Sparks JS, Smith WL. 2004a. Phylogeny and biogeography of cichlid fishes (Teleostei: Perciformes: Cichlidae). *Cladistics* 20:501–17

Sparks JS, Smith WL. 2004b. Phylogeny and biogeography of the Malagasy and Australasian rainbowfishes (Teleostei: Melanotaenioidei): Gondwanan vicariance and evolution in freshwater. *Mol. Phylogenet. Evol.* 33:719–34

Sparks JS, Smith WL. 2005. Freshwater fishes, dispersal ability, and nonevidence: "Gondwana life rafts" to the rescue. *Syst. Biol.* 54:158–65

Sparks JS, Stiassny MLJ. 2003. Introduction to the freshwater fishes. See Goodman & Benstead 2003, pp. 849–63

Stankiewicz J, Thiart C, Masters JC, de Wit MJ. 2006. Did lemurs have sweepstake tickets? An exploration of Simpson's model for the colonization of Madagascar by mammals. *J. Biogeogr.* 33:221–35

Storey M, Mahoney JJ, Saunders AD, Duncan RA, Kelley SP, Coffin MF. 1995. Timing of hot spot-related volcanism and the breakup of Madagascar and India. *Science* 267:852–55

Teissier O. 1859. Unpublished letters of Philibert Commerson. *Bull. Soc. Sci. Arts Belles-Lett. Toulon*, pp. 265–75

Thiel M, Gutow L. 2005. The ecology of rafting in the marine environment. II. The rafting organisms and community. *Oceanogr. Mar. Biol.* 43:279–418

Thorne JL, Kishino H. 2002. Divergence time and evolutionary rate estimation with multilocus data. *Syst. Biol.* 51:689–702

Thorne JL, Kishino H, Painter IS. 1998. Estimating the rate of evolution of the rate of evolution. *Mol. Biol. Evol.* 15:1647–57

Torres E, Lees DC, Vane-Wright RI, Kremen C, Leonard JA, Wayne RK. 2001. Examining monophyly in a large radiation of Madagascan butterflies (Lepidoptera: Satyrinae: Mycalesina) based on mitochondrial DNA data. *Mol. Phylogenet. Evol.* 20:460–73

Torsvik TH, Tucker RD, Ashwal LD, Carter LM, Jamtveit B, et al. 2000. Late Cretaceous India-Madagascar fit and timing of break-up related magmatism. *Terra Nova* 12:220–25

Torsvik TH, Tucker RD, Ashwal LD, Eide EA, Rakotosolofo NA. 1998. Late Cretaceous magmatism in Madagascar: paleomagnetic evidence for a stationary Marion hot spot. *Earth Planet. Sci. Lett.* 164:221–32

Valsangkar AB, Radhakrishnamurty K, Subbarao KV, Beckinsale RD. 1981. Paleomagnetism and potassium-argon age studies of acid igneous rocks from the St. Mary Islands. *Mem. Geol. Soc. India* 3:265–76

Vences M, Freyhof J, Sonnenberg R, Kosuch J, Veith M. 2001. Reconciling fossils and molecules: Cenozoic divergence of cichlid fishes and the biogeography of Madagascar. *J. Biogeogr.* 28:1091–99

Vences M, Kosuch J, Glaw F, Bohme W, Veith M. 2003a. Molecular phylogeny of hyperoliid treefrogs: biogeographic origin of Malagasy and Seychellean taxa. *J. Zool. Syst. Evol. Res.* 41:205–15

Vences M, Kosuch J, Rodel MO, Lotters S, Channing A, et al. 2004. Phylogeography of Ptychadena mascareniensis suggests transoceanic dispersal in a widespread African-Malagasy frog lineage. *J. Biogeogr.* 31:593–601

Vences M, Vieites DR, Glaw F, Brinkmann H, Kosuch J, et al. 2003b. Multiple overseas dispersal in amphibians. *Proc. R. Soc. London Ser. B* 270:2435–42

Warren BH, Bermingham E, Bowie RCK, Prys-Jones RP, Thebaud C. 2003. Molecular phylogeography reveals island colonization history and diversification of western Indian Ocean sunbirds (Nectarinia: Nectariniidae). *Mol. Phylogenet. Evol.* 29:67–85

Warren BH, Bermingham E, Prys-Jones RP, Thebaud C. 2005. Tracking island colonization history and phenotypic shifts in Indian Ocean bulbuls (Hypsipetes: Pycnonotidae). *Biol. J. Linn. Soc.* 85:271–87

Wells NA. 2003. Some hypotheses on the Mesozoic and Cenozoic paleoenvironmental history of Madagascar. See Goodman & Benstead 2003, pp. 16–34

Wiley EO. 1988. Vicariance biogeography. *Annu. Rev. Ecol. Syst.* 19:513–42

Yamagishi S, Honda M, Eguchi K, Thorstrom R. 2001. Extreme endemic radiation of the Malagasy vangas (Aves: Passeriformes). *J. Mol. Evol.* 53:39–46

Yang Z, Rannala B. 2006. Bayesian estimation of species divergence times under a molecular clock using multiple fossil calibrations with soft bounds. *Mol. Biol. Evol.* 23:212–26

Yang Z, Yoder AD. 2003. Comparison of likelihood and Bayesian methods for estimating divergence times using multiple gene loci and calibration points, with application to a radiation of cute-looking mouse lemur species. *Syst. Biol.* 52:705–16

Yoder AD. 1994. Relative position of the Cheirogaleidae in strepsirhine phylogeny: a comparison of morphological and molecular methods and results. *Am. J. Phys. Anthropol.* 94:25–46

Yoder AD. 1997. Back to the future: a synthesis of strepsirrhine systematics. *Evol. Anthropol.* 6:11–22

Yoder AD, Burns MM, Zehr S, Delefosse T, Veron G, et al. 2003. Single origin of Malagasy Carnivora from an African ancestor. *Nature* 421:734–37

Yoder AD, Cartmill M, Ruvolo M, Smith K, Vilgalys R. 1996. Ancient single origin of Malagasy primates. *Proc. Natl. Acad. Sci. USA* 93:5122–26

Yoder AD, Flynn JJ. 2003. Origin of Malagasy Carnivora. See Goodman & Benstead 2003, pp. 1253–56

Yoder AD, Yang Z. 2004. Divergence dates for Malagasy lemurs estimated from multiple gene loci: geological and evolutionary context. *Mol. Ecol.* 13:757–73

Zakharov EV, Smith CR, Lees DC, Cameron A, Vane-Wright RI, Sperling FAH, 2004. Independent gene phylogenies and morphology demonstrate a Malagasy origin for a wide-ranging group of swallowtail butterflies. *Evolution* 58:2763–82

Zhu B, Kidd WSF, Rowley DB, Currie BS, Shafique N. 2005. Age of initiation of the India-Asia collision in the east-central Himalaya. *J. Geol.* 113:265–85

Limits to the Adaptive Potential of Small Populations

Yvonne Willi,[1] Josh Van Buskirk,[2] and Ary A. Hoffmann[1]

[1]Centre for Environmental Stress and Adaptation Research, Department of Zoology and Department of Genetics, University of Melbourne, Parkville, VIC 3010 Australia; email: yvonne.willi@agrl.ethz.ch, ary@unimelb.edu.au

[2]Department of Zoology, University of Melbourne, Parkville, VIC 3010 Australia; and Institute of Zoology, University of Zürich, CH-8057 Zürich, Switzerland; email: josh.vanbuskirk@zool.unizh.ch

Key Words

adaptation, conservation biology, evolvability, genetic variation

Abstract

Small populations are predicted to have reduced capacity to adapt to environmental change for two reasons. First, population genetic models indicate that genetic variation and potential response to selection should be positively correlated with population size. The empirical support for this prediction is mixed: DNA markers usually reveal low heterozygosity in small populations, whereas quantitative traits show reduced heritability only in the smallest and most inbred populations. Quantitative variation can even increase in bottlenecked populations although this effect seems unlikely to increase the adaptive potential of populations. Second, individuals in small populations have lower fitness owing to environmental stress and genetic problems such as inbreeding, which can substantially increase the extinction probability of populations in changing environments. This second reason has not been included in assessments of critical population size assuring evolvability and makes it likely that many small threatened populations have a decreased potential for adaptation.

INTRODUCTION

The genetic and demographic consequences of small population size are of central concern in conservation biology. Threatened organisms often occur in demes that are small and isolated. Some threats to small populations are well known, such as demographic and environmental stochasticity (Lande 1988, Sæther et al. 2005) and declining individual fitness (Fischer & Matthies 1998, Reed 2005, Willi et al. 2005). This review focuses on a different threat to small populations that is widely assumed but poorly understood. Conservation biologists frequently assert that loss of genetic variation reduces the capacity of small populations to evolve in response to novel environmental conditions (e.g., Heschel & Paige 1995, Lande 1988, Reed & Frankham 2003, Templeton et al. 2001). This capacity is termed evolvability (Houle 1992) or adaptive potential. The danger of reduced evolvability is often raised in the context of global climate change, because extinction could be hastened by inability to respond to selection (Blows & Hoffmann 2005, Davis et al. 2005). But how good is the evidence that small populations have reduced evolvability? Our goal is to assess the theoretical and empirical basis for concluding that small and isolated populations have reduced capacity to respond to directional selection, and to evaluate whether reduced evolutionary potential should be a concern for conservation biologists.

The hypothesis of reduced evolvability in small populations makes sense, given certain assumptions. Small populations often harbor low neutral genetic variation and low heterozygosity at marker loci (Frankham 1996). Heterozygosity at marker loci may be proportional to heterozygosity at loci underlying phenotypically expressed traits, which influence additive genetic variance in the traits relevant for evolution. Hence, additive genetic variance should be reduced in small populations. The connection between additive genetic variance and evolutionary potential originates from the breeder's equation, which states that the short-term response to selection is proportional to the narrow-sense heritability (Falconer & Mackay 1996).

But effects of population size are more complex than this line of reasoning implies. First, population genetics theory since the 1960s shows that the amount of genetic variation depends on many factors beyond population size, such as the number of loci underlying a trait, the presence of dominance or epistasis, the effects and fixation probabilities of new mutations, selection intensity, and mode of selection. Second, factors other than genetic variation, including environmental or genetic stress, influence whether a small population adapts to a changing environment or declines toward extinction. Environmental stress refers to declining individual fitness owing to ecological factors, whereas genetic stress can be caused by inbreeding depression, genetic load, or reproductive incompatibility. So far, these well-documented consequences of habitat fragmentation and small population size have made little contribution to discussions about population size and evolvability.

The first section of this review examines models relating population size to heritable genetic variation for one-locus and polygenic traits, and the second section describes the empirical literature on the relationship between population size and genetic variation. The third section discusses mechanisms that may alter the evolutionary potential of small populations, without directly affecting genetic variation, arising from environmental stress, changes in local population processes, and reduced fitness. In the final section we argue that nongenetic mechanisms may be at least as important as standing levels of genetic variation in determining the critical population sizes guaranteeing potential for evolutionary change.

THEORY OF GENETIC VARIATION IN POPULATIONS OF FINITE SIZE

The capacity of a population to respond to selection depends on its level of genetic variation for the trait(s) undergoing selection (Falconer & Mackay 1996). Genetic variation within randomly mating populations is generally increased by mutation and dispersal, but decreased by drift and selection. Thus, we begin by reviewing theoretical findings concerned with equilibrium genetic variation under these four processes, focusing on their sensitivity to population size. The relevant evolutionary models are either single-locus or polygenic. Single-locus models may seem irrelevant because studies of quantitative trait loci (QTL) find that physiological and morphological traits are influenced by numerous loci (Otto & Jones 2000). However single-locus models are relevant for evolvability under some conditions, such as resistance to parasites and infectious diseases in plants (Bergelson et al. 2001) and QTLs of large effect (e.g., Schemske & Bradshaw 1999). Single-locus models are also relevant simply because marker variation is widely used to estimate population genetic variation.

Single-Locus Models

Neutral models. Single-locus models without selection usually measure genetic variation in terms of mean heterozygosity, because it reflects both the number of alleles in the population and their evenness in frequency. However, allelic diversity, a different measure of genetic variation, may better reflect adaptive potential under environmental change that favors a formerly neutral, possibly rare allele in a population, such as resistance to a new disease (Allendorf 1986).

When only drift and mutation occur, heterozygosity increases monotonically with population size (dashed lines in **Figure 1**). There are two explanations for this result. First, the magnitude of genetic drift is inversely proportional to N_e, the effective population size. Drift results in a decrease of heterozygosity at a rate of $1/[2N_e]$ per generation (Kimura 1955, Wright 1931). Drift is especially important when a population is currently declining, because each parent has fewer offspring than expected for a population of the same constant size, increasing the chance that rare alleles are lost (Otto & Whitlock 1997). The second reason, important only over long periods of time, is that fewer mutations appear in small populations. The expected number of generations until a mutation appears at any given locus is $1/(\mu N_e)$, where μ is the per-locus mutation rate (Muller 1964). Equilibrium heterozygosity depends to some degree on assumptions about mutation processes (e.g., see infinite alleles and stepwise mutation models in **Figure 1**), but larger populations are always predicted to have higher heterozygosity.

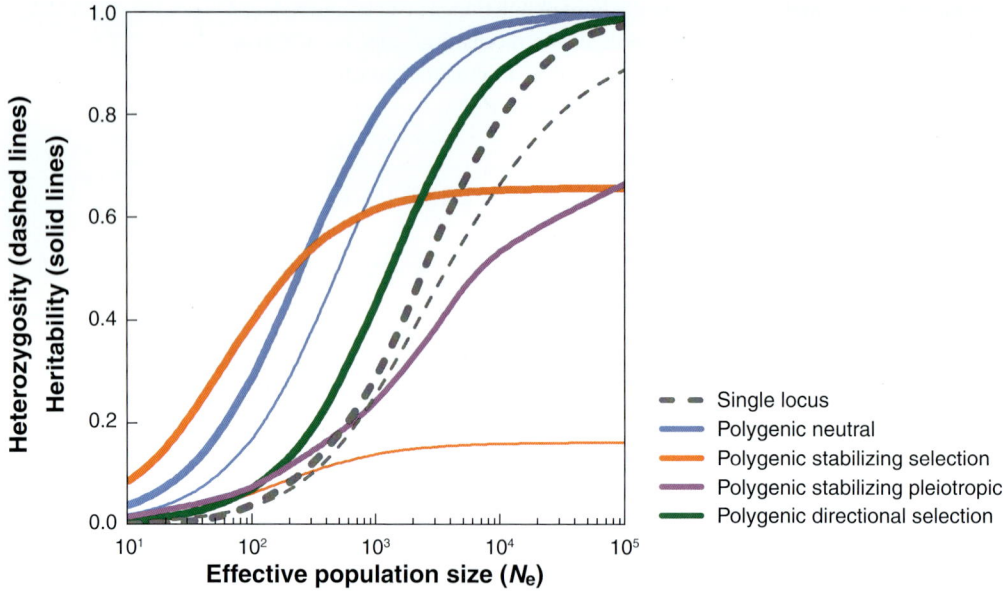

Figure 1
Single-locus and polygenic models predict sigmoidal increases in genetic variation with increasing population size. The dashed lines are heterozygosity predicted by single-locus neutral models (*narrow line*: stepwise mutation; *bold line*: infinite alleles), with mutation rate (μ) = 10^{-4} (Crow & Kimura 1970, Ohta & Kimura 1973). Neutral polygenic models are depicted in blue, assuming a ratio of mutational-to-environmental variance (V_m/V_E) of 0.001. The narrow blue line shows heritability in a single population without dispersal (Lynch & Hill 1986), and the bold blue curve is the situation within one population of size N_e that shares migrants with another population of equal size (island model, Whitlock 1999). Polygenic models with three alleles per locus and weak stabilizing selection are in orange (*narrow line*: 50 loci; *bold line*: 500 loci), with $\mu = 10^{-4}$ and other parameters chosen so that $V_m/V_E = 0.001$ when 100 loci are present (Houle 1989). The purple line is Zhang & Hill's (2002) polygenic model with pleiotropic effects of mutation, weak stabilizing selection, and 0.1 mutations affecting the character per generation. The green line depicts heritability under 50% truncational (directional) selection (Keightley & Hill 1987).

Models with selection. Single-locus models with selection and assuming additive gene effects predict that heterozygosity will decline at all population sizes, because selection increases the likelihood that the allele with the highest fitness will be fixed. But unlike drift, selection is more efficient in large populations rather than in small populations. Indeed, the probability of elimination of a deleterious mutation (whether recessive or not) shows a positive sigmoidal relationship with the product of N_e and the selection coefficient s (Kimura 1957, Robertson 1960). A locus under selection begins to behave like a neutral locus when $s \leq 1/[2N_e]$ (Wright 1931). Therefore as population size declines, stronger selection is required to prevent loss of rare beneficial mutations due to drift, and to ensure elimination of deleterious mutations. Consequently, genetic load becomes a problem.

Models with dispersal. There is a long tradition of modeling genetic structure under dispersal, with important implications for understanding how N_e and genetic variation are related (reviewed in Felsenstein 1976). Heterozygosity in these models depends on migration rate (m). If m exceeds a threshold value (in a two-dimensional rectangle without mutation, $N_e m = 1$), the local populations behave as if they were panmictic, and the rate of loss of heterozygosity within each is $1/[2N_T]$, where N_T is the effective size of the whole metapopulation. If migration rate is below the threshold, loss of heterozygosity within local demes is proportional to the inverse of the number of demes. The explanation is that one allele or the other always goes to fixation within each local deme when migration rate is sufficiently low, so the loss of an allele from the entire metapopulation depends on drift eliminating that allele from all local demes. The likelihood of this outcome decreases with the number of demes (Felsenstein 1976).

Recent work focuses on continuous spatial distributions and makes use of coalescence theory (reviewed in Charlesworth et al. 2003). One example is Barton et al.'s (2002) derivation of the probability that neutral alleles sampled from individuals at different distances apart in two-dimensional space are identical. One minus this probability, evaluated at short distances, is a measure of local genetic diversity. Interpreted this way, genetic diversity within demes is positively related to mutation rate, as expected, but also to average dispersal distance and population density (i.e., the population size of a local deme). In this model dispersal acts alongside mutation to maintain genetic variation in the face of drift, and there is no threshold above which genetic variation is unaffected by dispersal.

Polygenic Models

Neutral models. The development of theory for quantitative traits with polygenic inheritance has paralleled that for single loci. Quantitative traits can be modeled as neutral, exposed to selection, or influenced by dispersal, and genetic variance in quantitative traits is positively related to N_e and the amount of new genetic variance introduced by mutation.

Under assumptions of neutrality and additive gene effects, equilibrium additive genetic variance (V_A) of a quantitative trait increases linearly with effective population size (Chakraborty & Nei 1982, Lynch & Hill 1986). This occurs for the same reasons as in single-locus models: Variation is eroded in small populations because the loci that underlie quantitative traits are subject to drift, and variation is enhanced by new mutations, which are less frequent in small populations. The narrow-sense heritability of the trait, defined as the proportion of phenotypic variance that is genetically inherited by offspring (V_A/V_P; V_P is total phenotypic variance), increases sigmoidally with population size (*narrow blue curve* in **Figure 1**). Thus, according to these models, the signature of genetic drift should be visible not only in neutral marker loci, but also in the heritability of measurable quantitative traits.

Not every small population has low quantitative variation, because the outcome of neutral evolution becomes less predictable at low N_e. The expected variance of

V_A is inversely related to population size and the number of loci (Avery & Hill 1977, Chakraborty & Nei 1982). Small populations therefore should exhibit variable amounts of genetic variation, primarily because of differences among populations in the amount of linkage disequilibrium. Genetic covariances are just as susceptible to this effect as are genetic variances (Avery & Hill 1977, Jones et al. 2003): Small N_e leads to instability of the **G** matrix, which describes genetic variances and covariances for multiple traits. This suggests that the response of small populations to directional selection is reduced on average, and that both direct and correlated responses will be more variable.

Models with selection. Polygenic models that include selection predict that equilibrium genetic variance depends on factors apart from population size, including the type and intensity of selection, the rate and effect of mutations, and the number of loci involved (Houle 1989, Keightley & Hill 1987). As in neutral models, additive genetic variance is predicted to increase with population size under all forms of selection, because of the combined effects of more mutations and weaker drift. Under directional selection and free recombination there is no theoretical limit to the increase (Keightley & Hill 1987), whereas under stabilizing selection V_A reaches an asymptote in large populations (Houle 1989). This is because mutations with positive effect on the phenotype are favored under directional selection, whereas mutations having both positive and negative effects are disfavored under stabilizing selection. **Figure 1** illustrates that the maximum possible heritability under stabilizing selection depends on n, the number of loci underlying the trait (*thin* versus *thick red lines*) (Bulmer 1971, Bürger et al. 1989, Lande 1976). This assumes that all loci have the same mutational variance.

There are several reasons why directional and stabilizing selection usually reduce heritability relative to the neutral situation. First, selection directly eliminates genetic variation, although this effect is small when many loci are involved (Crow & Kimura 1970, Robertson 1960). Second, selection favors genotypes that carry beneficial alleles at multiple loci. This generates linkage disequilibrium, which introduces a negative component of variation that is subtracted from the total genetic variance (the Bulmer effect; Bulmer 1971). Linkage reduces the number of independent loci, which in turn reduces genetic variation because (under the assumption of additivity) V_A is roughly equal to the number of loci times the additive variance expected at equilibrium at each locus (Turelli 1984). The Bulmer effect can be substantial (5–25% of initial V_A; Verrier et al. 1991), but recombination lessens it. A third reason selection reduces heritability is that selection decreases the number of individuals that reproduce, increases the relatedness among them, and introduces variance in reproductive success, all of which reduce N_e. This increases the strength of drift and the frequency of inbreeding relative to the neutral situation, further eroding genetic variation (Robertson & Hill 1983, Verrier et al. 1991).

Models that relax the assumption of additive gene effects can produce high levels of V_G, at least at large population sizes (Barton 1990, Keightley & Hill 1990, Kondrashov & Turelli 1992). These models were motivated by the observed pleiotropic effects of most loci, and the fact that far more traits appear to be under stabilizing selection than can be justified under genetic load theory (Barton 1990). Mutations in these models contribute genetic variation to many traits at once, and cause apparent stabilizing selection on characters because they have negative pleiotropic effects on fitness. Heritabilities predicted at low N_e are no larger than those occurring under standard polygenic theory (*purple line*, **Figure 1**) (Zhang & Hill 2002).

Frequency-dependent selection may help maintain genetic variation (Endler 1986). Very few models explore the interaction between population size and frequency-dependent selection, and so far they suggest that this mechanism cannot maintain variation at small N_e (Roff 1998).

An important point from **Figure 1** is that the sensitivity of genetic variation to N_e depends on the range of population sizes considered (Foley 1992). Under stabilizing selection, there is almost no influence of N_e on heritability until populations decline below a few hundred individuals. In contrast, single-locus heterozygosity is sensitive to N_e at much higher population sizes (low thousands). Only models of two alleles of constant effect predict a proportional relationship between H and V_A (Foley 1992). This implies that empirical tests of single-locus and polygenic theory must sample a wide range of population sizes, from about ten to several hundred individuals.

Models with dispersal. Another class of polygenic models explores the impact of dispersal on genetic variation within local demes and metapopulations. For a broad range of population structures, any amount of movement among demes is predicted to increase V_A within each deme to the level expected for the metapopulation as a whole (Lande 1992, Lynch 1988, Whitlock 1999). Immigrant alleles counteract loss from drift much more rapidly than mutation alone. This means that V_A is not directly related to local population size, unless the size of the entire metapopulation is related to the size of its parts (*thick blue line*, **Figure 1**). Dispersal also reduces genetic differentiation among demes (Cockerham & Tachida 1987, Lande 1992), and can depress the effective population size of the entire metapopulation (Whitlock 1999). Thus, dispersal alters the evolutionary potential of the local deme in two conflicting ways, by increasing genetic variation beyond that expected for a population of its size, but also by decreasing variation within the larger metapopulation to which the deme belongs.

Response to Long-Term Directional Selection

The breeder's equation states that the response to one generation of selection is proportional to the strength of selection (selection differential) and the level of additive genetic variance of the trait under selection. As described above, standing levels of V_A are predicted to vary with N_e. For a given value of V_A, however, the long-term response to selection depends on population size in other ways. One reason is that drift decreases the chance of fixation of (rare) beneficial alleles under selection, and therefore counteracts the advance due to selection in small populations (Hill & Rasbash 1986, Robertson 1960). Also, new mutations become relatively less important, and inbreeding depression more important, as N_e decreases. Both factors decrease the response to selection (Hill & Rasbash 1986, Wei et al. 1996). Finally, the selection differential itself declines during long-term selection, because the population of selected individuals is highly related and therefore smaller than numbers alone would suggest (Verrier et al. 1991). Thus, small populations not only possess lower levels of genetic variation, but they are also less likely to reach the total response to selection predicted by their genetic variance. Moreover, the response to selection is less predictable in populations that are small or have low genetic variation (Hill 1982, Nomura 1997). The variance of the response is $2FV_A$, where F is the inbreeding coefficient and V_A is the initial additive genetic variance. Thus, the response becomes more variable as populations remain small for some time.

Bottlenecked Populations

A large literature addresses the effect of bottlenecks on genetic variance in quantitative traits. Large populations experience a bottleneck when they suddenly decline to small size for a short period. According to neutral theory, expected V_A should decline by a factor of $1/[2N_e]$ during a one-generation bottleneck of size N_e (Lynch & Hill 1986). But bottlenecks affect gene expression and evolutionary potential in complex ways, because, in addition to reducing additive variance due to drift, they also perturb gene frequencies and thereby change the genetic background of individual genes. This can cause V_A to increase immediately after the bottleneck. Goodnight (1987, 1988; Wade & Goodnight 1998) argued that epistatic variance in a large population can contribute to V_A if alleles at interacting loci suddenly change in frequency. However, Robertson (1952), Willis & Orr (1993), and Wang et al. (1998) concluded on theoretical grounds that changing frequencies of alleles exhibiting dominance account for most changes in V_A after a bottleneck. Recent work confirms that, even though both mechanisms are plausible, their effects are unpredictable and changes in V_A are probably based on increasing frequencies of deleterious alleles (Barton & Turelli 2004, Zhang et al. 2004). The outcome is also quite variable, so that a few populations show increasing V_A even if V_A decreases on average (Barton & Turelli 2004). In summary, a bottleneck may upset the prediction of declining V_A in small populations; but on average, and for traits with primarily additive inheritance, severe bottlenecks ($F > 0.25$) should decrease additive genetic variance.

GENETIC VARIATION IN SMALL AND LARGE POPULATIONS: EMPIRICAL STUDIES

The models discussed above predict that small populations show reduced variation at neutral marker loci, and reduced V_A and h^2 in quantitative traits as they become very small. All populations should exhibit increased genetic variance in response to gene flow. Heritabilities in very small populations are predicted to decline more strongly under directional selection than under stabilizing selection, and variability in genetic variance and selection response should increase at small population size. This section reviews empirical evidence relevant to these predictions.

Single-Locus Variation in Small Populations

Single-locus theory is well supported by comparative field studies on neutral markers and DNA sequence variation. The reduction in heterozygosity within small natural populations expected under neutral theory has been found repeatedly in comparisons among populations (Ellstrand & Elam 1993, Frankham 1996, Soulé 1976) and among species with differing abundance (Cole 2003, Spielman et al. 2004). In some studies of small populations the rate at which marker variation is lost matches the predictions of neutral theory almost exactly (e.g., Wayne et al. 1991). Small populations have proportionally greater temporal variation in population size than do large populations (Reed & Hobbs 2004), perhaps in part because they occur in more variable environments. This increases the importance of drift beyond that predicted from the absolute size of small populations, because N_e is related to the harmonic mean size through time (Crow & Kimura 1970).

Quantitative Genetic Variation in Small Populations

Experiments. The prediction that small populations contain reduced levels of quantitative genetic variation has been tested in the field and laboratory. Some experiments that allow populations to evolve at different sizes strongly support theory. A good example is Swindell & Bouzat's (2005) comparison between observed and expected rates at which quantitative genetic variation is lost through time. Replicate lines of *Drosophila melanogaster* were maintained at an effective size of 3.2 for 30 generations or 14 for 77 generations. At intervals of roughly 10 generations, the authors measured heritability by imposing selection for increased sternopleural bristle number. As expected, the response to selection declined with increasing time at small size. The data were quantitatively consistent with two alternative models, based on only drift and mutation or including stabilizing selection as well; both models predict similar levels of genetic variation at the population sizes tested (**Figure 1**).

We checked the generality of Swindell & Bouzat's (2005) findings by compiling data from 21 publications that reported heritability in experimental populations that were inbred for one or more generations (**Figure 2**). We used the inbreeding coefficient, F, as a surrogate of population size for this comparison because it can be derived from studies with very different designs. For example, an inbreeding coefficient of 0.25 can result from a one-generation bottleneck of two individuals or 30 generations of evolution at $N_e = 50$. In both cases, F gives the expected loss of heterozygosity at neutral loci, and V_A declines linearly with F under a neutral model with additive gene effects. To facilitate comparison of the experiments with studies of populations, **Figure 2** also depicts the N_e that would yield an equivalent level of inbreeding after a period of 10 generations. The figure should be interpreted cautiously because values do not indicate the level of replication, and some values represent different traits assessed on the same lines. Nevertheless, observed declines in heritability approach predictions from additive neutral theory for certain kinds of traits, and for all traits in populations that have been small for a long time or severely inbred. The main deviations from neutral predictions are for life-history traits closely associated with fitness.

Heritability may differ from predictions based on additive neutral theory for several reasons. Characters may experience weak fluctuating or stabilizing selection that opposes erosion due to drift (Kondrashov & Yampolsky 1996, Kristensen et al. 2005),

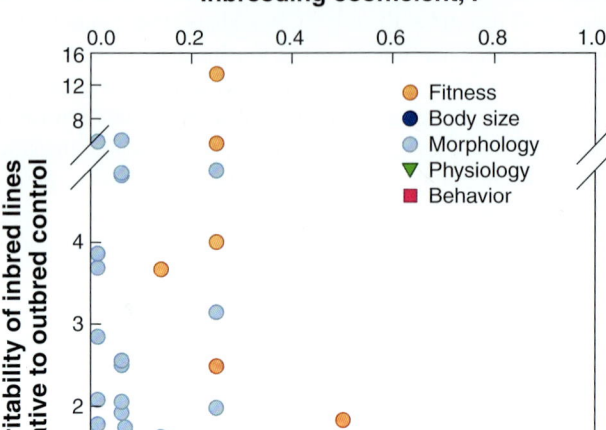

Figure 2
Change in heritability in inbred populations relative to outbred controls, measured in 21 studies that created small populations experimentally for one or several generations. The horizontal axis is expressed as either inbreeding coefficient (*upper axis*) or the value of N_e that would generate the same level of inbreeding over 10 generations. The figure is dominated by data on *Drosophila melanogaster* (10 studies) and *Musca domestica* (5 studies), but includes information from *Tribolium* beetles (2), two plant species (2), laboratory mice (1), and a butterfly (1). Each point on the figure refers to a character for which narrow-sense heritability was measured using a crossing design or artificial selection. The line shows the decline in heritability expected under an additive model with no selection, assuming $h^2 = 0.5$ in the outbred population. The trend for morphological traits ($n = 101$) is very close to the neutral additive expectation, whereas fitness-related life-history traits ($n = 12$) sometimes show increasing heritability at moderate inbreeding. Body size, physiology, and behavior have been studied too infrequently to support general conclusions.

or may be affected by alleles acting in a nonadditive way or linked to loci showing overdominance (Clegg et al. 1980). If traits are affected by epistasis or dominance, as fitness-related traits often are, genetic drift during short periods of small population size can inflate V_A above expectations from neutral additive theory (Barton & Turelli 2004, Robertson 1952). Nevertheless, the conclusion from experiments summarized in **Figure 2** is that the heritability of quantitative traits usually declines when population size is experimentally reduced.

Field observations. Comparison of population divergence in neutral markers and quantitative traits has been used to assess indirectly the effects of drift in nature.

Divergence between pairs of populations at marker loci, measured by Wright's (1943) F_{ST} or related measures, reflects the fraction of genetic variation that occurs between populations relative to within them. Genetic divergence in quantitative traits is measured by Q_{ST}, which theoretically equals F_{ST} if traits and markers are subject to the same evolutionary forces (Spitze 1993, Whitlock 1999). Indeed, many field studies report positive correlations between F_{ST} and Q_{ST} (Merilä & Crnokrak 2001), suggesting that quantitative traits are influenced by genetic drift. If drift accounts for this correlation, then F_{ST} and Q_{ST} should be negatively associated with population size, as found in one study (Willi et al. 2007).

In contrast, estimates of variation within natural populations often do not support the prediction that quantitative genetic variation declines at small N_e. Willi et al. (2007) directly estimated genetic variation in 13 populations of a plant, and found that small demes had significantly reduced narrow-sense heritability averaged over eight morphological and life-history characters. Heritability for two characters was significantly lower in small populations and tended downward in other characters. However, two studies comparing pairs of populations differing in size found no evidence for lower heritability in the smaller population, although these studies lacked power because of the number of populations involved (Waldmann 2001, Widén & Andersson 1993). Four studies included multiple populations, but found weak correlations between census size and broad sense genetic variation in a range of traits (Gravuer et al. 2005, Meyer & Allen 1999, Podolsky 2001, Waldmann & Andersson 1998). We combined data from the five studies of multiple populations in a meta-analysis (Rosenberg et al. 1997) and found that the mean effect size, measured by the correlation between genetic variation (V_G or h^2) and census size, was 0.0044 (95% CI: –0.159 to 0.167). Correlations involving traits related to fitness, overall plant size, and morphology or phenology did not differ ($P > 0.6$). In spite of these results, it may be premature to conclude that field data conflict with theory. Meyer & Allen (1999), Podolsky (2001), and Waldmann & Andersson (1998) observed no correlations between population size and neutral marker diversity, suggesting that current census size in these populations is not closely related to long-term N_e.

Reed & Frankham's (2001) meta-analysis of studies comparing estimates of quantitative and neutral marker variation also provides no support for a strong association between quantitative variation and population size. Although population size was not considered in this analysis, it can be inferred indirectly because of the close association between marker variation and N_e discussed above. Reed & Frankham found that heritability for many traits was essentially unrelated to neutral variation ($n = 15$ studies, average $r = -0.08$). The weakest correlations were for traits associated with fitness (i.e., life-history traits), perhaps because these are not neutral and are often governed by nonadditive loci (Crnokrak & Roff 1995).

Several factors may explain why field data do not support the predicted decline in small populations. One is suggested by the laboratory experiments summarized in **Figure 2**. Small populations may have declined in size only recently and still benefit from a temporary contribution of dominance or epistatic variance to additive variance (Barton & Turelli 2004). The time elapsed since population decline is important even under purely additive models, because the rate at which V_A drops under drift-mutation balance [$1/(2N_e)$ per generation] implies that substantial time is required to produce measurable differences among populations. Excursion to within 50% of asymptotic V_A requires $1.4N_e$ generations (Lynch & Hill 1986), which can amount to many decades for small populations in the field. Another factor is that traits may experience weak stabilizing selection, in which case V_A is sensitive to N_e only at very small population sizes (**Figure 1**). Although small populations in nature lose substantial marker heterozygosity (Frankham 1996), perhaps they are not yet small enough to have lost additive variance. A third explanation is that the smallest and most genetically impoverished populations are also most vulnerable to extinction prior to being sampled (Foley 1992), biasing field surveys with respect to genetic variation. Perhaps the most likely explanation is that gene flow among populations can decouple within-population V_A from local population size, according to polygenic models with dispersal (Lande 1992, Whitlock 1999). This could explain why experiments are more supportive of theory than field comparisons (gene flow is prevented in the lab) and why small populations have reduced heterozygosity at marker loci (H is correlated with local population density even under dispersal; Barton et al. 2002).

Consequences of dispersal. It has long been known that gene flow among inbred lines increases heritability (e.g., Tantawy 1957). However, the prediction that dispersal increases quantitative variation to the level expected for the size of the entire metapopulation remains untested. Also, little is known about the extent to which dispersal and population size are interconnected. If gametes disperse randomly and demes are equally spaced, then gametes will more likely encounter others from large demes. Indeed, dispersing pollinators and new recruits sometimes move preferentially among large demes, and immigration rates are often positively related to local population size (Hanski et al. 2000, Kunin 1993, Roland et al. 2000). Another mechanism that could exacerbate this pattern is that individuals in small populations often have reduced fitness because of ecological Allee effects and inbreeding (Stephens & Sutherland 1999), and therefore small populations may produce fewer propagules in the first place. Alternatively, sexual and natural selection can magnify the effects of gene flow into small and inbred populations by favoring immigrant genotypes (Ebert et al. 2002, Richards 2000, Willi & Fischer 2005). Clearly, it would be valuable to have more information on consequences of gene flow for genetic variation and responses to selection.

Response to Long-Term Directional Selection

Theory predicts that small populations show a reduced response to long-term selection because they contain lower levels of variation to begin with, they are more susceptible to drift during the selection process, and new mutations are less frequent (Hill 1982, Hill & Rasbash 1986, Robertson 1960). Experimental tests typically subject populations of different sizes to directional selection for several tens of generations by breeding a constant fraction of individuals that exhibit the most extreme phenotypes. These studies demonstrate that small populations show reduced response to selection in the short term (Frankham et al. 1968, Hanrahan et al. 1973, Silvela et al. 1989, Wade et al. 1996, Weyhrich et al. 1998) and in the long term (Eisen 1975, Jones et al. 1968, Weber 1990, Weber & Diggins 1990). The effect of population size can be substantial: Weber (1990) recorded a 57% lower selection response over 55 generations in *D. melanogaster* populations with $N_e \approx 8$ compared with $N_e \approx 200$.

Of the several mechanisms potentially contributing to reduced selection response, two are probably not important in these experiments. First, there can be little initial difference in the existing genetic variation between small and large populations, because populations of all sizes are founded by randomly drawing individuals from a large source pool. Even a founder population of several pairs contains the great majority of V_A present in the source. This mechanism would be important in nature, of course, if populations that had been small for some time were subjected to selection. Second, there is little opportunity for mutation to enhance the selection response of large populations, because these experiments last only tens of generations (Hill 1982). Weber & Diggins (1990) estimated that mutation contributed only 11% of the total selection response over their 65-generation experiment, and the contribution was similar in large and small populations.

The most likely interpretation is therefore that selection is less efficient in small populations because of drift counteracting selection. Weber & Diggins' (1990) summary of data from nine long-term experiments suggested that the selection response observed in small populations declines even more than predicted by a drift/selection model (Kimura 1957, Robertson 1960). The additional decline might be explained by assuming relatively few loci for the trait in question, or by invoking linkage disequilibrium generated by the Bulmer effect (Weber & Diggins 1990, Wei et al. 1996). Overall, the long-term selection studies strongly support the prediction that drift erodes the long-term adaptive potential of small populations.

These studies track genetic variation only in the trait that is the target of selection. From a conservation perspective, it would be valuable to know how selection acting in small and large populations affects genetic variation in a range of traits. Stated in quantitative genetic terms, evolutionary potential depends not simply on genetic variance for the character in question, but on the orientation of the **G** matrix relative to the direction of future selection.

Bottlenecked Populations

The first empirical study to describe the impact of bottlenecks on genetic variation was Bryant et al. (1986). The results contradicted purely additive predictions because there were increases in V_A and heritability, although effects varied among characters and bottleneck sizes. Bryant & Meffert (1993) subjected houseflies to multiple bottlenecks, and also found no evidence that V_A fell below that in nonbottlenecked populations. These results inspired a battery of follow-up investigations. A few have confirmed the capacity of bottlenecks to increase V_A (Fernández et al. 1995, García et al. 1994, López-Fanjul & Villaverde 1989, Ritchie & Kyriacou 1996, Wade et al. 1996), but for many traits the effect on variation is consistent with additive expectations (Saccheri et al. 2001, Swindell & Bouzat 2005, Wade et al. 1996, Whitlock & Fowler 1999).

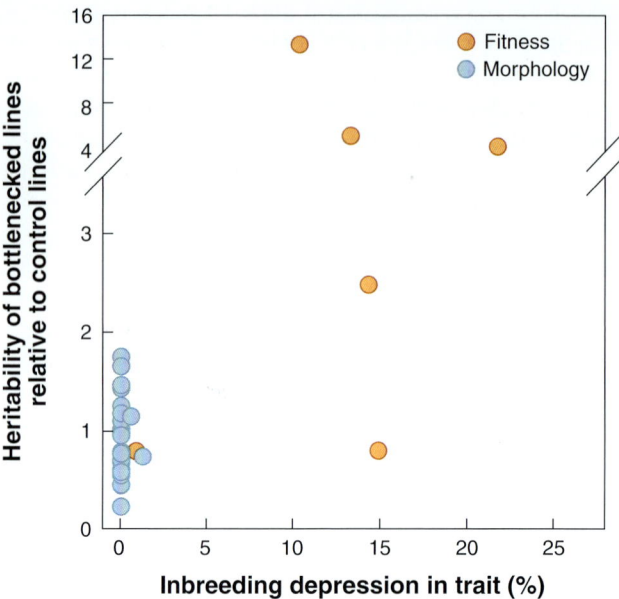

Figure 3
Relationship between the change in heritability after a bottleneck and inbreeding depression in the trait under study. Inbreeding depression is the change in trait value evaluated at $F = 0.25$ relative to the value of nonbottlenecked lines. The symbols represent characters that are either closely allied to fitness (life-history traits) or weakly connected to fitness (morphology). The data come from six studies that applied a one-generation bottleneck and assessed both heritability and inbreeding depression in the same traits (Fernández et al. 1995, 2003; García et al. 1994; Lopéz-Fanjul & Villaverde 1989; Saccheri et al. 2001; Tantawy 1957).

A pattern emerging from these studies is that traits showing increases in genetic variance are life-history traits, which experience inbreeding depression and perhaps strong selection (**Figure 3**). For traits without inbreeding depression (i.e., little change in mean trait value under inbreeding), heritability either follows additive expectations or shows extreme variation among replicate lines (Cheverud et al. 1999, Whitlock & Fowler 1999). This pattern is consistent with the hypothesis that bottlenecks increase V_A by changing nonadditive variance to additive variance, because traits closely connected to fitness and showing inbreeding depression tend to have high dominance variance (Crnokrak & Roff 1995).

Two observations lead us to question whether bottlenecks can increase the adaptive potential of a population. First, a bottleneck usually reduces mean fitness because of inbreeding depression and chance increases in the frequency of deleterious recessives. For example, López-Fanjul & Villaverde (1989) observed decreasing egg-to-pupal viability of *D. melanogaster* after a bottleneck. Even though the response to selection was 6.5 times higher in bottlenecked lines, due to higher V_A in viability, this was enough to overcome only about half the inbreeding depression resulting from the bottleneck. There is also evidence that bottlenecks can depress fitness in nature: The size of the founder population associated with bird introductions in New Zealand is inversely related to the hatchability of eggs (Briskie & Mackintosh 2004). In cases like these, bottlenecks can hardly be considered advantageous for adaptive potential. Although the bottleneck may increase heritability of a fitness trait, there is no guarantee that mean fitness can recover to the level of fitness in the initial population.

The second observation is that the net change in genetic variances of a collection of traits may always be negative. For every life-history trait showing enhanced V_A in a few lines, there are many other (usually morphological) traits showing decreased variance (Cheverud et al. 1999, Whitlock & Fowler 1999). Heritability in the majority of characters is negatively affected by bottlenecks in most lines, so any bottleneck will decrease V_A on average in a sample of characters. To properly evaluate the evolutionary significance of bottlenecks, experiments are needed that estimate genetic variance in a wide range of traits and the genetic correlations among them. New genetic variance must be sufficient to increase selection response in a bottlenecked population beyond that achieved by selection in a noninbred control population.

OTHER FACTORS AFFECTING ADAPTABILITY OF SMALL POPULATIONS

The adaptive potential of small populations can be indirectly influenced by factors that arise from altered environmental conditions and population processes. Two factors are likely to be important: environmental stress and reduced individual fitness. Both can reduce evolvability.

Environmental Stress

Small populations frequently occur in marginal environments, in habitat fragments, or near the edges of geographic distributions. These situations are associated with exposure to unfavorable conditions (Brown 1984, Hoffmann & Blows 1994), which influence both the response to selection and the selection regime itself.

Selection response is affected if genetic or environmental variances of characters, and hence heritabilities, change under stress. Although laboratory results are ambiguous on this question (Hoffmann & Merilä 1999), field data demonstrate that heritability decreases under stressful conditions, at least for morphological traits (Charmantier & Garant 2005). The decline is in some cases owing to lower V_A, and in other cases owing to increased residual or environmental variance (Charmantier & Garant 2005). Declining heritability under stressful conditions may be enhanced under inbreeding, which heightens the sensitivity of development to environmental differences (Lerner 1954). Whitlock & Fowler (1999) observed an 11% increase in environmental variance in inbred relative to outbred lines of *D. melanogaster*, and they cited other studies reporting similar findings. Though the interaction between environmental stress and inbreeding in nature is unclear, small populations may have reduced adaptability in part because they occupy unfavorable environments.

Ecological conditions in habitat fragments differ from those in continuous habitat (e.g., Laurance et al. 1998, Robinson et al. 1995). The selection regimes that prevail within small populations inhabiting marginal and fragmented environments will often differ from those in large populations. Environmental stress in fragments is likely to be stronger than in large populations and this could lead to rapid genetic divergence (Antonovics 1976). For instance, Karlsson & Van Dyck (2005) found that offspring of a woodland butterfly from a fragmented (agricultural) landscape and a closed forest differed in fecundity. Willi et al. (2007) compared genetic divergence among plant populations in quantitative traits (measured by Q_{ST}) and at neutral marker loci (measured by F_{ST}) and found that quantitative divergence exceeded neutral divergence, but only among small demes exposed to divergent selection. These studies suggest that the intensity and direction of selection can differ in small populations.

Reduced Individual Fitness

A second feature of small populations with consequences for adaptation is that individuals often have low mean fitness (Reed 2005) owing to inbreeding depression, genetic load, or nongenetic Allee effects (Stephens & Sutherland 1999, Willi et al. 2005). Crnokrak & Roff's (1999) review of inbreeding depression in natural populations indicated a decline in fitness ranging from 14% ± 3 SE (standard error) in plants (standardized for populations with $F = 0.25$) to 21% ± 12 SE in homeotherms. Lower individual fitness reduces the growth rate of the population, which in turn compromises its potential response to selection. Natural selection is costly for a population in the sense that it causes the death (or reduces fitness) of individuals that would otherwise have reproduced. The number of selective deaths per capita necessary to substitute a beneficial allele beginning at frequency p for an alternative less fit allele is $2\log_e (1/p)$ in diploids (Haldane 1957). This cost has been integrated with population growth rate and quantitative genetic variation in a model by Lynch & Lande (1993). If environmental stochasticity is disregarded, the model yields a critical rate of change of the character value conferring maximum fitness [k_c, in phenotypic standard deviation (SD) units] below which the population can respond by adaptation without extinction:

$$k_c = \frac{V_A}{\sigma_w}\sqrt{2r - \frac{1}{2N_e}},$$

where σ_w is the SD of the Gaussian fitness function underlying stabilizing selection, and r is the maximum possible population growth rate. This model illustrates how reduced individual fitness limits potential adaptive responses to environmental change. For instance, Fischer & Matthies (1998) estimated population growth rate in a plant, *Gentianella germanica*, in 23 local demes ranging in size from 21 to 3900 plants. Assuming that r is the maximum growth observed in any population ($r = 0.4$), then the largest populations are capable of adapting to reasonably rapid environmental change (0.28 phenotypic SD units per generation under weak stabilizing selection, $\sigma_w = 25$). However, in the smallest populations $k_c = 0.0015$ SD units, which suggests that the cost of selection will cause extinction of these populations even under mild environmental change. Even if these populations have sufficient genetic variation to respond, a slowly changing environment could exact a sufficient number of selective deaths to send them toward extinction. If anything, this calculation is optimistic because it assumes that only one character experiences selection, it includes a form of population growth favorable to persistence (logistic), and it equates N_e with N (Bürger & Lynch 1995). Indeed, Fischer & Matthies (1998) observed that all populations with fewer than around 500 individuals had negative growth rates. Low average fitness from inbreeding exacerbates the problem highlighted by Lynch & Lande (1993). But to the extent that ecological conditions in small populations also cause low fitness, results such as Fischer & Matthies (1998) may indicate that costly selection already occurs.

These two features of small populations—reduced average fitness and exposure to stressful environmental and genetic conditions—can act together to decrease the adaptability of small natural populations. A literature survey by Armbruster & Reed (2005) suggested that inbreeding depression is on average 69% more severe in stressful conditions. Also, the fitness cost of selection is likely to be amplified in environments that are extreme in multiple ways, exposing populations to selection on a wider range of characters. This decreases selection responses for specific traits because the potential rate of adaptation is limited by the number of characters under selection (Fisher 1930). The magnitude of the reduction could be substantial, because the rate of response to selection declines with the number of characters c at a rate greater than $1/c$ (Orr 2000).

REDUCED EVOLUTIONARY POTENTIAL AND CRITICAL POPULATION SIZES

There has been much discussion of the practical relevance of genetic erosion in conservation (Caughley 1994, Frankham 2005, Franklin 1980, Lande 1995). Franklin (1980; Franklin & Frankham 1998) first developed the idea that conservation biologists should agree upon a minimum critical population size below which the rate of loss of variation due to drift exceeds the rate of gain due to mutation. Franklin calculated that under mutation-drift balance an effective population size of ≈500 is sufficient to retain a heritability of 0.5 (**Figure 1**), a value commonly observed in natural populations. Lande (1995; Lynch & Lande 1998) disagreed, noting that many new mutations are not beneficial, and that models including selection require larger population sizes to maintain $h^2 = 0.5$. Lande suggested that a long-term effective size of 5000 would be more appropriate.

Both calculations begin with the idea of conserving heritability values as high as 0.5. However, field surveys indicate that h^2 can be considerably lower, especially for life-history and some physiological traits (Hoffmann 2000, Mousseau & Roff 1987), and in traits likely to limit species ranges (Blows & Hoffmann 2005). Target heritabilities in the range of 0.2–0.3 might make more sense, but would yield critical population sizes in the hundreds rather than thousands for most polygenic models (**Figure 1**).

In spite of this, we see two strong arguments against setting critical population sizes for evolvability in the hundreds. Foley (1992) and Lynch & Lande (1998), among other researchers, have pointed out that the potential to adapt depends on the type of environmental change and the type of traits that must respond to selection. In general, traits with a simple genetic basis lose variation rapidly at small population sizes.

Although quantitative characters can be influenced by many loci, mapping exercises often reveal small numbers of major genes associated with adaptive shifts (Lexer et al. 2005, Peichel et al. 2001, Schemske & Bradshaw 1999). Some environmental changes likely to affect populations in the future, such as introductions of diseases or toxins, may require adaptive response at rather few loci. In these cases, the population size required to maintain genetic variability is extremely large.

A second argument is that the evolutionary potential of small populations is compromised by processes that do not directly alter genetic variation. Individual fitness declines when populations experience Allee effects, suboptimal conditions, and inbreeding depression, and this will require upward revision of the minimum critical population size. Individual fitness is directly related to the growth rate of a population, which in turn influences the maximum rate at which the population can respond to selection without declining to extinction (Lynch & Lande 1993). If heritability is reduced under environmental stress or inbreeding, the cost of selection will be further exacerbated by the population's increasing lag behind the optimal phenotype.

The amount by which the critical population size should be revised upward depends on the extent to which inbreeding depression is present, as well as the extent to which individual fitness and genetic variance are influenced by marginal conditions where the population occurs. As an example, imagine that the largest *G. germanica* population studied by Fischer & Matthies (1998), containing 3900 plants, suffered a 20% decrease in V_A owing to genetic erosion. Lynch & Lande's (1993) model predicts that the maximum environmental change to which this population could adapt without decreasing to extinction (k_c) would decline from 0.28 to 0.22 phenotypic SD units per generation (20% change). A 20% decrease in r owing to inbreeding depression would reduce k_c from 0.28 to 0.25 SD units (11% change). Though these numbers may depend on assumptions specific to this example, the general point is that processes other than genetic erosion can have a quantitatively important impact on the adaptive potential of small populations.

We suggest that effective population sizes in the low thousands may be sufficient to maintain adequate adaptive potential in most cases. **Figure 1** illustrates that precise values differ depending on the number of underlying loci and the fitness consequences of the relevant traits. Further work on the expected distribution of environmental changes would be helpful, as would careful assessment of the importance of nongenetic threats to evolvability. In particular, we need to know more about the impact of inbreeding on fitness in nature, especially under suboptimal conditions caused by habitat fragmentation and habitat alteration. Meanwhile, many species that have sharply declined because of human activities now occur in such small populations that they currently experience decreased evolvability.

CONCLUSION

Neutral additive theory predicts that declining populations lose genetic variation for polygenic traits, just as they do for single-locus traits. The declines in marker and polygenic variation need not be highly correlated, depending on the range of population sizes involved. Theory also predicts that declines in population size lead to short-term increases in V_A when traits are affected by dominance or epistasis.

Laboratory studies generally confirm these predictions, although additive genetic variation generated by bottlenecks may not be of evolutionary importance. The few field studies that compare levels of quantitative genetic variation in populations of different size do not yet reveal a clear pattern. Selection and dispersal may account for the weak correlation between census size and V_A in nature.

We argue that factors other than standing levels of quantitative genetic variation influence the evolutionary potential of small populations. In particular, suboptimal conditions in the remaining habitat and reduced fitness owing to inbreeding depression will limit selection responses. Lower individual fitness within small natural populations suggests that evolvability is already reduced.

ACKNOWLEDGMENTS

We thank Markus Fischer, Torsten Kristensen, William Sherwin, Barry Sinervo, Anders Christian Sørensen, and Michael Turelli for ideas and comments on earlier drafts of the manuscript. Financial support was provided by the Swiss National Science Foundation and the Australian Research Council.

LITERATURE CITED

Allendorf FW. 1986. Genetic drift and the loss of alleles versus heterozygosity. *Zoo. Biol.* 5:181–90

Antonovics J. 1976. The nature of limits to natural selection. *Ann. Mo. Bot. Gard.* 63:224–47

Armbruster P, Reed DH. 2005. Inbreeding depression in benign and stressful environments. *Heredity* 95:235–42

Avery PJ, Hill WG. 1977. Variability in genetic parameters among small populations. *Genet. Res.* 29:193–13

Barton NH. 1990. Pleiotropic models of quantitative variation. *Genetics* 124:773–82

Barton NH, Depaulis F, Etheridge AM. 2002. Neutral evolution in spatially continuous populations. *Theor. Popul. Biol.* 61:31–48

Barton NH, Turelli M. 2004. Effects of genetic drift on variance components under a general model of epistasis. *Evolution* 58:2111–32

Bergelson J, Kreitman M, Stahl EA, Tian DC. 2001. Evolutionary dynamics of plant R-genes. *Science* 292:2281–85

Blows MW, Hoffmann AA. 2005. A reassessment of genetic limits to evolutionary change. *Ecology* 86:1371–84

Briskie JV, Mackintosh M. 2004. Hatching failure increases with severity of population bottlenecks in birds. *Proc. Natl. Acad. Sci. USA* 101:558–61

Brown JH. 1984. On the relationship between abundance and distribution of species. *Am. Nat.* 124:255–79

Bryant EH, McCommas SA, Combs LM. 1986. The effect of an experimental bottleneck upon quantitative genetic variation in the housefly. *Genetics* 114:1191–1211

Bryant EH, Meffert LM. 1993. The effect of serial founder-flush cycles on quantitative genetic variation in the housefly. *Heredity* 70:122–29

Bulmer MG. 1971. The effect of selection on genetic variability. *Am. Nat.* 105:201–11

Bürger R, Lynch M. 1995. Evolution and extinction in a changing environment: a quantitative genetic analysis. *Evolution* 49:151–63

Bürger R, Wagner GP, Stettinger F. 1989. How much heritable variation can be maintained in finite populations by mutation-selection balance? *Evolution* 43:1748–66

Caughley G. 1994. Directions in conservation biology. *J. Anim. Ecol.* 63:215–44

Chakraborty R, Nei M. 1982. Genetic differentiation of quantitative characters between populations or species. I. Mutation and random genetic drift. *Genet. Res.* 39:303–14

Charlesworth B, Charlesworth D, Barton NH. 2003. The effects of genetic and geographic structure on neutral variation. *Annu. Rev. Ecol. Evol. Syst.* 34:99–125

Charmantier A, Garant D. 2005. Environmental quality and evolutionary potential: lessons from wild populations. *Proc. R. Soc. London Ser. B* 272:1415–25

Cheverud JM, Vaughn TT, Pletscher LS, King-Ellison K, Bailiff J, et al. 1999. Epistasis and the evolution of additive genetic variance in populations that pass through a bottleneck. *Evolution* 53:1009–18

Clegg MT, Kidwell JF, Horch CR. 1980. Dynamics of correlated genetic systems. V. Rates of decay of linkage disequilibria in experimental populations of *Drosophila melanogaster*. *Genetics* 94:217–34

Cockerham CC, Tachida H. 1987. Evolution and maintenance of quantitative genetic variation by mutations. *Proc. Natl. Acad. Sci. USA* 84:6205–09

Cole CT. 2003. Genetic variation in rare and common plants. *Annu. Rev. Ecol. Evol. Syst.* 34:213–37

Crnokrak P, Roff DA. 1995. Dominance variance: associations with selection and fitness. *Heredity* 75:530–40

Crnokrak P, Roff DA. 1999. Inbreeding depression in the wild. *Heredity* 83:260–70

Crow JF, Kimura M. 1970. *An Introduction to Population Genetics Theory*. Minneapolis: Burgess

Davis MB, Shaw RG, Etterson JR. 2005. Evolutionary responses to changing climate. *Ecology* 86:1704–14

Ebert D, Haag C, Kirkpatrick M, Riek M, Hottinger JW, et al. 2002. A selective advantage to immigrant genes in a *Daphnia* metapopulation. *Science* 295:485–88

Eisen EJ. 1975. Population size and selection intensity effects on long-term selection response in mice. *Genetics* 79:305–23

Ellstrand NC, Elam DR. 1993. Population genetic consequences of small population-size: implications for plant conservation. *Annu. Rev. Ecol. Syst.* 24:217–42

Endler JA. 1986. *Natural Selection in the Wild*. Monographs in Population Biology 21. Princeton, NJ: Princeton Univ. Press

Falconer DS, Mackay TFC. 1996. *Introduction to Quantitative Genetics*, 4th edition. Essex, UK: Longman

Felsenstein J. 1976. The theoretical population genetics of variable selection and migration. *Annu. Rev. Ecol. Syst.* 10:253–80

Fernández A, Toro MA, López-Fanjul C. 1995. The effect of inbreeding on the redistribution of genetic variance of fecundity and viability in *Tribolium castaneum*. *Heredity* 75:376–81

Fernández J, Rodríguez-Ramilo ST, Pérez-Figueroa A, López-Fanjul C, Caballero A. 2003. Lack of nonadditive genetic effects on early fecundity in *Drosophila melanogaster*. *Evolution* 57:558–65

Fischer M, Matthies D. 1998. Effects of population size on performance in the rare plant *Gentianella germanica*. *J. Ecol.* 86:195–204

Fisher RA. 1930. *The Genetical Theory of Natural Selection*. Oxford, UK: Oxford Univ. Press

Foley P. 1992. Small population genetic variability at loci under stabilizing selection. *Evolution* 46:763–74

Frankham R. 1996. Relationship of genetic variation to population size in wildlife. *Conserv. Biol.* 10:1500–08

Frankham R. 2005. Genetics and extinction. *Biol. Conserv.* 126:131–40

Frankham R, Jones LP, Barker JSF. 1968. Effects of population size and selection intensity in selection for a quantitative character in *Drosophila*. I. Short-term response to selection. *Genet. Res.* 12:237–48

Franklin IR. 1980. Evolutionary change in small populations. In *Conservation Biology: an Evolutionary-Ecological Perspective*, ed. ME Soulé, BA Wilcox, pp. 135–50. Sunderland, MA: Sinauer

Franklin IR, Frankham R. 1998. How large must populations be to retain evolutionary potential? *Anim. Conserv.* 1:69–70

García N, López-Fanjul C, García-Dorado A. 1994. The genetics of viability in *Drosophila melanogaster*: effects of inbreeding and artificial selection. *Evolution* 48:1277–85

Goodnight CJ. 1987. On the effect of founder events on epistatic genetic variance. *Evolution* 41:80–91

Goodnight CJ. 1988. Epistasis and the effect of founder events on the additive genetic variance. *Evolution* 42:441–54

Gravuer K, von Wettberg E, Schmitt J. 2005. Population differentiation and genetic variation inform translocation decisions for *Liatris scariosa* var. *novae-angliae*, a rare New England grassland perennial. *Biol. Conserv.* 124:155–67

Haldane JBS. 1957. The cost of natural selection. *J. Genet.* 55:511–24

Hanrahan JP, Eisen EJ, LeGates JE. 1973. Effects of population size and selection intensity on short-term response to selection for postweaning gain in mice. *Genetics* 73:513–30

Hanski I, Alho J, Moilanen A. 2000. Estimating the parameters of survival and migration of individuals in metapopulations. *Ecology* 81:239–51

Heschel MS, Paige KN. 1995. Inbreeding depression, environmental stress, and population size variation in scarlet gilia, *Ipomopsis aggregata*. *Conserv. Biol.* 9:126–33

Hill WG. 1982. Predictions of response to artificial selection from new mutations. *Genet. Res.* 40:255–78

Hill WG, Rasbash J. 1986. Models of long-term artificial selection in finite population with recurrent mutation. *Genet. Res.* 48:125–31

Hoffmann AA. 2000. Laboratory and field heritabilities: some lessons from *Drosophila*. In *Adaptive Genetic Variation in the Wild*, ed. TA Mousseau, B Sinervo, J Endler, pp. 200–18. Oxford, UK: Oxford Univ. Press

Hoffmann AA, Blows MW. 1994. Species borders: ecological and evolutionary perspectives. *Trends Ecol. Evol.* 9:223–27

Hoffmann AA, Merilä J. 1999. Heritable variation and evolution under favourable and unfavourable conditions. *Trends Ecol. Evol.* 14:96–101

Houle D. 1989. The maintenance of polygenic variation in finite populations. *Evolution* 43:1767–80

Houle D. 1992. Comparing evolvability and variability of quantitative traits. *Genetics* 130:195–204

Jones AG, Arnold SJ, Borger R. 2003. Stability of the G-matrix in a population experiencing pleiotropic mutation, stabilizing selection, and genetic drift. *Evolution* 57:1747–60

Jones LP, Frankham R, Barker JSF. 1968. The effects of population size and selection intensity in selection for a quantitative character in *Drosophila*. II. Long-term response to selection. *Genet. Res.* 12:249–66

Karlsson B, Van Dyck H. 2005. Does habitat fragmentation affect temperature-related life-history traits? A laboratory test with a woodland butterfly. *Proc. R. Soc. London Ser. B* 272:1257–63

Keightley PD, Hill WG. 1987. Directional selection and variation in finite populations. *Genetics* 117:573–82

Keightley PD, Hill WG. 1990. Variation maintained in quantitative traits with mutation selection balance: pleiotropic side-effects on fitness traits. *Proc. R. Soc. London Ser. B* 242:95–100

Kimura M. 1955. Solution of a process of random genetic drift with a continuous model. *Proc. Natl. Acad. Sci. USA* 41:144–50

Kimura M. 1957. Some problems of stochastic processes in genetics. *Ann. Math. Stat.* 28:882–901

Kondrashov AS, Turelli M. 1992. Deleterious mutations, apparent stabilizing selection, and the maintenance of quantitative variation. *Genetics* 132:603–18

Kondrashov AS, Yampolsky LY. 1996. High genetic variability under the balance between symmetric mutation and fluctuating stabilizing selection. *Genet. Res.* 68:157–64

Kristensen TN, Sorensen AC, Sorensen D, Pedersen KS, Sorensen JG, et al. 2005. A test of quantitative genetic theory using *Drosophila*: effects of inbreeding and rate of inbreeding on heritabilities and variance components. *J. Evol. Biol.* 18:763–70

Kunin WE. 1993. Sex and the single mustard: population-density and pollinator behavior effects on seed-set. *Ecology* 74:2145–60

Lande R. 1976. The maintenance of genetic variability by mutation in a polygenic character with linked loci. *Genet. Res.* 26:221–35

Lande R. 1988. Genetics and demography in biological conservation. *Science* 241:1455–60

Lande R. 1992. Neutral theory of quantitative genetic variance in an island model with local extinction and colonization. *Evolution* 46:381–89

Lande R. 1995. Mutation and conservation. *Conserv. Biol.* 9:782–91

Laurance WF, Ferreira LV, Rankin-de Merona JM, Laurance SG. 1998. Rain forest fragmentation and the dynamics of Amazonian tree communities. *Ecology* 79:2032–40

Lerner IM. 1954. *Genetic Homeostasis*. New York: Wiley

Lexer C, Rosenthal DM, Raymond O, Donovan LA, Rieseberg LH. 2005. Genetics of species differences in the wild annual sunflowers, *Helianthus annuus* and *H. petiolaris*. *Genetics* 169:2225–39

López-Fanjul C, Villaverde A. 1989. Inbreeding increases genetic variance for viability in *Drosophila melanogaster*. *Evolution* 43:1800–4

Lynch M. 1988. The divergence of neutral quantitative characters among partially isolated populations. *Evolution* 42:455–66

Lynch M, Hill WG. 1986. Phenotypic evolution by neutral mutation. *Evolution* 40:915–35

Lynch M, Lande R. 1993. Evolution and extinction in response to environmental change. In *Biotic Interactions and Global Change*, ed. PM Kareiva, JG Kingsolver, RB Huey, pp. 234–50. Sunderland, MA: Sinauer

Lynch M, Lande R. 1998. The critical effective size for a genetically secure population. *Anim. Conserv.* 1:70–72

Merilä J, Crnokrak P. 2001. Comparison of genetic differentiation at marker loci and quantitative traits. *J. Evol. Biol.* 14:892–903

Meyer SE, Allen PS. 1999. Ecological genetics of seed germination regulation in *Bromus tectorum* L. I. Phenotypic variance among and within populations. *Oecologia* 120:27–34

Mousseau TA, Roff DA. 1987. Natural selection and the heritability of fitness components. *Heredity* 59:181–97

Muller HJ. 1964. The relation of recombination to mutational advance. *Mutat. Res.* 1:2–9

Nomura T. 1997. A simulation study on variation in response to selection and population size required for selection programmes. *J. Anim. Breed. Genet.* 114:13–21

Ohta T, Kimura M. 1973. A model of mutation appropriate to estimate the number of electrophoretically detectable alleles in a finite population. *Genet. Res.* 22:201–04

Orr HA. 2000. Adaptation and the cost of complexity. *Evolution* 54:13–20

Otto SP, Jones CD. 2000. Detecting the undetected: estimating the total number of loci underlying a quantitative trait. *Genetics* 156:2093–2107

Otto SP, Whitlock MC. 1997. The probability of fixation in populations of changing size. *Genetics* 146:723–33

Peichel CL, Nereng KS, Ohgi KA, Cole BLE, Colosimo PF, et al. 2001. The genetic architecture of divergence between threespine stickleback species. *Nature* 414:901–05

Podolsky RH. 2001. Genetic variation for morphological and allozyme variation in relation to population size in *Clarkia dudleyana*, an endemic annual. *Conserv. Biol.* 15:412–23

Reed DH. 2005. Relationship between population size and fitness. *Conserv. Biol.* 19:563–68

Reed DH, Frankham R. 2001. How closely correlated are molecular and quantitative measures of genetic diversity: a meta-analysis. *Evolution* 55:1095–1103

Reed DH, Frankham R. 2003. Correlation between fitness and genetic diversity. *Conserv. Biol.* 17:230–37

Reed DH, Hobbs GR. 2004. The relationship between population size and temporal variability in population size. *Anim. Conserv.* 7:1–8

Richards CM. 2000. Inbreeding depression and genetic rescue in a plant metapopulation. *Am. Nat.* 155:383–94

Ritchie MG, Kyriacou CP. 1996. Artificial selection for a courtship signal in *Drosophila melanogaster*. *Anim. Behav.* 52:603–11

Robertson A. 1952. The effect of inbreeding on the variation due to recessive genes. *Genetics* 37:189–207

Robertson A. 1960. A theory of limits in artificial selection. *Proc. R. Soc. London Ser. B* 153:234–49

Robertson A, Hill WG. 1983. Population and quantitative genetics of many linked loci in finite populations. *Proc. R. Soc. London Ser. B* 219:253–64

Robinson SK, Thompson FR, Donovan TM, Whitehead DR, Faaborg J. 1995. Regional forest fragmentation and the nesting success of migratory birds. *Science* 267:1987–90

Roff DA. 1998. The maintenance of phenotypic and genetic variation in threshold traits by frequency-dependent selection. *J. Evol. Biol.* 11:513–29

Roland J, Keyghobadi N, Fownes S. 2000. Alpine *Parnassius* butterfly dispersal: effects of landscape and population size. *Ecology* 81:1642–53

Rosenberg MS, Adams DC, Gurevitch J. 1997. *MetaWin: Statistical Software for Meta-Analysis with Resampling Tests*. Sunderland, MA: Sinauer

Saccheri IJ, Nichols RA, Brakefield PM. 2001. Effects of bottlenecks on quantitative genetic variation in the butterfly *Bicyclus anynana*. *Genet. Res.* 77:167–81

Sæther BE, Engen S, Møller AP, Visser ME, Matthysen E, et al. 2005. Time to extinction of bird populations. *Ecology* 86:693–700

Schemske DW, Bradshaw HD. 1999. Pollinator preference and the evolution of floral traits in monkeyflowers (*Mimulus*). *Proc. Natl. Acad. Sci. USA* 96:11910–15

Silvela L, Rodgers R, Barrera A, Alexander DE. 1989. Effect of selection intensity and population size on percent oil in maize, *Zea mays* L. *Theor. Appl. Gen.* 78:298–304

Soulé ME. 1976. Allozyme variation: its determinants in space and time. In *Molecular Evolution*, ed. FJ Ayala, pp. 60–77. Sunderland, MA: Sinauer

Spielman D, Brook BW, Frankham R. 2004. Most species are not driven to extinction before genetic factors impact them. *Proc. Natl. Acad. Sci. USA* 101:15261–64

Spitze K. 1993. Population structure in *Daphnia obtusa*: quantitative genetic and allozymic variation. *Genetics* 135:367–74

Stephens PA, Sutherland WJ. 1999. Consequences of the Allee effect for behavior, ecology, and conservation. *Trends Ecol. Evol.* 14:401–5

Swindell WR, Bouzat JL. 2005. Modeling the adaptive potential of isolated populations: experimental simulations using *Drosophila*. *Evolution* 59:2159–69

Tantawy AO. 1957. Heterosis and genetic variance in hybrids between inbred lines of *Drosophila melanogaster* in relation to the level of homozygosity. *Genetics* 42:535–43

Templeton AR, Robertson RJ, Brisson J, Strasburg J. 2001. Disrupting evolutionary processes: The effect of habitat fragmentation on collared lizards in the Missouri Ozarks. *Proc. Natl. Acad. Sci. USA* 98:5426–32

Turelli M. 1984. Heritable genetic-variation via mutation selection balance: Lerch's zeta meets the abdominal bristle. *Theor. Popul. Biol.* 25:138–93

Verrier E, Colleau JJ, Foulley JL. 1991. Methods for predicting response to selection in small populations under additive genetic models: a review. *Livest. Prod. Sci.* 29:93–114

Wade MJ, Goodnight CJ. 1998. The theories of Fisher and Wright in the context of metapopulations: when nature does many small experiments. *Evolution* 52:1537–53

Wade MJ, Shuster SM, Stevens L. 1996. Inbreeding: its effect on response to selection for pupal weight and the heritable variance in fitness in the flour beetle, *Tribolium castaneum*. *Evolution* 50:723–33

Waldmann P. 2001. Additive and nonadditive genetic architecture of two different-sized populations of *Scabiosa canescens*. *Heredity* 86:648–57

Waldmann P, Andersson S. 1998. Comparison of quantitative genetic variation and allozyme diversity within and between populations of *Scabiosa canescens* and *S. columbaria*. *Heredity* 81:79–86

Wang JL, Caballero A, Keightley PD, Hill WG. 1998. Bottleneck effect on genetic variance: a theoretical investigation of the role of dominance. *Genetics* 150:435–47

Wayne RK, Lehman N, Girman D, Gogan PJP, Gilbert DA, et al. 1991. Conservation genetics of the endangered Isle Royale gray wolf. *Conserv. Biol.* 5:41–51

Weber KE. 1990. Increased selection response in larger populations. I. Selection for wing-tip height in *Drosophila melanogaster* at three population sizes. *Genetics* 125:579–84

Weber KE, Diggins LT. 1990. Increased selection response in larger populations. II. Selection for ethanol vapor resistance in *Drosophila melanogaster* at two population sizes. *Genetics* 125:585–97

Wei M, Caballero A, Hill WG. 1996. Selection response in finite populations. *Genetics* 144:1961–74

Weyhrich RA, Lamkey KR, Hallauer AR. 1998. Effective population size and response to S-1-progeny selection in the BS11 maize population. *Crop Sci.* 38:1149–58

Whitlock MC. 1999. Neutral additive genetic variance in a metapopulation. *Genet. Res.* 74:215–21

Whitlock MC, Fowler K. 1999. The changes in genetic and environmental variance with inbreeding in *Drosophila melanogaster*. *Genetics* 152:345–53

Widén B, Andersson S. 1993. Quantitative genetics of life-history and morphology in a rare plant, *Senecio integrifolius*. *Heredity* 70:503–14

Willi Y, Fischer M. 2005. Genetic rescue in interconnected populations of small and large size of the self-incompatible *Ranunculus reptans*. *Heredity* 95:437–43

Willi Y, Van Buskirk J, Fischer M. 2005. A threefold genetic Allee effect: population size affects cross-compatibility, inbreeding depression, and drift load in the self-incompatible *Ranunculus reptans*. *Genetics* 169:2255–65

Willi Y, Van Buskirk J, Schmid B, Fischer M. 2007. Genetic isolation of fragmented populations is exacerbated by drift and selection. *J. Evol. Biol.* In press

Willis JH, Orr HA. 1993. Increased heritable variation following population bottlenecks: the role of dominance. *Evolution* 47:949–57

Wright S. 1931. Evolution in Mendelian populations. *Genetics* 16:97–159

Wright S. 1943. Isolation by distance. *Genetics* 28:114–38

Zhang XS, Hill WG. 2002. Joint effects of pleiotropic selection and stabilizing selection on the maintenance of quantitative genetic variation at mutation-selection balance. *Genetics* 162:459–71

Zhang XS, Wang J, Hill WG. 2004. Redistribution of gene frequency and changes of genetic variation following a bottleneck in population size. *Genetics* 167:1475–92

Resource Exchange in the Rhizosphere: Molecular Tools and the Microbial Perspective

Zoe G. Cardon[1] and Daniel J. Gage[2]

[1]Department of Ecology and Evolutionary Biology and Center for Integrative Geosciences, University of Connecticut, Storrs, Connecticut 06269; email: zoe.cardon@uconn.edu

[2]Department of Molecular and Cell Biology, University of Connecticut, Storrs, Connecticut 06269; email: daniel.gage@uconn.edu

Key Words

biosensor, diversity, food web, nutrient cycling, soil, symbiosis

Abstract

The interface between living plant roots and soils (the rhizosphere) is a central commodities exchange, where organic carbon flux from roots fuels decomposers that, in turn, can make nutrients available to roots. This ongoing exchange operates in the path of vast, transpiration-driven water flow. How the spatio-temporal patterning in resource availability around plant roots affects rhizosphere community composition, activity, and nutrient cycling remains unknown. This review considers how molecular approaches contribute to the exploration of rhizosphere resource exchange, highlighting several recently developed methods linking microbial identity with substrate uptake and gene expression. In particular, strengths and weaknesses of genetically engineered bioreporters are discussed, because currently they alone provide in situ spatio-temporal information at scales of rhizosphere organisms. The soil spatial context is an emerging frontier in ecological soils research. We conclude with parallels linking empirical investigation in the rhizosphere with the quest for understanding general rhizosphere function in Earth's diverse ecosystems.

1. INTRODUCTION

Four hundred million years ago, the first true plant roots evolved and grew downward into substrate—surrounded, from the beginning, by microbes and a soil faunal community. Today, plant roots pack or pepper the upper soil layers of diverse ecosystems over much of Earth's terrestrial surface. Because plant productivity is often limited by soil nutrient availability, the interface between living roots and soils (the rhizosphere, Lynch 1990) is a central commodities exchange, where organic carbon flux from roots fuels microbial and faunal decomposers that can, in turn, make nutrients available to those roots. Moreover, the majority of all water moving from soil to the atmosphere worldwide is transpired from leaves, and that transpired water passed first from soil into plants through the rhizospheres of innumerable roots. The ongoing exchange of carbon and nutrients in the rhizosphere thus operates in and is shaped by vast water flow toward roots. What functional commonalities underlie rhizosphere resource exchange, no matter the organismal players? And why ask such a question?

One practical reason to seek general rules is simply that, given the tremendous diversity of soil microbes, soil fauna, and plants, it is virtually impossible to investigate the intricacies of every potential rhizosphere interaction in every environmental circumstance. Yet, an understanding of controls over belowground function is becoming increasingly important as natural and agro-ecosystems around the globe are exposed to anthropogenic pressures (Drinkwater & Snapp 2006, Pregitzer et al. 2006, Wardle et al. 1998). In addition, the chemistry and development of soil present today have been strongly affected by the actions of rhizospheres over evolutionary time frames (Richter et al. 2006), and the evolution of true plant roots and their extension deep into substrate is hypothesized to have led to a revolution in planetary carbon and water cycling during the Devonian period (e.g., Algeo & Scheckler 1998, Beerling & Berner 2005). What is the biogeochemical function of the rhizosphere on Earth today? In what major ways is rhizosphere function belowground similar across terrestrial ecosystems, and in what fundamental ways can it differ?

These questions pursue common themes describing rhizosphere function, aiming to link mechanism at rhizosphere scales with pattern at ecosystem scales. Gathering empirical data belowground at both these spatial scales, however, is challenging. Key organisms in the rhizosphere, including microbes (bacteria and fungi) and their predators (e.g., protozoa, nematodes), inhabit microenvironments strongly influenced by soil structure, and they respond to conditions and resources at micrometer to millimeter scales that are difficult to characterize. Molecular methods are particularly promising for gathering empirical data at rhizosphere scales, and they are the focus of this review. The link to larger scales is challenging, not only because of inherent scaling issues and dramatic soil heterogeneity, but also because the assays of soil biogeochemistry used as estimators of belowground function at larger scales are often carried out on cored soil samples, where hyphae and roots are severed, and soil structure has been disturbed. Ultimately, the insights obtained by using molecular techniques to probe the microbial perspective of the environment and soil organismal response, at microscales, need to be coupled with improved nondestructive estimates of soil biogeochemical process rates at larger scales, through, for example, development of soil sensors capable of continuous measures of dynamic solute pools or gases in soil pore space.

In this review we consider how molecular approaches can contribute unique empirical data to the exploration of rhizosphere function. We first briefly discuss the major drivers behind molecular rhizosphere research and then describe a common framework in which to consider rhizosphere resource exchange, noting the recent call for model systems to support hypothesis-driven inquiry. We briefly review promising molecular methods, highlighting several that have been recently developed to link microbial identity with function (substrate uptake or gene expression). In particular, we emphasize the strengths and weaknesses of genetically engineered bioreporter bacteria (microbiosensors) that are unique in providing in situ temporal and spatial information at the scale of individual microbes. This spatial context for rhizosphere function—the 3D matrix in unsaturated soil—is an emerging frontier in soils and rhizosphere research. We conclude by exploring conceptual parallels linking empirical investigation in the rhizosphere itself with the quest for understanding general rhizosphere function in the diverse array of Earth's ecosystems.

PCR: polymerase chain reaction

1.1. Drivers of the Molecular Revolution in Rhizosphere Research

The recent explosion of soils research using molecular methods reveals astonishing diversity and fascinating relationships among organisms and their soil environment (e.g., Singh et al. 2004)—far more information than can be adequately reviewed in this space. Common techniques use PCR (polymerase chain reaction) to amplify selected fragments of DNA isolated from soil organisms, and most share an emphasis on fingerprinting the microbial community present (but not necessarily active) based on that DNA (see Section 2 and **Table 1**). These techniques are destructive; soil is harvested and DNA (sometimes RNA) extracted. Small-scale spatial information is lost; functional information (beyond phylogenetic) is often limited. Nevertheless, within these limitations, several major intellectual drivers have contributed to the rapid development and successful application of molecular methods to rhizosphere ecology. Environmental engineers are searching for specific rhizosphere organisms capable of degrading/transforming soil pollutants (see, e.g., Kuiper et al. 2004 for a review). Within agricultural research, efforts are ongoing to detect and biologically control specific disease organisms around roots (see, e.g., Compant et al. 2005 and Hawes et al. 2000 for reviews). In addition, soil microbial and ecosystems ecologists have documented remarkable soil biodiversity and the potential for individual plants, plant species, soil types, and management regimes to influence rhizosphere microbial diversity (see, e.g., Garbeva et al. 2004, Kent & Triplett 2002 and Lynch et al. 2004 for reviews).

In a recent meta-analysis, Hawkes et al. (2006) summarized 16S ribosomal DNA (bacterial rDNA) data from 13 papers examining rhizospheres of two woody dicots, nine herbaceous dicots, and three grasses. At least 35 taxonomic orders of bacteria were present, including abundant Proteobacteria (often γ-Proteobacteria,

Table 1 Strengths and limitations of methods commonly used for analyzing microbial diversity[a]

Strengths	Limitations
PCR-based methods: DGGE, RFLP, and related techniques (Acinas et al. 2005; Blackwood & Paul 2003; Marschner et al. 2001; Smalla et al. 1998, 2001; Spiegelman et al. 2005)	
-Amenable to high throughput analysis -Rapid and relatively easy to implement	-Subject to PCR biases -Relatively insensitive; phylotypes must be present at >1–5% in order to be reliably analyzed -A single band (phylotype) may represent more than one microbial species or type
PCR-based methods: sequencing of amplified genes (Hong et al. 2006, Thompson et al. 2005)	
-Can give a detailed picture of the diversity and contents of an amplified rhizosphere library	-Labor intensive; small numbers of samples analyzed -Subject to PCR biases
FISH (fluorescent in situ hybridization) (Spiegelman et al. 2005, Zwirglmaier 2005)	
-Can be used to visualize the location and prevalence of microbial species, or other taxonomic groups, at the single-cell level -Can be used to enumerate cells in samples	-Cells erroneously missed or labeled when probe is not universal or specific enough, respectively -Background fluorescence can make visualizing labeled cells difficult -Can be hard to detect labeled cells that constitute a small fraction of the community
Culturing (Ferrari et al. 2005, Joseph et al. 2003, Zengler et al. 2002)	
-Straightforward if methods for selective culturing of desired organisms exist	-Most rhizosphere microbes are currently not culturable (but methods development is underway)
Metagenomics (Daniel 2005, Handelsman 2004, Rondon et al. 2000)	
-Sequence information gained from large sections of environmental DNA, even whole genomes in some cases -Large cloned fragments can be used for the isolation of new enzymes or microbial products	-Time consuming and expensive -Cannot reconstruct large genomic contigs from organisms that are not highly represented in the environmental DNA -Subject to DNA isolation, cloning, and expression biases
PLFA (Bossio et al. 1998, Spiegelman et al. 2005)	
-Rapid and inexpensive	-Provides limited taxonomic information -Lipid profiles of single species can be influenced by environmental conditions

[a]References are generally to papers that give an overview of each technique or discuss its strengths or limitations. Also included are references that highlight soil or rhizosphere biology. Spiegelman et al. (2005) provide an excellent overview.

though α- and β-Proteobacteria also were present), gram-positive members of the CFB (*Cytophaga-Flavobacterium-Bacteroides*) group, and smatterings of Actinobacteria among others. Known capabilities of these groups hint at complex rhizosphere biogeochemistry; methanotrophy, nitrification, and diazotrophy are known in Proteobacteria, and the CFB group can use complex substrates (Hawkes et al. 2006). How biodiversity affects ecosystem function is a continuing debate (e.g., Coleman & Whitman 2005, Fitter et al. 2005 for comment and review); causal links between the diversity of soil organisms present and variation in soil function are very difficult to establish, in part because rhizosphere organisms detected using 16S or 18S rDNA are not necessarily active. As techniques to purify and amplify mRNA from soils improve, identification of the active members of the rhizosphere community, and their functional gene expression, will become more tractable.

Finally, molecular approaches have also contributed to understanding of the specific signaling systems within plant-microbe-faunal and symbiotic rhizosphere interactions (e.g., Graham & Miller 2005, Griffiths et al. 2006, Harrison 2005, Thies et al. 2001), as well as root interactions with specific rhizosphere bacteria causing disease (e.g., *Agrobacterium*, *Erwinia*) or having antibiotic properties (e.g., some *Pseudomonas*) (e.g., Brencic & Winans 2005).

1.2. Components of Rhizosphere Commonalities

Our focus is on finding common themes underlying rhizosphere organismal interactions in and with soil, with the ultimate goal of offering better understanding of controls over belowground processes across varied terrestrial ecosystems. Perhaps the most basic rhizosphere commonalities are the rhizosphere components themselves: a plant root (altering soil conditions and resource availability), substrate into which the root is growing, and a surrounding symbiotic, saprotrophic, and grazing soil community. Rhizosphere resource exchange is supported and constrained by these biotic and abiotic components. Crawford et al. (2005) advocate stepping back from the evolutionary and ecological macroscale to consider that (*a*) the soil environment is highly spatially structured and heterogeneous, leading to controls over belowground function that are linked with physical structure and location at very small scales, and (*b*) horizontal gene transfer is rampant so genetic information shuffles relatively easily among microbes, potentially muddying the correspondences between microbial functions and phylogenetic lineages. With these ideas in mind, we briefly discuss roots, the soil community, and the soil itself below, incorporating references to instances where molecular data have been, or promise to be, particularly useful for probing rhizosphere resource exchange.

1.2.1. Component 1: plant roots.
The rhizosphere is defined by the presence of a plant root, and all plant roots develop at the active tip with the same basic design, no matter their evolutionary lineage (Raven & Edwards 2001). Key developmental regions of roots are (**Figure 1a**) the tip zone (including root cap, meristem, and immature elongation region), the active uptake zone (where phloem and xylem are mature, and soil water and nutrients enter during transpiration), and the suberized zone (where a wax-like suberin barrier limits solute and water exchange with the surroundings). This ubiquitous developmental plan generates patterns of resource availability, soil water potential, and water flux in the rhizosphere that vary in space (e.g., around the three root zones) and in time (e.g., night versus day, **Figure 1a**). Some basic patterns, for example, rhizodeposition from the root tip during root growth, have long been described from roots growing in artificial liquid or sterilized solid media (e.g., see Whipps 1990 for a review). (Rhizodeposition, indicated in green in **Figure 1**, is the deposition of organic compounds to soil around the root, including secretion of mucilage, sloughing of root cap cells, and exudation of soluble compounds.) Exudation is likely enhanced at the tip because of basic root structure—xylem and phloem have not matured in the tissue immediately behind

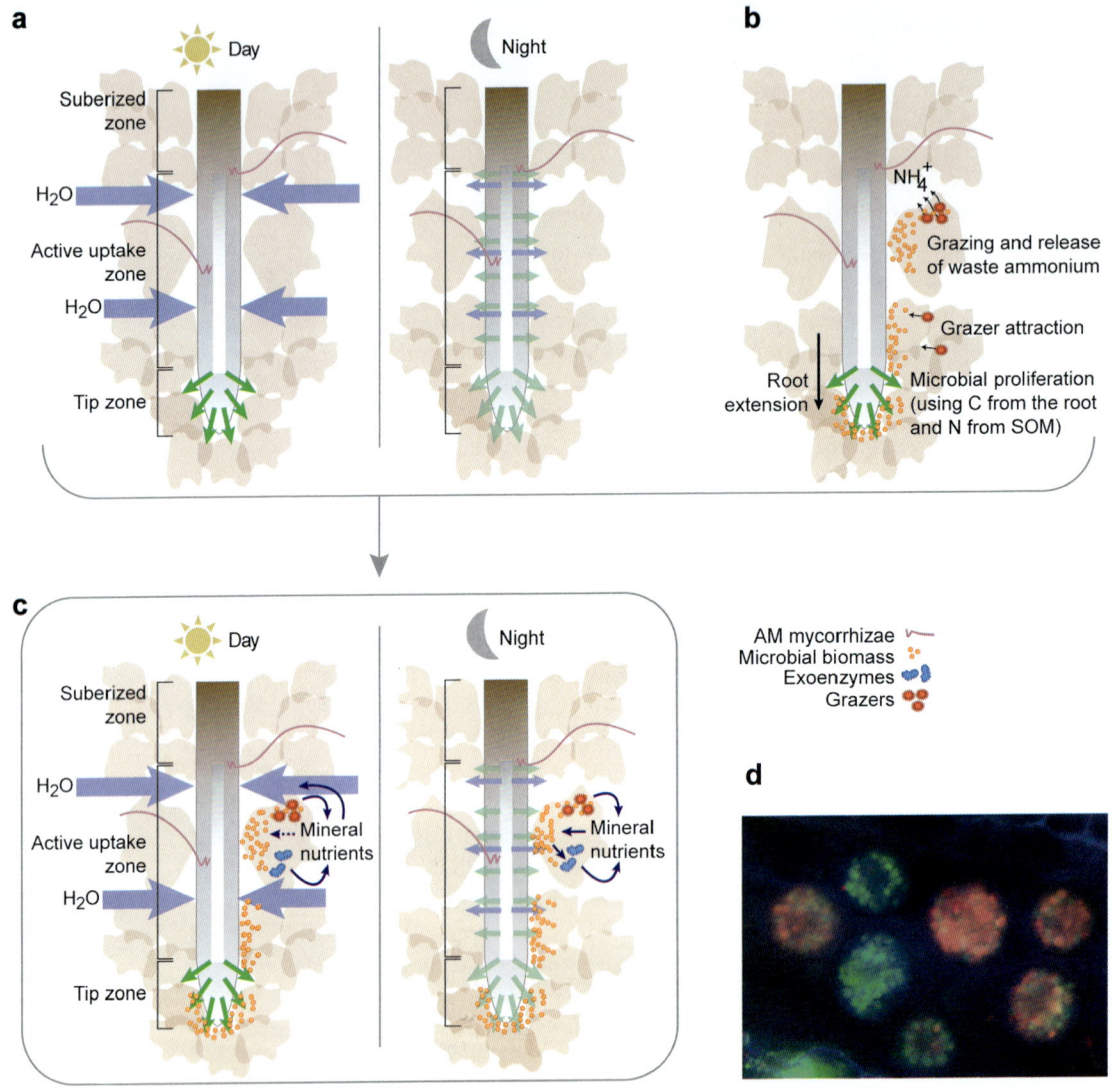

Figure 1

Organisms and interactions in the rhizosphere. (*a*) Root-driven water fluxes and rhizodeposition (including efflux of solution from roots during hydraulic redistribution at night). (*b*) Clarholm's (1985) model of tri-trophic resource exchange in the rhizosphere food web. (*c*) Combined influences of water flux, rhizodeposition, and community interactions in the rhizosphere. (*d*) Protozoa imaged in the rhizosphere of an alfalfa root. *Sinorhizobium meliloti* bioreporters that were genetically engineered to express red fluorescent and green fluorescent protein (Bringhurst et al. 2001) are visible in vacuoles after having been engulfed by protozoa.

the meristem, so some sugars and other cellular building blocks move from mature phloem to the meristem via an apoplastic route (through cell walls), where membranes and transporters are less able to control diffusion into soil (e.g., Bretharte & Silk 1994). How far various rhizodeposits can spread into nonsterile rhizosphere soil, and the complexities of their distributions, has remained unknown.

GFP: green fluorescent protein

The recent development of living bioreporter bacteria (microbiosensors, **Table 2**, Section 2) that can be deployed in nonsterile soil allows empirical investigation of such patterns. For example, Jaeger et al. (1999) genetically engineered *Erwinia herbicola* 299R to report (by producing ice nucleation protein) amounts of available sucrose and tryptophan in nonsterile soil around roots of *Avena barbata*. As expected, sucrose availability was maximal near the root tip and decreased dramatically with distance along the root axis, but, intriguingly, at 12–16 cm from the tip, sucrose report was still significantly above bulk soil levels. In contrast, tryptophan availability was minimal at the root tip back to 8 cm along the axis, then increased greatly until 12–16 cm from the tip (the most basipetal sampling point). Casavant et al. (2002) used an *Enterobacter cloacae* bioreporter to detect arabinose around barley seeds and the root/seed junction, but arabinose was not found at root tips. Further, Bringhurst et al. (2001) showed, using microbiosensors engineered from *Sinorhizobium meliloti*, that galactosides (common sugars in phloem sap) were present in nonsterile, rhizosphere soil around root hairs of a variety of grasses and legumes, and around locations where lateral roots emerged from the main root, but not at root tips. The microbiosensors' green fluorescent protein (GFP) report was detected by confocal microscopy up to 200 μm from the root surface. Such molecular data promise to provide a useful empirical envelope in which to examine mathematical predictions of solute distribution patterns around roots (e.g., Darrah 1991, Scott et al. 1995, and reviewed in Tinker & Nye 2000), particularly when the empirical data and modeling are developed in parallel.

Other microscale patterns in conditions and resource availability that result from the development and function of roots have not yet been as well explored. For example, microbiosensors indicate that common soluble organic compounds are not confined to soil around the root tip; transpiration should influence movements of solutes in soil around the root active uptake zone, with soil solution flow toward roots opposing diffusion of solutes away from the root (**Figure 1a**). At night, when transpiration stops, diffusion of any substances away from the active uptake zone could occur unimpeded, and if hydraulic redistribution is occurring, a flux of organic molecules out could even be aided by water flow out of the root (**Figure 1a**). [Hydraulic redistribution is the movement of water from regions of higher to regions of lower soil water potential via plant roots, and it is widespread (e.g., Caldwell et al. 1998).] Can this hypothesized diel rhythm in root-derived substrate availability around the active uptake zone (higher at night, lower during the day) be detected by rhizosphere microbiosensors? Moreover, because transpiration-driven water flux does not occur at the tip, is there an opposite rhythm driven by photosynthesis (higher exudate availability during the day, lower at night) in rhizosphere soil around the root tip? The existence and effects of such rhythms in single roots, and root systems of varying architecture, remain to be tested in situ in nonsterile soil.

Root water potential, too, varies in space and time in different root zones. Simple single-root models predict how far into rhizosphere soils of different textures and water contents the variation in root water potential propagates, and, e.g., Hillel (1998) calculates gradients in water potential developing out to more than several millimeters from the root over days. [See Tinker & Nye (2000) for more extensive discussion.] However, thermocouple psychrometers, time domain reflectometry, and tensiometers assay soil moisture at much larger scales. Recently, genetically engineered microbes have been developed that report total soil water potential at millimeter and micrometer scales in their surroundings, producing more GFP per cell as soil dries (Cardon & Herron 2005; P.M. Herron, D.J. Gage, Z.G. Cardon, submitted, building upon Axtell & Beattie 2002). This cellular-scale report can be viewed *in situ* using an epifluorescence microscope, without destructive harvest of rhizosphere soil. Expected transpiration-induced water potential gradients have been observed using these microbiosensors—rhizosphere water potential is least negative around the root tip, where xylem is not mature and transpiration-driven water uptake is minimal, and it becomes more negative around the active uptake zone where metaxylem has matured (P.M. Herron, D.J. Gage, Z.G. Cardon, submitted).

Water potentials and velocities of water movement toward roots in the active uptake zone have important implications for rhizosphere function, including organismal movements and resource exchanges (e.g., Scott et al. 1995). Protozoal grazers, for example, move more easily when water potentials are high (that is, when the soil is wet). Also, at the scales in which rhizosphere organisms operate, velocities of water movement strongly influence advective (bulk flow-linked) versus diffusion-linked delivery of nutrients through the rhizosphere, but in many current models the advective delivery to roots is ignored for most nutrients (Darrah et al. 2006). The behavior of roots as leaky pipes (Landsberg & Fowkes 1978) and the changes in axial and radial conductivity of roots during development behind the root tip (Doussan et al. 1998, Zwieniecki et al. 2002) suggest there should be dramatic variation in the radial velocity of transpiration-driven soil solution movement toward the root axis, increasing from zero at the tip to a maximum (but not constant) velocity in the active uptake zone (**Figure 1a,c**, *blue arrows* of different lengths). For example, Doussan et al.'s modeling, developed using Varney and Canny's dye-based data (cited in Doussan et al. 1998), finds a 30% increase in maximum radial water flow into roots between approximately 2–3 cm and 7–10 cm from the main root tip in corn, along with further variation possibly associated with maturation of early and late metaxylem elements and development of the endodermis and hypodermis, etc. (Doussan et al. 1998).

1.2.2. Component 2: rhizosphere communities.
Spatial and temporal heterogeneity in rhizosphere conditions and resources thus emerges from general root structure and function, and that heterogeneity is hypothesized to be important for community activity and resource exchange (e.g., Clarholm 1985). The rhizosphere microbial and soil faunal community has very diverse membership, including symbiotic and saprotrophic bacteria and fungi, grazing protozoa (e.g., heteroflagellates, ciliates, and amoebae), nematodes, mites, enchytraeid worms, and a host of other soil organisms (Griffiths et al. 2006, Hawkes et al. 2006, Johnson & Gehring 2006, Moore

et al. 2006). In 1985, Clarholm provided a now-classic, simple, function-oriented framework for exploring resource exchange among roots, microbes, and grazers in the rhizosphere. Very briefly, labile (easily used) carbon (*green arrows* in **Figure 1b**) is lost from the root tip as it extends through soil, and soil microbial activity increases as a result of the root carbon input. Building new microbial biomass (*yellow dots* in **Figure 1b**) requires not only carbon but also a source of nitrogen and other essential nutrients. (Use of the term nutrients varies between microbial and ecosystems literatures; here we use the word nutrients to represent essential, noncarbon elements such as nitrogen, phosphorus, and sulfur.) Focusing on nitrogen, if concentrations of mineral nitrogen in soil solution are not high, microbes can use exoenzymes to attack soil organic matter (SOM) for nitrogen to make their biomass, thus moving it into an actively cycling pool. Eventually, grazers (*red furry blobs*, **Figure 1b**; real protozoa in soil, **Figure 1d**) are attracted to and consume the microbial biomass. Meanwhile, the root tip continues to extend through soil, away from soil particles populated with now-thriving microbiota, and the root axis matures. Grazers consuming microbial biomass release waste ammonium, because of stoichiometric differences in carbon and nitrogen between biomasses of the two organism types or because of inefficiencies of assimilation, or both. If that mineral-nutrient pool is near the active uptake zone of the root (**Figure 1b**), ammonium can be taken up by the plant. In sum, through a multitrophic interaction, the root has exchanged carbon (which fed microbes initially, Cheng & Gershenson 2006) for nitrogen (released by grazers).

Clarholm's (1985) trophic interactions (**Figure 1b**) and the water fluxes into and out of roots (**Figure 1a**) interact (**Figure 1c**), but how the resulting spatio-temporal patterning in resource availability affects rhizosphere community biomass, gene expression, composition, trophic interactions, and nutrient cycling remains unknown. The balance of mineralization (release) of nutrients from SOM by microbes and immobilization (uptake) of nutrients into microbial biomass is key for nutrient availability to plants, because neither nutrients in SOM nor mineral nutrients immobilized in microbial biomass are immediately available for plant root uptake. Factors influencing the balance of mineralization and immobilization include stoichiometric constraints (C:N of soil community predators, prey, and chemical substrates), efficiencies of resource use, energy and nutrient status of microbes and, thus, likelihood of their producing exoenzymes to attack SOM, mineral nutrient availability in soil solution, and availability of SOM (e.g., Chen & Stark 2000, Clarholm 1985, Moore et al. 2004, Robinson et al. 1989, Schimel & Bennett 2004, Schimel & Weintraub 2003).

Of these factors, the energy and nutrient status of rhizosphere microbes clearly could be directly influenced by roots via rhizodeposition and transpiration-driven variation in water flow. Our combined model in **Figure 1c** shows, for example, that a diel rhythm of resource availability around the active uptake zones of roots (described in Section 1.2.1 and **Figure 1a**) could expose rhizosphere microbes to higher solute flux from roots at night, potentially providing carbon to rhizosphere microbes and increasing microbial growth demand for nutrients. As a result, depletion of the dissolved mineral nutrient pool in rhizosphere soil solution by microbes likely becomes more severe. This period of relative nutrient-limitation at night (caused by depletion of solutes from soil solution and potentially exacerbated by movement of organic compounds from the root into soil) could spur microbial production of long-lived exoenzymes (*blue blobs* in **Figure 1c**) that release mineral nutrients from SOM. Once transpiration starts again, relative carbon limitation of microbial growth could develop around the active uptake zone as transpiration again opposes diffusion of substances from the root surface. But, exoenzymes in soil, outside microbes, will still actively attack SOM and release nutrients that could be poached by the plant because the carbon-limited microbes are not growing (immobilizing nutrients) rapidly (**Figure 1c**, Day). Overall, average rhizodeposition from the entire root system (much of it at root tips) may be maximal during the day and lower at night (e.g., Murray et al. 2004), but organisms in the rhizosphere around the root's active uptake zone may experience exactly the opposite fluctuation. Over the course of days, the rhizosphere community around the active uptake zone may thus be pushed back and forth from carbon (daytime) to nutrient (nighttime) limitation on a diel cycle. Are the effects of such zone-specific rhythms on microbial gene expression detectable using functional gene arrays? Though the implications of such a hypothetical rhizosphere pump are not yet clear, preliminary mathematical modeling (Cardon & Rastetter 2004) suggests that persistence of a long-lived exoenzyme pool, releasing nutrients from SOM and replenished by periodic pushing of the rhizosphere microbial community into a state of moderate night-time nutrient limitation, enhances plant root access to nutrients in the long term.

Of course, interactions among organisms in the rhizosphere are governed by more than resources. Molecular and signal coordination among diverse members of the rhizosphere community extends beyond the well-known symbioses and is thought to be mediated through common active signaling compounds such as plant phenolics, phytohormones, and quorum-sensing signals (e.g., Bonkowski 2004, Hirsch et al. 2003, Loh et al. 2002). These signals can influence species-specific rhizosphere interactions and potentially can act much more broadly. For example, Bonkowski & Brandt (2002) found lateral root proliferation in *Lepidium sativum* seedlings was associated with the presence of protozoal grazers, perhaps because protozoa promote increased numbers of bacteria capable of producing auxin. The generality of such effects remains to be determined (Bonkowski 2004, Bonkowski & Brandt 2002, Clarholm 2005).

1.2.2.1. Perspectives on control of rhizosphere food webs. To conceptually organize such a dizzying array of potential signal molecules and interactions of rhizosphere organisms concentrated at the root surface, Phillips et al. (2003) proposed a framework founded on the idea of rhizosphere control points. These control points are defined as regulatory elements within organisms (e.g., genes coding for receptors binding rhizosphere signals) that can be acted upon by selective pressures in the rhizosphere, leading to altered fitness of those organisms and resulting in broader effects within the rhizosphere food web. Given this definition, there likely is a very large number of control points even within a very simple community. The strength of their effects on rhizosphere food web function can be examined using molecular techniques such as targeted mutation to knock out function. The framework is intriguing because it seeks to link molecular investigation of specific signals, receptors, and environmental responses within particular organisms to rhizosphere population and community ecology. The genetic nature of Phillips et al.'s control points leads them to suggest that "genes lie at the heart of rhizosphere food web regulation."

In contrast, Moore et al. (2003) argue that control points within rhizosphere food webs are embedded within the whole web structure itself, through which energy and resources flow in various paths (channels), in characteristic ways. For example, energy from plants can flow directly via a root channel to symbionts (e.g., mycorrhizae) and up the food chain that is using mycorrhizal hyphae as its base. Additionally, energy and resources can move from roots to upper trophic levels of rhizosphere food webs through bacterial (fast) or fungal (slow) saprotrophic channels with distinctly different dynamics. The relative importance of fast bacterial and slow fungal channels varies from ecosystem to ecosystem; the fungal channel dominates when more detritus, or more resistant detritus, is present (e.g., no-till agriculture, mor forest humus, Couteaux & Bottner 1994), and the bacterial channel dominates with soil disturbance and/or the prevalence of easily metabolized organic compounds (e.g., tilled agriculture, mull forest humus). Overall food web stability is enhanced by this flexible, multichannel structure (e.g., Moore et al. 2006). This view of control has the appeal of generality because rhizosphere stability and energy flow control are determined by community structural patterns, not particular species per se. Molecular techniques targeted toward describing community structure and capturing population dynamics could contribute to empirical tests of these ideas.

Finally, Young & Ritz (1998) advance a third perspective, that the soil habitat itself exerts significant control over function of rhizosphere food webs. Its heterogeneous three-dimensional soil structure and dynamic water content dramatically affect delivery paths for resources and mobility of organisms interacting in food webs, serving as a physical heart of rhizosphere function (see Section 1.2.3 below).

DGGE: denaturing gradient gel electrophoresis

1.2.2.2. The mycorrhizal menagerie. Plant roots have likely been involved in mycorrhizal symbioses and surrounded by saprotrophic microbes for 400 My (Harrison 2005). Today, the majority of plants are mycorrhizal, and for those plants, rhizosphere resource exchange occurs in two complementary domains: at the symbiotic physical interface between root and fungal tissues and in rhizosphere soil between plant roots and saprotrophic microbes. The various mycorrhizal types have differing capacities to attack SOM in order to release nutrients, with arbuscular mycorrhizal (AM) fungi (common, e.g., in grasslands) the least capable, and ericoid (ERM) and ectomycorrhizae (ECM; common in cold northern latitudes) very capable of saprotrophic activity (Chapman et al. 2006, Johnson & Gehring 2006, Read & Perez-Moreno 2003). Interestingly, it remains unclear whether, or to what extent, mycorrhizal hyphae can facilitate saprotrophic microbial attack on SOM in the rhizosphere (e.g., Fitter 2005, Jones et al. 2004), though Wamberg et al. (2003), using the molecular fingerprinting technique denaturing gradient gel electrophoresis (DGGE), detected that mycorrhizal hyphae can have significant effects on membership and perhaps activity in the soil bacterial community. Biogeographical patterns in mycorrhizal symbiosis and saprotrophic capability likely contribute to major differences in root-mycorrhizal-saprotrophic resource exchange in various terrestrial ecosystems. We return to these contrasts later.

The specific signals and responses involved in establishment of mycorrhizal symbioses have been the subject of intense molecular investigation (see Harrison 2005 and Graham & Miller 2005 for reviews), and the challenge is to weave this information into a better understanding of larger-scale phenomena. Graham & Miller (2005) note, for example, that an in-depth characterization of the mycorrhizal transporter responsible for uptake of a soil nutrient may not help with scaling up to understanding nutrient uptake in an ecosystem if release of that nutrient by saprotrophic microbiota is limiting. An ecosystem-level view of controls should help prioritize areas for molecular research.

1.2.3. Component 3: the soil structural stage for rhizosphere function.

The organisms driving essential ecosystem functions in soil through their gene expression, resulting physiology, and trophic interactions are operating at a scale that is extremely difficult to envision let alone examine empirically. Just as Earth's surface is heterogeneous at a scale we can perceive, soil particles' surfaces are heterogeneous for microbes, with water films of different thicknesses, and "mountains" and "valleys" of various resource availability. This soil structure strongly influences general rhizosphere resource exchange (e.g., Hillel 1998, Tinker & Nye 2000). The three-dimensional matrix is highly heterogeneous, with various-sized organic or inorganic particles piled together with space in between. If soils are saturated, that space is filled with water, constraining diffusion of gases relative to the unsaturated state, but facilitating diffusion of dissolved nutrients or carbon, and movements of swimming protozoal predators. As soil drains and dries, a number of changes take place. Spaces between particles become filled with gas, largest first and smallest last; water becomes confined to films on particles. Movements of swimming organisms are severely constrained; the tortuosity of the path for water movement in water films increases dramatically, as does the length of paths along which dissolved substances diffuse. Soil hydraulic conductivity plummets, and soil water potential drops, following characteristic trajectories of water potential as a function of soil moisture content that depend strongly on soil texture. Within this structural background, the chemical nature of soil also influences the potential for resource exchange; for example, clays and organic matter can provide cation exchange capacity that holds cations such as ammonium tightly in place. Crawford et al. (2005) suggest that a major current challenge within soil ecology is the need to monitor biological activity *in situ* within this physical structure, continuously, at multiple scales.

1.3. Model Systems for Function-Focused Rhizosphere Research?

Independently, both Phillips et al. (2003) and Crawford et al. (2005) call for development of model microcosms to facilitate hypothesis-driven, function-focused rhizosphere research. Crawford et al. champion the need for targeted investigation of soil structural and physical controls over soil ecology, and Phillips et al. emphasize the need for simplified soil communities of tractable size and genetic manipulability.

Such well-controlled biological systems would lend themselves well to integration of empirical results with mathematical modeling (e.g., see Darrah et al. 2006, Tinker & Nye 2000, Toal et al. 2000 and Wu et al. 2005 for reviews). They would also link beautifully to emerging function-focused molecular methods, including functional DNA arrays detecting expression of known genes among community members, stable isotope probing revealing use of ^{13}C-labeled rhizodeposits by identifiable microbes, and development of microbiosensors producing visual reports of rhizosphere conditions, resources, or microbial activity that can be assayed continuously (**Table 2**).

RFLP: restriction fragment length polymorphism

Ideally, both the expression of particular genes and a metric of numbers of organisms and/or groups of organisms (population and community biology, at the microscale) should be monitored, because both gene expression and trophic interactions contribute to biogeochemical cycling. Eventually, in more natural soil systems, rDNA genes from broad groups of soil prokaryotes, fungi, and predators might be sequenced and used to track changes in populations, but there are limitations still to such an approach, not least of which is the paucity of sequence information for soil organisms. Again, this effort would be greatly simplified at first in model microcosms, though even harvest design would be challenging, given the vast differences in sizes and numbers of such key soil organisms as nematodes and bacteria. To begin exploring genetic, trophic, and soil physical controls over rhizosphere resource exchange, not simply function in one type of microcosm, organisms should be included in various combinations, forming trophic webs of various structures and sizes (e.g., Ingham et al. 1985) in multiple soils with varied physical properties. The ultimate goal is to use simplified microcosms to test mechanistic hypotheses, and, from information gleaned, develop hypotheses testable in natural ecosystems.

2. MOLECULAR METHODS—COMMON AND CUTTING-EDGE

2.1. Diversity

Over the past 25 years, the development of molecular techniques has allowed microbial ecologists to explore microbial diversity far beyond that exposed by analysis of culturable microbes. There are 243,909 aligned and annotated bacterial 16S rRNA sequences alone in the current ribosome database at Michigan State University (Coleman & Whitman 2005), and, though large, this database represents a miniscule fraction of the total prokaryotic diversity estimated to exist worldwide (Curtis et al. 2002, Gans et al. 2005, Schloss & Handelsman 2004). The most powerful and convenient methods for investigating microbial diversity have been thoroughly reviewed elsewhere (**Table 1**); generally, they require the isolation of microbial DNA from the environment, followed by amplification of conserved regions (i.e., DNA regions with very similar or identical sequence) by PCR, and sequencing or other analyses [for example, DGGE separating fragments on gels, or T-RFLP (terminal-restriction fragment length polymorphism)] designed to characterize the diversity of the amplified fragments (Garbeva et al. 2004, Kent & Triplett 2002, Spiegelman et al. 2005). Resulting phylotypes from DGGE and T-RFLP are "operational taxonomic units"

SIP: stable isotope probing

FISH: fluorescent *in situ* hybridization

based on DNA fragment patterns; they do not necessarily correspond directly to single species. PCR-based methods ultimately underestimate diversity because of several inherent limitations. First, amplification of DNA in proportions matching those in complex environments is unlikely because microbes differ in their susceptibility to lysis and in the extractability of their nucleic acids. Second, primers often do not amplify DNA from all target organisms (Baker et al. 2003, Smalla et al. 1998). Third, genes are amplified at different efficiencies and their representation in the final pool of amplified products may not represent their environmental abundance. Finally, it is becoming clear that sampling of single or a small number of genes can vastly underestimate genome diversity at the species level (Thompson et al. 2005). Whether such genomic diversity within species confers functional diversity or contributes to differential selection is not always clear.

2.2. Function

Recent technical advances, described in more detail below, have moved microbial ecology beyond simply analyzing community diversity toward uncovering ecological activities of community members. Stable isotope probing (SIP) and mRNA-fluorescent in situ hybridization (mRNA-FISH) are newer nucleic acid-based methods that tag cells actively metabolizing an isotopically labeled compound or expressing certain genes, respectively. DNA arrays detect expression of selected functional genes in the environment; microbial bioreporters (**Table 2**) report their perception of, or response to, their environments. SIP and DNA arrays are just beginning to be applied to soil and rhizosphere environments, and mRNA-FISH to marine sediments and groundwater. Bioreporters have been used in rhizosphere studies to a limited extent for about 10 years (**Table 2**).

2.2.1. Stable isotope probing.
For SIP experiments, microbial communities are exposed to ^{13}C-labeled substrate, resulting in the biosynthesis of heavy, ^{13}C-labeled macromolecules by organisms using the substrate for growth. DNA with a high enough fraction of ^{13}C can be physically separated from unlabeled DNA by equilibrium density centrifugation (Meselson & Stahl 1958); this ^{13}C-labeled subpool provides genes to identify organisms that used the labeled substrate (Dumont & Murrell 2005; Lu & Conrad 2005; Radajewski et al. 2000, 2003; Saleh-Lakha et al. 2005; Singh et al. 2004; Wellington et al. 2003). Separation of ^{13}C-labeled nucleic acids from high G/C nucleic acids requires that at least 20% of the nucleic acid carbon be ^{13}C (Manefield et al. 2002). This requirement poses challenges for rhizosphere research in which photosynthetic fixation of ^{13}CO$_2$ provides the label; ^{13}C labeling of the DNA of rhizosphere organisms using rhizodeposits can be low (Ostle et al. 2003, Singh et al. 2004). Further challenges in interpretation of labeling patterns stem from cross feeding of organisms, either through conversion of labeled compounds by primary consumers or trophic interactions (Ostle et al. 2003, Radajewski et al. 2003).

RNA, in comparison with DNA, is more efficiently labeled because many RNA molecules are rapidly and continually synthesized in growing organisms. They are single stranded and short relative to the genomic DNA fragments, and thus are likely

Table 2 Representative bioreporter studies in the rhizosphere, grouped by report type

GFP Report

Induced by PCBs (Boldt et al. 2004) *Pseudomonas fluorescens* bioreporters in microcolonies were induced by PCBs in contaminated alfalfa rhizosphere soil. Comparison with growth rate bioreporters suggested that the PCB-degrading cells were not growing at high rates.

Induced by galactosides (Bringhurst et al. 2001) *Sinorhizobium meliloti* bioreporters were induced by galactosides in nonsterile rhizosphere soils around grasses and legumes. Protozoal grazing was detectable because engulfed bioreporter bacteria remained fluorescent.

Induced by arabinose (Casavant et al. 2002) An *Enterobacter cloacae* life history bioreporter (passing induction to all descendents) detected arabinose around barley seeds and the root/seed junction, but not at root tips, perhaps because arabinose is only cleaved from cell walls after elongation and differentiation of plant cells is complete.

Induced by toluene (Casavant et al. 2003) A *Pseudomonas fluorescens* life history bioreporter was induced by vapor-phase toluene, even at low or transient concentrations. In the absence of toluene, compounds in barley rhizosphere activated the bioreporter in localized areas.

Induced by nitrate (DeAngelis et al. 2005) An *Enterobacter cloacae* bioreporter detected lower levels of nitrate in the rhizosphere of wild oat than in bulk soil, suggesting detection of competition for nitrate between microbes and plant roots. (Ice nucleation report also used.)

Induced by low water potential (Cardon & Herron 2005; P.M. Herron, D.J. Gage & Z.G. Cardon, submitted) *Pseudomonas putida* and *Pantoea agglomerans* biosensors reported total soil water potential (Ψ); *P. agglomerans* reported more negative rhizosphere Ψ with increased axial distance from corn root tips, confirming the predicted gradient based on corn root anatomy.

Induced by high bioreporter growth rates (Ramos et al. 2000) *Pseudomonas putida* bioreporters host a growth rate–regulated rRNA promoter fused to unstable *gfp*. Bioreporters grew slowly in nonsterile barley rhizosphere, but rapidly around sloughing root sheath cells in sterile soils.

Induced by quorum sensing (Steidle et al. 2001) *Pseudomonas putida* bioreporters were induced by homoserine lactones, illustrating that quorum sensing–based gene expression occurs in nonsterile rhizosphere soil around tomato roots.

Lux Report

Constitutively expressed (Beauchamp & Kloepper 2003) Root colonization by bioluminescent, plant growth–promoting *Pseudomonas putida* inoculated onto soybean seeds was tracked in nonsterile soil; numbers were much lower on root sections near root tips.

Induced by root exudates (Darwent et al. 2003) *Pseudomonas fluorescens* bioreporters indicated nitrate limitation did not increase specific barley root exudation rates in sand microcosms, but did increase total root system length and thus overall root system exudation.

Induced by P. starvation (Kragelund et al. 1997) Luminescence from *Pseudomonas fluorescens* bioreporters was detected in the rhizosphere of barley plants in sterile, but not nonsterile, soils.

Constitutively expressed (Rattray et al. 1995) *Enterobacter cloacae* bioreporters luminesced throughout rhizospheres of wheat in sterile soils, but luminescence was restricted to shoot/root junctions in nonsterile wheat rhizospheres, perhaps owing to competition with native bacteria.

Constitutively expressed (Roberts et al. 1999) Bioluminescent *Enterobacter cloacae* migrated from cucumber seed to roots during root growth in nonsterile soil, but colonization was not even, and numbers of bioreporters were much lower near root tips.

Constitutively expressed (Sakai et al. 1997) *Pseudomonas fluorescens* bioreporters luminesced in sterile rhizosphere soil around spinach, but luminescence decreased in nonsterile soils.

Induced by copper (Tom-Petersen et al. 2001) Copper available to a *Pseudomonas fluorescens* bioreporter in soil amended with complex organic material was less than total copper content, suggesting that copper complexation with organic matter reduces bioavailability.

Ice Nucleation Protein Report

Induced by tryptophan and by sucrose (Jaeger et al. 1999) Two *Erwinia herbicola* biosensors reported increasing tryptophan and decreasing sucrose availabilities in nonsterile rhizosphere soil as a function of axial distance from root tips of *Avena barbata*.

PLFA: phospholipid fatty acid

to be labeled to higher specific activities. Using a short (6 hour) $^{13}CO_2$ pulse delivered to grassland plants in limed and unlimed field plots at the Sourhope Field Experimental Site, Rangel-Castro et al. (2005) successfully enriched RNA from rhizosphere fungi and bacteria that had consumed ^{13}C-labeled rhizodeposits. DGGE analysis of heavy and light RNA fractions identified distinct phylotypes that had preferentially used either carbon from the $^{13}CO_2$-labeled rhizodeposits or from other unlabeled (^{12}C) soil sources, respectively. Overall, the microbial community using plant rhizodeposits in more fertile, limed soils was more complex than that in unlimed soils. In another example, Lu & Conrad (2005) pulse-labeled rice growing in microcosms and found preferentially labeled RNA from rice cluster I archaea in rhizosphere soil, suggesting these methanogens may be particularly important in transforming root-derived carbon to methane. SIP can similarly be combined with phospholipid fatty acid analysis (PLFA; **Table 1**; e.g., Butler et al. 2003, Lu et al. 2004, Treonis et al. 2004). Lu et al. (2004) found rapid differential labeling of straight-chain and branched fatty acids extracted from rhizosphere microorganisms after pulse-labeling rice. Treonis et al. (2004) found preferential ^{13}C labeling of fatty acids indicative of fungi and gram-negative bacteria in the rhizosphere of pulse-labeled grasses; interestingly, labeling of PLFAs was similar in limed and unlimed treatments at Sourhope, perhaps because PLFA is less specific than DGGE analysis of extracted RNA.

2.2.2. DNA arrays. DNA microarrays are most often used to identify genes transcribed in a known organism grown in well-defined media. Arrays typically consist of thousands of different amplified gene fragments (or oligonucleotides) from a single species spotted onto a solid substrate such as a microscope slide (Schena et al. 1995). mRNA isolated from the target species is converted into fluorescently labeled cDNA and hybridized to the array, producing fluorescent spots that indicate genes that were transcriptionally active above a threshold level (DeRisi et al. 1997). For example, Mark et al. (2005) used DNA microarrays to identify unique *Pseudomonas aeruginosa* genes expressed during growth in artificial medium containing sugarbeet root exudates from two beet cultivars.

Functional gene arrays are similar to DNA arrays except they contain gene fragments or oligonucleotides from genes of functional interest (e.g., genes for enzymes important in nitrogen or carbon cycling), from a variety of organisms. RNA can be harvested from soil, converted to labeled cDNA, and hybridized to the array to identify which of the genes spotted on the array were transcriptionally active in soil (Saleh-Lakha et al. 2005, Sessitsch et al. 2002). Hybridization of harvested and labeled DNA from soil to the array identifies genes on the chip that were present in soil. Array design and fabrication are challenging. Genes of interest must be identified, and probes on the array must be designed to be specific enough to identify genes from particular organisms (or groups of organisms) among the diverse, often uncharacterized, soil community.

Schadt et al. (2005) recently described the design of an array targeting genes of soil bacteria involved in metal resistance, biodegradation, and carbon, nitrogen, and sulfur cycling. This array contained spots consisting of oligonucleotides of about 50 bases in length. Labeled genes would hybridize to spots if they and the spots were more than 86% identical, and given that the sequence identity between the same functional genes in different species is often less than 85%, the researchers suggest that the array should be able to provide species-level information about those genes represented on the array (Rhee et al. 2004, Schadt et al. 2005). Other functional gene arrays have also been described (Cho & Tiedje 2002, Dennis et al. 2003, Taroncher-Oldenburg et al. 2002). Using similar approaches, but with probes diagnostic for microbial species and higher taxonomic groups, DNA arrays are being developed to study microbial diversity in complex environments including soil (Bodrossy & Sessitsch 2004, Loy et al. 2002, Sanguin et al. 2006, Small et al. 2001).

2.2.3. mRNA-fluorescent in situ hybridization.
FISH is a standard technique in microbial ecology, often used to identify the phylogenetic affiliation of soil bacteria (Delong & Wickham 1989, Kobabe et al. 2004). Typically, fluorescently labeled RNA or DNA probes are designed to bind to the 16S ribosomal RNA of selected taxonomic groups. Following the fixation and permeabilization of target cells, probe is added and it hybridizes to rRNA within cells from those selected groups. Cells containing this specifically bound fluorescent probe can be visualized by epifluorescence microscopy or enumerated by flow cytometry. (In situ indicates that labeling of the rRNA occurs within cells, but those cells may or may not be in situ in their native environment.)

Recent advances have made it possible to visualize gene expression (mRNA-FISH) in individual cells, or to identify metabolically active cells, while simultaneously determining their taxonomic affiliations (Zwirglmaier 2005). To reliably detect gene expression, signal from bound probe needs to be amplified because target mRNA molecules are often present at only a few copies per cell, and they can turn over rapidly. Visualizing mRNAs by FISH has been achieved by tagging FISH probes with digoxigenin, and hybridizing the probe under conditions that optimize signal to noise. The bound probe is visualized using antibodies that are conjugated to horseradish peroxidase, which converts soluble substrates to insoluble fluorescent compounds. Using this method, marine sediment bacteria expressing methane monooxygenase, and cells in contaminated groundwater samples expressing a naphthalene dioxygenase gene, have been visualized (Bakermans & Madsen 2002, Pernthaler & Amann 2004). Combining FISH with microautoradiography, or with immunodetection of bromodioxyuridine, will further allow for the visualization and simultaneous taxonomic identification of individual cells that are capable of degrading particular growth substrates or that are in the process of actively synthesizing new DNA (Gray et al. 2000, Nielsen et al. 2003, Pernthaler et al. 2002, Zwirglmaier 2005).

2.2.4. Microbial bioreporters.
The function-focused methods described above require a destructive harvest of soil for assay. Bioreporters offer an alternative, often nondestructive method for investigating microbial distribution and function *in situ* in the rhizosphere (**Table 2**). A bioreporter is made by fusing DNA coding for an easily assayed reporter gene with a promoter of interest and then inserting the fusion into a living host bacterial strain. Common reporter genes include *lacZ*, *phoA*, and *inaZ* (all requiring destructive harvest for assay of gene products), *gfp* (coding for GFP), and genes that result in bacterial bioluminescence (*lux*). The visual reports from *gfp*

and *lux* can be assayed nondestructively, without supplying external cofactors or substrates to cells. GFP can be used to monitor gene expression in single cells using an epifluorescence microscope, providing information on micrometer to millimeter spatial scales in soil. Light from expression of *lux* genes is detectable using sensitive cameras, so a resident engineered population can report on conditions across entire imaged root systems (see **Table 2**). Use of bioreporters in complex environments is a relatively recent development (Larrainzar et al. 2005, Leveau & Lindow 2002, Prosser et al. 1996, van der Meer et al. 2004). Modeling and statistical treatment of data from bioreporters have added to the sophistication with which these tools can be used (Leveau & Lindow 2001, 2002; van der Meer et al. 2004).

Several studies have used bioreporters to explore rhizosphere biology (**Table 2**). In some cases the reporter gene was driven by a promoter induced by specific compounds in the soil (**Table 2**). In other cases, bioreporters were designed with promoters that fired without needing to be induced, thus constitutively expressing the report, and enabling detection of where the tagged bacteria were active. Such tagging is useful, for example, in testing whether growth-promoting bacteria introduced into soil on seeds can colonize the complete plant root system. However, bioreporters are living organisms, not electronic sensors, so the biology of the host organisms (e.g., substrate preferences, growth rates, promoter strengths) may influence the reporter response (e.g., Wright & Beattie 2004). Interpretation of reports must therefore be conservative. Even with this limitation, the information gained at microbial scales never before assayed is powerful. Challenges associated with interpretation of bioreports include the following:

1. In instances where production of report is to be assayed in individual bacteria, the strength of the report should be proportional to the strength, or concentration, of the signal or compound that activated the promoter driving the reporter gene (Axtell & Beattie 2002, Lee & Keasling 2005). However, in some cases the promoter is either switched on or off at some stimulus threshold in any particular cell (e.g., in the native *E. coli araC* and *lacZ* promoters; Novick & Weiner 1957, Siegele & Hu 1997), and the fraction of cells in which the promoter is on is proportional to the strength or concentration of the inducing signal. Bioreporter design and assay must take into account such differences.

2. As Leveau & Lindow's (2001) simple mathematical model of bioreports shows, factors beyond promoter activity can affect the amount of report (e.g., GFP molecules) per cell. For example, though all promoters in a population of cells may be firing at a constant rate and driving production of a constant stream of report molecules as a function of a stimulus, if various reporter cells are growing at different rates, the report molecules per cell are being divided more often into the daughter cells of bioreporters dividing quickly. Report molecules per cell could thus be lower in rapidly growing cells, even though promoter firing rates are the same in rapidly and slowly growing cells. Similarly, differential rates of degradation of report protein, or maturation of report protein, could also affect the concentration of report in cells (Leveau & Lindow 2001). Such patterns do not necessarily develop in all bioreporters; as Leveau & Lindow (2001)

point out, for example, promoter activity itself may be affected by growth rate, independent of the stimulus detected by the bioreporter. It is essential to test as much as possible for such effects under well-controlled conditions before microbiosensors are inoculated into complex environments. For example, P.M. Herron, D.J. Gage & Z.G. Cardon (submitted) developed water potential–sensing bacteria using *Pseudomonas putida*, and found that GFP per cell is a reliable report of water potential across a range of growth rates caused by varied carbon sources and salt types. Further, complex environments harbor zones of conditions and resources that may or may not be conducive to microbial activity. Kragelund et al. (1997) note, for example, that a high-level expression in a bioreporter cannot be expected from starved host cells, even if an inducing signal is present. Controls should be designed to reveal such resource limitations.

3. Finally, promoters in bioreporters may be induced by many compounds or conditions, some of which may not have been experimentally identified (e.g., Casavant et al. 2003; **Table 2**). For example, Bringhurst et al.'s (2001) galactoside bioreporter consisted of the *S. meliloti melA* promoter, normally induced by many α-galactosides, fused to *gfp*. But, the *melA* promoter has also been shown to be induced by galactose and some β-galactosides (Bringhurst & Gage 2000, 2002; Gage & Long 1998). Furthermore, carbon sources induced the *melA* promoter with varying efficiency, meaning bright bioreporter cells could have resulted from low levels of an efficient inducer (e.g., melibiose), or from high levels of an inefficient inducer (e.g., stachyose). For this reason, the inclination to view bioreporter results as absolutely quantifying levels of inducer should be avoided.

Good experimental design can diminish effects of these limitations. Bioreporters should be well-characterized in liquid medium before being deployed in complex environments. The influences of carbon source and growth rate should be investigated. Control strains that constitutively express a gene very similar to the reporter gene can be deployed in the environment under study to determine whether variations in environmental conditions (other than the resource or condition being assayed by the bioreporter) can cause variations in report levels independent of variation in promoter firing. For example, if the bioreporter gene is *gfp*, then red fluorescent protein (*rfp*) would be a good constitutively expressed control reporter (e.g., DeAngelis et al. 2005). Such control bacteria can also confirm that interesting spatial expression patterns are not due to uneven distribution or uneven growth of inoculated bioreporter cells (e.g., Bringhurst et al. 2001). If the constitutively expressed gene and the inducible bioreporter gene can be differentiated from each other, they can be engineered into the same cells (DeAngelis et al. 2005).

3. SEEKING BRIDGES ACROSS SCALES

Linking rhizosphere mechanisms to ecosystem function is a challenging goal, but already in the literature a variety of mathematical and conceptual models established at both scales have promisingly parallel components that recommend integration. At rhizosphere scales, a common functional framework should include spatially explicit, water-linked mixing of rhizosphere resources in soil hosting a community whose activity is constrained by trophic interactions, stoichiometry, and soil structure. There are recognized gross shifts in this framework ecosystem to ecosystem, including, for example, differences in saprotrophic abilities of mycorrhizae (and their biogeographical distribution), shifts in dominance of bacterial versus fungal saprotrophs, variation in soil structure and chemistry, seasonal extents of plant control over water fluxes, etc. These known, larger ecosystem-scale patterns can be woven into the developing conceptual rhizosphere framework and their consequences considered. As a simple example, shifts in dominance of fungi and bacteria among microbial decomposers have been well-documented across ecosystems and management regimes, and the implications of the simple morphological distinction between fungi and bacteria within the soil matrix are intriguing. The three-dimesional fungal hyphal structure can support integration of resources by a single organism from a much larger volume of rhizosphere soil, potentially diminishing the effect of spatial zonation of resource availability on microbes around tip, active uptake, and suberized root zones when fungi are dominant decomposers in ecosystems.

Models can be linked within as well as across scales. For example, Clarholm's (1985) model could be considered an extremely simple form of Moore et al.'s (2003) bacterial channel of energy flow in the rhizosphere food web, in which root-derived, labile carbon fuels bacterial growth and activity, leading to increased grazing by protozoa on that biomass (and concomitant nutrient mineralization). Recognizing mycorrhizae of various saprotrophic capabilities as part of the rhizosphere community, Clarholm's model is also paralleled by Chapman et al's (2006) conceptual ecosystems model describing open, leaky nutrient cycles. Such cycles operate in ecosystems where AM mycorrhizae (incapable of saprotrophic activities) are dominant, including grasslands (where almost all of the upper soil layer is rhizosphere soil) or rhizospheres of AM-infected trees. In this open ecosystem cycling scheme, availability of nutrient-rich, easily decomposed plant litter and rhizodeposits supports decomposition dominated by bacteria (as in Clarholm's model), release of mineral nutrients from that bacterial biomass (e.g., by grazing mortality, though that is not specified), and ultimate availability of those mineral nutrients to plants (taken up by roots directly from soil or delivered by the AM fungi). The cycle is open or leaky in the sense that nutrients are quite available in litter and are not being strongly conserved or passed tightly from organism to organism; the nutrients can pass through abiotic soil pools, where ammonium might be nitrified to nitrate, which could be lost by leaching, denitrification, or retained in the system through immobilization. AM hyphae operate in essence as extensions of the root system in this context, providing an enhanced surface area over which nutrients can be absorbed from soil. However, as noted previously, it remains unclear whether the AM fungal hyphae themselves can stimulate the activities of saprotrophs through either signaling or provision of resources, thus indirectly influencing decomposition rates. Nondestructive molecular approaches examining activity and gene expression in saprotrophic bacteria around intact AM fungal hyphae would be particularly interesting in this context.

In contrast to ecosystems dominated by AM mycorrhizae and bacterial saprotrophs, Chapman et al's (2006) conceptual model of closed or conservative ecosystem cycling characterizes ecosystems dominated by slow decomposition rates and

ECM-symbioses (e.g., northern boreal ecosystems). The closed ecosystem cycling scheme emphasizes the saprotrophic capability of the ECM fungi themselves in attacking litter that is naturally less rich, as well as the ECM fungal delivery of nutrients directly to the plant roots that directly supplied them carbon. This closed, tight cycling can be viewed as a variation on the rhizosphere fungal/slow and mycorrhizal channels described by Moore et al. (2006), in which the saprotrophic fungal community and the symbiotic fungal community are one and the same. The evolution of saprotrophic ECM symbionts greatly simplified the interorganismal rhizosphere transfers otherwise required to move carbon to decomposers and nutrients from litter or SOM to roots. In cold climates where microbial activity is limited and litter is recalcitrant, Chapman et al. (2006) suggest this streamlining represents an "uncorking of the microbial bottleneck" that otherwise could restrict nutrient availability to plants.

Discussion of open (leaky) and closed (conservative) recycling schemes and their shifts over time has a long tradition within ecosystems ecology, ranging from Odum's landmark paper in 1969 hypothesizing shifts from open to closed nutrient cycling during succession (see Vitousek & Reiners 1975 for contrast), to Bardgett et al.'s (2005) discussion of successional shifts under climate change. The rhizosphere is the microscopic stage in which these interactions are playing out, an empirical realm where molecular tools emphasizing function are invaluable for testing and inspiring understanding of belowground function in the diverse array of Earth's ecosystems.

ACKNOWLEDGMENTS

Thanks to Richard Ferrieri, Charles Harvey, Patrick Herron, N. Michelle Holbrook, Rebecca Neumann, Colin Orians, Edward Rastetter, Chris Schadt, John Stark, Michael Thorpe, Joe Zhou, and Maciej Zwieniecki for ongoing discussions, and Carolyn Malmstrom and Will Whitted for manuscript suggestions. Work was funded in part by grants from the Andrew W. Mellon Foundation (Z.G.C.), and NSF DEB-0415938 (Z.G.C., D.J.G., and John Stark). This is publication #8 from the Center for Integrative Geosciences at University of Connecticut.

LITERATURE CITED

Acinas SG, Sarma-Rupavtarm R, Klepac-Ceraj V, Polz MF. 2005. PCR-induced sequence artifacts and bias: Insights from comparison of two 16S rRNA clone libraries constructed from the same sample. *Appl. Environ. Microbiol.* 71:8966–69

Algeo TJ, Scheckler SE. 1998. Terrestrial-marine teleconnections in the Devonian: links between the evolution of land plants, weathering processes, and marine anoxic events. *Philos. Trans. R. Soc. London Ser. B* 353:113–28

Axtell CA, Beattie GA. 2002. Construction and characterization of a *proU-gfp* transcriptional fusion that measures water availability in a microbial habitat. *Appl. Environ. Microbiol.* 68:4604–12

Baker GC, Smith JJ, Cowan DA. 2003. Review and reanalysis of domain-specific 16S primers. *J. Microbiol. Methods* 55:541–55

Bakermans C, Madsen EL. 2002. Detection in coal tar waste-contaminated groundwater of mRNA transcripts related to naphthalene dioxygenase by fluorescent in situ hybridization with tyramide signal amplification. *J. Microbiol. Methods* 50:75–84

Bardgett RD, Bowman WD, Kaufmann R, Schmidt SK. 2005. A temporal approach to linking aboveground and belowground ecology. *Trends Ecol. Evol.* 20:634–41

Beauchamp CJ, Kloepper JW. 2003. Spatial and temporal distribution of a bioluminescent-marked *Pseudomonas putida* on soybean root. *Luminescence* 18:346–51

Beerling DJ, Berner RA. 2005. Feedbacks and the coevolution of plants and atmospheric CO_2. *Proc. Natl. Acad. Sci. USA* 102:1302–05

Blackwood CB, Paul EA. 2003. Eubacterial community structure and population size within the soil light fraction, rhizosphere, and heavy fraction of several agricultural systems. *Soil Biol. Biochem.* 35:1245–55

Bodrossy L, Sessitsch A. 2004. Oligonucleotide microarrays in microbial diagnostics. *Curr. Opin. Microbiol.* 7:245–54

Boldt TS, Sorensen J, Karlson U, Molin S, Ramos C. 2004. Combined use of different Gfp reporters for monitoring single-cell activities of a genetically modified PCB degrader in the rhizosphere of alfalfa. *FEMS Microbiol. Ecol.* 48:139–48

Bonkowski M. 2004. Protozoa and plant growth: the microbial loop in soil revisited. *New Phytol.* 162:617–31

Bonkowski M, Brandt F. 2002. Do soil protozoa enhance plant growth by hormonal effects? *Soil Biol. Biochem.* 34:1709–15

Bossio DA, Scow KM, Gunapala N, Graham KJ. 1998. Determinants of soil microbial communities: effects of agricultural management, season, and soil type on phospholipid fatty acid profiles. *Microb. Ecol.* 36:1–12

Brencic A, Winans SC. 2005. Detection of and response to signals involved in host-microbe interactions by plant-associated bacteria. *Microbiol. Mol. Biol. Rev.* 69:155–94

Bretharte MS, Silk WK. 1994. Nonvascular, symplasmic diffusion of sucrose cannot satisfy the carbon demands of growth in the primary root-tip of *Zea mays* L. *Plant Physiol.* 105:19–33

Bringhurst RM, Cardon ZG, Gage DJ. 2001. Galactosides in the rhizosphere: utilization by *Sinorhizobium meliloti* and development of a biosensor. *Proc. Natl. Acad. Sci. USA* 98:4540–45

Bringhurst RM, Gage DJ. 2000. An AraC-like transcriptional activator is required for induction of genes needed for α-galactoside utilization in *Sinorhizobium meliloti*. *FEMS Microbiol. Lett.* 188:23–27

Bringhurst RM, Gage DJ. 2002. Control of inducer accumulation plays a key role in succinate-mediated catabolite repression in *Sinorhizobium meliloti*. *J. Bacteriol.* 184:5385–92

Butler JL, Williams MA, Bottomley PJ, Myrold DD. 2003. Microbial community dynamics associated with rhizosphere carbon flow. *Appl. Environ. Microbiol.* 69:6793–6800

Caldwell MM, Dawson TE, Richards JH. 1998. Hydraulic lift: consequences of water efflux from the roots of plants. *Oecologia* 113:151–61

Cardon ZG, Herron PM. 2005. Sweeping water, oozing carbon: long distance transport and patterns of rhizosphere resource exchange. In *Vascular Transport in Plants*, ed. NM Holbrook, MA Zwieniecki, pp. 257–76. San Diego: Academic

Cardon ZG, Rastetter EB. 2004. What's running the rhizosphere—are both water and carbon at the root of it? *Proc. 89th Annu. Meet. Ecol. Soc. Am., Portland, Oregon*, p. 82 (Abstr.). Washington, DC: Ecol. Soc. Am.

Cardon ZG, Whitbeck JL. 2006. *The Rhizosphere: An Ecological Perspective*. San Diego: Elsevier. In press

Casavant NC, Beattie GA, Phillips GJ, Halverson LJ. 2002. Site-specific recombination-based genetic system for reporting transient or low-level gene expression. *Appl. Environ. Microbiol.* 68:3588–96

Casavant NC, Thompson D, Beattie GA, Phillips GJ, Halverson LJ. 2003. Use of a site-specific recombination-based biosensor for detecting bioavailable toluene and related compounds on roots. *Environ. Microbiol.* 5:238–49

Chapman SK, Langley JA, Hart SC, Koch GW. 2006. Plants actively control nitrogen cycling: uncorking the microbial bottleneck. *New Phytol.* 169:27–34

Chen J, Stark JM. 2000. Plant species effects and carbon and nitrogen cycling in a sagebrush-crested wheatgrass soil. *Soil Biol. Biochem.* 32:47–57

Cheng WX, Gershenson A. 2006. Carbon fluxes in the rhizosphere. See Cardon & Whitbeck 2006, In press

Cho JC, Tiedje TM. 2002. Quantitative detection of microbial genes by using DNA microarrays. *Appl. Environ. Microbiol.* 68:1425–30

Clarholm M. 1985. Interactions of bacteria, protozoa and plants leading to mineralization of soil-nitrogen. *Soil Biol. Biochem.* 17:181–87

Clarholm M. 2005. Soil protozoa: an under-researched microbial group gaining momentum. *Soil Biol. Biochem.* 37:811–17

Coleman DC, Whitman WB. 2005. Linking species richness, biodiversity and ecosystem function in soil systems. *Pedobiologia* 49:479–97

Compant S, Duffy B, Nowak J, Clement C, Barka EA. 2005. Use of plant growth-promoting bacteria for biocontrol of plant diseases: Principles, mechanisms of action, and future prospects. *Appl. Environ. Microbiol.* 71:4951–59

Couteaux MM, Bottner P. 1994. Biological interactions between fauna and the microbial community in soils. In *Beyond the Biomass*, ed. K Ritz, J Dighton, KE Giller, pp. 159–72. New York: Wiley

Crawford JW, Harris JA, Ritz K, Young IM. 2005. Towards an evolutionary ecology of life in soil. *Trends Ecol. Evol.* 20:81–87

Curtis TP, Sloan WT, Scannell JW. 2002. Estimating prokaryotic diversity and its limits. *Proc. Natl. Acad. Sci. USA* 99:10494–99

Daniel R. 2005. The metagenomics of soil. *Nat. Rev. Microbiol.* 3:470–78

Darrah PR. 1991. Models of the rhizosphere. 1. Microbial population dynamics around a root releasing soluble and insoluble carbon. *Plant Soil* 133:187–99

Darrah PR, Jones DL, Kirk GJD, Roose T. 2006. Modelling the rhizosphere: a review of methods for 'upscaling' to the whole-plant scale. *Eur. J. Soil Sci.* 57:13–25

Darwent MJ, Paterson E, McDonald AJS, Tomos AD. 2003. Biosensor reporting of root exudation from *Hordeum vulgare* in relation to shoot nitrate concentration. *J. Exp. Bot.* 54:325–34

DeAngelis KM, Ji PS, Firestone MK, Lindow SE. 2005. Two novel bacterial biosensors for detection of nitrate availability in the rhizosphere. *Appl. Environ. Microbiol.* 71:8537–47

DeLong EF, Wickham GS. 1989. Phylogenetic stains: Ribosomal RNA-based probes for the identification of single cells. *Science* 243:1360–63

Dennis P, Edwards EA, Liss SN, Fulthorpe R. 2003. Monitoring gene expression in mixed microbial communities by using DNA microarrays. *Environ. Microbiol.* 69:769–78

DeRisi JL, Iyer VR, Brown PO. 1997. Exploring the metabolic and genetic control of gene expression on a genomic scale. *Science* 278:680–86

Doussan C, Vercambre G, Pages L. 1998. Modeling of the hydraulic architecture of root systems: An integrated approach to water absorption—Distribution of axial and radial conductances in maize. *Ann. Bot. London* 81:225–32

Drinkwater LE, Snapp SS. 2006. Understanding and managing the rhizosphere in agroecosystems. See Cardon & Whitbeck 2006, In press

Dumont MG, Murrell JC. 2005. Stable isotope probing—linking microbial identity to function. *Nat. Rev. Microbiol.* 3:499–504

Ferrari BC, Binnerup SJ, Gillings M. 2005. Microcolony cultivation on a soil substrate membrane system selects for previously uncultured soil bacteria. *Appl. Environ. Microbiol.* 71:8714–20

Fitter AH. 2005. Darkness visible: reflections on underground ecology. *J. Ecol.* 93:231–43

Fitter AH, Gilligan CA, Hollingworth K, Kleczkowski A, Twyman RM, Pitchford JW. 2005. Biodiversity and ecosystem function in soil. *Funct. Ecol.* 19:369–77

Gage DJ, Long SR. 1998. α-galactoside uptake in *Rhizobium meliloti*: isolation and characterization of *agpA*, a gene encoding a periplasmic binding protein required for melibiose and raffinose utilization. *J. Bacteriol.* 180:5739–48

Gans J, Wolinsky M, Dunbar J. 2005. Computational improvements reveal great bacterial diversity and high metal toxicity in soil. *Science* 309:1387–90

Garbeva P, van Veen JA, van Elsas JD. 2004. Microbial diversity in soil: selection of microbial populations by plant and soil type and implications for disease suppressiveness. *Annu. Rev. Phytopathol.* 42:243–70

Graham J, Miller R. 2005. Mycorrhizas: gene to function. *Plant Soil* 274:79–100

Gray ND, Howarth R, Pickup RW, Jones JG, Head IM. 2000. Use of combined microautoradiography and fluorescence *in situ* hybridization to determine carbon metabolism in mixed natural communities of uncultured bacteria from the genus *Achromatium*. *Appl. Environ. Microbiol.* 66:4518–22

Griffiths BS, Christensen S, Bonkowski M. 2006. Microfaunal interactions in the rhizosphere—how nematodes and protozoa link above- and below-ground processes. See Cardon & Whitbeck 2006, In press

Handelsman J. 2004. Metagenomics: application of genomics to uncultured microorganisms. *Microbiol. Mol. Biol. Rev.* 68:669–85

Harrison MJ. 2005. Signaling in the arbuscular mycorrhizal symbiosis. *Annu. Rev. Microbiol.* 59:19–42

Hawes MC, Gunawardena U, Miyasaka S, Zhao XW. 2000. The role of root border cells in plant defense. *Trends Plant Sci.* 5:128–33

Hawkes CV, DeAngelis K, Firestone MK. 2006. Root interactions with soil microbial communities and processes. See Cardon & Whitbeck 2006, In press

Hillel D. 1998. *Environmental Soil Physics*, pp. 574–78. San Diego: Academic

Hirsch AM, Bauer WD, Bird DM, Cullimore J, Tyler B, Yoder JI. 2003. Molecular signals and receptors: controlling rhizosphere interactions between plants and other organisms. *Ecology* 84:858–68

Hong SH, Bunge J, Jeon SO, Epstein SS. 2006. Predicting microbial species richness. *Proc. Natl. Acad. Sci. USA* 103:117–22

Ingham RE, Trofymow JA, Ingham ER, Coleman DC. 1985. Interactions of bacteria, fungi and their nematode grazers: effects on nutrient cycling and plant growth. *Ecol. Monogr.* 55:119–40

Jaeger CH, Lindow SE, Miller S, Clark E, Firestone MK. 1999. Mapping of sugar and amino acid availability in soil around roots with bacterial sensors of sucrose and tryptophan. *Appl. Environ. Microbiol.* 65:2685–90

Johnson NC, Gehring CA. 2006. Mycorrhizas: symbiotic mediators of rhizosphere and ecosystem processes. See Cardon & Whitbeck 2006, In press

Jones DL, Hodge A, Kuzyakov Y. 2004. Plant and mycorrhizal regulation of rhizodeposition. *New Phytol.* 163:459–80

Joseph SJ, Hugenholtz P, Sangwan P, Osborne CA, Janssen PH. 2003. Laboratory cultivation of widespread and previously uncultured soil bacteria. *Appl. Environ. Microbiol.* 69:7210–15

Kent AD, Triplett EW. 2002. Microbial communities and their interactions in soil and rhizosphere ecosystems. *Annu. Rev. Microbiol.* 56:211–36

Kobabe S, Wagner D, Pfeiffer EM. 2004. Characterisation of microbial community composition of a Siberian tundra soil by fluorescence in situ hybridisation. *FEMS Microbiol. Ecol.* 50:13–23

Kragelund L, Hosbond C, Nybroe O. 1997. Distribution of metabolic activity and phosphate starvation response of *lux*-tagged *Pseudomonas fluorescens* reporter bacteria in the barley rhizosphere. *Appl. Environ. Microbiol.* 63:4920–28

Kuiper I, Lagendijk EL, Bloemberg GV, Lugtenberg BJJ. 2004. Rhizoremediation: a beneficial plant-microbe interaction. *Mol. Plant Microbe Interact.* 17:6–15

Landsberg JJ, Fowkes ND. 1978. Water movement through plant roots. *Ann. Bot.* 42:493–508

Larrainzar E, O'Gara F, Morrissey JP. 2005. Applications of autofluorescent proteins for in situ studies in microbial ecology. *Annu. Rev. Microbiol.* 59:257–77

Lee KL, Keasling JD. 2005. A propionate-inducible expression system for enteric bacteria. *Appl. Environ. Microbiol.* 71:6856–62

Leveau JHJ, Lindow SE. 2001. Predictive and interpretive simulation of green fluorescent protein expression in reporter bacteria. *J. Bacteriol.* 183:6752–62

Leveau JHJ, Lindow SE. 2002. Bioreporters in microbial ecology. *Curr. Opin. Microbiol.* 5:259–65

Loh J, Pierson EA, Pierson LS, Stacey G, Chatterjee A. 2002. Quorum sensing in plant-associated bacteria. *Curr. Opin. Plant Biol.* 5:285–90

Loy A, Lahner A, Lee N, Adamczyk J, Meier H, et al. 2002. Oligonucleotide microarray for 16S rRNA gene-based detection of all recognized lineages of sulfate reducing prokaryotes in the environment. *Appl. Environ. Microbiol.* 68:5064–81

Lu YH, Conrad R. 2005. In situ stable isotope probing of methanogenic archaea in the rice rhizosphere. *Science* 309:1088–90

Lu YH, Murase J, Watanabe A, Sugimoto A, Kimura M. 2004. Linking microbial community dynamics to rhizosphere carbon flow in a wetland rice soil. *FEMS Microbiol. Ecol.* 48:179–86

Lynch JM. 1990. *The Rhizosphere*. Chichester, NY: Wiley

Lynch JM, Benedetti A, Insam H, Nuti MP, Smalla K, et al. 2004. Microbial diversity in soil: ecological theories, the contribution of molecular techniques and the impact of transgenic plants and transgenic microorganisms. *Biol. Fertil. Soils* 40:363–85

Manefield M, Whiteley AS, Ostle N, Ineson P, Bailey MJ. 2002. Technical considerations for RNA-based stable isotope probing: an approach to associating microbial diversity with microbial community function. *Rapid Commun. Mass Spectrom.* 16:2179–83

Mark GL, Dow JM, Kiely PD, Higgins H, Haynes J, et al. 2005. Transcriptome profiling of bacterial responses to root exudates identifies genes involved in microbe–plant interactions. *Proc. Natl. Acad. Sci. USA* 102:17454–59

Marschner P, Yang CH, Lieberei R, Crowley DE. 2001. Soil and plant specific effects on bacterial community composition in the rhizosphere. *Soil Biol. Biochem.* 33:1437–45

Meselson M, Stahl FW. 1958. The replication of DNA in *Escherichia coli*. *Proc. Natl. Acad. Sci. USA* 44:671–82

Moore JC, Berlow EL, Coleman DC, de Ruiter PC, Dong Q, et al. 2004. Detritus, trophic dynamics and biodiversity. *Ecol. Lett.* 7:584–600

Moore JC, McCann K, de Ruiter PC. 2006. Soil rhizosphere food webs, their stability, and implications for soil processes and ecosystems. See Cardon & Whitbeck 2006, In press

Moore JC, McCann K, Setala H, De Ruiter PC. 2003. Top-down is bottom-up: Does predation in the rhizosphere regulate aboveground dynamics? *Ecology* 84:846–57

Murray P, Ostle N, Kenny C, Grant H. 2004. Effect of defoliation on patterns of carbon exudation from *Agrostis capillaris*. *J. Plant Nut. Soil Sci.* 167:487–93

Nielsen JL, Christensen D, Kloppenborg M, Nielsen PH. 2003. Quantification of cell-specific substrate uptake by probe-defined bacteria under in situ conditions by microautoradiography and fluorescence in situ hybridization. *Environ. Microbiol.* 5:202–11

Novick A, Weiner M. 1957. Enzyme induction as an all-or-none phenomenon. *Proc. Natl. Acad. Sci. USA* 42:553–66

Odum EP. 1969. The strategy of ecosystem development. *Science* 164:262–70

Ostle N, Whiteley AS, Bailey MJ, Sleep D, Ineson P, Manefield M. 2003. Active microbial RNA turnover in a grassland soil estimated using a (CO_2)-C-13 spike. *Soil Biol. Biochem.* 35:877–85

Pernthaler A, Amann R. 2004. Simultaneous fluorescence in situ hybridization of mRNA and rRNA in environmental bacteria. *Appl. Environ. Microbiol.* 70:5426–33

Pernthaler A, Pernthaler J, Schattenhofer M, Amann R. 2002. Identification of DNA-synthesizing bacterial cells in coastal North Sea plankton. *Appl. Environ. Microbiol.* 68:5728–36

Phillips DA, Ferris H, Cook DR, Strong DR. 2003. Molecular control points in rhizosphere food webs. *Ecology* 84:816–26

Pregitzer KS, Zak DR, Loya WM, Karberg NJ, King JS, Burton AJ. 2006. The contribution of root systems to biogeochemical cycles in a changing world. See Cardon & Whitbeck 2006, In press

Prosser JI, Killham K, Glover LA, Rattray EAS. 1996. Luminescence-based systems for detection of bacteria in the environment. *Crit. Rev. Biotechnol.* 16:157–83

Radajewski S, Ineson P, Parekh NR, Murrell JC. 2000. Stable-isotope probing as a tool in microbial ecology. *Nature* 403:646–49

Radajewski S, McDonald IR, Murrell JC. 2003. Stable-isotope probing of nucleic acids: a window to the function of uncultured microorganisms. *Curr. Opin. Biotechnol.* 14:296–302

Ramos C, Molbak L, Molin S. 2000. Bacterial activity in the rhizosphere analyzed at the single-cell level by monitoring ribosome contents and synthesis rates. *Appl. Environ. Microbiol.* 66:801–09

Rangel-Castro JI, Killham K, Ostle N, Nicol GW, Anderson IC, et al. 2005. Stable isotope probing analysis of the influence of liming on root exudate utilization by soil microorganisms. *Environ. Microbiol.* 7:828–38

Rattray EAS, Prossner JI, Glover LA, Killham K. 1995. Characterization of rhizosphere colonization by luminescent *Enterobacter cloacae* at the population and single-cell levels. *Appl. Environ. Microbiol.* 61:2950–57

Raven JA, Edwards D. 2001. Roots: evolutionary origins and biogeochemical significance. *J. Exp. Bot.* 52:381–401

Read DJ, Perez-Moreno J. 2003. Mycorrhizas and nutrient cycling in ecosystems—a journey towards relevance? *New Phytol.* 157:475–92

Rhee SK, Liu X, Wu L, Chong SC, Wan X, Zhou J. 2004. Detection of genes involved in biodegradation and biotransformation in microbial communities by using 50-mer oligonucleotide microarrays. *Appl. Environ. Microbiol.* 79:4303–17

Richter DD, Oh NH, Fimmen R, Jackson J. 2006. The rhizosphere and soil formation. See Cardon & Whitbeck 2006, In press

Roberts DP, Kobayashi DY, Dery PD, Short NMJ. 1999. An image analysis method for determination of spatial colonization patterns of bacteria in plant rhizosphere. *Appl. Environ. Microbiol.* 51:653–58

Robinson D, Griffiths B, Ritz K, Wheatley R. 1989. Root-induced nitrogen mineralization: a theoretical analysis. *Plant Soil* 117:185–93

Rondon MR, August PR, Bettermann AD, Brady SF, Grossman TH, et al. 2000. Cloning the soil metagenome: a strategy for accessing the genetic and functional diversity of uncultured microorganisms. *Appl. Environ. Microbiol.* 66:2541–47

Sakai M, Futamata H, Ozawa H, Sueguchi T, Kim JS, Matsuguchi T. 1997. Use of bacterial bioluminescence for monitoring the behavior of rhizobacteria introduced to plant rhizosphere. *Soil Sci. Plant Nutr.* 43:395–404

Saleh-Lakha S, Miller M, Campbell RG, Schneider K, Elahimanesh P, et al. 2005. Microbial gene expression in soil: methods, applications and challenges. *J. Microbiol. Methods* 63:1–19

Sanguin H, Herrera A, Oger-Desfeux C, Dechesne A, Simonet P, et al. 2006. Development and validation of a prototype 16S rRNA-based taxonomic microarray for Alphaproteobacteria. *Environ. Microbiol.* 8:289–307

Schadt CW, Liebich J, Chong SC, Gentry TJ, He ZL, et al. 2005. Design and use of functional gene microarrays (FGAs) for the characterization of microbial communities. *Methods Microbiol.* 34:331–68

Schena M, Shalon D, Davis RW, Brown PO. 1995. Quantitative monitoring of gene expression patterns with a complementary DNA microarray. *Science* 270:467–70

Schimel JP, Bennett J. 2004. Nitrogen mineralization: challenges of a changing paradigm. *Ecology* 85:591–602

Schimel JP, Weintraub MN. 2003. The implications of exoenzyme activity on microbial carbon and nitrogen limitation in soil: a theoretical model. *Soil Biol. Biochem.* 35:549–63

Schloss PD, Handelsman J. 2004. Status of the microbial census. *Microbiol. Mol. Biol. Rev.* 68:686–91

Scott EM, Rattray EAS, Prosser JI, Killham K, Glover LA, et al. 1995. A mathematical model for dispersal of bacterial inoculants colonizing the wheat rhizosphere. *Soil Biol. Biochem.* 27:1307–18

Sessitsch A, Gyamfi S, Stralis-Pavese N, Weilharter A, Pfeifer U. 2002. RNA isolation from soil for bacterial community and functional analysis: evaluation of different extraction and soil conservation protocols. *J. Microbiol. Methods* 51:171–79

Siegele DA, Hu JC. 1997. Gene expression from plasmids containing the *araBAD* promoter at subsaturating inducer concentrations represents mixed populations. *Proc. Natl. Acad. Sci. USA* 94:8168–72

Singh BK, Millard P, Whiteley AS, Murrell JC. 2004. Unravelling rhizosphere-microbial interactions: opportunities and limitations. *Trends Microbiol.* 12:386–93

Small J, Call DR, Brockman FJ, Straub TM, Chandler DP. 2001. Direct detection of 16S rRNA in soil extracts by using oligonucleotide microarrays. *Appl. Environ. Microbiol.* 67:4708–16

Smalla K, Wachtendorf U, Heuer H, Liu WT, Forney L. 1998. Analysis of BIOLOG GN substrate utilization patterns by microbial communities. *Appl. Environ. Microbiol.* 64:1220–25

Smalla K, Wieland G, Buchner A, Zock A, Parzy J, et al. 2001. Bulk and rhizosphere soil bacterial communities studied by denaturing gradient gel electrophoresis: plant-dependent enrichment and seasonal shifts revealed. *Appl. Environ. Microbiol.* 67:4742–51

Spiegelman D, Whissell G, Greer CW. 2005. A survey of the methods for the characterization of microbial consortia and communities. *Can. J. Microbiol.* 51:355–86

Steidle A, Sigl K, Schuhegger R, Ihring A, Schmid M, et al. 2001. Visualization of N-acylhomoserine lactone-mediated cell-cell communication between bacteria colonizing the tomato rhizosphere. *Appl. Environ. Microbiol.* 67:5761–70

Taroncher-Oldenburg G, Griner EM, Francis CA, Ward BB. 2002. Oligonucleotide microarray for the study of functional gene diversity in the nitrogen cycle in the environment. *Appl. Environ. Microbiol.* 69:1159–71

Thies JE, Holmes EM, Vachot A. 2001. Application of molecular techniques to studies in *Rhizobium* ecology: a review. *Aust. J. Exp. Agric.* 41:299–319

Thompson JR, Pacocha S, Pharino C, Klepac-Ceraj V, Hunt DE, et al. 2005. Genotypic diversity within a natural coastal bacterioplankton population. *Science* 307:1311–13

Tinker PB, Nye PH. 2000. *Solute Movement in the Rhizosphere.* New York: Oxford Univ. Press. 444 pp.

Toal ME, Yeomans C, Killham K, Meharg AA. 2000. A review of rhizosphere carbon flow modelling. *Plant Soil* 222:263–81

Tom-Petersen A, Hosbond C, Nybroe O. 2001. Identification of copper-induced genes in *Pseudomonas* fluorescens and use of a reporter strain to monitor bioavailable copper in soil. *FEMS Microbiol. Ecol.* 38:59–67

Treonis AM, Ostle NJ, Stott AW, Primrose R, Grayston SJ, Ineson P. 2004. Identification of groups of metabolically-active rhizosphere microorganisms by stable isotope probing of PLFAs. *Soil Biol. Biochem.* 36:533–37

van der Meer JJ, Tropel D, Jaspers M. 2004. Illuminating the detection chain of bacterial bioreporters. *Environ. Microbiol.* 6:1005–20

Vitousek PM, Reiners WA. 1975. Ecosystem succession and nutrient retention: a hypothesis. *Bioscience* 25:376–81

Wamburg C, Christensen S, Jakobsen I, Muller AK, Sorensen SJ. 2003. The mycorrhizal fungus *Glomus intraradices* affects microbial activity in the rhizosphere of pea plants (*Pisum sativum*). *Soil Biol. Biochem.* 35:1349–57

Wardle DA, Verhoef HA, Clarholm M. 1998. Trophic relationships in the soil microfood-web: predicting the response to a changing global environment. *Glob. Change Biol.* 4:713–27

Wellington EMH, Berry A, Krsek M. 2003. Resolving functional diversity in relation to microbial community structure in soil: exploiting genomics and stable isotope probing. *Curr. Opin. Microbiol.* 6:295–301

Whipps JM. 1990. Carbon economy. In *The Rhizosphere*, ed. JM Lynch, pp. 59–97. Chichester NY: Wiley

Wright CA, Beattie GA. 2004. Bacterial species specificity in *proU* osmoinducibility and *nptII* and *lacZ* expression. *J. Mol. Microbiol. Biotechnol.* 8:201–8

Wu L, McGechan MB, Watson CA, Baddeley JA. 2005. Developing existing plant root system architecture models to meet future, agricultural challenges. *Adv. Agron.* 85:181–219

Young IM, Ritz K. 1998. Can there be a contemporary ecological dimension to soil biology without a habitat? Discussion. *Soil Biol. Biochem.* 30:1229–32

Zengler K, Toledo G, Rappe M, Elkins J, Mathur EJ, et al. 2002. Cultivating the uncultured. *Proc. Natl. Acad. Sci. USA* 99:15681–86

Zwieniecki MA, Thompson MV, Holbrook NM. 2002. Understanding the hydraulics of porous pipes: Tradeoffs between water uptake and root length utilization. *J. Plant Growth Regul.* 21:315–23

Zwirglmaier K. 2005. Fluorescence in situ hybridisation (FISH)—the next generation. *FEMS Microbiol. Lett.* 246:151–58

RELATED REVIEWS

Pilon-Smits E. 2005. Phytoremediation. *Annu. Rev. Plant Biol.* 56:15–39

Larrainzar E, O'Gara F, Morrissey JP. 2005. Applications of autofluorescent proteins for in situ studies in microbial ecology. *Annu. Rev. Microbiol.* 59:257–77

Weller DM, Raaijmakers JM, Gardener BBM, Thomashow LS. 2002. Microbial populations responsible for specific soil suppressiveness to plant pathogens. *Annu. Rev. Phytopathol.* 40:309–48

The Role of Hybridization in the Evolution of Reef Corals

Bette L. Willis,[1] Madeleine J.H. van Oppen,[2] David J. Miller,[3] Steve V. Vollmer,[4] and David J. Ayre[5]

[1]Centre of Excellence for Coral Reef Studies, School of Marine Biology and Aquaculture, James Cook University, Townsville, Queensland 4811, Australia; email: Bette.Willis@jcu.edu.au

[2]Australian Institute of Marine Science, Townsville MC, Queensland 4810, Australia; email: m.vanoppen@aims.gov.au

[3]Comparative Genomics Center, James Cook University, Townsville, Queensland 4811, Australia; email: David.Miller@jcu.edu.au

[4]Smithsonian Tropical Research Institute, Balboa, Panama: email: vollmers@si.edu

[5]School of Biological Sciences, University of Wollongong, New South Wales 2522, Australia; email: dja@uow.edu.au

Key Words

Acropora, hybridfitness, introgression, reticulate evolution, sperm choice crosses

Abstract

The importance of hybridization in the evolution of plant species is widely accepted, but its contributions to animal species evolution remain less recognized. Here we review evidence that hybridization has contributed to the evolution of reef corals, a group underpinning the coral reef ecosystem. Increasingly threatened by human and climate-related impacts, there is need to understand the evolutionary processes that have given rise to their diversity and contribute to their resilience. Reticulate evolutionary pathways among the ecologically prominent, mass-spawning genus *Acropora* suggest that hybridization, although rare on ecological timescales, has been instrumental in their diversification on evolutionary timescales. Evidence that coral hybrids colonize marginal habitats distinct from those of parental species' and that hybridization may be more frequent at peripheral boundaries of species' ranges supports a role for hybridization in range expansion and adaptation to changing environments. We conclude that outcomes of hybridization are significant for the future resilience of reef corals and warrant inclusion in conservation strategies.

INTRODUCTION

Hybridization has long been considered a major creative force in the evolution of plant species (e.g., Anderson 1949, Anderson & Stebbins 1954, Grant 1981, Stebbins 1959), but traditionally, its role in the evolution of animal species has been thought to be inconsequential, or at most, significant only in terms of reinforcing barriers to interspecific fertilization (Dobzhansky 1937, Mayr 1963). However, as predicted in a comparatively recent review of animal hybridization (Dowling & Secor 1997), evidence is mounting that hybridization is more common among animal species than previously thought and occurs in the majority of phyla in both terrestrial and marine habitats (reviewed in Arnold 1997, Gardner 1997, Mallet 2005). Phylogenetic hotspots, where (as in British vascular plants) approximately 25% of animal species hybridize, include British duck and game bird species, and American warblers and butterflies, but overall, 6–12% of bird, butterfly, and mammal species hybridize naturally (Mallet 2005). There are comparatively few studies of hybridization in modular, sessile marine animals (Gardner 1997), whose modes of speciation are more likely to resemble those of plants given similarities in their life histories. For example, both have sessile modular adults, frequent hermaphroditism, absence of mating behavior, broadcast spawning of gametes, passive dispersal of juveniles, and extensive asexual reproduction. Further studies of animals with plant-like life histories will enhance current understanding of the evolutionary significance of hybridization in animals.

On tropical coral reefs, the simultaneous mass spawning of many species of stony corals (Harrison et al. 1984) represents a unique breeding strategy among animals and suggests that hybridization might have played a role in the evolution of this functionally important group, the cornerstone of the coral reef ecosystem. In the past two decades, an upsurge of studies on the reproduction of scleractinian corals has shown that synchronized spawning among more than two species (i.e., mass spawning sensu Willis et al. 1985) occurs in the majority of reef regions (e.g., Baird et al. 2002, Carroll et al. 2006, de Graaf et al. 1999, Gittings et al. 1992, Guest et al. 2002, Hayashibara et al. 1993, Simpson 1991). In highly synchronized events, up to 35 species in sympatry may spawn within two hours of each other (e.g., mass spawning of corals on Australia's Great Barrier Reef (GBR); Babcock et al. 1986, Willis et al. 1985). In combination, the apparent absence of temporal barriers to interspecific breeding provided by mass spawning events, the co-occurrence of large numbers of coral species on reefs where currents mix positively buoyant gametes in a thin layer at the sea surface, and the reliance of the mate recognition system on interactions among gametes for assortative fertilization (Palumbi 1994) provide outstanding opportunities for hybridization. Concurrence in the timing of spawning among many coral species and the global nature of the mass-spawning phenomenon suggest that the Scleractinia provide a novel system for gaining insights into the role that hybridization has played in the evolution of animal species.

Understanding the evolutionary processes that have shaped modern reef corals and given rise to their diversity has become urgent as the number and intensity of threats to the biodiversity and resilience of coral reef ecosystems escalate against a backdrop of global climate change (e.g., Harvell et al. 2002, Hoegh-Guldberg 1999, Hughes et al. 2003). Estimates that recent declines in coral cover and colony numbers have been greater than 97% for two of the three Atlantic species of the once-common coral genus *Acropora* and predictions that the downward trajectory may continue (*Acropora* Biological Review Team 2005) have led to their listing as threatened and endangered wildlife under the U.S. Endangered Species Act (ESA). However, evidence that the third coral, *Acropora prolifera*, is a hybrid and thus fails to meet the ESA definition of a species disqualifies it from consideration for protection under the act. If appropriate management strategies are to be developed for the Scleractinia, greater understanding of the evolutionary processes giving rise to their biodiversity is needed, particularly in relation to the role that hybridization has played in adaptive radiation and evolution within the group. The recent publication of a body of reproductive, morphological, ecological, and molecular data for mass-spawning corals in both the Indo-Pacific and Caribbean (see references in sections below) provides new perspectives on coral evolution and has prompted the present review of the evolutionary importance of hybridization within the Scleractinia.

HYBRIDIZATION AS AN EVOLUTIONARY PROCESS

Introgressive hybridization, here defined as the exchange of genes between genetically differentiated species, may have a variety of contrasting outcomes. Extensive and prolonged episodes of introgressive hybridization may lead to the merging of species and ultimately to the extinction of pure parental species along with their morphological, behavioral and/or ecological distinctions. This process of genetic mixing may provide increased genetic diversity, new traits, and heterosis for the emerging species, but may carry the cost of a net loss of species diversity. Regardless of contrasting perspectives on the fate of species following merging and homogenization of gene pools (i.e., the loss of current species versus the emergence of new ones), the comparative speed with which new traits and species can arise suggests that, in some cases, there may be positive selective pressure for hybridization and introgressive events [e.g., among diversifying species of sunflowers (Ungerer et al. 1998) and cyprinid fish (Rosenfield & Kodric-Brown 2003)].

Hybridization may also give rise to new species that are reproductively isolated from parental species through either polyploid speciation or recombinational speciation, both processes representing mechanisms for the rapid evolution of genetic novelty without the loss of parental species. Polyploid speciation involves the production of allopolyploids, which by definition contain three or more sets of chromosomes from two different species and thus achieve immediate reproductive isolation from parental species. Allopolyploid species are common in plants (Grant 1981, Ramsey & Schemske 1998, Soltis & Soltis 1999) and account for 2–4% of angiosperm species and 7% of fern species (Otto & Whitton 2000). In studies so far, animal allopolyploids are much rarer and are predominantly parthenogenic rather than sexually reproducing (Bullini 1994, Dowling & Secor 1997, Otto & Whitton 2000, White 1978). However, because uniparental species may contribute sexual propagules capable of introgressing with other species (e.g., diploid, hemiclonal species in the hybridogenetic fish

Poeciliopsis; Mateos & Vrijenhoek 2002) and may have greater distributional ranges than their bisexual ancestors (e.g., *Bacillus* stick insects, Bullini 1994), lack of sexual viability is not grounds for dismissing the evolutionary potential of asexual hybrid lineages.

Recombinational species (Grant 1981), also called homoploid hybrid species, are hybrid species that have the same number of chromosomes as their parent species but are reproductively isolated from them and are true breeding (Grant 1981, Rieseberg 1997). The number of clear-cut cases of recombinational speciation identified suggests that it may be rare in both plants and animals (Coyne & Orr 2004, Rieseberg 1997). For example, Rieseberg (1997) identified only eight well-documented cases of recombinational speciation in plants, three of which are *Helianthus* sunflower species. However, by its very nature, recombinational speciation, which requires that fit, interfertile recombinant hybrid genotypes be produced that are reproductively isolated from their sympatric parent species, is difficult to identify (Rieseberg 1997, Coyne & Orr 2004). The discovery of increasing numbers of diploid hybrid species with the advent of improved molecular techniques suggests that the real number of cases may be much higher than generally thought (Seehausen 2004). Seehausen argues that hybridization is particularly significant where ecological conditions favor adaptive radiations, such as where there are underutilized niches following disturbance or at the ecological or geographical peripheries of species ranges, and based on the prevalence of hybridization in studies of more recent adaptive radiations [e.g., Darwin's finches (Freeland & Boag 1999), African cichlids (Salzburger et al. 2002), Hawaiian crickets (Shaw 2002)], its contributions to past species diversifications have been underestimated. Evidence of transgressive segregation, in which phenotypic variation in hybrid populations is greater than the combined variation of parent populations (e.g., enhanced salt tolerance of hybrid *Helianthus* sunflowers through acquisition of different parental loci with additive effects; Lexer et al. 2003), provides a mechanism by which recombinational species can have greater fitness than parental species. When recombinational speciation does occur, empirical and theoretical evidence suggests that it does so rapidly and is often accompanied by ecological divergence that promotes reproductive isolation (Buerkle et al. 2000, McCarthy et al. 1995, Rieseberg et al. 2003). This further impedes the detection of past hybridization events in the origin and transfer of novel adaptations. Although hybrid zones provide valuable microcosms in which current, ongoing evolutionary processes can be studied (Hewitt 2001), the lack of a similar window on past hybridization events makes more distant contributions of hybridization events difficult to detect.

There is wide acceptance among botanists that the transfer of genes between taxa through introgressive hybridization potentially provides more raw material for evolution than can be produced directly by mutation (Anderson 1949, Anderson & Stebbins 1954, Arnold 1997), and hybridization is estimated to have produced a significant fraction (∼11%) of flowering plant lineages (Rieseberg 1997). Even though cases of hybridization in animals continue to be described, the prevalent view is that hybridization has not played a major creative role in the evolution of animal species. This may be partly due to difficulties in identifying the products of hybrid speciation in past events. Even in contemporary studies, hybrid taxa or lineages may be difficult to detect because they may be much closer genetically or morphologically to one parental species than the other (Jiggins & Mallet 2000, Naisbit et al. 2003). Moreover, establishing that hybrid lineages constitute distinct evolutionary entities that are reproductively isolated from parent species requires multiple lines of evidence including morphological, reproductive, molecular, and chromosomal karyotype data. Without such evidence of effective reproductive isolation, hybrid lineages are hybrids and not hybrid species. In summary, the variety of ways in which hybridization may contribute to the evolution of lineages attests to its scope as an evolutionary process, but evidence of its past contributions may be elusive.

MATING SYSTEMS OF MASS-SPAWNING CORALS

The mating systems of corals are based on either broadcast spawning of gametes for external fertilization or internal fertilization followed by larval brooding, with approximately two thirds of species ($n = 227$) utilizing the former, external mode of development (Harrison & Wallace 1990). The breeding periods of broadcast-spawning corals may be temporally isolated on scales that range from hours (Fukami et al. 2003, Knowlton et al. 1997, Levitan et al. 2004, van Oppen et al. 2002) to weeks or months (Wolstenholme 2004; B.L. Willis, unpublished data), but most corals spawn in tightly synchronized breeding events that overlap with those of a number of species, most commonly congeners (Harrison & Wallace 1990, Richmond & Hunter 1990). Opportunities for interspecific breeding in the mating systems of corals thus vary in response to breeding times that range from complete temporal isolation to complete overlap, with the dominant breeding strategy of synchronized mass spawning providing unparalleled potential for hybridization. Interpreting this potential for hybridization, however, requires knowledge of both prezygotic (including fine-scale temporal barriers and gamete incompatibility) and postzygotic isolating barriers.

Next we review isolating barriers for the dominant reef-building genera in two of the major ocean basins: the *Acropora* in the Indo-Pacific and the *Montastraea* in the Western Atlantic. The Indo-Pacific *Acropora* constitute a highly diverse group of more than 100 species, up to 76 of which can occur in sympatry (Wallace 1999; **Figure 1a**) and at least 35 of which are known to participate in synchronized mass-spawning events (Babcock et al. 1986, Willis et al. 1985). This genus has undergone the greatest adaptive radiation of any scleractinian genus in the Indo-Pacific, where it commonly dominates coral communities. In contrast, there are only 3 extant species of *Acropora* in the Western Atlantic (**Figure 1b**), indicating that different processes have shaped the evolution of this genus in these two biogeographic regions.

Prezygotic Isolating Barriers: Mating Precedence in Sperm Choice Experiments

Tests of gamete compatibility in laboratory crosses between mass-spawning species have had variable outcomes. Although the data are as yet fragmentary, there are some basic differences in interspecific gamete compatibility between those Indo-Pacific and Caribbean corals for which data are available. Among species of *Acropora*

Figure 1

Comparison of *Acropora* (Order: Scleractinia) assemblages on (*a*) Indo-Pacific reefs, where the genus attains the greatest diversity (more than 100 species) of all extant corals and typically dominates coral communities (almost all corals in the assemblage pictured are species of *Acropora*); and (*b*) Caribbean reefs, where there are only 3 extant species: *A. cervicornis* (*top right*); *A. palmata* (*bottom right*); and the hybrid *A. prolifera* (*bottom left*).

from the central GBR, outcomes of colony crosses between 38 pairs of species have spanned the complete spectrum of fertilization success from high (50–100%) between 8 pairs, through moderate (10–50%) between 7 pairs, to low (3–10%) between a further 3 pairs of species (van Oppen et al. 2002, Willis et al. 1997). Overall, one third of species pairs of Indo-Pacific *Acropora* crossed experimentally (n = 73 species combinations) have resulted in greater than 10% fertilization in some colony pairings, based on studies combined from the central GBR (van Oppen et al. 2002, Willis et al. 1997), northern GBR (Wolstenholme 2004), and Japan (Fukami et al. 2003, Hatta et al. 1999). The fraction rises to more than 45% if species pairs with low gamete compatibility (3–10% interspecific fertilization success) in at least some colony crosses are included. Thus the capacity to hybridize appears to be a common feature of mating systems in the Indo-Pacific *Acropora*, and this feature is consistent over large geographic areas. Similarly, breeding trials between seven morphospecies in the Indo-Pacific genus *Platygyra* have shown high levels of interspecific fertilization on the GBR (Miller & Babcock 1997, Willis et al. 1997). The lack of genetic distinctiveness among morphospecies of *Platygyra* in allozyme studies (Miller & Benzie 1997) suggests that hybridization may be more common in this genus than in the Indo-Pacific *Acropora*.

Although laboratory crosses in which eggs are provided only with the opportunity to mate with heterospecific sperm demonstrate that gametes are compatible, significant variation in the relative frequency of inter- and intraspecific fertilization success and dramatic variation among individual pairs of colonies (from 0% to 100% in some species crosses; Fukami et al. 2003, van Oppen et al. 2002, Willis et al. 1997) imply that prezygotic isolating barriers may exist among the *Acropora*. In order to test for evidence of premating isolation under conditions that more closely resemble in situ mass-spawning events, we carried out a series of sperm-choice breeding trials. We incubated eggs with both conspecific and heterospecific sperm in lab crosses, where sperm came from parents (*Acropora millepora* and *A. pulchra*) that were alternately homozygous for allozyme genotypes. Some 14 different sperm-choice crosses were performed, involving 3 cases in which *A. millepora* was the mother and 11 cases in which *A. pulchra* was the mother. Electrophoretic analyses of individual larval offspring demonstrated that, in the 11 trials in which *A. pulchra* was the mother, all offspring (n = 475 larvae) tested were the result of intraspecific fertilizations. Similarly, in 2 of 3 sperm-choice trials involving *A. millepora* as the mother, all offspring (n = 117 larvae) tested were purebreds. However, in one trial where *A. millepora* was the mother, all offspring (n = 120) tested were *A. millepora* × *A. pulchra* hybrids. It is conceivable that the colony used as the source of intraspecific sperm in this latter trial (which resulted in hybrid offspring) had been misidentified and was actually *A. spathulata*, a cryptic species that has recently been separated from *A. millepora* (Wallace 1999) on the basis of incompatibility in breeding trials (Willis et al. 1997). However, either hybridization was favored over conspecific fertilization (assuming that the fathers were correctly identified as *A. millepora* and *A. pulchra*) or, given a choice between sperm from two different species (*A. spathulata* and *A. pulchra*), *A. millepora* eggs were fertilized by the species with the more compatible sperm. These results suggest that, although eggs of many Indo-Pacific species of *Acropora* are compatible with heterospecific sperm in no-choice lab crosses, prezygotic isolating mechanisms exist that favor conspecific mating in these species at least most of the time. Species specificity in gamete recognition and binding proteins has been found in other marine invertebrates, for example, in abalone (Vacquier et al. 1990) and sea urchins (Metz & Palumbi 1996) and would be a fruitful area of research to further understand the mating systems of mass-spawning corals.

Because many species of *Acropora* are able to hybridize in no-choice experimental crosses, the lack of hybrid offspring produced in the presence of conspecific sperm in the few sperm-choice trials so far performed suggests that prezygotic isolating barriers are semipermeable based on a hierarchy in gamete compatibility. Decreasing gamete compatibility, in rank order from intraspecific to closely related interspecific, and lastly to more distantly related interspecific matings, offers an alternative avenue for reproductive success if mates are scarce, such as in marginal or disturbed environments. At the same time, precedence of conspecific mating limits gene flow between species, thus maintaining species boundaries and morphologically recognizable species. Regardless of whether the capacity for hybridization under these circumstances is an artifact of their mating systems or a selected feature, the net result is a form of bet-hedging. Although interspecific pairings can lead to gamete wastage, there is ample evidence that they may also result in offspring that, if not sexually viable, are nevertheless able to found clonal lineages that may be ecologically successful (Bullini 1994, Kearney 2005, Spolsky et al. 1992).

In the Caribbean, there are no published studies of interspecific gamete compatibility for the genus *Acropora*, but studies of the *Montastraea* species complex have found generally low interspecific gamete compatibility between species that have overlapping spawning times. In Panama, evidence of gamete incompatibility in laboratory crosses provides corroborative evidence that *M. faveolata*, previously thought to be a morphological variant of *M. annularis*, is a cryptic species (Knowlton et al. 1997, Levitan et al. 2004). A second cryptic species (*M. franksi*) had moderate (~40–60%) gamete compatibility with *M. annularis*, but gametes age rapidly so that fine-scale (two hours) temporal isolation in spawning times provides a reasonably effective reproductive barrier. Significantly, the two species that have the more compatible gametes and are most closely related genetically (i.e., *M. franksi* and *M. annularis*) show strong temporal isolation, whereas the two species that spawn together (*M. annularis* and *M. faveolata*) have incompatible gametes and are genetically most different (Fukami et al. 2004, Levitan et al. 2004). Thus a suite of isolating barriers maintains species boundaries in this group in Panama.

Contrasting patterns of gamete compatibility were found at a site near the northern geographical periphery of the ranges of these western Atlantic species. In the Florida Keys, *M. faveolata* hybridized well with both *M. franksi* and *M. annularis* (Szmant et al. 1997). Genetic differences among these three taxa were also weaker in the Bahamas than in Panama (Fukami et al. 2004), potentially reflecting the occasional moderate to high numbers of viable larvae found in some experimental crosses, particularly with aged eggs, between pairs of these three species (Fukami et al. 2004). Similarly, consistent differences in patterns of corallite morphology were detected among these three species across their distributional ranges. Distinct corallite morphologies among all three species in Panama, in contrast to overlapping corallite morphologies in the Bahamas, led Fukami et al. (2004) to speculate that a gradient in hybridization exists; the strongest introgression occurred at northern sites. Differences in gamete compatibility and morphologies among the same species at different geographic locations provide support for Veron's (1995) suggestion that species may interact differently through hybridization across their ranges. Morphological analyses of fossil specimens from Panama and the Bahamas suggest that present patterns of greatest introgressive hybridization and least morphological differentiation among *Montastraea* species in the Bahamas have persisted throughout the geological history of this species complex (Budd & Pandolfi 2004). Such persistent differences in the extent of interspecific breeding over broad geographic ranges throughout evolutionary histories are consistent with the notion that semipermeable barriers at the extremities of species' ranges afford increased opportunities for mating.

Patterns of temporal barriers to interbreeding in relation to species relatedness within the Caribbean *Montastraea* are the inverse of those within the Indo-Pacific *Acropora*; temporal isolating barriers are strongest between the most closely related species of Caribbean *Montastraea* (i.e., *M. franksi* and *M. annularis*; Levitan et al. 2004) but weakest between closely related species of Indo-Pacific *Acropora*, where temporal barriers reflect the greatest genetic divergence among species (Fukami et al. 2003; van Oppen et al. 2001, 2002; Wolstenholme 2004). Similarly, patterns of synchronous spawning with gamete compatibility differ between the two reef regions; synchronous spawning in Panama occurs between *Montastraea* species with the least compatible gametes (Levitan et al. 2004), whereas many Indo-Pacific *Acropora* and *Platygyra* species that spawn synchronously have compatible gametes (Miller & Babcock 1997, Fukami et al. 2003, Willis et al. 1997). Gametes of at least some Indo-Pacific species of *Platygyra* and *Acropora* remain competent for a significantly longer period of time (6–8 hours post spawning; Miller & Babcock 1997, Willis et al. 1997) than do those of their Caribbean counterparts (1–2 hours, Levitan et al. 2004), contributing to differences between the two regions in the degree of temporal isolation among species that spawn on the same night.

Postzygotic Isolating Barriers: Developmental Competence of Coral Hybrids

Studies of postzygotic isolating barriers among coral species are scarce, largely because of their long pre-reproductive periods (Harrison & Wallace 1990). In the Indo-Pacific, hybrid offspring from two different species crosses (*Acropora millepora* × *A. pulchra* and *A. hyacinthus* × *A. cytherea*) are developmentally competent and have been successfully reared for three years (**Figure 2**). Moreover, in a large-scale grow-out program on Orpheus Island reefs in the central GBR, patterns in the survival and growth of three-month-old hybrids of *A. millepora* × *A. pulchra* in three habitats (i.e., the intertidal, inner reef-flat habitat of *A. pulchra*, the reef-slope habitat of *A. millepora*, and an intermediate reef-crest habitat) indicate that, at this early life history stage, their fitness does not differ from that of purebred offspring reared in the reef-flat and reef-crest habitats (**Figure 3**). Generally, purebred offspring survived best in the habitat where the corresponding parental species was most abundant. Thus offspring

Figure 2

Three-year-old coral offspring produced from experimental crosses demonstrate that coral hybrids are developmentally competent. (*a*) *Acropora millepora* purebred juvenile; (*b*) *A. pulchra* purebred juvenile; (*c*) *A. pulchra* (mother) × *A. millepora* (father) hybrid juvenile; (*d*) adult *A. millepora* showing typical corymbose (pillow) morphology and subtidal, reef-slope habitat; and (*e*) adult *A. pulchra* showing typical arborescent (branching) morphology and intertidal, reef-flat habitat. Offspring were reared in flow-through tanks at Orpheus Island Research Station, Central Great Barrier Reef, but died in a bleaching event before reproductive competency could be assessed.

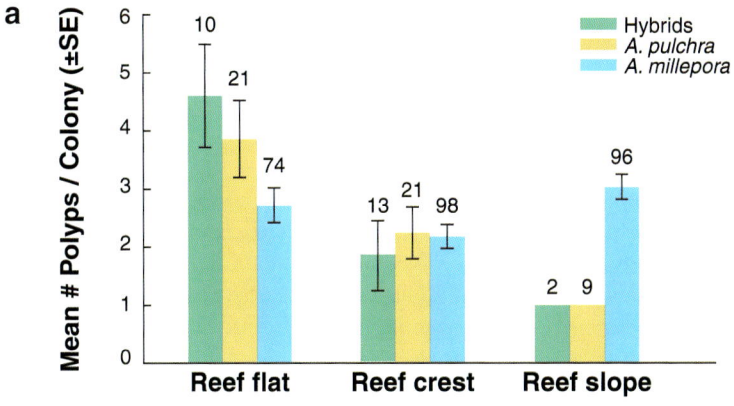

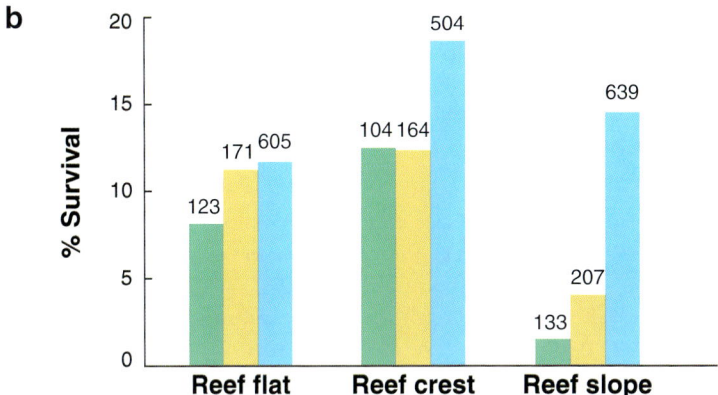

Figure 3

Comparative mean growth ± SE (*a*) and survival (*b*) at 3 months among purebred and hybrid offspring of *Acropora millepora* and *A. pulchra* produced in experimental crosses and outplanted to three habitats (reef flat, crest, and slope) on northeast Orpheus Island reef, central Great Barrier Reef (see text for details of parental habitats). Numbers above histograms represent sample sizes: n = # offspring at census 2 (growth); n = # offspring at time 0 (survival).

of *A. pulchra* survived best on the inner reef flat and reef crest, which are the habitats to which it is largely restricted as an adult. Moreover, *A. pulchra* offspring grew fastest on the inner reef flat (**Figure 3**), where the species is most abundant and typically forms monospecific stands. There was a trend toward poorest survival of *A. millepora* offspring in the intertidal reef-flat habitat, where adults are typically not found, although their growth in the first three months was not markedly different among the three habitats. Interestingly, patterns of hybrid offspring survival and growth were most similar to those of *A. pulchra*. Hybrids grew faster than *A. millepora* on the inner reef flat (**Figure 3a**) and also survived better there and on the reef crest than on the reef slope. Although results for F_2-generation hybrids are unknown, results for F_1 hybrids are consistent with a review of hybrid fitness by Arnold & Hodges (1995), who found that existing evidence does not support a general pattern of reduced hybrid fitness and concluded that the role of hybridization in the process of evolutionary diversification has been underestimated. The greater growth and survival of hybrids in the more environmentally variable reef-flat and reef-crest habitats suggest a potential role for hybrids in providing a source of variation for adaptation to new or extreme environments.

In summary, there are few apparent pre- or postzygotic isolating mechanisms that would preclude hybridization from contributing to the evolutionary diversification of at least some of the Indo-Pacific mass-spawning *Acropora*, particularly when mating opportunities with conspecifics are scarce. However, parallel studies suggest that hybridization has played a different role in the evolution of the dominant Caribbean coral genera. Within the Caribbean *Acropora*, it has been suggested that postzygotic selection against hybrid genotypes (reviewed below) may be acting as a strong filter against genetic mixing and hence the potential for further evolutionary diversification among the three Caribbean species (Vollmer & Palumbi 2002). Similarly, a variety of mechanisms acting in concert are thought to render hybridization among the three species in the *Montastraea* complex unlikely in Panama (Levitan et al. 2004), although it appears to have played a greater role in the Bahamas (Fukami et al. 2004). Studies of the fitness of hybrid offspring would shed further light on the nature of postzygotic isolating mechanisms within these two Caribbean species complexes.

MOLECULAR EVIDENCE OF HYBRIDIZATION IN CORALS

Hybrid Origin of Coral Species in the Caribbean *Acropora*: *A. prolifera* Case Study

The only scleractinian coral known to be of hybrid origin, so far, is the Caribbean species *Acropora prolifera*. Three species of *Acropora* are known from the Caribbean, all of which are endemic: *A. cervicornis*, *A. palmata* and *A. prolifera* [collectively referred to as the *A. cervicornis* group (Wallace 1999)]. *A. cervicornis* and *A. palmata* are sister species with good fossil records, the earliest fossils being approximately 6.6 (Budd & Johnson 1999) and 3.6–2.6 (McNeill et al. 1997) Myr old respectively. In contrast, *A. prolifera* is of recent (Holocene) origin and has no fossil record (Budd et al. 1994). The three species are differentiated by colony growth form (**Figure 1b**) and habitat preference (Cairns 1982, Rützler & Macintyre 1982). *Acropora cervicornis* has an arborescent, staghorn morphology and occurs in comparatively sheltered fore- and back-reef habitats. *Acropora palmata* has a robust, elkhorn morphology and occurs primarily in more exposed, reef-crest habitats. The third described species, *Acropora prolifera*, has a fused-branched morphology that is intermediate between the two species, and has been further differentiated into co-occurring palmate and bushy morphs at one location (in Puerto Rico, Vollmer & Palumbi 2002). It is rarer than the other two Caribbean *Acropora* species and tends to occur in marginal, shallow-water, back-reef and reef-crest habitats that the other species do not occupy. All three species are sympatric throughout the Caribbean (Adey et al. 1977, Budd et al. 1994, Goreau 1959, Rützler & Macintyre 1982, Wallace 1999) and, although they generally occupy different reef zones, their depth distributions may overlap (Adey et al. 1977, Goreau 1959). Despite their distinct morphologies and ecological niches, the taxonomic status of these three species has been debated for over a century (Cairns 1982, Gregory 1895, Vaughan 1901, Wells 1973). The intermediate morphology of *A. prolifera* and the restriction of its distribution to reefs on which both *A. palmata* and *A. cervicornis* co-occur suggest that it may be a hybrid of the other two species.

The putative parent species, *Acropora palmata* and *A. cervicornis*, spawn synchronously once or twice a year in August and/or September (de Graaf et al. 1999, Szmant 1986; S.V. Vollmer, personal observation) and have compatible gametes in experimental crosses (B.L. Willis, unpublished data); thus opportunities for interspecific hybridization exist. Genetic data confirm that *A. prolifera* has indeed had a hybrid origin (van Oppen et al. 2000, Vollmer & Palumbi 2002). Allele frequencies of a nuclear intron (*PaxC* 46/47 intron) were significantly different between sympatric colonies of the parental species; *A. cervicornis* carried two alleles that were not present in *A. palmata* at a frequency of 0.769 and 0.039, respectively, confirming that they are distinct species (van Oppen et al. 2000). All colonies of *A. prolifera* were heterozygous for this locus, the expected outcome if two species with very distinct allele complements and frequencies (i.e., *A. palmata* and *A. cervicornis*) hybridize. Ribosomal DNA ITS sequence types were shared among all three species, confirming that *A. prolifera* is a hybrid between the other two species that backcrosses at low frequency. At two additional nuclear loci, all *A. prolifera* were also heterozygous for both species' alleles, indicating that *A. prolifera* colonies are most likely first generation (F_1) hybrids (Vollmer & Palumbi 2002, 2004). Moreover, maternally inherited mitochondrial DNA demonstrated that hybridization occurs in both directions (Vollmer & Palumbi 2002). Three mtDNA haplotypes were observed (A, B, and C), with haplotypes A and C occurring only in *A. cervicornis* and *A. prolifera*, whereas haplotype B occurs in all three species. Using phylogenies and Bayesian coalescent models, Vollmer & Palumbi (2002) demonstrated that backcrossing occurs with only one of the parental species, *A. cervicornis*, thereby passing genes from *A. palmata* to *A. cervicornis*. Even limited unidirectional flow of genes may represent a significant mechanism for evolutionary change (Mallet 2005), thus by providing a conduit for gene flow; *A. prolifera* may provide a mechanism for the acquisition of novel genes for at least one (*A. cervicornis*) of the Caribbean *Acropora*.

Interestingly, *A. prolifera* lacks a fossil record (Budd et al. 1994) suggesting that either hybridization between *A. palmata* and *A. cervicornis* is a relatively recent phenomenon or that environmental conditions have only recently favored survival of the hybrid *A. prolifera*. The current Caribbean-wide distribution of *A. prolifera* (Veron 2000, Wallace 1999) suggests that whatever triggers hybridization between its parental species has operated over a large geographic scale.

A Polyploid or Recombinational Hybrid Species?

Karyotype data for corals are few (Heyward 1985, Kenyon 1997); however, there is evidence of polyploidy in multiple species groups of Pacific *Acropora* (Kenyon 1997). In several scleractinian genera (*Acropora, Fungia, Gonipora, Lobophyllia,* and *Montipora*), the diploid chromosomal number is 28 (Heyward 1985, Kenyon 1997). Of 22 *Acropora* species sampled, 3 apparent cases of polyploidy were identified—*A. elseyi, A. ocellata,* and *A. valida*. The last of these 3 species is a probable triploid, whereas the others are likely tetraploids (Kenyon 1997). Without additional molecular or morphological data, it is not clear if these 3 species are allopolyploid hybrid species or autopolyploids (i.e., polyploids produced by the same species). However, the number of polyploid species found in the single study and aforementioned evidence for hybridization implies the existence of polyploid hybrids among Indo-Pacific *Acropora* species and suggests that this phenomenon may not be uncommon among corals.

Although no karyotype data are yet available for the three Caribbean *Acropora*, *A. prolifera* is unlikely to be polyploid because hybrids never had more than two alleles at single copy nuclear loci (S.V. Vollmer, unpublished data). Although known to be of hybrid origin, it is unclear whether *A. prolifera* is reproductively isolated from its parent species. In laboratory crosses, *A. prolifera* is both self-fertile and cross-fertile with *A. cervicornis* (backcrosses with *A. palmata* have not yet been completed; S.V. Vollmer, unpublished data). However, genetic evidence showing limited one-way gene flow from *A. palmata* to *A. cervicornis* implies that hybrids backcross with *A. cervicornis* at only low levels (Vollmer & Palumbi 2002). Both lines of evidence suggest that *A. prolifera* is not reproductively isolated from *A. cervicornis*. Moreover, the fact that all colonies of the hybrid *A. prolifera* so far tested appear to be F_1 hybrids (Vollmer & Palumbi 2002) suggests that hybrids are not interbreeding at high frequency. Thus, the available genetic and reproductive evidence suggest that *A. prolifera* is a hybrid and not a hybrid species.

Although *A. prolifera* may not strictly constitute a recombinational hybrid species, its ability to propagate clonally via asexual fragmentation allows these coral hybrids to persist locally, potentially over long periods of time (Vollmer & Palumbi 2002). The ability to persist through asexual propagation makes coral hybrids similar to clonal parthenogenic taxa (Miller & van Oppen 2003) that have been shown to be ecologically successful in a variety of animal groups (Bullini 1994, Vrijenhoek 1984). Indeed, *A. prolifera*, although generally rare, is most abundant in marginal and shallow water environments where neither of the parental species occurs (Cairns 1982; S.V. Vollmer, personal observation). Thus, these hybrids appear to be successful in an ecological niche that is distinct from that of the parent species. Similar habitat diversification is shown to be important in the evolution of plant hybrid species (Arnold 1997, Rieseberg et al. 2003). The recent detection of *A. prolifera*'s hybrid status raises the possibility that further molecular studies will reveal other coral morphospecies to be either hybrids or hybrid species that have diversified to occupy new habitats (Vollmer & Palumbi 2002). Recent evidence that the European temperate soft coral *Alcyonium hibernicum* harbors the ITS variants of two congeners, *A. coralloides* and *A.* sp. *M2*, and consequently is a hybrid of the two species (McFadden & Hutchinson 2004) supports this view. Despite its exclusively asexual mode of reproduction, *A. hibernicum* has a considerable distribution beyond the northern margins of the putative parent species' distributions, highlighting the potential role that hybrid lineages may have for range expansion beyond the limits of parental species' distributions. If climate change and other human-related pressures on Caribbean reefs continue at present levels or escalate, the survival of coral species may be increasingly dependent on their ability to colonize marginal habitats, highlighting the potentially increasing significance of hybrids and hybrid species.

Para- and Polyphyly in Molecular Phylogenies of Indo-Pacific Mass-Spawning Species of *Acropora*

In contrast to the low diversity of the Caribbean *Acropora*, there are more than 100 species of Indo-Pacific *Acropora* and up to 76 of these have been recorded from a single area (Wallace 1999). This complexity effectively precludes direct extrapolation of conclusions drawn on the basis of the much simpler, three-species Caribbean system to the Indo-Pacific *Acropora*. In particular, unlike the Caribbean fauna, there are no clear cases of species with morphologies intermediate between two putative parent species and, so far, molecular approaches have identified only two likely first- or early-generation hybrids (van Oppen et al. 2002, Wolstenholme 2004). The lack of obvious morphological hybrids may be partly explained by high levels of morphological variation within many coral species, which makes it difficult to locate species borders within morphological space, a prerequisite for distinguishing potential hybrids from intraspecific variation. Recently, however, morphs that have affinities to two species have been identified within the *A. humilis* group (Wolstenholme 2004, Wolstenholme et al. 2003). Evidence from molecular and reproductive studies suggests that species boundaries may be at various stages of formation among a group of five species and seven intermediate morphs and that some of this morphological variation may have arisen through hybridization (e.g., morphs intermediate between *A. digitifera* and *A. gemmifera*; Wolstenholme 2004). In contrast to the Caribbean, where opportunities for hybridization are limited to two species, in the Indo-Pacific, introgression between any pair of species can theoretically occur through backcrossing with a range of other species (and intermediate morphs). The high number of potentially interbreeding species, combined with intraspecific morphological variation, poses difficulties for unraveling molecular phylogenies within the Indo-Pacific *Acropora*.

Ribosomal DNA ITS phylogenies for Indo-Pacific *Acropora* spp. are consistent with interspecific hybridization and introgression in that they show sharing of highly divergent sequence types between a wide range of species (Marquez et al. 2003, Odorico & Miller 1997, van Oppen et al. 2002). A divergence of more than 20% has been observed between ITS1-5.8S-ITS2 sequences within species and even within individual corals, while at the same time some sequence variants show high similarity between species. In the *A. aspera* species group, one species, *A. aspera*, differs at least in some years in the timing of gamete release and harbors mainly a distinct and unique ITS type (van Oppen et al. 2002). The other four species that have been examined in this group spawn simultaneously every year and also share ITS types, supporting the view that synchronized spawning has led to hybridization and the sharing of ITS types. Furthermore, there is some evidence that some *A. pulchra* individuals may be recent generation hybrids (van Oppen et al. 2002). As ITS homogenization may proceed extremely slowly when divergent sequence types are combined in a single genome (Modrich & Lahue 1996), rDNA data cannot readily distinguish between ancient or recent hybridization events. Incomplete lineage sorting may further mimic hybridization signatures (Vollmer & Palumbi 2004). Nevertheless, phylogenies based on single-copy nuclear DNA (scnDNA) and mtDNA are broadly consistent with the rDNA analyses, also showing extended para- and polyphyly of a wide range of species and resolving *A. aspera* from other members of the *A. aspera* species group (Hatta et al. 1999, van Oppen et al. 2001).

Within the *A. humilis* group, sequence data from the partial nuclear ribosomal large subunit DNA regions and the mitochondrial control region suggest that sequence types are shared between some species through occasional introgression without disrupting morphological boundaries (Wolstenholme 2004, Wolstenholme et al. 2003). Morphs intermediate between true species are possibly of hybrid origin based on fertilization tests, their morphological affinities, and their phylogenetic position in molecular phylogenies (Wolstenholme 2004, Wolstenholme et al. 2003). Paraphyly has also been observed for several other *Acropora* species at scnDNA and mtDNA markers (Hatta et al. 1999, Márquez et al. 2002b, van Oppen et al. 2001). In addition, some aspects of mtDNA and scnDNA phylogenies are inconsistent. Such patterns could be explained by interspecific hybridization and introgression, incomplete lineage sorting (i.e., shared ancestral polymorphism), or a combination of both. Two lines of evidence, however, lead us to believe that hybridization is at least partly responsible. First, as enumerated above, many mass-spawning species of Indo-Pacific *Acropora* are capable of successful cross-fertilization (Fukami et al. 2003, Hatta et al. 1999, van Oppen et al. 2002, Willis et al. 1997), and second, the genetic distinctiveness of a nominal species appears to be directly correlated with the extent of temporal or other reproductive barriers (Fukami et al. 2003, van Oppen et al. 2001, 2002, Wolstenholme 2004), with species having the most effective temporal isolating mechanisms containing distinct rDNA ITS1 variants, scnDNA alleles, and mtDNA haplotypes. For example, *A. donei* and *A. yongei* in Japan, which spawn 1–3 hours before other sympatric species, constitute a genetically distinct cluster based on *Mini-C* intron 2 sequences (Fukami et al. 2003, Hatta et al. 1999). *A. digitifera* on the GBR, which spawns three months after the main mass-spawning event, constitutes a separate clade from the other four species in the *A. humilis* group based on analyses of the mtDNA intergenic region (Wolstenholme 2004). Spawning of *A. yongei* in the central GBR was correctly inferred to coincide with that of *A. tenuis*, which similarly spawns 2–3 hours before congeners (Babcock et al. 1986) based on its phylogenetic position in the *PaxC* intron and control region trees (M.J.H. van Oppen, B.L. Willis, D.J. Miller, unpublished data). Moreover, ITS diversity correlates with the permeability of isolating mechanisms; ITS1 variability within and between most Indo-Pacific *Acropora* species is much higher than that observed among the three-species Caribbean *A. cervicornis* group. Thus, ITS1 p-distances range from 0% to ~13% in the *A. cervicornis* group, but are up to four times greater in the Indo-Pacific *A. aspera* group. *A. tenuis* has even lower ITS1 diversity (0–5.3%; M.J.H. van Oppen, unpublished data) and is genetically distinct based on scnDNA and mtDNA, which is consistent with its earlier spawning time and low interspecies fertilization success in vitro (Babcock et al. 1986, Willis et al. 1997). Similarly, *A. latistella* typically (but not always) spawns two weeks out of phase with most congeners (Babcock et al. 1986, Willis et al. 1985), and this species as well as *A. aspera* (which may spawn earlier than the mass spawning) both form distinct clusters in the molecular analyses. Thus transient reproductive barriers imposed by year-to-year variation in the lunar night of spawning may also contribute to the relative genetic distinction of *Acropora* species. In summary, a clear pattern linking lack of temporal barriers to breeding with nonmonophyly in molecular phylogenies helps to build the case that hybridization has contributed to the evolution of the mass-spawning, Indo-Pacific *Acropora*.

Population Genetic Approaches Show that Some Nonmonophyletic Species are Genetically Distinct

Molecular phylogenies are extremely powerful tools for identifying likely cases of natural hybridization, but additional approaches are required to corroborate the occurrence and extent of introgressive hybridization. Although allele sharing in phylogenetic trees of the Indo-Pacific *Acropora* is strongly indicative of natural hybridization and introgression within the genus, such evidence alone cannot distinguish occasional hybridization between true species from the existence of a single, morphologically polymorphic species or from cases of incomplete lineage sorting. Because differences in allele frequencies occur more rapidly than mutational changes underlying phylogenetic analyses, population genetic approaches provide further insights into whether closely related cross-fertile species share the same gene pool or are connected through limited hybridization and introgression.

Population genetic studies of two corals, *A. cytherea* and *A. hyacinthus*, which have highest interspecific fertility in experimental crosses (Willis et al. 1997) and share alleles extensively in phylogenetic analyses (Márquez et al. 2002b), demonstrate that these two taxa do not represent a single morphologically plastic species (Márquez et al. 2002a). Although very low, levels of genetic differentiation between these two species in sympatry were significant at eight polymorphic allozyme loci. Higher levels of gene flow between conspecific, widely allopatric (eastern versus western Australian) populations of these species in comparison to interspecific gene flow between them in sympatry support the conclusion that *A. cytherea* and *A. hyacinthus* constitute distinct entities (Márquez et al. 2002a). The question of whether incomplete lineage sorting is responsible for the small genetic divergences is more difficult to address. Based on the absence of a fossil record for either *A. cytherea* or *A. hyacinthus* prior to the Pleistocene (Wallace 1999), both species are assumed to be of relatively recent origin. The lack of fixed allelic differences and low allele frequency differences between these two species are both consistent with natural hybridization and introgression (i.e., interspecific gene flow following secondary contact) occurring infrequently, but could also be explained by them being incipient species with incomplete reproductive barriers and retention of ancestral polymorphisms (Márquez et al. 2002b).

Low, but distinct genetic differentiation in population genetic studies of the *A. nasuta* group (MacKenzie 2005) provides a second example of infrequent hybridization events potentially contributing to the evolution of species in the Indo-Pacific *Acropora*. Three species in this group, *A. nasuta*, *A. valida*, and *A. secale*, showed highly significant indices of pairwise genetic differentiation, both in sympatry and in allopatry (F_{st} values based on two nuclear introns and one microsatellite locus ranged from 0.08–0.09, 0.24–0.37, and 0.38–0.42 between *A. secale* and *A. valida*, *A. nasuta* and *A. valida*, and *A. nasuta* and *A. secale*, respectively) (MacKenzie 2005). Of the three species studied, two (*A. nasuta* and *A. valida*) were included in the *PaxC* intron phylogeny (van Oppen et al. 2001) and showed para- and polyphyly, respectively.

These examples illustrate that nominal *Acropora* species constitute genetically distinct entities, some of which are likely to exchange genes with congeneric species (as recognized within the current taxonomic framework; Wallace 1999) at low frequencies through introgressive hybridization. Thus, although no-choice cross-fertilization trials demonstrate mating compatibilities (Hatta et al. 1999, Miller & Babcock 1997, Szmant et al. 1997, van Oppen et al. 2002, Willis et al. 1997), sperm-choice trials are a much better predictor of the extent of hybridization occurring under natural conditions, where eggs are exposed to complex mixtures of con- and heterospecific sperm that compete to fertilize eggs. Nevertheless, even rare hybridization events on ecological timescales are likely to be significant on evolutionary timescales.

Hybridization in Other Mass-Spawning and Brooding Corals

There have been few genetic studies of mating systems in other mass-spawning coral genera on Indo-Pacific reefs, but the one study that has explored the potential for hybridization in a nonacroporid coral found evidence of extensive interspecific breeding. The genus *Platygyra* (Faviidae) on the GBR comprises seven recognized morphospecies, and estimates of pairwise genetic differences among species based on allozymes are low (Miller & Benzie 1997). Although corroborative evidence from additional molecular markers is required, low allozyme divergence (Nei's D range from 0.032 to 0.057) combined with the fact that the species show overlap in spawning times and are reproductively compatible in experimental breeding trials (Miller & Babcock 1997) suggest that introgressive hybridization has also played a role in the evolution of this genus. In contrast to the Indo-Pacific *Acropora*, however, natural hybridization may be operating to homogenize morphospecies within the *Platygyra*. Alternatively, recent speciation has occurred with or without ongoing hybridization (Miller & Benzie 1997).

Molecular studies of Caribbean species in the genus *Madracis* suggest that hybridization may also occur among brooding corals. Mixed paraphyly and site-specific polymorphisms, including additivity at nine sites, were interpreted as provisional evidence of hybridization among *M. decactis*, *M. pharensis*, and *M. formosa* (Diekmann et al. 2001). Interestingly, a recent allozyme study in the Indo-Pacific of two brooding scleractinian corals, *Pocillopora damicornis* and *Stylophora pistillata*, at the edge of their ranges on high latitude reefs of Lord Howe Island (beyond the southern extremity of the GBR), revealed a small proportion of apparently introgressed hybrids (Miller & Ayre 2004). Such hybridization among brooding coral species may yet prove to be as widespread as for mass-spawning species. Alternatively, such events may be more prevalent in areas such as Lord Howe Island, where conspecific sperm may occur at unusually low densities.

COMPARATIVE BIOGEOGRAPHIC PATTERNS AND THE GEOGRAPHIC SCALE OF HYBRIDIZATION

Molecular studies reviewed in the above sections demonstrate that hybridization events have occurred throughout the evolutionary history of corals on both Caribbean and Indo-Pacific reefs; however, the frequency and outcomes of hybridization appear to vary between the two regions and between genera. In the Indo-Pacific *Acropora*, allele and haplotype sharing indicate that hybridization occurs between many pairs of species on evolutionary timescales, but these hybridization events appear to be rare on ecological timescales. Because a few hybrids can provide a means for transferring alleles between species, even rare events on ecological timescales may represent a significant mechanism for evolutionary change (Mallet 2005). Such events may have contributed significantly to adaptive radiation of the Indo-Pacific *Acropora*, particularly as they colonized new shallow-water habitats on continental shelves following frequent sea-level transgressions during the Pleistocene (Veron 1995). In contrast, hybridization between two of the three Caribbean *Acropora* species appears to be common, resulting in a hybrid that is recognized as the separate morphospecies, *A. prolifera*, although the outcome of hybridization appears to be limited to the production of F_1 hybrids. One factor that may contribute significantly to apparent differences in the frequency of hybridization is the comparative species richness of these two biogeographic regions. In the Caribbean, there are only three species within the genus *Acropora* and the coral fauna overall is quite small; around 30 scleractinian genera are present (Veron 2000), but each characteristically contains few species. This contrasts markedly with the Indo-Pacific fauna, where there are more than 100 species of *Acropora* and more than 80 genera overall (Veron 2000). When many congeneric species spawn simultaneously in sympatry, there may be stronger selection for efficient gamete recognition, whereas in locations where gametes of fewer species have an opportunity to interact, selection may be less stringent. However, as discussed above, even on Indo-Pacific reefs, isolating mechanisms are not absolute; many species of *Acropora* retain options for opportunistic interspecific mating.

Limited evidence suggests that there may also be variation in the frequency of hybridization within biogeographic regions. Hybridization appears to be more frequent at the edges of some coral species' distributional ranges, for example, at the northern periphery of the distribution of the Caribbean genus *Montastraea* (Fukami et al. 2004). There is also limited evidence of intergeneric hybridization between *Pocillopora* and *Stylophora* at the southern extremity of their ranges on the GBR (Miller & Ayre 2004). In such peripheral locations, densities of mates are likely to be low, few species are present, and hybrids may be able to exploit nonparental niches, as shown for plant species (Lexer et al. 2003, Rieseberg et al. 2003).

The broad geographic scales over which corals interact through introgressive hybridization are unparalleled in terrestrial animal groups. The distribution of the coral hybrid *Acropora prolifera* throughout the Caribbean (Veron 2000, Wallace 1999) suggests that hybridization occurs over the entire biogeographic ranges of the two parental species, *A. palmata* and *A. cervicornis*. Similarly, species of Indo-Pacific *Acropora* are sympatric over hundreds to thousands of kilometers and have consistent patterns of low interspecific gene flow on both the eastern and western coasts of Australia (Márquez et al. 2002a). Other pairs of *Acropora* species have a reticulate evolutionary history on reefs in Japan (Hatta et al. 1999). Interbreeding on this scale differs radically from the norm for terrestrial animals, where hybrid zones are usually narrow and maintained by a balance between dispersal and selection (reviewed in Arnold 1997, Barton & Hewitt 1985). Although concepts of hybrid zones have expanded to encompass mosaic hybrid zones and attention is now focusing on geographical distributions of genealogical lineages (reviewed in Hewitt 2001), in general, animal hybrids have been assumed to be limited to contact zones where neither parent species is particularly fit. The closest parallels to potential geographic scales of hybridization among corals are found in plants. For example, oak species in the genus *Quercus* in the eastern United States (Whittemore & Schaal 1991) and *Eucalyptus* species in Australia (Potts & Reid 1988) are accepted as syngameons of interbreeding species across broad distributional ranges. Recent evidence that hybridizing species of coral reef fish are also broadly sympatric but ecologically partitioned by reef habitat (van Herwerden et al. 2006) suggests that broad scales of interbreeding may not be uncommon among broadcast-spawning animals.

SPECIES CONCEPTS FOR HYBRIDIZING CORALS

Coral species are traditionally recognized on the basis of morphological characters, particularly skeletal structures, but evolutionary relationships implied by morphology are not always reflected in molecular phylogenies (Fukami et al. 2004, van Oppen et al. 2001, Wolstenholme et al. 2003). Different rates of evolution for morphological and molecular characters (Wolstenholme et al. 2003), phenotypic plasticity, and convergent evolution may partly explain these mismatches, but reticulate evolutionary relationships detected among mass-spawning corals of Indo-Pacific *Acropora* (Hatta et al. 1999; van Oppen et al. 2001, 2002) are also likely to have contributed, particularly in cases of closely related species that interbreed. Regardless of such mismatches, it is clear that discrete coral species with distinct ecological characteristics do exist and define stable and cohesive entities despite occasional interspecific gene flow (Márquez et al. 2002a, Miller & Babcock 1997, Wolstenholme et al. 2003, as described above for *A. millepora* and *A. pulchra*).

Examples of interspecific gene exchange occurring between corals that are nonetheless morphologically and ecologically distinct suggest that, just as in plants, cohesive mechanisms override occasional gene flow to maintain recognizable species boundaries. Spatial heterogeneity on reefs, particularly as caused by steep environmental gradients (e.g., light, hydrodynamic and exposure gradients with depth), leads to dramatically different habitats over scales measured in tens of meters and provides a mechanism by which disruptive selection could maintain morphologically discrete taxa with distinct physiological requirements over distances small enough to allow mixing of gametes. Thus, even when adults of mass-spawning species differ in their benthic habitats along depth-defined environmental gradients, gametes from these species are able to interact and fertilize at the sea surface, continually renewing the potential for interspecific gene flow. Disruptive selection combined with disjunction between the benthic habitat of adults and the sea surface location of fertilization may partly explain how well-defined morphological and ecological coral species persist while maintaining opportunities for some interspecific gene flow.

Because many coral species share genetic variation with closely related species, phylogenetic species concepts based on species monophyly (e.g., Baum & Shaw 1995, Cracraft 1989) should not be applied to corals, especially because genes can have different histories of separation and gene flow (e.g., within the Caribbean *Acropora*; Vollmer & Palumbi 2002). A more appropriate species concept for corals is one that embodies the spirit of the biological species concept (sensu Dobzansky 1937, Mayr 1963) in which species are viewed as entities that are effectively reproductively isolated from other such groups (i.e., the cohesion species concept of Templeton 1989), but acknowledges that complete reproductive isolation among species does not reflect biological reality. The approach of plant biologists, who have always tolerated a degree of hybridization among well-defined species (sensu Grant 1981, Rieseberg 1997), is particularly relevant to coral species. Similarly, the botanically derived concept of a syngameon (Grant 1981), which links species that are capable of interbreeding and exchanging genes in a reproductive community, usefully describes many groups of closely related coral species. The possession of a mating system that tolerates and potentially takes advantage of a degree of interspecific gene exchange adds a new dimension to the similarities between the life histories of corals and plants. In view of this, delimiting good coral species requires multiple lines of evidence (e.g., as for the Caribbean *Montastraea* and Indo-Pacific *Acropora*) to interpret how sometimes contradictory morphological, ecological, reproductive, and genetic characters best characterize cohesive coral species (Wallace & Willis 1994). Although the extent to which the consequences of gene flow are potentially adaptive versus negative requires further exploration, this propensity of corals adds to current understanding of the role of hybridization in animal speciation.

SIGNIFICANCE OF HYBRIDIZATION FOR TROPICAL CORAL SPECIES AND THE RESILIENCE OF CORAL REEFS

Tropical reef corals are facing a suite of challenges, the cumulative impact of which may jeopardize the persistence of some (e.g., Caribbean) coral reef ecosystems as they have been known in recent geological times. The threat of ocean warming is potentially catastrophic to animals that live within 1–2°C of their upper thermal limits (Hoegh-Guldberg 1999, Hughes et al. 2003). Moreover, exploitation of coral reef communities has disturbed their trophic balance, in some cases causing community phase shifts from coral- to algal-dominated reefs (Hughes et al. 2003). The unprecedented rise of coral disease on Caribbean reefs has decimated populations of the dominant framework-building corals in the genera *Acropora* and *Montastraea*, as well as major gorgonian species (Harvell et al. 2002). To respond to these challenges, coral species need a variety of mechanisms for rapid evolutionary change.

Hybridization events are an important source of raw material for rapid evolutionary change in a variety of plant and animal groups and also have the potential to facilitate adaptive radiation when new adaptive zones are invaded (Seehausen 2004). A study of Darwin's finches in the Galapagos Islands highlights the ability of hybrids to provide a rapid response to environmental change, in this case through novel beak morphologies when climatological changes abruptly reduced the seed types available (Grant & Grant 1996). Moreover, backcrossing facilitated the persistence of the parental species' genetic diversity when finches failed to breed successfully in purebred matings. Hybridization may also significantly increase resilience to novel disease challenges. For example, hybrids of frog species arising through allopolyploid speciation had increased disease resistance to two different parasites that each infected one of the parent species (Jackson & Tinsley 2003). Similarly, hybrid parakeets had significantly higher measures of immune function than inbred parental species, through enrichment of depauperate gene pools (Tompkins et al. 2006). Such evidence that hybridization can mediate the response of species to rapid climatological change and the development of disease resistance highlights the potential significance hybridization may hold for the persistence of coral species, which face similar environmental challenges.

In conclusion, hybridization has contributed in diverse and significant ways to the evolution of coral species in the ecologically prominent Indo-Pacific *Acropora*, as well as to the major Caribbean framework-building genera *Montastraea* and *Acropora*. We assert that the evolutionary potential of hybridization is important to conserve, thus hybrids like *A. prolifera* represent important reservoirs of novel genetic diversity that may facilitate adaptive radiation under changed environmental circumstances. In addition, their ability to pass genes between species represents a potentially significant mechanism for rapid evolutionary change of parental species. As advocated by Ennos et al. (2005) in a recent review of conservation principles for taxonomically complex groups, efforts should be directed toward the conservation of evolutionary processes that generate such biodiversity rather than the current focus on the conservation of species, which are ambiguous entities in taxonomically complex groups. Recommendations against the need to protect hybridizing species because they fall outside the framework of conventional species definitions or are interpreted as jeopardizing the persistence of pure native species (e.g., Allendorf et al. 2001) ignore the dynamic nature of species and the potentially integral role that hybridization may play in the founding of lineages. Given the implications of ocean warming for increasing frequency and severity of mass bleaching events and emerging coral diseases, it is important to realize that hybridization has contributed to the evolutionary diversification of corals and that it has the potential to contribute significantly to the resilience of coral species and the coral reef ecosystem into the future. Conservation strategies that protect the evolutionary processes that have given rise to modern reef corals and are likely to contribute to their future resilience should be paramount.

SUMMARY POINTS

1. Hybridization has contributed to the evolution of many ecologically dominant and structurally important corals in diverse and significant ways.

2. Interspecific gamete compatibility in no-choice crosses and developmental competence of hybrids suggest there may be few absolute pre- or postzygotic barriers to interspecific breeding among many mass-spawning species of Indo-Pacific *Acropora*. However, in sperm choice experiments, conspecific matings take precedence, providing a mechanism for maintaining morphospecies boundaries when prezygotic barriers are semi-permeable.

3. Although decoupling of adult (benthic) and natal (sea surface) habitats in mass-spawning corals continually renews opportunities for hybridization, spatial heterogeneity and steep, depth-related environmental gradients on reefs provide a mechanism by which disruptive selection could also contribute to the maintainance of morphologically and ecologically discrete taxa despite occasional gene flow.

4. Juvenile coral hybrids are not less fit than purebreds, as indicated by comparative survival and growth in a large-scale grow-out program. Greater growth and survival of juvenile hybrids in environmentally variable and extreme habitats suggest a role for hybrids in adaptation to new environments.

5. In the Indo-Pacific, molecular phylogenies of mass-spawning species in the genus *Acropora* are consistent with reticulate evolutionary pathways. Although population genetic studies indicate hybridization events are rare on ecological timescales, on evolutionary timescales, they are likely to have facilitated adaptive radiations leading to their current high diversity.

6. In the comparatively depauperate Caribbean, molecular studies reveal that one of only three exant Atlantic *Acropora* species is a hybrid, providing a conduit for one-way gene flow from *A. palmata* to *A. cervicornis* and a mechanism for the introduction of novel genetic material.

7. The capacity of the hybrid *A. prolifera* to colonize marginal habitats distinct from its parent species and evidence of hybridization at geographical boundaries of the Caribbean *Montastraea* support an evolutionary role for hybridization in range expansion and adaptation to changing environments.

8. The distribution of the hybrid *A. prolifera* throughout the entire Caribbean ranges of its parent species highlights the broad geographic scales of hybridization possible in marine environments when the location of gamete interactions is spatially segregated from adult habitats. Such scales differ from the narrow hybrid zones typical of most animals and provide new insights into the process of hybridization in animal species evolution.

9. In combination, outcomes of hybridization are likely to be significant for the future resilience of reef corals, for example, by providing options for rapid response to changing environments and climatologies as well as increasing resilience to novel disease challenges. Hybridization warrants consideration when developing conservation strategies for reef corals.

ACKNOWLEDGMENTS

We thank D. Thomson, A. Beer, and B. Miller for field assistance; F. van der Leest for data management in the hybrid and purebred outplanting study; R. Standish for assistance with allozyme studies; J. Veron, H. Ruiz, and F. Seneca for use of images in **Figures 1a** and **1b**, respectively; D. Harvell and L. van Herwerden for useful comments on the manuscript; and R. Willis for editorial assistance. This work was supported by funds from the Australian Research Council and James Cook University.

LITERATURE CITED

Acropora Biological Review Team. 2005. Atlantic *Acropora* status review document. Report to National Marine Fisheries Service, Southeast Regional Office. March 3, 2005. 152 pp. + Appendix

Adey WH, Gladfelter W, Odgen J, Dill R. 1977. *Field guidebook to the reefs and reef communities of St. Croix, Virgin Islands. Third Int. Symp. Fla: Coral Reefs*, The Atlantic Reef Comm., Univ. Miami

Allendorf FW, Leary RF, Spruell P, Wenburg JK. 2001. The problems with hybrids: setting conservation guidelines. *Trends Ecol. Evol.* 16:613–22

Anderson E. 1949. *Introgressive Hybridization*. New York: Wiley

Anderson E, Stebbins GL. 1954. Hybridization as an evolutionary stimulus. *Evolution* 8:378–88

Arnold ML. 1997. *Natural Hybridization and Evolution*. New York: Oxford Univ. Press. 215 pp.

Arnold ML, Hodges SA. 1995. Are natural hybrids fit or unfit relative to their parents? *Trends Ecol. Evol.* 10:67–71

Babcock RC, Bull GD, Harrison PL, Heyward AJ, Oliver JK, et al. 1986. Mass spawning of 105 scleractinian coral species on the Great Barrier Reef. *Marine Biol.* 90:379–94

Barton NH, Hewitt GM. 1985. Analysis of hybrid zones. *Annu. Rev. Ecol. Syst.* 16:113–48

Baird AH, Marshall PA, Wolstenholme J. 2002. Mass spawning of *Acropora* in the Coral Sea. *Proc. Ninth Int. Coral Reef Symp. Bali.* 1:385–89

Baum DA, Shaw KL. 1995. Genealogical perspectives on the species problem. In *Experimental and Molecular Approaches to Plant Biosystematics*, ed. PC Hoch, AG Stevenson, pp. 289–303. St. Louis: Missouri Bot. Gard.

Budd AF, Johnson KG. 1999. Origination preceding extinction during late Cenozoic turnover of Caribbean reefs. *Paleobiology* 25:188–200

Budd AF, Pandolfi JM. 2004. Overlapping species boundaries and hybridization within the *Montastraea "annularis"* reef coral complex in the Pleistocene of the Bahama Islands. *Paleobiology* 30:396–425

Budd AF, Stemann TA, Johnson KG. 1994. Stratigraphic distributions of genera and species of Neogene to recent Caribbean reef corals. *J. Paleontol.* 68:951–77

Buerkle CA, Morris RJ, Asmussen MA, Rieseberg LH. 2000. The likelihood of homoploid hybrid speciation. *Heredity* 84:441–51

Bullini L. 1994. Origin and evolution of animal hybrid species. *Trends Ecol. Evol.* 9:422–26

Cairns SD. 1982. Stony corals (Cnidaria; Hydrozoa, Scleractinia) of Carrie Bow Cay, Belize. See Rützler & Macintyre 1982, pp. 271–302

Carroll A, Harrison P, Adjeroud M. 2006. Sexual reproduction of *Acropora* reef corals at Moorea, French Polynesia. *Coral Reefs* 25:93–97

Coyne JA, Orr HA. 2004. *Speciation*. Sunderland: Sinauer. 545 pp.

Cracraft J. 1989. Speciation and its ontology: the empirical consequences of alternative species concepts for understanding patterns and processes of differentiation. In *Speciation and its Consequences*, ed. D Otte, JA Endler, pp. 28–59. New York: Sinauer

de Graaf M, Geertjes GJ, Videler JJ. 1999. Observations on spawning of scleractinian corals and other invertebrates on the reefs of Bonaire (Netherlands Antilles, Caribbean). *Bull. Mar. Sci.* 64:189–94

Diekmann OE, Bak RPM, Stam WT, Olsen JL. 2001. Molecular genetic evidence for probable reticulate speciation in the coral genus *Madracis* from a Caribbean fringing reef slope. *Mar. Biol.* 139:221–33

Dobzhansky T. 1937. *Genetics and the Origin of Species*. New York: Columbia Univ. Press. 364 pp.

Dowling TE, Secor CL. 1997. The role of hybridization and introgression in the diversification of animals. *Annu. Rev. Ecol. Syst.* 28:593–619

Ennos RA, French GC, Hollingsworth PM. 2005. Conserving taxonomic complexity. *Trends Ecol. Evol.* 20:164–68

Freeland JR, Boag PT. 1999. The mitochondrial and nuclear genetic homogeneity of the phenotypically diverse Darwin's ground finches. *Evolution* 53:1553–63

Fukami H, Budd AF, Levitan DR, Jara J, Kersanach R, Knowlton N. 2004. Geographic differences in species boundaries among members of the *Montastraea annularis* complex based on molecular and morphological markers. *Evolution* 58:324–37

Fukami H, Omori M, Shimoike K, Hayashibara T, Hatta M. 2003. Ecological and genetic aspects of reproductive isolation by different spawning times in *Acropora* corals. *Mar. Biol.* 142:679–84

Gardner JPA. 1997. Hybridization in the sea. *Adv. Mar. Biol.* 31:1–78

Gittings SR, Boland GS, Deslarzes KJ, Combs CL, Holland BS, Brights TJ. 1992. Mass spawning and reproductive variability of reef corals at the East Flower Garden Bank, north-west Gulf of Mexico. *Bull. Mar. Sci.* 51:420–28

Goreau TF. 1959. The ecology of Jamaican coral reefs I: Species composition and zonation. *Ecology* 40:67–90

Grant V. 1981. *Plant Speciation*. New York: Columbia Univ. Press. 563 pp. 2nd ed.

Grant PR, Grant R. 1996. High survival of Darwin's finch hybrids: effects of beak morphology and diet. *Ecology* 77:500–9

Gregory JW. 1895. Contributions to the palaeontology and physical geology of the West Indies. *Q. J. Geol. Soc. London* 51:255–312

Guest JR, Chou LM, Baird AM, Goh BPL. 2002. Multispecific, synchronous coral spawning in Singapore. *Coral Reefs* 21:422–23

Harrison PL, Babcock RC, Bull GD, Oliver JK, Wallace CC, Willis BL. 1984. Mass spawning in tropical reef corals. *Science* 223:1186–89

Harrison PL, Wallace CC. 1990. Reproduction, dispersal and recruitment of scleractinian corals. In *Ecosystems of the World, 25: Coral Reefs*, ed. Z Dubinsky, pp 133–207. Amsterdam: Elsevier

Harvell CD, Mitchell CE, Ward JR, Altizer S, Dobson AP, et al. 2002. Climate warming and disease risks for terrestrial and marine biota. *Science* 296:2158–62

Hatta M, Fukami H, Wang W, Mori M, Shimoike K, et al. 1999. Reproductive and genetic evidence for a reticulate evolutionary history of mass-spawning corals. *Mol. Biol. Evol.* 16:1607–13

Hayashibara T, Shimoike K, Kimura T, Hosaka S, Heyward A, et al. 1993. Patterns of coral spawning at Akajima Island, Okinawa, Japan. *Mar. Ecol. Prog. Ser.* 10:253–62

Hewitt GM. 2001. Speciation, hybrid zones and phylogeography—or seeing genes in space and time. *Mol. Ecol.* 10:537–49

Heyward AJ. 1985. Comparative coral karyology. *Proc. Fifth Int. Coral Reef Cong. Tahiti* 6:47–51

Hoegh-Guldberg O. 1999. Coral bleaching, climate change and the future of the world's coral reefs. *Mar. Freshw. Res.* 50:839–66

Hughes TP, Baird AH, Bellwood DR, Card M, Connolly SR, et al. 2003. Climate change, human impacts and the resilience of coral reefs. *Science* 301:929–33

Jackson JA, Tinsley RC. 2003. Parasite infectivity to hybridizing host species: a link between hybrid resistance and allopolyploid speciation? *Int. J. Parasitol.* 33:137–44

Jiggins CD, Mallet J. 2000. Bimodal hybrid zones and speciation. *Trends Ecol. Evol.* 15:250–55

Kearney M. 2005. Hybridization, glaciation and geographical parthenogenesis. *Trends Ecol. Evol.* 20:495–502

Kenyon JC. 1997. Models of reticulate evolution in the coral genus *Acropora* based on chromosome numbers: parallels with plants. *Evolution* 51:756–67

Knowlton N, Mate JL, Guzman HM, Rowan R, Jara J. 1997. Direct evidence for reproductive isolation among the three species of the *Montastraea annularis* complex in Central America (Panama and Honduras). *Mar. Biol.* 127:705–11

Lexer C, Welch ME, Raymond O, Rieseberg LH. 2003. The origin of ecological divergence in *Helianthus paradoxus* (Asteraceae): selection on transgressive characters in a novel hybrid habitat. *Evolution* 57(9):1989–2000

Levitan DR, Fukami H, Jara J, Kline D, McGovern TM, et al. 2004. Mechanisms of reproductive isolation among sympatric broadcast-spawning corals of the *Montastraea annularis* species complex. *Evolution* 58:308–23

MacKenzie JB. 2005. Population-species architecture of corals in the *Acropora nasuta* group. PhD thesis. James Cook Univ., Townsville, Aust.

Mallet J. 2005. Hybridization as an invasion of the genome. *Trends Ecol. Evol.* 20:229–37

Márquez LM, Miller DJ, MacKenzie JB, van Oppen MJH. 2003. Pseudogenes contribute to the extreme diversity of nuclear ribosomal DNA in the hard coral *Acropora*. *Mol. Biol. Evol.* 20:1077–86

Márquez LM, van Oppen MJH, Willis BL, Miller DJ. 2002a. Sympatric populations of the highly cross-fertile coral species *Acropora hyacinthus* and *A. cytherea* are genetically distinct. *Proc. R. Soc. London Ser. B* 269:1289–94

Márquez LM, van Oppen MJH, Willis BL, Reyes A, Miller DJ. 2002b. The highly cross-fertile coral species, *Acropora hyacinthus* and *A. cytherea*, constitute statistically distinguishable lineages. *Mol. Ecol.* 11:1339–49

Mateos M, Vrijenhoek RC. 2002. Ancient versus reticulate origin of a hemiclonal lineage. *Evolution* 56:985–92

Mayr E. 1963. *Animal Species and Evolution*. Cambridge: Harvard Univ. Press. 797 pp.

McCarthy EM, Asmussen MA, Anderson WW. 1995. A theoretical assessment of recombinational speciation. *Heredity* 74:502–09

McFadden CS, Hutchinson MB. 2004. Molecular evidence for the hybrid origin of species in the soft coral genus Alcyonium (Cnidaria: Anthozoa: Octocorallia). *Mol. Ecol.* 13:1495–1505

McNeill DF, Budd AF, Borne FP. 1997. An earlier (late Pliocene) first appearance of the reef-building coral *Acropora palmata*: stratigraphic and evolutionary implications. *Geology* 25:891–94

Metz EC, Palumbi SR. 1996. Positive selection and sequence rearrangements generate extensive polymorphism in the gamete recognition protein bindin. *Mol. Biol. Evol.* 13:397–406

Miller DJ, van Oppen MJH. 2003. A "fair go" for coral hybridization. *Mol. Ecol.* 12:805–7

Miller KJ, Ayre DJ. 2004. The role of sexual and asexual reproduction in structuring high latitude populations of the reef coral *Pocillopora damicornis*. *Heredity* 92:557–68

Miller KJ, Babcock RC. 1997. Conflicting morphological and reproductive species boundaries in the coral genus *Platygyra Biol. Bull.* 192:98–110

Miller KJ, Benzie JAH. 1997. No clear genetic distinction between morphological species within the coral genus *Platygyra Bull. Mar. Sci.* 61:907–17

Modrich P, Lahue R. 1996. Mismatch repair in replication fidelity, genetic recombination and cancer biology. *Annu. Rev. Biochem.* 65:101–33

Naisbit RE, Jiggins CD, Mallet J. 2003. Mimicry: developmental genes that contribute to speciation. *Evol. Dev.* 5:269–80

Odorico DM, Miller DJ. 1997. Variation in the ribosomal internal transcribed spacers and 5.8S rDNA among five species of *Acropora* (Cnidaria; Scleractinia): patterns of variation consistent with reticulate evolution. *Mol. Biol. Evol.* 14:465–73

Otto SP, Whitton J. 2000. Polyploid incidence and evolution. *Annu. Rev. Genet.* 34:401–37

Palumbi SR. 1994. Genetic divergence, reproductive isolation, and marine speciation. *Annu. Rev. Ecol. Syst.* 25:547–72

Potts BM, Reid JB. 1988. Hybridization as a dispersal mechanism. *Evolution* 42:1245–55

Ramsey J, Schemske DW. 1998. Pathways, mechanisms, and rates of polyploid formation in flowering plants. *Annu. Rev. Ecol. Syst.* 29:467–501

Richmond RH, Hunter CL. 1990. Reproduction and recruitment of corals: comparisons among the Caribbean, the Tropical Pacific, and the Red Sea. *Mar. Ecol. Prog. Ser.* 60:185–203

Rieseberg LH. 1997. Hybrid origins of plant species. *Annu. Rev. Ecol. Syst.* 28:359–89

Rieseberg LH, Raymond O, Rosenthal DM, Lai Z, Livingstone K, et al. 2003. Major ecological transitions in wild sunflowers facilitated by hybridization. *Science* 301(5637):1211–16

Rosenfield JA, Kodric-Brown A. 2003. Sexual selection promotes hybridization between Pecos pupfish, *Cyprinodon pecosensis*, and sheepshead minnow, *C-variegatus*. *J. Evol. Biol.* 16:595–606

Rützler K, Macintyre IG. 1982. Habitat distribution and community structure of the barrier reef complex at Carrie Bow Cay, Belize. In *The Atlantic Barrier Reef Ecosystem at Carrie Bow Bay, Belize, I. Structure and Communities*, K Rützler, IG Macintyre, eds., pp. 9–45. Washington: Smithsonian Inst. Press. 539 pp.

Salzburger W, Baric S, Sturmbauer C. 2002. Speciation via introgressive hybridization in east African cichlids? *Mol. Ecol.* 11:619–25

Seehausen O. 2004. Hybridization and adaptive radiation. *Trends Ecol. Evol.* 19:198–207

Shaw KL. 2002. Conflict between nuclear and mitochondrial DNA phylogenies of a recent species radiation: what mtDNA reveals and conceals about modes of speciation in Hawaiian crickets. *Proc. Natl. Acad. Sci. USA* 99:16122–27

Simpson CJ. 1991. Mass spawning of corals on Western Australian reefs and comparisons with the Great Barrier Reef. *J. R. Soc. West. Aust.* 74:85–91

Soltis DE, Soltis PS. 1999. Polyploidy: recurrent formation and genome evolution. *Trends Ecol. Evol.* 14:348–52

Spolsky CM, Phillips CA, Uzzell T. 1992. Antiquity of clonal salamander lineages revealed by mitochondrial DNA. *Nature* 356:706–10

Stebbins GL. 1959. The role of hybridization in evolution. *Proc. Am. Philos. Soc.* 103:231–51

Szmant AM. 1986. Reproductive ecology of Caribbean reef corals. *Coral Reefs* 5:43–54

Szmant AM, Weil E, Miller MW, Colon DE. 1997. Hybridization within the species complex of *Montastrea annularis*. *Mar. Biol.* 129:561–72

Templeton AR. 1989. In *Speciation and its Consequences*, ed. D Otte, JA Endler. New York: Sinauer. 670 pp.

Tompkins DM, Mitchell RA, Bryant DM. 2006. Hybridization increases measures of innate and cell-mediated immunity in an endangered bird species. *J. Anim. Ecol.* 75:559–64

Ungerer MC, Baird SJE, Pan J, Rieisberg LH. 1998. Rapid hybrid speciation in wild sunflowers. *Proc. Natl. Acad. Sci. USA* 95:11757–762

Vacquier VD, Corner KR, Stout CD. 1990. Species specific sequences of abalone lysine, the sperm protein that creates a hole in the egg envelope. *Proc. Natl. Acad. Sci. USA* 87:5792–96

van Herwerden L, Choat JH, Dudgeon CD, Carlos G, Newman SJ, et al. 2006. Contrasting patterns of genetic structure in two species of the coral trout *Plectropomus* (Serranidae) from east and west Australia: introgressive hybridization or ancestral polymorphisms. *Mol. Phyl. Evol.* In press

van Oppen MJH, McDonald BJ, Willis BL, Miller DJ. 2001. The evolutionary history of the coral genus *Acropora* (Scleractinia, Cnidaria) based on a mitochondrial and a nuclear marker: Reticulation, incomplete lineage sorting or morphological convergence? *Mol. Biol. Evol.* 18:1315–29

van Oppen MJH, Willis BL, van Rheede T, Miller DJ. 2002. Spawning times, reproductive compatibilities and genetic structuring in the *Acropora aspera* group:

evidence for natural hybridization and semipermeable species boundaries in corals. *Mol. Ecol.* 11:1363–76

van Oppen MJH, Willis BL, van Vugt H, Miller DJM. 2000. Examination of species boundaries in the *Acropora cervicornis* group (Scleractinia, Cnidaria) using nuclear DNA sequence analysis. *Mol. Ecol.* 9:1363–73

Vaughan TW. 1901. The stony corals of the Porto Rican waters. *Bull. US Fish Commiss. 1900* 20:290–320

Veron JEN. 1995. *Corals in Space and Time. The Biogeography and Evolution of the Scleractinia.* Sydney: UNSW Press. 321 pp.

Veron JEN. 2000. *Corals of the World.* Vol. 1. Townsville, Aust.: AIMS. 463 pp.

Vollmer SV, Palumbi SR. 2002. Hybridization and the evolution of reef coral diversity. *Science* 296:2023–25

Vollmer SV, Palumbi SR. 2004. Testing the utility of internally transcribed spacer sequences in coral phylogenetics. *Mol. Ecol.* 13:2763–72

Vrijenhoek RC. 1984. Ecological differentiation among clones: The frozen niche variation model. In *Population Biology and Evolution*, ed. K Wohrmann, V Loeschecke. Berlin: Springer-Verlag

Wallace CC. 1999. *Staghorn Corals of the World: A Revision of the Genus Acropora.* Collingwood, Aust.: CSIRO Publishing. 421 pp. 1st ed.

Wallace CC, Willis BL. 1994. The systematics of *Acropora*: The effect of new biological findings on species concepts. *Annu. Rev. Ecol. Syst.* 25:237–62

Wells JW. 1973. New and old scleractinian corals from Jamaica. *Bull. Mar. Sci.* 23:16–58

White MJ. 1978. *Modes of Speciation.* San Francisco: Freeman

Whittemore AT, Schaal BA. 1991. Interspecific gene flow in sympatric oaks. *Proc. Natl. Acad. Sci. USA* 88:2540–44

Willis BL, Babcock RC, Harrison PL, Oliver JK, Wallace CC. 1985. Patterns in the mass spawning of corals on the Great Barrier Reef. *Proc. Fifth Int. Coral Reef Symp. Tahiti* 4:343–48

Willis BL, Babcock R, Harrison P, Wallace C. 1997. Experimental hybridization and breeding incompatibilities within the mating systems of mass spawning corals. *Coral Reefs* 16:S53–65

Wolstenholme JK. 2004. Temporal reproductive isolation and gametic compatibility are evolutionary mechanisms in the *Acropora humilis* species group (Cnidaria; Scleractinia). *Mar. Biol.* 144:567–82

Wolstenholme JK, Wallace CC, Chen CA. 2003. Species boundaries within the *Acropora humilis* species group (Cnidaria; Scleractinia): a morphological and molecular interpretation of evolution. *Coral Reefs* 22:155–66

The New Bioinformatics: Integrating Ecological Data from the Gene to the Biosphere

Matthew B. Jones,[1] Mark P. Schildhauer,[1] O.J. Reichman,[1] and Shawn Bowers[2]

[1]National Center for Ecological Analysis and Synthesis, University of California, Santa Barbara, California 93101; email: jones@nceas.ucsb.edu, schild@nceas.ucsb.edu, reichman@nceas.ucsb.edu

[2]Genome Center, University of California, Davis, California 95616; email: sbowers@ucdavis.edu

Key Words

ecoinformatics, data integration, data sharing, metadata, ontology, scientific workflows, semantics

Abstract

Bioinformatics, the application of computational tools to the management and analysis of biological data, has stimulated rapid research advances in genomics through the development of data archives such as GenBank, and similar progress is just beginning within ecology. One reason for the belated adoption of informatics approaches in ecology is the breadth of ecologically pertinent data (from genes to the biosphere) and its highly heterogeneous nature. The variety of formats, logical structures, and sampling methods in ecology create significant challenges. Cultural barriers further impede progress, especially for the creation and adoption of data standards. Here we describe informatics frameworks for ecology, from subject-specific data warehouses, to generic data collections that use detailed metadata descriptions and formal ontologies to catalog and cross-reference information. Combining these approaches with automated data integration techniques and scientific workflow systems will maximize the value of data and open new frontiers for research in ecology.

1. INTRODUCTION

Bioinformatics: the use of computational and statistical techniques to more effectively manage and analyze biological data

Ecoinformatics: a field of research and development focused on the interface between ecology, computer science, and information technology

In 2003, *Science* and *Nature* simultaneously published cover stories about the demise of gorillas in Africa (Kaiser 2003, Whitfield 2003). The title of the story in *Science* ("Ebola, Hunting Push Ape Populations to the Brink of Extinction") revealed that the ecological issues were quite broad. Ebola, a pernicious hemorrhagic disease that can spread to humans, is affected by its local environment. Understanding this disease requires knowledge of epidemiology, genetics, and transmission modes, along with their ecological contexts. Hunting pressure engendered by the need for bushmeat relates to the nutritional status of local humans and sociological features of their culture. With regard to the apes, information about their population dynamics (birth and death rates, longevity, social interactions) is needed to take timely, effective action to save the species. Virtually every ecological question, whether this dire or not, requires access to a similarly diverse array of data and information in order to develop robust analyses. Integrating ecologically pertinent data into the chain of information from the gene to the biosphere will significantly enhance our understanding of the natural world and promote wise management strategies for natural resources. In this review, we examine challenges and solutions relative to locating, accessing, integrating, and analyzing data from ecology and allied disciplines.

2. THE NEED FOR A NEW BIOINFORMATICS

Bioinformatics is the application of techniques from computer science and statistics to manage and analyze biological data. The initial focus within bioinformatics has been on tools and analytical techniques that operate on genetic and protein sequence data (many useful databases have emerged in this context, including GenBank; Benson et al. 2005) and there is ongoing discussion about the need to further integrate these resources with higher systems levels such as data describing biological processes at the metabolic level (Thomas & Ganji 2006).

Ecology as a discipline grew out of a natural history tradition with a strong emphasis on observation in the field. By the late nineteenth century, ecology was becoming a more quantitative science with fewer purely descriptive studies. Later studies increasingly moved toward mathematically derived models that focused on assessments of the distribution and abundance of organisms along with related information about the abiotic environment (Real & Brown 1991). Since the 1960s there has been a strong emphasis on experimental manipulation to elucidate causal relationships (e.g., Brown & Munger 1985, Connell 1961, Lubchenco & Real 1991, Paine 1966). Experiments are typically designed to test a particular set of hypotheses and therefore the types of manipulations performed and the formats of data collected vary tremendously across studies. These factors contribute to making ecological data highly heterogeneous.

The most significant challenge in ecological informatics (ecoinformatics) is dealing with the inherent complexity and breadth of data used in ecological studies. Ecological data do not only document entities of interest—such as the numbers of individuals, or sequences of nucleotides. Rather, they frequently contain measurements of processes (e.g., rates of competition, or herbivory) or surrogates for these (extent of shading, or assessment of leaf damage) that often require specialized expertise to accurately document and interpret. Ecological data also occur in many forms (text, numbers, images, videos), and numerous legacy data that are important for dealing with scientific and environmental issues remain undigitized. These characteristics make the access, interpretation, analysis, and modeling of ecological data especially challenging.

Ecologists have recognized the need for integrated data systems to support cross-disciplinary collaboration to understand the basic ecological principles that govern the biosphere (Green et al. 2005). With the rapid growth of human populations and their impacts, it becomes critically important to better describe and understand natural processes. The increasing demands within ecology for greater access to more types of data emphasize the need for integrated data-management solutions that span biological subdisciplines from the gene to the biosphere.

3. CASE STUDIES IN SYNTHESIS

Although data from the observations and experiments of individual investigators remain at the core of ecological and evolutionary research, their value increases substantially when they are integrated and synthesized to reveal important patterns and to generate broad generalities. Data synthesis allows a broader perspective over time and space, and across many disciplines, than is possible from one or a few studies. Even more important in the long run, synthesis allows data to be used for purposes other than those for which they were originally intended, to address questions that were unknown or unapproachable at the time the data were collected (e.g., Andelman et al. 2004).

Regardless of whether synthetic and integrative research is undertaken in collaborative frameworks (such as is sponsored by various synthesis centers) or by individuals, these efforts will almost certainly depend on access to complementary data that were collected by other individuals or under the auspices of other projects. Once the data needs extend beyond local data collection efforts specifically tailored for a given analysis, efficient access to data will be severely hampered for a variety or reasons, ranging from difficulties owing to sociological and legal reasons (e.g., lack of permission to use data) to technical issues such as data contained within incompatible management systems, or integration challenges resulting from variable spatial and temporal scales of sampling, taxonomic irregularities in the identification of specimens, and idiosyncratic labeling of variables and their units of measurement.

A project comparing the effects of grazers and fire on grasslands in North America (Konza Prairie) and South Africa (Kruger Park; Knapp et al. 2004) provides an excellent example of both the difficulties of synthesizing data from multiple sources and creative solutions for dealing with them. The researchers dealt with incongruous variables that quantified the effects of treatments by using those that were similar, and correlating surrogates for other variables that were dissimilar. Many plant taxa were sampled in one study, whereas only trees and grasses were sampled in the other, so the researchers had to use growth form rather than taxon, per se, to compare data. Plot size and methodology differed so data from each location were transformed into relative abundances. Because the fire and grazing regimes were imposed differently at

Relational database: the prevalent software method for storing tabular information that uses named 2-dimensional tables

Data model (data schema): how information should be conceptually and logically described to optimize storage, access, and interpretation in some computer application

the two locations (experimental versus natural), the treatments had to be reduced to ordinal rankings (e.g., high or low fire frequency) to analyze how each affected plant productivity and community-level characteristics. Significant amounts of data had to be discarded because of incompatibilities among variables in the source data sets. By comparing the consolidated measures within a site (e.g., converting actual plant abundance counts to relative abundance, species to growth form, and actual fire frequencies to relative frequencies) the researchers conducted a robust synthetic analysis using the data. Still, the process of integrating the data was arduous and inefficient, and the strength of the analyses was impacted by the inability to incorporate all the available data and the need to use ordinal rankings instead of the original numeric values.

4. CURRENT METHODS FOR STORING AND ACCESSING ECOLOGICAL DATA

The most common method that scientists currently use to manage ecological data is to enter it in an ad-hoc manner in spreadsheet-based software tools. Spreadsheets are flexible, easy to learn, and allow scientists to quickly enter, review, and get summaries of data in a simple although statistically unreliable manner (McCullough & Wilson 1999). Spreadsheets do not provide the tools to promote good data management practices, however, because they lack sufficient structure to adequately describe and constrain the data. Spreadsheets often contain multiple data tables on a single page, and easily permit intermixing of raw data values with statistical summarizations, and marginal sums and annotations. For the scientist, using a spreadsheet in this way is often convenient over the short term. But unless these usages are well-documented, such informal data management practices cause difficulty for scientists looking back at their own data and inhibit reuse of the data by other scientists that may be unfamiliar with the organization of the spreadsheet. Although having data available in any format is probably more valuable than not having it at all, automated data processing approaches will always be constrained by the unstructured way in which spreadsheets store data.

Scientists seeking a more robust way to store their data frequently learn to use relational database systems such as Microsoft Access or Filemaker Pro. Alternatively, many ecologists store and analyze their data in a statistical package such as SAS (**http://www.sas.com**) or the "R Statistical Package" (**http://www.r-project.org**). In all these cases, a researcher typically models the data beforehand by deciding how to separate them into tables (usually with columns representing variables, and rows equaling observations). The researcher then specifies how these tables, which each contain some distinct conceptual type of information, or entity, can be joined or merged (Brunt 2000, Pascal 2000, Porter 2000). For example, one might join a table containing observations of species abundance on a given day with a table of observations of the temperature taken at that same location through time to create a new table that has information from both. Ecological researchers usually learn these skills on their own, because data modeling and implementation in a database management system (DBMS) are not currently standard parts of a biologist's academic training, despite their relevance for better understanding how to collect and manage data.

Owing to limitations in desktop DBMSs and the computer operating systems on which they run, it is relatively difficult to share databases and spreadsheets with colleagues. This problem grows with time as the software on which the data depends becomes obsolete and is replaced by newer tools, leaving the older proprietary databases and spreadsheets in an inaccessible state. Moreover, the data in these systems is typically structured to serve the specific needs of the project. The practice of creating project-specific data sets prevents standardization of approaches among otherwise similar studies and ultimately leads to data integration challenges for synthetic analyses.

Data integration: matching up and combining information from different sources, ideally in ways that are meaningful and useful

4.1. Data Warehouses (Vertically Integrated Databases)

One solution to the problem of project-specific databases complicating data integration is to develop vertically integrated databases that store data collected by many different investigators, but all following a common theme. These data systems are often called data warehouses, and usually have a Web-based interface for querying and downloading data. They are thus broadly accessible to individuals, without the need to locally store the data, or install specialized DBMS software. Examples of vertically integrated data warehouses include centralized data archives such as GenBank, VegBank (Jennings et al. 2004), and TreeBase (Morell 1996); others provide network-distributed access to data from a number of compatible servers (e.g., access to biodiversity and specimen collection data in the GBIF portal; Canhos et al. 2004). These databases typically are more complex than desktop databases because they attempt to reconcile the differences in the data models among existing independent research projects. The resulting data model is more general than its project-specific counterparts, and usually represents a least-common denominator approach that only allows some data from each of the contributing projects to be integrated in a useful way. Consequently, data warehouses usually cannot suffice for project-specific data management tasks, because they do not accommodate all of the information contained in any project-specific database.

As an example, consider the VegBank data model. This model allows federation of vegetation plot data for quantifying the composition of plant communities in space and time. The VegBank database contains raw data for vegetation plots, which can be collected in a number of different ways, but is focused on describing the floral composition within an area, along with the environmental context. Plots data form the basis for the classification of vegetation communities, which are associations of co-occurring plant taxa. Data about community classifications are contained within VegBank along with the plots data.

Plots stored in VegBank can have an optional value for the "disturbanceType" to which a plot may have been exposed. The values for "disturbanceType" must be chosen from a controlled list that includes "Animal, general," "Grazing, domestic stock," "Grazing, native ungulates," "Herbivory, invertebrate," and "Herbivory, vertebrates." Individual project data collection methods may not all map precisely into this breakdown of disturbance types, because the categories are not mutually exclusive and allow for differences in interpretation. Consequently, the data in VegBank may contain less detailed information than the original data sources as a byproduct of being transformed into a more standardized and broadly accessible form. This is a common and unavoidable trade-off for data warehouses. Nevertheless, there are great benefits from integrating the plot data into a common data model because it allows efficient searching for relevant records across a much larger collection and the ability to analyze the data in a common format.

Another limitation of data warehouses is that contributing project-specific data to one warehouse does not automatically make the information available in others. As an example, part of the VegBank plot observation includes a characterization of its soil profile. Contributing these soil data to VegBank, however, does not relate them in any way to the Natural Resources Conservation Service Soil Data Mart from the U.S. Department of Agriculture (**http://soildatamart.nrcs.usda.gov**). Thus, the approach of using vertically integrated databases does not address the issue of integrating with data resources outside the scope of its own data model. Data warehouses essentially have the same data integration problems as project-specific databases, but at a higher level.

Because of this mismatch between the information management needs of individual projects and those of vertically integrated databases, there will always be a cost to contributing data to the latter. In addition, despite the highly integrated nature of data warehouses, they still require extensive documentation of the data to permit their reasonable interpretation. For example the term location in one data set might refer to a very proximate observation such as "on a tree" whereas in another data set it might refer to an entire region, such as "Delaware." Additional metadata—precise, structured descriptions of what a variable is referring to—is needed to resolve these types of issues.

Finally, whether a given scientist is willing to bear the cost in time and effort to contribute their own project-specific data to a data warehouse is driven primarily by their expectations of utilizing the warehouse for their own research purposes. As there is no reward system in place that recognizes the contributions made by sharing data, it is difficult to convince researchers to build and contribute to these data systems simply because there will be a benefit to their scientific discipline.

Metadata: information used to document and interpret data

4.2. Metadata-Driven Databases (Data Collections)

Metadata is the contextual information needed to understand and use a set of data (i.e., data about data). The importance of metadata has been emphasized both within ecology (Jones et al. 2001; Michener 2000, 2006; Michener et al. 1997) and in other communities (Attig et al. 2004, Daniel et al. 1998, Dekkers & Weibel 2003, Nair & Jeevan 2004, Theile 1998, Weibel 1995). Detailed human-readable metadata about the context of data collection, the protocols used to collect the data, and the structure and format of the data objects are a necessary prerequisite to the long-term preservation and interpretation of data. Michener et al. (1997) illustrate this point as a decline in information content with increasing time from when the results of the data are published. They emphasize that particular events, such as retirement, career change, or death of the original investigator, can have a dramatic impact on the availability of metadata and therefore the utility of data.

An alternative, more robust approach to the highly structured, vertically integrated data warehouse is a more loosely structured collection of project-specific data sets accompanied by structured metadata about each of the data sets. Advantages of this approach include: (*a*) data represented using different data models can be stored together in a single uniform storage system; (*b*) metadata-based data collection is familiar to scientists because it focuses on the same project-level data model as their typical spreadsheet data management approaches; (*c*) the metadata collected is much more detailed than the metadata used in a typical relational database, thus promoting long-term utility of the data; and, (*d*) the metadata is typically more concise than the raw data and can be used as a proxy when searching for data of interest. Each of the data sets is stored in a manner that is opaque to the data system in that the data themselves cannot be directly queried; rather, the structured metadata describing the data is queried in order to locate data sets of interest. After data sets of interest are located, more detailed information (such as the detailed data model that specifies, e.g., the definitions of the variables) can be extracted from the metadata and used to load, query, and manipulate individual data sets.

EML: Ecological Metadata Language

LTER: Long-term Ecological Research Network

BDP: Biological Data Profile

Some researchers have pointed out that there are limits to what can be captured in metadata and that any attempt to document all aspects of data will necessarily be incomplete and subject to the biases of the original investigators (Bowker 2000). However, this criticism points to the fact that data often reflect implicit assumptions and subtleties in meaning that will be challenging to capture in any structured framework. The increased emphasis on experimental approaches in ecology has certainly resulted in a need for more extensive documentation of methods and procedures in order to reasonably interpret data from these studies. Nevertheless, with the increasing recognition of the value of existing data, even partial descriptions of data that facilitates its reuse beyond that of its original purpose will be important for synthetic studies.

Although, loosely speaking, metadata refers to any information that provides additional context for interpreting data, in practice the term typically has connotations of structured, well-defined categories for systematically documenting critical aspects of a data set, which would be amenable to storage in a data format rather than natural language. A consistent and rigorous set of definitions for metadata categories is called a metadata content specification and, when broadly adopted or endorsed by some community, becomes a metadata standard. Several metadata content specifications could be used for documenting ecological and biological data. Within ecology, the Ecological Metadata Language (EML), developed through the efforts of ecologists and information managers, has garnered support from a variety of institutions such as the National Center for Ecological Analysis and Synthesis (NCEAS), and the Long-Term Ecological Research Network (LTER; Jones et al. 2001). The National Biological Information Infrastructure (NBII) uses the Biological Data Profile (BDP) from the Federal Geographic Data Committee (FGDC; Frondorf et al. 1999, Parr & Cummings 2005). Other metadata standards are in use and development, including

GML: Geographic Markup Language

ISO: International Standards Organization

KNB: Knowledge Network for Biocomplexity

the Geographic Markup Language (GML), ISO (International Standards Organization) Geospatial Metadata (ISO 19115), the Directory Interchange Format (DIF), and several taxonomic standards such as the Taxonomic Concept Schema (TCS) from the Taxonomic Databases Working Group (Goodchild 2003). The library community has many metadata standards, including the Dublin Core Element Set for use in providing basic bibliographic information about digital objects (Dekkers & Weibel 2003, Nair & Jeevan 2004). Each of these metadata specifications addresses metadata at differing levels of detail (granularity) that is relevant to biological data.

The multitude of metadata standards creates problems for interoperability. Data providers wanting to make their data available in various national archives often need to comply with multiple metadata standards, which increases the burden on data providers. Partial mappings between metadata standards can be accomplished, but there is no mechanism for automatically synonymizing, or crosswalking metadata concepts among the multiple standards. Within ecology, the Knowledge Network for Biocomplexity (KNB) is a distributed metadata-driven data repository (see Section 4.3 below), which provides a translation from EML to the BDP, but the reverse mapping is not yet available (partly because the BDP metadata is not as fine-grained as EML in several key areas). Another problem with the diversity of metadata standards is that metadata repositories typically support searching via only a single metadata standard, and an integrated search mechanism that spans multiple content standards is still elusive (but see the discussion on EcoGrid in Section 4.3 below).

The level of detail (granularity) and comprehensiveness are important factors for an effective metadata standard. Although several metadata standards provide for basic descriptions of bibliographic information, only a few attempt to fully describe the structure and content of scientific data sets. For example, EML provides a detailed and machine-readable description of the logical structure of data tables (i.e., the data model) as well as their physical structure (data format). This information can be used to automatically parse and load the data sets into analytical systems and relational databases (such as R, SAS, or PostgreSQL), which can significantly improve efficiency of data handling for larger synthesis studies.

Ultimately the choice of a metadata standard for ecological data should depend on the set of capabilities that the metadata provides in terms of facilitating data discovery, access, and reuse for future research, although the relative importance of these might vary depending on institutional priorities and budgets (e.g., federal agencies, NGOs, university researchers). It is possible that ecological metadata may well become accessible in several standards, especially if technology facilitates the interchange among these via automatic translations.

4.3. Current Data Collections

Several national data collections have been created to promote data sharing and long-term preservation of scientific data that are relevant to ecology and biology (**Table 1**). Many are metadata-driven data collections based on one or a few metadata standards. The KNB Metacat system (Berkley et al. 2001) provides a distributed data archive with search facilities that are customized for the specific needs of an

Table 1 Metadata and data collections in widespread use within ecology

Collection name	Standards supported	Archives data
Knowledge Network for Biocomplexity	EML, BDP, others	yes
NBII Metadata Clearinghouse	BDP	no[a]
NSDI Metadata Clearinghouse	CSDGM	no[a]
GBIF Taxonomic Collections	Darwin Core	no[a]
TOPP	EML	yes
Kruger National Park/SAEON	EML	yes
Open-Source Project for a Network Data Access Protocol (OPeNDAP) / Integrated Ocean Observing System (IOOS)	—	yes
VegBank	US NVC	yes
TreeBase	—	yes
Storage Resource Broker	various	yes
ORNL DAAC	BDP, other	yes
Global Change Master Directory	DIF	no[a]
ESA Data Registry	EML	no[a]
Ecological Archives (data papers)	various	yes
Ocean Biogeographic Information System (OBIS)	Darwin Core[b]	no[a]
Global Population Dynamics Database (GPDD)	various	yes

[a]Although this is a metadata clearinghouse, the data are often archived in other systems.
[b]OBIS uses an extension of Darwin Core, see Grassle (2000) for details.

GBIF: Global Biodiversity Information Facility (Taxonomic Names and Specimen Collections)

organization (**Table 1**). This interlinked set of data archives provides a mechanism for investigators to publicly share data or to share only with a limited set of colleagues. Since its inception in 2000, the data holdings in the KNB have been growing rapidly to the current level of over 12,000 databases described in the system (**Figure 1**). In the museum community, the Global Biodiversity Information Facility (GBIF; Canhos et al. 2004) is providing a centralized portal for searching over 450 museum collections via a federated subset of the data that is held in the collections. Both of these resources and others in **Table 1** enable new types of collaborative studies and meta-analysis that were previously impossible.

Several of the systems in **Table 1** only store metadata entries that describe the data and do not store the actual data, whereas others permit archiving the actual data along with the metadata. Registries, which house only metadata, are helpful in increasing the awareness of data holdings, and promote data sharing while bypassing some difficult cultural issues associated with asking scientists to archive (contribute) their data. For example, the Ecological Society of America (ESA) Data Registry (**http://data.esa.org**) was created to better document the data used in articles published in the ESA's journals. The society plans to promote data sharing and preservation by registering (through metadata) the data used for a publication. In future versions of the ESA registry, the society plans to add a data archive feature and integrate its *Ecological Archives* journal into a single integrated system (Peet 1998). The data registries for the Organization of Biological Field Stations (OBFS),

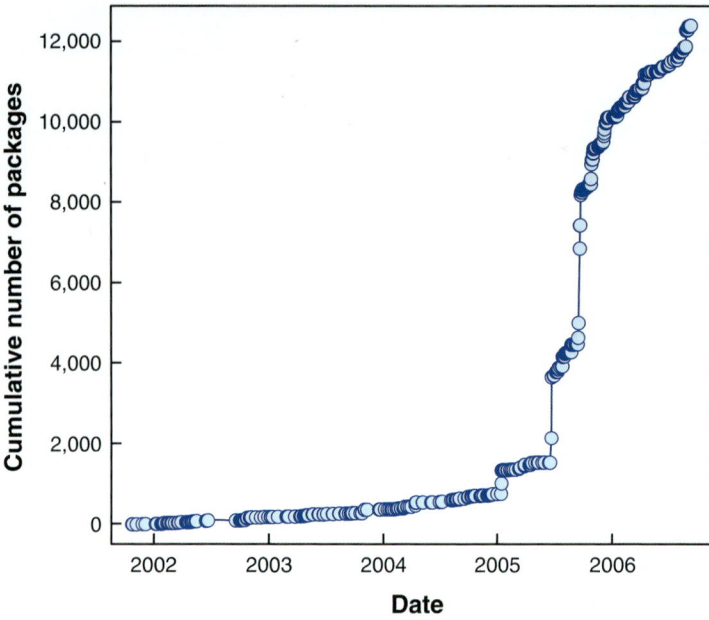

Figure 1
Cumulative number of data packages deposited in the Knowledge Network for Biocomplexity (KNB) over time. Recent advances in data sharing networks such as the KNB have promoted a surge in the number of ecological data sets available. Other systems such as the National Biological Information Infrastructure Metadata Clearinghouse are also growing rapidly (see Parr & Cummings 2005).

U.C. Natural Reserve System (UCNRS), and the NBII are adopting similar strategies of first encouraging scientists to catalog their data and later introducing the idea of archiving the raw data. The OBFS, UCNRS, and ESA data registries are all currently based on the KNB framework.

Data collections must ultimately provide easy access to the actual data, and not just the metadata, if they are to be highly useful in broad-scale synthesis studies. The KNB provides access to data through the use of EML metadata, which describes where to download the data and what format the data will be in. Because the metadata is in a machine-readable format, the process of locating, downloading, and reloading the data can be automated. This automation can significantly increase efficiency for scientists who want to utilize many data sources in a single integrated analysis or model. Other metadata systems use different mechanisms for archiving or associating the data with metadata entries.

Development of a single informatics solution to enable ecologists to gain reliable, long-term data access across these various systems is still a challenge. Recent efforts by the Science Environment for Ecological Knowledge (SEEK) project to create a uniform access interface that works with widely different data systems has seen some success. The SEEK EcoGrid interface currently provides ecologists with unified access to several of the data collections listed in **Table 1**, including the KNB Metacats, the GBIF taxonomic data portal, and the Storage Resource Broker, as well as to the Geosciences Network (GEON) data portal (Michener et al. 2005; http://seek.ecoinformatics.org). This integrated data access system has proved to be extremely useful in the Kepler analytical environment for building analyses and models that utilize heterogeneous data from one or more of these different data systems (Altintas et al. 2004; and see below).

GEON: Geosciences Network

ISBN: International Standard Book Number

ISSN: International Standard Serial Number

LSID: life science identifier

DOI: digital object identifier

4.4. Identification and Versioning of Data

Many of the data collections described in **Table 1** contain entries for the same or overlapping data sets. For example, some of the data found in the KNB are also cataloged in the NBII metadata clearinghouse. Any given data set also likely exists as multiple different versions that represent successive generations of additions, error corrections, or other changes. In addition, integrated data products used in meta-analysis and synthesis studies contain records from multiple primary data sets. Consequently, identifying unique, nonduplicated data objects within and across data repositories is labor intensive.

Standardized data identifiers that are recognized across data collections provide one potential solution for definitively identifying a data set. Standardized identifiers are used by publishers for books (ISBN) and periodicals (ISSN), but this practice is less well established within the electronic data publishing community. Several approaches are emerging within the Internet community for providing unique, location-independent identification of digital data and metadata, including the Life Science Identifier (LSID), Digital Object Identifier (DOI), and other specifications (Clark et al. 2004). These types of identifiers can be used to label unique snapshots of data so that they can be referenced permanently and unambiguously.

The ability to unambiguously reference a specific data set is especially important to the scientific ideal of repeatability. To replicate an analysis, one must reassemble the same data used in the original analysis, which is typically impossible because of current data management practices with respect to error correction and handling of revisions and updates to the data. Relational database systems, which currently provide the main repositories for large amounts of archived biological data, can be complex and dynamic frameworks that are dependent on expert support for continued operation. This can cause problems for future repeatability because these systems may not be available years later when a researcher needs to replicate an analysis. Consequently, good scientific practice requires permanently archiving snapshot versions of a data set in nonproprietary formats with an unambiguous identifier that clearly differentiates that data set from others (see Buneman et al. 2004 for another approach).

4.5. Data Curation

Several technical and nontechnical issues must be dealt with to ensure that data are adequately preserved or curated for future use (National Research Council 1997, Olson & McCord 2000). One of the most pressing issues is establishing a cyberinfrastructure that supports reliable and long-term data archives (Atkins et al. 2003). Data curation does not come free—it requires appropriate technical infrastructure to which data can be contributed, supporting data providers in the use of this infrastructure, and maintaining the day-to-day operations of the data archives.

The prevalent model for funding of scientific research overlooks the need for long-term preservation of data, with a strong focus on the production of scientific results, and their presentation in the scientific literature. The genomic and molecular biology community is an exception, with well-known and well-supported repositories for the data in the National Center for Biotechnology Information (NCBI). As noted elsewhere, the success of this effort has been facilitated by the relative uniformity and simplicity of sequence data. But the budget to sustain such a data curation center is substantial, amounting to tens of millions of dollars per year. Natural history museums are another notable exception, where digital curation of specimen data is a high priority task (Krishtalka & Humphrey 2000).

Current efforts toward curating data in other areas of biological science are less focused. There is general recognition of the need to preserve the data, but no official or authorized repositories exist for most research data. Despite interest from the digital library community in extending its archiving functions to that of scientific data preservation, scientific data archives are not considered core to their mission (Fox et al. 2002).

The NBII (**http://www.nbii.gov**) provides a registry for metadata that focuses on documenting biological data arising out of federally supported research, as well as from a growing list of agency and academic partners. As part of the United States Geological Survey (USGS), the NBII is subject to the vagaries of budgeting within the USGS, but holds great promise for constituting a robust and persistent archive for biological information. Unfortunately, the current NBII metadata clearinghouse is not a data archive. Moreover, whether the data curated by the NBII contains adequately detailed metadata for future interpretation and scientific reuse needs to be carefully evaluated.

A persistent and reliable data archive, based on well-established standards for metadata and data provisioning, especially for organismal, ecological, and environmental data, remains largely undeveloped. This deficit is also apparent throughout other scientific disciplines, including the physical and social sciences (Atkins et al. 2003, Lord & MacDonald 2003).

The technical needs regarding data curation are similar for all sciences (Freeman et al. 2005, Lord & MacDonald 2003, Raven et al. 1998). The peculiar challenge for the ecological realm is that the field is so broad that diverse types of data are potentially useful for future work (Gross & Pake 1995). Also, although computing and storage capabilities are becoming more affordable, the volume of potentially relevant data is growing even faster. This growth in data volumes is likely to accelerate as new technologies such as dense ground-based sensor networks (e.g., National Ecological Observatory Network; also see Porter et al. 2005) enable exciting new forms of science to be accomplished.

Data curation is not only technically challenging, it can also be expensive. The ongoing costs of data curation break down into three areas: hardware and software, networking, and staffing. It is difficult to clearly separate costs associated with necessary aspects of cyberinfrastructure (e.g., having a fast Internet connection, or staff to maintain databases and Web servers), from those specifically dedicated to data curation and provisioning. One recent study (Lord & MacDonald 2003) found that staffing was the most significant cost component at several data archiving sites, ranging from 69% to 82% of the total budget.

Maximizing the utility of any investment in data archiving will depend on providing adequate outreach and support so that the data can be effectively discovered and reused by scientists. Finally, clear determination of the cost/benefit of data archiving is difficult owing to the lack of clear metrics for assessing this ratio (Lord & MacDonald 2003). Still, it is undeniable that vast funds are expended on data creation and acquisition. It is false economy, and poor scientific practice, not to ensure that the data are present and useful to all users in the future.

The most cost-effective and accurate way to document data may often involve having the researcher document their own data at the time when they are collected. It will be challenging to find the balance of responsibility for documenting data between individual researchers and trained data stewards who have advanced expertise with appropriate metadata standards and technologies. One hopes that standard approaches to data curation will become common practice once appropriate tools and frameworks are in place (Jones et al. 2001, Michener et al. 1997).

5. DATA INTEGRATION, ANALYSIS, AND MODELING

The previous section described a number of informatics problems—the paucity of metadata and other documentation, incompatibility among different data management systems, the lack of persistent archives, and the need to promote sound data curation practices within the research community. In this section we introduce the possibilities of several emerging informatics technologies for providing major new capabilities to ecological researchers.

5.1. Bridging Data Islands

Much biological information is collected by researchers working relatively autonomously, carefully designing experiments and gathering data that address specific, predetermined hypotheses. This leads to many bioinformatics resources being distributed in data islands, bounded by subdisciplines with their specific focus of interests, specialized vocabularies, and entrenched traditions with regards to informatics (Davis et al. 2005). These data islands present a challenge for recently created synthesis centers in fields such as ecology, evolution, and hydrology, which depend on using existing data for their analyses. There is also a growing realization that more integrative work in biology is not only needed, but increasingly possible owing to emerging informatics solutions that will enable researchers to reach across these islands of data (**http://www.nsf.gov/od/lpa/forum/colwell/rc010324aibs.htm**) to achieve exciting new synthetic results.

Ultimately the rationale for preserving data lies in their potential reuse for addressing new hypotheses, and this usually entails significant data integration challenges.

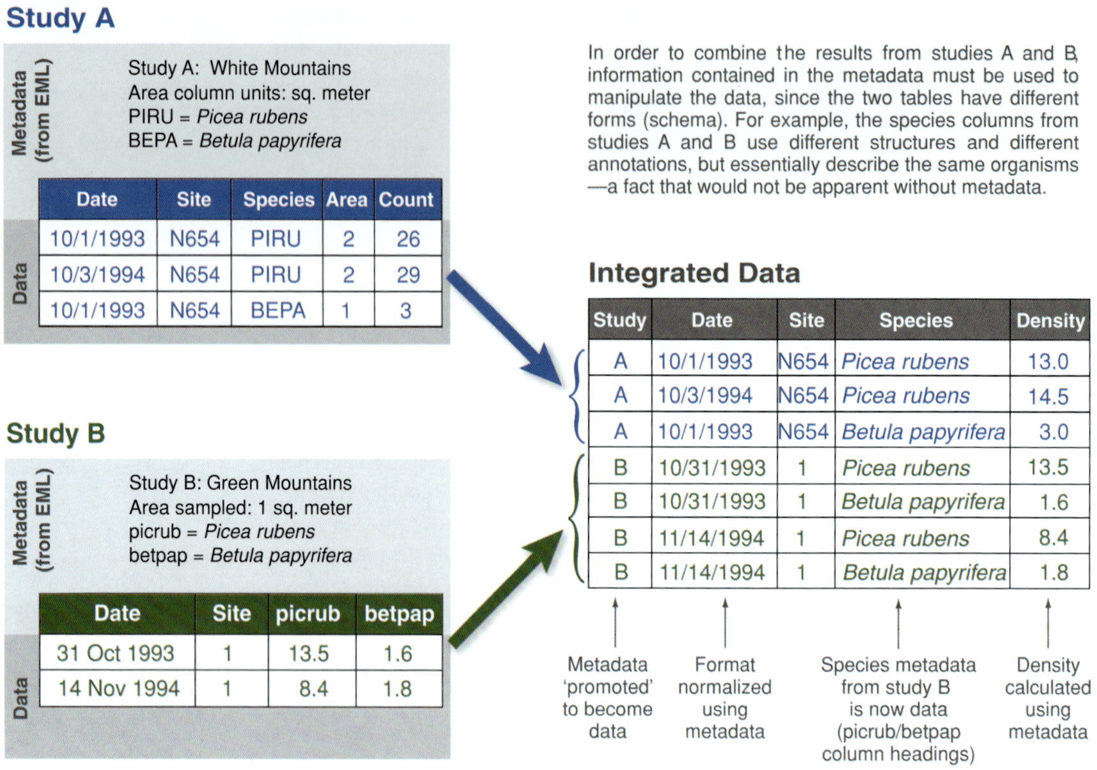

Figure 2

Data integration involves combining two or more heterogeneous but compatible data sets into a uniform product that resolves differences among the source data. Integration requires metadata about each source data set that can provide bridging information, but more importantly requires an understanding of the underlying semantics of the data in order to make reasonable decisions regarding correspondences among the source data sets.

Figure 2 shows two hypothetical source data tables (*left*) that a researcher might want to integrate for use in an analysis (*right*). Performing this integration requires resolving differences in the data format, logical data model, and semantic meaning of data. Although this can be accomplished manually through painstaking expert evaluation of the data sources, such approaches are not practical when investigators need to integrate tens or hundreds of data sources. Thus, automating data integration as much as possible is critical to advances in broad-scale synthesis studies.

Semantics: in computer science, refers to making computers capable of interacting powerfully and appropriately using familiar and meaningful concepts for humans

5.2. Traditional Data Integration Approaches

Data integration involves determining whether and how two or more data resources can be effectively combined. The issue has been widely studied in computer science, resulting in many approaches and systems for managing data differences at the system, format, data model, and semantic levels (e.g., see Haas et al. 2002, Hammer &

McLeod 1999, Ludäscher et al. 2006). Systems-level integration involves reconciling differences in network protocols (e.g., HTTP versus FTP for file transfer), operating systems, and data management applications (e.g., Oracle, Excel, and R). Systems-level integration is necessary for providing low-level support for accessing and transferring data (e.g., data transfers between Windows and Unix), but does not guarantee that an integrated data product is useful or interpretable from a scientific perspective. Format-level integration is similar, but deals with differences in data representation schemes, such as whether the data is stored in a relational (i.e., tabular) or hierarchical database system, or clarifying whether the data object is a raster file or table. Detailed metadata can often enable software applications to cope with formatting issues.

Data-model integration begins with information on how data sources are logically structured. We use the term data model (also called a schema) to refer to a number of essential and concrete features of a data set, e.g., in the case of tabular data, the definitions and data types (integer, string) of the variables (columns) of the table and how those columns might match up with columns from other tables. Structural integration focuses on matching up simple features of data sets, such as like-named variables and checking for consistent data types (e.g., integer or character string). Such information is often explicitly stored within a relational DBMS, but can be lacking in less structured data such as spreadsheets.

One traditional approach to data integration that requires working with data models is data federation (see Haas et al. 2002). In this approach, data integration requires defining a single, global data model (or global schema). Once clarified, the schemas of the local data sources are mapped to the global schema so that one can issue a query against the global schema. The system then uses those mappings to retrieve the associated data from the local databases. This process is a formal description of what one does when creating the vertically integrated database warehouses described above.

The utility of the federation approach is hampered by the difficulty of defining a useful global schema (Batini et al. 1992). Moreover, the federation process cannot be easily extended, because every additional data set to be integrated may require significant modification to the global schema, or lead to major compromises in developing the local schema. An additional potential impediment to federating data via a global schema is that biological databases are often structured for specific usages and have closer affinities with on-line analytical processing (OLAP) and multidimensional databases than a typical relational database (Gray et al. 1997, Pedersen & Jensen 2001, Shoshani 2003). Automated matching of data models (Leser & Naumann 2005, Rahm & Bernstein 2001) as well as peer-to-peer-based integration methods (Bernstein et al. 2002) are also under development to assist in data integration. All of these approaches are likely to benefit from incorporating semantics (Bowers et al. 2004a, Ludäscher et al. 2003, Paton et al. 1999).

5.3. Semantic Approaches to Data Integration

One of the most challenging hurdles to integrating data is uncertainty about its precise meaning—for the data set as a whole, its individual variables (in the case of a table),

Ontology: a formal model of knowledge in a particular subject area useful in making inferences about data

OWL: Web Ontology Language

or even the broader context that motivated its collection. This occurs because the semantics or meaning of some aspects of the data often are unclear. For example, column labels of *wt*, *bm*, and *LL* in separate tables might all refer to a measure of "biomass of leaf litter," but this biologically meaningful concept is not explicit in the abbreviated labels and might not be in any of the metadata. The areas within computer science that deal with clarifying these semantic issues include conceptual data modeling, knowledge representation and semantic mediation, and the Semantic Web (Antoniou & van Harmelen 2004, Berners-Lee et al. 2001, Ludäscher et al. 2003).

Semantic integration involves clarifying data content in ways that are similar to controlled vocabularies, but using more powerful formal structures known as ontologies. Ontologies establish a set of well-defined concepts or terms of interest within a domain and clearly specify how these terms are interrelated (Baker et al. 1999, Brilhante 2003, Horridge et al. 2004, Noy & Hafner 2000, Rector et al. 2004). Owing to the formal logical structure of ontologies, computer-based reasoning systems can use them to draw inferences or conclusions (Gali et al. 2004, Horridge et al. 2004, Sowa 2000). This enables ontologies to help identify important aspects of a data set that were hidden, or implicit. For example, a data set might contain information for a count of organisms and an area over which this measurement was taken. An ontology could be used to identify that density (count divided by area) is implicit in this data set. Other examples of the use of ontologies for integrating ecological data can be found in Bowers et al. (2004b).

The genomics community has made significant advances using ontologies to facilitate data integration by unifying terminologies among communities of genetic researchers working with different model systems (originally fruit fly, mouse, and yeast). These groups of researchers recognized that their work had much in common, but that communication was hindered owing to subdisciplinary variations in terminology. The Gene Ontology project arose as a collaborative effort to develop a structured, controlled vocabulary for associating gene products with their cellular location, molecular function, and biological process, regardless of the model system. The resulting Gene Ontology facilitates a better understanding of the structure and function of genes across taxa (Ashburner et al. 2000).

The creation of robust and useful ontologies requires understanding of formal logic (Baader et al. 2003, Rector et al. 2004), as well as proficiency with specialized software tools (e.g., Protégé; Horridge et al. 2004). The general approach today involves using the OWL Web Ontology Language (McGuinness & Van Harmelen 2004) along with Description Logic reasoners (e.g., Pellet or Racer; Sirin et al. 2005). In ecology, the SEEK (http://seek.ecoinformatics.org) and SPIRE (http://spire.umbc.edu) projects are developing ontologies to investigate how semantic approaches can assist with data integration (Bowers et al. 2004a,b). These are challenging tasks involving knowledge engineers and computer scientists working closely with biologists to identify and explicate the concepts and relationships that are meaningful for interpreting their data. Although constructing robust ontologies requires considerable skill (Guarino & Welty 2002, Pinto & Martins 2004), and these capabilities are still nascent within the ecological community, we believe that semantic approaches to data integration are likely to remain an active frontier in bioinformatics for the near future and could be extremely important for a diverse, complicated discipline like ecology.

5.4. Analysis and Modeling Using Scientific Workflow Systems

Scientific workflow: formal, executable model of a quantitative process linking discrete analytical modules; used to document analytical processing, data flow and provenance

As more data become available and techniques improve for integrating datasets, the bottleneck in synthesis increasingly becomes analysis and modeling support. Using spreadsheets and scripted analysis frameworks (e.g., R, SAS, and Perl), scientists are often limited to accomplishing their analyses using the techniques and tools available within the particular software package. For many analyses, however, various techniques and tools are desired across multiple software packages, which invariably results in the process of exporting data from one package and importing it into another. Combining software packages in this way is often challenging and makes tracing and reconstructing the flow of data among applications difficult. As a consequence, information about the analytical procedures that were applied to data is frequently lost. Thus, the process used to obtain a particular analytical result can be difficult to replicate.

Several projects are working to solve these issues by explicitly modeling the flow of data through an entire analytical process (Ludäscher et al. 2005, Osterweil et al. 2006). These scientific workflow systems typically support multiple analytical frameworks and components and have been successfully used in a variety of disciplines, including ecology, the geosciences, molecular biology, and other areas where data access, modeling, and visualization are complex and multistaged (Altintas et al. 2004, Deelman et al. 2004, Ellison et al. 2006, McPhillips & Bowers 2005, Oinn et al. 2000, Pennington & Michener 2005, Taylor et al. 2003).

There are several advantages of scientific workflow systems. First, they provide a formal description of the analytical steps used in a process. Second, they often provide direct access to data sources that would otherwise require significant effort to gather and collate (Shankar et al. 2005). For example, the Kepler scientific workflow system (**Figure 3**; http://www.kepler-project.org) is being developed by several scientific communities (Altintas et al. 2004). For ecologists, Kepler provides direct access to ecological data from hundreds of field stations, collections data from various natural history museums, and molecular biology data through services such as GenBank. Third, scientific workflow systems provide a number of tools for managing data including query, discovery, and integration support (Berkley et al. 2005). Fourth, scientific workflow systems typically provide high-level graphical user interfaces for constructing complex analytical processes (see **Figure 3**) and can display analytical results in a clear and intuitive way to scientists.

Workflows are similar to scripted systems like R in that they allow a formal process to be completely specified, but they have the added advantage of being readily understandable to nonexperts. Scientific workflow systems also focus on reusability, e.g., analytical tools from multiple software packages can be incorporated into a single workflow, and workflows can typically be separated into reusable modules,

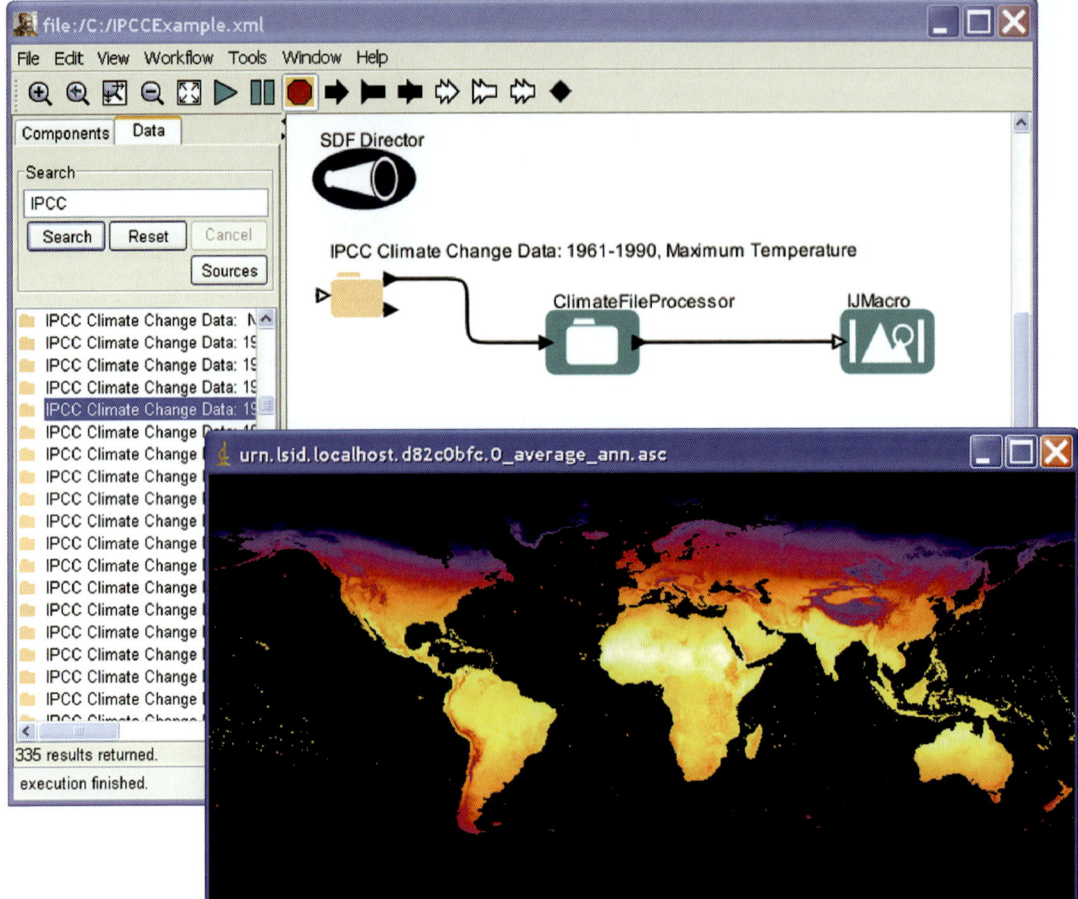

Figure 3

The Kepler scientific workflow system provides a visual model of the flow of data and access to diverse data sources. This workflow shows data from the Intergovernmental Panel on Climate Change (IPCC) that was retrieved from the EcoGrid, processed through a workflow, and visualized as a map.

making complex workflows simpler to understand. Finally, scientific workflows are themselves a form of metadata that can be easily archived and shared with colleagues.

6. CULTURAL ISSUES

There are also nontechnical, cultural, and sociological hurdles to making ecological data broadly available (Palmer et al. 2005). One problem is the reluctance to incur additional operational overhead by learning unfamiliar tools to manage data. This is difficult to overcome even for one's own data, but may be especially onerous when directed toward enhancing the ability of others to use one's data. Many scientists are reluctant to make their data freely available for fear that others will use it before they themselves have extracted as much as possible for their own analyses and publications. This reluctance is understandable in some cases—imagine conducting a 10-year project during which one makes the data publicly available as it is gathered. It is possible that someone outside the project might use the data available for the first seven or eight years to scoop the primary investigators. However, the opportunities that arise from new collaborations due to data sharing will likely outnumber any unethical use of data, which should be quickly self-correcting owing to the offenders being ostracized and cut out of the flow of funding and scholarship.

Perhaps the most pervasive cultural factor stalling access to digital data is our reward system. Publishing in respectable journals, securing extramural funding, and training the next generation of scholars to do the same are the primary criteria by which we gain the respect of peers and are formally evaluated. Ecological researchers lack an incentive to invest in making their data more broadly available to other scientists because it is neither rewarded nor appreciated and reduces time for respected activities (Olson & McCord 2000). Although it will ultimately benefit science to make data openly accessible, the personal advantages to doing so are not clear in the short term.

Despite these circumstances, Cech et al. (2003) list five principles and 10 recommendations that revolve around the obligation of authors to make data and other materials publicly available. The report stresses that funding entities should require that data gathered under their auspices be made available, that the funding entity provide the resources to do so, and that scientists have a responsibility to the science community to make all data, algorithms, and other information associated with a publication publicly available.

These are admirable goals that have been met, to some extent, in a few disciplines. For example, authors publishing gene sequences must submit their data to GenBank and publish an accession number. GenBank is, however, a relatively simple database of uniform entries compared to the heterogeneity of ecological data. Furthermore, the requirement to publish accession numbers was not an altruistic action on the part of authors or publishers. Rather, publishers could not invest the time and money required to proof and print the massive gene sequences in journals. Currently, there is no such driver to force ecologists to do the same.

Parr & Cummings (2005) comment on how data sharing has transformed other fields and address the points raised above about what has inhibited ecologists from fully participating in the data-sharing revolution. They suggest that it is short-sighted to restrict access to one's data and that the technical barriers to sharing information are illusory. Although not as sanguine about these issues, we believe there are solutions to the cultural hurdles to data sharing.

Ecologists must begin to appreciate efforts made to make data openly available and reward these efforts accordingly. Peers, department heads, deans, and others who evaluate and fund research must recognize that making data openly available is an integral component of research, similar in value to the highly regarded service performed by reviewing journals and grant proposals. We must place a similar value on data sharing.

7. A NEW WORLD OF ONLINE DATA

Major technology advances in ecology have included vast increases in the computing power available to analyze and model patterns and processes, enhanced means to measure and record events and observations, and new methods for exchanging information one-to-one or over the Web. It is ironic then, that the vast majority of ecological data still remain virtually undetectable and inaccessible.

As this situation improves through technological and cultural advances, we can expect changes similar to those occurring with on-line journal publications. For example, citations of articles available electronically are two to five times higher than for those available only in print (T.C. Bergstrom, personal communication). Furthermore, Tenopir et al. (2003) show that the number of articles read and the time spent reading by scientists significantly increases as journals become available electronically. We expect that the changes will be even more dramatic for electronic access to data, because paper journals have been widely available for centuries whereas most ecological data are not accessible in any format. We can also be certain that entirely new ways of using data will emerge, some imaginable (e.g., data analysis might replace reading abstracts or even scientific articles as a means to test ideas for a dissertation or research project) and others beyond our current imagination.

As ecological research becomes more complicated with the addition of critical information from adjacent disciplines (Palmer et al. 2005), access to an ever broader array of data will be indispensable. Although such a capability will advance knowledge about natural systems, it is absolutely essential for the wise conservation and management of natural resources, to the extent that it is unconscionable not to move rapidly toward open access to data. Consider again the headlines of recent articles in *Science* (Kaiser 2003) and *Nature* (Whitfield 2003) about the demise of gorilla populations—"Ebola, Hunting Push Ape Populations to the Brink" and "Ape Populations Decimated by Hunting and Ebola Virus" Although the underlying issue is ecological in nature, the information needed to address this concern comes from many domains, including economics, local culture, disease, and complex population dynamics. It will only be through fundamental advances in informatics that researchers will be able to efficiently access and analyze the diversity of data needed to address such questions in a holistic and comprehensive way. With the clear advantages of online, open access to data, to both our discipline and our planet, we cannot delay in the development and adoption of advanced bioinformatics solutions for enhancing digital access to ecological information.

SUMMARY POINTS

1. There is a critical need for more synthetic and integrative analyses in ecology to better understand and wisely manage the Earth's biological resources.
2. Synthetic analyses in ecology require information from multiple disciplines and perspectives, which raises significant challenges in accessing and integrating relevant data.

3. Although a number of ecological data archives exist, much ecological data is still unavailable.

4. Advances in ecoinformatics are addressing issues regarding data access and integration by adopting semantic approaches and building scientific workflow systems to assist researchers in locating and documenting their data and analyses.

ACKNOWLEDGMENTS

We thank Eric Seabloom, Jim Regetz, and Josh Madin for valuable comments on the manuscript, Leslie Allfree for her administrative support, and Matthew Brooke for assistance with the graphics. Jones, Schildhauer and Reichman are at the National Center for Ecological Analysis and Synthesis, a Center funded by the National Science Foundation (DEB-0072909 and DBI-0225676), the University of California, and the Santa Barbara campus. Bowers is supported by NSF Grant DBI-0533368.

LITERATURE CITED

Altintas I, Berkley C, Jaeger E, Jones M, Ludäscher B, Mock S. 2004. Kepler: an extensible system for design and execution of scientific workflows. *Proc. 16th Int. Conf. Sci. Stat. Database Manag.*, *Santorini Island, Greece*

Andelman SJ, Bowles CM, Willig MR, Waide RB. 2004. Understanding environmental complexity through a distributed knowledge network. *BioScience* 54(3):240–46

Antoniou G, van Harmelen F. 2004. *A Semantic Web Primer*. Cambridge, MA: MIT Press. 272 pp.

Ashburner M, Ball CA, Blake JA, Botstein D, Butler H, et al. 2000. Gene ontology: Tool for the unification of biology. *Nat. Genet.* 25:25–29

Atkins DE, Droegemeier KK, Feldman SI, Garcia-Molina H, Klein ML, et al. 2003. Revolutionizing science and engineering through cyberinfrastructure. *Rep. Natl. Sci. Found. Blue-Ribbon Advis. Panel Cyberinfrastructure* **http://www.communitytechnology.org/nsf_ci_report**

> Provides a blueprint of the information technology needs for U.S. scientists and outlines the importance of access to data archives.

Attig J, Copeland A, Pelikan M. 2004. Context and meaning: The challenges of metadata for a digital image library within the university. *Coll. Res. Libr.* 65(3):251–61

Baader F, Calvanese D, McGuinness D, Nardi D, Patel-Schneider P. 2003. *The Description Logic Handbook: Theory, Implementation, and Applications*. Cambridge, UK: Cambridge Univ. Press. 574 pp.

Baker P, Goble C, Bechhofer S, Paton N, Stevens R, Brass A. 1999. An ontology for bioinformatics applications. *Bioinformatics* 15(6):510–20

Batini C, Ceri S, Navathe SB. 1992. *Conceptual Database Design: An Entity-Relationship Approach*. Redwood City, CA: Benjamin Cummings

Benson DA, Karsch-Mizrachi I, Lipman DJ, Ostell J, Wheeler DL. 2005. GenBank. *Nucleic Acids Res.* 33:D34–38

Berkley C, Bowers S, Jones MB, Ludäscher B, Schildhauer M, Tao J. 2005. Incorporating semantics in scientific workflow authoring. *Proc. 17th Int. Conf. Sci. Stat. Database Manag.*, IEEE Comput. Soc.

> Describes the Kepler scientific workflow system and how it uses ontologies to provide advanced data discovery and analysis.

Berkley C, Jones MB, Bojilova J, Higgins D. 2001. Metacat: a schema-independent XML database system. *Proc. 13th Int. Conf. Sci. Stat. Database Manag.*, IEEE Comput. Soc.

Berners-Lee T, Hendler J, Lassila O. 2001. The Semantic Web. *Sci. Am.* 284(5):34–43

> General introduction to the approaches and expected advantages of using ontologies to enhance Web resources.

Bernstein PA, Guinchiglia F, Kementsietsidis A, Mylopoulos J, Serafini L, Zaihrayeu I. 2002. Data management for peer-to-peer computing: a vision. *Proc. Int. Workshop Web Databases (WebDB)*, pp. 89–94

Bowers S, Lin K, Ludäscher B. 2004a. On integrating scientific resources through semantic registration. *Proc. 16th Int. Conf. Sci. Stat. Database Manag.*, pp. 349–52. IEEE Comput. Soc.

Bowers S, Thau D, Williams R, Ludäscher B. 2004b. Data procurement for enabling scientific workflows: on exploring interant parasitism. *Proc. 2nd Int. Workshop Semantic Web Databases (SWDB), LNCS*

Bowker GC. 2000. Biodiversity datadiversity. *Soc. Stud. Sci.* 30(5):643–83

Brilhante VB. 2003. *Ontology and reuse in model synthesis*. PhD thesis. Univ. Edinburgh

Brown JH, Munger JC. 1985. Experimental manipulation of a desert rodent community: Food addition and species removal. *Ecology* 66(5):1545–63

Brunt JW. 2000. Data management principles, implementation, and administration. See Michener & Brunt 2000, 2:25–47

Buneman P, Khanna S, Tajima K, Tan W. 2004. Archiving scientific data. *ACM Trans. Database Syst. (TODS)* 29(1):2–42

Canhos VP, Souza S, Giovanni R, Canhos DAL. 2004. Global biodiversity informatics: setting the scene for a "new world" of ecological modeling. *Biodiver. Inf.* 1:1–13

> Articulates the need for access to more data for effective global modeling of species distributions.

Cech TR, Eddy SR, Eisenberg D, Hersey K, Holtzman SH, et al. 2003. Sharing publication-related data and materials: responsibilities of authorship in the life sciences. *Natl. Res. Counc. Rep. Comm. Responsib. Authorship Biol. Sci.* Washington, DC: Natl. Acad. Press

> Presents a thorough overview of the cultural and ethical issues surrounding data sharing in the life sciences.

Clark T, Martin S, Liefeld T. 2004. Globally distributed object identification for biological knowledgebases. *Brief. Bioinform.* 5(1):59–70

Connell JH. 1961. The influence of interspecific competition and other factors on the distribution of the barnacle *Chthamalus stellatus*. *Ecology* 42(4):710–23

Daniel R, Lagoze C, Payette SD. 1998. A metadata architecture for digital libraries. *Proc. IEEE Int. Forum Res. Technol. Adv. Digital Libr.*, pp. 276–88. IEEE Comput. Soc., Los Alamitos, CA

Davis MA, Pergl J, Truscott AM, Kollmann J, Bakker JP, et al. 2005. Vegetation change: a reunifying concept in plant ecology. *Perspect. Plant Ecol. Evol. Syst.* 7:69–76

Deelman E, Blythe J, Gil Y, Kesselman C, Mehta G, et al. 2004. Pegasus: Mapping scientific workflows onto the grid. *Proc. Eur. Grids Conf., 2nd, Nicosia, Cyprus*

Dekkers M, Weibel S. 2003. State of the Dublin Core Metadata initiative. *D-Lib Mag.*, 9 (April) No. 4

Ellison AM, Osterweil LJ, Hadley JL, Wise A, Boose E, et al. 2006. Analytic webs support the synthesis of ecological datasets. *Ecology* 87:1345–58

Fox EA, Moore RW, Larsen RL, Myaeng SH, Kim SH. 2002. Toward a global digital library: generalizing US-Korea collaboration on digital libraries. *D-Lib Mag.*, 8(Oct.) No. 10

Freeman PA, Crawford DL, Kim S, Muñoz JL. 2005. Cyberinfrastructure for Science and engineering: promises and challenges. *Proc. IEEE* 93(3):682–91

Frondorf A, Jones MB, Stitt S. 1999. Linking the FGDC geospatial metadata content standard to the biological/ecological sciences. *Proc. 3rd IEEE Comput. Soc. Metadata Conf., Bethesda, MD*, April 6–7

Gali A, Chen C, Claypool K, Uceda-Sosa R. 2004. From ontology to relational databases. *Int. Workshop Concept.-Model Driven Web Inf. Integr. Min., Shanghai, China*

Goodchild MF. 2003. Geographic information science and systems for enviromental management. *Annu. Rev. Environ. Resourc.* 28:493–519

Grassle FJ. 2000. The Ocean Biogeographic Information System (OBIS): an on-line, worldwide atlas for accessing, modeling and mapping marine biological data in a multidimensional geographic context. *Oceanography* 13:5–7

Gray J, Chaudhuri S, Bosworth A, Layman A, Reichart D, et al. 1997. DataCube: A relational aggregation operator generalizing group-by, cross-tab, and subtotals. *Data Min. Knowl. Discov.* 1:29–53

Green JL, Hastings A, Arzberger P, Ayala FJ, Cottingham KL, et al. 2005. Complexity in ecology and conservation: mathematical, statistical, and computational challenges. *BioScience* 55(6):501–10

Good summary of the current possibilities and future needs for more effectively using technology in ecology and conservation.

Gross KL, Pake CE, eds. 1995. *The Future of Long-term Ecological Data. A Report to the Ecological Society of America.* Vol. I: *Text of the Report.* 123 pp. Vol. II: *Directories to Sources of Long-term Ecological Data.* 114 pp. http://intranet.lternet.edu/archives/documents/other/fledvol2.pdf

Guarino N, Welty C. 2002. Evaluating ontological decisions with ONTOCLEAN. *Commun. ACM* 45(2):61–65

Haas LM, Lin ET, Roth MT. 2002. Data integration through database federation. *IBM Syst. J.* 41(4):578–96

Hammer J, McLeod D. 1999. Resolution of representational diversity in multi-database systems. In *Management of Heterogeneous and Autonomous Database Systems*, ed. A Elmagarmid, M Rusinkiewicz, A Sheth, 4:91–117. San Francisco: Morgan Kaufmann. 413 pp.

Horridge M, Knublauch H, Rector A, Stevens R, Wroe C. 2004. *Practical Guide to Building OWL Ontologies Using the Protégé-OWL Plugin and CO-ODE Tools Edition 1.0.* Manchester, UK: Univ. Manchester. 118 pp. http://www.co-ode.org/resources/tutorials/ProtegeOWLTutorial.pdf

Jennings M, Faber-Langendoen D, Peet R, Loucks O, Glenn-Lewin D, et al. 2004. Guidelines for describing associations and alliances of the U.S. National Vegetation Classification. Version 4.0. Washington, DC: Ecol. Soc. Am.

Jones MB, Berkley C, Bojilova J, Schildhauer M. 2001. Managing scientific metadata. *IEEE Internet Comput.* 5(5):59–68

Kaiser J. 2003. Conservation biology - Ebola, hunting push ape populations to the brink. *Science* 300:232

Knapp AK, Smith MD, Collins SL, Zambatis N, Peel M, et al. 2004. Generality in ecology: testing North American grassland rules in South African savannas. *Front. Ecol. Environ.* 2(9):483–91

Krishtalka L, Humphrey PS. 2000. Can natural history museums capture the future? *BioScience* 50:611–17

Leser U, Naumann F. 2005. (Almost) Hands-off information integration for the life sciences. *Conf. Innovative Database Syst. Res.*, pp. 131–43

Lord P, MacDonald A. 2003. Data curation for e-science in the UK: an audit to establish requirements for future curation and provision. *JISC Comm. Support Res. (JCSR).* **http://www.jisc.ac.uk/uploaded_documents/e-ScienceReportFinal.pdf**

Lubchenco J, Real LA. 1991. Manipulative experiments as tests of ecological theory. See Real & Brown 1991, pp. 715–33

Ludäscher B, Altintas I, Berkley C, Higgins D, Jaeger-Frank E, et al. 2005. Scientific workflow management and the Kepler system. *Concurr. Comput.: Pract. Exp.* Special issue Scientific Workflow

Ludäscher B, Gupta A, Martone ME. 2003. A model-based mediator system for scientific data management. In *Bioinformatics: Managing Scientific Data*, ed. Z Lacroix, T Critchlow, 12:335–70. San Francisco: Morgan Kaufman

Ludäscher B, Lin K, Bowers S, Jaeger-Frank E, Brodaric B, Baru C. 2006. Managing scientific data: from data integration to scientific workflows. *GSA Special Paper 397, Geoinformatics: From Data to Knowledge*, ed. A. Krishna Sinha, pp. 109–30. Boulder, CO: Geol. Soc. Am.

McCullough BD, Wilson B. 1999. On the accuracy of statistical procedures in Microsoft Excel 97. *Comput. Stat. Data Anal.* 31:27–37

McGuinness DL, Van Harmelen F. 2004. OWL Web ontology language overview. *W3C Recommendation*, Feb. 10. **http://www.w3.org/TR/owl-features**

McPhillips TM, Bowers S. 2005. An approach for pipelining nested collections in scientific workflows. *SIGMOD Rec.* 34:12–17

Michener WH. 2006. Meta-information concepts for ecological data management. *Ecol. Inform.* 1:3–7

Michener WK. 2000. Metadata. See Michener & Brunt 2000, pp. 5:92–116

Michener WK, Brunt JW, eds. 2000. *Ecological Data: Design, Management and Processing*. Oxford: Blackwell. 180 pp.

Michener WK, Brunt JW, Helly JJ, Kirchner TB, Stafford SG. 1997. Non-geospatial metadata for the ecological sciences. *Ecol. Appl.* 7:330–42

The seminal paper articulating the need to better preserve ecological data through metadata.

Michener WK, Reichman OJ, Beach J, Jones MB, Ludäscher B, et al. 2005. Creating and providing data management services for the biological and ecological sciences: Science environment for ecological knowledge. *Proc. 17th Int. Conf. Sci. Stat. Database Manag.*, IEEE Comput. Soc.

Morell V. 1996. TreeBASE: The roots of phylogeny. *Science* 273:569

Nair SS, Jeevan VKJ. 2004. A brief overview of metadata formats. *DESIDOC Bull. Inf. Technol.* 24(4):3–11

Natl. Res. Counc., Comm. Issues Transborder Flow. 1997. *Bits of Power: Issues in Global Access to Scientific Data*. Washington, DC: Natl. Acad. Press. 235 pp.

Noy N, Hafner C. 2000. Ontological foundations for experimental science knowledge bases. *Appl. Artif. Intell.* 14(6):565–618

Oinn T, Greenwood M, Addis M, Alpdemir MN, Ferris J, et al. 2005. Taverna: Lessons in creating a workflow environment for the life sciences. *Concurr. Comput.: Pract. Exp.* 18:1067–1100

Olson RJ, McCord RA. 2000. Archiving ecological data and information. See Michener & Brunt 2000, 6:117–41

Osterweil LJ, Wise A, Clarke LA, Ellison AM, Hadley JL, et al. 2006. Process technology to facilitate the conduct of science. *Lect. Notes Comput. Sci.* 3840:403–15

Paine RT. 1966. Food web complexity and species diversity. *Am. Nat.* 100:65–75

Palmer M, Bernhardt ES, Chornesky EA, Collins SL, Dobson AP, et al. 2005. Ecological science and sustainability for the 21st century. *Front. Ecol. Environ.* 3:4–11

Parr CS, Cummings MP. 2005. Data sharing in ecology and evolution. *Trends Ecol. Evol.* 20(7):362–63

Pascal F. 2000. *Practical Issues in Database Management: A Reference for the Thinking Practitioner*. Boston: Addison-Wesley. 256 pp.

> One of many good introductions to the conceptual issues underlying data modeling and management.

Paton NW, Stevens R, Baker P, Goble CA, Bechhofer S, Brass A. 1999. Query processing in the TAMBIS bioinformatics source integration system. *Proc. 11th Int. Conf. Sci. Stat. Database Manag.*, pp. 138–47

Pedersen TB, Jensen CS. 2001. Multidimensional database technology. *IEEE Comput.* 34(12):40–46

Peet RK. 1998. ESA journals: Evolution and revolution. *Bull. Ecol. Soc. Am.* 79:177–81

Pennington DD, Michener WK. 2005. The EcoGrid and the Kepler Workflow System: A new platform for conducting ecological analyses. *Bull. Ecol. Soc. Am.* 86(3):169–76

Pinto HS, Martins JP. 2004. Ontologies: how can they be built? *Knowl. Inf. Syst.* 6:441–64

Porter J, Arzberger P, Braun HW, Bryant P, Gage S, et al. 2005. Wireless sensor networks for ecology. *BioScience* 55:561–72

Porter JH. 2000. Scientific databases. See Michener & Brunt 2000, 3:48–69

Rahm E, Bernstein PA. 2001. A survey of approaches to automatic schema matching. *VLDB J.* 10(1):334–50

Raven P, Ayala FJ, Bowker GC, Colwell RR, Cracraft JL, et al. 1998. *Teaming with Life: Investing in Science to Understand and Use America's Living Capital*. President's Comm. Advis. Sci. Technol., Panel Biodivers. Ecosyst., Washington, D.C.

Real LA, Brown JH, eds. 1991. *Foundations of Ecology: Classic Papers with Commentaries*. Chicago: Univ. Chicago Press. 920 pp.

Rector A, Drummond N, Horridge M, Roger J, Knublauch H, et al. 2004. OWL Pizzas: Practical experience of teaching OWL-DL: Common errors and common patterns. *Int. 14th Conf. Knowl. Eng. Knowl. Manag. (EKAW)*. Northamptonshire, UK: Whittlebury Hall

Shankar S, Kini A, DeWitt DJ, Naughton J. 2005. Integrating databases and workflow systems. *ACM SIGMOD Rec.* 34(3):5–11

Shoshani A. 2003. Multidimensionality in statistical, OLAP, and scientific databases. In *Multidimensional Databases: Problems and Solutions*, ed. M Rafanelli, pp. 46–68. Hershey, PA: Idea Group

Sirin E, Parsia B, Grau BC, Kalyanpur A, Katz Y. 2005. *Pellet: a practical OWL-DL reasoner.* **http://www.mindswap.org/papers/PelletJWS.pdf**

Sowa JF. 2000. *Knowledge Representation: Logical, Philosophical, Conceptual Foundations*. Pacific Grove, CA: Brooks Cole. 594 pp.

Taylor I, Shields M, Wang I, Rana O. 2003. Triana applications within grid computing and peer to peer environments. *J. Grid Comput.* 1:199–217

Tenopir C, King DW, Boyce P, Grayson M, Zhanf Y, Ebuen M. 2003. Patterns of journal use by scientists through three evolutionary phases. *D-Lib Mag.* 9(May) No. 5

Theile H. 1998. The Dublin Core and Warwick Framework. *D-Lib Mag.*, Jan. 4. **http://www.dlib.org**

Thomas CE, Ganji G. 2006. Integration of genomic and metabonomic data in systems biology: are we 'there' yet? *Curr. Opin. Drug Discov. Dev.* 9(1):92–100

Weibel S. 1995. Metadata: the foundations of resource description. *D-Lib Mag.*, July 1. **http://www.dlib.org**

Whitfield J. 2003. Ape populations decimated by hunting and Ebola virus. *Nature* 422:551

Incorporating Molecular Evolution into Phylogenetic Analysis, and a New Compilation of Conserved Polymerase Chain Reaction Primers for Animal Mitochondrial DNA

Chris Simon,[1,2] Thomas R. Buckley,[3] Francesco Frati,[4] James B. Stewart,[5,6] and Andrew T. Beckenbach[5]

[1]Ecology and Evolutionary Biology, University of Connecticut, Storrs, Connecticut 06269; email: chris.simon@uconn.edu

[2]School of Biological Sciences, Victoria University of Wellington, Wellington 6014, New Zealand

[3]Landcare Research, Auckland 1142, New Zealand; email: BuckleyT@landcareresearch.co.nz

[4]Department of Evolutionary Biology, University of Siena, 53100 Siena, Italy; email: frati@unisi.it

[5]Department of Molecular Biology and Biochemistry, Simon Fraser University, Burnaby, British Columbia V5A 1S6, Canada; email: jbs@alumni.sfu.ca, beckenba@sfu.ca

[6]Department of Laboratory Medicine, Division of Metabolic Diseases, Karolinska Institutet, Norvum 141 86, Stockholm, Sweden

Key Words

among-site rate variation, covarion-like evolution, molecular clocks, mtDNA genomes, nodal support, PCR primers

Abstract

DNA data has been widely used in animal phylogenetic studies over the past 15 years. Here we review how these studies have used advances in knowledge of molecular evolutionary processes to create more realistic models of evolution, evaluate the information content of data, test phylogenetic hypotheses, attach time to phylogenies, and understand the relative usefulness of mitochondrial and nuclear genes. We also provide a new compilation of conserved polymerase chain reaction (PCR) primers for mitochondrial genes that complements our earlier compilation.

mtDNA: mitochondrial DNA

Gene order rearrangement: an evolutionary change in the location and/or direction of transcription of a gene with respect to other genes

PCR: polymerase chain reaction

rRNA: ribosomal RNA

INTRODUCTION

The properties of the genes used and our ability to accommodate these properties have a much larger influence on the outcome of a molecular phylogenetic analysis than the particular method chosen to build a tree [although there are good reasons for preferring some phylogenetic methods over others (Holder & Lewis 2003; Swofford et al. 1996, 2001)]. It is from a careful study of data and their properties that empiricists can gain insight into the type of analyses needed (Simon 1991, Simon et al. 1994). Here we update our previous review of the evolution, weighting, and phylogenetic utility of mitochondrial genes and expand the focus from insects to all animals and from mitochondria to all DNA—although many of our examples still come from mitochondrial DNA (mtDNA). We summarize advances that have been made in the past 12 years especially in the area of (*a*) accommodating rate variation among sites, among data partitions, and among lineages; (*b*) understanding the information content of data; and (*c*) taking advantage of the relative phylogenetic usefulness of mitochondrial and nuclear genes. We do not include a section on the phylogenetic usefulness of different mtDNA genes because this has been updated for animals by others (e.g., Lin & Danforth 2004, Meyer & Zardoya 2003). Similarly, useful reviews have appeared recently that focus on animal mitochondrial genome evolution (Boore et al. 2005), cytonuclear coevolution (Burger et al. 2003, Rand et al. 2004), mechanisms of gene order rearrangement (Boore 2000), the use of mtDNA in phylogeographic/species-level studies (Funk & Omland 2003), and the population biology of mtDNA (Ballard & Rand 2005). Finally, we include as a web-resource an updated compilation of conserved mtDNA polymerase chain reaction (PCR) primers (see the Supplemental Appendix; follow the Supplemental Material link from the Annual Reviews home page at **http://www.annualreviews.org/**) using the standardized naming system of Simon et al. (1994). This compilation contains 70 new primers that are useful for sequencing large sections of the mitochondrial genome.

The Beginnings of Molecular Systematics and the Rapid Pace of Change: The Influence of Molecular Technology

In 2003, the world celebrated the 50th anniversary of the discovery of the structure of DNA. Since 1953, DNA sequences have been incorporated into every aspect of biology. The development of molecular technology and subsequent production of data have dictated the direction of molecular phylogenetics. Despite the advances introduced by chain termination sequencing (Sanger et al. 1977), DNA sequencing was difficult and slow before the advent of PCR (Saiki et al. 1985); molecular phylogenetic analysis was therefore largely based on amino acid sequences, immunological distances, DNA-DNA hybridization, allozymes, and mitochondrial DNA restriction site mapping (reviewed in Simon 1991). Before the development of large batteries of conserved PCR primers for mitochondrial DNA (e.g., Kocher et al. 1989, Simon et al. 1994), direct sequencing of RNA was easier than sequencing of DNA, and large data sets of 18S ribosomal RNA (rRNA) accumulated and grew for comparative purposes. Thus was set into motion the collection of a large amount of sequence data for a severely problematic—thus very interesting—macromolecule simply because of sequencing technology. In the early to mid-1990s, PCR primers and the development of fast and efficient automated sequencing machines greatly increased the rate of collection of DNA data. The sequencing of complete organelle and nuclear genomes has greatly facilitated the development of additional PCR primers and the selection of genes. Despite the promise of nuclear genes (Zhang & Hewitt 2003), mtDNA still remains the most used genome in animal phylogenetics for studies of mid- to late Cenozoic-age divergences because of its faster rate of evolution, ease of sequencing, paucity of visible recombination, and conserved gene content (Caterino et al. 2000, Lin & Danforth 2004, Simon et al. 1994). A greater understanding of mitochondrial evolution (Ballard & Rand 2005, Funk & Omland 2003) allows potential pitfalls in data interpretation to be recognized and avoided.

The Importance of Phylogenetic Computer Applications

The presence of user-friendly tree-building programs has heavily influenced the choice of phylogenetic methods made by most systematists. The value of model-based methods such as maximum likelihood (ML) became apparent as more was learned about the mechanisms of evolution of DNA sequences, and empiricists and theoreticians began to consider the necessity of accommodating the peculiarities of molecular evolution in increasingly realistic models (Swofford et al. 1996; see **Figure 1**). In the 1980s and 1990s tree-building programs that implemented maximum likelihood facilitated model-based analyses (e.g., Felsenstein 1981; Swofford 1998, beta version in 1993; Yang 1997, beta version 1993). In 2001, the first version of the program, MrBayes, became available (Huelsenbeck & Ronquist 2001). Because these programs allow a strong focus on models of evolution, building realistic models and choosing among them are two of the most active areas of research in systematics today (Posada & Buckley 2004, Sullivan & Joyce 2005). User-friendly programs like Modeltest (Posada & Crandall 1998) have facilitated the selection of models.

In the precursor to this review (Simon et al. 1994), we pointed out that likelihood and spectral-analysis methods "show great promise for phylogenetic analysis but are computationally intensive and currently work well only for a limited number of taxa." For this reason, models of evolution were discussed in terms of distance corrections and parsimony weighting. Between 1996 and 2001, as computers and algorithms picked up speed, likelihood became the method of choice. Bayesian phylogenetic analysis (Huelsenbeck et al. 2001, Larget & Simon 1999, Yang & Rannala 1997) was rapidly embraced once MrBayes became available (Huelsenbeck & Ronquist 2001). The advantage of Bayesian analysis lies in its ability to reveal phylogenetic uncertainty in trees directly constructed using probabilistic models (Holder & Lewis 2003, Huelsenbeck & Imennov 2002). Leache & Reeder (2002) were the first to compare parsimony, likelihood, and Bayesian phylogenetic analyses for a large mitochondrial data set and discuss the comparative advantages of these procedures.

Today, nucleotide sequence data are accumulating faster than they can be analyzed. Better and better models of evolution are being developed. Still, it is not apparent whether current Bayesian tree-building and fast maximum likelihood (e.g., Zwickl

```
                    ┌─────────────────────────────────────────┐
                    │      1969: Jukes-Cantor (JC)            │
                    │                                         │
                    │  Equal base frequencies: $\pi_A=\pi_C=\pi_G=\pi_T$ │
                    │  All substitutions        $\alpha = \beta$        │
                    │    equally likely                       │
                    └─────────────────────────────────────────┘
```

Allows for transition/transversion bias → **1980: Kimura 2 parameter (K2P)**

Equal base frequencies: $\pi_A=\pi_C=\pi_G=\pi_T$
Two substitution types: $\alpha \neq \beta$
  Transitions and transversions have different substitution rates

Allows base frequencies to vary → **1981: Felsenstein (F81)**

Unequal base frequencies: $\pi_A \neq \pi_C \neq \pi_G \neq \pi_T$
One substitution type: $\alpha = \beta$
  All substitutions equally likely

Allows three substitution types → **1981: Kimura 3 parameter (K3P)**

Equal base frequencies: $\pi_A=\pi_C=\pi_G=\pi_T$
Three substitution types: $\alpha_1 \neq \beta_1 \neq \beta_2$
  One transition class
  Two tranversion classes

Allows for transition/transversion bias → **1985: Hasegawa et al. (HKY85)**

Unequal base frequencies: $\pi_A \neq \pi_C \neq \pi_G \neq \pi_T$
Two substitution types: $\alpha \neq \beta$
  Transitions and transversions have different substitution rates

Allows six substitution types → **1994: Zharkikh symmetrical (SYM)**

Equal base frequencies: $\pi_A=\pi_C=\pi_G=\pi_T$
All six substitution types have different rates

Allows three substitution types → **1993: Tamura-Nei (TrN)**

Unequal base frequencies: $\pi_A \neq \pi_C \neq \pi_G \neq \pi_T$
Three substitution types: $\alpha_1 \neq \alpha_2 \neq \beta$
  Two transition classes
  One transversion class

Allows base frequencies to vary / *Allows six substitution types* → **1984–1990: General time reversible (GTR)**

Unequal base frequencies: $\pi_A \neq \pi_C \neq \pi_G \neq \pi_T$
All six substitution types have different rates

2006) programs that incorporate complex models can handle the very large numbers of taxa to be analyzed in future data sets. One major question is whether large numbers of genes and especially enormous numbers of finely sampled taxa (sequences) can rescue distance analyses that do not make full use of the character information in the data and nonmodel-based methods such as evenly weighted parsimony that ignore complex substitution patterns. The remainder of this review explores substitution patterns and their significant effects on phylogenetic analyses and data interpretation.

Molecular clock: the assumption that the rate of molecular substitutions is constant per unit time and can be used to date divergences

HOW MOLECULES EVOLVE

Evolution and Weighting of Molecular Data

Our previous review (Simon et al. 1994) describes how model-based corrections and analogous phylogenetic weighting schemes were devised to correct for the fact that nucleotide substitutions at single sites are obscured by later substitutions that can mislead phylogenetic and molecular clock analyses. Beginning with the Jukes-Cantor (1969) model, we traced the parallel development of weighting schemes and models of evolution that relax each of Jukes-Cantor's unrealistic assumptions: (*a*) all bases are found in equal proportions within a sequence, (*b*) every base changes to every other base with equal probability, and (*c*) the rate of substitution at every site is the same.

To incorporate molecular realism, parsimony tree-building methods rely on (*a*) weighting (e.g., Cunningham 1997), or (*b*) conversion to distances, correction with a model of evolution and conversion back to character data using a Hadamard transformation (Penny et al. 1996). Because many proponents of parsimony insist on even weighting of bases (which makes the same unrealistic assumptions as the Jukes-Cantor model) and because complex weighting takes away one of parsimony's greatest advantages relative to maximum likelihood and Bayesian analyses—speed—little research progress has been made in data weighting. So, the discussion below focuses on models of evolution. Although some models will always fit data better than others, data

Figure 1

The figure shows models of evolution from simplest (most unrealistic) at top to the most complex (most general) at bottom (modified from Swofford et al. 1996, their figure 11, and Page & Holmes 1998, their figure 5.14). $\alpha =$ transition rate; $\beta =$ transversion rate. All models are symmetrical (probability of changing from base X to base Y is the same as changing from base Y to base X). HKY85 is similar to the Felsenstein 1984 (F84) model (formally described by Kishino & Hasegawa 1989) in that both allow for unequal base frequencies and unequal transition and transversion rates. The general time reversible (GTR) model was developed several times between 1984 and 1990 (Felsenstein 2004), but not implemented until 1993 (e.g., Swofford 1993) for computational reasons. Note that none of the models described above include an accommodation for among-site rate variation (ASRV) but this can be added by attaching a Γ, invariant sites correction, and/or by partitioning data. Standard ASRV corrections all assume that the pattern of ASRV does not change over time; violation of this assumption is addressed by covarion-like models. Another factor not accommodated by the models shown is correlation among sites. Note also that although some of these models accommodate nucleotide bias, this accommodation assumes that the bias is the same in all taxa.

might not fit any model very well (Bollback 2002). Still, Sullivan & Swofford (2001) convincingly argue that it is better to use a poorly fitting model than no model at all. Posada & Buckley (2004) and Sullivan & Joyce (2005) discuss the need for models in science and phylogenetics, review the properties of molecular evolutionary models, and/or compare and contrast methods of model selection. Below we discuss models from the point of view of the data.

Nucleotide and Amino Acid Substitution Models

The models of evolution described in **Figure 1** show the progressive incorporation of realism from top to bottom including accommodation of biased base composition within a sequence and biased substitution patterns from one nucleotide to another. As stated in the legend, none of these models incorporate among-site rate variation (ASRV) per se, but ASRV can be added as an additional parameter(s). Accommodating ASRV has more of an effect on phylogenetic analyses than accommodating nucleotide and base substitutional biases within sequences (Sullivan & Swofford 2001, Yang et al. 1994). Adding ASRV to amino acid evolution models has also resulted in substantial improvement in model fit (e.g., Susko et al. 2003, contra Yang et al. 1994). Among-lineage rate variation (ALRV; covarion-like evolution) is a more complex process to model and, as a result, models of evolution that address these processes are less well developed. Below we discuss among-site and ALRV and their effects on phylogenetic analysis.

Among-site rate variation (ASRV): a ubiquitous property of molecules where different numbers of substitutions per unit time occur at different DNA or amino acid sites

Among-lineage rate variation (ALRV): a property of molecules where the number of substitutions per unit time occurring at any given position varies among taxa

Covarion-like evolution: nucleotide or amino acid substitutions whose pattern of ASRV varies across lineages; encompasses heterotachy, mosaic evolution, and covarion/covariatide evolution

Among-Site Rate Variation

History. Ideally, slowly evolving genes would be most useful for deep-level phylogenetics whereas rapidly evolving genes would be necessary for reconstructing recent divergences. Unfortunately, most genes are not well characterized by a single average rate of evolution. With the unraveling of the genetic code between 1961 and 1966, it became immediately obvious that substitutions at different codon and amino acid positions would be accepted at different rates owing to the degeneracy of first and third positions. Shortly thereafter, evolutionary biologists began to examine how this ASRV might affect genetic distances among taxa and phylogenetic analyses based on them (e.g., Fitch & Margoliash 1967). Simon et al. (1994) reviewed early studies that attempted to incorporate ASRV into models of evolution and weighting schemes. They also demonstrated how knowledge of molecular structure and function can help to understand the constraints that create rate variation among sites. Yang (1996) produced an exceptionally complete review of the discovery of ASRV and its incorporation as both discrete rate classes and continuous distributions into usable models of DNA sequence evolution including methods for calculating the α-shape parameter of the Γ-distribution of rates across sites. It is now well established that even genes that are considered to be strongly conserved contain rapidly evolving sites (e.g., Simon et al. 1996), with the converse also being true. Therefore it is inadequate to characterize a gene as fast or slow and expect that categorization to hold across all sites.

Detrimental effects of ignoring ASRV. In his 1996 review, Yang summarized the detrimental effects of ignoring ASRV: severe underestimation of genetic distances, incorrect estimation of transition-transversion rate ratios, and confounding of phylogenetic tree-building algorithms. On the population level, Yang (1996) pointed out that the infinite alleles model ignores ASRV, which in turn invalidates Tajima's D statistic for testing neutrality and causes the distribution of pairwise sequence differences to mimic patterns of population expansion. Revell et al. (2005) found that incorrect inferences of rapid cladogenesis early in a group's history due to bias in tree shape was caused by model underparameterization, especially the omission or misestimation of ASRV. Buckley et al. (2001a) and Buckley & Cunningham (2002) showed that ignoring ASRV has a strong effect on estimates of nonparametric bootstrap support. Similarly, Lemmon & Moriarty (2004) found that when ASRV is ignored in simulations model misspecification has a strong effect on Bayesian posterior probability estimates of nodal support. Although Kumar et al. (1993) warned that correcting sequences that differ by less than 5% may result in overparameterization (trading an increased error variance for realism), Yang (1996) argued that this point of view is a misconception. Sullivan & Joyce (2005) present an extensive discussion of the impact of model misspecification and overparameterization.

SSR: site-specific rates

Topology: the branching order of a phylogenetic tree

Among-site rate variation in amino acid sequences. Like nucleotides, amino acids show considerable variability in rate of substitution among sites. Although the earliest studies of ASRV focused on amino acids (e.g., Fitch & Markowitz 1970), later phylogenetic studies of amino acids ignored ASRV. Recently, there has been a rebirth of interest in ASRV and the beneficial effects of its incorporation into models of amino acid evolution (e.g., Susko et al. 2003, 2004). As noted above, ASRV is caused by structural and functional constraints. Mitochondria-specific models of evolution that use information on the probability of particular amino acid replacements (e.g., Adachi & Hasegawa 1996) and secondary structure (e.g., Lió & Goldman 2002) also improve phylogeny construction, especially for deeper-level phylogenetic studies where adaptive shifts in molecules among lineages (see below) are more likely to have taken place.

Different methods of accommodating among-site rate variation vary in their effects. ASRV can be addressed by partitioning the data into different rate classes and assigning each rate class its own rate (site-specific rates or SSR models). However, SSR models (where each site in a rate class is assumed to evolve at the same rate) give much lower branch-length estimates than Γ models, invariant sites models, and partitioned-Γ models (**Figure 2**). Although ignoring ASRV may not in all circumstances affect topology, it is more likely to have an effect on nodal support (Buckley et al. 2001a). Buckley & Cunningham (2002) evaluated the effect of different ASRV models using six real data sets for taxa whose relationships are supported strongly by other data. By examining the match of trees constructed from a variety of models to the well-supported trees, they found that SSR models and evenly weighted parsimony performed poorly in recovering the topology and produced lower branch supports than models that incorporated a gamma (Γ) or invariant sites (I) correction.

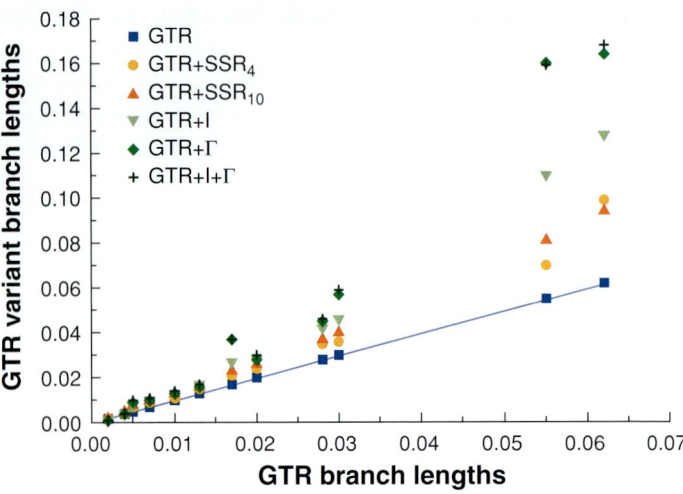

Figure 2

Maximum-likelihood estimates of branch lengths under a range of variants of the general time reversible (GTR) substitution model (*colored symbols*). The diagonal line connecting the GTR points on the x- and y-axes illustrates the deviation of the ASRV-corrected branch lengths from the equal-rates estimates. Models of evolution are described in **Figure 1**. Subscripts indicate number of partitions; e.g., in the GTR + SSR$_4$ model the data is partitioned into four SSR classes (first, second, and third positions of proteins plus transfer RNA sites). Redrawn with permission from Buckley et al. 2001a.

Among-Partition Rate Variation

Although partitioning data does not work well via SSR models because of the usually unrealistic assumption that all sites within the partition evolve at the same rate, partitioning strategies that estimate the distribution of ASRV separately for each partition (SSR+Γ) avoid this problem. Yang's PAML program has allowed partitioned models for many years but worked slowly for large numbers of taxa. Recently MrBayes has introduced partitioned models for Bayesian Markov chain Monte Carlo (MCMC) analyses that work well for many taxa. Using partitioned models on combined data with heterogeneous rates addresses many of the concerns about data combinability (e.g., Barker & Lutzoni 2002, Buckley et al. 2002, Bull et al. 1993)—assuming that the same topology (history) underlies different data partitions. Independently modeling data partitions improves branch support and tree likelihood (Brandley et al. 2005, Castoe et al. 2004, Nylander et al. 2004), but must be done with care because partitioning data without properly accommodating among-partition rate variation (APRV) and/or using branch-length priors that are too diffuse can seriously distort branch lengths (Marshall et al. 2006).

Mixture Models

There are many ways any one data set can be partitioned, and the method by which this is done is usually arbitrary. Indeed, for many genes it is not always clear how

APRV: among-partition rate variation

Mixture models: models in which data points are viewed as generated by one of a number of distributions, each contributing to the likelihood best to partition data. For protein coding genes, first, second, and third positions are common partitions; however, within each of these categories there is considerable variation in functional constraint. At third positions, sites can be twofold, threefold, or fourfold degenerate with the fourfold degenerate sites displaying the most freedom to vary (although these sites may be somewhat constrained by codon usage bias). In addition, messenger RNA secondary structure, protein secondary/tertiary structure, and other functional considerations all influence the rate and pattern of evolution. In rRNA, partitioning stems (paired) versus loops (unpaired) regions is tempting, but there is heterogeneity within each of these partitions. Although some authors have partitioned rRNA data into stems versus loops (e.g., Springer & Douzery 1996), this does not make sense in terms of partition variability because some stems are slowly evolving whereas others evolve rapidly (**Figure 3**). Furthermore, within a stem the pattern of variation may differ depending on the identity of the base; tertiary structure and protein interactions also add complexity (Gutell et al. 2002, Hickson et al. 1996, Pagel & Meade 2004). A more informative method of accommodating rate and pattern variation among data subsets is to use mixture models that do not require a priori specification of partitions. In mixture models, characters are viewed as having been generated by one of a number of distributions, and each of these distributions contributes to the likelihood during an analysis. The parameters of each model distribution and the weights assigned to them are estimated from the data. In the end of the analysis, each character can be assigned a probability of membership in each model. Pagel & Meade (2004) developed a Bayesian mixture model and characterized pattern variation in published protein (EF1α and DDC) and 12S small subunit (SSU) mitochondrial rRNA data. For the rRNA data, their program converged on four rate matrices to describe the patterns of variation in the data but these matrices only partially corresponded to stems and loops. In fact, the sites weighted heaviest for one of the matrices were evenly divided between stems and loops. Similarly, for the protein coding genes, although each Q matrix specialized on a particular codon position, each matrix also provided the best fit to some other codon positions. Lartillot & Philippe (2004) developed a mixture model for amino acid sequence evolution. Their use of a Bayesian Dirichlet process prior allowed the association of each amino acid site to a given Q matrix to be determined during the analysis. Using Bayes factors they demonstrated that the mixture model outperformed standard amino acid rate matrices.

rRNA secondary structure: the pattern of helices and unpaired regions formed when a single-stranded ribosomal RNA primary sequence folds and pairs with itself

Covariation: two bases or two amino acids whose rates of evolution are correlated usually due to functional or structural constraints

Mixture models do not solve all problems for ribosomal RNA.

In rRNA there is clearly covariation among sites that is related to base-pairing in helices, long-distance base-pairing interactions across domains, and ribosomal-protein-rRNA interactions (e.g., Hickson et al. 1996). In fact, covariation analysis has been particularly important in devising elegant models of rRNA secondary structure that have now been completely verified using experimental methods and X-ray crystallography (Gutell et al. 2002). Accommodating correlation among sites in models of rRNA evolution is important for phylogenetic analysis (e.g., Huelsenbeck & Nielsen 1999, Smith et al. 2004) and needs further study. Finally, rRNA molecules often include large numbers of variable length indels that can cause alignment difficulties and provide information that is difficult to objectively incorporate into phylogenetic reconstruction. In general,

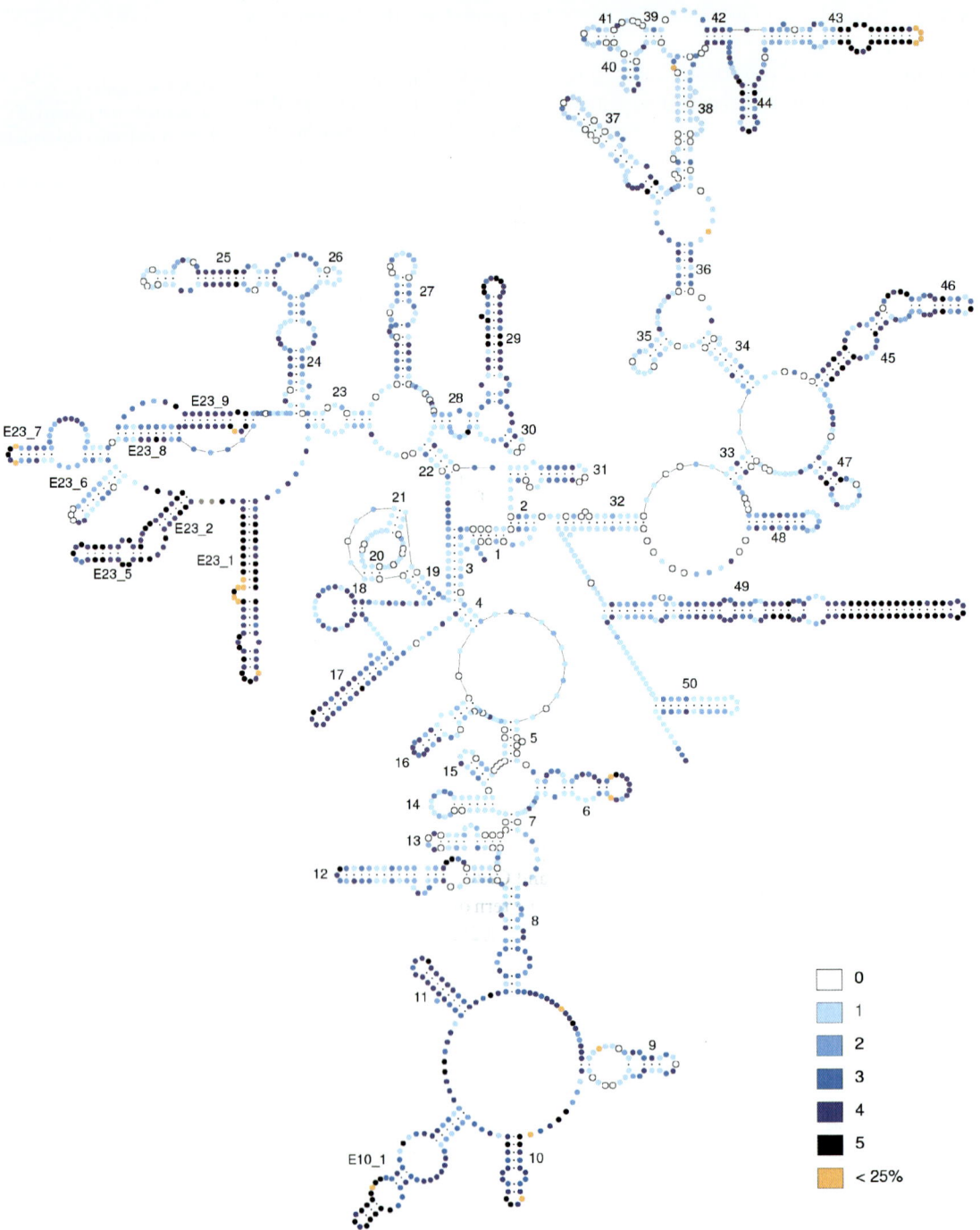

information in indels has been quantified separately and added into phylogenetic analyses later (Kjer et al. 2001, Lutzoni et al. 2000). A growing number of models explicitly describe covariation between bases in RNA molecules (e.g., Smith et al. 2004), but these are only rarely employed in phylogenetic reconstructions (e.g., Kjer 2004).

Among-Lineage Rate Variation

Covarions, covariatides, heterotachy.
Fitch & Markowitz (1970) proposed that adaptive shifts in protein function over time would result in a change in the probability of substitution of a particular amino acid site over time (across lineages). Based on earlier ideas by Margoliash & Smith (1965) and Fitch & Margoliash (1967), they proposed the covarion hypothesis, which speculated that a certain proportion of sites in a protein were not free to vary but could become variable if other sites assumed their function. This was later extended to nucleotides and called covariotide evolution (Fitch 1986). Miyamoto & Fitch (1995) reviewed the development of the covarion hypothesis and showed that the distribution of variable and invariant positions is different in seven mammal and seven plant sorbitol dehydrogenase amino acid sequences and thus follows a covarion model. Miyamoto & Fitch (1995) and others (e.g., Steel et al. 2000) pointed out that a constant-sized covarion class is unrealistic because the number of variable positions can differ among lineages. Newer covarion models (Galtier 2001, Tuffley & Steel 1998) do not require this restriction.

Because nucleotide functional/rate shifts occur in rRNA as well as in protein coding genes (Simon et al. 1994, 1996), a more general name for this phenomenon would be helpful. Philippe & Lopez (2001) coined the useful term heterotachy to describe positions that evolve at different rates in different lineages. Earlier this type of evolution had been called "independent and episodic" (Johannes & Berger 1993), "mosaic evolution" (Simon et al. 1996), or "covarion-like evolution" (Lockhart et al. 1998, Lopez et al. 1999). Demonstrations that evolutionary rate of a given position is not always constant throughout time, apart from discussions of codon evolution, include those by Johannes & Berger (1993), Philippe et al. (1996), Simon et al. (1996), Lockhart et al. (1996, 1998), Lopez et al. (1999), and Gaucher et al. (2001).

Functional shifts in molecules cause shifts in the pattern of ASRV and are expected over the course of long-term evolution. Lopez et al. (2002) studied more than 2000 vertebrate cytochrome-b sequences from 32 large monophyletic groups and found

Figure 3

Rates of variability of individual nucleotide positions contingent on nucleotide variabilities were based on 500 sequences of species belonging to the eukaryotic crown taxa small subunit rRNA molecules, superimposed on the secondary structure of *Saccharomyces cerevisiae*. The most variable positions are in black, the most conserved in light blue, and invariable positions are in white. Sites containing a nucleotide in *S. cerevisiae* but vacant in more than 75% of sequences, which were not considered for the variability calculations, are indicated in orange. Areas that could not be aligned with confidence are also indicated in orange. All rRNA molecules have similar patterns of variability. Redrawn with permission from Yves Van de Peer (http://www.psb.ugent.be/rRNA/varmaps/Scer_ssu.html).

Nonstationarity: any process characterized by statistical properties that vary over time; in phylogenetics most often discussed with reference to nucleotide bias

that all variable positions are heterotachous but that, surprisingly, there was no obvious relationship between variability, function and three-dimensional structure. Gribaldo et al. (2003) found that not all sites in vertebrate hemoglobin show a relationship to structure and function; there is a class of sites that are constant (within groups) but different (among groups) and these CBD sites show "signatures of functional specialization." Lockhart et al. (2000) showed that evolving distributions of variable sites alone provide support for deep-branching patterns in eubacterial phylogenies. If these shifts exhibit convergent evolution, Lockhart and colleagues pointed out, the tree or parts of it could be artifactual (bad covarion evolution). Reassuringly, different genes recovered similar patterns of evolution (good covarion evolution). Misof et al. (2002) showed how the Γ-distributed rate at particular sites in the insect mitochondrial 16S rRNA gene varied extensively among different insect orders, indicating that covariotide evolution operates in mtDNA at deep levels of divergence.

Progress has been made in developing covarion-like models of sequence evolution. In addition to the Tuffley & Steel (1998) and Galtier (2001) models cited above, Yang et al. (2000) created a model that allows selection constraints to change across the sequence. One drawback of this model is that the branches with different selection pressures must be specified a priori. Guindon et al. (2004) created a model that not only allows selection to change across the sequence, but also allows selective constraints to change over time without a priori specification. Recently, Huelsenbeck et al. (2006) applied a Dirichlet process prior in a fully Bayesian approach to model variation in nonsynonymous sites and allow selection to vary across a sequence. These models are close in spirit to the original covarion models in which an amino acid substitution at one position in a gene changes the selective constraints elsewhere.

Nucleotide bias among lineages. Shifts in patterns of ASRV can cause changes in the nucleotide bias of taxa across the tree. This is because the nucleotide bias of an organism's genome is most evident at the most variable sites (Simon et al. 1994). Earlier it had been shown that substitutional bias can seriously affect phylogenetic trees (e.g., Lockhart et al. 1992, Weisburg et al. 1989). If patterns of nucleotide bias have changed over lineages, models that assume a stationary distribution of nucleotide bias among taxa can cause systematic error in phylogeny construction. LogDet-type models were designed to incorporate nucleotide bias nonstationarity (Steel 1994). Because ASRV and ALRV (covarion-like evolution) can occur simultaneously the LogDet model should be combined with an invariant sites model because the LogDet does not correct for ASRV. Haddrath & Baker (2001) showed that ratite bird phylogenies based on complete mitochondrial genomes were consistent with traditional expectations only when they were corrected for variation in nucleotide bias among lineages. Similarly, using complete mitochondrial genomes corrected for nonstationarity, Paton et al. (2002) refute the controversial conclusions based on fossils/morphology that modern birds are descended from shorebirds or passerines. Jermiin et al. (2004) review the literature on nucleotide nonstationarity, discuss methods to detect it and, using simulations, examine the effects of nucleotide nonstationarity on tree building.

Although the biasing effects of nucleotide nonstationarity are well known, they are often forgotten. For example, in a recent paper, Rokas et al. (2003) used data from eight complete yeast nuclear genomes to discover the minimum number of genes necessary to build a robust and well-supported phylogeny. They concluded that approximately 20 genes (average length 1198 nucleotides) were needed and that there were no discernable characteristics of these genes that "predicted the performance" of one over the other. There are several problems with these conclusions. First—trivial but often overlooked—it is not the number of genes that is important but the number of informative sites, and this differs among genes and at different depths in the tree. Second and more important, there are many characteristics of genes that can make them less useful for phylogenetic analysis. One is high levels of ASRV; another is high levels of ALRV. Collins et al. (2005) demonstrated that many of the genes that Rokas et al. (2003) employed contained nonstationary nucleotide frequencies. When these genes were excluded, it was concluded that on average, eight yeast genes were required to recover the underlying phylogeny. Whereas, as Collins and colleagues point out, genes do not come in two classes (stationary and nonstationary), the greater the variation in levels of base compositional bias among lineages, the more problematic the gene is likely to be for phylogenetic reconstruction. A recent paper by Hedtke et al. (2006) demonstrated that the Rokas et al. (2003) phylogeny also suffered from poor taxon sampling leading to long branch problems.

Codon models. Codon models, reviewed by Yang (2003), operate at the level of the codon as a unit. They are an advancement over amino acid substitution models that only consider the probability of changing from one particular amino acid to another. Codon models take into account the fact that a switch from one type of codon family (e.g., twofold versus fourfold versus sixfold degeneracy) changes the probability of substitution of individual bases within the codon. Although earlier codon models assumed that the rate of synonymous substitution is constant among sites within genes (e.g., Muse & Gaut 1994) newer codon models reflect the fact that there is significant variability of synonymous rates in the majority of genes. This is observed, for example, in complete mitochondrial genome sequences of 111 animal taxa sampled from 10 disparate clades (F.V. Mannino & S.V. Muse, in review). Because of their computational complexity, codon models are only rarely used for inferring tree topology (Ren & Yang 2005).

INTERPRETING TREES AND DATA SUPPORT

A thorough knowledge of the properties of molecular data and how they evolve can also aid in the design and intrepretation of phylogenetic studies. Taxon sampling, measuring nodal support, testing alternative phylogenetic hypotheses and attaching time to phylogenies are four areas where understanding molecular evolution can improve the process.

Taxon Sampling

Theoretical and simulation studies have shown that trees can be very hard to reconstruct when branch lengths are unequal and rates of change vary over the tree, although this is partly dependent on the method of phylogenetic inference employed (e.g., Felsenstein 1978, Swofford et al. 2001). Adding taxa to a tree in order to break up long branches improves the accuracy of topology estimation by better reconstructing the history of character-state changes, an implicit step in parsimony and likelihood inference. The more nodes in a phylogenetic tree, the more information on character-state change. The result is that more densely sampled subsections of a tree tend to have longer total branch lengths (Venditti et al. 2006). Another factor that can reduce sampling bias is increased sequence length, which also leads to more accurate estimates of branch lengths and tree topology as long as the added nucleotides have properties similar to the original sample. The above observations have led to a debate on whether it is preferable to sample more characters or more taxa in order to increase phylogenetic accuracy (e.g., Pollock et al. 2002, Rosenberg & Kumar 2001, Zwickl & Hillis 2002). The extensive simulation studies of Zwickl & Hillis (2002) and Pollock et al. (2002) showed how adding taxa to a phylogenetic analysis was more effective than adding more characters once a certain sequence length was reached (Hillis et al. 2003). Maximum-likelihood simulations showed consistent improvement with the addition of taxa, probably owing to the improved estimation of model parameters (Pollock & Bruno 2000). However, taxon addition should be done strategically to avoid adding taxa that increase the depth of the tree or that create long branches, which can introduce further biases into the reconstruction (e.g., Mitchell et al. 2000, Poe 2003, Hedtke et al. 2006).

The second debate surrounding the density of taxon sampling concerns the efficacy of evenly weighted parsimony in reconstructing large trees with many taxa. Hillis (1996) suggested that densely sampled phylogenies were actually easier to reconstruct accurately than were sparsely sampled phylogenies, even for simple models of evolution such as evenly weighted parsimony. He showed that highly variable rates of evolution and ASRV were much less of a problem for densely sampled trees than for trees of only a few taxa. Dense sampling of taxa decreases the number of superimposed changes of characters that must be reconstructed along lineages and therefore decreases the reliance on accurate and complex models of evolutionary change. DeBry (2005) summarized the ensuing debate and conducted new simulations, which agreed that overall accuracy for parsimony increases with increased taxon sampling. Similarly, Salamin et al. (2005) showed that evenly weighted parsimony performed quite well in reconstructing large phylogenies. However, although it is inevitable that increased taxon sampling will help to reconstruct superimposed changes, the conclusions about the efficacy of evenly weighted parsimony may be less applicable to empirical sequence data, which tend to fit models less well than simulated data and have more ASRV and more nucleotide bias (e.g., Holder 2001). Empirical data may also show variation in patterns of substitution among lineages as described above.

Measures of Nodal Support

Measures of nodal support are generally more satisfying than whole tree measures of information content because for most phylogenetic trees some clades are better supported than others. Measures of nodal support provide a useful summary of how well data support the relationships defined by a tree. Poorly supported relationships are of little use in evolutionary studies other than to illustrate where more data are needed before conclusions can be drawn. Of course, high support values do not mean that a node is accurate, only that it is well supported by the data; model misspecification and taxon sampling can mislead the analysis (e.g., Hedtke et al. 2006).

Currently, the nonparametric bootstrap (Felsenstein 1985) is still the most widely used method for assessing nodal support, despite a long-running debate as to the validity and interpretation of the bootstrap in phylogenetics (e.g., Efron et al. 1996, Felsenstein & Kishino 1993, Hillis & Bull 1993, Holmes 2005, Sanderson 1995). Perhaps the best interpretation is that the bootstrap quantifies the sensitivity of a node to perturbations in the data (Holmes 2005). However, as commonly implemented, the bootstrap gives a biased estimate of accuracy (Hillis & Bull 1993, Holmes 2005), where accuracy is defined as the probability of obtaining a correct phylogenetic reconstruction (Penny et al. 1992). The reason for this bias is related to the complex geometry of tree space and site-pattern space (Efron et al. 1996, Holmes 2005), which is described as follows. All possible site patterns (i.e., sets of sites that show identical states across all taxa) for a given data set can be divided into regions, each one separated by a boundary. Each region has an optimal topology associated with it. An observed data set lies within the sampling distribution of what we can consider to be the truth. Because bootstrap replicates are generated from the observed data rather than the truth, the proportion of replicate data sets that lie within each region can become distorted, which in turn can bias the bootstrap (Sanderson & Wojciechowski 2000). More sophisticated bootstrap techniques are available to correct for this bias (e.g., Efron et al. 1996, Shimodaira 2002); unfortunately, these are rarely implemented to measure nodal support.

The well-known bias of the bootstrap has led researchers to seek other methods of estimating nodal support, and perhaps the most popular alternative is Bayesian posterior probability (e.g., Larget & Simon 1999, Yang & Rannala 1997). The increasing reliance on posterior probabilities as measures of nodal support, as opposed to the bootstrap, has initiated a debate as to the merits of the two approaches (e.g., Huelsenbeck & Rannala 2004, Suzuki et al. 2002). This debate arose from early observations of Bayesian inference in phylogenetics that demonstrated a tendency for posterior probabilities to be more extreme than ML nonparametric bootstrap proportions, although the two tended to be correlated. This observation was made from both empirical (e.g., Buckley et al. 2002, Wilcox et al. 2002) and simulated data (e.g., Cummings et al. 2003, Suzuki et al. 2002, Wilcox et al. 2002). Here we address the following questions: Why are bootstrap proportions and posterior probabilities different? Is this really a problem? If so what can be done about it?

Comparing posterior probabilities and bootstrap proportions is difficult because they represent fundamentally different quantities. A nodal posterior probability is the probability that a given node is found in the true tree, conditional on the observed data, and the model (including both the prior model and the likelihood model). Some researchers argue that posterior probabilities are superior to bootstrap proportions because the former give a more direct measure of confidence in a node (e.g., Huelsenbeck & Rannala 2004). A bootstrap proportion is based on the concept of resampling, and its exact interpretation depends on how it was calculated, as discussed above. Furthermore, posterior probabilities are calculated by assuming a prior distribution on all model parameters, including the branch lengths and topology, and these priors will influence the posterior in many cases (e.g., Yang & Rannala 2005). This dependence on prior distributions also complicates the comparison of the two measures of support. Yang & Rannala (2005) demonstrated how some prior distributions on branch length can cause nodal posterior probabilities to become extreme. Lewis et al. (2005) also showed how posterior probabilities can be biased if a prior that excludes zero-length branches is applied to data generated from a topology that, in fact, includes polytomies, an observation first made by Suzuki et al. (2002). Lewis et al. (2005) demonstrated that if a polytomy exists but is not accommodated in the prior, resolution of the polytomy will be arbitrary and the nodal support indicated by the posterior probability will appear unusually high compared to ML bootstraps. As with the problems noted by Yang & Rannala (2005), this can be circumvented by applying more appropriate prior distributions or by using reversible jump MCMC to permit internal branches not supported by the data to collapse to polytomies (Lewis et al. 2005).

Another observation from simulated (Huelsenbeck & Rannala 2004, Lemmon & Moriarity 2004, Suzuki et al. 2002) and empirical data (Buckley 2002, Waddell et al. 2002) is how model misspecification affects posterior probabilities relative to bootstrap proportions. The simulations by Huelsenbeck & Rannala (2004) and the empirical study of Buckley (2002) show how posterior probabilities respond in a more extreme fashion to model misspecification than the bootstrap or bootstrap-based topology tests. This problem is likely to be exacerbated by branch-length heterogeneity (Felsenstein and inverse-Felsenstein zone problems) and a high rate of change across the tree, both of which typify many mtDNA data sets. This problem can obviously be rectified by implementing a more complex substitution model; however, there is no guarantee that the models as implemented in available software packages will be sufficient. Because we have little knowledge of the goodness of fit between data and model in typical phylogenetic studies (although goodness of fit tests do exist), we have little idea of the seriousness of the problem of model misspecification in current implementations of Bayesian phylogenetic inference. Finally, failure of convergence of the MCMC algorithm is an underappreciated problem, especially for large data sets. Failure to diagnose a lack of convergence of the algorithm will lead to incorrect posterior probabilities (Huelsenbeck et al. 2002).

Given these issues, which method is best for quantifying phylogenetic support? This is a difficult question to answer because it partly depends on one's philosophical approach to statistical inference. However, if the desired measure of support is the probability that a node is correct given the data set and the model, then the only way to calculate this is by Bayes' theorem. Some researchers have attempted to reconcile Bayesian and bootstrap approaches by merging multiple Bayesian analyses from bootstrapped data sets, the so-called Bayesian bootstrap (Douady et al. 2003,

Waddell et al. 2002). However, the exact statistical interpretation of these values is not at all obvious, and this combination of distinct statistical paradigms has yet to be justified. In terms of practical advice, a review of the current Bayesian phylogenetic literature indicates that much more emphasis needs to be placed on developing more realistic models, checking the effects of the priors, and monitoring the convergence of posterior distributions.

Tests of Topology

In many phylogenetic studies it is important to assess the information content of the data with respect to the entire tree relative to an alternative hypothesis in addition to understanding support for individual nodes. A variety of tests of topology have been designed to achieve this goal. The currently used tests of topology can be divided into two types: frequentist and Bayesian. The frequentist tests can in turn be divided into parametric tests and nonparametric tests. The most widely used parametric test is the Swofford-Olsen-Waddell-Hillis (SOWH) test (Swofford et al. 1996), which uses parametric bootstrapping to simulate replicate data sets that are in turn used to obtain the null distribution. The SOWH test has been applied in a wide variety of studies ranging from comparative phylogeography (e.g., Carstens et al. 2005b) to deeper phylogenetics (e.g., Rokas et al. 2002). The nonparametric tests use the nonparametric bootstrap to generate replicates that are then used to construct the null distribution. The Kashino-Hasegawa (KH) test (Kishino & Hasegawa 1989) is a nonparametric test designed to compare pairs of topologies selected before a phylogenetic analysis and may become too liberal when the maximum-likelihood topology (selected a posteriori) is tested against another topology (Goldman et al. 2000). The Shimodaira-Hasegawa (SH) test (Shimodaira & Hasegawa 1999) and the approximately-unbiased (AU) test (Shimodaira 2002) simultaneously compare sets of topologies and incorporate more complex bootstrap procedures to correct for the bias associated with multiple comparisons and inclusion of the maximum-likelihood topology. The study by Buckley et al. (2001b) demonstrates the effect of these assumptions on the SH test relative to the KH test. For these reasons, and because of the nature of the null hypotheses employed by the nonparametric tests, the SH and KH tests are generally more conservative than the parametric tests (e.g., Aris-Brosou 2003, Buckley 2002, Goldman et al. 2000). The more explicit reliance on models of evolution by the parametric tests makes them very powerful tests, yet they are also more susceptible to model misspecification (e.g., Buckley 2002, Huelsenbeck et al. 1996, Shimodaira 2002).

Bayesian tests of topology (e.g., Aris-Brosou 2003) are much less commonly implemented than the frequentist tests. The Bayesian tests generally rely on Bayes factors (Kass & Raftery 1995) to compare marginal likelihoods generated under two hypotheses corresponding to different topologies. The use of Bayes factors in testing topologies will likely receive much greater attention in the future (Huelsenbeck & Imennov 2002, Suchard et al. 2005). One example of a Bayesian test of topology is that of Carstens et al. (2005a), who assessed whether posterior distributions of trees contained topologies consistent with a priori demographic hypotheses.

Attaching Time to Phylogenies

The past several years have seen a renewed interest in methods for attaching time estimates to phylogenetic trees (Arbogast et al. 2002, Welch & Bromham 2005). Early attempts to estimate divergence times when the molecular clock was violated often involved removing taxa with aberrant rates of evolution (Hedges & Kumar 2003), decomposing the procedure into quartets of taxa that conformed to the molecular clock (e.g., Rambaut & Bromham 1998), or fitting local clocks to different regions of the phylogeny (Yoder & Yang 2000). More recently, the focus has shifted to explicitly accounting for rate changes over the tree with a growing emphasis on using models to quantify the uncertainty. The first attempts to correct for changing rates with time were the nonparametric and semiparametric rate-smoothing methods described by Sanderson (1997, 2002). The methods based on explicit models, known as relaxed-clock models, used Bayesian estimation to obtain posterior distributions of node times (Aris-Brosou & Yang 2002, Thorne et al. 1998). For example, the method of Kishino et al. (2001) assumes that the rate of evolution along a descendant branch is a random variable drawn from a log-normal distribution, whose mean is the rate of evolution of the parent branch. This approach has been modified by other researchers who use different distributions in relaxed-clock models (e.g., Aris-Brosou & Yang 2002). Given the expanding range of relaxed-clock models it is becoming increasingly important to justify the use of one model over another. However, model selection procedures are rarely applied to relaxed-clock models. Aris-Brosou & Yang (2002) first applied Bayesian model selection to different relaxed-clock models, and we expect these methods to be more commonly applied in the future, especially as different relaxed-clock models are incorporated into user-friendly software packages. We currently have little information as to how well DNA data sets fit the various relaxed-clock models, although simulation and empirical studies show that these methods can be misleading when the relaxed-clock model deviates strongly from the actual process of changing rates over the tree (Welch et al. 2005), as is likely to be true for comparisons of closely related populations experiencing slightly deleterious mutations (Ho et al. 2005). Another serious source of error in dating studies is the manner in which dates are calibrated. It is desirable to have as many calibration points as possible and new methods that improve the ability to incorporate uncertainty into fossil dates are an important step forward (Yang & Rannala 2006).

A further complication for dating divergences using mitochondrial DNA is model misspecification compounded by the typically rapid rate of evolution even for divergences that are only a few million years old. Buckley et al. (2001b) observed large differences in branch-length estimates among substitution models for a group of New Zealand cicada genera that began to radiate 10 Mya (Arensburger et al. 2004). These results indicate that, even for divergences a few million years old, the substitution model can be very important for obtaining reliable divergence times. Furthermore, if data are partitioned, then APRV must be properly accommodated and suitable branch-length priors employed (Marshall et al. 2006).

Finally, when attaching estimates of divergence time to recent speciation events, the well-known discordance between species and gene divergence times must be taken into account. Coalescent theory predicts that mtDNA haplotype divergence dates will often predate the speciation event by a substantial amount if ancestral effective population size was large (Edwards & Beerli 2000). Jennings & Edwards (2005) showed how up to 10 loci were required before stable estimates of species divergence times were obtained for three closely related species of Australian finches. Although large numbers of loci are rarely available for most empirical studies, this uncertainty can be adequately captured if coalescent times are accounted for in divergence-time estimation (Edwards & Beerli 2000). Various methods have been described to accommodate this potential bias (e.g., Edwards & Beerli 2000); however, these methods and the relaxed-clock models have yet to be implemented in a single framework of divergence-time estimation.

THE PHYLOGENETIC USEFULNESS OF MITOCHONDRIAL DNA

Mitochondrial Genes and Phylogeny

At the species level, mitochondrial gene data are by far the most widely used marker for assessing phylogenetic relationships. They offer some advantages over nuclear data for practical reasons (ease of actually obtaining the sequences, direct comparisons across different studies, and higher levels of variability). In addition, mtDNA genes have faster coalescent times owing to the smaller effective population size of their haploid, maternally inherited genomes. Thus, through genetic drift, their gene trees achieve species-level reciprocal monophyly sooner after speciation than do gene trees generated from nuclear substitutions (Sunnucks 2000). The extensive use of mitochondrial genes in phylogenetic reconstructions (e.g., Caterino et al. 2000) has generated an overwhelmingly greater amount of data for mitochondrial genes compared to nuclear genes for all taxa of Metazoa (e.g., over 800 complete, or nearly complete, mitochondrial genomes were available in GenBank as of July 2006). Historically, the mitochondrial genes most often used for phylogenetic purposes are *co1*, *co2*, *ssu* (small subunit) and *lsu* (large subunit) *rRNA*, *cytb*, and the control region (Caterino et al. 2000, Meyer & Zardoya 2003), but most regions of the mitochondrial genome are similarly useful; Simon et al. (1994) discuss the relative usefulness of the various mitochondrial genes at different levels of divergence (see especially their table 1).

Comparison of Substitution Rates in Mitochondrial versus Nuclear Genes

It has long been known that mitochondrial genes evolve faster than the majority of genes encoded in the nuclear genome (Brown et al. 1982). Although the number of nuclear genes is much greater, and the variance of the average nonsynonymous substitution rates among them is reasonably expected to be much larger, synonymous substitutions of mitochondrial genes have been empirically estimated to accumulate 1.7–3.4 times as fast as in the most rapidly evolving nuclear genes, and 4.5–9 times as fast if one averages across all nuclear genes studied (Moriyama & Powell 1997). These estimates may be biased, however, because genes chosen for analysis are usually the most conserved in order to facilitate primer design. Nuclear introns, although faster than nuclear coding regions, are still slower than mtDNA (Zhang & Hewitt 2003). Heterogeneity in substitution rate divergence between mitochondrial and nuclear genes has been observed as a function of the genes and taxa studied (Lin & Danforth 2004). Faster rates of evolution in mitochondrial genes have been related to higher rates of transition mutations (Brown et al. 1982) and stronger constraints in nuclear genes due to selection for codon usage (Moriyama & Powell 1997). Other factors believed to influence the rates of evolution of mitochondrial genes include thermal adaptation, mitochondrial-nuclear interactions, and infection with *Wolbachia* (Ballard & Rand 2005). The faster evolution of mitochondrial genes implies higher levels of multiple substitutions than nuclear genes, especially at synonymous sites (Goto & Kimura 2001, Overton & Rhoads 2004), with an obvious effect on levels of homoplasy when genes are used for phylogenetic inference (Lin & Danforth 2004). Another major difference between nuclear and mitochondrial genes that could influence phylogenetic inference is the greater nucleotide compositional bias of mtDNA, especially in insects where A+T bias can be very extreme (Simon et al. 1994). As a further complication, each of the two strands of mtDNA may exhibit different patterns of base compositional bias, and these patterns can change because of gene rearrangements (Hassanin et al. 2005).

Performance of Mitochondrial versus Nuclear Genes for Phylogeny

The combination of these factors affects phylogenetic inference and has led to many attempts to evaluate the differential performance of nuclear and mitochondrial genes in phylogenetic reconstructions. Mitochondrial and nuclear genes have often been found to differ significantly in phylogenetic signal (Overton & Rhoads 2004), a pattern that is to be expected at shallow phylogenetic levels, where different genes may present truly different allelic histories (gene trees) owing to as yet incomplete sorting of ancestral mtDNA haplotype polymorphisms under drift. Nuclear genes are suspected to outperform mitochondrial genes in phylogenetic inference when the depth of the tree is such that the nuclear genes possess sufficient variability (Lin & Danforth 2004), whereas more rapidly evolving mitochondrial genes will have experienced more multiple substitutions and associated homoplasy. Obviously, the faster-evolving mitochondrial genes provide more resolving power for the phylogeny of closely related taxa and for phylogeographic and population genetic studies (Avise 2000, Zhang & Hewitt 2003), but are more problematic for resolving the deepest nodes of a phylogenetic tree (distantly related taxa), because of the extreme compositional biases, the asymmetry of transformation-rate matrices, the higher amount of homoplasy, and the higher levels of ASRV (Lin & Danforth 2004, Springer et al. 2001). This has led to the conclusion that mitochondrial genes should be used largely for the phylogenetics of closely related taxa (Cenozoic divergences) and that they require highly parameterized models that correct for some of the best known evolutionary anomalies if they are to be used for the phylogenetic analysis of more ancient divergences (Lin & Danforth 2004). Nevertheless, as discussed earlier, rates of evolution are not the only important parameters; other features, such as the heterogeneity of rates of variation across sites and across lineages, may be more important. In addition, the use of structural information from the sequence of mtDNA-encoded proteins has proven valuable for the resolution of deep phylogenetic relationships, such as those among different eukaryotic lineages (reviewed in Bullerwell & Gray 2004).

However, mitochondrial genes have a number of advantages over nuclear genes that have stimulated their extensive use for phylogenetic inference at all taxonomic levels. In addition to faster coalescence times (discussed above), these features include haploidy, the absence (or reduced rates) of recombination (Birky 2001), maternal inheritance, and more neutral patterns of evolution. Although some of these features have been challenged by evidence of recombination (Guo et al. 2006, Shao et al. 2005), doubly uniparental inheritance (Passamonti et al. 2003), heteroplasmy with paternal leakage (Bromham et al. 2003), and selective sweeps (Ballard & Rand 2005), these phenomena are likely to have little effect on mitochondrial phylogenetics above the species level, especially if rapid phylogenetic radiations need not be resolved (an expensive proposition anyway). Even assuming that recombination occurs, its frequency is certainly much lower in mtDNA than in the nuclear genome, and it most likely occurs between more similar haplotypes. In terms of phylogenetic utility this means that nuclear genes are more likely to be a mosaic of parts having a different evolutionary history and to be telling a different phylogenetic tale, either because of recombination or because of sorting of ancestral polymorphisms (e.g., Pääbo 2003). The mitochondrial genome is much less affected by this problem, and it tends to evolve as a single, rarely recombinant locus. Nuclear genes are most effective if multiple unlinked loci can be sequenced and if these loci are variable enough to answer the questions posed.

The combination of nuclear and mitochondrial data is useful because incongruence between nuclear and mitochondrial phylogenies can reveal important aspects of species histories such as introgression, hybridization, direct or indirect selection, and incomplete lineage sorting (reviewed by Funk & Omland 2003 and Ballard & Rand 2005). Because mtDNA can introgress faster than nuclear DNA and can sometimes entirely replace the mtDNA of the species invaded (Ballard & Rand 2005, Machado & Hey 2003), the combined use of nuclear traits is an important check. It is difficult to list all the different studies where markers from the two genomes have been used in conjunction, but examples can be found of either perfect congruence (e.g., Overton & Rhoads 2004) or sharp conflict (e.g., Goto & Kimura 2001) of the phylogenetic signal. They have helped to reveal recent hybridization events (e.g., Alexandrino et al. 2005, Buckley et al. 2006, Jordan et al. 2003), and cases of peculiar reproductive systems, such as hybridogenesis and androgenesis (e.g., Mantovani et al. 2001).

Nuclear Pseudogenes

One serious problem with the use of mitochondrial genes for phylogenetic analysis is the occurrence of copies of mitochondrial genes translocated to the nucleus (numts) where they become pseudogenes (Bensasson et al. 2001, Thalmann et al. 2004, Zhang & Hewitt 1996). Identifying pseudogenes by reading-frame disruption is not sufficient, as functional mitochondrial genes containing translational frameshifts have been observed (see the Supplemental Appendix for discussion;

follow the Supplemental Material link from the Annual Reviews home page at
http://www.annualreviews.org/). A way to overcoming this problem would be to
use gene sequences retrieved from longer sections of the mitochondrion or to take
advantage of the fact that the mtDNA is a circle to devise clever long PCR strategies
for testing the integrity of the molecule (Thalmann et al. 2004).

Usefulness of Complete Mitochondrial Genome Sequences

The collection of complete mitochondrial genomes and the analysis of information
from all or most of the 13 protein-coding genes have increased dramatically in the
past decade (Boore et al. 2005) and the use of this information for phylogenetic
purposes has therefore intensified (reviewed in Carapelli et al. 2006 for hexapods,
and Meyer & Zardoya 2003 for vertebrates). This has allowed the investigation of
the phylogenetic usefulness of complete mitochondrial genomes for deep levels and
the testing of different approaches to phylogeny reconstruction (Cook et al. 2005,
Delsuc et al. 2003, Hassanin et al. 2005, Nardi et al. 2003). A detailed knowledge
of the misleading effects of nucleotide compositional and substitutional biases of
mitochondrial protein-coding genes, as well as accelerated rates of evolution owing
to genomic rearrangements (e.g., Shao et al. 2003), parasitic lifestyles (e.g., Dowton &
Austin 1995), and metabolic rate (e.g., Gillooly et al. 2005), allows the development
of more realistic models of DNA evolution and the estimation of well-supported
phylogenetic reconstructions (Hassanin 2006).

The gene rearrangements of entire mitochondrial genomes provide another potential source of phylogenetic information, especially for the deepest nodes, because
shared gene-order changes are assumed to be evolutionarily rare events (and therefore less likely to occur convergently), and may represent synapomorphic characters
useful for reconstructing phylogenetic relationships (Boore et al. 2005, Larget et al.
2005). The best known example was provided by Boore et al. (1998), who suggested
an insect-crustacean (Pancrustacea) relationship based on the shared translocation of
one transfer RNA (tRNA)-encoding gene, but other examples have been obtained
from other taxa (reviewed in Boore et al. 2005). However, the discovery of taxa with
extensive variation in gene order (Dowton et al. 2003 for hymenopterans; Shao et al.
2003 for hemipteroids; Scouras et al. 2004 for echinoderms) and convergent gene arrangements (Macey et al. 2004 in amphiesbaenian reptiles) suggests caution when interpreting gene order changes in a phylogenetic framework. Several mechanisms have
been proposed by which gene order rearrangements may occur, including intramitochondrial recombination (Dowton & Campbell 2001), and duplication followed by
random or nonrandom loss of duplicated genes (Boore 2000, Lavrov et al. 2002).

CONCLUDING REMARKS

Despite recent challenges, mitochondrial data remain at the forefront of phylogenetic
studies at all levels of divergence. When used with care (Ballard & Rand 2005, Funk &
Omland 2003), mtDNA is particularly valuable in studies of taxa that have evolved in
the past 15 million years, where most nuclear genes show very little change. mtDNA

data have played a major role in studies of molecular evolution that have helped to refine models of evolution and understand how rates and patterns of substitution vary across sequences and over time. The combination of mitochondrial and nuclear data provides useful insights that cannot be obtained with either type of data alone. To facilitate the sequencing of mtDNA, we provide an online appendix of conserved PCR primers that complement those presented in the compilation of Simon et al. (1994).

SUMMARY POINTS

1. Virtually all nucleotide and amino acid sequence data display ASRV within sequences; accommodating this bias will have more of an effect on phylogenetic analyses than any of the other commonly used substitution model parameters.
2. Mixture models can accommodate among-partition rate and pattern variability without the need to pre-assign partitions.
3. Changes in patterns of ASRV over time (covarion-like evolution) can result in nucleotide bias among taxa that can introduce systematic error if ignored.
4. Many sequences show covarion-like evolution. This is a serious problem for tree building because most models of evolution assume that it does not exist.
5. Measures of nodal support reflect the relative information content of the data for different groupings on the tree. A phylogenetic tree with no support values is meaningless.
6. In Bayesian phylogenetic analyses, it is important to check the effects of the priors and to monitor the convergence of posterior distributions of all parameters.
7. Complete mitochondrial genome sequences contain significant phylogenetic information when analyzed using concatenated nucleotide and amino acid sequences or gene order rearrangements.
8. The combination of nuclear and mitochondrial data is important because incongruence between nuclear and mitochondrial phylogenies can reveal important aspects of species histories.

FUTURE RESEARCH DIRECTIONS

1. The further development of models of rRNA evolution that incorporate correlation among sites, variable length insertions and deletions, and strong differences in rate variation among sites.
2. The development of better models of covarion-like evolution and continued development of user-friendly programs that incorporate the newest models of evolution.

3. Further exploration of the effects of priors and the convergence of posterior distributions in Bayesian phylogenetic analyses.

4. The exploration of new methods for model selection.

ACKNOWLEDGMENTS

We thank the following people for information, helpful discussions, or comments on all or part of the manuscript: Jeff Boore, Tim Collins, Dan Funk, Francesco Nardi, Mark Pagel, Paul Lewis, David Hillis, John Huelsenbeck, Mark Holder, Pete Lockhart, Hervé Philippe, Dave Rand, Ziheng Yang, Joe Felsenstein, Spencer Muse, Andrew Roger, Brad Schaffer, Jack Sullivan, and Yves Van de Peer. For unselfishly devoting countless hours designing important computer programs that allow progress in the kind of molecular systematics studies presented here we thank Keith Crandall, Alexei Drummond, Joe Felsenstein, Mark Holder, John Huelsenbeck, Daniel Huson, Sudhir Kumar, Paul Lewis, David & Wayne Maddison, Rod Page, Mark Pagel, David Posada, Fredrik Ronquist, David Swofford, Jeff Thorne, Ziheng Yang, Derrick Zwickl, and others. We also thank Eliza Jewitt and Virge Kask for help with the illustrations. Funding from the following grants aided in the production of this manuscript: NSF DEB 0089946 (C.S.), NSF DEB -04-22386 (C.S., T.B.), NSF DEB 05-29679 (C.S.) and funds from the University of Connecticut (C.S.) and Victoria University of Wellington Visiting Professorial Fellowship (C.S.); the New Zealand Foundation for Research, Science and Technology (T.B.), the New Zealand Marsden Fund (T.B., C.S.), the University of Siena (F.F.), and the Italian Ministry for University and Research (F.F.); and an NSERC of Canada Discovery Grant (A.T.B.).

LITERATURE CITED

Adachi J, Hasegawa M. 1996. Model of amino acid substitution in proteins encoded by mitochondrial DNA. *J. Mol. Evol.* 42:459–68

Alexandrino J, Baird SJ, Lawson L, Macey JR, Moritz C, Wake DB. 2005. Strong selection against hybrids at a hybrid zone in the *Ensatina* ring species complex and its evolutionary implications. *Evol. Int. J. Org. Evol.* 59:1334–47

Arbogast BS, Edwards SV, Wakeley J, Beerli P, Slowinski JB. 2002. Estimating divergence times from molecular data on phylogenetic and population genetic timescales. *Annu. Rev. Ecol. Syst.* 33:707–40

Arensburger P, Buckley TR, Simon C, Moulds M, Holsinger KE. 2004. Biogeography and phylogeny of the New Zealand cicada genera (*Hemiptera: Cicadidae*) based on nuclear and mitochondrial DNA data. *J. Biogeogr.* 31:557–69

Aris-Brosou S. 2003. How Bayes tests of molecular phylogenies compare with frequentist approaches. *Bioinfomatics* 19:618–24

Aris-Brosou S, Yang Z. 2002. The effects of models of rate evolution on estimation of divergence dates with a special reference to the metazoan 18S rRNA phylogeny. *Syst. Biol.* 51:703–14

Avise JC. 2000. *Phylogeography: The History and Formation of Species*. Cambridge, MA: Harvard Univ. Press

Ballard JWO, Rand DM. 2005. The population biology of mitochondrial DNA and its phylogenetic implications. *Annu. Rev. Ecol. Evol. Syst.* 36:621–42

Barker FK, Lutzoni FM. 2002. The utility of the incongruence length difference test. *Syst. Biol.* 51:625–37

Bensasson D, Zhang X, Hartl DL, Hewitt GM. 2001. Mitochondrial pseudogenes: evolution's misplaced witnesses. *Trends Ecol. Evol.* 16:314–21

Birky CWJ. 2001. The inheritance of genes in mitochondria and chloroplasts: laws, mechanisms, and models. *Annu. Rev. Genet.* 35:125–48

Bollback JP. 2002. Bayesian model adequacy and choice in phylogenetics. *Mol. Biol. Evol.* 19:1171–80

Boore JL. 2000. The duplication/random loss model for gene rearrangement exemplified by mitochondrial genomes of deuterostome animals. In *Comparative Genomics*, ed. D Sankoff, JH Nadeau, pp. 133–47. Netherlands: Kluwer

Boore JL, Lavrov D, Brown WM. 1998. Gene translocation links insects and crustaceans. *Nature* 393:667–68

Boore JL, Macey JR, Medina M. 2005. Sequencing and comparing whole mitochondrial genomes of animals. *Methods Enzymol.* 395:311–48

Brandley MC, Schmitz AS, Reeder TW. 2005. Partitioned Bayesian analyses, partition choice, and the phylogenetic relationships of scincid lizards. *Syst. Biol.* 54:373–90

Bromham L, Eyre-Walker A, Smith NH, Maynard Smith J. 2003. Mitochondrial Steve: paternal inheritance of mitochondria in humans. *Trends Ecol. Evol.* 18:2–4

Brown WM, Prager EM, Wang A, Wilson AC. 1982. Mitochondrial DNA sequences of primates: tempo and mode of evolution. *J. Mol. Evol.* 18:225–39

Buckley TR. 2002. Model misspecification and probabilistic tests of topology: evidence from empirical data sets. *Syst. Biol.* 51:509–23

Buckley TR, Arensburger P, Simon C, Chambers GK. 2002. Combined data, Bayesian phylogenetics, and the origin of the New Zealand cicada genera. *Syst. Biol.* 51:4–18

Buckley TR, Cordeiro M, Marshall DC, Simon C. 2006. Differentiating between hypotheses of lineage sorting and introgression in New Zealand alpine cicadas (*Maoricicada* Dugdale). *Syst. Biol.* 55:411–25

Buckley TR, Cunningham CW. 2002. The effects of nucleotide substitution model assumptions on estimates of nonparametric bootstrap support. *Mol. Biol. Evol.* 19:394–402

Buckley TR, Simon C, Chambers GK. 2001a. Exploring among-site rate variation models in a maximum likelihood framework using empirical data: effects of model assumptions on estimates of topology, branch lengths and bootstrap support. *Syst. Biol.* 50:67–86

Show that different methods of correcting for ASRV have varying effects on branch lengths and nodal support.

Buckley TR, Simon C, Shimodaira H, Chambers GK. 2001b. Evaluating hypotheses on the origin and evolution of the New Zealand alpine cicadas (*Maoricicada*) using multiple comparison tests of tree topology. *Mol. Biol. Evol.* 18:223–34

Bull JJ, Huelsenbeck JP, Cunningham CW, Swofford DL, Waddell PJ. 1993. Partitioning and combining data in phylogenetic analysis. *Syst. Biol.* 42:384–97

Bullerwell CE, Gray MW. 2004. Evolution of the mitochondrial genome: protist connections to animals, fungi and plants. *Curr. Opin. Microbiol.* 7:528–34

Burger G, Gray MW, Lang BF. 2003. Mitochondrial genomes: Anything goes. *Trends Genet.* 19:709–16

Carapelli A, Nardi F, Dallai R, Frati F. 2006. A review of molecular data for the phylogeny of basal hexapods. *Pedobiologia* 50:191–204

Carstens BC, Brunsfeld SJ, Demboski JR, Good JM, Sullivan J. 2005a. Investigating the evolutionary history of the Pacific Northwest mesic forest ecosystem: hypothesis testing within a comparative phylogeographic framework. *Evolution* 59:1639–52

Carstens BC, Demboski JR, Good JM, Brunsfeld SJ, Sullivan J. 2005b. The evolutionary history of the northern Rocky Mountain mesic forest ecosystem. *Evolution* 59:1639–52

Castoe TA, Doan TM, Parkinson CL. 2004. Data partitions and complex models in Bayesian analysis: the phylogeny of gymnophthalmid lizards. *Syst. Biol.* 53:448–59

Caterino MS, Cho S, Sperling FAH. 2000. The current state of insect molecular systematics: a thriving tower of Babel. *Annu. Rev. Entomol.* 45:1–54

Collins T, Fedrigo O, Naylor GJP. 2005. Choosing the best genes for the job: the case for stationary genes in genome-scale phylogenetics. *Syst. Biol.* 54:493–500

> Demonstrates that nucleotide bias nonstationarity has a strong effect on the phylogenetic usefulness of genes.

Cook CE, Yue Q, Akam M. 2005. Mitochondrial genomes suggest that hexapods and crustaceans are mutually paraphyletic. *Proc. R. Soc. London Ser. B* 272:1295–304

Cummings MP, Handley SA, Myers DS, Reed RL, Rokas A, Winka K. 2003. Comparing bootstrap and posterior probability values in the four-taxon case. *Syst. Biol.* 52:477–87

Cunningham CW. 1997. Is congruence between data partitions a reliable predictor of phylogenetic accuracy? Empirically testing an iterative procedure for choosing among phylogenetic methods. *Syst. Biol.* 46:464–78

DeBry RW. 2005. The systematic component of phylogenetic error as a function of taxonomic sampling under parsimony. *Syst. Biol.* 54:432–40

Delsuc F, Phillips MJ, Penny D. 2003. Comment on "Hexapod origins: monophyletic or paraphyletic?" *Science* 301:1482d

Douady CJ, Delsuc F, Boucher Y, Doolittle WF, Douzery EJP. 2003. Comparison of Bayesian and maximum likelihood bootstrap measures of phylogenetic reliability. *Mol. Biol. Evol.* 20:248–54

Dowton M, Austin AD. 1995. Increased genetic diversity in mitochondrial genes is correlated with the evolution of parasitism in the Hymenoptera. *J. Mol. Evol.* 41:958–65

Dowton M, Campbell NJH. 2001. Intramitochondrial recombination—is it why some mitochondrial genes sleep around? *Trends Ecol. Evol.* 16:269–71

Dowton M, Castro LR, Campbell SL, Bargon SD, Austin AD. 2003. Frequent mitochondrial gene rearrangements at the hymenopteran nad3-nad5 junction. *J. Mol. Evol.* 56:517–26

Edwards SV, Beerli P. 2000. Perspective: gene divergence, population divergence, and the variation in coalescence time in phylogeographic studies. *Evolution* 54:1839–54

Efron B, Halloran E, Holmes S. 1996. Bootstrap confidence levels for phylogenetic trees. *Proc. Natl. Acad. Sci. USA* 93:13429–34

Felsenstein J. 1978. Cases in which parsimony and compatibility methods will be positively misleading. *Syst. Zool.* 27:401–10

Felsenstein J. 1981. Evolutionary trees from DNA sequences: a maximum likelihood approach. *J. Mol. Evol.* 17:368–76

Felsenstein J. 1985. Confidence limits on phylogenies: an approach using the bootstrap. *Evolution* 39:783–91

Felsenstein J. 2004. *Inferring Phylogenies*. Sunderland, MA: Sinauer. 664 pp

Felsenstein J, Kishino H. 1993. Is there something wrong with the bootstrap on phylogenies? A reply to Hillis and Bull. *Syst. Biol.* 42:193–200

Fitch WM. 1986. The estimate of total nucleotide substitutions from pairwise differences is biased. *Philos. Trans. R. Soc. London Ser. B* 312:317–24

Fitch WM, Margoliash E. 1967. A method for estimating the number of invariant amino acid coding positions in a gene using cytochrome *c* as a model case. *Biochem. Genet.* 1:65–71

Fitch WM, Markowitz E. 1970. An improved method for determining codon variability in a gene and its application to the rate of fixation of mutations in evolution. *Biochem. Genet.* 4:579–93

Funk DJ, Omland KE. 2003. Species-level paraphyly and polyphyly: Frequency, causes, consequences, with insights from animal mitochondrial DNA. *Annu. Rev. Ecol. Evol. Syst.* 34:397–423

Galtier N. 2001. Maximum-likelihood phylogenetic analysis under a covarion-like model. *Mol. Biol. Evol.* 18:866–73

Gaucher EA, Miyamoto MM, Benner SA. 2001. Function-structure analysis of proteins using covarion-based evolutionary approaches: elongation factors. *Proc. Natl. Acad. Sci. USA* 98:548–52

Gillooly JF, Allen AP, West GB, Brown JH. 2005. The rate of DNA evolution: effects of body size and temperature on the molecular clock. *Proc. Natl. Acad. Sci. USA* 102:140–45

Goldman N, Anderson JP, Rodrigo AG. 2000. Likelihood-based tests of topologies in phylogenetics. *Syst. Biol.* 49:652–70

Goto SG, Kimura MT. 2001. Phylogenetic utility of mitochondrial *COI* and nuclear *Gpdh* genes in *Drosophila*. *Mol. Phylogenet. Evol.* 18:404–22

Gribaldo S, Casane D, Lopez P, Philippe H. 2003. Functional divergence prediction from evolutionary analysis: a case study of vertebrate hemoglobin. *Mol. Biol. Evol.* 20:1754–59

Guindon S, Rodrigo AG, Dyer KA, Huelsenbeck JP. 2004. Modeling the site-specific variation of selection patterns along lineages. *Proc. Natl. Acad. Sci. USA* 35:12957–62

Guo X, Liu SJ, Liu Y. 2006. Evidence for recombination of mitochondrial DNA in triploid crucian carp. *Genetics* 172:1745–49

> Created a covarion-like model that allows selective constraints to change over time without a priori specification of the branches on which the changes occur.

Gutell RR, Lee JC Cannone JJ. 2002. The accuracy of ribosomal RNA comparative structure models. *Curr. Opin. Struct. Biol.* **12:301–10**

> Provides elegant, proven, rRNA secondary-structure diagrams expandable in the PDF.

Haddrath OP, Baker AJ. 2001. Complete mitochondrial DNA genome sequences of extinct birds: ratite phylogenetics and the vicariance biogeography hypothesis. *Proc. R. Soc. London Ser. B* 268:939–45

Hassanin A. 2006. Phylogeny of Arthropoda inferred from mitochondrial sequences: strategies for limiting the misleading effects of multiple changes in pattern and rates of substitutions. *Mol. Phylogenet. Evol.* 38:100–16

Hassanin A, Léger N, Deutsch J. 2005. Evidence for multiple reversals of asymmetric mutational constraints during the evolution of the mitochondrial genome of Metazoa, and consequences for phylogenetic inferences. *Syst. Biol.* **54:277–98**

> Studied the effects of gene translocations on strand-specific mutational bias and developed a phylogenetic method to accommodate reversed mutational constraints.

Hedges SB, Kumar S. 2003. Genomic clocks and evolutionary timescales. *Trends Genet.* 19:200–6

Hedtke SM, Townsend TM, Hillis DM. 2006. Resolution of phylogenetic conflict in large data sets by increased taxon sampling. *Syst. Biol.* 55:522–29

Hickson RE, Simon C, Cooper A, Spicer GS, Sullivan J, Penny D. 1996. Conserved sequence motifs, alignment, and secondary structure for the third domain of animal 12S rRNA. *Mol. Biol. Evol.* 13:150–69

Hillis DM. 1996. Inferring complex phylogenies. *Nature* 383:130–31

Hillis DM, Bull JJ. 1993. An empirical test of bootstrapping as a method for assessing confidence in phylogenetic analysis. *Syst. Biol.* 42:182–92

Hillis DM, Pollock D, McGuire JA, Zwickl DJ. 2003. Is sparse taxon sampling a problem for phylogenetic inference? *Syst. Biol.* 52:124–26

Ho SY, Phillips MJ, Drummond AJ, Cooper A. 2005. Accuracy of rate estimation using relaxed-clock models with a critical focus on the early metazoan radiation. *Mol. Biol. Evol.* 22:1355–63

Holder M, Lewis PO. 2003. Phylogeny estimation: traditional and Bayesian approaches. *Nat. Rev. Genet.* **4:275–84**

> Provides an excellent overview of the advantages, disadvantages, and functioning of different phylogenetic methods.

Holder MT. 2001. *Using a complex model of sequence evolution to evaluate and improve phylogenetics methods*. PhD thesis. Univ. Texas, Austin. 165 pp.

Holmes S. 2005. Statistical approach to tests involving phylogenetic trees. In *Mathematics of Evolution and Phylogeny*, ed. O Gascuel, pp. 91–121. New York: Oxford Univ. Press

Huelsenbeck JP, Hillis DM, Nielsen R. 1996. A likelihood ratio test of monophyly. *Syst. Biol.* 45:546–58

Huelsenbeck JP, Imennov NS. 2002. Geographic origin of human mitochondrial DNA: accommodating phylogenetic uncertainty and model comparison. *Syst. Biol.* 51:155–65

Huelsenbeck JP, Jain S, Frost SWD, Kosakovsky Pond SL. 2006. A Dirichlet process model for detecting positive selection in protein-coding DNA sequences. *Proc. Natl. Acad. Sci. USA* 103:6263–68

Huelsenbeck JP, Larget B, Miller RE, Ronquist F. 2002. Potential applications and pitfalls of Bayesian inference of phylogeny. *Syst. Biol.* 51:673–88

Huelsenbeck JP, Nielsen R. 1999. Effect of nonindependent substitution on phylogenetic accuracy. *Syst. Biol.* 48:317–28

Huelsenbeck JP, Rannala B. 2004. Frequentist properties of Bayesian posterior probabilities of phylogenetic trees under complex and simple substitution models. *Syst. Biol.* 53:904–13

Huelsenbeck JP, Ronquist F. 2001. MrBayes: Bayesian inference of phylogenetic trees. *Bioinformatics* 17:754–55

Huelsenbeck JP, Ronquist F, Nielsen R, Bollback JP. 2001. Bayesian inference of phylogeny and its impact on evolutionary biology. *Science* 294:2310–14

Jennings WB, Edwards SV. 2005. Speciational history of Australian grass finches (*Poephila*) inferred from 30 gene trees. *Evol. Int. J. Org. Evol.* 59:2033–47

Jermiin L, Ho SYW, Ababneh F, Robinson J, Larkum AWD. 2004. The biasing effect of compositional heterogeneity on phylogenetic estimates may be underestimated. *Syst. Biol.* 53:638–43

Johannes GJ, Berger EG. 1993. Domains within the mammalian ornithine decarboxylase messenger RNA have evolved independently and episodically. *J. Mol. Evol.* 36:555–67

Jordan S, Simon C, Polhemus D. 2003. Molecular systematics and adaptive radiation of Hawaii's endemic damselfly genus *Megalagrion* (Odonata: Coenagrionidae). *Syst. Biol.* 52:89–109

Jukes TH, Cantor CR. 1969. Evolution of protein molecules. In *Mammalian Protein Metabolism*, ed. N Munro, pp. 21–132. New York: Academic

Kass RE, Raftery AE. 1995. Bayes factors. *J. Am. Stat. Assoc.* 90:773–95

Kishino H, Hasegawa M. 1989. Evaluation of the maximum likelihood estimate of the evolutionary tree topologies from DNA sequence data, and the branching order in Hominoidea. *J. Mol. Evol.* 29:170–79

Kishino H, Thorne JL, Bruno WJ. 2001. Performance of a divergence time estimation method under a probabilistic model of rate evolution. *Mol. Biol. Evol.* 18:352–61

Kjer K. 2004. Aligned 18S and insect phylogeny. *Syst. Biol.* 53:506–14

Kjer KM, Blahnik RJ, Holzenthal RW. 2001. Phylogeny of Trichoptera (Caddisflies): characterization of signal within multiple datasets. *Syst. Biol.* 50:781–816

Kocher TD, Thomas WK, Meyer A, Edwards SV, Pääbo S, et al. 1989. Dynamics of mitochondrial DNA evolution in animals: amplification and sequencing with conserved primers. *Proc. Natl. Acad. Sci. USA* 86:6196–200

Kumar S, Tamura K, Nei M. 1993. MEGA: molecular evolutionary genetics analysis. Version 1.0. University Park, Pennsylvania: Pennsylvania State University

Larget B, Simon DL. 1999. Markov chain Monte Carlo algorithms for the Bayesian analysis of phylogenetic trees. *Mol. Biol. Evol.* 16:750–59

Larget B, Simon DL, Kadane JB, Sweet D. 2005. A Bayesian analysis of metazoan mitochondrial genome rearrangements. *Mol. Biol. Evol.* 22:486–95

Lartillot N, Philippe H. 2004. A Bayesian mixture model for across-site heterogeneities in the amino-acid replacement process. *Mol. Biol. Evol.* 21:1095–1109

Lavrov DV, Boore JL, Brown WM. 2002. Complete mtDNA sequences of two millipedes suggest a new model for mitochondrial gene rearrangement: duplication and nonrandom loss. *Mol. Biol. Evol.* 19:163–69

Leache A, Reeder T. 2002. Molecular systematics of the eastern fence lizard (*Sceloporus undulatus*): a comparison of parsimony, likelihood, and Bayesian approaches. *Syst. Biol.* 51:44–68

Lemmon AR, Moriarty EC. 2004. The importance of proper model assumption in Bayesian phylogenetics. *Syst. Biol.* 53:265–77

Lewis PO, Holder MT, Holsinger KE. 2005. Polytomies and Bayesian phylogenetic inference. *Syst. Biol.* 54:241–53

> Demonstrates that reversible jump MCMC analysis can improve estimation of posterior probabilities for polytomies.

Lin C-P, Danforth BN. 2004. How do insect nuclear and mitochondrial gene substitution patterns differ? Insights from Bayesian analyses of combined datasets. *Mol. Phylogenet. Evol.* 30:686–702

Lió P, Goldman N. 2002. Modeling mitochondrial protein evolution using structural information. *J. Mol. Evol.* 54:519–29

Lockhart PJ, Howe CJ, Bryant DA, Beanland TJ, Larkum AWD. 1992. Substitutional bias confounds inference of cyanelle origins from sequence data. *J. Mol. Evol.* 34:153–62

Lockhart PJ, Huson D, Maier U, Fraunholz MJ, Van de Peer Y, et al. 2000. How molecules evolve in Eubacteria. *Mol. Biol. Evol.* 17:835–38

Lockhart PJ, Larkum AWD, Steel MA, Waddell PJ, Penny D. 1996. Evolution of chlorophyll and bacteriochlorophyll: the problem of invariant sites in sequence analysis. *Proc. Natl. Acad. Sci. USA* 93:1930–34

Lockhart PJ, Steel MA, Barbrook AC, Huson DH, Charleston MA, Howe CJ. 1998. A covariotide model explains apparent phylogenetic structure of oxygenic photosynthetic lineages. *Mol. Biol. Evol.* 15:1183–88

Lopez P, Casane D, Philippe H. 2002. Heterotachy, an important process of protein evolution. *Mol. Biol. Evol.* 19:1–7

Lopez P, Forterre P, Philippe H. 1999. The root of the tree of life in the light of the covarion model. *J. Mol. Evol.* 49:496–508

Lutzoni F, Wagner P, Reeb V, Zoller S. 2000. Integrating ambiguously aligned regions of DNA sequences in phylogenetic analyses without violating positional homology. *Syst. Biol.* 49:628–51

Macey JR, Papenfuss TJ, Kuhel JV, Fourcade HM, Boore JL. 2004. Phylogenetic relationships among amphisbaenian reptiles based on complete mitochondrial genomic sequences. *Mol. Phylogenet. Evol.* 33:22–31

Machado CA, Hey J. 2003. The causes of phylogenetic conflict in a classic *Drosophila* species group. *Proc. R. Soc. London Ser. B* 270:1193–202

Mantovani B, Passamonti M, Scali V. 2001. The mitochondrial cytochrome oxidase II gene in *Bacillus* stick insects: ancestry of hybrids, androgenesis, and phylogenetic relationships. *Mol. Phylogenet. Evol.* 19:157–63

Margoliash E, Smith E. 1965. Structural and functional aspects of cytochrome *c* in relation to evolution. In *Evolving Genes and Proteins*, ed V Bryson, HJ Vogel. New York: Academic. 221 pp.

Marshall DC, Simon C, Buckley TR. 2006. Accurate branch length estimation in partitioned Bayesian analyses requires accommodation of among-partition rate variation and attention to branch length priors. *Syst. Biol.* In press

Meyer A, Zardoya R. 2003. Recent advances in the (molecular) phylogeny of vertebrates. *Annu. Rev. Ecol. Evol. Syst.* 34:311–38

Misof B, Anderson CL, Buckley TR, Erpenbeck D, Rickert A, Misof K. 2002. An empirical analysis of mt 16S rRNA covarion-like evolution in insects: site-specific rate variation is clustered and frequently detected. *J. Mol. Evol.* 55:460–69

Mitchell A, Mitter C, Regier JC. 2000. More taxa or more characters revisited: combining data from nuclear protein-encoding genes for phylogenetic analyses of Noctuoidea (Insecta: Lepidoptera). *Syst. Biol.* 49:202–24

Miyamoto MM, Fitch WM. 1995. Testing the covarion hypothesis of molecular evolution. *Mol. Biol. Evol.* 12:503–13

Moriyama EN, Powell JR. 1997. Synonymous substitution rates in *Drosophila*: mitochondrial versus nuclear genes. *J. Mol. Evol.* 45:378–91

Muse SV, Gaut BS. 1994. A likelihood approach for comparing synonymous and nonsynonymous nucleotide substitutions rates, with application to the chloroplast genome. *Mol. Biol. Evol.* 11:1139–51

Nardi F, Spinsanti G, Boore JL, Carapelli A, Dallai R, Frati F. 2003. Hexapod origins: monophyletic or polyphyletic? *Science* 299:1887–89

Nylander JAA, Ronquist F, Huelsenbeck JPP, Nieves-Aldrey JL. 2004. Bayesian phylogenetic analysis of combined data. *Syst. Biol.* 53:47–67

Overton LC, Rhoads DD. 2004. Molecular phylogenetic relationships based on mitochondrial and nuclear gene sequences for the Todies (*Todus*, Todidae) of the Caribbean. *Mol. Phylogenet. Evol.* 32:524–38

Pääbo S. 2003. The mosaic that is our genome. *Nature* 421:409–12

Page RDM, Holmes EC. 1998. *Molecular Evolution. A Phylogenetic Approach*. Oxford, UK: Blackwell Sci. 352 pp.

Pagel M, Meade A. 2004. A phylogenetic mixture model for detecting pattern-heterogeneity in gene sequences or character-state data. *Syst. Biol.* 53:571–81

> Describes a general mixture model within a Bayesian MCMC framework that can simultaneously detect rate and pattern heterogeneity.

Passamonti M, Boore JL, Scali V. 2003. Molecular evolution and recombination in gender-associated mitochondrial DNAs of the Manila clam *Tapes philippinarum*. *Genetics* 164:603–11

Paton T, Haddrath O, Baker AJ. 2002. Complete mitochondrial DNA genome sequences show that modern birds are not descended from transitional shorebirds. *Proc. R. Soc. London Ser. B* 269:839–46

Penny D, Hendy MD, Lockhart PJ, Steel MA. 1996. Corrected parsimony, minimum evolution and Hadamard conjugations. *Syst. Biol.* 45:593–603

Penny D, Hendy MD, Steel MA. 1992. Progress with evolutionary trees. *Trends Ecol. Evol.* 7:73–79

Philippe H, Lecointre G, Van Le HL, Le Guyader H. 1996. A critical study of homoplasy in molecular data with the use of a morphologically based cladogram. *Mol. Biol. Evol.* 13:1174–86

Philippe H, Lopez P. 2001. On the conservation of protein sequences in evolution. *Trends Biochem. Sci.* 26:414–16

Poe S. 2003. Evaluation of the strategy on long-branch subdivision to improve the accuracy of phylogenetic methods. *Syst. Biol.* 52:423–28

Pollock DD, Bruno WJ. 2000. Assessing an unknown evolutionary process: effect of increasing site-specific knowledge through taxon addition. *Mol. Biol. Evol.* 17:1854–58

Pollock DD, Zwickl DJ, McGuire JA, Hillis DM. 2002. Increased taxon sampling is advantageous for phylogenetic inference. *Syst. Biol.* 51:664–71

Posada D, Buckley TR. 2004. Model selection and model averaging in phylogenetics: advantages of Akaike information criterion and Bayesian approaches over likelihood ratio tests. *Syst. Biol.* 53:793–808

Posada D, Crandall KA. 1998. Modeltest: testing the model of DNA substitution. *Bioinformatics* 14:817–18

Rambaut A, Bromham L. 1998. Estimating divergence dates from molecular sequences. *Mol. Biol. Evol.* 15:442–48

Rand DM, Haney RA, Fry AJ. 2004. Cytonuclear coevolution: the genomics of cooperation. *Trends Ecol. Evol.* 19:645–53

Ren F, Yang Z. 2005. An empirical examination of the utility of codon-substitution models in phylogeny reconstruction. *Syst. Biol.* 54:808–18

Revell LJ, Harmon LJ, Glor RE. 2005. Underparameterized model of sequence evolution leads to bias in the estimation of diversification rates from molecular phylogenies. *Syst. Biol.* 54:973–83

Rokas A, Nylander JAA, Ronquist F, Stone GA. 2002. A maximum-likelihood analysis of eight phylogenetic markers in gallwasps (Hymenoptera: Cynipidae): implications for insect phylogenetic studies. *Mol. Phylogenet. Evol.* 22:206–19

Rokas A, Williams BL, King N, Carroll SB. 2003. Genome-scale approaches to resolving incongruence in molecular phylogenies. *Nature* 425:798–804

Rosenberg MS, Kumar S. 2001. Incomplete taxon sampling is not a problem for phylogenetic inference. *Proc. Natl. Acad. Sci. USA* 98:10751–56

Saiki RK, Scharf S, Faloona F, Mullis KB, Horn GT, et al. 1985. Enzymatic amplification of B-globin genomic sequences and restriction site analysis for diagnosis of sickle cell anemia. *Science* 230:1350–54

Salamin N, Hodkinson TR, Savolainen V. 2005. Towards building the tree of life: a simulation study for all angiosperm genera. *Syst. Biol.* 54:183–96

Sanderson MJ. 1995. Objections to bootstrapping phylogenies: a critique. *Syst. Biol.* 44:299–320

Sanderson MJ. 1997. A nonparametric approach to estimating divergence times in the absence of rate constancy. *Mol. Biol. Evol.* 14:1218–31

Sanderson MJ. 2002. Estimating absolute rates of molecular evolution and divergence times: a penalized likelihood approach. *Mol. Biol. Evol.* 19:101–9

Sanderson MJ, Wojciechowski MF. 2000. Improved bootstrap confidence limits in large-scale phylogenies, with an example from Neo-*Astragalus* (Leguminosae). *Syst. Biol.* 49:671–85

Sanger F, Nicklen S, Coulson AR. 1977. DNA sequencing with chain-termination inhibitors. *Proc. Natl. Acad. Sci. USA* 74:5463–67

Scouras A, Beckenbach K, Arndt A, Smith MJ. 2004. Complete mitochondrial genome DNA sequence for two ophiuroids and a holothuroid: the utility of protein gene sequence and gene maps in the analyses of deep deuterostome phylogeny. *Mol. Phylogenet. Evol.* 31:50–65

Shao R, Dowton M, Murrell A, Barker SC. 2003. Rates of gene rearrangement and nucleotide substitution are correlated in the mitochondrial genomes of insects. *Mol. Biol. Evol.* 20:1612–19

Shao R, Mitani H, Barker SC, Takahashi M, Fukunaga M. 2005. Novel mitochondrial gene content and gene arrangement indicate illegitimate intermtDNA recombination in the chigger mite, *Leptotrombidium pallidum*. *J. Mol. Evol.* 60:764–73

Shimodaira H. 2002. An approximately unbiased test of phylogenetic tree selection. *Syst. Biol.* 51:492–508

Shimodaira H, Hasegawa M. 1999. Multiple comparisons of log-likelihoods with applications to phylogenetic inference. *Mol. Biol. Evol.* 16:1114–16

Simon C. 1991. Molecular systematics at the species boundary: exploiting conserved and variable regions of the mitochondrial genome of animals via direct sequencing from enzymatically amplified DNA. In *Molecular Techniques in Taxonomy*, ed. GM Hewitt, AWB Johnston, JPW Young, H57:33–71. New York: Springer Verlag, NATO Advanced Studies Institute

Simon C, Frati F, Beckenbach A, Crespi B, Liu H, Flook P. 1994. Evolution, weighting, and phylogenetic utility of mitochondrial gene sequences and a compilation of conserved polymerase chain reaction primers. *Ann. Entomol. Soc. Am.* 87:651–701

Simon C, Nigro L, Sullivan J, Holsinger K, Martin A, et al. 1996. Large differences in substitutional pattern and evolutionary rate of 12S ribosomal RNA genes. *Mol. Biol. Evol.* 13:923–32

Smith AD, Lui TWH, Tillier ERM. 2004. Empirical models for substitution in ribosomal RNA. *Mol. Biol. Evol.* 21:419–27

Springer MS, DeBry RW, Douady C, Amrine HM, Madsen O, et al. 2001. Mitochondrial versus nuclear gene sequences in deep-level mammalian phylogeny reconstruction. *Mol. Biol. Evol.* 18:132–43

Springer MS, Douzery E. 1996. Secondary structure and patterns of evolution among mammalian mitochondrial 12S rRNA molecules. *J. Mol. Evol.* 43:357–73

Steel MA. 1994. Recovering a tree from the leaf colorations it generates under a Markov model. *Appl. Math. Lett.* 7:19–24

Steel MA, Huson D, Lockhart PJ. 2000. Invariable sites models and their use in phylogeny reconstruction. *Syst. Biol.* 49:225–32

Suchard MA, Weiss RE, Sinsheimer JS. 2005. Models for estimating Bayes factors with applications to phylogeny and tests of monophyly. *Biometrics* 61:665–73

Sullivan J, Joyce P. 2005. Model selection in phylogenetics. *Annu. Rev. Ecol. Evol. Syst.* 36:445–66

Sullivan J, Swofford DL. 2001. Should we use model-based methods for phylogenetic inference when we know assumptions about among-site rate variation and nucleotide substitution pattern are violated? *Syst. Biol.* 50:723–29

> Show that ASRV has more of an effect than other model parameters on the accuracy of phylogenetic trees.

Sunnucks P. 2000. Efficient genetic markers for population biology. *Trends Ecol. Evol.* 15:199–203

Susko E, Field C, Blouin C, Roger AJ. 2003. Estimation of rates-across-sites distributions in phylogenetic substitution models. *Syst. Biol.* 52:594–603

Susko E, Inagaki Y, Roger AJ. 2004. On inconsistency of the neighbor-joining, least squares, minimum evolution estimation when substitution processes are incorrectly modeled. *Mol. Biol. Evol.* 21:1629–42

Suzuki Y, Glazko GV, Nei M. 2002. Overcredibility of molecular phylogenies obtained by Bayesian phylogenetics. *Proc. Natl. Acad. Sci. USA* 99:15138–43

Swofford DL. 1993/1998. *PAUP*: Phylogenetic Analysis Using Parsimony (*and Other Methods)*. Sunderland, MA: Sinauer

Swofford DL, Olsen GJ, Waddell PJ, Hillis DM. 1996. Phylogenetic inference. In *Molecular Systematics*, ed. DM Hillis, CM Moritz, BK Mable, pp. 407–514. Sunderland, MA: Sinauer

Swofford DL, Waddell PJ, Huelsenbeck JP, Foster PG, Lewis PO, Rogers JS. 2001. Bias in phylogenetic estimation and its relevance to the choice between parsimony and likelihood methods. *Syst. Biol.* 50:525–39

> Clearly explain the long-branch attraction problem and how it affects different phylogenetic tree-building methods.

Thalmann O, Hebler J, Poinar N, Pääbo S, Vigilant L. 2004. Unreliable mtDNA data due to nuclear insertions: a cautionary tale from analysis of humans and other great apes. *Mol. Ecol.* 13:321–35

Thorne JL, Kishino H, Painter IS. 1998. Estimating the rate of evolution of the rate of molecular evolution. *Mol. Biol. Evol.* 15:1647–57

Tuffley C, Steel M. 1998. Modeling the covarion hypothesis of nucleotide substitution. *Math. Biosci.* 147:63–91

Venditti C, Meade A, Pagel M. 2006. Detecting the node-density artifact in phylogeny reconstruction. *Syst. Biol.* 55:637–43

Waddell PJ, Kishino H, Ota R. 2002. Very fast algorithms for evaluating the stability of ML and Bayesian phylogenetic trees from sequence data. *Genome Inform.* 13:82–92

Weisburg WG, Giovannoni SJ, Woese CR. 1989. The *Deinococcus* and *Thermum* phylum and the effect of ribosomal RNA composition on phylogenetic tree construction. *Syst. Appl. Microbiol.* 11:128–34

Welch JJ, Bromham L. 2005. Molecular dating when rates vary. *Trends Ecol. Evol.* 20:320–27

Welch JJ, Fontanillas E, Bromham L. 2005. Molecular dates for the "Cambrian explosion": the influence of prior assumptions. *Syst. Biol.* 54:672–78

Wilcox TP, Zwickl DJ, Heath TA, Hillis DM. 2002. Phylogenetic relationships of the dwarf boas and a comparison of Bayesian and bootstrap measures of phylogenetic support. *Mol. Phylogenet. Evol.* 25:361–71

Yang Z. 1996. Among-site rate variation and its impact on phylogenetic analysis. *Trends Ecol. Evol.* 11:367–72

Yang Z. 1997. PAML: a program for phylogenetic analysis by maximum likelihood. *Comput. Appl. Biosci.* 13:555–56

Yang Z. 2003. Adaptive molecular evolution. In *Handbook of Statistical Genetics*, ed. DJ Balding, M Bishop, C Cannings, pp. 229–54. New York: Wiley. 2nd ed.

Yang Z, Goldman N, Friday AE. 1994. Comparison of models for nucleotide substitution used in maximum likelihood phylogenetic estimation. *Mol. Biol. Evol.* 11:316–24

Yang Z, Nielsen R, Goldman N, Pedersen AMK. 2000. Codon-substitution models for heterogeneous selection pressure at amino acid sites. *Genetics* 155:431–49

Yang Z, Rannala B. 1997. Bayesian phylogenetic inference using DNA sequences: a Markov chain Monte Carlo method. *Mol. Biol. Evol.* 14:717–24

Yang Z, Rannala B. 2005. Branch-length prior influences Bayesian posterior probability of phylogeny. *Syst. Biol.* 54:455–70

Yang Z, Rannala B. 2006. Bayesian estimation of species divergence times under a molecular clock using fossil calibrations with soft bounds. *Mol. Biol. Evol.* 23:212–26

Yoder AD, Yang Z. 2000. Estimation of primate speciation dates using local molecular clocks. *Mol. Biol. Evol.* 17:1081–90

Zhang D-X, Hewitt GM. 1996. Nuclear integrations: challenges for mitochondrial DNA markers. *Trends Ecol. Evol.* 11:247–51

Zhang D-X, Hewitt GM. 2003. Nuclear DNA analyses in genetic studies of populations: practice, problems and prospects. *Mol. Ecol.* 12:563–84

Zwickl DJ. 2006. GARLI version 0.94. **http://www.zo.utexas.edu/faculty/antisense/Garli.html**

Zwickl DJ, Hillis DM. 2002. Increased taxon sampling greatly reduces phylogenetic error. *Syst. Biol.* 51:588–98

RELATED RESOURCE

Sanderson MJ, Shaffer HB. 2002. Troubleshooting molecular phylogenetic analyses. *Annu. Rev. Ecol. Syst.* 33:49–72

The Developmental, Physiological, Neural, and Genetical Causes and Consequences of Frequency-Dependent Selection in the Wild

Barry Sinervo[1] and Ryan Calsbeek[2]

[1]Department of Ecology and Evolutionary Biology, University of California, Santa Cruz, California 95064; email: sinervo@biology.ucsc.edu

[2]Department of Biological Sciences, Dartmouth College, Hanover, New Hampshire, 03755; email: Ryan.G.Calsbeek@Dartmouth.edu

Key Words

cognition, correlational selection, interspecific interactions, intraspecific, perception, rock-paper-scissors

Abstract

We outline roles of frequency-dependent selection (FDS) in coadaptation and coevolutionary change. Coadaptation and coevolution occur because correlational selection (CS) and correlated evolution couple many traits. CS arises from causal interactions between traits expressed in two or more interactors, which invariably involve different traits (signalers-receivers). Thus, the causes of CS are due to FDS acting on trait interactions. Negative FDS, a rare advantage, is often coupled to positive FDS generating complex dynamics and FD cycles. Neural mechanisms of learning and perception create analogous routes by which traits are reinforced in cognitive and perceptual systems of interactors, substituting for positive FDS. FDS across all levels of biological organization is thus best understood as proximate causes that link interactors and shape genetic correlations within and among interactors on long timescales, or cognitive trait correlations within interactors on short timescales. We find rock-paper-scissors dynamics are common in nature.

INTRODUCTION

FDS: frequency-dependent selection

Meme: (and related terms memetic and memotype) reflects information storage and transmission forms in cultural evolution (analogous to the terms gene, genetic, and genotype)

Correlational selection (CS): nonlinear force of selection that serves to couple traits within individuals, or forces that correlate traits among individuals and species

We address causes and consequences of frequency-dependent selection (FDS) in nature. All forms of FDS arise from biotic interactions; therefore, FDS is involved in coevolutionary interactions, including those acting on clones, mating systems, social systems, and ecosystems. FDS links causes intrinsic to interactors with causes extrinsic to interactors in a parsimonious fashion, thereby bridging what Mayr (1993) refers to as proximate and ultimate causation.

We highlight FDS in mating systems, given well-characterized genetics and methods (i.e., payoff matrix) for diagnosing forms of FDS. Analysis of clonal prokaryotes requires pinpointing single base-pair changes that alter invasion dynamics (Elena & Lenski 1997, Remold & Lenski 2004). Molecular change is difficult to track in the context of confounding macroscopic effects of natural systems. For this reason, causes of FDS in clonal prokaryotes are best studied in laboratory settings and we review a few of these studies. We also link higher levels of biotic interactions, such as ecosystems, with a set of principles for positive and negative FDS caused by either genetic changes of Darwinian evolution or memetic changes of cultural evolution.

Characterizing FDS requires analysis of how relative or absolute frequency (e.g., density) of one form, arising from genetic or memetic causes, impacts fitness of other interactors. Fitness gains of replicators, copies of single alleles, often come at the expense of the fitness of individual phenotypes in which replicators reside (Nowak & Sigmund 2004). Thus, elucidation of frequency-dependent (FD) causes are intimately related to levels of selection such as genic, individual, kin, group, and species as well as concepts of altruism, antagonism, and mutualism (Hamilton 1964).

FD causes arise from proximate interactions of traits between interactors. Because traits need not be the same between interactors and often involve different traits for obvious reasons in signaler-receiver interactions, the proximate causes of FDS will generate correlational selection (CS) within individuals (Brodie 1992, Chevrud 1984, Sinervo & Svensson 2002) as well as between individuals (Sinervo & Clobert 2003). CS couples the fates of multiple traits and their underlying genetic or memetic control in a nonlinear fashion, generating correlations at the level of genotype and/or memotype where none existed before.

Given that underlying causes of trait interactions involve behavior or physiology, our pursuit of FD causes delves into neural processes such as perception and learning, the routes by which interactors interpret extrinsic biotic environments, comprised of interactors. Likewise, at a molecular level, consideration of physiology is required to understand how interactions arise in clonal, floral, fungal, and ecosystem dynamics. Physiological causes homologize with neural processes in that molecule and molecular action under FDS are parallel to signaler-receiver interactions. Consideration of behavior and physiology allows generalizations to be made at all biological levels of organization, unifying FD causes and the ultimate consequences of FDS such as CS that drives intraspecific coadaptation or interspecies coevolution.

THE FORMS OF FREQUENCY-DEPENDENT SELECTION AND NATURAL SELECTION

We begin our discussion by defining and referencing forms of FDS to forms of natural selection. Conceptual understanding of the causes of selection (Bock 1977, Lande & Arnold 1983, Wade & Kalisz 1990) has been advanced via a multivariate treatment of traits and cascading effects on fitness (Lande & Arnold 1983). Phenotypic selection analysis decomposes fitness into directional, stabilizing, and disruptive components. CS gradients, which describe synergisms among phenotypic traits, are added to multitrait models (Brodie 1992, Chevrud 1984, Lande & Arnold 1983, Sinervo & Svensson 2002). In the case of mating interactions, CS is substituted by special forms of assortative or dissassortative mating, the nonrandom union of gametes driven by phenotypic traits of signalers and receivers.

These key generalizations, homologies among CS on traits within individuals, CS on traits between individuals, and special forms of CS due to nonrandom mating (sexual selection), will be useful in generalizing FDS to ecosystem interactions. We generalize relations further by noting that social causes of CS and FDS, social selection, arise from trait interactions between individuals, even though efforts to date have ignored CS on traits between individuals (but see Sinervo & Clobert 2003), other than the case of nonrandom mating. This is because a focus on individual selection defines the unit of selection as phenotype, but in FDS the unit of selection is interaction and its effects on phenotype(s). CS on traits between individuals drives coadaptation of trait combinations within species and coevolution between species. CS shapes and intensifies tradeoffs linking multivariate suites of traits (**Figure 1**), which reinforces and refines the FDS acting on subsequent generations. Tradeoffs are implicit to CS and FDS and we point out key tradeoffs.

The CS gradient, which acts on two or more traits, is a higher-order description of stabilizing or disruptive selection on a single trait. Forms of FD nonrandom mating arise from either linear or quadratic interactions of two or more traits. Sexual selection generates CS between the sexes via mate preference and within the sexes due to mate competition (McGlothlin et al. 2005) or male cooperation (Sinervo & Clobert 2003). Forms of FD mate success that preserve diversity are possible (termed balancing selection, see Hedrick, this volume), including forms associated with the major histocompatibility complex (MHC) of vertebrates, analogues in invertebrates, in which diversifying selection favors mate preference for nonself (Aeschlimann et al. 2003, Grosberg & Hart 2000, Landry et al. 2001, Potts et al. 1991, Reusch et al. 2001) or self-incompatibility systems of plants (Weller & Sakai 1999). Other forms of nonrandom mating based on MHC exert stabilizing selection, in which self-mate preference may arise at contact zones where hybrid unfitness promotes evolution of premating isolation (Howard 1999).

Forms of FDS in predator-prey interactions are referred to as apostatic and antiapostatic selection. Apostatic selection, a rare morph advantage, reflects negative FDS. In other contexts, negative FDS can arise from self-poisoning effects of common forms. Antiapostatic selection, a common morph advantage, reflects positive

Social selection: a form of FDS in which mating or social systems are coupled via signaler and receiver coevolution

MHC: major histocompatibility complex

Apostatic selection: a form of FDS in which rare types have advantages during positive and negative reinforcement of predator-prey detection and learning

Antiapostatic selection: a form of FDS in which the common type has advantages during positive and negative reinforcement of predator-prey detection and learning

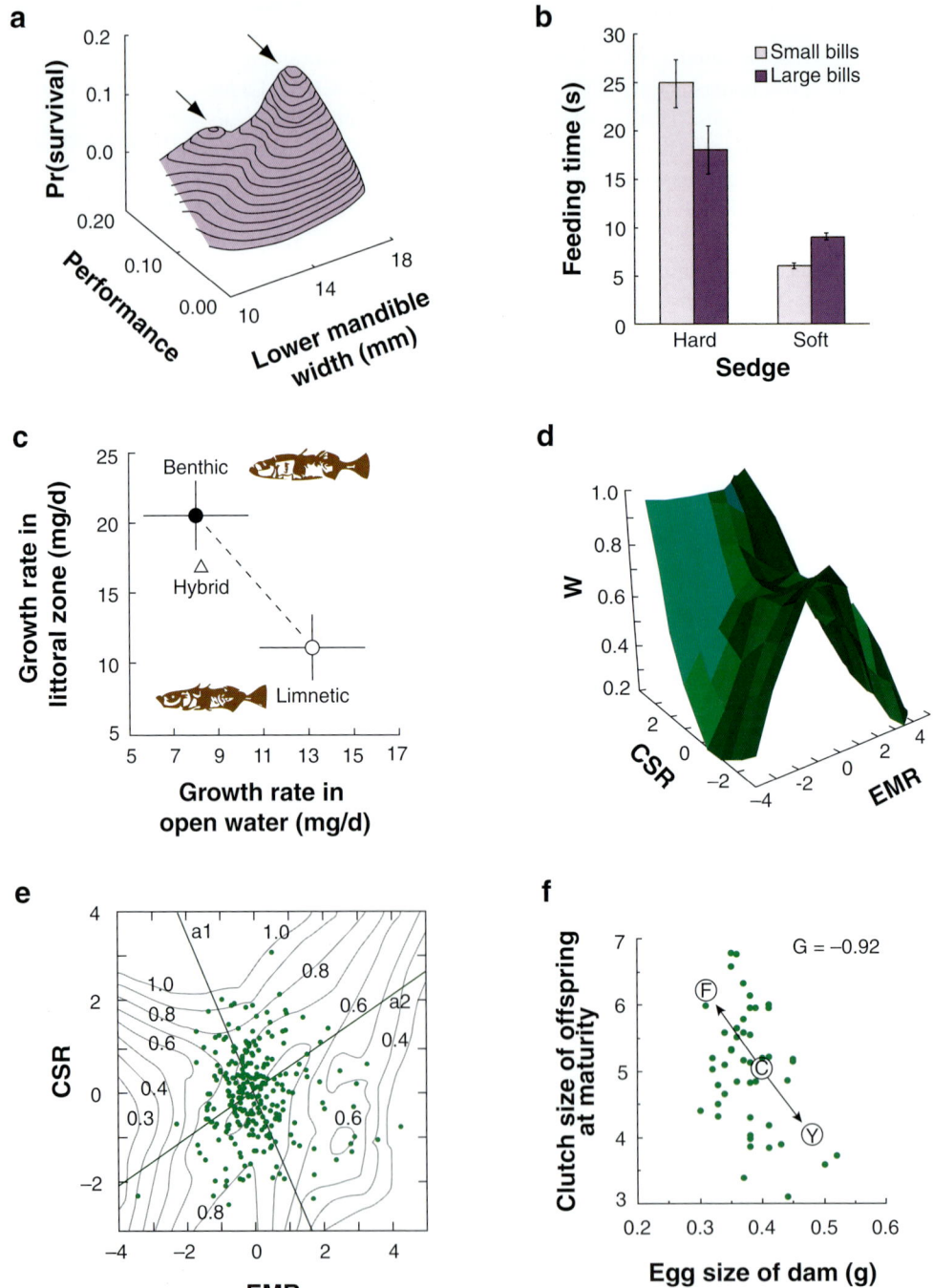

FDS. Positive FDS can be due to Allee effects (Greene & Stamps 2001) in which rarity limits and high density permits interaction, or from aposematic effects where aggregations enhance fitness (Endler & Mappes 2004).

Distinctions between positive and negative FDS can be further subdivided into open-ended versus closed outcomes (cycles: discussed below). Open-ended positive FDS moves trait distributions rapidly in one concerted direction; higher frequencies of traits enhance change in subsequent generations. Depletion of genetic variation places ultimate limits on such processes. Open-ended positive FDS does not arise from unity of FD cause acting on distributions of single traits, but from unity of FD cause that shapes trait frequency distributions in two interactors (signaler-receiver) or, more generally, from any interaction force that links interactors.

The distinction is made clear by Fisherian runaway sexual selection (Fisher 1930), the most celebrated example of open-ended positive FDS (see Kokko et al., this volume). Mate preference for traits in the opposite sex generates positive feedback on genetic correlations. The proximate cause of this runaway is thought to be open-ended perceptual mapping of preference and preferred traits (Lande 1981). Positive FDS occurs because of this invariance of signaler-receiver relations that link preferences and preferred traits. In contrast, balancing selection at the MHC arises from negative FDS in which rare progeny genotypes are advantageous in the context of disease resistance (Reusch et al. 2001), an open-ended diversity-generating mechanism.

Red Queen dynamics: (van Valen 1976) coevolutionary inter- or intraspecific interactions that arise from FDS and generate never-ending cycles of adaptation and counteradaptation

CYCLES AND NEGATIVE FREQUENCY-DEPENDENT SELECTION BETWEEN THE SEXES

Other forms of sexual selection involve evolutionary cycles and Red Queen dynamics. Interlocus sexual conflict, for example, arises when traits have alternative fitness optima in the sexes (Rice & Chippindale 2001). Alleles that move one sex toward one optimum moves the other sex away; this is a rare advantage of strategy and counterstrategy.

Figure 1

(*a*) Survival selection on bill morphology and feeding performance of African seed crackers, *Pyrenestes ostrinus*, will promote correlational selection and alternative fitness peaks (Smith & Girman 2000). (*b*) Tradeoffs in handling time of hard and soft seeds from the sedge *Scleria verrucosa* as a function of bill morph (table 1, Smith 1993). Distribution of seeds varies with rainfall. The tradeoff plays an important role in maintaining the bill polymorphism (with permission from T.B. Smith). (*c*) Tradeoffs in growth of limnetic, benthic, and hybrid forms of the stickleback, *Gasterosteus* spp., in two different habitats (from Schluter 1995). (*d, e*) Fitness surface and contour plot of survival of adult female side-blotched lizards, *Uta stansburiana*, from first to second clutch (residuals after removing year are depicted). Disruptive selection on clutch mass and stabilizing selection on egg mass shape a tradeoff between clutch size and egg mass. (*f*) Negative correlation of daughter's clutch size and dam's egg mass is indicative of a genetic correlation (G = –0.92, Sinervo et al. 2000a). Experiments [follicle stimulating hormone (F), follicle ablation (Y), relative to controls (C)] elucidate the physiological basis of the tradeoff.

RPS: rock-paper-scissors

Sequestering genes on sex chromosomes, evolving sex-specific modifiers or sex steroid control of traits (Rice 1984) can ameliorate such cyclical FDS. Experiments on *Drosophila* support the importance of X chromosomes for harboring sexually antagonistic variation, given that in each generation two copies are under selection in females whereas only a single copy is under selection in males (Gibson et al. 2002). Thus, sexually antagonistic selection may explain a large proportion of traits that are sexually dimorphic, if dimorphism evolves to limit sex-specific costs of sexually antagonistic alleles (Rice 1984). However, sex limitation (dimorphism) of traits may rely on the evolution of modifiers to alter expression of traits initially expressed in both sexes.

Moreover, some sex-limited traits are thought to promote sexual conflict. For example, in *Drosophila* seminal fluid proteins are hypothesized to reduce female remating rates and enhance paternity assurance (Rice 2000). Physiologically naïve females, which have been selected under monogyny and thus not previously exposed to male strains that have evolved under polygyny, suffer higher rates of mortality when mated to polygamous strains compared with polyandrous females that have evolved under polygyny (Holland & Rice 1999). Evolved female resistance to male proteins may be countered by increased efficiency of seminal fluid proteins.

THREE-PLAYER MATING SYSTEM DYNAMICS

Much generalization is afforded from a study of FD mating systems in which genetics of alternative strategies have been elucidated. In mating systems analysis, we highlight the role of CS between individuals in shaping FD interactions (**Figure 2**, and see the **Figure 2** Supplementary Materials; follow the Supplemental Material link from the Annual Reviews home page at http://www.annualreviews.org/). Trimorphisms are common in animal and plant mating systems. We argue that mating systems with dimorphisms, sexual dimorphisms included, involve similar forces of FDS, but a reduced subset of those acting in trimorphic mating systems. In animal mating systems, spatial tradeoffs associated with territoriality (despotic strategy) are counterbalanced by paternity assurance afforded by mate guarding (Sinervo & Svensson 2002). Mating system tradeoffs exert survival costs, thereby invoking life history tradeoffs. Tradeoffs of social interactions (Sinervo et al. 2006b) generate a third axis along which CS acts. Trimorphic mating systems exhibit sets of multivariate tradeoffs that sustain CS. Increased dimensionality in tradeoffs, such as in mating systems that involve life history tradeoffs, sexually-selected tradeoffs, and socially-selected tradeoffs provide enough dimensionality to sustain cyclical dynamics such as the rock-paper-scissors dynamic.

The trimorphism referred to as the rock-paper-scissors game (RPS) (Maynard Smith 1982) is named for intransitive fitness interactions in which rock beats scissors, paper beats rock, and scissors beats paper. In each pair-wise interaction, players are involved in negative FDS of their own type that enhances fitness of another type. However, RPS cycles in biological systems may necessarily involve positive FDS within forms (Sinervo et al. 2006b). When mixtures of negative and positive FDS interact, a system can become destabilized and oscillate. Forces of positive FDS cause RPS cycles to spiral outward from the attractor, whereas forces of negative FDS cause the RPS to spiral inwards from the absorbing boundaries and toward the attractor.

The simplicity of the lizard RPS mating system (**Figure 2**) also involves complex dynamics of female morphs that sustain density cycles via female life history tradeoffs (**Figure 1**). FDS on female morphs have been addressed with experimental manipulations on large-scale plots of both morph allele frequency and egg-size phenotype frequency (Sinervo et al. 2000a, 2001). Experiments are critical in elucidating causes of FDS and CS (Sinervo & Basolo 1996) and in elucidating the proximate causes of genetic correlations underlying tradeoffs (**Figure 1e,f**).

In other mating systems, pure apostasis among three forms is common. For example, trimorphisms of female damselflies found in Europe and North America are maintained by pure apostasis among three female morphs based on equations for FD fitness gains of the damselfly *Ischnura elegans* (Svensson et al. 2005), which we have re-expressed as a payoff matrix (**Figure 2**). Apostatic FDS is thought to arise from search image formation in males for common female morphs (Fincke 2004). Given intense sexual conflict arising from male harassment, rare cryptic female morphs gain higher fecundity, thus involving previously discussed tradeoffs for costs of reproduction, fecundity, polyandry, and sexually antagonistic selection owing to polygyny.

Complex dynamics involving both negative and positive FDS (**Figure 2**) are also exhibited by the marine isopod *Paracerceis sculpta* (Shuster & Wade 1991a,b). Apostatic (female mimic or cryptic) male mating strategies can invade despotic male strategies that guard female harems, however, invasion is synergistically dependent on density of female harems and negative FDS acting on male strategies. Reanalysis of data (Shuster & Wade 1991a) in the form of a payoff matrix (**Figure 2**) indicates that stability of the system is in part determined by apostatic selection.

A tristylous arrangement of anthers and styles in some plant mating systems (Barrett et al. 2004) is thought to promote outcrossing (Eckert et al. 1996), because the rare anther form more efficiently transfers pollen to common stigma types (Oneil & Schmitt 1993). This negative FDS is often not strong enough to preserve all three types and many populations are fixed for combinations of dystyly, a response to genetic drift or founder events (Eckert & Barrett 1992, Eckert et al. 1996, Husband & Barrett 1992). However, reanalysis of frequency and fitness data from Ågren & Ericson (1996) in the form of a payoff matrix (**Figure 2d**) indicates that stability of tristyly in *Lythrum salicaria* arises from more complex FD epistatic interactions (and three-way interactions) among plants and their neighbors (**Figure 2d**, *right panel*), not merely a rare advantage.

ESS analyses (**Figure 2**) indicate that three of the four mating systems exhibit RPS fitness intransitivity, which was first characterized in nature for lizards (Sinervo & Lively 1996). In **Figure 2**, we define evolutionarily stable strategy (ESS) conditions for an apostatic RPS, based on the damselfly mating system:

$$W_{IO,A} > W_{I,A} > W_{A,A}, W_{A,I} > W_{IO,I} > W_{I,I} \quad \text{and}$$
$$W_{I,IO} > W_{A,IO} > W_{IO,IO}. \qquad 1.$$

Evolutionarily stable strategy (ESS) analysis: (Maynard Smith 1982) used to analyze the invasion conditions of forms. Often, systems exhibit ESStates of many interactors and cyclical dynamics, maintained by interplay of negative and positive FDS

Payoff matrix

Evolutionary dynamic

Correlational FDS

a

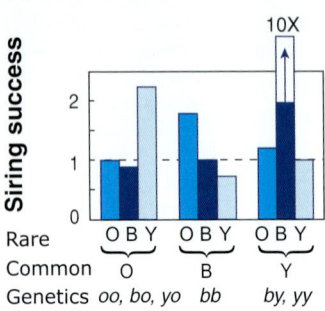

True RPS:

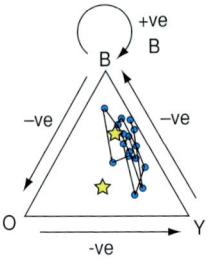

b

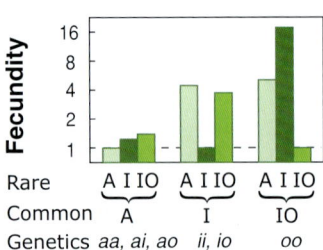

Apostatic RPS:

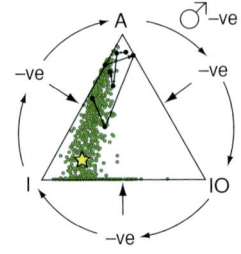

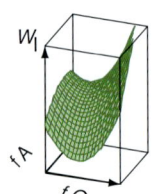

c

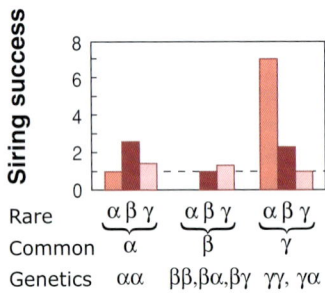

True & apostatic RPS:

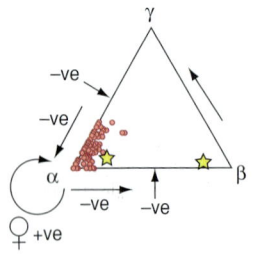

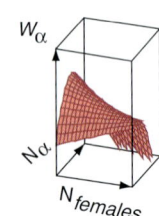

d

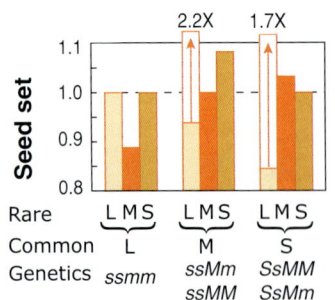

Fisherian sex ratio apostasis:

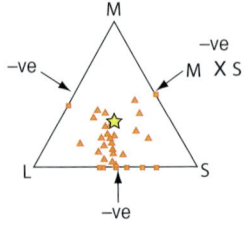

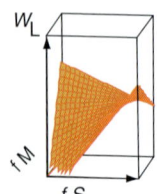

The isopod mating system exhibits a mix of a true RPS and an apostatic RPS:

$$W_{\beta,\alpha} > W_{\gamma,\alpha} > W_{\alpha,\alpha}, W_{\gamma,\beta} > W_{\beta,\beta} > W_{\alpha,\beta} \text{ and } W_{\alpha,\gamma} > W_{\beta,\gamma} > W_{\gamma,\gamma}. \qquad 2.$$

In an apostatic RPS, common strategies lose to rare strategies. In a true RPS, each rare strategy loses to one common strategy and each beats one rare strategy (Sinervo 2001), which generates more intense FDS among types relative to RPS apostasis. Nevertheless, RPS fitness intransitivity is common to both the true and apostatic RPS. Prevalence of RPS intransitivity in trimorphic taxa provides a simple avenue for lineage diversification. Loss of morphs during RPS excursions may rapidly fix the systems on degenerate mating systems that lack one or more morphs. These new degenerate mating systems can diverge rapidly, forming new mating systems and species (discussed below).

Figure 2

Diagnosing frequency-dependent selection (FDS) in mating systems with payoff matrices (*left panels*: $W_{rare,common}$), evolutionary dynamics (*center panels*: ternary plots with calculated or theorized attractors, *star*), and correlational FDS (*right panels*: nonlinearity that deviates from linear FD relations assumed if payoffs were purely due to linear and negative FDS). Fitness (W) is shown on the vertical axis. In payoff matrices *a* and *d*, supplementary fitness is illustrated (histograms with *arrows*), and arise from nonlinear and/or three-way interactions that generate correlational FDS. Supplementary Materials (follow the Supplemental Material link from the Annual Reviews home page at **http://www.annualreviews.org/**) provide: methods of payoff matrix construction, derivation of nonlinear equations for FDS, discussion of FD correlational selection, and emergent properties of genetics. (*a*) Data on siring success of side-blotched lizards, *Uta stansburiana* (Sinervo 2001), are nearly consistent with true rock-paper-scissors (RPS) fitness intransitivity in which one rare strategy has high fitness while the other rare strategy has low fitness relative to the common strategy $W_{Y,O} > W_{O,O} > W_{B,O}$; $W_{O,B} > W_{B,B} > W_{Y,B}$ and $W_{B,Y} > W_{Y,Y} > W_{O,Y}$ (it only deviates from a true RPS in that $W_{O,Y} > W_{Y,Y}$, however, payoffs based on overlap show $W_{Y,Y} > W_{O,Y}$, Sinervo 2001). Strong positive FDS on blue male cooperation (positive loop, *center panel*; CS surface, *right panel*) enhances the ability of B to invade the Y sneaker male strategy (Sinervo et al. 2006b). (*b*) FDS (Svensson et al. 2005) on fecundity of the damselfly, *Ischnura elegans*, is consistent with a purely apostatic mating system, which is diagnosed by both rare strategies having higher fitness than each common strategy. However, the payoff matrix also exhibits RPS intransitivity (Equation 1) (*c*) Reanalysis of data (Shuster & Wade 1991a) on siring success and female harems in the isopod, *Paracerceis sculpta*. Isopods exhibit a mixture of true RPS and apostatic RPS (Equation 2). The despotic α morph (R = rock) controls harems, but is self-limiting (*right panel*) at high harem densities, when two cryptic (P = paper) morphs (β, γ) invade. Correlational and stabilizing selection on α–male frequency and female density is due to female copying (Shuster & Wade 1991b). (*d*) Reanalysis of data (Ågren & Ericson 1996) on the tristylous purple loosestrife, *Lythrum salicaria*. A payoff matrix based on linear FD relations indicates that two morphs, M and S, will exist in a stable evolutionarily stable strategy state (ESState), but the third morph, L, cannot invade M or S. However, the negative FDS assumed for tristyly (Barrett et al. 2004) does not entirely stabilize the system. Stability is also conferred by epistatic (nonlinear and three-way) interactions on L, which exhibits two optima, when either M or S is fixed. At intermediate values of M and S, L has low fitness (fitness saddle).

THE ROLE OF COGNITION, PERCEPTION, AND INNATE SIGNALS IN FREQUENCY-DEPENDENT SELECTION

Cognition constitutes three steps (Roitblat 1987):
1. perception—units of information are collected and stored,
2. processing—these data, stored in memory, are analyzed with computational rules built into nervous systems,
3. environmental representations form from data processing—adaptive behaviors are based on these "pictures".

Though cognition promotes negative FDS (e.g., uncommon "picture" advantage), other noncognitive processes can influence FDS. Innate recognition is a genetically codified signaler-receiver interaction. Habituation, sensitization, and input matching supply critical filtering systems in a messy world. For example, habituation of lizards to sinusoidal waveforms (blowing branches) favors mimicry to evolve in snake movement and morphology, thereby thwarting antipredator detection systems of lizards (Fleishman 1986). Learning builds correlations among many interrelated "pictures" of the environment. Cognition and innate neural processes thus contribute to correlations that build among the pictures or memotypes of neural systems (plastic: imprinted, habituated, sensitized, or learned signals; or genetic: innate), much like correlational selection on traits, but through positive and negative reinforcement or through signaler-receiver coevolution.

LEARNING MECHANISMS AND FREQUENCY-DEPENDENT CYCLES

Cognition: neural process involving learning; breaks down into the three phases of perception, data manipulation, and formation of a representation of the external environment

Our goal in this section is to link correlative forces of cognition, perception, and learning (i.e., cognitive representations of traits or memotypes) directly to CS and FDS on genotypes. Apostasis in predator-prey interactions (Cook & Kenyon 1991, Mallet & Joron 1999), analogous to mating system apostasis, promotes evolutionary cycles of highly variable forms.

Learning experiments on *Cyanocitatta cristata* in a virtual-reality environment (Bond & Kamil 1998, 2002), which used many alternative cryptic forms, produced cycles in frequency of computer-generated cryptic prey morphotypes. As *C. cristata* switched between common type, learning preserved and cyclically generated new variation (**Figure 3**), experimentally confirming the role of prey learning in driving cycles of apostatic selection. Analogous effects with dorsal pattern manipulations have been demonstrated in nature (Forsman & Appelqvist 1998).

Handedness is a common form of FDS, which is hypothesized to become fixed owing to advantages of a bias from bilateral symmetry that allows for rapid stereotyped actions to always commence within the same hemisphere of the brain and propagate through a dominant-handed motor pattern (Propper et al. 2005). Handedness is ancient in origin. Handed attack patterns appear on trilobite prey, owing to their

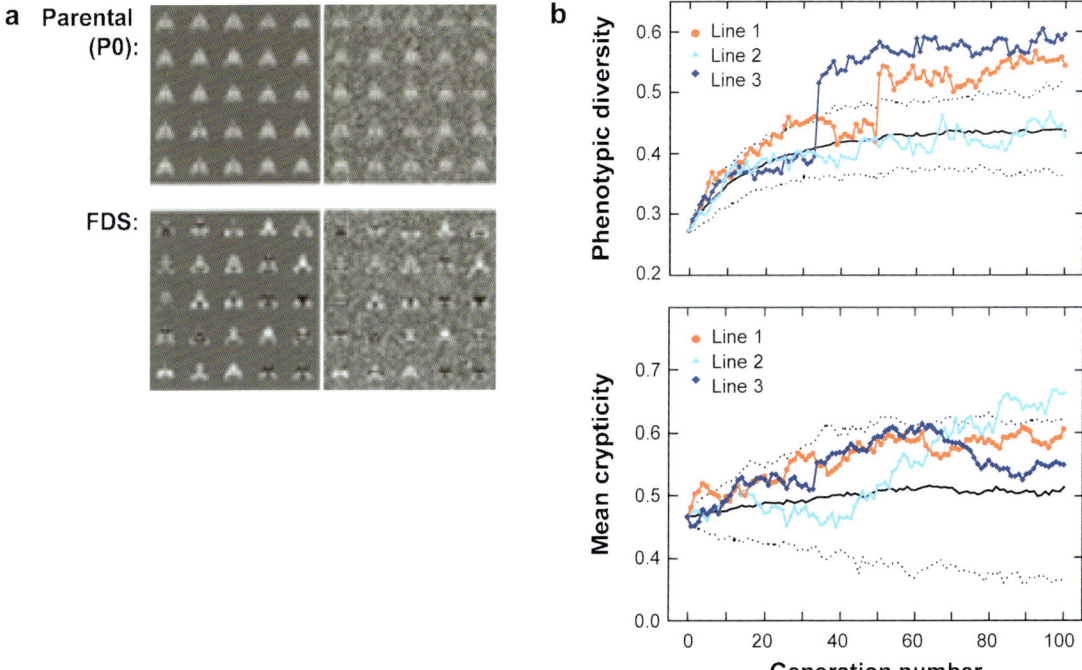

Figure 3

(*a*) Samples of virtual prey (digital moths) for blue jays, *Cyanocitatta cristata*, shown on uniform gray (*left*) and cryptically textured (*right*) backgrounds. Panels show prey items from parental prey population, P0, and from prey populations after 100 generations of frequency-dependent selection (FDS) by jays. (*b*) Moths from the FDS lines were more cryptic than those in the nonselected lines, and more variable in appearance than those in the lines subjected to frequency-independent selection (not shown). Changes in mean crypticity and phenotypic variance across successive generations in three experimental lines (plotted in *color*), contrasted with distribution of values from two sets of control lines (plotted in *black*) (from Bond & Kamil 2002, with permission).

handedly biased predators that hunted Cambrian ecosystems (Babcock 1993). Attack handedness of the scale-eating cichlid, *Perissodus microlepis*, which exhibits right- and left-jawed morphs, drives FD cycles via either learning or sensitization of their prey cichlid species (Hori 1993). Handed feeding polymorphism in crossbills, *Loxia curvirostra*, generates FD advantages (Benkman 1996, Benkman & Lindholm 1991) in opening cones that are either sinistral or dextral with respect to the spiral orientation of bracts.

In humans, the advantage of handedness and other rare behaviors is common knowledge in sports (switch hitting, a form of ambidexterity; south paw in boxing; or regular versus goofy, a footedness advantage to surfing the rare left- or right-hand wave depending on breaking surf). The advantage of left-handedness (Billiard et al. 2005), as judged by handedness frequencies in sports (Raymond et al. 1996), is most prominent in close contact sports (e.g., fencing or boxing compared to tennis). A

Aposematic selection: a form of signal-receiver evolution in which noxious or deadly prey gain protection by evolving a conspicuous signal advertised to predators

rare left-handed advantage may have first arisen in close contact fighting (Faurie & Raymond 2005). Handedness in humans has a genetic basis (Klar 2005, McKeever 2004). However, learning is involved, because most training partners and contestants are right-handed. Defensive and offensive strategies will always be reinforced in real contests with right-handed opponents regardless of novel training regimes, such as sparring with south-paws.

Invasion of cheaters is also driven by rare advantage. For example, rewardless orchid species evolve conspicuous and colorful flowers but forgo provisioning them with nectar (Gigord et al. 2004). CS on floral morphology has been observed in rewardless orchids (O'Connell & Johnston 1998). As bumblebees emerge to feed in spring, they often visit rewardless orchids, which have evolved color mutations with rare advantage. Bumblebees visiting common floral types are negatively reinforced, and subsequently avoid that color (Smithson & Macnair 1997).

Aposematic selection, which favors conspicuous prey signals coupled to noxious or lethal defenses, involves learning or innate recognition. Fisher (1930) realized that a kin benefit in prey would favor evolution of aposematism if kin were aggregated, a form of positive FDS (Endler & Mappes 2004). Death of an individual that reinforced predator learning would benefit nearby kin. A constraint on studying origins of aposematism is a universal innate predator aversion to certain colors, which in avian systems are usually yellow or red (Brodie & Janzen 1995). Thus, extant bird species share innate aversion to feeding on certain colors, reinforced over eons of interactions with prey that have all converged on yellow or red aposematic signals.

To circumvent this constraint, ingenious learning environments involving novel worlds (Lindstrom et al. 2001) allow abstract geometrical shapes to be substituted for aposematic colors (**Figure 4**). Early experiments (Mappes & Alatalo 1997) confirmed Fisher's idea that gregarious aposematic prey gain an advantage through single-trial learning of predators, in which clustered and obvious but noxious forms have a survival advantage over dispersed noxious forms. CS makes origin of warning signals contingent on prey behaviors or life history traits that aggregate signals. Aposematism is common in butterflies (Langham 2004), which often evolve to lay eggs in batches, thus establishing kin aggregations. Extensions of the novel world approach (Mappes et al. 2005) demonstrated advantages of Mullerian forms that are conspicuous but not necessarily aggregated. Other studies demonstrated that aposematic forms are antiapostatic (Lindstrom et al. 2001), further underscoring impacts of rarity on origin of aposematism. Use of three morphs also confirmed the role of cryptic forms in maintaining imperfect Batesian mimicry (**Figure 4**). Batesian mimics evolve to resemble a noxious or toxic aposematic model and gain FD protection (rare advantage) from attack even though they lack defense, provided that cryptic forms are common.

Additional constraints and tradeoffs involved in perceptual systems of predators can drive CS and FDS on alternative antipredator prey traits such as escape behavior and dorsal patterns (Brodie 1992, Niskanen & Mappes 2005). For example, predators attacking a moving snake with stripes often miss because moving stripes appear stationary. The alternative tactic, freeze, becomes coupled to cryptic patterns. Though speed is often coupled to stripes, differential crypsis can arise through either pattern matching a background (e.g., spots on fine backgrounds, bars on bark),

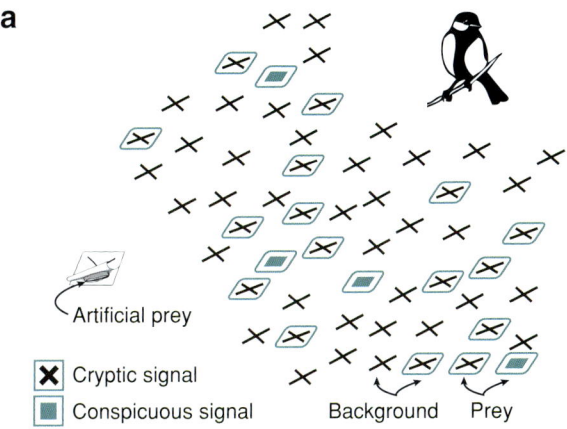

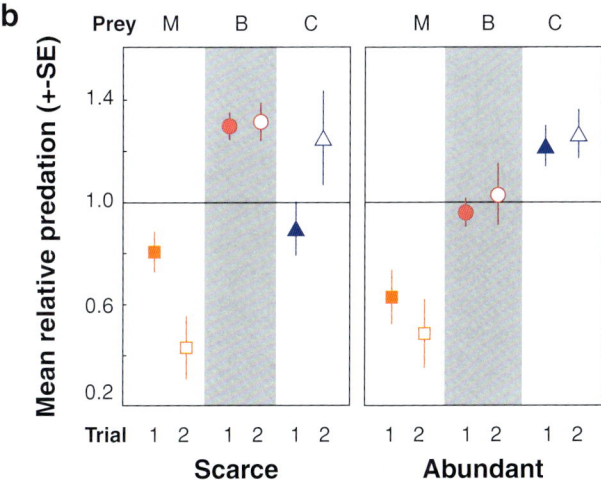

Figure 4

Predator learning, aposematic, and apostatic selection on prey. (*a*) A great tit inspects the floor of a novel world aviary during learning trials. (*b*) Data from novel worlds involving an aposematic model (M), weakly Batesian mimics, (B) and cryptic prey (C), which were presented at two frequencies, scarce and common (Lindstrom et al. 2004). Relative predation of models (*squares*), mimics (*circles*), and cryptic (*triangles*) prey in the two alternative prey treatments. Filled symbols indicate mean relative predation (with standard error bars) in the first trial, and open symbols in the second trial. A line indicates the expectation based on random predation. When alternative prey were scarce, imperfect Batesian mimics were selected against, but abundantly available alternative prey caused selection against imperfect mimics to be relaxed (*Top panel* with permission of M. Joron and *bottom panel* with permission of L. Lindstrom).

or disruptive patterns like large spots that break up the prey outline (Ruxton et al. 2004).

Coevolved strategies of hosts and brood parasites are also coupled by FDS (Soler et al. 2003). Brood parasites lay eggs in nests of conspecifics or other species, to be cared for by hosts. To reduce reproductive costs of brood parasitism (Payne et al. 2001), hosts evolve defenses including refined recognition of parasitic eggs and chicks, nest defense (Amundsen et al. 2002), or egg counting (Lyon 2003). Brood parasites, counterselected to combat host defenses, evolve refined egg mimicry (Soler et al. 2003) and signals to enhance host feeding (Lyon et al. 1994).

Viduine finches provide a dramatic example of mimetic evolution. Indigo birds, *Vidua chalybeata*, learn songs of host species, and as adults, males attract females with songs of foster parents (Payne & Payne 1994). Mutual production and preference of mimetic host songs in both sexes reflect sexually and parasitically-selected traits.

Host-song imprinting of female brood parasites as chicks attracts them to mates and back to the nests of specific hosts in a culturally selected runaway (Payne et al. 2000). Positive FDS may be held in check by negative FDS and counterstrategies of specific hosts that limit specific memetic lineages of Viduine finches. Other brood parasites such as cowbirds have innate songs (divergent from their hosts), which results in a more generalized niche in which they parasitize diverse species (Garamszegi & Aviles 2005).

GENERALIZING THE CORRELATING MECHANISMS OF COGNITION AND PERCEPTION

In preceding examples, neural processes are potent correlating agents by which cues or signals become coupled to preference or performance. Positive- or negative-reinforcement learning couples signals and conditioned response in a FD fashion, thereby coupling memotypes. Likewise, perceptual biases or innate behavior couples alternative attack behaviors in predators with alternative dorsal patterns and escape behavior of prey, thereby coupling genotypes. In other cases, signals attain universal meaning among predator guilds via powerful aposematism.

A second point merits mention. The examples of neural processes (either innate or conditioned), including mating system apostasis, involve negative (rather than purely positive) FDS. This is no coincidence. Observing FDS requires variation in behaviors, which is only preserved by negative FDS. As noted above, positive FDS (like runaway) rapidly fixes populations on single behavioral types. In the mating system examples (**Figure 2**), negative FDS maintains types and in those cases with positive FDS, cycles arise through interplay of these two forces. Thus, negative FDS in one interactor (prey, predators, or hosts) is maintained in the context of the FD of learning or cultural processes of another interactor (predators, prey, or parasites). Learning and imitation have positive correlating effects on memotypes, analogous to CS on multitrait genotypes. Thus, negative FDS on variation is held in check by positively conditioned signals underlying self-reinforcing neural processes such as reinforcement learning, rehearsal, imitative learning, innate recognition, habituation, and/or sensitization.

Are there examples of pure positive FDS that involve learning? Many cases of positive FDS are present and of great import for theories of memetic evolution. Consider populations of naïve birds in which novel behaviors arise, which have net benefits on all interactors. As the behavior spreads through imitation and learning, there are more teachers available to spread the behavior in a culturally selected runaway process. The clearest example is a rapid spread of milk bottle cap opening within and then among species of passerines in Britain (Lefebvre 1995).

Analogies between the positive FD of learning and imitation and sexual selection are germane. In light of this analogy, links between perceptual bias and other theories of FD sexual selection, such as sensory bias (Ryan & Rand 1999) are obvious. Positive FDS has made both processes difficult to study with extant forms because endpoints are rapidly attained. For converse reasons, we often observe negative FDS in nature. A reciprocal rare advantage of two forms or RPS fitness intransitivity of three forms preserves variation. In mating systems, positive FDS would fix morphs were it not for the fact that positive FDS is cyclically held in check by negative FDS. Phylogenetic solutions provide recourse to the dilemma of identifying cases of pure positive FDS. The phylogenetic approach used to study positive FDS underlying receiver perceptual (sensory) bias (Ryan & Rand 1999) could be applied to test whether ancestral perceptual bias of receivers has driven positive FDS in culturally selected systems. Paradigms of novel worlds could also be used to identify rules for neural processes in a potential case of pure positive FDS. In light of these analogies, learning polymorphisms and animal personalities (Sih et al. 2004), which involve observational (imitative) types, provide key linkage to FDS on culture.

INTERPLAY BETWEEN NEGATIVE AND POSITIVE FREQUENCY-DEPENDENT SELECTION, SELF-RECOGNITION, AND SPECIATION

RPS mating systems (**Figure 2**) have important implications for speciation. Interplay among positive FDS in both sexes and negative FDS is clear in the context of RPS mating systems (**Figures 2, 5**). Positive FDS in the lizard RPS is the result of cooperative behavior (Sinervo & Clobert 2003) not merely positive FD mate preference. Signaler-receiver relations of cooperation and sexual selection share many runaway properties because both promote a buildup of genetic correlations among signalers and receivers. Recent gene mapping studies (Sinervo et al. 2006b) indicate that genetic factors for male self-recognition and the OBY color signal locus, named for the throat colors orange, blue, and yellow, generate true greenbeard altruism (Sinervo et al. 2006b). When orange-throated male despots invade the cooperative blue-throated strategy, the blue allele confers an altruistic benefit among unrelated blue males: blue males next to orange lose competitions for paternity to the orange, but this altruistic male buffers his unrelated territorial partner from orange male aggression. In contrast, when cooperative blue males invade yellow the blue alliance is mutualistic. Thus, RPS social dynamics drive an evolutionary cycle of altruism and mutualism (**Figure 5a** and **5b**).[1]

Altruism exhibited in RPS systems may also predispose such mating systems to socially mediated speciation (Hochberg et al. 2003), a theory of speciation involving FDS and social behaviors like true altruism (**Figure 5a**). This is because the same genetic factors governing blue male self-recognition are also expressed in females and drive mate preference (Sinervo et al. 2006b), a form of positive FDS for self-mating. The dual action of positive FDS in mating and male-male cooperation may rapidly eliminate strategies other than blue (**Figure 5a**). Speciation requires both assortative mating and hybrid unfitness. Hybrid unfitness arises under social selection because self-recognition loci and loci that promote cooperation are distributed across the

---

[1]The concepts of altruism, mutualism, competition, and spite form a key framework for Hamilton's (1964) ideas on genic selection. Altruism $(-,+)$, mutualism $(+,+)$, competition $(+,-)$, and spite $(-,-)$ form fitness relations among interactors expressed in terms of the costs and benefits of interaction (Hamilton 1964).

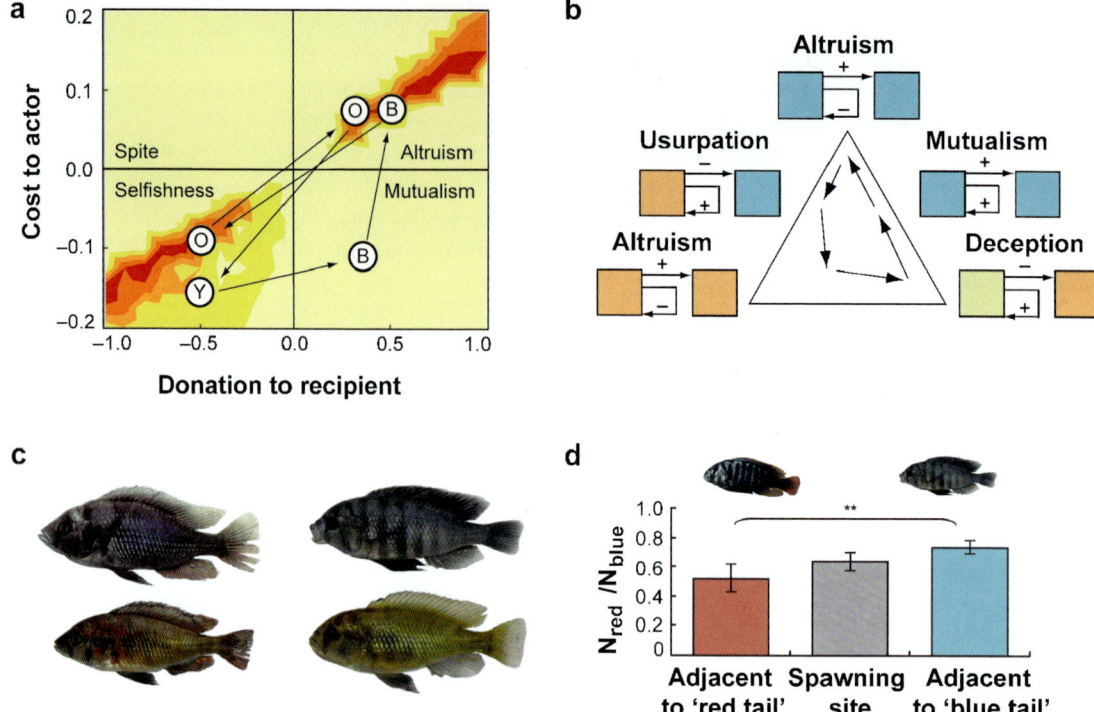

Figure 5

(a) Results from a simulation model of socially mediated speciation (Hochberg et al. 2003) with fitness payoffs (measured in nature, Sinervo et al. 2006b; Sinervo & Clobert 2003) from rock-paper-scissor (RPS) cycles of side-blotched lizards superimposed. The speciation model indicates that reproductive isolation is likely (*red*) when alternatively tagged altruists that donate to self, or usurpers that take from nonself, interact on spatially structured landscapes. Speciation is unlikely (*yellow*) for mutualistic or spiteful social interactions. (*b*) The RPS mating system traverses these social domains during the five-year cycle (superimposed on *panel a*): B altruists are invaded by O (Sinervo et al. 2006a); when O becomes common, O altruists must disperse and pay costs of dispersal (Sinervo & Clobert 2003; Sinervo et al. 2006a,b); Y selfishness invades O, and B mutualists invade Y, driving the system to a point where O can reinvade. Thus, the lizard RPS mating system is susceptible to processes of socially mediated speciation: B females exhibit self-preference for B sires (Sinervo et al. 2006b) and B male cooperation also generates positive FDS. For example, elimination of Y should fix the system on B and O or on B alone, a system of true altruism. (*c*) Socially mediated speciation may have arisen from negative FDS on cichlids. *Left panel*: Representative "blue" and "yellow–red" male nuptial-color types of two species of Lake Victoria cichlids: *Lithochromis rubripinnis* (*top*) and *Lithochromis* spp. "red dorsum" (*bottom*). *Right panel*: two nuptial-color morphs from a single population (Makobe Island) of *Neochromis omnicaeruleus*. (*d*) Under-representation of a territory owner's own coloration among the males that occupy adjacent territories. White bar, the mean ratio *n* ("red tail")/*n* ("blue tail") among territorial males on a spawning site (256 m^2) at Makobe Island (Lake Victoria). Colored bars, the same ratios among territories adjacent to territory owners with a "red tail" (*Neochromis rufocaudalis*) or a "blue tail" (*N. omnicaeruleus*) (*c–d* from Seehausen & Schluter 2004).

genome (Sinervo et al. 2006b); any mixing of this variation with alternative strategies such as usurpation (orange) or parasitism (yellow) breaks up the coadapted gene complex of cooperation.

In this regard, self-recognition of either true or kin altruism may parsimoniously arise from the MHC (Aeschlimann et al. 2003, Grosberg & Hart 2000, Landry et al. 2001, Potts et al. 1991, Reusch et al. 2001). Thus, MHC self-recognition may, in some social systems, serve as a key signal-recognition complex that maintains coadapted gene complexes of social behavior. However, selection favoring self-mate preference produces self-similar progeny, which should tradeoff with the nonself mating benefits of MHC in the context of disease resistance (see above).

Although links between the positive FD of sexual selection and speciation have been made (Lande 1981), the explicit role of negative and positive FDS of social systems and speciation has received little empirical attention (however, see Seehausen & Schluter 2004) (**Figure 5c**), even though mating system examples (**Figure 2**) and theory (Dieckmann & Doebeli 1999, Hochberg et al. 2003) suggest an important role in diversification. Self- versus nonself-recognition systems like MHC may underlie self-mate preference at hybrid zones between species where reproductive isolation is forming (Howard 1999). Similarly, memetic forms of self-recognition, such as song imprinting among male and female viduine finches and their host species, provide analogous social avenues for speciation that involve interplay between cultural and biological evolution.

ECOLOGICAL INTERACTIONS AND FREQUENCY-DEPENDENT SELECTION

The FDS of RPS mating systems (**Figure 2**) and social interactions like altruism, mutualism and antagonism (**Figure 5a** and **5b**) have properties directly related to ecosystem-level interactions. The roles of FD learning in ecosystem interactions (see above) make these homologies explicit. Indeed, aposematism can be thought of as a greenbeard (Guilford 1988) in which reinforcement learning of a signal benefits unrelated individuals that merely share alleles for the conspicuous signal. Links between RPS cycles and altruism are further exemplified by a bacterial RPS. Colicins are bacterial plasmids carried by some strains of *Escherichia coli*. Col plasmids code for production of both a toxic protein called colicin, and one of several immunity factors against that toxin (Kerr et al. 2002). Col plasmids are costly, reducing growth rates of col strains relative to those lacking the plasmid. However, col bacteria can invade most noncol strains; cells of the col strain that contact other strains lyse and release colicin, a kin altruistic act clearing the way for self. Other mutant strains evolve colicin resistance, even in the absence of col plasmids (Feldgarden & Riley 1999, Kerr et al. 2002). When the three strains are grown in spatially structured environments, theory suggests (Durrett & Levin 1997), and in vitro (Kerr et al. 2002) and in vivo studies (Kirkup and & 2004) show, that they are maintained by an RPS. This example reinforces the analogous roles of physiology in mediating negative FDS. Molecules that can be detoxified by specialists are of great significance to interspecific allelopathy in plants (Thijs et al. 1994) as well as for plant-fungal mutualism and parasitism (Bruns et al. 2000, Taylor et al. 2004) (**Figure 6a**).

At the ecosystem level (**Figure 6b**), fitness surfaces are shaped by interactions among many species [e.g., host-parasite (Ewald 1983), competition, and predator-prey (Bolker et al. 2003), intra- and interspecific brood parasitism (Lyon 2003)]. Interspecific dynamics often reflect Red-Queen escalation (van Valen 1976), in which selection acting among interactors amplifies the variance in traits as individuals evolve counteradaptations. For example, interactions of hosts and pathogens involve disruptive selection on both interactors (**Figure 6b**). Host genotypes are under selection for rare immune responses that have no countertactic in pathogens, whereas pathogens are under FD selection to evolve counteradaptations to common resistance genotypes of hosts. In the case of MHC molecules at Class I and II loci, there is a perfect physiological analogy between evolution of novel surface proteins that escape detection by the evolving host, and signaler and receiver evolution (Bernatchez & Landry 2003, Piertney & Oliver 2006). The phenomenon of immune system memory forges the analogy with neural mechanisms of learning.

Multispecies interactions and proximate mechanisms that generate FDS, when embedded in rich ecosystem interactions, characterize FDS at higher levels of biological organization. Thompson (2005) has formalized these ideas in the tripartite

Figure 6

Summary of two-way, three-way, and multiway interactions discussed in the review. Arrows with straight lines indicate negative frequency-dependent selection (FDS), whereas circular arrows indicate positive FDS or analogues for positive FDS in learning, innate behaviors, or immune recognition and memory in the major histocomopatibility complex (MHC). (*a*) The bacterial rock-paper-scissor (RPS) is hypothesized (Kerr et al. 2002) to generalize to allelopathy in higher ecosystem interactions. If so, additional frequency-dependent FD detoxification interactions or cheater dynamics might involve coevolutionary interactions and positive and negative FDS between plants and fungal mutualists or parasites (Bruns et al. 2000, Taylor et al. 2004). (*b*) The simple coevolutionary dynamic involving a host's immune system (e.g., MHC) and a foreign pathogen. The immune system, which retains immunity from cross reactivity, is susceptible to invasion by rare mutant pathogens that beat the immune system. (*c*) Coevolutionary arms race between a toxic newt, *Taricha torosa*, and a snake predator, *Thamnophis elegans* (Brodie et al. 2005), generates a coevolutionary hotspot in Northern California that coincides with a Batesian mimic, *Ensatina eschscholtzii xanthoptica* (Kuchta 2005). In addition, alternative cryptic prey found in the ring species of *Ensatina* (Wake 1997) are hypothesized to be maintained by either background matching or two alternative forms of disruptive coloration. This complex of mimicry and crypsis may serve to exert reinforcement learning on avian predators (**Figures 3** and **4**) in an RPS dynamic. The complete predator-prey dynamic is actually best visualized as a tetrahedron, rather than two RPS triangles (*c, lower left*). The predator resides at the apex, aposematic model (and Mullerian forms) at one vertex, Batesian mimics at the other vertex, and cryptic forms at the third vertex. (*d*) Indirect effects are hypothesized to govern three-way interactions between a predator and two-prey species (Bolker et al. 2003). An RPS may arise from competition among three competitors, which is set up by tradeoffs among exploitative and interference competition, and additional tradeoffs from either cooperative competitors or competitors under other forms of positive FDS. Alternative predator behaviors such as individual versus group foragers might generate tradeoffs and negative and positive FDS, respectively.

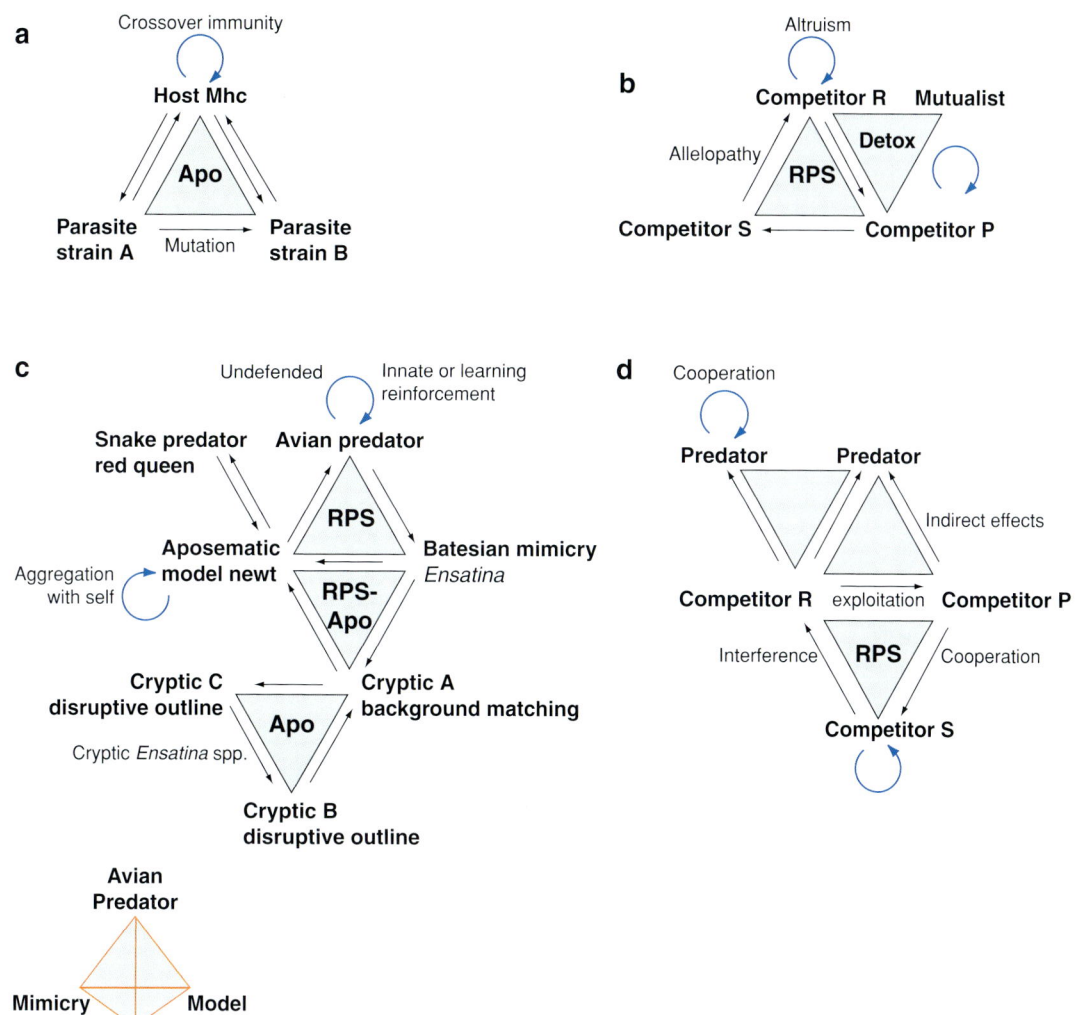

Geographic Mosaic theory of ecovolution. In brief, populations are geographically structured on landscapes and forms of FDS vary in space and time creating hotspots of reciprocal selection acting between species, and evolutionary coldspots, where selection is not reciprocal. Finally, connectivity among spatially structured populations (i.e., gene flow) mixes traits that are locally adaptive in one population but not in another. A geographical hodgepodge of locally adapted/maladapted populations results, constantly evolving alternative solutions to similar problems under FDS.

For example, the moth *Greya pollitella* pollinates flowers of the woodland star, *Lithophragma*, as an incidental byproduct of oviposition behavior. *Greya* larvae feed on some seeds produced by the flower, imposing a reproductive cost. A geographic mosaic across both species' ranges arises from parasitic costs and mutualistic benefits

TTX: tetrodotoxin

of *Greya* moths pollinating *Lithophragma*, relative to more neutral pollinators like flies (Thompson & Pellmyr 1992).

Coevolution between garter snakes and toxic newts in western North America provides another example of a geographic mosaic (**Figure 6c**). Garter snakes in the genus *Thamnophis* are predators of newts in the genus *Taricha*. These newts produce a paralytic toxin (tetrodotoxin; TTX) whose potency is an effective deterrent against all predation except that by *Thamnophis* (Brodie et al. 2005). Though garter snakes can detoxify TTX, resistance comes with locomotor tradeoffs (Brodie & Brodie 1991). Newt toxicity and snake TTX resistance are geographically variable and closely matched in nature. Moreover, a Batesian aposematic mimic (Kuchta 2005), *Ensatina eschscholtzii xanthoptica*, straddles two coevolutionary hotspots in the newt-garter snake dynamic—the San Francisco Bay Area and central Sierras of California—intensifying the arms race. Other cryptic *Ensatina* salamanders complete a ring species of western North America (Wake 1997). The contact zones and hybrid unfitness in the *Ensatina* ring species are thought to arise from CS on alternative antipredator coloration in *Ensatina* subspecies. If so, *Ensatina* salamanders provide another example of negative FDS in speciation, discussed above. Thus, an arms race (**Figure 6c**) with the attributes of strong FDS has shaped a geographic mosaic of coevolution between predator and prey, perhaps intensified by invasion of mimics.

ON THE NUMBERS TWO VERSUS THREE IN SPECIES INTERACTIONS

Rather than focusing on two-way interactions, it will become necessary to shift our focus to three-way interactions of FDS, arising from interplay between positive and negative FDS (**Figure 6c**). Three players and RPS dynamics may be a common Red Queen dynamic in coevolutionary hotspots. We suggest that a system with model, mimic, and cryptic forms is RPS if ($W_{rare,common}$):

$$W_{mimic,model} > W_{model,model}; W_{cryptic,mimic} > W_{mimic,mimic};$$
$$W_{model,cryptic} > W_{cryptic,cryptic}. \quad\quad 3.$$

Such RPS conditions are entirely plausible given results of learning experiments in novel worlds (**Figure 4**), and if the model pays costs of defense, which is likely in the case of chemical defense like TTX. Costs of defense, a tradeoff ignored in learning experiments, could reduce growth or delay maturation, thereby allowing mimics to invade, particularly at low frequency and when alternative cryptic prey are common. At high frequency, Batesian mimics should lose against rare cryptic forms, particularly because crypsis is under apostatic selection. To invade, models with weak defense should be aggregated, and common cryptic and edible prey must also be present.

In the case of a model, a mimic, and generalist predator (Kokko et al. 2003), it is difficult to equate fitness of predator and alternative prey. We propose a modification to standard game theoretic approaches (e.g., Equation 3) and introduce the idea of rare versus common cognitive representations developed in predators. A common aposematic form in a predator's search image depresses a predator's fitness, which would otherwise be able to feed on mimetic forms. Conversely, predator fitness is elevated when Batesian mimics are common in its search image. Thus, this system is an RPS in which (*a*) mimic beats model and applies selection to model—mimic pays no costs of defense but gains signal benefits, (*b*) predators pay costs of mistakes and learn quickly to recognize and eat imperfect mimics or evolve refined discrimination or learning, (*c*) ongoing predator mistakes with models cause models to chase away from mimics and refine the signal or add greater defense. The RPS cycle repeats endlessly (**Figure 6***c*), further refining each player via powerful CS in a runaway cycle, or when the Batesian mimic itself evolves defense and is converted in a more mutualistic Mullerian form. We substitute a cognitive representation of frequency (i.e., memotype), predator{mimic} versus predator{model}, for genotype frequency, model versus mimic, to define ESS conditions under which the three-player dynamic is RPS:

$$W_{mimic,model} > W_{model,model}; W_{predator\{mimic\},mimic} > W_{mimic,mimic};$$
$$W_{mimic,predator\{model\}} > W_{model,predator\{model\}}. \qquad 4.$$

Recent advances in our understanding of FD ecological interactions highlights a role for three players even if such interactions are not RPS players. Trait-mediated interactions, in which the outcome of competition and trophic interactions depend on individual traits and density (Bolker et al. 2003), begin to address FDS on ecosystem interactors. For example, apparent competition is an ecological interaction that is couched in terms of three players: two competitors (or herbivores) that compete for the same resource, which is controlled by FD predation (**Figure 6***d*), provided that predators exhibit functional responses to prey frequency (Bolker et al. 2003). Competition often involves a behaviorally dominant species that exerts interference competition, whereas the other competitor exerts effects via exploitative competition, each type carrying a set of tradeoffs.

Inferences from FDS in mating system dynamics and the dimensionality of tradeoffs implies that a true RPS competition might be similarly structured in terms of a third tradeoff owing to social interactions. For example, group foraging (cooperative) or learned foraging strategies of the third competitor (Sih et al. 2004) may generate additional social tradeoffs that are necessary for RPS intransitivity (**Figure 6***d*). Moreover, our broader definition of RPS intransitivity developed for mating systems (**Figure 2**), which involves either a true RPS or RPS apostasis, broadens this theory of FDS. Many other ecological interactions may involve RPS intransitivity that generates ESS conditions for a three-player Red Queen dynamic, which sustains self-reinforcing CS.

Coevolution in ecological interactions might not involve genetic change, but may involve induced plasticity or learning polymorphisms (Sih et al. 2004). Theoretical models on the value of induced predator defenses, costs of defense, and reliability of inducing cues can be found in Lively (2005). For example, inducible feeding morphs under FDS within a species can lead to ecological character displacement (Dayan & Simberloff 2005, Pfennig 1992) and divergent selection pressures that alleviate competition for common resources (Pfennig & Murphy 2003).

For example, divergent selection favors a limnetic form of sticklebacks, *Gasterosteus* spp., which is specialized on zooplankton, and a benthic form, which is specialized for the edges of postglacial lakes in North America. Experimental studies (Schluter 2003) demonstrate that fitness costs paid by different morphs (**Figure 1c**) are greatest when competitors are similar in phenotype, a form of FDS. New genomic methods have begun to pinpoint genetic factors involved in divergence of sticklebacks (Colosimo et al. 2005). Alleles responsible for alternative phenotypes are selected differentially in the benthic form relative to the ancestral marine form, suggesting roles for CS and negative FDS in driving this radiation during the last 20,000 years.

Linkage disequilibrium (LD): nonrandom association of alleles at loci that form via selection and physical linkage, or via correlational selection on linked or unlinked loci

CORRELATIONAL SELECTION AND LINKAGE DISEQUILIBRIUM

It is often assumed that because different species do not exchange genes, no linkage disequilibrium (LD) can arise from interspecific FD interactions. However, LD forms among longitudinally transmitted parasites found in alternate hosts that vary genetically in susceptibility (Lively et al. 2005). Likewise, if host specificity or parasite preference arises from culturally heritable behaviors, this sets up analogues of LD (memotypes with genotypes) owing to CS on interactor traits, even if interactors reside in different species (e.g., viduine finches and host lineages). Coevolutionary relations arise from these analogues of LD, byproducts of correlational FDS and neural systems. CS refines traits in each player generating coevolutionary runaway.

In this regard, genomic approaches for understanding correlated trait evolution (Colisimo et al. 2005) and CS (Sinervo & Clobert 2003, Sinervo et al. 2006b) will become more important in expanding these ideas to other systems. In the case of genetic variation within species, strong CS owing to FDS will generate LD among even unlinked loci (Chevrud 1984, Lynch &Walsh 1998). LD equilibrates when a balance is struck between the force of CS that builds LD and recombination and segregation that destroys LD. Cycles, driven by strong FDS (**Figure 2**), generate LD that is detectable with marker loci. For example, LD associated with FD male cooperation in the lizard RPS was used to map the color signal locus and loci for self-recognition (Sinervo et al. 2006b). LD analysis of genes controlling alternative behaviors in humans (Ding et al. 2002) has likewise identified genes likely to be under FDS (Harpending & Cochran 2002). Similarly, the approach to detecting balancing selection on the MHC is explicitly marker based (see Hedrick, this volume). Though direct measures of CS gradients would be very useful in studying evolution in ecological processes like character displacement, such methods may be impractical. In this regard, genomic approaches to studying consequences of FDS on CS and correlated trait evolution alleviate this difficulty at higher levels of biological organization.

SUMMARY POINTS

1. FDS often involves CS given that it arises from interactions between interactors. Different traits in interactors are coupled by functional relationships (e.g., signaler-receiver, etc.).

2. Tradeoffs are often, if not always, a component of CS associated with FDS. Identifying tradeoffs is useful in elucidating proximate causes of FDS and functional axes along which CS modifies trait correlations to generate functional integration.

3. The payoff matrix provides a succinct way to diagnose the forms of FDS (apostasis, RPS, negative or positive FDS) and these analogies are useful at diverse levels of biological organization, particularly those involving Red Queen dynamics.

4. In learning or perception, correlations between cognitive representations in one interactor may serve as positive agents, which can substitute for genetic CS. In other cases, such representations are codified by selection as innate signal recognition.

5. Thus, roles of rare and common cognitive representations, memotypes, in many interactions (e.g., predator) substitute for rare and common genotypes in ESS analysis.

6. Systems exhibiting FDS across all levels of biological organization often involve a complex interplay between both negative (apostatic, rare advantage, common disadvantage) and positive FDS (aposematic, cooperation, learning, sexual selection).

7. Ecosystem dynamics and stability are often governed by three-way interactions that arise from FDS among three interactors, owing to homologous sets of the rules elaborated for mating system and social system evolution such as the true RPS or apostatic RPS.

FUTURE ISSUES

1. Much research is based on the premise that simple pair-wise relationships can describe FDS. Lessons from ESS analyses of mating systems (with well-characterized genetics), FDS, and payoff matrices indicate this assumption is rarely met. Nonlinear and three-way interaction drives FDS. Given the prevalence of RPS in trimorphic mating systems, is the RPS common in systems that even lack morphs?

2. Though positive FDS is built into some speciation models, the pervasive role of negative and positive FDS in mating systems implies that explicit models are required to capture dynamics of evolving mating systems, at contact zones between diverging species or in sympatry. Self-recognition loci should be common in the origin of reproductive isolation.

3. Few experiments link FDS in ecological systems. Experiments will be required to understand the complex biological interactions of FDS in ecosystem interactions.

4. Though spatial dynamics are an implicit or explicit component of most ecological systems with FDS, few studies are replicated across a broad geographic framework to test whether interaction strength changes from the center to the edge of a species range or between contact zones of diverging species. The critique also holds for mating and social systems.

LITERATURE CITED

Aeschlimann PB, Haberli MA, Reusch TBH, Boehm T, Milinski M. 2003. Female sticklebacks *Gasterosteus aculeatus* use self-reference to optimize MHC allele number during mate selection. *Behav. Ecol. Sociobiol.* 54:119–26

Ågren J, Ericson L. 1996. Population structure and morph-specific fitness differences in tristylous *Lythrum salicaria*. *Evolution* 50:126–39

Amundsen T, Brobakken PT, Moksnes A, Roskaft E. 2002. Rejection of common cuckoo Cuculus canorus eggs in relation to female age in the bluethroat *Luscinia svecica*. *J. Avian Biol.* 33:366–70

Babcock LE. 1993. Trilobite malformations and the fossil record of behavioral asymmetry. *J. Paleontol.* 67:217–29

Barrett SCH, Harder LD, Cole WW. 2004. Correlated evolution of floral morphology and mating-type frequencies in a sexually polymorphic plant. *Evolution* 58:964–75

Benkman CW. 1996. Are the ratios of bill crossing morphs in crossbills the result of frequency-dependent selection? *Evol. Ecol.* 10:119–26

Benkman CW, Lindholm AK. 1991. The advantages and evolution of a morphological novelty. *Nature* 349:519–20

Bernatchez L, Landry C. 2003. MHC studies in nonmodel vertebrates: what have we learned about natural selection in 15 years? *J. Evol. Biol.* 16:363–77

Billiard S, Faurie C, Raymond M. 2005. Maintenance of handedness polymorphism in humans: a frequency-dependent selection model. *J. Theor. Biol.* 235:85–93

Bock WJ. 1977. Toward an ecological morphology. *Vogelwarte* 29:127–35

Bolker B, Holyoak M, Krivan V, Rowe L, Schmitz O. 2003. Connecting theoretical and empirical studies of trait-mediated interactions. *Ecology* 84:1101–14

Bond AB, Kamil AC. 1998. Apostatic selection by blue jays produces balanced polymorphism in virtual prey. *Nature* 395:594–96

Bond AB, Kamil AC. 2002. Visual predators select for crypticity and polymorphism in virtual prey. *Nature* 415:609–13

Brodie ED III. 1992. Correlational selection for color pattern and antipredator behavior in the garter snake *Thamnophis ordinoides*. *Evolution* 46:1284–98

Brodie ED III, Brodie EDJ. 1991. Evolutionary response of predators to dangerous prey: reduction of toxicity of newts and resistance of garter snakes in island populations. *Evolution* 45:221–24

Brodie ED III, Feldman CR, Hanifin CT, Motychak JE, Mulcahy DG, Williams BL. 2005. Parallel arms races between garter snakes and newts involving tetrodotoxin as the phenotypic interface of coevolution. *J. Chem. Ecol.* 31:343–56

Brodie ED III, Janzen FJ. 1995. Experimental studies of coral snake mimicry—Generalized avoidance of ringed snake patterns by free-ranging avian predators. *Funct. Ecol.* 9:186–90

Bruns TD, Bidartondo MI, Taylor DL. 2000. Interactions of ectomycorrhizal fungi and ectomycorrhizal epiparasites. *Am. Zool.* 40:956–59

Chevrud JM. 1984. Quantitative genetics and developmental constraints on evolution by selection. *J. Theor. Biol.* 110:155–71

Colosimo PF, Hosemann KE, Balabhadra S, Villarreal G, Dickson M, et al. 2005. Widespread parallel evolution in sticklebacks by repeated fixation of ectodysplasin alleles. *Science* 307:1928–33

Cook LM, Kenyon G. 1991. Frequency-dependent selection with background heterogeneity. *Heredity* 66:67–73

Dayan T, Simberloff D. 2005. Ecological and community-wide character displacement: the next generation. *Ecol. Lett.* 8:875–94

Dieckmann U, Doebeli M. 1999. On the origin of species by sympatric speciation. *Nature* 400:354–57

Ding YC, Chi HC, Grady DL, Morishima A, Kidd JR, et al. 2002. Evidence of positive selection acting at the human dopamine receptor D4 gene locus. *Proc. Natl. Acad. Sci. USA* 99:309–14

Durrett R, Levin S. 1997. Allelopathy in spatially distributed populations. *J. Theor. Biol.* 185:165–71

Eckert CG, Barrett SCH. 1992. Stochastic loss of style morphs from populations of tristylous *Lythrum salicaria* and *Decodon verticillatus* (Lythraceae). *Evolution* 46:1014–29

Eckert CG, Manicacci D, Barrett SCH. 1996. Frequency-dependent selection on morph ratios in tristylous *Lythrum salicaria* (Lythraceae). *Heredity* 77:581–88

Elena SF, Lenski RE. 1997. Test of synergistic interactions among deleterious mutations in bacteria. *Nature* 390:395–98

Endler JA, Mappes J. 2004. Predator mixes and the conspicuousness of aposematic signals. *Am. Nat.* 163:532–47

Ewald PW. 1983. Host-parasite relations, vectors, and the evolution of disease severity. *Annu. Rev. Ecol. Evol. Syst.* 14:465–85

Faurie C, Raymond M. 2005. Handedness, homicide and negative frequency-dependent selection. *Proc. R. Soc. London Ser. B* 272:25–28

Feldgarden M, Riley MA. 1999. The phenotypic and fitness effects of colicin resistance in *Escherichia coli* K-12. *Evolution* 53:1019–27

Fincke OM. 2004. Polymorphic signals of harassed female odonates and the males that learn them support a novel frequency-dependent model. *Anim. Behav.* 67:833–45

Fisher R. 1930. *The Genetical Theory of Natural Selection*. Oxford, UK: Clarendon, 268 pp.

Fleishman LJ. 1986. Motion detection in the presence and absence of background motion in an *Anolis* lizard. *J. Comp. Physiol. A* 159:711–20

Forsman A, Appelqvist S. 1998. Visual predators impose correlational selection on prey color pattern and behavior. *Behav. Ecol.* 9:409–13

Garamszegi LZ, Aviles JM. 2005. Brood parasitism by brown-headed cowbirds and the expression of sexual characters in their hosts. *Oecologia* 143:167–77

Gibson JR, Chippindale AK, Rice WR. 2002. The X chromosome is a hotspot for sexually antagonistic fitness variation. *Proc. R. Soc. London Ser. B* 269:499–505

Gigord LDB, Macnair MR, Smithson A. 2004. Negative frequency-dependent selection maintains a dramatic flower color polymorphism in the rewardless orchid *Dactylorhiza sambucina* (L.) *Proc. Natl. Acad. Sci. USA* 101:7839

Greene CM, Stamps JA. 2001. Habitat selection at low population densities. *Ecology* 82:2091–2100

Grosberg RK, Hart MW. 2000. Mate selection and the evolution of highly polymorphic self/nonself recognition genes. *Science* 289:2111–14

Guilford T. 1988. The evolution of conspicuous coloration. *Am. Nat.* 131:S7–21

Hamilton WD. 1964. The evolution of social behavior. *J. Theor. Biol.* 7:1–52

Harpending H, Cochran G. 2002. In our genes. *Proc. Natl. Acad. Sci. USA* 99:10–12

Hochberg ME, Sinervo B, Brown SP. 2003. Socially mediated speciation. *Evolution* 57:154–58

Holland B, Rice WR. 1999. Experimental removal of sexual selection reverses intersexual antagonistic coevolution and removes a reproductive load. *Proc. Natl. Acad. Sci. USA* 96:5083–88

Hori M. 1993. Frequency-dependent natural selection in the handedness of scale-eating cichlid fish. *Science* 260:216–19

Howard DJ. 1999. Conspecific sperm and pollen precedence and speciation. *Annu. Rev. Ecol. Evol. Syst.* 30:109–32

Husband BC, Barrett SCH. 1992. Effective population-size and genetic drift in tristylous *Eichhornia paniculata* (Pontederiaceae). *Evolution* 46:1875–90

Kerr B, Riley MA, Feldman MW, Bohannan BJM. 2002. Local dispersal promotes biodiversity in a real-life game of rock-paper-scissors. *Nature* 418:171–74

Kirkup BC, Riley MA. 2004. Antibiotic-mediated antagonism leads to a bacterial game of rock-paper-scissors in vivo. *Nature* 428:412–14

Klar AJS. 2005. A 1927 study supports a current genetic model for inheritance of human scalp hair-whorl orientation and hand-use preference traits. *Genetics* 170:2027–30

Kokko H, Mappes J, Lindstrom L. 2003. Alternative prey can change model-mimic dynamics between parasitism and mutualism. *Ecol. Lett.* 6:1068–76

Kuchta SR. 2005. Experimental support for aposematic coloration in the salamander *Ensatina eschscholtzii xanthoptica*: Implications for mimicry of Pacific newts. *Copeia* 2005:265–71

Lande R. 1981. Models of speciation by sexual selection on polygenic traits. *Proc. Natl. Acad. Sci. USA* 78:3721–25

Lande R, Arnold SJ. 1983. The measurement of selection on correlated characters. *Evolution* 37:1210–26

Landry C, Garant D, Duchesne P, Bernatchez L. 2001. 'Good genes as heterozygosity': the major histocompatibility complex and mate choice in Atlantic salmon (*Salmo salar*). *Proc. R. Soc. London Ser. B* 268:1279–85

Langham GM. 2004. Specialized avian predators repeatedly attack novel color morphs of *Heliconius* butterflies. *Evolution* 58:2783–87

Lefebvre L. 1995. The opening of milk bottles by birds: evidence for accelerating learning rates, but against the wave-of-advance model of cultural transmission. *Behav. Process* 34:43–54

Lindstrom L, Alatalo RV, Lyytinen A, Mappes J. 2001. Strong antiapostatic selection against novel rare aposematic prey. *Proc. Natl. Acad. Sci. USA* 98:9181–84

Lindstrom L, Alatalo RV, Lyytinen A, Mappes J. 2004. The effect of alternative prey on the dynamics of imperfect Batesian and Mullerian mimicries. *Evolution* 58:1294–1302

Lively CM, Clay K, Wade MJ, Fuqua C. 2005. Competitive co-existence of vertically and horizontally transmitted parasites. *Evol. Ecol. Res.* 7:1183–90

Lynch M, Walsh B. 1998. *Genetics and Analysis of Quantitative Traits*. Sunderland, MA: Sinauer. 980 pp.

Lyon BE. 2003. Egg recognition and counting reduce costs of avian conspecific brood parasitism. *Nature* 422:495–99

Lyon BE, Eadie JM, Hamilton LD. 1994. Parental choice selects for ornamental plumage in American coot chicks. *Nature* 371:240–43

Mallet J, Joron M. 1999. Evolution of diversity in warning color and mimicry: Polymorphisms, shifting balance, and speciation. *Annu. Rev. Ecol. Evol. Syst.* 30:201–33

Mappes J, Alatalo RV. 1997. Effects of novelty and gregariousness in survival of aposematic prey. *Behav. Ecol.* 8:174–77

Mappes J, Marples N, Endler JA. 2005. The complex business of survival by aposematism. *Trends Ecol. Evol.* 20:598–603

Maynard Smith J. 1982. *Evolution and Theory of Games*. Cambridge, MA: Cambridge Univ. Press. 224 pp.

Mayr E. 1993. Proximate and ultimate causations. *Biol. Philos.* 8:93–94

McGlothlin JW, Parker PG, Nolan V, Ketterson ED. 2005. Correlational selection leads to genetic integration of body size and an attractive plumage trait in dark-eyed juncos. *Evolution* 59:658–71

McKeever WF. 2004. An X-linked three allele model of hand preference and hand posture for writing. *Laterality* 9:149–73

Niskanen M, Mappes J. 2005. Significance of the dorsal zigzag pattern of *Vipera latastei gaditana* against avian predators. *J. Anim. Ecol.* 74:1091–1101

Nowak MA, Sigmund K. 2004. Evolutionary dynamics of biological games. *Science* 303:793–99

O'Connell LM, Johnston MO. 1998. Male and female pollination success in a deceptive orchid, a selection study. *Ecology* 79:1246–60

Oneil P, Schmitt J. 1993. Genetic constraints on the independent evolution of male and female reproductive characters in the tristylous plant *Lythrum salicaria*. *Evolution* 47:1457–71

Payne RB, Payne LL. 1994. Song mimicry and species associations of West-African indigobirds *Vidua* with quail-finch *Ortygospiza atricollis*, goldbreast *Amandava subflava* and brown twinspot *Clytospiza monteiri*. *Ibis* 136:291–304

Payne RB, Payne LL, Woods JL, Sorenson MD. 2000. Imprinting and the origin of parasite-host species associations in brood-parasitic indigobirds, *Vidua chalybeata*. *Anim. Behav.* 59:69–81

Payne RB, Woods JL, Payne LL. 2001. Parental care in estrildid finches: experimental tests of a model of Vidua brood parasitism. *Anim. Behav.* 62:473–83

Pfennig DW. 1992. Polyphenism in spadefoot toad tadpoles as a locally adjusted evolutionarily stable strategy. *Evolution* 46:1408–20

Pfennig DW, Murphy PJ. 2003. A test of alternative hypotheses for character divergence between coexisting species. *Ecology* 84:1288–97

Piertney SB, Oliver MK. 2006. The evolutionary ecology of the major histocompatibility complex. *Heredity* 96:7–21

Potts WK, Manning CJ, Wakeland EK. 1991. Mating patterns in seminatural populations of mice influenced by MHC genotype. *Nature* 352:619–21

Propper RE, Christman SD, Phaneuf KA. 2005. A mixed-handed advantage in episodic memory: A possible role of interhemispheric interaction. *Mem. Cogn.* 33:751–57

Raymond M, Pontier D, Dufour AB, Moller AP. 1996. Frequency-dependent maintenance of left-handedness in humans. *Proc. R. Soc. London Ser. B* 263:1627–33

Remold SK, Lenski RE. 2004. Pervasive joint influence of epistasis and plasticity on mutational effects in *Escherichia coli*. *Nat. Genet.* 36:423–26

Reusch TBH, Haberli MA, Aeschlimann PB, Milinski M. 2001. Female sticklebacks count alleles in a strategy of sexual selection explaining MHC polymorphism. *Nature* 414:300–2

Rice WR. 1984. Sex chromosomes and the evolution of sexual dimorphism. *Evolution* 38:735–42

Rice WR. 2000. Dangerous liaisons. *Proc. Natl. Acad. Sci. USA* 97:12953–55

Rice WR, Chippindale AK. 2001. Intersexual ontogenetic conflict. *J. Evol. Biol.* 14:685–93

Roitblat HL. 1987. *Introduction to Comparative Cognition*. New York: Freeman. 377 pp.

Ruxton GD, Sherratt TN, Speed MP. 2004. *Avoiding Attack: the Evolutionary Ecology of Crypsis, Warning Signals, and Mimicry*. Oxford, UK: Oxford Univ. Press. 260 pp.

Ryan MJ, Rand AS. 1999. Phylogenetic influence on mating call preferences in female tungara frogs, *Physalaemus pustulosus*. *Anim. Behav.* 57:945–56

Schluter D. 1995. Adaptive radiation in sticklebacks: Trade-offs in feeding performance and growth. *Ecology* 76:82–90

Schluter D. 2003. Frequency dependent natural selection during character displacement in sticklebacks. *Evolution* 57:1142–50

Seehausen O, Schluter D. 2004. Male-male competition and nuptial-color displacement as a diversifying force in Lake Victoria cichlid fishes. *Proc. R. Soc. London Ser. B* 271:1345–53

Shuster SM, Wade MJ. 1991a. Equal mating success among male reproductive strategies in a marine isopod. *Nature* 350:606–61

Shuster SM, Wade MJ. 1991b. Female copying and sexual selection in a marine isopod crustacean, *Paracerceis sculpta*. *Anim. Behav.* 41:1071–78

Sih A, Bell AM Johnson JC. 2004. Behavioral syndromes: an ecological and evolutionary overview. *Trends Ecol. Evol.* 19:372–78

Sinervo B. 2001. Runaway social games, genetic cycles driven by alternative male and female strategies, and the origin of morphs. *Genetica* 112:417–34

Sinervo B, Basolo AL. 1996. Testing adaptation using phenotypic manipulations. In *Adaptation*, ed. MR Rose, G Lauder, pp. 148–85. New York: Academic

Sinervo B, Bleay C, Adamopoulou C. 2001. Social causes of correlational selection and the resolution of a heritable throat color polymorphism in a lizard. *Evolution* 55:2040–52

Sinervo B, Calsbeek R, Comendant T, Adamopoulou C, Both C, Clobert J. 2006a. Genetic and maternal determinants of dispersal. *Am. Nat.* 168:88–99

Sinervo B, Chaine A, Clobert J, Calsbeek R, Hazard L, et al. 2006b. Color morphs and genetic cycles of greenbeard mutualism and transient altruism. *Proc. Natl. Acad. Sci. USA.* 103:7372–77

Sinervo B, Clobert J. 2003. Morphs, dispersal behavior, genetic similarity, and the evolution of cooperation. *Science* 300:1949–51

Sinervo B, Lively C. 1996. The rock-paper-scissors game and the evolution of alternative male strategies. *Nature* 380:240–43

Sinervo B, Svensson E. 2002. Correlational selection and the evolution of genomic architecture. *Heredity* 89:329–38

Sinervo B, Svensson E, Comendant T. 2000a. Density cycles and an offspring quantity and quality game driven by natural selection. *Nature* 406:985–88

Smith TB. 1993. Disruptive selection and the genetic-basis of bill size polymorphism in the African finch *Pyrenestes*. *Nature* 363:618–20

Smith TB, Girman DJ. 2000. Reaching new adaptive peaks: Evolution of alternative bill forms in an African finch. In *Adaptive Genetic Variation in the Wild*, ed. TA Mousseau, B Sinervo, J Endler, pp. 139–56. New York: Oxford Univ. Press

Smithson A, Macnair MR. 1997. Negative frequency-dependent selection by pollinators on artificial flowers without rewards. *Evolution* 51:715–23

Soler JJ, Aviles JM, Soler M, Moller AP. 2003. Evolution of host egg mimicry in a brood parasite, the great spotted cuckoo. *Biol. J. Linn. Soc.* 79:551–63

Svensson EI, Abbott J, Hardling R. 2005. Female polymorphism, frequency dependence, and rapid evolutionary dynamics in natural populations. *Am. Nat.* 165:567–76

Taylor DL, Bruns TD, Hodges SA. 2004. Evidence for mycorrhizal races in a cheating orchid. *Proc. R. Soc. London Ser. B* 271:35–43

Thijs H, Shann JR, Weidenhamer JD. 1994. The effect of phytotoxins on competitive outcome in a model system. *Ecology* 75:1959–64

Thompson JN. 2005. *The Geographic Mosaic of Coevolution*. Chicago: Univ. Chicago Press. 443 pp.

Thompson JN, Pellmyr O. 1992. Mutualism with pollinating seed parasites amid co-pollinators—constraints on specialization. *Ecology* 73:1780–91

van Valen L. 1976. Red Queen lives. *Nature* 260:575

Wade MJ, Kalisz S. 1990. The causes of natural selection. *Evolution* 44:1947–55

Wake DB. 1997. Incipient species formation in salamanders of the *Ensatina* complex. *Proc. Natl. Acad. Sci. USA* 94:7761–67

Weller SG, Sakai AK. 1999. Using phylogenetic approaches for the analysis of plant breeding system evolution. *Annu. Rev. Ecol. Evol. Syst.* 30:167–99

Carbon-Nitrogen Interactions in Terrestrial Ecosystems in Response to Rising Atmospheric Carbon Dioxide

Peter B. Reich,[1] Bruce A. Hungate,[2] and Yiqi Luo[3]

[1]Department of Forest Resources, University of Minnesota, St. Paul, Minnesota 55108; email: preich@umn.edu

[2]Department of Biological Sciences and Merriam-Powell Center for Environmental Research, Northern Arizona University, Flagstaff, Arizona 86011; email: Bruce.Hungate@NAU.EDU

[3]Department of Botany and Microbiology, University of Oklahoma, Norman, Oklahoma 73019; email: yluo@ou.edu

Key Words

biogeochemistry, global change, nitrogen, photosynthesis, stoichiometry

Abstract

Interactions involving carbon (C) and nitrogen (N) likely modulate terrestrial ecosystem responses to elevated atmospheric carbon dioxide (CO_2) levels at scales from the leaf to the globe and from the second to the century. In particular, response to elevated CO_2 may generally be smaller at low relative to high soil N supply and, in turn, elevated CO_2 may influence soil N processes that regulate N availability to plants. Such responses could constrain the capacity of terrestrial ecosystems to acquire and store C under rising elevated CO_2 levels. This review highlights the theory and empirical evidence behind these potential interactions. We address effects on photosynthesis, primary production, biogeochemistry, trophic interactions, and interactions with other resources and environmental factors, focusing as much as possible on evidence from long-term field experiments.

1. INTRODUCTION

Carbon (C) and nitrogen (N) are critical to many aspects of plant, herbivore, and microbial metabolism. Given rising levels of atmospheric CO_2 (hereafter eCO_2), the coupled cycling of C and N is also critical to ecosystem function today and in the future. Interactions involving C and N that might influence the global C cycle are of great importance to atmosphere-biosphere interactions, and thus to human society, because changes in eCO_2 impact global climate.

This importance is highlighted by several kinds of studies suggesting that C-N interactions may substantially constrain the CO_2 fertilization effect at local and global scales (Hungate et al. 2003, Luo et al. 2004, Oren et al. 2001, Reich et al. 2006). The mechanisms involve responses to resource supply rates, modifications of resource supply rates, or both (**Figure 1**). First, if plants experience multiple resource limitations, interactions of CO_2 and N supply could limit the CO_2 fertilization effect on biomass and C accumulation (Oren et al. 2001, Reich et al. 2006, Schneider et al. 2004). Such interactions occur at ecophysiological to ecosystem scales, and involve plant-microbial, plant-consumer, and plant-plant interactions, or all of these. If responses to eCO_2 are generally larger when N supply is high rather than low, as shown by contrasting response curves in a simple multiple limitation framework (**Figure 1**), then N supply rate would routinely influence eCO_2 response.

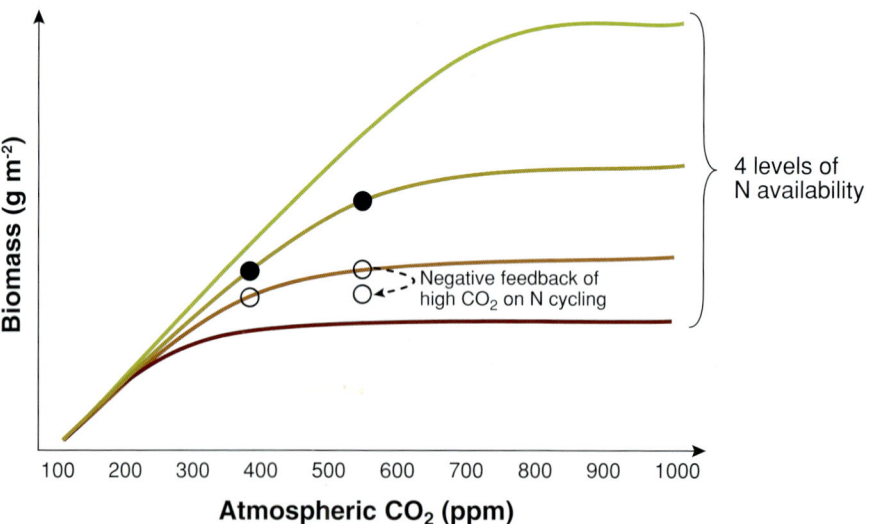

Figure 1

Simplified theoretical model of the way multiple resource limitation theory can lead to significant CO_2-N interactions (i.e., greater-than-additive) and of the way in which elevated CO_2-induced progressive N limitation of soil net N supply is superimposed on such multiple limitation responses. In this example, biomass increase with a 50% increase in CO_2 concentration is greater at higher than lower N supply rates (compare filled circles to open circles on the middle two N supply lines). Moreover, elevated CO_2 can diminish N supply rates, further suppressing biomass at elevated CO_2 (compare the two open circles connected with the dotted line).

Second, rising eCO_2 may result in feedbacks that lead to a suppression of plant N availability that limits the CO_2 fertilization effect, called progressive nitrogen limitation (PNL) (Luo et al. 2004). PNL could occur even if responses to eCO_2 were not influenced by N-supply level (i.e., effects of CO_2 and N availability were strictly additive) and would reduce CO_2 stimulation of biomass by reducing N supply, especially in N-limited conditions (note the downward shift in N supply owing to eCO_2; **Figure 1**). Finally, multiple resource limitations and PNL can work in tandem over time, exacerbating the extent to which lack of N availability may influence ecosystem responses to eCO_2.

The main C-N interactions proposed to influence responses to eCO_2 include the following: down-regulation of leaf N concentration and, hence, of net photosynthetic capacity; altered rates of herbivory, disease, or fungal mutualism due to changes in plant chemistry and stoichiometry; alterations of biogeochemical cycling; compositional change in plant or soil microbial communities; and increased N fixation rates or abundances of fixers, or both. Other drivers of environmental change (e.g., temperature or water) may also potentially influence C-N interactions under eCO_2.

Summarizing and synthesizing information about C-N interactions under eCO_2 is the central goal of this review. Unfortunately, empirical evidence of long-term C-N interactions under eCO_2 is still rare for either managed or unmanaged ecosystems. In a strange twist of fate, in contrast to the thousands of publications on laboratory eCO_2 experiments, we are severely limited, and will remain so over the next 25 years, in our ability to generalize about CO_2-N interactions over meaningful ecosystem time frames, by the low number of realistic long-term experiments [such as those conducted without chambers using free-air CO_2 enrichment (FACE)]. This limitation is especially true for experiments using communities that are realistic approximations of natural or managed mixed species communities, as well as those examining CO_2 responses across a range of nutrient supply conditions (see **Table 1**). Thus, herein we focus on C-N interactions in a set of long-term studies (at least three years in length) of eCO_2 in near-natural conditions, emphasizing especially those that manipulate both CO_2 and N.

2. ABIOTIC AND BIOTIC CARBON-NITROGEN INTERACTIONS UNDER RISING ATMOSPHERIC CARBON DIOXIDE

The stoichiometry of tissue C:N ratios and the relative supply of carbon vis-à-vis nitrogen influence an enormous number of biotic and abiotic processes and relationships in terrestrial ecosystems. In this section we review some of these, organized by hierarchical and trophic scale.

2.1. Carbon-Nitrogen Interactions at the Point of Carbon Capture: The Leaf

How might C-N interactions down-regulate photosynthetic capacity under eCO_2 and therefore limit the sustainability of potential productivity response? Growth at

Table 1 Long-term (three or more field seasons) studies of elevated CO_2 under contrasting replicated N treatments. Studies listed by longevity of treatments

Study & System	Location	Ecosystem	Treatment period	References
Swiss FACE	Eschikon, Switzerland	Managed grassland	1993–present	Zanetti et al. 1996, Schneider et al. 2004
Jasper Ridge Mini FACE	California, USA	Grassland (annual)	1997–present	Shaw et al. 2002, Dukes et al. 2005
BioCON FACE	Minnesota, USA	Grasslands	1998–present	Reich et al. 2001, 2004, 2006
California pine OTC	California, USA	Pine juveniles	1991–1997	Johnson et al. 1997, Haile-Mariam et al (2000)
Swiss model ecosystem OTC	Birmensdorf, Switzerland	Juvenile trees	1995–1998	Spinnler et al. (2002)
Japan FACE	Shizukuizhi, Japan	Rice	1998–2000	Okada et al. 2001, Kim et al. 2003
Swiss OTC	Central Alps, Switzerland	Alpine grassland	1991–1994	Schäppi & Körner 1996
Iceland OTC	Gunnarsholt, Iceland	Poplar juveniles	1994–1996	Sigurdsson et al. 2001

eCO_2 can lead to acclimation (i.e., down-regulation) of carboxylation capacity (V_{cmax}) driven by reduced ribulose bisphosphate carboxylase (Rubisco) amount and activity. This can occur by either of two ways, or by both. Down-regulation can occur by a general decrease in leaf N (especially on a mass basis) owing to enhanced C uptake under eCO_2. Additionally, down-regulation of Rubisco amount and activity could occur owing to decreased expression of specific photosynthetic genes as a result of increased sucrose cycling in mesophyll cells that occurs when net C uptake exceeds the capacity for carbohydrate export and utilization (Ainsworth & Long 2005). Thus, if eCO_2 decreases soil N availability (see Section 2.4) and/or increases C pools in biomass, this should lead to lower leaf N (mass or area basis), leading to down-regulation of net photosynthetic capacity (Ellsworth et al. 2004, Nowak et al. 2004).

In a meta-analysis of FACE studies (n = 3 to 11), eCO_2 decreased mass-based leaf N by 13% (range of 95% confidence interval, 10% to 17%), and area-based N, V_{cmax} and Rubisco, by 5% (2% to 7%), 13% (11% to 16%), and 19% (6% to 31%), respectively (Ainsworth & Long 2005). For 104 studies [predominantly open-top chambers (OTC)], the eCO_2-induced reduction in mass-based N was 11% on average (Luo et al. 2006). The repeated experimental observations of down-regulation of N, Rubisco, and V_{cmax} strongly suggest that eCO_2 is likely having similar effects globally, although species responses vary substantially.

Across all FACE studies, eCO_2 increased light-saturated photosynthetic capacity by 31% (Ainsworth & Long 2005), comparable to earlier meta-analyses based largely on chamber studies (Curtis & Wang 1998, Wand et al. 1999). Does the long-term eCO_2-enhancement of net photosynthesis vary with N supply? Surprisingly, our only data come from two FACE studies, the managed pasture Swiss FACE (*Lolium perenne*) and the tallgrass prairie BioCON study in Minnesota (**Table 1, Figure 2**). Leaf-level

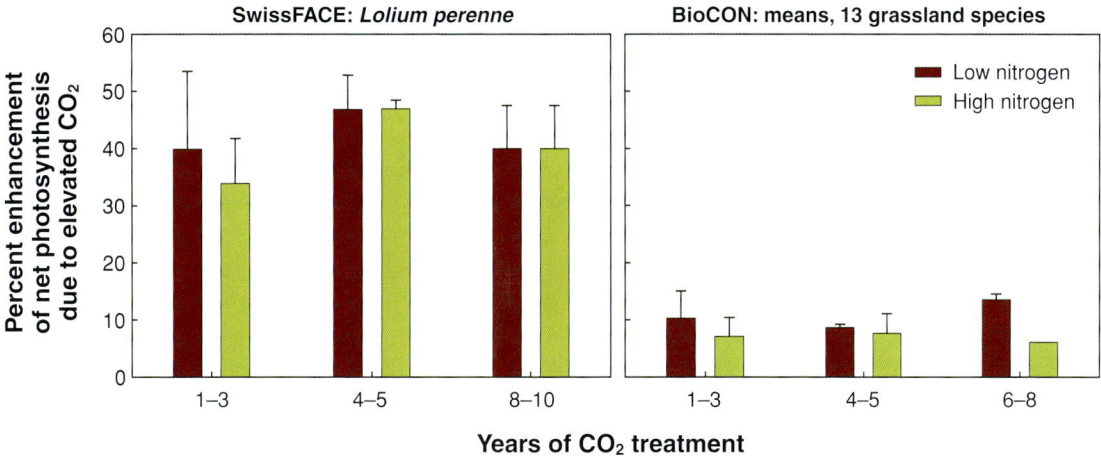

Figure 2

Effects of elevated CO₂ and N supply on light-saturated net photosynthesis in two long-term FACE studies, the managed pasture study in Eschikon, Switzerland (Swiss FACE) and the temperate grassland study in Cedar Creek, Minnesota (BioCON). For BioCON, data are the means of 11–13 species each year in monocultures. Data from Lee et al. (2001), Ainsworth & Long (2005), and T.D. Lee & P.B. Reich (unpublished data).

photosynthesis was increased by eCO_2 in both studies, but much more so for *Lolium* in the managed pasture than for the dozen grassland species in Minnesota. However, despite the different magnitude of photosynthetic enhancement, in neither study did this vary across contrasting N treatments at any point in time (**Figure 2**). Given that in both studies, biomass responses to eCO_2 differ at contrasting N supply rates (Schneider et al. 2004, Reich et al. 2006), the results suggest that plants are more prone to C-N interactions at the system scale than at the leaf scale, and these may be manifest in the size, organization or turnover rates of canopies and root systems.

2.2. Productivity and Biomass Accumulation

Both multiple resource limitation theory and PNL provide mechanisms by which C-N interactions can influence biomass and productivity responses to eCO_2. Direct evidence of such interactions can only be obtained from experiments with contrasting soil N supply rates, which we focus on herein (**Table 1**, plus additional shorter-term, multiyear studies).

In OTC studies with CO_2 and N manipulations, poplar seedlings and saplings (≈1–5 years old) had greater aboveground growth stimulation by eCO_2 at high rather than at low N supply after two years in Michigan (Zak et al. 2000) and Italy (Liberloo et al. 2005) and three years in Iceland (Sigurdson et al. 2001), as did young North Carolina pine plantation trees (≈15 years old) after two years of CO_2 and N manipulations in a FACE system (Oren et al. 2001). In contrast, juvenile (1–7 years old) ponderosa pine in California and spruce-beech in Switzerland had similar responses to eCO_2 at high and low N supply over six and four years in OTC, respectively (Haile-Mariam et al. 2000, Johnson et al. 1997, Spinnler et al. 2002). Results of these six studies suggest young trees may often, but not always, be more responsive to eCO_2 under enriched rather than ambient N conditions. There have been no such experiments with mature forests though, nor any long-term (more than three years) CO_2 and N studies with woody plants grown in natural settings. Thus, whether mature forests will be responsive to eCO_2 only at high N supply remains an open question, as does the general response of young forests over 5-, 10-, or 20-year time frames.

In managed agricultural and unmanaged herbaceous ecosystems with both FACE and N manipulations, there are aboveground biomass data for wheat (2 years), rice (3 years), annual grassland (5 years), perennial grassland (7 years), and ryegrass (10 years) (**Table 1**). In wheat the eCO_2 effect was no different at high N than at low N (Ainsworth & Long 2005). In rice there was greater eCO_2 enhancement of tiller number and biomass at higher rather than lower N fertilization rates (15 versus 9 versus 4 g N m^{-2} years^{-1}) during early- and mid-season stages (such as panicle initiation and anthesis), which disappeared, however, by the end of the growing season (Kim et al. 2003). In the ryegrass study in Switzerland, there was a consistent ($\approx$26%) enhancement over 10 years of harvestable aboveground biomass by eCO_2 at high N fertilization (56 g N m^{-2} years^{-1}), and no enhancement with lower (14 g N m^{-2} years^{-1}) fertilization rates (Schneider et al. 2004; **Figure 3**). The two natural system FACE studies (Jasper Ridge, California and BioCON, Minnesota) had divergent responses. In California annual grasslands there was no effect by five years of eCO_2 on aboveground biomass regardless of N supply (Dukes et al. 2005). In Minnesota perennial grassland N addition (4 g N m^{-2} years^{-1}) caused a slightly smaller eCO_2 stimulation of total biomass in years 1–3 and then a significantly larger eCO_2 stimulation of biomass after year 3 of the study (Reich et al. 2006; **Figure 3**). Additionally, a three-year OTC study in Swiss alpine grasslands reported no statistically significant interactions of CO_2 with nutrients treatments (Schäppi & Körner 1996); however, after three years total biomass was 13% greater under eCO_2 in fertilized plots and 5% lower under eCO_2 in unfertilized plots, consistent with the patterns of the Minnesota grassland and Swiss-managed pasture results.

Summarizing these 12 agricultural, grassland, and woody plant studies that manipulated both CO_2 and N for at least two years, greater biomass accumulation under eCO_2 at high rather than low N availability was observed in two thirds of the cases (with neutral findings in the others). Thus, the evidence suggests that a general N limitation to the eCO_2 effect is common, although not ubiquitous.

Measurements of leaf area index (LAI), soil CO_2 flux, and root biomass have been made in too few of the studies, and in insufficient intensity within studies, to come to any meaningful general conclusion regarding the interactions of CO_2 with N on these properties and processes, despite their general stimulation by eCO_2 (Ainsworth & Long 2005, Luo et al. 2006, Nowak et al. 2004). However, in the Japanese FACE study, eCO_2 influenced the seasonal trajectory of LAI in a manner consistent with a progressive within-season N limitation. LAI was enhanced early by eCO_2 but declining relative N availability eventually eliminated the LAI enhancement (Kim et al. 2003). Kim et al. (2003) state that management of rice under eCO_2 must

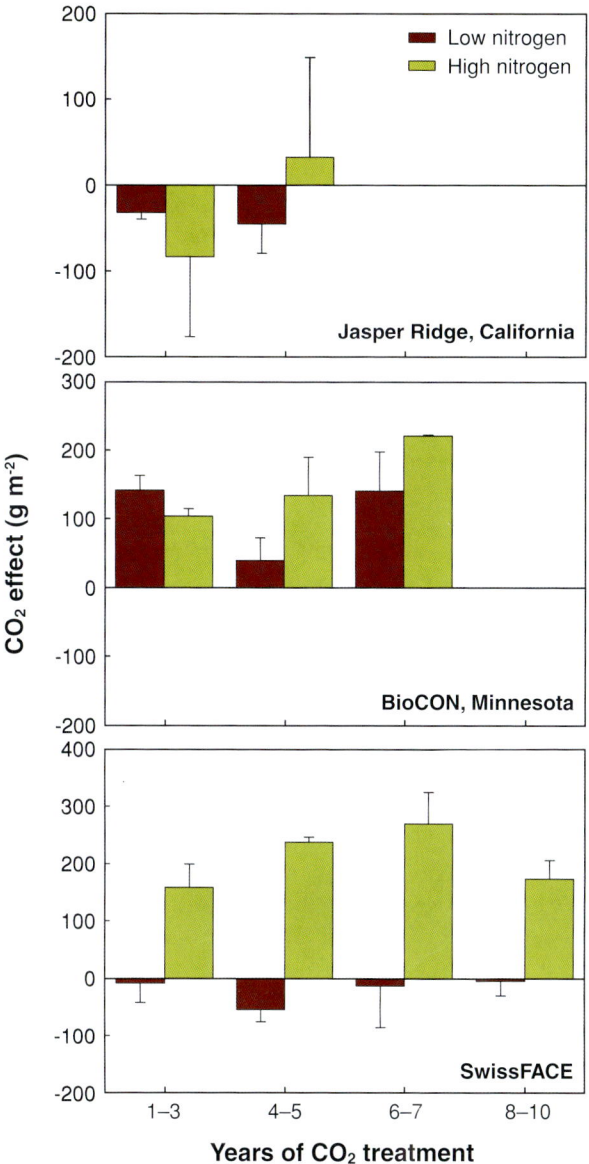

Figure 3

Effects of elevated CO_2 and N supply on biomass accumulation in three long-term studies, the annual grassland study at Jasper Ridge, California; managed pasture study in Eschikon, Switzerland (Swiss FACE) and the temperate grassland study in Cedar Creek, Minnesota (BioCON). The effect of CO_2 on biomass is defined as the biomass at elevated CO_2 minus biomass at ambient CO_2, estimated separately at contrasting low and high N supply. For the BioCON study, the data are for total biomass (above and belowground, 0–20 cm). For the other two studies, only aboveground biomass data are available. Data from Dukes et al. (2005), Reich et al. (2006), Schneider et al. (2004), and P.B. Reich (unpublished data).

therefore include a ramping up of N fertilization sufficient to maintain enhanced LAI and stable tissue N concentrations in order for the eCO_2 supply to be converted into increased C uptake.

2.3. Net Ecosystem Production and Carbon Sequestration

Net ecosystem production (NEP), the sum of net primary production (NPP) minus total heterotrophic respiration, is one explicit measure of ecosystem C uptake. Does response of NEP to eCO_2 depend on N supply? Given that plant biomass production is one key component of NEP, the larger eCO_2 enhancement of plant biomass with added N (see above) indicates the potential for added N to stimulate the response of NEP to eCO_2. Whether this potential is realized depends on how eCO_2 affects soil heterotrophic respiration from standing and soil surface detritus, and heterotrophic respiration from microbial breakdown of soil organic carbon stocks, and whether this response is sensitive to added N (van Groenigen et al. 2006). In the Swiss FACE pasture, mean soil C pools were higher with eCO_2, and the difference was slightly larger with added N; none of the changes were statistically significant, however, even after 10 years of experimental treatment (Xie et al. 2005). In the cold perennial grassland in Minnesota (BioCON), soil C also did not change significantly with eCO_2, but any tendency was toward a loss of C, with the loss being greater when N was also added (Dijkstra et al. 2005). Overall, changes in NEP due to eCO_2 in these (and other) experiments are apparently small on an annual basis and appear to exhibit little sensitivity to N additions. However, our ability to detect such effects is limited by the magnitude of the changes in soil C and the scarcity of ongoing experiments—only long-term experimentation in more systems than currently being studied will adequately characterize the sign and magnitude of CO_2-N effects on C storage.

Assuming that we are years, if not decades, from having adequate direct tests of long-term C sequestration under contrasting CO_2-N regimes, data syntheses are an alternative tool. A synthesis that divided up data (total n = 80) from a broad array of indoor, open-top chamber, and FACE experiments into contrasting N fertilization levels suggests that soil C is insensitive to eCO_2 in the absence of N supplements and that exogenous N is needed for eCO_2 to increase soil carbon (van Groenigen et al. 2006). A different synthesis (Luo et al. 2006) that directly contrasted only the limited number (n = 6) of studies that included contrasting N treatments (i.e., with versus without N additions) also found that soil C only increased in response to eCO_2 when high N was added. However, based on the same data set (but not shown in Luo et al. 2006), the larger sample (n = 28) of studies without N addition did show a significant increase (of 6.5%) in soil C in response to eCO_2. Thus, these recent meta-analyses (Luo et al. 2006, van Groenigen et al. 2006) suggest that evidence is still equivocal regarding whether eCO_2 increases soil C under ambient soil conditions (i.e., without added N). However, both analyses indicate that the effects of eCO_2 on soil C depends on N supply; in both cases, responses of soil C to eCO_2 were larger with added N than without.

These findings underscore the importance of considering N supply and availability when projecting eCO_2-induced changes in soil C sequestration to the global scale

(Hungate et al. 2003). The following sections discuss processes that are involved in such C-N interactions.

2.4. Carbon Dioxide Effects on Soil Organic Matter Turnover and N Cycling

Nitrogen mineralization largely controls N availability to plants, so effects of eCO_2 on N mineralization have the potential to feed back to N-limited plant growth, and thus C gain in ecosystems. The effects of eCO_2 on organic matter mineralization are also important because, although eCO_2 usually increases photosynthesis, changes in respiration are equally important for total system C balance. Elevated CO_2 has been postulated to decrease litter quality, increase C input to soil, and increase soil water content, and these in turn are expected to alter organic matter mineralization, plant N uptake, and whole-system carbon balance. In this section, we summarize findings from field experiments examining these effects.

Despite early speculation (Strain & Bazzaz 1983) based on the common observation that eCO_2 reduces the N concentration of live plant leaves (see above), exhaustive tests suggest little influence of eCO_2 on the C:N ratio of litter or on the rate of plant litter decomposition (Norby et al. 2001), nor any dependence thereof on soil N supply (Henry et al. 2005; M.A. de Graaff, K.J. van Groenigen, J. Six, B. Hungate, C. van Kessel, in review). Because of large interspecific differences in decomposition rates between plant species (Dijkstra et al. 2006), eCO_2 is likely to have larger effects on litter decomposition by altering the composition of the plant assemblage (Dukes et al. 2005), though this indirect mechanism is likely to be very system specific (Henry et al. 2005).

In contrast to litter C:N ratios, eCO_2 has been shown to influence soil organic matter decomposition through several mechanisms. Plant growth affects mineralization of soil organic matter via the priming effect (Kuzyakov 2002). In the context of eCO_2 and soil N availability, there are three hypotheses describing mechanisms modulating the priming effect: competition, microbial activation, and preferential substrate use (Cheng 1999, Kuzyakov 2002). The microbial activation hypothesis (Kuzyakov 2002) predicts that eCO_2 increases the input of C-rich organic matter from the growth and death of roots to which soil microorganisms respond initially by immobilizing available nutrients, later by mining older soil organic matter. The preferential substrate use hypothesis holds that soil microorganisms prefer the higher-quality substrates from plant roots, such that when eCO_2 enhances C input to soil, decomposition of older soil organic matter declines; this hypothesis assumes that soil nutrients do not limit decomposition. Where nutrients do limit decomposition, the competition hypothesis is relevant, whose predictions vary depending on the presumed winner in plant-microbe competition for soil nutrients: When plants triumph, greater plant uptake of nutrients reduces decomposition (Hu et al. 2001), whereas when microorganisms emerge victorious, the reduced nutrient availability caused by immobilization restricts plant growth responses to eCO_2 (Gill et al. 2002). These hypotheses are broad, and collectively explain a wide range of experimental results, but we still lack a general reconciliation of when each is dominant.

Does greater plant growth in eCO_2 affect decomposition of soil organic matter, and does this depend on soil nutrient status? Experiments have assessed this directly by measuring gross and net rates of soil N transformations. There is little evidence that eCO_2 alters gross N mineralization: No overall significant effect was found via meta-analysis (M.A. de Graaff, K.J. van Groenigen, J. Six, B. Hungate, C. van Kessel, in review), nor do the field experiments assessed here show any strong response or any dependence on soil N supply (**Table 2**). In a synthesis including pot and greenhouse studies, eCO_2 enhanced gross immobilization (M.A. de Graaff, K.J. van Groenigen, J. Six, B. Hungate, C. van Kessel, in review), yet this generalization does not reflect the field experiments considered here, where eCO_2 effects on gross consumption of inorganic N are small to nonexistent (**Table 2**). In summary, effects of eCO_2 on soil gross N mineralization and immobilization appear to be small compared to background variation, and there is no clear pattern revealing a dependence on soil N supply. Nevertheless, it is difficult to state with certainty that such effects are

Table 2 Summary of gross N transformation measurements conducted in elevated CO_2 field experiments*

Ecosystem	Years	Gross NH_4^+ mineralization ($g N g^{-1} d^{-1}$)		Gross NH_4^+ consumption ($g N g^{-1} d^{-1}$)		Seasonal gross NH_4^+ flux
		Ambient	Elevated	Ambient	Elevated	$g N m^{-2} years^{-1}$
Tallgrass prairie[1]	7.5	2.5 ± 0.4	2.3 ± 0.4	3.1 ± 0.8	2.7 ± 0.6	72
Lolium pasture[2]	7.2	6.2 ± 0.9	7.7 ± 1.3	8.0 ± 0.8	10.2 ± 1.2	217
Trifolium pasture[2]	7.2	6.9 ± 0.5	5.6 ± 1.1	8.8 ± 0.6	7.4 ± 0.7	194
Loblolly pine[3]	3.5	1.6 ± 0.5	1.6 ± 0.5	2.8 ± 0.4	2.9 ± 0.4	60
Sweetgum[4]	1.5	0.7 ± 0.2	0.6 ± 0.1	0.7 ± 0.3	0.9 ± 0.3	20
Aspen[5]	2.5	0.9 ± 0.3	1.3 ± 0.3	1.7 ± 0.8	2.3 ± 0.8	42
Scrub-oak[6]	1.0	2.9 ± 0.4	2.0 ± 0.2	2.9 ± 0.5	2.4 ± 0.6	68
Perennial grassland[7]	5.0	2.0 ± 0.1	2.1 ± 0.1			55
Perennial grassland (+N)	5.0	2.2 ± 0.2	2.3 ± 0.2			61

*Gross NH_4^+ mineralization and consumption were calculated using the ^{15}N isotope dilution technique over 24- to 48-h periods. As a first approximation of seasonal flux rates, values were expressed on a real basis for a 180-day period using the average rates of both fluxes (mineralization and immobilization) for both CO_2 treatments, and assuming a bulk density of $1 g cm^{-3}$. The largest change in plant uptake reported to date is for the tallgrass prairie site, where eight years of CO_2 enrichment caused an increment in plant N content of $9.8 g N m^{-2}$, or $1.2 g N m^{-2} years^{-1}$ (Jastrow et al. 2000 for belowground, and calculated for aboveground from Owensby et al. 1999 and personal communication). After four years of CO_2 enrichment in the loblolly pine plantation, total plant N increased by $1.91 g N m^{-2}$, or $0.4 g N m^{-2} years^{-1}$ (Finzi & Schlesinger 2003). If driven by increases in gross NH_4^+ transformations, such changes could easily go undetected.
[1] Average over depths and times (Williams et al. 2001).
[2] Average over depths and times for the low N treatment ($14 g N m^{-2} years^{-1}$) (Richter et al. 2003).
[3] Finzi & Schlesinger (2003).
[4] Zak et al. (2003).
[5] Average of high and low O_3 treatments (Zak et al. 2003).
[6] Average over times (Hungate et al. 1999).
[7] Average across diversity treatments (West et al. 2006).

unimportant, because very small changes in gross N transformations (empirically undetectable using isotope dilution) would be more than adequate to explain changes in plant N acquisition caused by eCO_2. Scaling gross N cycling rates measured in the field as a first approximation of annual gross fluxes, in cases where eCO_2 has been shown to increase plant N uptake, gross N mineralization need increase (or immobilization decrease) by only 2% or less to explain the differences observed (**Table 2**). Thus, even if only a small portion of gross N turnover is available to plants, still only small changes in gross N mineralization would be needed to explain the differences observed.

Net N mineralization is typically on the same scale as plant N uptake, and thus less challenged by a low signal-to-noise ratio compared to gross N transformation measurements. eCO_2 often has no detectable effect on net N mineralization or on proxies of soil N availability, such as resin bags or resin sticks (Finzi & Schlesinger 2003, Matamala & Drake 1999, D.W. Johnson et al. 2003). In contrast, in the absence of N supplements eCO_2 has been found to reduce net N mineralization under field conditions in a cold perennial grassland (Reich et al. 2001, 2006) and in a warm perennial grassland (Gill et al. 2002). Also in the absence of N supplements, eCO_2 has been found to reduce N availability as estimated by extractable inorganic nitrogen (Hu et al. 2001, Hungate et al. 1999, Matamala & Drake 1999), in some cases despite having no effect on net N mineralization. In total, these findings indicate that under typical ambient soil conditions, eCO_2 has neutral or negative effects on net N mineralization rates.

Do these effects of eCO_2 on net N mineralization and plant available N, in turn, depend on soil N supply? Added N gradually (by years 5–6) reversed the depressing effect of eCO_2 on net N mineralization in the cold perennial grassland (**Figure 4**) and resulted in a substantially greater stimulation of total plant N pools under eCO_2 at high rather than low N, likely contributing to the greater plant biomass response to eCO_2 with added N in that ecosystem (**Figure 3**) that began only after several years (Reich et al. 2006). Similarly, in the Swiss FACE study, eCO_2 consistently reduced total aboveground N pools in ryegrass over 10 years at low N fertilization (14 g N m^{-2} years^{-1}), with no temporal trend, and for the first 5 years at high N fertilization (56 g N m^{-2} years^{-1}), but with a gradual switch from suppression to stimulation at high N over 10 years (Schneider et al. 2004). Thus, these two long-term grassland studies with radically different N cycles (ambient versus 4 g N m^{-2} years^{-1} in BioCON; 14 versus 56 g N m^{-2} years^{-1} in Swiss FACE) have similar results (**Figure 4**). In both studies there was an interactive effect of N supply on the eCO_2 effect on total plant available N pools that had a significant temporal trajectory, with stimulation of net N mineralization and of total plant available N by eCO_2 developing over time in N-rich conditions, first showing up in the fifth year in BioCON and the sixth year in Swiss FACE. It is unfortunate that, to our knowledge, only three grassland studies (**Table 1**) provide long-term direct evidence of the degree to which eCO_2 effects on the N cycle are dependent upon (the collective set of factors that regulate) the overall level of plant available N.

What can we conclude about the possible importance of eCO_2-induced changes in organic matter mineralization? Although a far cry from a mechanistic understanding

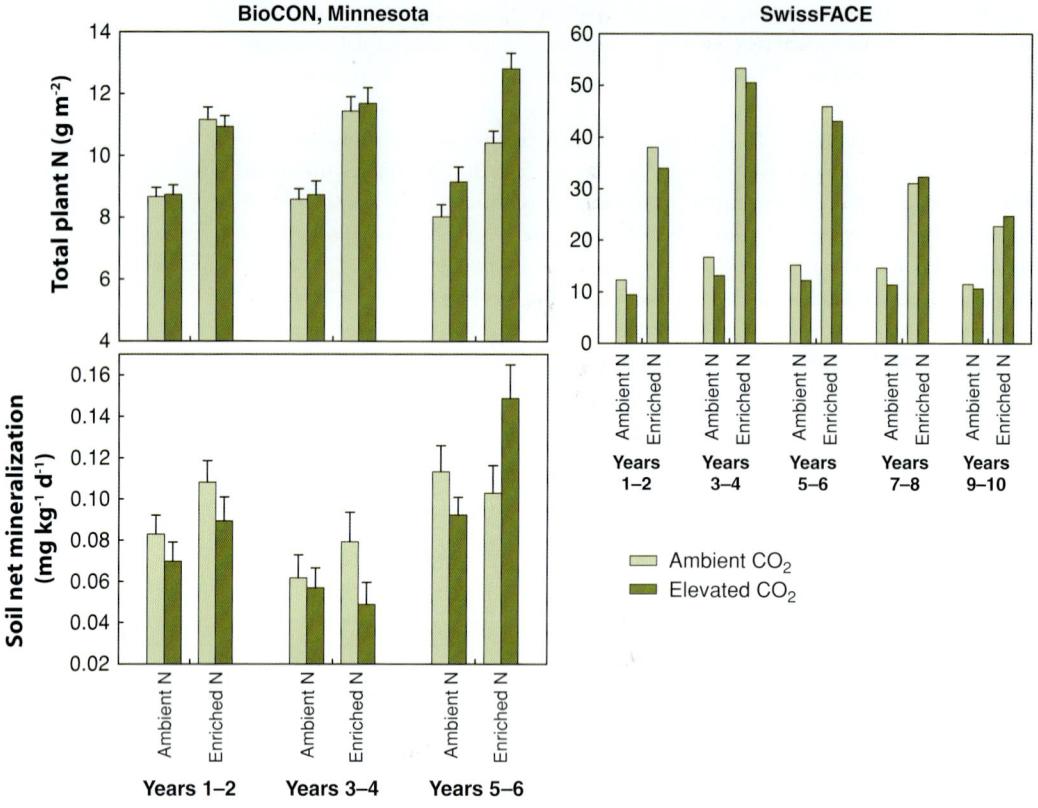

Figure 4

Mid-summer soil net N mineralization rates and plant N pools at elevated and ambient CO_2, at ambient and enriched soil N for a cold temperate grassland in Minnesota (BioCON) and a managed pasture in Switzerland (Swiss FACE). Using annual data in BioCON, there was significant interaction between CO_2 and N, because elevated CO_2 suppressed net N mineralization at both ambient and enriched N supply in the first four years of the study, but in year 5–6 CO_2 stimulated net N mineralization at enriched N supply. A similar interaction over time was noted for plant N pools in both studies. Data from Reich et al. (2006) and Schneider et al. (2004).

of the importance of the priming effect, two conclusions emerge from the finding that eCO_2 increases soil organic C consistently only when N is also added. First, even if N addition augments the priming effect with eCO_2, any loss of soil C caused by priming is insufficient to offset the increased C input to soil caused by eCO_2-enhanced plant production with added N. Thus, the priming effect, if it occurs, does not appear to be large enough to dominate soil C balance response to eCO_2 with added N. Second, a eCO_2-induced enhancement of the priming effect may partly explain why eCO_2 often does not increase soil organic carbon in the absence of N additions, even when plant production increases, for example, in the Swiss FACE experiment (Xie et al. 2005). Quantifying the importance of the priming effect under field conditions, and thereby assessing its importance compared to other, simpler mechanisms (e.g., the simple fact that plant responses to eCO_2 are larger with added nutrients) remains an important challenge to global change research.

2.5. Nitrogen Fixation: Interactions with Other Nutrients

Symbiotic biological N_2 fixation often increases with eCO_2 (Wilson 1933, Soussana & Hartwig 1996). Fixation of atmospheric N_2 requires reduced C, which higher rates of photosynthesis in response to eCO_2 can supply. Bacterial symbionts use this C surplus to fix N_2, providing needed N to the plants (Hartwig 1998).

Is this expected increase in N_2 fixation observed in ecosystems exposed to eCO_2? Sometimes. In experiments where phosphorus (P), potassium (K), and/or other non-N nutrients have been added, N_2 fixation often shows a positive response to eCO_2 (van Groenigen et al. 2006). For example, in a long-term experiment of pasture receiving annual supplements of P, K, and magnesium (Mg), eCO_2 increased N_2 fixation (Zanetti et al. 1996, Lüscher et al. 2000). But in the absence of such nutritional supplements, N_2 fixation is often unresponsive to eCO_2 (van Groenigen et al. 2006; see Section 3.3 below). In the cases where significant increases in N_2 fixation have been observed (e.g., Lee et al. 2003a), they may occur as a short-term response to eCO_2 (e.g., Dakora & Drake 2000, Hungate et al. 1999), a response that can subsequently decline (Hungate et al. 2004, van Groenigen et al. 2006) (**Figure 5**). In the cold temperate grassland experiment in Minnesota, evidence from 1998–2004 was mixed: two of four species showed some evidence of a decline over time in eCO_2 stimulation of N_2 fixation and the other two did not (J.B. West, T.D. Lee, S. Hobbie, and P.B. Reich, unpublished data). Unfortunately, little other evidence is available to further evaluate this question.

Nitrogen fixation often declines as soil N supply increases (e.g., Hartwig 1998, Lee et al. 2003b), because N uptake from soil is less costly than fueling N_2-reduction, and thus N addition should depress the proportion of N derived from N_2 fixation, as shown experimentally by Lee et al. (2003b). N addition did reduce the response of N_2 fixation to exposure to eCO_2 in managed pasture (Zanetti et al. 1996, Lüscher et al. 2000) and in rice crops (Hoque et al. 2001). In each of these experiments, nonnitrogenous fertilizers were applied to both treatments, likely promoting the N_2 fixation response to eCO_2 with low N additions. In a more global analysis of all available observations, there was no effect of N addition on the response of N_2 fixation to eCO_2 (van Groenigen et al. 2006).

Elevated CO_2 can increase or decrease the relative abundance of N_2-fixing plants, suggesting that some of the effects of eCO_2 on N_2 fixation may be manifest as changes in plant communities rather than changes in N_2 fixation on a per plant basis. For example, in the New Zealand FACE pasture experiment, legumes responded positively to eCO_2, constituting an increasing proportion of the total community biomass and productivity (Ross et al. 2004). Two leguminous understory trees increased biomass production with eCO_2 in a loblolly pine plantation (Mohan et al. 2006), but there have been no assessments of whether these species fix N under field conditions. Positive competitive responses of legumes to eCO_2 are by no means universal. Legumes

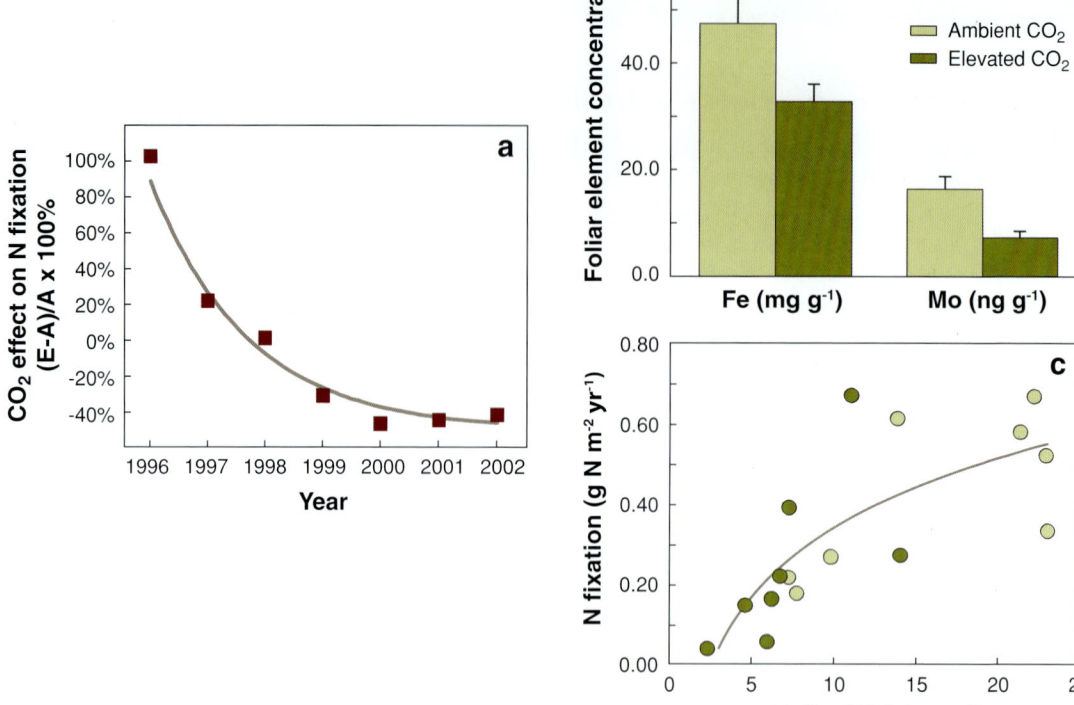

Figure 5

(*a*) The relative effect of elevated CO_2 on N fixation of *G. elliottii*, (*b*) the concentration of Fe and Mo, for ambient (*light green*) and elevated (*dark green*) treatments, and (*c*) correlation between foliar Mo concentration in *G. elliottii* and N fixation rate for plots exposed to ambient or elevated CO_2. From Hungate et al. 2004.

exhibited no significant abundance responses to eCO_2 in a calcareous grassland after six years of exposure to eCO_2 (Niklaus & Körner 2004), nor in an alpine grassland after four years (Arnone 1999). Similarly, over eight years in diverse assemblages in the Minnesota grassland, legumes made up a much larger fraction of plant cover in ambient than enriched N, but a similar fraction at eCO_2 as at ambient CO_2 at both N levels (P.B. Reich, unpublished data; **Figure 6**). In addition, eCO_2 can even depress productivity of N-fixing legumes as observed over time in the Florida scrub-oak ecosystem (Hungate et al. 2006).

What about free-living and associate N fixation in soil? In salt marsh, associative N_2 fixation increased after four months of eCO_2 (Dakora & Drake 2000), but in longer-term field experiments, eCO_2 did not appear to alter N_2 fixation by free-living heterotrophs. During years three to six of experimental treatment, soil N fixers were unresponsive to eCO_2 in a loblolly pine plantation (K. Hofmackel &

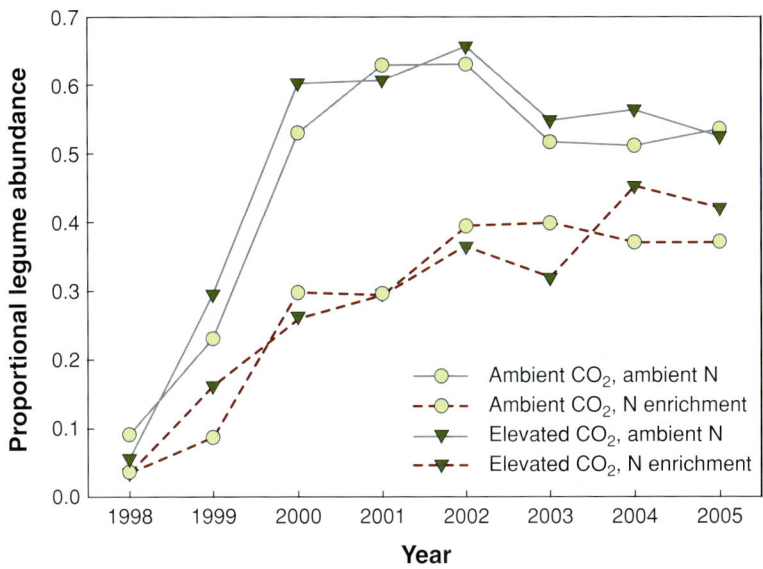

Figure 6

Legume abundance (total for four species) as a proportion of total abundance (based on aboveground biomass) in 16-species assemblages in the BioCON Minnesota FACE study over 8 years of treatment. From P.B. Reich, unpublished data.

W.H. Schlesinger, unpublished data). Bacterial N fixers were also unresponsive to eCO_2 in a desert ecosystem (Billings et al. 2003).

2.6. Carbon-Nitrogen Interactions: Other Trophic Dimensions

Elevated CO_2 usually increases plant growth rates, leaf C:N ratios, and secondary compounds, such as lignin and phenolics, and decreases N-based metabolites (Ainsworth & Long 2005, Matros et al. 2006). The increased C-based compounds and decreased N-based metabolites consequently impact trophic interactions with pathogens, herbivores, and mycorrhizal symbionts (Chakraborty & Datta 2003, N.C. Johnson et al. 2003, Matros et al. 2006, Mitchell et al. 2003).

2.6.1. Plant pathogens.
The shift in the balance of C- and N-based secondary metabolites can enhance resistance to pathogen invasion under eCO_2. In inoculation experiments with *potato virus Y*, for example, the titer of viral coat-protein was markedly reduced in leaves under eCO_2 (Matros et al. 2006). Also, oats infected with the Barley Yellow Dwarf Virus showed a greater biomass response to eCO_2 than did uninfected oat plants (Malmström & Field 1997). eCO_2 also reduced disease incidence and severity of a red maple fungal pathogen due to reduced stomatal opening, reduced leaf N, and increased defensive chemistry (McElrone et al. 2005).

Increased plant resistance to pathogens and leaf C:N ratios under eCO_2 could, in theory, result in increased plant photosynthesis and production (e.g., Strengbom

& Reich 2006) and reduced decomposition. However, some studies have shown the opposite results, that eCO_2 promoted foliar diseases such as rusts, leaf spots, and blights owing to increased canopy size and density, decreased water stress, and increased canopy spore-trapping (Mitchell et al. 2003, Wand et al. 1999). In addition, some fungal pathogens produce more spores on host tissues under eCO_2 because of increased fecundity (Chakraborty & Datta 2003, Hibberd et al. 1996). The effects of eCO_2 and N on foliar fungal disease severity may also depend on the plant photosynthetic pathway (Mitchell et al. 2003). Given the diverse observed responses, it is difficult to draw general conclusions about the effects of eCO_2 on the interaction of plants and pathogens, let alone whether it systematically varies with N.

2.6.2. Herbivory. Plant-herbivore interactions are likely altered under eCO_2 because eCO_2-induced changes in plant chemistry affect the quality of herbivore diets and can alter host plant preferences (Lindroth 1996). Decreased leaf N concentration and increased secondary compounds under eCO_2 reduce palatability and nutritional quality of foliage to some insect pests, especially at early larval stages (Agrell et al. 2000, Lindroth 1996). The detrimental effects may decrease herbivore growth rates, performance, and fecundity and increase insect developmental time and mortality (Lindroth 1996). However, some insect herbivores fed with CO_2-enriched foliage increased their consumption rate to accumulate requisite amounts of N (i.e., compensatory consumption), especially at late larval stages (Agrell et al. 2000, Knepp et al. 2005). Nonetheless, overall herbivory may decrease under eCO_2 owing to decreased abundance of insects (Knepp et al. 2005).

However, N availability can exert considerable effects on these processes. For instance, survival rate and longevity of the silkworm (*Erisan*, a generalist herbivore) were lower when they were fed with birch, oak, and maple leaves under eCO_2 and infertile soil conditions than under more fertile conditions (Koike et al. 2006). Responses to increased N deposition can counteract and mitigate the effects of eCO_2 on insect performance (Hättenschwiler & Schafellner 1999, Kerslake et al. 1998). However, the heterogeneity of changes in the chemistry (types of compounds) as well as the stoichiometry of plant tissues under eCO_2 makes it difficult to generalize.

2.6.3. Soil mutualists. Unlike pathogens and insect herbivores, mycorrhizal fungi [including arbuscular mycorrhizal (AM) and ectomycorrhizal (ECM) fungi] form mutualistic association with most plant roots and link C transfers from plant to soil. Generally, stimulated supply of carbohydrate to roots under eCO_2 promotes the growth and colonization of mycorrhizae, resulting in increased uptake of limiting nutrients (i.e., N and P), which in turn facilitate plant growth and resistance to drought and pathogens (Allen et al. 2005, Olsrud et al. 2004, Treseder & Allen 2000). However, mycorrhizal responses to eCO_2 decline with added N (Treseder & Allen 2000). Additionally, mycorrhizae can be important mediators of plant community responses to eCO_2, which can be further adjusted by soil N availability (N.C. Johnson et al. 2003). Under a mycocentric view, increased fungal biomass could increase competition for nutrients between plants and mycorrhizae, shifting the fungal community toward less nutrient-limited species (Alberton et al. 2005).

3. ADDITIONAL MULTIPLE FACTOR INTERACTIONS

All of the processes and interactions discussed above are potentially sensitive to other environmental factors. Hence, the occurrence or strength of specific C-N interactions under eCO_2 may also depend on environmental factors such as soil pH, air or soil temperature, moisture availability, and/or other resources. In the following section, we briefly discuss a few potentially important multiple factor interactions, but note that empirical evidence about such interactions is generally quite scarce.

3.1. Interactive Effects of Temperature and Elevated Carbon Dioxide on Carbon-Nitrogen Interactions

Climate warming and eCO_2 could interactively alter plant and soil N cycling, which in turn could influence response to eCO_2. Experiments found altered eCO_2 fertilization response due to temperature only at temperatures substantially different than typical or optimal thermal environments (e.g., Dukes et al. 2005, Tjoelker et al. 1998). Such studies are rare, however, especially in the field, and typically can only detect dramatic interactions, given their limited replication (Norby & Luo 2004). Thus, we must use our understanding of interactions to deduce potential effects of temperature on coupled C-N cycling under eCO_2.

Temperature has both direct physiological (Tjoelker et al. 1999) and indirect biogeochemical (e.g., Shaw & Harte 2001, Wan et al. 2005) impacts on tissue-N concentrations. In cold regions, warming often increases leaf N concentrations because of enhanced soil N mineralization (Rustad et al. 2001, Shaw & Harte 2001), which in turn could enhance NPP responses to eCO_2 (e.g., **Figure 3**).

For instance, warming increased soil N availability and thereby enhanced relative responses of photosynthesis to eCO_2 (Kellomäki & Wang 1996) and a 3°C temperature increase alleviated the suppressive eCO_2 effects on N cycling (Loiseau & Soussana 2000). In a warmer region though, the initial stimulation of N mineralization by warming disappeared quickly (Wan et al. 2005), resulting in progressively larger decreases in green leaf N concentrations over time in warmed rather than in control plots (An et al. 2005). Whether interactive effects of eCO_2 and temperature on C and N dynamics are common or consistent requires further study, especially for long-term processes.

3.2. Influence of Water on Carbon-Nitrogen Interactions under Elevated CO_2

Intersite or interannual variation in water supply has been long hypothesized to be a potential regulator of eCO_2 effects, with positive responses expected in drier years or drier microenvironments. The main rationale for this hypothesis is that plants under eCO_2 have lower stomatal conductance (by far the most consistent eCO_2 effect out of dozens studied), which should ameliorate soil water deficits, all else being equal (Morgan et al. 2004).

Most studies do support the notion of modestly increased soil water under eCO_2. However, despite this, based on comparisons of responses in relatively dry versus wet years, there is weak support for the hypothesized greater eCO_2 stimulation of productivity in dry versus wet years. Studies of aboveground biomass responses of grasslands and deserts supported this hypothesis in Kansas, Colorado, and Switzerland, but not in Texas, Minnesota, California, or New Zealand, and results from Nevada were opposite (Dukes et al. 2005, Hammerlynk et al. 2002, Morgan et al. 2004, Niklaus & Körner 2004, Owensby et al. 1999; also P.B. Reich, unpublished data). Unfortunately, it is difficult to draw a specific conclusion from these studies. Years that are relatively wet versus dry can vary in many other ways that could potentially confound or complicate the water availability effects, as can complex water-C-N interactions.

By reducing evapotranspiration, eCO_2 has been found to increase soil water content, and this has been invoked as a possible driver of increased plant N uptake and soil N turnover in some grassland ecosystems (Hungate et al. 1997, Rice et al. 1994). However, such effects will be modest or nonexistent if soil water enhancement under eCO_2 is modest or nonexistent, as can occur because of increased LAI (Ainsworth & Long 2005, Hungate et al. 2006).

We propose an alternative hypothesis. When water shortage is a predominant limitation compared with C or N, little positive response to eCO_2 or added N might occur, because the relative supply of each is high compared to water. According to such a hypothesis, we would expect a three-way interaction among C, N, and water availability: Response to high levels of all three would be greater than additive, but additionally eCO_2 might compensate for mild water limitation but be of little impact when water limitations were severe.

Direct evidence would come from experimental manipulations of water and CO_2, and given widespread water limitations, one might expect reports of many such experiments. However, to our knowledge there is only one such experiment in the world, at Jasper Ridge, California, which involves a largely annual plant community in a strongly Mediterranean system. In the first five years of that study there was no evidence of a water-CO_2 interaction on productivity (Dukes et al. 2005), but there was also, surprisingly, no evidence of either a CO_2 effect or a water limitation on total NPP, and no water-N or water-CO_2-N interactions. Dukes et al. (2005) hypothesized that a number of factors could be responsible for the lack of CO_2 or water effects as well as the lack of interactions, including phosphorus limitation.

In summary, elevated CO_2 does often slightly increase soil water, and higher soil water is usually associated with higher rates of net N mineralization, so moister conditions could ameliorate both the synergistic CO_2-N interactions and PNL. This effect will likely occur only in systems where water and N colimit plant production, because the interaction requires a convergence of N-limited growth and water-limited soil activity. Also, this mechanism is self-limiting, because it will likely decline to the extent eCO_2 enhances growth and leaf area, minimizing changes in soil water content.

3.3. Multiple Element Interactions

The finding that positive responses of N fixation to eCO_2 depend on the availability of other nutrients illustrates the importance of element interactions beyond C and N. For example, the reduction in N_2-fixation in the vine, *Galactia elliottii*, in the scrub-oak ecosystem was accompanied by reduced foliar molybdenum (Mo) and iron (Fe), essential elements for N_2 fixation (**Figure 5**). After nine years of experimental treatment, eCO_2 tended to reduce extractable Mo concentrations in the soil (P = 0.069), implicating Mo limitation of N_2 fixation in this ecosystem (B. Duval & B.A. Hungate, unpublished data). Limitation of N fixation by Mo may not be a general phenomenon in terrestrial ecosystems, but the finding illustrates the importance of element interactions that may lurk lower in the periodic table than terrestrial biogeochemists are accustomed to looking.

Phosphorus limitation of the response of N_2 fixation to eCO_2 may be more common. In controlled-environment and short-term studies, low P availability often restricts the responses of N_2 fixers to eCO_2 (Edwards et al. 2006, Sa & Israel 1998). The small biomass and NPP responses to eCO_2 in the long-term calcareous grassland experiment in Switzerland were attributed to limitation of N_2 fixation by P (Niklaus & Körner 2004). In CO_2-P experiments using calcareous grassland mesocosms, growth of N fixers only responded to eCO_2 when P was also added (e.g., Niklaus & Körner 2004). Across all field experiments examined (van Groenigen et al. 2006), N fixation was far more responsive to eCO_2 in ecosystems receiving inputs of P (often in combination with K and Mg).

It is not known whether and to what extent interactions among the cycles of other elements influence ecosystem responses to eCO_2, in part because few experiments have documented responses or potential drivers. A survey of crop and short-term experiments showed that eCO_2 reduced plant concentrations of other elements, including P, K, Ca, S, Mg, Fe, Zn, Mn, and Cu (Loladze 2002). Consistent with findings from these short-term experiments, eCO_2 significantly reduced concentrations of many of these elements in long-term field studies (Hagedorn et al. 2001, D.W. Johnson et al. 2003; also S. Natali, unpublished data). Fe and Mo are involved in biological N transformations, so changes in their concentrations and availabilities with eCO_2 have implications for N cycling (e.g., Hungate et al. 2003).

Although eCO_2 caused no change in foliar concentrations of Zn in the Florida scrub-oak ecosystem, eCO_2 caused a 30% increase in Zn concentrations in the surface litter layer and a 17% reduction in Zn availability in the underlying mineral soil. Thus, eCO_2 can alter trace-element cycling through mechanisms other than changes in foliar chemistry. One such mechanism involves carbonic acid formation and base cation availability, both of which were found to increase in the loblolly pine experiment after three years of CO_2 enrichment (Andrews & Schlesinger 2001); such changes could exacerbate base cation losses in ecosystems subject to acid deposition.

In cases where eCO_2 does alter foliar element concentrations, reductions are by no means the universal response. Elevated CO_2 increased foliar concentrations of Mn by nearly 40% in the Florida scrub-oak site (D. Johnson et al. 2003), and increased concentrations of P and Zn in spruce-beech community exposed to eCO_2 for four years (Hagedorn et al. 2001) contrary to results from short-term controlled environment studies (Loladze 2002). Significant reductions and increases in element concentrations underscores the potential for global changes like rising CO_2 to alter the stoichiometries of plant-soil systems, which could in some cases shift or exacerbate nutrient limitations. For example, the 30% reduction in the N:P ratio of vegetation in an acidic loam soil for spruce-beech suggests P limitation. Limitation by elements other than N could also contribute to negative or small responses to elevated CO_2 and other global changes in annual grasslands (Dukes et al. 2005, Shaw et al. 2002). These results show that eCO_2 has the potential to alter ecosystem stoichiometries. For the most part, however, the consequences of such changes are not well understood.

SUMMARY POINTS

1. Evidence from long-term field studies suggests that both progressive N limitation under eCO_2 and a significant interaction between CO_2 and plant available N supply that constrains NPP responses to eCO_2 are likely to be common, although not ubiquitous, in many natural and managed ecosystems. The combination of progressive N limitation and an interaction of CO_2 and plant available N supply will likely lead to smaller NPP enhancement under eCO_2 than widely anticipated.

2. Surprisingly, given the importance implications to Earth's C balance and future climate, there will likely be only a very small number (less than five) of long-term experiments that manipulate both CO_2 and N from which to extrapolate to the globe.

LITERATURE CITED

Agrell J, McDonald EP, Lindroth RL. 2000. Effects of CO_2 and light on tree phytochemistry and insect performance. *Oikos* 88:259–72

Ainsworth EA, Long SP. 2005. What have we learned from 15 years of free-air CO_2 enrichment (FACE)? A meta-analytic review of the responses of photosynthesis, canopy properties and plant production to rising CO_2. *New Phytol.* 165:351–72

Alberton O, Kuyper TW, Gorissen A. 2005. Taking mycocentrism seriously: mycorrhizal fungi and plant responses to elevated CO_2. *New Phytol.* 167:859–68

Allen MF, Klironomos JN, Treseder KK, Oechel W. 2005. Responses of soil biota to elevated CO_2 in a chaparral ecosystem. *Ecol. Appl.* 15(5):1701–11

An Y, Wan S, Zhou X, Subeda AA, Wallace LL, Luo Y. 2005. Plant nitrogen concentration, use efficiency, and contents in a tallgrass prairie ecosystem under experimental warming. *Glob. Change Biol.* 11:1733–44

Andrews JA, Schlesinger WH. 2001. Soil CO_2 dynamics, acidification, and chemical weathering in a temperate forest with experimental CO_2 enrichment. *Glob. Biogeochem. Cycles* 15:149–62

Arnone JA. 1999. Symbiotic N-2 fixation in a high Alpine grassland: effects of four growing seasons of elevated CO_2. *Funct. Ecol.* 13:383–87

Billings SA, Zitzer SF, Weatherly H, Schaeffer SM, Charlet T, et al. 2003. Effects of elevated carbon dioxide on green leaf tissue and leaf litter quality in an intact Mojave Desert ecosystem. *Global Change Biol.* 9:729–35

Chakraborty S, Datta S. 2003. How will plant pathogens adapt to host plant resistance at elevated CO_2 under a changing climate? *New Phytol.* 159:733–42

Cheng WX. 1999. Rhizosphere feedbacks in elevated CO_2. *Tree Physiol.* 194:313–20

Curtis PS, Wang X. 1998. A meta-analysis of elevated CO_2 effects on woody plant mass, form, and physiology. *Oecologia* 113:299–313

Dakora FD, Drake BG. 2000. Elevated CO_2 stimulates associative N_2 fixation in a C_3 plant of the Chesapeake Bay wetland. *Plant Cell Environ.* 23:943–53

Dijkstra FA, Hobbie SE, Reich P. 2006. Soil processes affected by sixteen grassland species grown under different environmental conditions. *Soil Sci. Soc. Am. J.* 70:770–77

Dijkstra FA, Hobbie SE, Reich P, Knops J. 2005. Divergent effects of elevated CO_2, N fertilization, and plant diversity on soil C and N dynamics in a grassland field experiment. *Plant Soil* 272:41–52

Dukes JS, Chiariello NR, Cleland EE, Moore LA, Shaw MR, et al. 2005. Responses of grassland production to single and multiple global environmental changes. *PLoS Biol.* 3:1829–37

Edwards EJ, McCaffery S, Evans JR. 2006. Phosphorus availability and elevated CO_2 affect biological nitrogen fixation and nutrient fluxes in a clover-dominated sward. *New Phytol.* 169:157–67

Ellsworth DS, Reich PB, Naumburg ES, Koch GW, Kubiske M, Smith S. 2004. Photosynthesis, carboxylation and leaf nitrogen responses of 16 species to elevated $pCO2$ across four free-air CO_2 enrichment experiments in forest, grassland and desert. *Glob. Change Biol.* 10:2121–38

Finzi AC, Schlesinger WH. 2003. Soil-nitrogen cycling in a pine forest exposed to 5 years of elevated carbon dioxide. *Ecosystems* 6:444–56

Gill RA, Polley HW, Johnson HB, Anderson LJ, Maherali H, Jackson RB. 2002. Nonlinear grassland responses to past and future atmospheric CO_2. *Nature* 417:279–82

Hagedorn F, Maurer S, Egli P, Blaser P, Bucher JB, Siegwolf R. 2001. Carbon sequestration in forest soils: effect of soil type, atmospheric CO_2 enrichment, and N deposition. *Eur. J. Soil Sci.* 52:619–28

Haile-Mariam S, Cheng W, Johnson DW, Ball JT, Paul ES. 2000. Use of carbon-13 and carbon-14 to measure the effects of carbon dioxide and nitrogen fertilization on carbon dynamics in ponderosa pine. *Soil Sci. Soc. Am. J.* 64:1984–93

Hammerlynk EP, Huxman TE, Charlet TN, Smith SD. 2002. Effects of elevated CO_2 (FACE) on the functional ecology of the drought-deciduous Mojave Desert shrub, *Lycium andersonii*. *Environ. Exp. Bot.* 48:93–106

Hartwig UA. 1998. The regulation of symbiotic N_2 fixation: a conceptual model of N feedback from the ecosystem to the gene expression level. *Persp. Plant Ecol. Evol. Syst.* 1:92–120

Hättenschwiler S, Schafellner C. 1999. Opposing effects of elevated CO_2 and N deposition on Lymantria monacha larvae feeding on spruce trees. *Oecologia* 118(2):210–17

Henry HAL, Cleland EE, Field CB, Vitousek PM. 2005. Interactive effects of elevated CO_2, N deposition and climate change on plant litter quality in a California annual grassland. *Oecologia* 142:465–73

Hibberd JM, Whitbread R, Farrar JF. 1996. Effect of elevated concentrations of CO_2 on infection of barley by *Erysiphe graminis*. *Physiol. Mol. Plant Pathol.* 48:37–53

Hoque MM, Inubushi K, Miura S, Kobayashi K, Kim HY, et al. 2001. Biological dinitrogen fixation and soil microbial biomass carbon as influenced by free-air carbon dioxide enrichment (FACE) at three levels of nitrogen fertilization in a paddy field. *Biol. Fertil. Soils* 34:453–59

Hu S, Chapin FS, Firestone MK, Field CB, Chiariello NR. 2001. Nitrogen limitation of microbial decomposition in a grassland under elevated CO_2. *Nature* 409:188–91

Hungate BA, Dijkstra P, Johnson DW, Hinkle CR, Drake BG. 1999. Elevated CO_2 increases nitrogen fixation and decreases soil nitrogen mineralization in Florida scrub oak. *Glob. Change Biol.* 5:781–90

Hungate BA, Dukes JS, Shaw MR, Luo Y, Field CB. 2003. Nitrogen and climate change. *Science* 302:1512–13

Hungate BA, Stiling P, Dijkstra P, Johnson DW, Ketterer M, et al. 2004. CO_2 elicits long-term decline in nitrogen fixation. *Science* 304:1291

Hungate BA, Holland EA, Jackson RB, Chapin III FS, Mooney HA, Field CB. 1997. The fate of carbon in grasslands under carbon dioxide enrichment. *Nature* 388:576–79

Hungate BA, Johnson DW, Dijkstra P, Hymus GJ, Stiling P, et al. 2006. Nitrogen cycling during seven years of atmospheric CO_2 enrichment in a scrub oak woodland. *Ecology* 87:26–40

Jastrow JD, Miller RM, Owensby CE. 2000. Long-term effects of elevated atmospheric CO_2 on below-ground biomass and transformations to soil organic matter in grassland. *Plant Soil* 224:85–97

Johnson DW, Hungate BA, Dijkstra P, Hymus G, Hinkle CR, et al. 2003. The effects of elevated CO_2 on nutrient distribution in a fire-adapted scrub oak forest. *Ecol. Appl.* 13:1388–99

Johnson NC, Wolf J, Koch GW. 2003. Interactions among mycorrhizae, atmospheric CO_2 and soil N impact plant community composition. *Ecol. Lett.* 6:532–40

Johnson NC, Graham JH, Smith FA. 1997. Functioning of mycorrhizal associations along the mutualism-parasitism continuum. *New Phytol.* 135:575–85

Kellomäki S, Wang KY. 1996. Photosynthetic responses to needle water potentials in Scots pine after a four-year exposure to elevated CO_2 and temperature. *Tree Phys.* 16:765–72

Kerslake JE, Woodin SJ, Hartley SE. 1998. Effects of carbon dioxide and nitrogen enrichment on a plant-insect interaction: the quality of *Calluna vulgaris* as a host for *Operophtera brumata*. *New Phytol.* 140(1):43–53

Kim HY, Lieffering M, Kobayashi K, Okada M, Miura S. 2003. Seasonal changes in the effects of elevated CO_2 on rice at three levels of nitrogen supply: a free air CO_2 enrichment (FACE) experiment. *Glob. Change Biol.* 9:826–37

Knepp RG, Hamilton JG, Mohan JE, Zangerl AR, Berenbaum MR, DeLucia EH. 2005. Elevated CO_2 reduces leaf damage by insect herbivores in a forest community. *New Phytol.* 167:207–18

Koike T, Tobita H, Shibata T, Matsuki S, Konno K, et al. 2006. Defense characteristics of seral deciduous broad-leaved tree seedlings grown under differing levels of CO_2 and nitrogen. *Pop. Ecol.* 48(1):23–29

Kuzyakov Y. 2002. Review: Factors affecting rhizosphere priming effects. *J. Plant Nutr. Soil Sci.* 165:382–96

Lee TD, Reich PB, Tjoelker MG. 2003a. Legume presence increases photosynthesis and N concentrations of co-occurring nonfixers but does not modulate their responsiveness to carbon dioxide enrichment. *Oecologia* 137:22–31

Lee TD, Tjoelker MG, Russelle MP, Reich PB. 2003b. Contrasting response of an N-fixing and non N-fixing forb to elevated CO_2: dependence on soil N supply. *Plant Soil* 255:475–86

Liberloo M, Dillen SY, Calfapietra C, Marinari S, Luo ZB, et al. 2005. Elevated CO_2 concentration, fertilization and their interaction: growth stimulation in a short-rotation poplar coppice (EUROFACE). *Tree Phys.* 25:179–89

Lindroth RL. 1996. CO_2-mediated changes in tree chemistry and tree-Lepidoptera interactions. In *Carbon Dioxide and Terrestrial Ecosystem*, ed. GW Koch, HA Mooney, pp. 105–20. San Diego, CA: Academic

Loiseau P, Soussana JF. 2000. Effects of elevated CO_2, temperature and N fertilization on nitrogen fluxes in a temperate grassland ecosystem. *Glob. Change Biol.* 6:953–65

Loladze I. 2002. Rising atmospheric CO_2 and human nutrition: toward globally imbalanced plant stoichiometry? *Trends Ecol. Evol.* 17:457–61

Luo Y, Hui H, Zhang D. 2006. Elevated CO_2 stimulates net accumulations of carbon and nitrogen in land ecosystems: a meta-analysis. *Ecology* 87:53–63

Luo Y, Su B, Currie WS, Dukes JS, Finzi A, et al. 2004. Progressive nitrogen limitation of ecosystem responses to rising atmospheric carbon dioxide. *BioScience* 54:731–39

Lüscher A, Hartwig UA, Suter D, Nosberger J. 2000. Direct evidence that symbiotic N-2 fixation in fertile grassland is an important trait for a strong response of plants to elevated atmospheric CO_2. *Glob. Change Biol.* 6:655–62

Malmström CM, Field CB. 1997. Virus-induced differences in the response of oat plants to elevated carbon dioxide. *Plant Cell Environ.* 20:178–88

Matamala R, Drake BG. 1999. The influence of atmospheric CO_2 enrichment on plant-soil nitrogen interactions in a wetland plant community on the Chesapeake Bay. *Plant Soil* 210:93–101

Matros A, Amme S, Kettig B, Buck-Sorlin GH, Sonnewald U, Mock HP. 2006. Growth at elevated CO_2 concentrations leads to modified profiles of secondary metabolites in tobacco cv. SamsunNN and to increased resistance against infection with *potato virus Y*. *Plant Cell Environ.* 29:126–37

McElrone AJ, Reid CD, Hoye KA, Hart E, Jackson RB. 2005. Elevated CO_2 reduces disease incidence and severity of a red maple fungal pathogen via changes in host physiology and leaf chemistry. *Glob. Change Biol.* 11:1828–36

Mitchell CE, Reich PB, Tilman D, Groth JV. 2003. Effects of elevated CO_2, nitrogen deposition, and decreased species diversity on foliar fungal plant disease. *Glob. Change Biol.* 9:438–51

Mohan JE, Clark JS, Schlesinger WH. 2006. Long-term CO_2 enrichment of a forest ecosystem: implications for forest regeneration and succession. *Ecol. Appl.* In press

Morgan JA, Pataki DE, Körner C, Clark H, Del Grosso SJ, et al. 2004. Water relations in grassland and desert ecosystems exposed to elevated atmospheric CO_2. *Oecologia* 140:11–25

Niklaus P, Körner C. 2004. Synthesis of a six-year study of calcareous grassland responses to in situ CO_2 enrichment. *Ecol. Monogr.* 74:491–511

Norby RJ, Cotrufo MF, Ineson P, O Neill EG, Canadell JG. 2001. Elevated CO_2, litter chemistry, and decomposition: a synthesis. *Oecologia* 127:153–65

Norby RJ, Luo Y. 2004. Evaluating ecosystem responses to rising atmospheric CO_2 and global warming in a multi-factor world. *New Phytol.* 162:281–93

Nowak RS, Zitzer SF, Babcock D, Smith-Longozo V, Charlet TN, et al. 2004. Elevated atmospheric CO_2 does not conserve soil water in the Mojave Desert. *Ecology* 85(1):93–99

Okada M, Lieffering M, Nakamura H, Yoshimoto M, Kim HY, Kobayashi K. 2001. Free-air CO_2 enrichment (FACE) using pure CO_2 injection: system description. *New Phytol.* 150:251–60

Olsrud M, Melillo JM, Christensen TR, Michelsen A, Wallander H. 2004. Response of ericoid mycorrhizal colonization and functioning to global change factors. *New Phytol.* 162:459–69

Oren R, Ellsworth DS, Johnsen KH, Phillips N, Ewers BE, et al. 2001. Soil fertility limits carbon sequestration by forest ecosystems in a CO_2 enriched world. *Nature* 411:469–72

Owensby CE, Ham JM, Knapp AK, Auen LM. 1999. Biomass production and species composition change in a tall grass prairie ecosystem after long-term exposure to elevated atmospheric CO_2. *Glob. Change Biol.* 5:497–506

Reich PB, Hobbie SE, Lee T, Ellsworth D, West JB, et al. 2006. Nitrogen limitation constrains sustainability of ecosystem response to CO_2. *Nature* 440:922–25

Reich PB, Knops J, Tilman D, Craine J, Ellsworth D, et al. 2001. Plant diversity enhances ecosystem responses to elevated CO_2 and nitrogen deposition. *Nature* 410:809–12

Reich PB, Tilman D, Naeem S, Ellsworth D, Knops J, et al. 2004. Species and functional group diversity independently influence biomass accumulation and its response to CO_2 and N. *Proc. Natl. Acad. Sci. USA* 101:10101–6

Rice CW, Garcia FO, Hampton CO, Owensby CE. 1994. Soil microbial response in tallgrass prairie to elevated CO_2. *Plant Soil* 165:67–74

Richter M, Cadisch G, Fear J, Frossard E, Nösberger J, Hartwig UA. 2003. Gross nitrogen fluxes in grassland soil exposed to elevated atmospheric pCO_2 for seven years. *Soil Biol. Biochem.* 35:1325–35

Ross DJ, Newton PCD, Tate KR. 2004. Elevated CO_2 effects on herbage production and soil carbon and nitrogen pools and mineralization in a species-rich, grazed pasture on a seasonally dry sand. *Plant Soil* 260:183–96

Rustad LE, Campbell JL, Marion GM, Norby RJ, Mitchell MJ, et al. 2001. A meta-analysis of the response of soil respiration, net nitrogen mineralization, and aboveground plant growth to experimental ecosystem warming. *Oecologia* 126:543–62

Sa TN, Israel DW. 1998. Phosphorus-deficiency effects on response of symbiotic N-2 fixation and carbohydrate status in soybean to atmospheric CO_2 enrichment. *J. Plant Nutr.* 21:2207–18

Schäppi B, Körner C. 1996. Growth responses of an alpine grassland to elevated CO_2. *Oecologia* 105:43–52

Schneider MK, Luscher A, Richter M, Aeschlimann U, Hartwig UA, et al. 2004. Ten years of free-air CO_2 enrichment altered the mobilization of N from soil in *Lolium perenne* L. swards. *Glob. Change Biol.* 10:1377–88

Shaw MR, Harte J. 2001. Response of nitrogen cycling to simulated climate change: differential responses along a subalpine ecotone. *Glob. Change Biol.* 7:193–210

Shaw MR, Zavaleta ES, Chiariello NR, Cleland EE, Mooney HA, Field CB. 2002. Grassland responses to global environmental changes suppressed by elevated CO_2. *Science* 298:1987–90

Sigurdsson BD, Thorgeirsson H, Linder S. 2001. Growth and dry-matter partitioning of young *Populus trichocarpa* in response to carbon dioxide concentration and mineral nutrient availability. *Tree Phys.* 21:941–50

Soussana JF, Hartwig UA. 1996. The effects of elevated CO_2 on symbiotic N2 fixation: a link between the carbon and nitrogen cycles in grassland ecosystems. *Plant Soil* 187:321–32

Spinnler D, Egli P, Körner C. 2002. Four-year growth dynamics of beech-spruce model ecosystems under CO_2 enrichment on two different forest soils. *Trees* 16:423–36

Strain BR, Bazzaz FA. 1983. Terrestrial plant communities. In *CO_2 and Plants*, ed. ER Lemon, pp. 177–222. Boulder, CO, USA: Westview

Strengbom J, Reich PB. 2006. Elevated CO_2 and increased N supply reduce leaf disease and related photosynthetic impacts on *Solidago rigida*. *Oecologia*. In press

Tjoelker MG, Oleksyn J, Reich PB. 1998. Temperature and ontogeny mediate growth response to elevated CO_2 in seedlings of five boreal tree species. *New Phytol.* 140:197–210

Tjoelker MG, Reich PB, Oleksyn J. 1999. Changes in leaf nitrogen and carbohydrates underlie temperature and CO_2 acclimation of dark respiration in five boreal tree species. *Plant Cell Environ.* 22:767–78

Treseder KK, Allen MF. 2000. Mycorrhizal fungi have a potential role in soil carbon storage under elevated CO_2 and nitrogen deposition. *New Phytol.* 147:189–200

van Groenigen KJ, Six J, Hungate BA, de Graaf M, van Breeman N, van Kessel C. 2006. Element interactions limit soil carbon storage. *Proc. Natl. Acad. Sci. USA* 103:6571–74

Wan S, Hui DF, Wallace LL, Luo Y. 2005. Direct and indirect effects of experimental warming on ecosystem carbon processes in a tallgrass prairie. *Glob. Biogeochem. Cycles* 19:GB2014, doi:10.1029/2004GB002315

Wand SE, Midgley GF, Jones MH, Curtis PS. 1999. Responses of wild C_4 and C_3 grasses (Poaceae) species to elevated atmospheric CO_2 concentration: a meta-analytic test of current theories and perceptions. *Glob. Change Biol.* 5:723–41

West JB, Hobbie SE, Reich PB. 2006. Effects of plant species diversity, atmospheric [CO_2], and N addition on gross rates of inorganic N release from soil organic matter. *Glob. Change Biol.* 12:1400–8

Williams MA, Rice CW, Owensby CE. 2001. Nitrogen competition in a tallgrass prairie ecosystem exposed to elevated carbon dioxide. *Soil Sci. Soc. Am. J.* 65:340–46

Wilson PW, Fred EB, Salmon MR. 1933. Relation between carbon dioxide and elemental nitrogen assimilation in leguminous plants. *Soil Sci.* 35:145–65

Xie ZB, Cadisch G, Edwards G, Baggs EM, Blum H. 2005. Carbon dynamics in a temperate grassland soil after 9 years exposure to elevated CO_2 (Swiss FACE). *Soil Biol. Biochem.* 37:1387–95

Zak DR, Holmes WE, Finzi AC, Norby RJ, Schlesinger WH. 2003. Soil nitrogen cycling under elevated CO_2: a synthesis of forest face experiments. *Ecol. Appl.* 13:1508–14

Zak DR, Pregitzer KS, King JS, Holmes WE. 2000. Elevated atmospheric CO_2, fine roots and the response to soil microorganisms: a review and hypothesis. *New Phytol.* 147:201–22

Zanetti S, Hartwig UA, Luscher A, Hebeisen T, Frehner M, et al. 1996. Stimulation of symbiotic N-2 fixation in *Trifolium repens* L. under elevated atmospheric pCO_2 in a grassland ecosystem. *Plant Phys.* 112:575–83

Ecological and Evolutionary Responses to Recent Climate Change

Camille Parmesan

Section of Integrative Biology, University of Texas, Austin, Texas 78712;
email: parmesan@mail.utexas.edu

Key Words

aquatic, global warming, phenology, range shift, terrestrial, trophic asynchrony

Abstract

Ecological changes in the phenology and distribution of plants and animals are occurring in all well-studied marine, freshwater, and terrestrial groups. These observed changes are heavily biased in the directions predicted from global warming and have been linked to local or regional climate change through correlations between climate and biological variation, field and laboratory experiments, and physiological research. Range-restricted species, particularly polar and mountaintop species, show severe range contractions and have been the first groups in which entire species have gone extinct due to recent climate change. Tropical coral reefs and amphibians have been most negatively affected. Predator-prey and plant-insect interactions have been disrupted when interacting species have responded differently to warming. Evolutionary adaptations to warmer conditions have occurred in the interiors of species' ranges, and resource use and dispersal have evolved rapidly at expanding range margins. Observed genetic shifts modulate local effects of climate change, but there is little evidence that they will mitigate negative effects at the species level.

INTRODUCTION

Historical Perspective

Climate change is not a new topic in biology. The study of biological impacts of climate change has a rich history in the scientific literature, since long before there were political ramifications. Grinnell (1917) first elucidated the role of climatic thresholds in constraining the geographic boundaries of many species, followed by major works by Andrewartha & Birch (1954) and MacArthur (1972). Observations of range shifts in parallel with climate change have been particularly rich in northern European countries, where observational records for many birds, butterflies, herbs, and trees date back to the mid-1700s. Since the early part of the twentieth century, researchers have documented the sensitivity of insects to spring and summer temperatures (Bale et al. 2002, Dennis 1993, Uvarov 1931). Ford (1945) described northward range shifts of several butterflies in England, attributing these shifts to a summer warming trend that began around 1915 in Britain. Ford noted that one of these species, *Limenitis camilla*, expanded to occupy an area where attempted introductions prior to the warming had failed. Kaisila (1962) independently documented range shifts of Lepidoptera (primarily moths) in Finland, using historical data on range boundaries dating back to 1760. He showed repeated instances of southward contractions during decades of "harsh" climatic conditions (cold wet summers), followed by northward range expansion during decades with climate "amelioration" (warm summers and lack of extreme cold in winter). Further corroboration came from the strong correlations between summer temperatures and the northern range boundaries for many butterflies (Dennis 1993).

Similar databases exist for northern European birds. A burst of papers documented changed abundances and northerly range shifts of birds in Iceland, Finland, and Britain associated with the 1930s–1940s warming period (Gudmundsson 1951; Harris 1964; Kalela 1949, 1952; Salomonsen 1948). A second wave of papers in the 1970s described the subsequent retreats of many of these temperate bird and butterfly species following the cool, wet period of the 1950s–1960s (Burton 1975, Heath 1974, Severnty 1977, Williamson 1975).

Complementing this rich observational database is more than 100 years of basic research on the processes by which climate and extreme weather events affect plants and animals. As early as the 1890s, Bumpus (1899) noted the differential effects of an extreme winter storm on the introduced house sparrow (*Parus domesticus*), resulting in stabilizing selection for intermediate body size in females and directional selection for large body size in males (Johnston et al. 1972). The first extensive studies of climate variability as a powerful driver of population evolution date back to the 1940s, when Dobzhansky (1943, 1947) discovered repeated cycles of seasonal evolution of temperature-associated chromosomal inversions within *Drosophila pseudoobscura* populations in response to temperature changes from spring through summer.

In summary, the history of biological research is rich in both mechanistic and observational studies of the impacts of extreme weather and climate change on wild species: Research encompasses impacts of single extreme weather events; experimental studies of physiological tolerances; snapshot correlations between climatic variables and species' distributions; and correlations through time between climatic trends and changes in distributions, phenologies, genetics, and behaviors of wild plants and animals.

Anthropogenic Climate Change

In spite of this wealth of literature on the fundamental importance of climate to wild biota, biologists have been reluctant to believe that modern (greenhouse gas-driven) climate change is a cause of concern for biodiversity. In his introduction to the 1992 *Annual Review of Ecology, Evolution, and Systematics* volume on "Global Environmental Change," Vitousek wrote, "ultimately, climate change probably has the greatest potential to alter the functioning of the Earth system.... nevertheless, the major effects of climate change are mostly in the future while most of the others are already with us." Individual authors in that volume tended to agree—papers were predominantly concerned with other global change factors: land use change, nitrogen fertilization, and the direct effects of increased atmospheric CO_2 on plant ecophysiology.

Just 14 years later, the direct impacts of anthropogenic climate change have been documented on every continent, in every ocean, and in most major taxonomic groups (reviewed in Badeck et al. 2004; Hoegh-Guldberg 1999, 2005b; Hughes 2000; IPCC 2001a; Parmesan 2005b; Parmesan & Galbraith 2004; Parmesan & Yohe 2003; Peñuelas & Filella 2001; Pounds et al. 2005; Root & Hughes 2005; Root et al. 2003; Sparks & Menzel 2002; Thomas 2005; Walther et al. 2002, 2005). The issue of whether observed biological changes can be conclusively linked to anthropogenic climate change has been analyzed and discussed at length in a plethora of syntheses, including those listed above. Similarly, complexity surrounding methodological issues of detection (correctly detecting a real trend) and attribution (assigning causation) has been explored in depth (Ahmad et al. 2001; Dose & Menzel 2004; Parmesan 2002, 2005a,b; Parmesan & Yohe 2003; Parmesan et al. 2000; Root et al. 2003, Root & Hughes 2005, Schwartz 1998, 1999; Shoo et al. 2006). The consensus is that, with proper attention to sampling and other statistical issues and through the use of scientific inference, studies of observed biological changes can provide rigorous tests of climate-change hypotheses. In particular, independent syntheses of studies worldwide have provided a clear, globally coherent conclusion: Twentieth-century anthropogenic global warming has already affected Earth's biota.

Detection: ability to discern long-term trends above yearly variability and real changes from apparent changes brought about by changes in sampling methodology and/or sampling intensity

Attribution: teasing out climate change as the causal driver of an observed biological change amid a backdrop of potential confounding factors

Globally coherent: a common term in economics, a process or event is globally coherent when it has similar effect across multiple systems spread across different locations throughout the world

Scope of This Review

This review concentrates on studies of particularly long time series and/or particularly good mechanistic understanding of causes of observed changes. It deals exclusively with observed responses of wild biological species and systems to recent, anthropogenic climate change. In particular, agricultural impacts, human health, and ecosystem-level responses (e.g., carbon cycling) are not discussed. Because they have been extensively dealt with in previous publications, this review does not repeat discussions of detection and attribution, nor of the conservation implications of climate change. Rather, some of the best-understood cases are presented to illustrate the complex ways in which various facets of climatic change impact wild biota. The choice of studies for illustration attempts to draw attention to the taxonomic and geographic breadth of climate-change impacts and to the most-recent literature not already represented in prior reviews.

Researchers have frequently associated biological processes with indices of ocean-atmosphere dynamics, such as the El Niño Southern Oscillation and the North Atlantic Oscillation (Blenckner & Hillebrand 2002, Holmgren et al. 2001, Ottersen et al. 2001). However, the nature of the relationship between atmospheric dynamics, ocean circulation, and temperature is changing (Alley et al. 2003, IPCC 2001b, Karl & Trenberth 2003, Meehl et al. 2000). Therefore, there is large uncertainty as to how past relationships between biological systems and ocean indices reflect responses to ongoing anthropogenic climate change. Although I use individual examples where appropriate, this complex topic is not fully reviewed here.

OVERVIEW OF IMPACTS LITERATURE

An extensive, but not exhaustive, literature search revealed 866 peer-reviewed papers that documented changes through time in species or systems that could, in whole or in part, be attributed to climate change. Some interesting broad patterns are revealed. Notably, the publication rate of climate-change responses increases sharply each year. The number of publications between 1899 and January 2003 (the date of two major syntheses) was 528. Therefore, approximately 40% of the 866 papers compiled for this review were published in the past three years (January 2003 to January 2006).

The studies are spread broadly across taxonomic groups. Whereas distributional studies concentrated on animals rather than plants, the reverse is true of phenological time series. This may simply be because historical data on species range boundaries have higher resolution for animals than for plants. Conversely, local records of spring events are much more numerous for plants (e.g., flowering and leaf out) than for animals (e.g., nesting).

Although there is still a terrestrial bias, studies in marine and freshwater environments are increasing in proportional representation. The largest gaps are geographic rather than taxonomic. In absolute numbers, most biological impact studies are from North America, northern Europe and Russia. Few biological studies have come from South America, and there are large holes in Africa and Asia, with most of the studies from these two continents coming from just two countries: South Africa and Japan. In past decades, Australia's impact studies have stemmed predominantly from the coral reef community, but in recent years scientists have dug deep to find historical data, and terrestrial impact studies are now emerging. Similarly, the Mediterranean/North African region (Spain, France, Italy, and Israel) has recently spawned a spate of studies. Antarctica stands out as a region where impacts (or lack of impacts) on most species and systems have been documented, even though data often have large geographic or temporal gaps.

Few studies have been conducted at a scale that encompasses an entire species' range (i.e., a continental scale), with only a moderate number at the regional scale (e.g., the United Kingdom or Germany). Most have been conducted at local scales, typically at a research station or preserve. Continental-scale studies usually cover most or all of a species' range in terrestrial systems (Both et al. 2004, Burton 1998a,b, Dunn & Winkler 1999, Menzel & Fabian 1999, Parmesan 1996, Parmesan et al. 1999). However, even a continental scale cannot encompass the entire ranges of many oceanic species (Ainley & Divoky 1998, Ainley et al. 2003, Beaugrand et al. 2002, Croxall et al. 2002, Hoegh-Gulberg 1999, McGowan et al. 1998, Reid et al. 1998, Spear & Ainley 1999). Terrestrial endemics, in contrast, can have such small ranges that regional, or even local, studies may represent impacts on entire species (Pounds et al. 1999, 2006).

Meta-analysis: set of statistical techniques designed to synthesize quantitative results from similar and independent experiments

Meta-Analyses and Syntheses: Globally Coherent Signals of Climate-Change Impacts

A handful of studies have conducted statistical meta-analyses of species' responses or have synthesized independent studies to reveal emergent patterns. The clear conclusion across global syntheses is that twentieth-century anthropogenic global warming has already affected the Earth's biota (IPCC 2001a; Parmesan 2005a,b; Parmesan & Galbraith 2004; Parmesan & Yohe 2003; Peñuelas & Filella 2001; Pounds et al. 2005; Root & Hughes 2005; Root et al. 2003; Thomas 2005; Walther et al. 2002, 2005).

One study estimated that more than half (59%) of 1598 species exhibited measurable changes in their phenologies and/or distributions over the past 20 to 140 years (Parmesan & Yohe 2003). Analyses restricted to species that exhibited change documented that these changes were not random: They were systematically and predominantly in the direction expected from regional changes in the climate (Parmesan & Yohe 2003, Root et al. 2003). Responding species are spread across diverse ecosystems (from temperate grasslands to marine intertidal zones and tropical cloud forests) and come from a wide variety of taxonomic and functional groups, including birds, butterflies, alpine flowers, and coral reefs.

A meta-analysis of range boundary changes in the Northern Hemisphere estimated that northern and upper elevational boundaries had moved, on average, 6.1 km per decade northward or 6.1 m per decade upward ($P < 0.02$) (Parmesan & Yohe 2003). Quantitative analyses of phenological responses gave estimates of advancement of 2.3 days per decade across all species (Parmesan & Yohe 2003) and 5.1 days per decade for the subset of species showing substantive change (>1 day per decade) (Root et al. 2003).

A surprising result is the high proportion of species responding to recent, relatively mild climate change (global average warming of 0.6°C). The proportion of wild species impacted by climate change was estimated at 41% of all species (655 of 1598) (Parmesan & Yohe 2003). This estimate was derived by focusing on multispecies studies that reported stable as well as responding species. Because responders and stable species were often sympatric, variation of response is not merely a consequence of differential magnitudes of climate change experienced.

PHENOLOGICAL CHANGES

By far, most observations of climate-change responses have involved alterations of species' phenologies. This is partly a result of the tight links between the seasons and agriculture: Planting and harvest dates (and associated climatic events such as day of last frost) have been well recorded, dating back hundreds of years for some crops. But the plethora of records also stems from the strong sociological significance of the change of the seasons, particularly in high-latitude countries. Peoples of Great Britain, the Netherlands, Sweden, and Finland have been keen on (some might say even obsessed with) recording the first signs of spring—the first leaf on an oak, the first peacock butterfly seen flying, the first crocus in bloom—as a mark that the long, dark winter is finally over. Fall has not captured as much enthusiasm as spring, but some good records exist, for example, for the turning of leaf color for trees.

The longest records of direct phenological observations are for flowering of cherry trees *Prunus jamasakura* and for grape harvests. Menzel & Dose (2005) show that timing of cherry blossom in Japan was highly variable among years, but no clear trends were discerned from 1400 to 1900. A statistically significant change point is first seen in the early 1900s, with steady advancement since 1952. Recent advancement exceeds observed variation of the previous 600 years. Menzel (2005) analyzed grape-harvest dates across Europe, for which April-August temperatures explain 84% of the variation. She found that the 2003 European heat wave stands out as an extreme early harvest (i.e., the warmest summer) going back 500 years. Although such lengthy observational records are extremely rare, these two unrelated plants on opposite sides of the world add an important historical perspective to results from shorter time series.

Several lines of evidence indicate a lengthening of vegetative growing season in the Northern Hemisphere, particularly at higher latitudes where temperature rise has been greatest. Summer photosynthetic activity (normalized difference vegetation index estimates from satellite data) increased from 1981–1991 (Myneni et al. 1997), concurrent with an advance and increase in amplitude of the annual CO_2 cycle (Keeling et al. 1996). White et al. (1999) modeled meteorological and satellite data to estimate actual growing season length each year from 1900–1987 in the United States. Growing season was unusually long during the warm period of the 1940s at all 12 sites. However, patterns have recently diverged. Since 1966, growing season length has increased only in four of the coldest, most-northerly zones (42°–45° latitude), not in the three warmest zones (32°–37° latitude). Across the European Phenological Gardens (experimental clones of 16 species of shrubs and trees at sites across Europe), a lengthening of the growing season by 10.8 days occurred from 1959–1993 (Menzel 2000, Menzel & Fabian 1999). Analysis of climatological variables (e.g., last frost date of spring and first frost date of fall) mirrors this finding, with an estimated lengthening of the growing season of 1.1–4.9 days per decade since 1951 (Menzel et al. 2003).

Bradley et al. (1999) built on Aldo Leopold's observations from the 1930s and 1940s on the timing of spring events on a Wisconsin farm. Of 55 species resurveyed in the 1980s and 1990s, 18 (35%) showed advancement of spring events, whereas the rest showed no change in timing (with the exception of cowbirds arriving later). On average, spring events occurred 7.3 days earlier by the 1990s compared with 61 years before, coinciding with March temperatures being 2.8°C warmer.

Another long-term (100-year) study by Gibbs & Breisch (2001) compared recent records (1990–1999) of the calling phenology of six frog species in Ithaca, New York, with a turn-of-the-century study (1900–1912). They showed a 10–13-day advance associated with a 1.0–2.3°C rise in temperature during critical months. Amphibian breeding has also advanced in England, by 1–3 weeks per decade (Beebee 1995). Ecophysiological studies in frogs have shown that reproduction is closely linked to both nighttime and daytime temperatures (Beebee 1995).

In the United Kingdom, Crick et al. (1997), analyzing more than 74,000 nest records from 65 bird species between 1971 and 1995, found that the mean laying dates of first clutches for 20 species had advanced by an average 8.8 days. Brown et al. (1999) found a similar result for the Mexican jay (*Aphelocoma ultramarina*) in the mountains of southern Arizona. In the North Sea, migrant birds have advanced their passage dates by 0.5–2.8 days per decade since 1960, with no significant difference between short- and long-distance migrants (Hüppop & Hüppop 2003). In contrast, Gordo et al. (2005) found that three of six long-distance migrant birds had significantly delayed arrival to breeding grounds in Spain, with arrival date highly correlated with climatic conditions in their overwintering grounds in the southern Sahara.

Butterflies frequently show a high correlation between dates of first appearance and spring temperatures, so it is not surprising that their first appearance has advanced for 26 of 35 species in the United Kingdom (Roy & Sparks 2000) and for all 17 species analyzed in Spain (Stefanescu et al. 2003). Seventy percent of 23 species of butterfly in central California have advanced their first flight date over 31 years, by an average of 24 days (Forister & Shapiro 2003). Climate variables explained 85% of variation in flight date in the California study, with warmer, drier winters driving early flight.

There are only two continental-scale studies of bird phenology. Dunn & Winkler (1999) analyzed changes in breeding for tree swallows (*Tachycineta bicolor*) from 1959 to 1991 over the entire breeding range in the contiguous United States and Canada. Laying date was significantly correlated with mean May temperature and had advanced by an average of nine days over the 32-year period. In a complementary study, Both et al. (2004) analyzed the pied flycatcher (*Ficedula hypoleuca*) at 23 sites across Europe and found a significant advance in laying date for nine of the populations, which also tended to be those with the strongest warming trends. Continental-scale studies of both lilac (*Syringa vulgaris*) and honeysuckle (*Lonicera tatarica* and *L. korolkowii*) in the western United States have shown an advance in mean flowering dates of 2 and 3.8 days per decade, respectively (Cayan et al. 2001).

Aquatic systems exhibit similar trends to those of terrestrial systems. In a lake in the northwestern United States, phytoplankton bloom has advanced by 19 days from

1962 to 2002, whereas zooplankton peak is more varied, with some species showing advance and others remaining stable (Winder & Schindler 2004). The Arctic seabird Brunnich's guillemot, *Uria lomvia*, has advanced its egg-laying date at its southern boundary (Hudson Bay) with no change at its northern boundary (Prince Leopold Island); both trends are closely correlated with changes in sea-ice cover (Gaston et al. 2005).

Roetzer et al. (2000) explicitly quantified the additional impacts of urban warming by comparing phenological trends between urban and rural sites from 1951 to 1995. Urban sites showed significantly stronger shifts toward earlier spring timing than nearby rural sites, by 2–4 days. An analysis of greening across the United States via satellite imagery also concluded that urban areas have experienced an earlier onset of spring compared with rural areas (White et al. 2002).

Researchers generally report phenological changes as a separate category from changes in species' distributions, but these two phenomena interplay with each other and with other factors, such as photoperiod, to ultimately determine how climate change affects each species (Bale et al. 2002, Chuine & Beaubien 2001).

INTERACTIONS ACROSS TROPHIC LEVELS: MATCHES AND MISMATCHES

Species differ in their physiological tolerances, life-history strategies, probabilities of population extinctions and colonizations, and dispersal abilities. These individualistic traits likely underlie the high variability in strength of climate response across wild species, even among those subjected to similar climatic trends (Parmesan & Yohe 2003). For many species, the primary impact of climate change may be mediated through effects on synchrony with that species' food and habitat resources. More crucial than any absolute change in timing of a single species is the potential disruption of coordination in timing between the life cycles of predators and their prey, herbivorous insects and their host plants, parasitoids and their host insects, and insect pollinators with flowering plants (Harrington et al. 1999, Visser & Both 2005). In Britain, the butterfly *Anthocharis cardamines* has accurately tracked phenological shifts of its host plant, even when bud formation came two to three weeks early (Sparks & Yates 1997). However, this may be the exception rather than the rule.

Visser & Both (2005) reviewed the literature and found only 11 species' interactions in which sufficient information existed to address the question of altered synchrony. Nine of these were predator-prey interactions, and two were insect–host plant interactions. In spite of small sample size, an important trend emerged from this review: In the majority of cases (7 of 11), interacting species responded differently enough to climate warming that they are more out of synchrony now than at the start of the studies. In many cases, evidence for negative fitness consequences of the increasing asynchrony has been either observed directly or predicted from associated studies (Visser & Both 2005).

In one example, Inouye et al. (2000) reported results of monitoring between 1975 and 1999 at Rocky Mountain Biological Laboratory in Colorado, where there has been a 1.4°C rise in local temperature. The annual date of snowmelt and plant flowering did not change during the study period, but yellow-bellied marmots (*Marmota flaviventris*) advanced their emergence from hibernation by 23 days, changing the relative phenology of marmots and their food plants. In a similar vein, Winder & Schindler (2004) documented a growing asynchrony between peak phytoplankton bloom and peak zooplankton abundances in a freshwater lake.

More complex phenomena resulting from trophic mismatches have also been documented. For example, phenological asynchrony has been linked to a range shift in the butterfly *Euphydryas editha*. Warm and/or dry years alter insect emergence time relative to both the senescence times of annual hosts and the time of blooming of nectar sources (Singer 1972, Singer & Ehrlich 1979, Singer & Thomas 1996, Thomas et al. 1996, Weiss et al. 1988). Field studies have documented that butterfly-host asynchrony has resulted directly in population crashes and extinctions. Long-term censuses revealed that population extinctions occurred during extreme droughts and low snowpack years (Ehrlich et al. 1980, Singer & Ehrlich 1979, Singer & Thomas 1996, Thomas et al. 1996), and these extinctions have been highly skewed with respect to both latitude and elevation, shifting mean location of extant populations northward and upward (Parmesan 1996, 2003, 2005a).

Van Nouhuys & Lei (2004) showed that host-parasitoid synchrony was influenced signficantly by early spring temperatures. Warmer springs favored the parasitoid wasp *Cotesia melitaearum* by bringing it more in synchrony with its host, the butterfly *Melitaea cinxia*. Furthermore, they argue that because most butterfly populations are protandrous (i.e., males pupating earlier than females), temperature-driven shifts in synchrony with parasitoids may affect butterfly sex ratios.

> **Intergovernmental Panel on Climate Change:** a scientific panel formed under the auspices of the United Nations and the World Meteorological Organization for the purpose of synthesizing literature and forming scientific consensus on climate change and its impacts

OBSERVED RANGE SHIFTS AND TRENDS IN LOCAL ABUNDANCE

Expected distributional shifts in warming regions are poleward and upward range shifts. Studies on these shifts fall mainly into two types: (*a*) those that infer large-scale range shifts from small-scale observations across sections of a range boundary (with the total study area often determined by a political boundary such as state, province, or country lines) and (*b*) those that infer range shifts from changes in species' composition (abundances) in a local community. Studies encompassing the entire range of a species, or at least the northern and southern (or lower and upper) extremes, are few and have been concentrated on amphibians (Pounds et al. 1999, 2006), a mammal (Beever et al. 2003), and butterflies (Parmesan 1996, Parmesan et al. 1999). The paucity of whole-range studies likely stems from the difficulties of gathering data on the scale of a species' range—often covering much of a continent.

Shifts at Polar Latitudes

Broad impacts of climate change in polar regions—from range shifts to community restructuring and ecosystem functioning—have been reviewed by the Intergovernmental Panel on Climate Change (Anisimov et al. 2001), the Arctic Climate Impact Assessment (2004) and the Millenium Ecosystem Assessment (Chapin et al. 2006).

SST: sea surface temperature

Antarctic. Plant, bird, and marine life of Antarctica have exhibited pronounced responses to anthropogenic climate change. These responses have been largely attributed to extensive changes (mostly declines) in sea-ice extent, which in turn appears to have stimulated a trophic cascade effect in biological systems. Declines in sea-ice extent and duration since 1976 have apparently reduced abundances of ice algae, in turn leading to declines in krill (from 38%–75% per decade) in a large region where they have been historically concentrated, the southwest Atlantic (Atkinson et al. 2004). Krill (*Euphausia superba*) is a primary food resource for many fish, seabirds, and marine mammals. Interestingly, McMurdo Dry Valleys, which actually cooled between 1990 and 2000, also showed declines in lake phytoplankton abundances and in soil invertebrate abundances (Doran et al. 2002).

Penguins and other seabirds in Antarctica have shown dramatic responses to changes in sea-ice extent over the past century (Ainley et al. 2003, Croxall et al. 2002, Smith et al. 1999). The sea-ice dependent Adélie and emperor penguins (*Pygoscelis adeliae* and *Aptenodytes forsteri*, respectively) have nearly disappeared from their northernmost sites around Antarctica since 1970. Emperors have declined from 300 breeding pairs down to just 9 in the western Antarctic Peninsula (Gross 2005), with less severe declines at Terre Adélie (66° S), where they are now at 50% of pre-1970s abundances (Barbraud & Weimerskirch 2001). Adélies have declined by 70% on Anvers Island (64°–65° S along the Antarctic peninsula (Emslie et al. 1998, Fraser et al. 1992), whereas they are thriving at the more-southerly Ross Island at 77° S (Wilson et al. 2001)—effectively shifting this species poleward. In the long-term, sea-ice-dependent birds will suffer a general reduction of habitat as ice shelves contract [e.g., as has already occured in the Ross Sea (IPCC 2001b)] or collapse [e.g., as did the Larsen Ice Shelves along the Antarctic Peninsula in 2002 (Alley et al. 2005)].

In contrast, open-ocean feeding penguins—the chinstrap and gentoo—invaded southward along the Antarctic Peninsula between 20 and 50 years ago, with paleological evidence that gentoo had been absent from the Palmer region for 800 years previously (Emslie et al. 1998, Fraser et al. 1992). Plants have also benefited from warming conditions. Two Antarctic vascular plants (a grass, *Deschampsia antarctica*, and a cushion plant, *Colobanthus quitensis*) have increased in abundance and begun to colonize novel areas over a 27-year period (Smith 1994).

Arctic. Nearly every Arctic ecosystem shows marked shifts. Diatom and invertebrate assemblages in Arctic lakes have shown huge species' turnover, shifting away from benthic species toward more planktonic and warm-water-associated communities (Smol et al. 2005). Across northern Alaska, Canada, and parts of Russia, shrubs have been expanding into the tundra (Sturm et al. 2005). Field studies, experimentation, and modeling link this major community shift to warming air temperatures, increased snow cover, and increased soil microbial activity (Chapin et al. 1995; Sturm et al. 2001, 2005). Populations of a pole-pole migrant, the sooty shearwater (*Puffinus griseus*), have shifted their migration routes by hundreds of kilometers in concert with altered sea surface temperature (SST) in the Pacific (Spear & Ainley 1999).

Sea-ice decline in the Arctic has been more evenly distributed than in the Antarctic. Because of differing geology, with an ocean at the pole rather than land, Arctic species that are sea-ice dependent are effectively losing habitat at all range boundaries. Polar bears have suffered significant population declines at opposite geographic boundaries. At their southern range boundary (Hudson Bay), polar bears are declining both in numbers and in mean body weight (Stirling et al. 1999). Climate change has caused a lengthening of ice-free periods on Hudson Bay, periods during which the bears starve and live on their reserves because an ice shelf is necessary for feeding. Furthermore, researchers have also linked warming trends to reductions of the bears' main food, the ringed seal (Derocher et al. 2004, Ferguson et al. 2005). At the bears' northern range boundaries off Norway and Alaska, sea ice has also been reduced, but poorer records make it is less clear whether observed declines in body size and the number of cubs per female are linked to climate trends or to more basic density-dependent processes (Derocher 2005, Stirling 2002).

Shifts in Northern-Hemisphere Temperate Species

On a regional scale, a study of the 59 breeding bird species in Great Britain showed both expansions and contractions of northern range boundaries, but the average boundary change for 12 species that had not experienced overall changes in density was a mean northward shift of 18.9 km over a 20-year period (Thomas & Lennon 1999). For a few well-documented bird species, their northern U.K. boundaries have tracked winter temperatures for over 130 years (Williamson 1975). Physiological studies indicate that the northern boundaries of North American songbirds may generally be limited by winter nighttime temperatures (Burger 1998, Root 1988).

Analogous studies exist for Lepidoptera (butterflies and moths), which have undergone an expansion of northern boundaries situated in Finland (Marttila et al. 1990, Mikkola 1997), Great Britain (Hill et al. 2002, Pollard 1979, Pollard & Eversham 1995, Warren 1992), and across Europe (Parmesan et al. 1999). Depending on the study, some 30% to 75% of northern boundary sections had expanded north; a smaller portion (<20%) had contracted southward; and the remainder were classified as stable. In a study of 57 nonmigratory European butterflies, data were obtained from both northern and southern range boundaries for 35 species (Parmesan et al. 1999). Nearly two thirds (63%) had shifted their ranges to the north by 35–240 km, and only two species had shifted to the south (Parmesan et al. 1999). In the most-extreme cases, the southern edge contracted concurrent with northern edge expansion. For example, the sooty copper (*Heodes tityrus*) was common in the Montseny region of central Catalonia in the 1920s, but modern sightings are only from the Pyrenees, 50 km to the north. Symmetrically, *H. tityrus* entered Estonia for the first time in 1998, by 1999 had established several successful breeding populations, and by 2006 had reached the Baltic Sea (Parmesan et al. 1999; T. Tammaru, personal communication).

Another charismatic insect group with good historical records is Odonata (dragonflies and damselflies). In a study of all 37 species of resident odonates in the United Kingdom, Hickling et al. (2005) documented that 23 of the 24 temperate species had expanded their northern range limit between 1960–1995, with mean northward shift of 88 km.

Nondiapausing (i.e., active year-round) butterfly species are also moving northward with warmer winters. The northern boundary of the sachem skipper butterfly has expanded from California to Washington State (420 miles) in just 35 years (Crozier 2003, 2004). During a single year—the warmest on record (1998)—it moved 75 miles northward. Laboratory and field manipulations showed that individuals are killed by a single, short exposure to extreme low temperatures (−10°C) or repeated exposures to −4°C, indicating winter cold extremes dictate the northern range limit (Crozier 2003, 2004). The desert orange tip (*Colotis evagore*), which historically was confined to northern Africa, has established resident populations in Spain while maintaining the same ecological niche. Detailed ecological and physiological studies confirm that *C. evagore* has remained a specialist of hot microclimates, needing more than 164 days at greater than 12°C to mature. It has not undergone a host switch in its new habitat, and it has not evolved a diapause stage (Jordano et al. 1991).

In the Netherlands between 1979 and 2001, 77 new epiphytic lichens colonized from the south, nearly doubling the total number of species for that community (van Herk et al. 2002). Combined numbers of terrestrial and epiphytic lichen species increased from an average of 7.5 per site to 18.9 per site. An alternate approach to documenting colonizations is to document extinction patterns. Comparing recent censuses across North America (1993–1996) with historical records (1860–1986), Parmesan (1996) documented that high proportions of population extinctions along the southern range boundary of Edith's checkerspot butterfly (*E. editha*) had shifted the mean location of living populations 92 km farther north (Parmesan 1996, 2003, 2005a).

Shifts of Tropical Species Ranges

Warming trends at lower latitudes are associated with movements of tropical species into more-temperate areas. The rufous hummingbird has undergone a dramatic shift in its winter range (Hill et al. 1998). Thirty years ago it wintered mainly in Mexico, and between 1900 and 1990, there were never more than 30 winter sightings per year along the Gulf Coast of the United States. In the early 1990s, sightings increased to more than 100 per year in the southern United States. The number of sightings has increased steadily since then—up to 1,643 by 1996, with evidence that, by 1998, resident populations had colonized 400 km inland (Howell 2002). Over this same period, winter temperatures rose by approximately 1°C (IPCC 2001b). In Florida, five new species of tropical dragonfly established themselves in 2000, an apparently natural invasion from Cuba and the Bahamas (Paulson 2001).

Similarly, North African species are moving into Spain and France, and Mediterranean species are moving up into the continental interior. The African plain tiger butterfly (*Danaus chrysippus*) established its first population in southern Spain in 1980 and by the 1990s had established multiple, large metapopulations (Haeger 1999).

Elevational Shifts

Montane studies have generally been scarcer and less well documented (lower sampling resolution), but a few good data sets show a general movement of species upward in elevation. By comparing species compositions in fixed plots along an elevational gradient in Monteverde National Park, Costa Rica, Pounds et al. (1999, 2005) documented that lowland birds have begun breeding in montane cloud-forest habitat over the past 20 years. A similar study across 26 mountains in Switzerland documented that alpine flora have expanded toward the summits since the plots were first censused in the 1940s (Grabherr et al. 1994, Pauli et al. 1996). Upward movement of treelines has been observed in Siberia (Moiseev & Shiyatov 2003) and in the Canadian Rocky Mountains, where temperatures have risen by 1.5°C (Luckman & Kavanagh 2000).

The few studies of lower elevational limits show concurrent contractions upward of these warm range boundaries. Because warm boundaries generally have data gaps through time, these studies have conducted recensuses of historically recorded (sedentary) populations and looked for nonrandom patterns of long-term population extinctions.

A 1993–1996 recensus of Edith's checkerspot butterfly (*E. editha*) populations recorded 1860–1986 throughout its range (Mexico to Canada) documented that more than 40% of populations from 0–2400 m were extinct (in spite of having suitable habitat), whereas less than 15% were extinct at the highest elevations (2400–3500 m) (Parmesan 1996). Over the past 50–100 years, snowpack below 2400 m has become lighter by 14% and melts 7 days earlier, whereas higher elevations (2400–3500 m) have 8% heavier snowpack and no change in melt date (Johnson et al. 1999). In concert with altered snow dynamics, the mean location of *E. editha* populations has shifted upward by 105 m (Parmesan 1996, 2003, 2005a).

In southern France, metapopulations of the cool-adapted Apollo butterfly (*Parnassius apollo*) have gone extinct over the past 40 years on plateaus less than 850 m high but have remained healthy where plateaus were greater than 900 m high (Descimon et al. 2006). The data suggest that dispersal limitation was important, and this strong flyer can persist when nearby higher elevation habitats exist to colonize. In Spain, the lower elevational limits of 16 species of butterfly have risen an average of 212 m in 30 years, concurrent with a 1.3°C rise in mean annual temperatures (Wilson et al. 2005).

In the Great Basin of the western United States, 7 out of 25 recensused populations of the pika (*Ochotona princeps*, Lagomorpha) were extinct since being recorded in the 1930s (Beever et al. 2003). Human disturbance is minimal because pika habitat is high-elevation talus (scree) slopes, which are not suitable for ranching or recreational activities. Extinct populations were at significantly lower elevations than those still present (Parmesan & Galbraith 2004). Field observations by Smith (1974) documented that adult pika stopped foraging in the midday heat in August at low elevation sites. Subsequent experiments showed that adults were killed within a half hour at more than 31°C (Smith 1974).

Marine Community Shifts

Decades of ecological and physiological research document that climatic variables are primary drivers of distributions and dynamics of marine plankton and fish (Hays et al. 2005, Roessig et al. 2004). Globally distributed planktonic records show strong shifts of phytoplankton and zooplankton communities in concert with regional oceanic climate regime shifts, as well as expected poleward range shifts and changes in timing of peak biomass (Beaugrand et al. 2002, deYoung et al. 2004, Hays et al. 2005, Richardson & Schoeman 2004). Some copepod communities have shifted as much as 1000 km northward (Beaugrand et al. 2002). Shifts in marine fish and invertebrate communities have been been particularly well documented off the coasts of western North America and the United Kingdom. These two systems make an interesting contrast (see below) because the west coast of North America has experienced a 60-year period of significant warming in nearshore sea temperatures, whereas much of the U.K. coast experienced substantial cooling in the 1950s and 1960s, with warming only beginning in the 1970s (Holbrook et al. 1997, Sagarin et al. 1999, Southward et al. 2005).

Sagarin et al. (1999) related a 2°C rise of SST in Monterey Bay, California, between 1931 and 1996 to a significant increase in southern-ranged species and decrease of northern-ranged species. Holbrook et al. (1997) found similar shifts over the past 25 years in fish communities in kelp habitat off California.

Much of the data from the North Atlantic, North Sea, and coastal United Kingdom have exceptionally high resolution and long time series, so they provide detailed information on annual variability, as well as long-term trends. Over 90 years, the timing of animal migration (e.g., veined squid, *Loligo forbesi*, and flounder *Platichthys flesus*) followed decadal trends in ocean temperature, being later in cool decades and up to 1–2 months earlier in warm years (Southward et al. 2005).

In the English Channel, cold-adapted fish (e.g., herring *Clupea harengus*) declined during both warming periods (1924 to the 1940s, and post-1979), whereas warm-adapted fish did the opposite (Southward et al. 1995, 2005). For example, pilchard *Sardina pilchardus* increased egg abundances by two to three orders of magnitude during recent warming. In the North Sea, warm-adapted species (e.g., anchovy *Engraulis encrasicolus* and pilchard) have increased in abundances since 1925 (Beare et al. 2004), and seven out of eight have shifted their ranges northward (e.g., bib, *Trisopterus luscus*) by as much as 100 km per decade (Perry et al. 2005). Records dating back to 1934 for intertidal invertebrates show equivalent shifts between warm- and cold-adapted species (e.g., the barnacles *Semibalanus balanoides* and *Chthamalus* spp., respectively), mirroring decadal shifts in coastal temperatures (Southward et al. 1995, 2005).

Pest and Disease Shifts

Pest species are also moving poleward and upward. Over the past 32 years, the pine processionary moth (*Thaumetopoea pityocampa*) has expanded 87 km at its northern range boundary in France and 110–230 m at its upper altitudinal boundary in Italy (Battisti et al. 2005). Laboratory and field experiments have linked the feeding behavior and survival of this moth to minimum nighttime temperatures, and its expansion has been associated with warmer winters. In the Rocky Mountain range of the United States, mountain pine beetle (*Dendroctonus ponderosae*) has responded to warmer temperatures by altering its life cycle. It now only takes one year per generation rather than its previous two years, allowing large increases in population abundances, which, in turn, have increased incidences of a fungus they transmit (pine blister rust,

Cronartium ribicola) (Logan et al. 2003). Increased abundance of a nemotode parasite has also occurred as its life cycle shortened in response to warming trends. This has had associated negative impacts on its wild musk oxen host, causing decreased survival and fecundity (Kutz et al. 2005).

In a single year (1991), the oyster parasite *Perkinsus marinus* extended its range northward from Chesapeake Bay to Maine—a 500 km shift. Censuses from 1949 to 1990 showed a stable distribution of the parasite from the Gulf of Mexico to its northern boundary at Chesapeake Bay. The rapid expansion in 1991 has been linked to above-average winter temperatures rather than human-driven introduction or genetic change (Ford 1996). A kidney disease has been implicated in low-elevation trout declines in Switzerland. High mortality from infection occurs above 15°–16°C, and water temperatures have risen in recent decades. High infection rates (27% of fish at 73% of sites) at sites below 400 m have been associated with a 67% decline in catch; mid-elevation sites had lower disease incidence and only moderate declines in catch; and the highest sites (800–3029 m) had no disease present and relatively stable catch rates (Hari et al. 2006).

Changes in the wild also affect human disease incidence and transmission through alterations in disease ecology and in distributions of their wild vectors (Parmesan & Martens 2006). For example, in Sweden, researchers have documented marked increases in abundances of the disease-transmitting tick *Ixodes ricinus* along its northernmost range limit (Lindgren & Gustafson 2001). Between the early 1980s and 1994, numbers of ticks found on domestic cats and dogs increased by 22%–44% along the tick's northern range boundary across central Sweden. In the same time period, this region had a marked decrease in the number of extremely cold days ($<-12°C$) in winter and a marked increase in warm days ($>10°C$) during the spring, summer, and fall. Previous studies on temperature developmental and activity thresholds indicated the observed warmer temperatures cause decreased tick mortality and longer growing seasons (Lindgren & Gustafson 2001).

Trees and Treelines: Complex Responses

A complex of interacting factors determines treeline, often causing difficulties in interpretation of twentieth-century trends. Some species are "well behaved" in that they show similar patterns of increased growth at treeline during the early warming in the 1930s and 1940s as during the recent warming of the past 20 years. In recent decades, treelines have shifted northward in Sweden (Kullman 2001) and eastern Canada (Lescop-Sinclair & Payette 1995), and upward in Russia (Meshinev et al. 2000, Moiseev & Shiyatov 2003) and New Zealand (Wardle & Coleman 1992).

However, in other studies, researchers saw a strong response to warming in the late 1930s and 1940s but a weaker (or absent) response in recent warm decades (Innes 1991, Jacoby & D'Arrigo 1995, Lescop-Sinclair & Payette 1995, Briffa et al. 1998a,b), possibly resulting from differences in rainfall between the two warm periods. In Alaska, recent decades have been relatively dry, which may have prevented trees from responding to current warming as they did before (Barber et al. 2000, Briffa et al. 1998b). In contrast, treelines in the arid southwest United States, which has had increased rainfall, have shown unprecedented increased tree-ring growth at high elevations (Swetnam & Betancourt 1998).

An impressive study across all of northern Russia from 1953–2002 showed a shift in tree allometries. In areas where summer temperatures and precipitation have both increased, a general increase in biomass (up 9%) is primarily a result of increased greenery (33% more carbon in leaves and needles), rather than woody parts (roots and stem). In areas that have experienced warming and drying trends, greenery has decreased, and both roots and stems have increased (Lapenis et al. 2005).

EXTINCTIONS

Amphibians

Documented rapid loss of habitable climate space makes it no surprise that the first extinctions of entire species attributed to global warming are mountain-restricted species. Many cloud-forest-dependent amphibians have declined or gone extinct on a mountain in Costa Rica (Pounds et al. 1999, 2005). Among harlequin frogs in Central and South American tropics, an astounding 67% have disappeared over the past 20–30 years. Pounds et al. (2006) hypothesised that recent trends toward warmer nights and increased daytime cloud cover have shifted mid-elevation sites (1000–2400 m), where the preponderance of extinctions have occurred, into thermally optimum conditions for the chytrid fungus, *Batrachochytrium dendrobatidis*.

Tropical Coral Reefs

Elevated sea temperatures as small as 1°C above long-term summer averages lead to bleaching (loss of coral algal symbiont), and global SST has risen an average of 0.1°–0.2°C since 1976 (Hoegh-Guldberg 1999, IPCC 2001b). A more acute problem for coral reefs is the increase in extreme temperature events. El Niño events have been increasing in frequency and severity since records began in the early 1900s, and researchers expect this trend to continue over coming decades (Easterling et al. 2000, IPCC 2001b, Meehl et al. 2000). A particularly strong El Niño in 1997–1998 caused bleaching in every ocean (up to 95% of corals bleached in the Indian Ocean), ultimately resulting in 16% of corals rendered extinct globally (Hoegh-Guldberg 1999, 2005b; Wilkinson 2000).

Recent evidence for genetic variation among the obligate algal symbiont in temperature thresholds suggests that some evolutionary response to higher water temperatures may be possible (Baker 2001, Rowan 2004). Changes in genotype frequencies toward increased frequency of high-temperature-tolerant symbiont appear to have occurred within some coral populations between the mass bleaching events of 1997–1998 and 2000–2001 (Baker et al. 2004). However, other studies indicate that many entire reefs are already at their thermal tolerance limits (Hoegh-Guldberg 1999). Coupled with poor dispersal of symbiont between reefs, this has led several researchers to conclude that local evolutionary responses are unlikely to mitigate the negative impacts of future temperature rises (Donner et al. 2005, Hoegh-Guldberg et al. 2002).

One optimistic result suggests that corals, to some extent, may be able to mirror terrestrial range shifts. Two particularly cold-sensitive species (staghorn coral, *Acropora ceervicornis*, and elkhorn coral, *Acropora palmata*) have recently expanded their ranges into the northern Gulf of Mexico (first observation in 1998), concurrent with rising SST (Precht & Aronson 2004). Although continued poleward shift will be limited by light availability at some point (Hoegh-Guldberg 1999), small range shifts may aid in developing new refugia against extreme SST events in future.

Although impacts have not yet been observed, the fate of coral reefs may be as, or more, affected in coming decades by the direct effects of CO_2 rather than temperature rise. Increased atmospheric CO_2 since industrialization has significantly lowered ocean pH by 0.1. The more dire projections (a doubling to tripling of current CO_2 levels) suggest that, by 2050, oceans may be too acidic for corals to calcify (Caldeira & Wickett 2003, Hoegh-Guldberg 2005a, Orr et al. 2005).

Population Extinctions Leading to Range Contractions

Many species have suffered reduced habitable area due to recent climate change. For those species that have already been driven extinct at their equatorial or lower range boundaries, some have either failed to expand poleward or are unable to expand due to geographic barriers. Such species have suffered absolute reductions in range size, putting them at greater risk of extinction in the near future.

This is particularly evident in polar species, as these are already pushed against a geographical limit. Researchers have seen large reductions in population abundances and general health along the extreme southern populations of Arctic polar bears (Derocher 2005, Derocher et al. 2004, Stirling et al. 1999) and the extreme northern populations of Antarctic Adélie and emperor penguins (Ainley et al. 2003, Croxall et al. 2002, Emslie et al. 1998, Fraser et al. 1992, Smith et al. 1999, Taylor & Wilson 1990, Wilson et al. 2001). In the United Kingdom, four boreal odonates have contracted northward by an average of 44 km over 40 years (Hickling et al. 2005).

Similarly, high numbers of population extinctions have occurred along the lower elevational boundaries of mountaintop species, such as pikas in the western United States (Beever et al. 2003) and the Apollo butterfly in France (Descimon et al. 2006). For 16 mountain-restricted butterflies in Spain, warming has already reduced their habitat by one third in just 30 years (Wilson et al. 2005). Warming and drying trends on Mt. Kiliminjaro have increased fire impacts, which have caused a 400-m downward contraction of closed (cloud) forest, now replaced by an open, dry alpine system (Hemp 2005). Temperate low-elevation species are not immune: Twenty-five percent of temperate butterflies in Europe contracted northward by 35–50 km over a 30–70-year period. For one of these, its northern range boundary had not expanded, so it suffered an overall contraction of range size (Parmesan et al. 1999).

EVOLUTION AND PLASTICITY

Species ranges are dynamic. Historically, ecologists have viewed species' niches as static and range shifts over time as passive responses to major environmental changes (global climate shifts or geological changes in corridors and barriers).

There is no doubt that climate plays a major role in limiting terrestrial species' ranges (Andrewartha & Birch 1954; Bale et al. 2002; Parmesan et al. 2000, 2005; Precht et al. 1973; Webb & Bartlein 1992; Weiser 1973; Woodward 1987). Recent physiological and biogeographic studies in marine systems also implicate temperature as a primary driver of species' ranges (Hoegh-Guldberg 1999, 2005b; Hoegh-Guldberg & Pearse 1995).

However, evolutionary processes clearly can substantially influence the patterns and rates of response to climate change. Theoretically, evolution can also drive range shifts in the absence of environmental change (Holt 2003). A prime example of this is the hybridization of two species of Australian fruit fly that led to novel adaptations, allowing range expansion with no concomitant environmental change (Lewontin & Birch 1966).

The problem of estimating the relative roles of evolution and plasticity is tractable with extensive, long-term ecological and genetic data. For example, genetic analysis of a population of red squirrels in the Arctic indicated that 62% of the change in breeding dates occurring over a 10-year period was a result of phenotypic plasticity, and 13% was a result of genetic change in the population (Berteaux et al. 2004, Réale et al. 2003).

Geneticists in the 1940s noticed that certain chromosomal inversions in fruit flies (*Drosophila*) were associated with heat tolerance (Dobzhansky 1943, 1947). These "hot" genotypes were more frequent in southern than in northern populations and increased within a population during each season, as temperatures rose from early spring through late summer. Increases in the frequencies of warm-adapted genotypes have occurred in wild populations of *Drosophila ssp* in Spain between 1976 and 1991 (Rodríguez-Trelles & Rodriguez 1998, Rodríguez-Trelles et al. 1996, 1998), as well as in the United States between 1946 and 2002 (Levitan 2003). The change in the United States was so great that populations in New York in 2002 were converging on genotype frequencies found in Missouri in 1946.

In contrast, red deer in Norway show completely plastic responses. Their body size responds rapidly to yearly variability of winter temperatures. Warmer winters cause developing males to become larger while females become smaller (Post et al. 1999). In consequence, the end result of a gradual winter warming trend has been an increase in sexual dimorphism.

A surprising twist is that species whose phenology is under photoperiodic control have also responded to temperature-driven selection for spring advancement or fall delay. Bradshaw & Holzapfel (2001) showed that the pitcher plant mosquito, *Wyeomyia smithii*, has evolved a shorter critical photoperiod in association with a longer growing season. Northern populations of this mosquito now use a shorter day-length cue to enter winter diapause, doing so later in the fall than they did 24 years ago.

The Role of Evolution in Shaping Species' Impacts

Increasing numbers of researchers use analyses of current intraspecific genetic variation for climate tolerance to argue for a substantive role of evolution in mitigating negative impacts of future climate change (Baker 2001, Baker et al. 2004, Davis & Shaw 2001, Rowan 2004). However, in spite of a plethora of data indicating local adaptation to climate change at specific sites, the fossil record shows little evidence for the evolution of novel phenotypes across a species as a whole. Pleistocene glaciations represent shifts 5–10 times the magnitude of twentieth-century global warming. These did not result in major evolution at the species level (i.e., appearance of new forms outside the bounds of known variation for that species), nor in major extinction or speciation events. Existing species appeared to shift their geographical distributions as though tracking the changing climate, rather than remaining stationary and evolving new forms (Coope 1994, Davis & Zabinski 1992, Huntley 1991).

Most of the empirical evidence for rapid adaptation to climate change comes from examples of evolution in the interiors of species' ranges toward higher frequencies of already existing heat-tolerant genotypes. In studies that focus on dynamics at the edge of a species' range or across an entire range, a different picture emerges. Several studies suggest that the effects of both genetic constraints and asymmetrical gene flow are intensified close to species' borders (Antonovics 1976, Garcia-Ramos & Kirkpatrick 1997, Hoffmann & Blows 1994). It is expected that a warming climate strengthens climate stress at equatorial range boundaries and reduces it at poleward boundaries. Equatorial boundary populations are often under natural selection for increased tolerance to extreme climate in the absence of climate change, but may be unable to respond due to lack of necessary genetic variance. Furthermore, gene flow from interior populations may stifle response to selection at the range limits, even when sufficient genetic variation exists (Kirkpatrick & Barton 1997).

Because of strong trade-offs between climate tolerance and resource/habitat preferences, a relaxation of selection on climate tolerance at northern boundaries may cause rapid evolution of these correlated traits. This process has been investigated in the European butterfly *Aricia agestis*, in which populations near the northern range boundary had previously adapted to cool conditions by specializing on the host genus, *Helianthemum*, which grows in hot microclimates and hence supports fast larval growth. Climate warming did not initially cause range expansion because *Helianthemum* was absent to the immediate north of the range limit. However, warming did permit rapid evolution of a broader diet at the range limit, to a host used in more southern populations, *Geranium*, which grows in cooler microclimates. Once this local diet evolution occurred, the boundary expanded northward across the band from which *Helianthemum* was absent but *Geranium* was present (Thomas et al. 2001).

This example shows how a complex interplay may occur between evolutionary processes and ecological responses to extreme climates and climate change. However, these evolutionary events did not constitute alternatives to ecological responses to climate change; they modulated those changes. Adaptive evolution of host preference occurred at the northern range boundary in response to temperature rise, but genetic variation for host use already existed within the *A. agestis* butterfly. In this case, evolutionary processes are not an alternative to range movement, but instead modulate the magnitude and dynamics of the range shift. This is not likely to be an isolated example because populations of other species near poleward boundaries are known to specialize on resources that mitigate the effects of cool climate. Such resources either support rapid growth or occur in the hottest available microclimates (Nylin 1988, Scriber & Lederhouse 1992, Thomas et al. 2001).

In addition to resource choice, dispersal tendency evolves at range margins in response to climate change. In nonmigratory species, the simplest explanation of northward range expansions is that individuals have always crossed the species' boundary, and with climate warming, some of these emigrants are successful at founding new populations outside the former range. When dispersal tendency is heritable, these new populations contain dispersive individuals and higher rates of dispersal will soon evolve at the expanding boundary.

Evolution toward greater dispersal has indeed been documented in several species of insect. Two species of wing-dimorphic bush crickets in the United Kingdom have evolved longer wings at their northern range boundary, as mostly long-winged forms participated in the range expansion and short-winged forms were left behind (Thomas et al. 2001). Adults of newly colonized populations of the speckled wood butterfly (*Pararge aegeria*) in the United Kingdom have larger thoraces and greater flight capability than historical populations just to the south (Hill et al. 1999). Variation in dispersal abilities can be cryptic. Newly founded populations of the butterfly *M. cinxia* contained females that were genetically superior dispersers due to increased production of ATP (Hanski et al. 2004).

Overall, empirical evidence suggests that evolution can complement, rather than supplant, projected ecological changes. However, there is little theoretical or experimental support to suggest that climate warming will cause absolute climatic tolerances of a species to evolve sufficiently to allow it to conserve its geographic distribution in the face of climate change and thereby inhabit previously unsuitable climatic regimes (Donner et al. 2005; Hoegh-Guldberg 1999, 2005b; Hoegh-Guldberg et al. 2002; Jump & Peñuelas 2005).

CONCLUDING THOUGHTS ON EVOLUTION AND CLIMATE CHANGE

For species-level evolution to occur, either appropriate novel mutations or novel genetic architecture (new gene complexes) would have to emerge to allow a response to selection. Lynch & Lande (1993) used a genetic model to infer rates of environmental change that would allow populations to respond adaptively. However, Travis & Futuyma (1993)—discussing the same question from broad paleontological, population, genetic, and ecological perspectives—highlighted the complexity of predicting future responses from currently known processes. Fifteen years later, answers still lie very much in empirical observations. These observations indicate that, although local evolutionary responses to climate change have occurred with high frequency, there is no evidence for change in the absolute climate tolerances of a species. This view is supported by the disproportionate number of population extinctions documented along southern and low-elevation range edges in response to recent climate warming, resulting in contraction of species' ranges at these warm boundaries, as well as by extinctions of many species.

SUMMARY POINTS

1. The advance of spring events (bud burst, flowering, breaking hibernation, migrating, breeding) has been documented on all but one continent and in all major oceans for all well-studied marine, freshwater, and terrestrial groups.

2. Variation in phenological response between interacting species has already resulted in increasing asynchrony in predator-prey and insect-plant systems, with mostly negative consequences.

3. Poleward range shifts have been documented for individual species, as have expansions of warm-adapted communities, on all continents and in most of the major oceans for all well-studied plant and animal groups.

4. These observed changes have been mechanistically linked to local or regional climate change through long-term correlations between climate and biological variation, experimental manipulations in the field and laboratory, and basic physiological research.

5. Shifts in abundances and ranges of parasites and their vectors are beginning to influence human disease dynamics.

6. Range-restricted species, particularly polar and mountaintop species, show more-severe range contractions than other groups and have been the first groups in which whole species have gone extinct due to recent climate change. Tropical coral reefs and amphibians are the taxonomic groups most negatively impacted.

7. Although evolutionary responses have been documented (mainly in insects), there is little evidence that observed genetic shifts are of the type or magnitude to prevent predicted species extinctions.

FUTURE ISSUES

1. Ocean-atmosphere processes are dynamically changing in response to anthropogenic forcings. Indices such as the El Niño Southern Oscillation and the North Atlantic Oscillation may be a poor basis for projecting future biological impacts.

2. Projections of impacts will be aided by a better mechanistic understanding of ecological, behavioral, and evolutionary responses to complex patterns of climate change, and in particular to impacts of extreme weather and climate events.

ACKNOWLEDGMENTS

I would like to give many thanks to C. Britt, M. Butcher, J. Mathews, C. Metz, and P. van der Meer for help with the literature search and to D. Simberloff and D. Futuyma for helpful comments on earlier versions. I also want to give special appreciation to M.C. Singer for his critique and editorial assistance.

LITERATURE CITED

Ahmad QK, Warrick RA, Downing TE, Nishioka S, Parikh KS, et al. 2001. Methods and tools. See IPCC 2001a, pp. 105–43

Ainley DG, Ballard G, Emslie SD, Fraser WR, Wilson PR, Woehler EJ. 2003. Adelie penguins and environmental change. *Science* 300:429–30

Ainley DG, Divoky GJ. 1998. Climate change and seabirds: a review of trends in the eastern portion of the Pacific Basin. *Pac. Seab.* 25:20

Alley RB, Clark PU, Huybrechts P, Joughin I. 2005. Ice-sheet and sea-level changes. *Science* 310:456–60

Alley RB, Marotzke J, Nordhaus WD, Overpeck JT, Peteet DM, et al. 2003. Abrupt climate change. *Science* 299:2005–10

Andrewartha HG, Birch LC. 1954. *The Distribution and Abundance of Animals*. Chicago, IL: Univ. Chicago Press

Anisimov O, Fitzharris B, Hagen JO, Jefferies R, Marchant H, et al. 2001. Polar regions (Arctic and Antarctic). See IPCC 2001a, pp. 801–47

Antonovics J. 1976. The nature of limits to natural selection. *Ann. Mo. Bot. Gard.* 63:224–47

Arctic Climate Impact Assessment. 2004. *Impacts of a Warming Arctic*. Cambridge, UK: Cambridge Univ. Press

Atkinson A, Siegel V, Pakhomov E, Rothery P. 2004. Long-term decline in krill stock and increase in salps within the Southern Ocean. *Nature* 432:100–3

Badeck FW, Bondeau A, Böttcher K, Doktor D, Lucht W, et al. 2004. Responses of spring phenology to climate change. *New Phytol.* 162:295–309

Baker AC. 2001. Reef corals bleach to survive change. *Nature* 411:765–66

Baker AC, Starger CJ, McClanahan TR, Glynn PW. 2004. Coral reefs: corals' adaptive response to climate change. *Nature* 430:741

Bale JS, Masters GJ, Hodkinson ID, Awmack C, Bezemer TM, et al. 2002. Herbivory in global climate change research: direct effects of rising temperature on insect herbivores. *Glob. Change Biol.* 8:1–16

Barber VA, Juday GP, Finney BP. 2000. Reduced growth of Alaskan white spruce in the twentieth century from temperature-induced drought stress. *Nature* 405:668–73

Barbraud C, Weimerskirch H. 2001. Emperor penguins and climate change. *Nature* 411:183–86

Battisti A, Stastny M, Netherer S, Robinet C, Schopf A, et al. 2005. Expansion of geographic range in the pine processionary moth caused by increased winter temperatures. *Ecol. Appl.* 15:2084–96

Beare DJ, Burns F, Greig A, Jones EG, Peach K, et al. 2004. Long-term increases in prevalence of North Sea fishes having southern biogeographic affinities. *Mar. Ecol. Prog. Ser.* 284:269–78

Beaugrand G, Reid PC, Ibanez F, Lindley JA, Edwards M. 2002. Reorganization of North Atlantic marine copepod biodiversity and climate. *Science* 296:1692–94

Beebee TJC. 1995. Amphibian breeding and climate. *Nature* 374:219–20

Beever EA, Brussard PF, Berger J. 2003. Patterns of apparent extirpation among isolated populations of pikas (*Ochotona princeps*) in the Great Basin. *J. Mammal.* 84:37–54

Berteaux D, Reale D, McAdam AG, Boutin S. 2004. Keeping pace with fast climate change: Can Arctic life count on evolution? *Integr. Comp. Biol.* 44:140–51

Blenckner T, Hillebrand H. 2002. North Atlantic Oscillation signatures in aquatic and terrestrial ecosystems: a meta-analysis. *Glob. Change Biol.* 8:203–12

Both C, Artemyev AV, Blaauw B, Cowie RJ, Dekhuijzen AJ, et al. 2004. Large-scale geographical variation confirms that climate change causes birds to lay earlier. *Proc. R. Soc. London Ser. B* 271:1657–62

Bradley NL, Leopold AC, Ross J, Wellington H. 1999. Phenological changes reflect climate change in Wisconsin. *Proc. Natl. Acad. Sci. USA* 96:9701–4

Bradshaw WE, Holzapfel CM. 2001. Genetic shift in photoperiodic response correlated with global warming. *Proc. Natl. Acad. Sci. USA* 98:14509–11

Briffa KR, Schweingruber FH, Jones PD, Osborn TJ, Harris IC, et al. 1998a. Trees tell of past climates, but are they speaking less clearly today? *Philos. Trans. R. Soc. London Ser. B* 353:65–73

Briffa KR, Schweingruber FH, Jones PD, Osborn TJ, Shiyatov SG, Vaganov EA. 1998b. Reduced sensitivity of recent tree-growth to temperature at high northern latitudes. *Nature* 391:678–82

Brown JL, Li SH, Bhagabati N. 1999. Long-tem trend toward earlier breeding in an American bird: a response to global warming? *Proc. Natl. Acad. Sci. USA* 96:5565–69

Bumpus HC. 1899. The elimination of the unfit as illustrated by the introduced sparrow, *Passer domesicus*. In *Biological Lectures Delivered at the Marine Biological Laboratory of Wood's Holl*, 1896–97, pp. 209–26. Boston: Ginn & Co

Burger M. 1998. *Physiological mechanisms limiting the northern boundary of the winter range of the northern cardinal* (Cardinalis cardinalis). PhD diss. Univ. Mich., Ann Arbor

Burton J. 1975. The effects of recent climatic change on British insects. *Bird Study* 22:203–4

Burton JF. 1998a. The apparent effects of climatic changes since 1850 on European lepidoptera. *Mém. Soc. R. Belge Entomol.* 38:125–44

Burton JF. 1998b. The apparent responses of European lepidottera to the climate changes of the past hundred years. *Atropos* 5:24–30

Caldeira K, Wickett ME. 2003. Anthropogenic carbon and ocean pH. *Nature* 425:365

Cayan DR, Kammerdiener SA, Dettinger MD, Caprio JM, Peterson DH. 2001. Changes in the onset of spring in the western United States. *Bull. Am. Meteorol. Soc.* 82:399–415

Chapin FS III, Berman M, Callaghan TV, Convey P, Crepin AS, et al. 2006. Polar systems. In *Millennium Ecosystem Assessment, Ecosystems and Human Well-Being, Volume 1: Current State and Trends*, ed. R Hassan, R Scholes, N Ash, pp. 717–43. Washington, DC: Island Press

Chapin FS III, Shaver GR, Giblin AE, Nadelhoffer KG, Laundre JA. 1995. Response of Arctic tundra to experimental and observed changes in climate. *Ecology* 76:694–711

Chuine I, Beaubien E. 2001. Phenology is a major determinant of tree species range. *Ecol. Lett.* 4:500–10

Coope GR. 1994. The response of insect faunas to glacial-interglacial climatic fluctuations. *Philos. Trans. R. Soc. London Ser. B.* 344:19–26

Crick HQ, Dudley C, Glue DE. 1997. UK birds are laying eggs earlier. *Nature* 388:526

Croxall JP, Trathan PN, Murphy EJ. 2002. Environmental change and Antarctic seabird populations. *Science* 297:1510–14

Crozier L. 2003. Winter warming facilitates range expansion: cold tolerance of the butterfly *Atalopedes campestris*. *Oecologia* 135:648–56

Crozier L. 2004. Warmer winters drive butterfly range expansion by increasing survivorship. *Ecology* 85:231–41

Davis MB, Shaw RG. 2001. Range shifts and adaptive responses to quaternary climate change. *Science* 292:673–79

Davis MB, Zabinski C. 1992. Changes in geographical range resulting from greenhouse warming: effects on biodiversity in forests. In *Global Warming and Biological Diversity*, ed. TEL Peters, R Lovejoy, pp. 297–308. New Haven, CT: Yale Univ. Press

Dennis RLH. 1993. *Butterflies and Climate Change*. Manchester, UK: Manchester Univ. Press

Derocher AE. 2005. Population ecology of polar bears at Svarlbad, Norway. *Popul. Ecol.* 47:267–75

Derocher AE, Lunn NJ, Stirling I. 2004. Polar bears in a warming climate. *Integr. Comp. Biol.* 44:163–76

Descimon H, Bachelard P, Boitier E, Pierrat V. 2006. Decline and extinction of *Parnassius apollo* populations in France—continued. In *Studies on the Ecology and Conservation of Butterflies in Europe (EBIE)*, ed. E Kühn, R Feldmann, J Settele. Bulgaria: PENSOFT. In press

deYoung B, Harris R, Alheit J, Beaugrand G, Mantua N, Shannon L. 2004. Detecting regime shifts in the ocean: data considerations. *Prog. Oceanogr.* 60:143–64

Dobzhansky TH. 1943. Genetics of natural populations. IX. Temporal changes in the composition of populations of *Drosophila pseudoobscura*. *Genetics* 28:162–86

Dobzhansky TH. 1947. A response of certain gene arrangements in the third chromosome of *Drosophila pseudoobscura* to natural selection. *Genetics* 32:142–60

Donner SD, Skirving WJ, Little CM, Oppenheimer M, Hoegh-Guldberg O. 2005. Global assessment of coral bleaching and required rates of adaptation under climate change. *Glob. Change Biol.* 11:2251–65

Doran PT, Priscu JC, Lyons WB, Walsh JE, Fountain AG, et al. 2002. Antarctic climate cooling and terrestrial ecosystem response. *Nature* 415:517–22

Dose V, Menzel A. 2004. Bayesian analysis of climate change impacts in phenology. *Glob. Change Biol.* 10:259–72

Dunn PO, Winkler DW. 1999. Climate change has affected the breeding date of tree swallows throughout North America. *Proc. R. Soc. London Ser. B* 266:2487–90

Easterling DR, Meehl J, Parmesan C, Chagnon S, Karl TR, Mearns LO. 2000. Climate extremes: observations, modeling, and impacts. *Science* 289:2068–74

Ehrlich PR, Murphy DD, Singer MC, Sherwood CB, White RR, Brown IL. 1980. Extinction, reduction, stability and increase: the responses of checkerspot butterfly populations to the California drought. *Oecologia* 46:101–5

Emslie SD, Fraser W, Smith RC, Walker W. 1998. Abandoned penguin colonies and environmental change in the Palmer Station area, Anvers Island, Anatarctic Peninsula. *Antarct. Sci.* 10:257–68

Ferguson SH, Stirling I, McLoughlin P. 2005. Climate change and ringed seal (*Phoca hispida*) recruitment in western Hudson Bay. *Mar. Mamm. Sci.* 21:121–35

Ford EB. 1945. *Butterflies*. London: Collins

Ford SE. 1996. Range extension by the oyster parasite *Perkinsus marinus* into the northeastern United States: response to climate change? *J. Shellfish Res.* 15:45–56

Forister ML, Shapiro AM. 2003. Climatic trends and advancing spring flight of butterflies in lowland California. *Glob. Change Biol.* 9:1130–35

Fraser WR, Trivelpiece WZ, Ainley DC, Trivelpiece SG. 1992. Increases in Antarctic penguin populations: reduced competition with whales or a loss of sea ice due to environmental warming? *Polar Biol.* 11:525–31

Garcia-Ramos G, Kirkpatrick M. 1997. Genetic models of adaptation and gene flow in peripheral populations. *Evolution* 51:21–28

Gaston AJ, Gilchrist HG, Hipfner M. 2005. Climate change, ice conditions and reproduction in an Arctic nesting marine bird: Brunnich's guillemot (*Uria lomvia* L.). *J. Anim. Ecol.* 74:832–41

Gibbs JP, Breisch AR. 2001. Climate warming and calling phenology of frogs near Ithaca, New York, 1900–1999. *Conserv. Biol.* 15:1175–78

Gordo O, Brotons L, Rerrer X, Comass P. 2005. Do changes in climate patterns in wintering areas affect the timing of the spring arrival of trans-Saharan migrant birds? *Glob. Change Biol.* 11:12–21

Grabherr G, Gottfried M, Pauli H. 1994. Climate effects on mountain plants. *Nature* 369:448

Grinnell J. 1917. Field tests of theories concerning distributional control. *Am. Nat.* 51:115–28

Gross L. 2005. As the Antarctic ice pack recedes, a fragile ecosystem hangs in the balance. *PLoS Biol.* 3(4):e127

Gudmundsson F. 1951. The effects of the recent climatic changes on the bird life of Iceland. *Proc. 10th Int. Ornithol. Congr., Uppsala, June 1950*, pp. 502–14

Gurevitch J, Hedges LV. 1999. Statistical issues in ecological meta-analyses. *Ecology* 80:1142–49

Haeger JF. 1999. *Danaus chrysippus* (Linnaeus 1758) en la Península Ibérica: migraciones o dinámica de metapoblaciones? *Shilap* 27:423–30

Hanski I, Erälahti C, Kankare M, Ovaskainen O, Sirén H. 2004. Variation in migration propensity among individuals maintained by landscape structure. *Ecol. Lett.* 7:958–66

Hari RE, Livingstone DM, Siber R, Burkhardt-Holm P, Güttinger H. 2006. Consequences of climatic change for water temperature and brown trout population in Alpine rivers and streams. *Glob. Change Biol.* 12:10–26

Harrington R, Woiwod I, Sparks T. 1999. Climate change and trophic interactions. *Trends Ecol. Evol.* 14:146–50

Harris G. 1964. Climatic changes since 1860 affecting European birds. *Weather* 19:70–79

Hays GC, Richardson AJ, Robinson C. 2005. Climate change and marine plankton. *Trends Ecol. Evol.* 20:337–44

Heath J. 1974. A century of changes in the lepidoptera. In *The Changing Flora and Fauna of Britain*, ed. DL Hawkesworth, 6:275–92. London: Syst. Assoc.

Hemp A. 2005. Climate change-driven forest fires marginalize the impact of ice cap wasting on Kilimanjaro. *Glob. Change Biol.* 11:1013–23

Hickling R, Roy DB, Hill JK, Thomas CD. 2005. A northward shift of range margins in British Odonata. *Glob. Change Biol.* 11:502–6

Hill GE, Sargent RR, Sargent MB. 1998. Recent change in the winter distribution of Rufous Hummingbirds. *Auk* 115:240–45

Hill JK, Thomas CD, Fox R, Telfer MG, Willis SG, et al. 2002. Responses of butterflies to twentieth century climate warming: implications for future ranges. *Proc. R. Soc. London Ser. B* 269:2163–71

Hill JK, Thomas CD, Lewis OT. 1999. Flight morphology in fragmented populations of a rare British butterfly, *Hesperia comma*. *Biol. Conserv.* 87:277–84

Hoegh-Guldberg O. 1999. Climate change, coral bleaching and the future of the world's coral reefs. *Mar. Freshw. Res.* 50:839–66

Hoegh-Guldberg O. 2005a. Low coral cover in a high-CO_2 world. *J. Geophys. Res.* 110:C09S06

Hoegh-Guldberg O. 2005b. Marine ecosystems and climate change. See Lovejoy & Hannah 2005, pp. 256–71

Hoegh-Guldberg O, Jones RJ, Ward S, Loh WK. 2002. Is coral bleaching really adaptive? *Nature* 415:601–2

Hoegh-Guldberg O, Pearse JS. 1995. Temperature, food availability, and the development of marine invertebrate larvae. *Am. Zool.* 35:415–25

Hoffmann AA, Blows MW. 1994. Species borders: ecological and evolutionary perspectives. *Trends Ecol. Evol.* 9:223–27

Holbrook SJ, Schmitt RJ, Stephens JS Jr. 1997. Changes in an assemblage of temperate reef fishes associated with a climatic shift. *Ecol. Appl.* 7:1299–310

Holmgren M, Scheffer M, Ezcurra E, Gutierrez JR, Mohren GMJ. 2001. El Niño effects on the dynamics of terresrial ecosystems. *Trends Ecol. Evol.* 16:89–94

Holt RD. 2003. On the evolution ecology of species' ranges. *Evol. Ecol. Res.* 5:159–78

Howell SNG. 2002. *Hummingbirds of North America*. San Diego, CA: Academic

Hughes L. 2000. Biological consequences of global warming: Is the signal already apparent? *Trends Ecol. Evol.* 15:56–61

Huntley B. 1991. How plants respond to climate change: migration rates, individualism and the consequences for the plant communities. *J. Bot.* 67:15–22

Hüppop O, Hüppop K. 2003. North Atlantic Oscillation and timing of spring migration in birds. *Proc. R. Soc. London Ser. B* 270:233–40

Innes JL. 1991. High-altitude and high-latitude tree growth in relation to past, present, and future global climate change. *Holocene* 1:168–73

Inouye DW, Barr B, Armitage KB, Inouye BD. 2000. Climate change is affecting altitudinal migrants and hibernating species. *Proc. Natl. Acad. Sci. USA* 97:1630–33

IPCC (Intergovernmental Panel Climate Change). 2001a. *Climate Change 2001: Impacts, Adaptation, and Vulnerability, Contribution of Working Group II to the Intergovernmental Panel on Climate Change Third Assessment Report*, ed. JJ McCarthy, OF Canziani, NA Leary, DJ Dokken, KS White. Cambridge, UK: Cambridge Univ. Press

IPCC (Intergovernmental Panel Climate Change). 2001b. *Climate Change 2001: The Science of Climate Change, Contribution of Working Group I to the Intergovernmental Panel on Climate Change Third Assessment Report*, ed. JT Houghton, Y Ding, DJ Griggs, M Noguer, PJ van der Linden, X Dai, K Maskell, CA Johnson. Cambridge, UK: Cambridge Univ. Press

Jacoby GC, D'Arrigo RD. 1995. Tree ring width and density evidence of climatic and potential forest change in Alaska. *Glob. Biogeochem. Cycles* 9:227–34

Johnson T, Dozier J, Michaelsen J. 1999. Climate change and Sierra Nevada snowpack. *IAHS Publ.* 256:63–70

Johnston RF, Niles DM, Rohwer SA. 1972. Hermon Bumpus and natural selection in the house sparrow *Passer domesticus*. *Evolution* 26:20–31

Jordano D, Retamosa EC, Fernandez H. 1991. Factors facilitating the continued presence of *Colotis evagore* (Klug 1829) in southern Spain. *J. Biogeogr.* 18:637–46

Jump AS, Peñuelas J. 2005. Running to stand still: adaptation and the response of plants to rapid climate change. *Ecol. Lett.* 8:1010–20

Kaisila J. 1962. Immigration und Expansion der Lepidopteren in Finnland in den Jahren 1869–1960. *Acta Entomol. Fenn.* 18:1–452

Kalela O. 1949. Changes in geographic ranges in the avifauna of northern and central Europe in response to recent changes in climate. *Bird-Band.* 20:77–103

Kalela O. 1952. Changes in the geographic distribution of Finnish birds and mammals in relation to recent changes in climate. In *The Recent Climatic Fluctuation in Finland and its Consequences: A Symposium*, ed. I Hustichi, pp. 38–51. Helsinki: Fennia

Karieva PM, Kingsolver JG, Huey RB, eds. 1993. *Biotic Interactions and Global Change*. Sunderland, MA: Sinauer

Karl TR, Trenberth KE. 2003. Modern global climate change. *Science* 302:1719–23

Keeling CD, Chin JFS, Whorf TP. 1996. Increased activity of northern vegetation inferred from atmospheric CO_2 measurements. *Nature* 382:146–49

Kirkpatrick M, Barton NH. 1997. Evolution of a species' range. *Am. Nat.* 150:1–23

Kullman L. 2001. 20th century climate warming and tree-limit rise in the southern Scandes of Sweden. *Ambio* 30:72–80

Kutz SJ, Hoberg EP, Polley L, Jenkins EJ. 2005. Global warming is changing the dynamics of Arctic host-parasite systems. *Proc. R. Soc. London Ser. B* 272:2571–76

Lapenis A, Shvidenko A, Shepaschenko D, Nilsson S, Aiyyer A. 2005. Acclimation of Russian forests to recent changes in climate. *Glob. Change Biol.* 11:2090–102

Lescop-Sinclair K, Payette S. 1995. Recent advance of the Arctic treeline along the eastern coast of Hudson Bay. *J. Ecol.* 83:929–36

Levitan M. 2003. Climatic factors and increased frequencies of 'southern' chromosome forms in natural populations of *Drosophila robusta*. *Evol. Ecol. Res.* 5:597–604

Lewontin RC, Birch LC. 1966. Hybridization as a source of variation for adaptation to new environments. *Evolution* 20:315–36

Lindgren E, Gustafson R. 2001. Tick-borne encephalitis in Sweden and climate change. *Lancet* 358:16–18

Logan JA, Regniere J, Powell JA. 2003. Assessing the impacts of global warming on forest pest dynamics. *Front. Ecol. Environ.* 1:130–37

Lovejoy T, Hannah L, eds. 2005. In *Climate Change and Biodiversity*. New Haven, CT: Yale Univ. Press

Luckman B, Kavanagh T. 2000. Impact of climate fluctuations on mountain environments in the Canadian Rockies. *Ambio* 29:371–80

Lynch M, Lande R. 1993. Evolution and extinction in response to environmental change. See Karieva et al. 1993, pp. 234–50

MacArthur RM. 1972. *Geographical Ecology*. New York: Harper & Row

Marttila O, Haahtela T, Aarnio H, Ojalainen P. 1990. *Suomen Päiväperhoset*. Helsinki: Kirjayhtymä

McGowan JA, Cayan DR, Dorman LM. 1998. Climate-ocean variability and ecosystem response in the Northeast Pacific. *Science* 281:210–17

Meehl GA, Zwiers F, Evans J, Knutson T, Mearns LO, Whetton P. 2000. Trends in extreme weather and climate events: issues related to modeling extremes in projections of future climate change. *Bull. Am. Meteorol. Soc.* 81:427–36

Menzel A. 2000. Trends in phenological phases in Europe between 1951 and 1996. *Int. J. Biometeorol.* 44:76–81

Menzel A. 2005. A 500 year pheno-climatological view on the 2003 heatwave in Europe assessed by grape harvest dates. *Meteorol. Z.* 14:75–77

Menzel A, Dose V. 2005. Analysis of long-term time-series of beginning of flowering by Bayesian function estimation. *Meteorol. Z.* 14:429–34

Menzel A, Fabian P. 1999. Growing season extended in Europe. *Nature* 397:659

Menzel A, Jakobi G, Ahas R, Scheifinger H, Estrella N. 2003. Variations of the climatological growing season (1951–2000) in Germany compared with other countries. *Int. J. Climatol.* 23:793–812

Meshinev T, Apostolova I, Koleva E. 2000. Influence of warming on timberline rising: a case study on *Pinus peuce* Griseb. in Bulgaria. *Phytocoenologia* 30:431–38

Mikkola K. 1997. Population trends of Finnish Lepidoptera during 1961–1996. *Entomol. Fenn.* 3:121–43

Moiseev PA, Shiyatov SG. 2003. The use of old landscape photographs for studying vegetation dynamics at the treeline ecotone in the Ural Highlands, Russia. In *Alpine Biodiversity in Europe*, ed. L Nagy, G Grabherr, C Körner, DBA Thompson, pp. 423–36. Berlin: Springer-Verlag

Myneni RB, Keeling CD, Tucker CJ, Asrar G, Nemani RR. 1997. Increased plant growth in the northern high latitudes from 1981 to 1991. *Nature* 386:698–702

Nylin S. 1988. Host plant specialization and seasonality in a phytophagous butterfly, *Polygonia c-album* (Nymphalidae). *Oikos* 53:381–86

Orr JC, Fabry VJ, Aumont O, Bopp L, Doney SC, et al. 2005. Anthropogenic ocean acidification over the twenty-first century and its impact on calcifying organisms. *Nature* 437:681–86

Ottersen G, Planque B, Belgrano A, Post E, Reid PC, Stenseth NC. 2001. Ecological effects of the North Atlantic Oscillation. *Oecologia* 128:1–14

Parmesan C. 1996. Climate and species' range. *Nature* 382:765–66

Parmesan C. 2002. Detection of range shifts: general methodological issues and case studies using butterflies. In *Fingerprints of Climate Change: Adapted Behaviour and Shifting Species' Ranges*, ed. G-R Walther, CA Burga, PJ Edwards, pp. 57–76. Dordrecht, Netherlands: Kluwer Acad./Plenum

Parmesan C. 2003. Butterflies as bio-indicators of climate change impacts. In *Evolution and Ecology Taking Flight: Butterflies as Model Systems*, ed. CL Boggs, WB Watt, PR Ehrlich, pp. 541–60. Chicago: Univ. Chicago Press

Parmesan C. 2005a. Detection at multiple levels: *Euphydryas editha* and climate change. Case study. See Lovejoy & Hannah 2005, pp. 56–60

Parmesan C. 2005b. Range and abundance changes. See Lovejoy & Hannah 2005, pp. 41–55

Parmesan C, Gaines S, Gonzalez L, Kaufman DM, Kingsolver J, et al. 2005. Empirical perspectives on species' borders: environmental change as challenge and opportunity. *Oikos* 108:58–75

Parmesan C, Galbraith H. 2004. *Observed Ecological Impacts of Climate Change in North America*. Arlington, VA: Pew Cent. Glob. Clim. Change

Parmesan C, Martens P. 2006. Climate change. In *Biodiversity, Health and the Environment: SCOPE/Diversitas Rapid Assessment Project*, ed. O Sala, L Meyerson, C Parmesan. Washington, DC: Island Press. In press

Parmesan C, Root TL, Willig MR. 2000. Impacts of extreme weather and climate on terrestrial biota. *Bull. Am. Meteorol. Soc.* 81:443–50

Parmesan C, Ryrholm N, Stefanescu C, Hill JK, Thomas CD, et al. 1999. Poleward shifts in geographical ranges of butterfly species associated with regional warming. *Nature* 399:579–83

Parmesan C, Yohe G. 2003. A globally coherent fingerprint of climate change impacts across natural systems. *Nature* 421:37–42

Pauli H, Gottfried M, Grabherr G. 1996. Effects of climate change on mountain ecosystems: upward shifting of mountain plants. *World Res. Rev.* 8:382–90

Paulson DR. 2001. Recent odonata records from southern Florida: effects of global warming? *Int. J. Odonatol.* 4:57–69

Peñuelas J, Filella I. 2001. Responses to a warming world. *Science* 294:793–94

Perry AL, Low PJ, Ellis JR, Reynolds JD. 2005. Climate change and distribution shifts in marine fishes. *Science* 308:1912–15

Pollard E. 1979. Population ecology and change in range of the white admiral butterfly *Ladoga camilla* L. in England. *Ecol. Entomol.* 4:61–74

Pollard E, Eversham BC. 1995. Butterfly monitoring 2: interpreting the changes. In *Ecology and Conservation of Butterflies*, ed. AS Pullin, pp. 23–36. London: Chapman & Hall

Post E, Langvatn R, Forchhammer MC, Stenseth NC. 1999. Environmental variation shapes sexual dimorhism in red deer. *Proc. Natl. Acad. Sci. USA* 96:4467–71

Pounds JA, Bustamente MR, Coloma LA, Consuegra JA, Fogden MPL, et al. 2006. Widespread amphibian extinctions from epidemic disease driven by global warming. *Nature* 439:161–67

Pounds JA, Fogden MPL, Campbell JH. 1999. Biological response to climate change on a tropical mountain. *Nature* 398:611–15

Pounds JA, Fogden MPL, Masters KL. 2005. Responses of natural communities to climate change in a highland tropical forest. Case study. See Lovejoy & Hannah 2005, pp. 70–74

Precht H, Christophersen J, Hensel H, Larcher W. 1973. *Temperature and Life*. New York: Springer-Verlag

Precht WF, Aronson RB. 2004. Climate flickers and range shifts of reef corals. *Front. Ecol. Environ.* 2:307–14

Réale D, McAdam A, Outin G, Berteaux S. 2003. Genetic and plastic response of a northern mammal to climate change. *Proc. R. Soc. London Ser. B* 270:591–96

Reid PC, Edwards M, Hunt HG, Warner AJ. 1998. Phytoplankton change in the North Atlantic. *Nature* 391:546

Richardson AJ, Schoeman DS. 2004. Climate impact on plankton ecosystems in the Northeast Atlantic. *Science* 305:1609–12

Rodríguez-Trelles F, Álvarez G, Zapata C. 1996. Time-series analysis of seasonal changes of the O inversion polymorphism of *Drosophila subobscura*. *Genetics* 142:179–87

Rodríguez-Trelles F, Rodriguez MA. 1998. Rapid micro-evolution and loss of chromosomal diversity in *Drosophila* in response to climate warming. *Evol. Ecol.* 12:829–38

Rodríguez-Trelles F, Rodriguez MA, Scheiner SM. 1998. Tracking the genetic effects of global warming: *Drosophila* and other model systems. *Conserv. Ecol.* 2:2

Roessig JM, Woodley CM, Cech JJ, Hansen LJ. 2004. Effects of global climate change on marine and estuarine fishes. *Rev. Fish Biol. Fish.* 14:215–75

Roetzer T, Wittenzeller M, Haeckel H, Nekovar J. 2000. Phenology in central Europe: difference and trends of spring phenophases in urban and rural areas. *Int. J. Biometeorol.* 44:60–66

Root TL. 1988. Energy constraints on avian distributions and abundances. *Ecology* 69:330–39

Root TL, Hughes L. 2005. Present and future phenological changes in wild plants and animals. See Lovejoy & Hannah 2005, pp. 61–69

Root TL, Price JT, Hall KR, Schneider SH, Rosenzweig C, Pounds JA. 2003. Fingerprints of global warming on wild animals and plants. *Nature* 421:57–60

Rowan R. 2004. Thermal adaptation in reef coral symbionts. *Nature* 430:742

Roy DB, Sparks TH. 2000. Phenology of British butterflies and climate change. *Glob. Change Biol.* 6:407–16

Sagarin RD, Barry JP, Gilman SE, Baxter CH. 1999. Climate-related change in an intertidal community over short and long time scales. *Ecol. Monogr.* 69:465–90

Salomonsen F. 1948. The distribution of birds and the recent climatic change in the North Atlantic area. *Dansk. Orn. Foren. Tidsskr.* 42:85–99

Schwartz MD. 1998. Green-wave phenology. *Nature* 394:839–40

Schwartz MD. 1999. Advancing to full bloom: planning phenological research for the 21st century. *Int. J. Biometeorol.* 42:113–18

Scriber JM, Lederhouse RC. 1992. The thermal environment as a resource dictating geographic patterns of feeding specialization of insect herbinores. In *Effects of Resource Distribution on Plant-Animal Interactions*, ed. MR Hunter, T Ohgushi, PW Price, pp 429–66. New York: Academic

Severnty DL. 1977. The use of data on the distribution of birds to monitor climatic changes. *Emu* 77:162–66

Shoo LP, Williams SE, Hero JM. 2006. Detecting climate change induces range shifts: Where and how should we be looking? *Aust. Ecol.* 31:22–29

Singer MC. 1972. Complex components of habitat suitability within a butterfly colony. *Science* 176:75–77

Singer MC, Ehrlich PR. 1979. Population dynamics of the checkerspot butterfly *Euphydryas editha*. *Fortschr. Zool.* 25:53–60

Singer MC, Thomas CD. 1996. Evolutionary responses of a butterfly metapopulation to human and climate-caused environmental variation. *Am. Nat.* 148:S9–39

Smith AT. 1974. The distribution and dispersal of pikas: influences of behavior and climate. *Ecology* 55:1368–76

Smith RC, Ainley D, Kaber K, Domack E, Emslie S, et al. 1999. Marine ecosystem sensitivity to historical climate change in the Antarctic Peninsula. *BioScience* 49:393–404

Smith RIL. 1994. Vascular plants as bioindicators of regional warming in Antarctica. *Oecologia* 99:322–28

Smol JP, Wolfe AP, Birks HJB, Douglas MSV, Jones VJ, et al. 2005. Climate-driven regime shifts in the biological communities of Arctic lakes. *Proc. Natl. Acad. Sci. USA* 102:4397–402

Southward AJ, Hawkins SJ, Burrows MT. 1995. Seventy years' observations of changes in distribution and abundance of zooplankton and intertidal organisms in the western English Channel in relation to rising sea temperature. *J. Therm. Biol.* 20:127–55

Southward AJ, Langmead O, Hardman-Mountford NJ, Aiken J, Boalch GT, et al. 2005. Long-term oceanographic and ecological research in the western English Channel. *Adv. Mar. Biol.* 47:1–105

Sparks TH, Menzel A. 2002. Observed changes in seasons, an overview. *Int. J. Climatol.* 22:1715–26

Sparks TH, Yates TJ. 1997. The effect of spring temperature on the appearance dates of British butterflies 1883–1993. *Ecography* 20:368–74

Spear LB, Ainley DG. 1999. Migration routes of sooty shearwaters in the Pacific Ocean. *Condor* 101:205–18

Stefanescu C, Peñuelas J, Filella I. 2003. Effects of climatic change on the phenology of butterflies in the northwest Mediterranean Basin. *Glob. Change Biol.* 9:1494

Stirling I. 2002. Polar bears and seals in the eastern Beaufort Sea and Amundsen Gulf: a synthesis of population trends and ecological relationships over three decades. *Arctic* 55:59–76

Stirling I, Lunn NJ, Iacozza J. 1999. Long-term trends in the population ecology of polar bears in western Hudson Bay in relation to climatic change. *Arctic* 52:294–306

Sturm M, Racine C, Tape K. 2001. Increasing shrub abundance in the Arctic. *Nature* 411:546–47

Sturm M, Schimel J, Mechaelson G, Welker JM, Oberbauer SF, et al. 2005. Winter biological processes could help convert Arctic tundra to shrubland. *BioScience* 55:17–26

Swetnam TW, Betancourt JL. 1998. Mesoscale disturbance and ecological response to decadal climatic variability in the American Southwest. *J. Clim.* 11:3128–47

Taylor RH, Wilson PR. 1990. Recent increase and southern expansion of Adelie penguin populations in the Ross Sea, Antarctica, related to climatic warming. *N.Z. J. Ecol.* 14:25–29

Thomas C. 2005. Recent evolutionary effects of climate change. See Lovejoy & Hannah 2005, pp. 75–90

Thomas CD, Bodsworth EJ, Wilson RJ, Simmons AD, Davies ZG, et al. 2001. Ecological and evolutionary processes at expanding range margins. *Nature* 411:577–81

Thomas CD, Lennon JJ. 1999. Birds extend their ranges northwards. *Nature* 399:213

Thomas CD, Singer MC, Boughton D. 1996. Catastrophic extinction of population sources in a butterfly metapopulation. *Am. Nat.* 148:957–75

Travis J, Futuyma DJ. 1993. Global change: lessons from and for evolutionary biology. See Karieva et al. 1993, pp. 234–50

Uvarov BP. 1931. Insects and climate. *R. Entomol. Soc. London* 79:174–86

van Herk CM, Aptroot A, van Dobben HF. 2002. Long-term monitoring in the Netherlands suggests that lichens respond to global warming. *Lichenologist* 34:141–54

Van Nouhuys S, Lei G. 2004. Parasitoid-host metapopulation dynamics: the causes and consequences of phenological asynchrony. *J. Anim. Ecol.* 73:526–35

Visser ME, Both C. 2005. Shifts in phenology due to global climate change: the need for a yardstick. *Proc. R. Soc. B* 272:2561–69

Walther GR, Hughes L, Vitousek P, Stenseth NC. 2005. Consensus on climate change. *Trends Ecol. Evol.* 20:648–49

Walther GR, Post E, Convery P, Menzel A, Parmesan C, et al. 2002. Ecological responses to recent climate change. *Nature* 416:389–95

Wardle P, Coleman MC. 1992. Evidence for rising upper limits of four native New Zealand forest trees. *N.Z. J. Bot.* 30:303–14

Warren MS. 1992. The conservation of British butterflies. In *The Ecology of Butterflies in Britain*, ed. RLH Dennis, pp. 246–74. Oxford, UK: Oxford Univ. Press

Webb T III, Bartlein PJ. 1992. Global changes during the last 3 million years: climatic controls and biotic responses. *Annu. Rev. Ecol. Syst.* 23:141–73

Weiser W, ed. 1973. *Effects of Temperature on Ectothermic Organisms*. New York: Springer-Verlag

Weiss SB, Murphy DD, White RR. 1988. Sun, slope and butterflies: topographic determinants of habitat quality for *Euphydryas editha*. *Ecology* 69:1486–96

White MA, Nemani RR, Thornton PE, Running SW. 2002. Satellite evidence of phenological differences between urbanized and rural areas of the eastern United States deciduous broadleaf forest. *Ecosystems* 5:260–73

White MA, Running SW, Thornton PE. 1999. The impact of growing-season length variability on carbon assimilation and evapotranspiration over 88 years in the eastern US deciduous forest. *Int. J. Biometeorol.* 42:139–45

Wilkinson CR, ed. 2000. *Global Coral Reef Monitoring Network: Status of Coral Reefs of the World in 2000*. Townsville, Qld: Aust. Inst. Mar. Sci.

Williamson K. 1975. Birds and climatic change. *Bird Study* 22:143–64

Wilson JW, Gutiérrez D, Martinez D, Agudo R, Monserrat VJ. 2005. Changes to the elevational limits and extent of species ranges associated with climate change. *Ecol. Lett.* 8:1138–46

Wilson PR, Ainley DG, Nur N, Jacobs SS, Barton KJ, et al. 2001. Adélie penguin population change in the pacific sector of Antarctica: relation to sea-ice extent and the Antarctic Circumpolar Current. *Mar. Ecol. Prog. Ser.* 213:301–9

Winder M, Schindler DE. 2004. Climate change uncouples trophic interactions in an aquatic ecosystem. *Ecology* 85:2100–6

Woodward FI. 1987. *Climate and Plant Distribution*. Cambridge, UK: Cambridge Univ. Press

RELATED RESOURCES

Alford, RA, Richards SJ. 1999. Global amphibian declines: a problem in applied ecology. *Annu. Rev. Ecol. Syst.* 30:133–65

Glynn PW. 1988. El Niño-Southern Oscillation 1982–1983: nearshore population, community, and ecosystem responses. *Annu. Rev. Ecol. Syst.* 19:309–45

Ludwig D, Marc M, Haddad B. 2001. Ecology, conservation, and public policy. *Annu. Rev. Ecol. Syst.* 32:481–517

Smith SV, Buddemeier RW. 1992. Global change and coral reef ecosystems. *Annu. Rev. Ecol. Syst.* 23:89–118

Stevens GC, Fox JF. 1991. The causes of treeline. *Annu. Rev. Ecol. Syst.* 22:177–91

Cumulative Indexes

Contributing Authors, Volumes 33–37

A

Aanen DK, 36:563–95
Ackerly DD, 33:475–505
Agrawal AA, 33:641–64
Alfaro ME, 37:19–42
Allan JD, 35:257–84
Allen JE, 36:373–97
Altizer S, 34:517–47
Antonovics J, 34:517–47
Arbogast BS, 33:707–40
Armbruster WS, 35:375–403
Ashman T-L, 36:467–97
Ayre DJ, 37:489–517

B

Baker AC, 34:661–89
Ballard JWO, 36:621–42
Barnosky AD, 37:215–50
Barton NH, 34:99–125
Basset Y, 36:597–620
Battin J, 35:491–522
Bazzaz FA, 35:323–45
Beckenbach AT, 37:545–79
Beerli P, 33:707–40
Behrensmeyer AK, 35:285–322
Benard MF, 35:651–73
Bininda-Emonds OR, 33:265–89
Bjørnstad ON, 35:467–90
Blackburn TM, 34:71–98
Bobe R, 35:285–322
Bowers S, 37:519–44
Boyce MS, 37:317–42

Bradshaw HD Jr, 36:243–66
Brady KU, 36:243–66
Brakefield PM, 34:633–60
Brinkmann H, 36:541–62
Brooks R, 37:43–66
Buckley TR, 37:545–79
Burd M, 36:467–97
Burkepile DE, 35:175–97
Byers JE, 36:643–89

C

Calsbeek R, 37:581–610
Campbell DR, 36:467–97
Cardon ZG, 37:459–88
Carpenter S, 35:557–81
Case TJ, 33:181–233
Caudill CC, 35:175–97
Chapin FS III, 34:455–85
Charlesworth B, 34:99–125
Charlesworth D, 34:99–125
Chave J, 34:575–604
Chequer AD, 35:175–97
Chetkiewicz C-LB, 37:317–42
Chiappe LM, 33:91–124
Cole CT, 34:213–37
Cory JS, 34:239–72
Côté SD, 35:113–47
Cotner JB, 36:219–42
Cowan JH Jr, 34:127–51
Crease TJ, 36:219–42
Cunnningham AA, 34:517–47
Cushman FA, 36:499–518

D

Daehler C, 34:183–211
Davic RD, 35:405–34
Dawson TE, 33:507–59
Day TA, 33:371–96
Dayan T, 34:153–81
Dayton PK, 33:449–73
DeAngelis DL, 36:147–68
Dearing MD, 36:169–89
Delsuc F, 36:541–62
Demuth JP, 37:289–316
D'Hondt S, 36:295–317
DiMichele WA, 35:285–322
Dixon AFG, 36:345–72
Dobson AP, 34:517–47
Donoghue MJ, 33:475–505
Dudash MR, 35:375–403; 36:467–97
Dukas R, 35:347–74
Duncan RP, 34:71–98
Dussault C, 35:113–47
Dyke GJ, 33:91–124
Dynesius M, 33:741–78

E

Eckert CG, 36:47–79
Edwards SV, 33:707–40
Elmqvist T, 35:557–81
Elser JJ, 36:219–42
Emmerson M, 36:419–44
Evans EW, 37:95–122
Eviner VT, 34:455–85
Ezenwa V, 34:517–47

F

Fahrig L, 34:487–515
Falkowski PG, 35:523–56
Falster DS, 33:125–59
Finkel ZV, 35:523–56
Fletcher RJ Jr, 35:491–522
Foley WJ, 36:169–89
Folke C, 35:557–81
Ford SE, 35:31–54
Foster MS, 37:343–72
Frati F, 37:545–79
French V, 34:633–60
Funk DJ, 34:397–423

G

Gage DJ, 37:459–88
Gerardo NM, 36:563–95
Gilbert GS, 35:675–700
Gioia P, 35:623–50
Gittleman JL, 33:265–89; 34:517–47
Goodwillie C, 36:47–79
Goulson D, 34:1–26
Graham AL, 36:373–97
Graham CH, 36:519–39
Grove SJ, 33:1–23
Grzebyk D, 35:523–56
Gunderson L, 35:557–81

H

Halanych KM, 35:229–56
Hallinan ZP, 35:175–97
Hampe A, 37:187–214
Hansen TF, 37:123–57
Harvell CD, 37:251–88
Hättenschwiler S, 36:191–218
Hauser MD, 36:499–518
Hawkins S, 37:373–404
Hay ME, 35:175–97
Hedrick PW, 37:67–93
Heil M, 34:425–53
Heiman KW, 36:643–89
Helmuth B, 37:373–404
Hill GE, 34:27–49
Hoffman AA, 37:433–58
Holder MT, 37:19–42
Holland SM, 33:561–88
Holling CS, 35:557–81
Holway D, 33:181–233
Hopper SD, 35:623–50

Huber SK, 35:55–87
Hungate BA, 37:611–36

I

Irwin RE, 35:435–66

J

Jansson R, 33:741–78
Jennions MD, 37:43–66
Johannes RE, 33:317–40
Johnson JB, 33:665–706
Johnston MO, 36:467–97
Jones KE, 34:517–47
Jones LE, 37:251–288
Jones MB, 37:519–44
Joyce P, 36:445–66

K

Kalisz S, 36:47–79
Karban R, 33:641–64
Kareiva PM, 33:665–706
Katz ME, 35:523–56
Kaufman DM, 34:273–309
Kay MC, 36:643–89
Kelly D, 33:427–47
Kidwell SM, 33:561–88
Klingenberg CP, 36:1–21
Knight TM, 36:467–497
Knoll AH, 35:523–56
Koch PL, 37:215–50
Koenig WD, 35:467–90
Kokko H, 37:43–66
Kronfeld-Schor N, 34:153–81
Kruckeberg AR, 36:243–66

L

Labandeira CC, 35:285–322
Lach L, 33:181–233
Lafferty KD, 35:31–54
Lartillot N, 36:541–62
Leamy LJ, 36:1–21
Lenihan HS, 36:643–89
Levin P, 33:665–706
Levin SA, 34:575–604
Levine JM, 34:549–74
Lewinsohn TM, 36:597–620
Lieberman BS, 34:51–69
Liebhold A, 35:467–90

Linder HP, 36:107–24
Liow LH, 35:323–45
Luo Y, 37:611–36

M

Mambelli S, 33:507–59
Markow TA, 36:219–42
Marshall JC, 35:199–227
Mateos M, 36:219–42
Mazer SJ, 36:467–97
McDonald ME, 35:89–111
McKey D, 34:425–53
McLean S, 36:169–89
McPeek MA, 33:475–505
Meyer A, 34:311–38
Micheli F, 36:643–89
Mieszkowska N, 37:373–404
Milinski M, 37:159–86
Miller DJ, 37:489–517
Mitchell RJ, 36:467–97
Moles AT, 33:125–59
Mooij WM, 36:147–68
Moore P, 37:373–404
Mueller UG, 36:563–95
Muller-Landau HC, 34:575–604
Murrell DJ, 34:549–74
Mydlarz LD, 37:251–88
Myers JH, 34:239–72

N

Nathan R, 34:575–604
Neale PJ, 33:371–96
Nee S, 37:1–17
Niemi GJ, 35:89–111
Noor MAF, 34:339–64
Novotny V, 36:597–620
Nowak MD, 37:405–31
Nunn CL, 34:517–47

O

Ohgushi T, 36:81–105
Olivera B, 33:25–47
Olszewski TD, 35:285–322
Omland KE, 34:397–423

P

Pandolfi JM, 35:285–322
Pannell JR, 33:397–425

Parker IM, 35:675–700
Parker JD, 35:175–97
Parmesan C, 37:637–69
Pedersen AB, 34:517–47
Petit RJ, 37:187–214
Philippe H, 36:541–62
Pilson D, 35:149–74
Plamboeck AH, 33:507–59
Podos J, 35:55–87
Porter JW, 35:31–54
Poss M, 34:517–47
Prendeville HR, 35:149–74
Pulliam JRC, 34:517–47

R

Rabalais NN, 33:235–63
Rahel FJ, 33:291–315
Ramsey J, 33:589–639
Rand DM, 36:621–42
Read AF, 36:373–97
Reich PB, 37:611–36
Reichman OJ, 37:519–44
Ries L, 35:491–522
Rooney TP, 35:113–47
Roopnarine PD, 34:605–32
Rose KA, 34:127–51
Ruckelshaus M, 33:665–706
Rudall PJ, 36:107–24
Ruesink JL, 36:643–89

S

Sanderson MJ, 33:49–72
Scheffer M, 35:557–81
Schemske DW, 33:589–639
Scheu S, 36:191–218
Schiel DR, 37:343–72
Schildhauer MP, 37:519–44
Schlupp I, 36:399–417
Schultz TR, 36:563–95
Servedio MR, 34:339–64

Shaffer H, 33:49–72
Shine R, 36:23–46
Simmons LW, 36:125–46
Simon C, 37:545–79
Sinervo B, 37:581–610
Sisk TD, 35:491–522
Sites JW Jr, 35:199–227
Six DL, 36:563–95
Slowinski JB, 33:707–40
Snyder WE, 37:95–122
Sodhi NS, 35:323–45
Sol D, 34:71–98
Sork VL, 33:427–47
Srivastava DS, 36:267–94
Stadler B, 36:345–72
St. Clair CC, 37:317–42
Steel MA, 33:265–89
Steets JA, 36:467–97
Stevens JR, 36:499–518
Stevens RD, 34:273–309
Stewart IRK, 34:365–96
Stewart JB, 37:545–79
Strauss SY, 35:435–66
Suarez AV, 33:181–233
Sullivan J, 36:445–66
Swanson WJ, 33:161–79

T

Taft B, 35:55–87
Templer PH, 33:507–59
Thewissen J, 33:73–90
Thomson JD, 35:375–403
Thrall PH, 34:517–47
Thrush S, 33:449–73
Tiffney BH, 35:1–29
Tiunov AV, 36:191–218
Tremblay J-P, 35:113–47
Trimble AC, 36:643–89
Tsutsui ND, 33:181–233
Tu KP, 33:507–59
Turner MG, 36:319–44
Turner RE, 33:235–63

V

Vacquier VD, 33:161–79
Vamosi JC, 36:467–97
Van Buskirk J, 37:433–58
Vanni MJ, 33:341–70
van Oppen MJH, 37:489–517
Vellend M, 36:267–94
Vesk PA, 33:125–59
Vollmer SV, 37:489–517

W

Wade MJ, 37:289–316
Wakeley J, 33:707–40
Walker B, 35:557–81
Waller DM, 35:113–47
Webb CO, 33:475–505
Weider LJ, 36:219–42
Welsh HH Jr, 35:405–34
Westneat DF, 34:365–96
Westoby M, 33:125–59
Whigham DF, 35:583–621
Wiens JJ, 36:519–39
Willi Y, 37:433–58
Williams EM, 33:73–90
Willig MR, 34:273–309
Willis BL, 37:489–517
Wilson AE, 35:175–97
Wilson P, 35:375–403
Wing SL, 35:285–322
Wiseman WJ Jr, 33:235–63
Wootton JT, 36:419–44
Wright IJ, 33:125–59

Y

Yoder AD, 37:405–31

Z

Zardoya R, 34:311–38
Zwaan BJ, 34:633–60

Chapter Titles, Volumes 33–37

Volume 33 (2002)

Saproxylic Insect Ecology and the Sustainable Management of Forests	SJ Grove	33:1–23
Conus Venom Peptides: Reflections from the Biology of Clades and Species	BM Olivera	33:25–47
Troubleshooting Molecular Phylogenetic Analyses	MJ Sanderson, HB Shaffer	33:49–72
The Early Radiations of Cetacea (Mammalia): Evolutionary Pattern and Developmental Correlations	JGM Thewissen, EM Williams	33:73–90
The Mesozoic Radiation of Birds	LM Chiappe, GJ Dyke	33:91–124
Plant Ecological Strategies: Some Leading Dimensions of Variation Between Species	M Westoby, DS Falster, AT Moles, PA Vesk, IJ Wright	33:125–59
Reproductive Protein Evolution	WJ Swanson, VD Vacquier	33:161–79
The Causes and Consequences of Ant Invasions	DA Holway, L Lach, AV Suarez, ND Tsutsui, TJ Case	33:181–233
Gulf of Mexico Hypoxia, a.k.a. "The Dead Zone"	NN Rabalais, RE Turner, WJ Wiseman Jr	33:235–63
The (Super)Tree of Life: Procedures, Problems, and Prospects	ORP Bininda-Emonds, JL Gittleman, MA Steel	33:265–89

Homogenization of Freshwater Faunas	FJ Rahel	33:291–315
The Renaissance of Community-Based Marine Resource Management in Oceania	RE Johannes	33:317–40
The Role of Animals in Nutrient Cycling in Freshwater Ecosystems	MJ Vanni	33:341–70
Effects of UV-B Radiation on Terrestrial and Aquatic Primary Producers	TA Day, PJ Neale	33:371–96
The Evolution and Maintenance of Androdioecy	JR Pannell	33:397–425
Mast Seeding in Perennial Plants: Why, How, Where?	D Kelly, VL Sork	33:427–47
Disturbance to Marine Benthic Habitats by Trawling and Dredging–Implications for Marine Biodiversity	S Thrush, PK Dayton	33:449–73
Phylogenies and Community Ecology	CO Webb, DD Ackerly, MA McPeek, MJ Donoghue	33:475–505
Stable Isotopes in Plant Ecology	TE Dawson, S Mambelli, AH Plamboeck, PH Templer, KP Tu	33:507–59
The Quality of the Fossil Record: Implications for Evolutionary Analyses	SM Kidwell, SM Holland	33:561–88
Neopolyploidy in Flowering Plants	DW Schemske, J Ramsey	33:589–639
Herbivore Offense	R Karban, AA Agrawal	33:641–64
The Pacific Salmon Wars: What Science Brings to the Challenge of Recovering Species	M Ruckelshaus, P Levin, JB Johnson, PM Kareiva	33:665–706
Estimating Divergence Times from Molecular Data on Phylogenetic and Population Genetic Timescales	BS Arbogast, SV Edwards, J Wakeley, P Beerli, JB Slowinski	33:707–40
The Fate of Clades in a World of Recurrent Climatic Change - Milankovitch Oscillations and Evolution	R Jansson, M Dynesius	33:741–78

Volume 34 (2003)

| Effects of Introduced Bees on Native Ecosystems | D Goulson | 34:1–26 |

Avian Sexual Dichromatism in Relation to Phylogeny and Ecology	AV Badyaev, GE Hill	34:27–49
Paleobiogeography: The Relevance of Fossils to Biogeography	BS Lieberman	34:51–69
The Ecology of Bird Introductions	RP Duncan, TM Blackburn, D Sol	34:71–98
The Effects of Genetic and Geographic Structure on Neutral Variation	B Charlesworth, D Charlesworth, NH Barton	34:99–125
Data, Models, and Decisions in U.S. Marine Fisheries Management: Lessons for Ecologists	KA Rose, JH Cowan Jr.	34:127–51
Partitioning of Time as an Ecological Resource	N Kronfeld-Schor, T Dayan	34:153–81
Performance Comparisons of Co-Occuring Native and Alien Invasive Plants: Implications for Conservation and Restoration	CC Daehler	34:183–211
Genetic Variation in Rare and Common Plants	CT Cole	34:213–37
The Ecology and Evolution of Insect Baculoviruses	JS Cory, JH Myers	34:239–72
Latitudinal Gradients of Biodiversity: Pattern, Process, Scale, and Synthesis	MR Willig, DM Kaufman, RD Stevens	34:273–309
Recent Advances in the (Molecular) Phylogeny of Vertebrates	A Meyer, R Zardoya	34:311–38
The Role of Reinforcement in Speciation: Theory and Data	MR Servedio, MAF Noor	34:339–64
Extra-Pair Paternity in Birds: Causes, Correlates, and Conflict	DF Westneat, IRK Stewart	34:365–96
Species-Level Paraphyly and Polyphyly: Frequency, Causes, and Consequences, with Insights from Animal Mitochondrial DNA	DJ Funk, KE Omland	34:397–423
Protective Ant-Plant Interactions as Model Systems in Ecological and Evolutionary Research	M Heil, D McKey	34:425–53
Functional Matrix: A Conceptual Framework for Predicting Plant Effects on Ecosystem Processes	VT Evinerm, FS Chapin III	34:455–85
Effects of Habitat Fragmentation on Biodiversity	L Fahrig	34:487–515

Social Organization and Disease Risk in Mammals: Integrating Theory and Empirical Studies	S Altizer, CL Nunn, PH Thrall, JL Gittleman, J Antonovics, AA Cunningham, AP Dodson, V Ezenwa, KE Jones, AB Pedersen, M Poss, JRC Pulliam	34:517–47
The Community-Level Consequences of Seed Dispersal Patterns	JM Levine, DJ Murrell	34:549–74
The Ecology and Evolution of Seed Dispersal: A Theoretical Perspective	SA Levin, HC Muller-Landau, R Nathan, J Chave	34:575–604
Analysis of Rates of Morphologic Evolution	PD Roopnarine	34:605–32
Development and the Genetics of Evolutionary Change Within Insect Species	PM Brakefield, V French, BJ Zwaan	34:633–60
Flexibility and Specificity in Coral-Algal Symbiosis: Diversity, Ecology, and Biogeography of *Symbiodinium*	AC Baker	34:661–89

Volume 35 (2004)

Vertebrate Dispersal of Seed Plants Through Time	BH Tiffney	35:1–29
Are Diseases Increasing in the Ocean?	KD Lafferty, JW Porter, SE Ford	35:31–54
Bird Song: The Interface of Evolution and Mechanism	J Podos, SK Huber, B Taft	35:55–87
Application of Ecological Indicators	GJ Niemi, ME McDonald	35:89–111
Ecological Impacts of Deer Overabundance	SD Côté, TP Rooney, J-P Tremblay, C Dussault, DM Waller	35:113–47
Ecological Effects of Transgenic Crops and the Escape of Transgenes into Wild Populations	D Pilson, HR Prendeville	35:149–74

Mutualisms and Aquatic Community Structure: The Enemy of My Enemy Is My Friend	ME Hay, JD Parker, DE Burkepile, CC Caudill, AE Wilson, ZP Hallinan, AD Chequer	35:175–97
Operational Criteria for Delimiting Species	JW Sites Jr, JC Marshall	35:199–227
The New View of Animal Phylogeny	KM Halanych	35:229–56
Landscapes and Riverscapes: The Influence of Land Use on Stream Ecosystems	JD Allan	35:257–84
Long-Term Stasis in Ecological Assemblages: Evidence from the Fossil Record	WA DiMichele, AK Behrensmeyer, TD Olszewski, CC Labandeira, JM Pandolfi, SL Wing, R Bobe	35:285–322
Avian Extinctions from Tropical and Subtropical Forests	NS Sodhi, LH Liow, FA Bazzaz	35:323–45
Evolutionary Biology of Animal Cognition	R Dukas	35:347–74
Pollination Syndromes and Floral Specialization	CB Fenster, WS Armbruster, P Wilson, MR Dudash, JD Thomson	35:375–403
On the Ecological Roles of Salamanders	RD Davic, HH Welsh Jr.	35:405–34
Ecological and Evolutionary Consequences of Multispecies Plant-Animal Interactions	SY Strauss, RE Irwin	35:435–66
Spatial Synchrony in Population Dynamics	A Liebhold, WD Koenig, ON Bjørnstad	35:467–90
Ecological Responses to Habitat Edges: Mechanisms, Models, and Variability Explained	L Ries, RJ Fletcher Jr, J Battin, TD Sisk	35:491–522
Evolutionary Trajectories and Biogeochemical Impacts of Marine Eukaryotic Phytoplankton	ME Katz, ZV Finkel, D Grzebyk, AH Knoll, PG Falkowski	35:523–56

Regime Shifts, Resilience, and Biodiversity in Ecosystem Management	C Folke, S Carpenter, B Walker, M Scheffer, T Elmqvist, L Gunderson, CS Holling	35:557–81
Ecology of Woodland Herbs in Temperate Deciduous Forests	DF Whigham	35:583–621
The Southwest Australian Floristic Region: Evolution and Conservation of a Global Hot Spot of Biodiversity	SD Hopper, P Gioia	35:623–50
Predator-Induced Phenotypic Plasticity in Organisms with Complex Life Histories	MF Benard	35:651–73
The Evolutionary Ecology of Novel Plant-Pathogen Interactions	IM Parker, GS Gilbert	35:675–700

Volume 36 (2005)

The Genetics and Evolution of Fluctuating Asymmetry	LJ Leamy, CP Klingenberg	36:1–21
Life-History Evolution in Reptiles	R Shine	36:23–46
The Evolutionary Enigma of Mixed Mating Systems in Plants: Occurrence, Theoretical Explanations, and Empirical Evidence	C Goodwillie, S Kalisz, CG Eckert	36:47–79
Indirect Interaction Webs: Herbivore-Induced Effects Through Trait Change in Plants	T Ohgushi	36:81–105
Evolutionary History of Poales	HP Linder, PJ Rudall	36:107–24
The Evolution of Polyandry: Sperm Competition, Sperm Selection, and Offspring Viability	LW Simmons	36:125–46
Individual-Based Modeling of Ecological and Evolutionary Processes	DL DeAngelis, WM Mooij	36:147–68
The Influence of Plant Secondary Metabolites on the Nutritional Ecology of Herbivorous Terrestrial Vertebrates	MD Dearing, WJ Foley, S McLean	36:169–89
Biodiversity and Litter Decomposition in Terrestrial Ecosystems	S Hättenschwiler, AV Tiunov, S Scheu	36:191–218
The Functional Significance of Ribosomal (r)DNA Variation: Impacts on the Evolutionary Ecology of Organisms	LJ Weider, JJ Elser, TJ Crease, M Mateos, JB Cotner, TA Markow	36:219–42

Evolutionary Ecology of Plant Adaptation to Serpentine Soils	KU Brady, AR Kruckeberg, HD Bradshaw Jr.	36:243–66
Biodiversity-Ecosystem Function Research: Is it Relevant to Conservation?	DS Srivastava, M Vellend	36:267–94
Consequences of the Cretaceous/Paleogene Mass Extinction for Marine Ecosystems	S D'Hondt	36:295–317
Landscape Ecology: What Is the State of the Science?	MG Turner	36:319–44
Ecology and Evolution of Aphid-Ant Interactions	B Stadler, AFG Dixon	36:345–72
Evolutionary Causes and Consequences of Immunopathology	AL Graham, JE Allen, AF Read	36:373–97
The Evolutionary Ecology of Gynogenesis	I Schlupp	36:399–417
Measurement of Interaction Strength in Nature	JT Wootton, M Emmerson	36:419–44
Model Selection in Phylogenetics	J Sullivan, P Joyce	36:445–66
Pollen Limitation of Plant Reproduction: Pattern and Process	TM Knight, JA Steets, JC Vamosi, SJ Mazer, M Burd, DR Campbell, MR Dudash, MO Johnston, RJ Mitchell, T-L Ashman	36:467–97
Evolving the Psychological Mechanisms for Cooperation	JR Stevens, FA Cushman, MD Hauser	36:499–518
Niche Conservatism: Integrating Evolution, Ecology, and Conservation Biology	JJ Wiens, CH Graham	36:519–39
Phylogenomics	H Philippe, F Delsuc, H Brinkmann, N Lartillot	36:541–62
The Evolution of Agriculture in Insects	UG Mueller, NM Gerardo, DK Aanen, DL Six, TR Schultz	36:563–95
Insects on Plants: Diversity of Herbivore Assemblages Revisited	TM Lewinsohn, V Novotny, Y Basset	36:597–620
The Population Biology of Mitochondrial DNA and Its Phylogenetic Implications	JWO Ballard, DM Rand	36:621–42

Introduction of Non-Native Oysters: Ecosystem Effects and Restoration Implications	JL Ruesink, HS Lenihan, AC Trimble, KW Heiman, F Micheli, JE Byers, MC Kay	36:643–89

Volume 37 (2006)

Birth-Death Models in Macroevolution	S Nee	37:1–17
The Posterior and the Prior in Bayesian Phylogenetics	ME Alfaro, MT Holder	37:19–42
Unifying and Testing Models of Sexual Selection	H Kokko, MD Jennions, R Brooks	37:43–66
Genetic Polymorphism in Heterogeneous Environments: The Age of Genomics	PW Hedrick	37:67–93
Ecological Effects of Invasive Arthropod Generalist Predators	WE Snyder, EW Evans	37:95–122
The Evolution of Genetic Architecture	TF Hansen	37:123–57
The Major Histocompatibility Complex, Sexual Selection, and Mate Choice	M Milinski	37:159–86
Some Evolutionary Consequences of Being a Tree	RJ Petit, A Hampe	37:187–214
Late Quaternary Extinctions: State of the Debate	PL Koch, AD Barnosky	37:215–50
Innate Immunity, Environmental Drivers, and Disease Ecology of Marine and Freshwater Invertebrates	LD Mydlarz, LE Jones, CD Harvell	37:251–88
Experimental Methods for Measuring Gene Interactions	JP Demuth, MJ Wade	37:289–316
Corridors for Conservation: Integrating Pattern and Process	C-LB Chetkiewicz, C St. Clair, MS Boyce	37:317–42
The Population Biology of Large Brown Seaweeds: Ecological Consequences of Multiphase Life Histories in Dynamic Coastal Environments	DR Schiel, MS Foster	37:343–72
Living on the Edge of Two Changing Worlds: Forecasting the Responses of Rocky Intertidal Ecosystems to Climate Change	B Helmuth, N Mieszkowska, P Moore, SJ Hawkins	37:373–404

Has Vicariance or Dispersal Been the Predominant Biogeographic Force in Madagascar? Only Time will Tell	AD Yoder, MD Nowak	37:405–31
Limits to the Adaptive Potential of Small Populations	Y Willi, J Van Buskirk, AA Hoffmann	37:433–58
Resource Exchange in the Rhizosphere: Molecular Tools and the Microbial Perspective	ZG Cardon, DJ Gage	37:459–88
The Role of Hybridization in the Evolution of Reef Corals	BL Willis, MJH van Oppen, DJ Miller, SV Vollmer, DJ Ayre	37:489–517
The New Bioinformatics: Integrating Ecological Data from the Gene to the Biosphere	MB Jones, MP Schildhauer, OJ Reichman, S Bowers	37:519–44
Incorporating Molecular Evolution into Phylogenetic Analysis, and a New Compilation of Conserved Polymerase Chain Reaction Primers for Animal Mitochondrial DNA	C Simon, TR Buckley, F Frati, JB Stewart, AT Beckenbach	37:545–79
The Developmental, Physiological, Neural, and Genetical Causes and Consequences of Frequency-Dependent Selection in the Wild	B Sinervo, R Calsbeek	37:581–610
Carbon-Nitrogen Interactions in Terrestrial Ecosystems in Response to Rising Atmospheric Carbon Dioxide	PB Reich, BA Hungate, Y Luo	37:611–36
Ecological and Evolutionary Responses to Recent Climate Change	C Parmesan	37:637–69